**Advances in
Carbohydrate Chemistry and Biochemistry**

Volume 68

Advances in Carbohydrate Chemistry and Biochemistry

Editor
DEREK HORTON

Board of Advisors

DAVID C. BAKER
DAVID R. BUNDLE
STEPHEN HANESSIAN
YURIY A. KNIREL
TODD L. LOWARY

SERGE PÉREZ
PETER H. SEEBERGER
J.F.G. VLIEGENTHART
ARNOLD A. STÜTZ

Volume 68

AMSTERDAM • BOSTON • HEIDELBERG • LONDON
NEW YORK • OXFORD • PARIS • SAN DIEGO
SAN FRANCISCO • SINGAPORE • SYDNEY • TOKYO
Academic Press is an imprint of Elsevier

Academic Press is an imprint of Elsevier
The Boulevard, Langford Lane, Kidlington, Oxford, OX51GB, UK
32, Jamestown Road, London NW1 7BY, UK
Radarweg 29, PO Box 211, 1000 AE Amsterdam, The Netherlands
225 Wyman Street, Waltham, MA 02451, USA
525 B Street, Suite 1900, San Diego, CA 92101-4495, USA

First edition 2012

Copyright © 2012 Elsevier Inc. All rights reserved

No part of this publication may be reproduced, stored in a retrieval system or transmitted in any form or by any means electronic, mechanical, photocopying, recording or otherwise without the prior written permission of the publisher

Permissions may be sought directly from Elsevier's Science & Technology Rights Department in Oxford, UK: phone (+44) (0) 1865 843830; fax (+44) (0) 1865 853333; email: permissions @elsevier.com. Alternatively you can submit your request online by visiting the Elsevier web site at http://elsevier.com/locate/permissions, and selecting *Obtaining permission to use Elsevier material*

Notice
No responsibility is assumed by the publisher for any injury and/or damage to persons or property as a matter of products liability, negligence or otherwise, or from any use or operation of any methods, products, instructions or ideas contained in the material herein. Because of rapid advances in the medical sciences, in particular, independent verification of diagnoses and drug dosages should be made

ISBN: 978-0-12-396523-3
ISSN: 0065-2318

British Library Cataloguing in Publication Data
A catalogue record for this book is available from the British Library

Library of Congress Cataloging-in-Publication Data
A catalog record for this book is available from the Library of Congress

For information on all Academic Press publications
visit our website at store.elsevier.com

Printed and bound in USA

12 13 14 15   11 10 9 8 7 6 5 4 3 2 1

Working together to grow
libraries in developing countries

www.elsevier.com | www.bookaid.org | www.sabre.org

ELSEVIER   BOOK AID International   Sabre Foundation

CONTENTS

CONTRIBUTORS ix
PREFACE xi

Molecular Architecture and Therapeutic Potential of Lectin Mimics
YU NAKAGAWA AND YUKISHIGE ITO

I. Introduction. 2
II. Synthetic Lectin Mimics 4
   1. Molecular Architecture of Boronic Acid-Dependent Lectin Mimics. 4
   2. Antiviral Potential of Boronic Acid-Dependent Lectin Mimics 11
   3. Molecular Architecture of Boronic Acid-Independent Lectin Mimics. 15
   4. Antiviral and Antimicrobial Potential of Boronic Acid-Independent Lectin Mimics 27
III. Naturally Occurring Lectin Mimics 33
   1. Antimicrobial and Carbohydrate-Binding Profiles of Pradimicins and Benanoimicins 33
   2. Molecular Basis of Carbohydrate Recognition by PRMs 40
   3. Antiviral Profile and Mode of Action of Pradimicins 45
IV. Conclusion and Future Prospects 48
Acknowledgments. 49
References. 49

Enzymatic Conversions of Starch
PIOTR TOMASIK AND DEREK HORTON

I. Introduction 61
   1. Introduction and General Remarks. 61
   2. Historical Background 62
   3. Former Reviews 64
II. Enzymes and Microorganisms for Conversion of Starch 65
   1. Introduction 65
   2. Alpha Amylases (EC 3.2.1.1). 80
   3. Beta Amylases (EC 3.2.1.2) 98
   4. Glucoamylase (EC 3.2.1.3) 103
   5. Other Amylases 111
   6. α-Glucosidase (EC 3.2.1.20) 111
   7. Pullulanase (EC 3.2.1.41) 114
   8. Neopullulanase (EC 3.2.1.135). 117
   9. Isoamylase (EC 3.2.1.68) 117
   10. Other Hydrolases. 119

	11. Enzymatic Cocktails	120
	12. Glycosyltransferases (EC 2.4.1)	123
	13. Microorganisms	124
III.	Hydrolysis Pathways and Mechanisms	127
	1. Role of Adsorption	127
	2. Mechanism of Inhibition	130
	3. Mathematical Models of Enzymatic Hydrolysis	132
	4. Effect of Light, Microwaves, and External Electric Field	135
	5. Kinetics	136
IV.	Amylolytic Starch Conversions	144
	1. Introduction	144
	2. Pulping	145
	3. Malting	146
	4. Mashing	146
	5. Liquefaction	148
	6. Saccharification	151
	7. Effect of the Botanical Origin of Starch	159
	8. Role of Starch Pretreatment	174
	9. Role of Temperature	180
	10. Role of the Substrate Concentration	181
	11. Role of Water	181
	12. Role of Elevated Pressure	182
	13. Role of pH	182
	14. Role of Admixed Inorganic Salts	183
	15. Role of Inhibitors	186
	16. Stimulators of Hydrolysis	188
	17. Engineering Problems	189
	18. Applications of the Enzymatic Processes	191
V.	Starch as a Feedstock for Fermentations	208
	1. General Remarks on Fermentation	208
	2. Alcohol and Alcohol–Acetone Fermentations	209
	3. Carboxylic Acid Fermentations	232
VI.	Nonamylolytic Starch Conversions	238
	1. Glycosylation	238
	2. Esterification and Hydrolysis	241
	3. Methanogenic and Biosulfidogenic Conversions	243
	4. Isomerization	243
	5. Hydrogen Production	245
	6. Trehalose	246
	7. Bacterial Polyester Formation	247
	8. Branching of Starch	247
	9. Oxidation	247
	10. Polymerization	247
	11. Cyclodextrins	249
VII.	Starch Metabolism in Human and Animal Organisms	258
	1. Digestible Starch	258
	2. Resistant Starch	261

VIII.	Starch Analytics Involving Enzymes	262
	1. Starch Evaluation and Analysis	262
	2. Enzyme Evaluation	266
	References	268

AUTHOR INDEX 437
SUBJECT INDEX 525

CONTRIBUTORS

Derek Horton, Chemistry Department, Ohio State University, Columbus, Ohio, USA

Yukishige Ito, Synthetic Cellular Chemistry Laboratory, RIKEN Advanced Science Institute, and ERATO, Japan Science and Technology Agency, Glycotrilogy Project, Wako, Saitama, Japan

Yu Nakagawa, Synthetic Cellular Chemistry Laboratory, RIKEN Advanced Science Institute, Wako, Saitama, Japan

Piotr Tomasik, Krakow College of Health Promotion, Krakow, Poland

PREFACE

This volume of *Advances* constitutes a departure from the customary format in that it features one article on a rapidly evolving new field of research in its early stage of development, together with another very large article that presents a broad overview of a mature area of carbohydrate science.

The contribution from Nakagawa and Ito (Saitama, Japan) focuses on small molecules that mimic the carbohydrate-binding properties of lectins, a class of proteins commonly derived from plant sources and which are used extensively as research tools in glycobiology. Lectins offer interesting potential in drug applications but are expensive to manufacture, have low chemical stability, and may be immunogenic.

Consequently, there is high current interest in molecules that behave as lectin mimics, emulating the strong carbohydrate-binding properties of natural lectins. Some of these occur naturally (pradimicins and benanomicins) and others are readily accessible by synthesis.

The synthetic examples fall into two broad categories, the boronic acid-dependent and boronic acid-independent lectin mimics, and they lend themselves readily to chemical modification for "tuning" the architecture of the carbohydrate-binding cavity and optimizing the strength of binding.

Although the carbohydrate-binding ability of most of the lectin mimics studied thus far does not reach the level of natural lectins, their therapeutic potential as antiviral and antimicrobial agents clearly warrants active pursuit toward goals in medicine. In the broader context, these studies will further our understanding of the molecular basis of carbohydrate recognition.

The most abundant carbohydrates in the biosphere are cellulose, chitin, and starch. A recorded procedure for the isolation of starch from cereal grains, by the forerunner of today's wet-milling process, dates back two millennia to the writings of Pliny The Elder. The manifold applications to which starch has been put during those 2000 years, by conversion processes involving enzymatic, chemical, and physical transformations, long predates most modern scientific documentation.

In previous volumes of *Advances*, Tomasik (Cracow) has contributed comprehensive surveys of the conversions of starch by chemical and physical methods. The huge volume of work on the enzymatic transformations of starch published during the past 200 years, much of it the subject of claims in the patent literature, greatly exceeds the usual scope of an article in *Advances*, but there has been no obvious way to subdivide this body of work into a series of shorter articles. The large report contributed here by Tomasik, in conjunction with this writer, surveys methods for the enzymatic conversion of starch by hydrolases and other enzymes, together with the role of microorganisms producing such enzymes and applications of these enzymes in the food, pharmaceutical, pulp, textile, and other branches of industry.

An effort has been made to coordinate the descriptions of the enzymes and their action patterns within the framework of the modern EC classification system. However, much early work, as well as some contemporary technology, has involved the use of crude enzyme isolates that are mixtures of individual enzymes, along with starches from different plant sources that exhibit distinctive differences in their behavior toward enzymes.

The two largest classes of starch-degrading enzymes are named uniformly in this article as alpha amylases and beta amylases. Although the terms α-amylase and β-amylase are very frequently encountered in the literature, the Greek designators are properly reserved to designate stereochemical configuration at the anomeric center in a formal sugar name. Applying the Greek designators to the enzymes has led some instructors and students to the erroneous notion that α-amylases cleave alpha linkages and β-amylases cleave beta linkages.

The products of enzyme-catalyzed starch degradation, mixtures of variously depolymerized glucosaccharides, are frequently described by the ill-defined term "dextrins," and this term may have a multiplicity of meanings in different situations. Likewise, many different names have been used for various enzymes and their plant, fungal, and microbial sources. As part of a comprehensive treatment, the article incorporates much information on starch enzymology recorded in the patent literature. Such sources have clearly not been subject to the peer-review process of the major journals, and they must be judged in this light; inconsistencies and contradictions remain evident.

Future issues of *Advances* are expected to revert to the regular format of five or six individual articles on a variety of carbohydrate topics.

The death is noted with regret of Malcolm B. Perry, on June 25, 2012. Dr. Perry worked for many years in Ottawa at the National Research Council of Canada, where he made extensive contributions in the carbohydrate field, especially on the structures of lipopolysaccharide O-antigens of gram-negative bacteria.

DEREK HORTON

Washington, DC
October 2012

MOLECULAR ARCHITECTURE AND THERAPEUTIC POTENTIAL OF LECTIN MIMICS

Yu Nakagawa[a] and Yukishige Ito[a,b]

[a]Synthetic Cellular Chemistry Laboratory, RIKEN Advanced Science Institute, Wako, Saitama, Japan
[b]ERATO, Japan Science and Technology Agency, Glycotrilogy Project, Wako, Saitama, Japan

I. Introduction	2
II. Synthetic Lectin Mimics	4
1. Molecular Architecture of Boronic Acid-Dependent Lectin Mimics	4
2. Antiviral Potential of Boronic Acid-Dependent Lectin Mimics	11
3. Molecular Architecture of Boronic Acid-Independent Lectin Mimics	15
4. Antiviral and Antimicrobial Potential of Boronic Acid-Independent Lectin Mimics	27
III. Naturally Occurring Lectin Mimics	33
1. Antimicrobial and Carbohydrate-Binding Profiles of Pradimicins and Benanoimicins	33
2. Molecular Basis of Carbohydrate Recognition by PRMs	40
3. Antiviral Profile and Mode of Action of Pradimicins	45
IV. Conclusion and Future Prospects	48
Acknowledgments	49
References	49

Abbreviations

2D-DARR, two-dimensional dipolar-assisted rotational resonance; AIDS, acquired immunodeficiency syndrome; BAMP, 2,5-bis-(aminomethyl)pyrrole; CA, *Cymbidium* sp. agglutinin; CP/MAS, cross-polarization/magic angle spinning; CV-N, cyanovirin-N; EGTA, ethylene glycol-bis(2-aminoethyl ether)-N,N,N',N'-tetraacetic acid; ELISA, enzyme-linked immunosorbent assay; Gal, D-galactose; Glc, D-glucose; GlcNAc, *N*-acetyl-D-glucosamine; GNA, *Galanthus nivalis* agglutinin; HCV, hepatitis C virus; HEA, *Epipactis helleborine* agglutinin; HHA, *Hippeastrum hybrid* agglutinin; HIV, human immunodeficiency virus; ICD, induced circular dichroism;

LOA, *Listera ovata* agglutinin; MALDI-MS, matrix-assisted laser desorption/ionization mass spectroscopy; Man, D-mannose; Me-α-Glc, methyl α-D-glucopyranoside; Me-β-Glc, methyl β-D-glucopyranoside; Me-α-GlcNAc, methyl 2-acetamido-2-deoxy-α-D-glucopyranoside; Me-β-GlcNAc, methyl 2-acetamido-2-deoxy β-D-glucopyranoside; Me-α-Man, methyl α-D-mannopyranoside; Me$_2$SO, dimethyl sulfoxide; MIC, minimum inhibitory concentration; NMR, nuclear magnetic resonance; NPA, *Narcissus pseudonarcissus* agglutinin; Oct-β-Gal, *n*-octyl β-D-galactopyranoside; Oct-α-Gal, *n*-octyl α-D-galactopyranoside; Oct-α-Glc, *n*-octyl α-D-glucopyranoside; Oct-β-Glc, *n*-octyl β-D-glucopyranoside; Oct-β-GlcNAc, *n*-octyl 2-acetamido-2-deoxy-β-D-glucopyranoside; Oct-β-Man, *n*-octyl β-D-mannopyranoside; Oct-α-Man, *n*-octyl α-D-mannopyranoside; PET, photoinduced electron transfer; PRM, pradimicin; ROS, reactive oxygen species; sLex, sialyl Lewis X; SPR, surface plasmon resonance; STD, saturation-transfer difference; TF, Thomsen–Friedenreich; TIMS, targeted imaging mass spectroscopy; UDA, *Urtica dioica* agglutinin; WGA, wheat-germ agglutinin

I. INTRODUCTION

Carbohydrates regulate a variety of biochemical pathways and disease processes, including protein folding, fertilization, embryogenesis, neuronal development, cell proliferation, microbial and viral infections, and cancer metastasis.[1–5] Coupled with advances in understanding of the functions of carbohydrates in biological and pathological processes, the importance of carbohydrate-binding molecules has been rapidly growing as chemical tools in glycobiology. The most extensively used carbohydrate-binding molecules are lectins, which are defined as proteins of non-immune origin that bind specific carbohydrates without modifying them.[6,7] They have proved to be extremely useful tools not only for the isolation and characterization of glycoproteins but also for detection of dynamic spatiotemporal changes that are associated with pathological processes.[8–10] In addition, several lectins derived from prokaryotic, plant, invertebrate, or vertebrate species having binding specificity for mannose, galactose, or *N*-acetylglucosamine (GlcNAc) have been found to inhibit the infection of enveloped viruses, such as human immunodeficiency virus (HIV) and hepatitis C virus (HCV).[11–13] Representative lectins having antiviral activity are the mannose-specific agglutinins from *Galanthus nivalis* (GNA), *Hippeastrum hybrid* (HHA), *Cymbidium* sp. (CA), *Narcissus pseudonarcissus* (NPA), *Epipactis helleborine* (HEA), and *Listera ovata* (LOA), the GlcNAc-specific agglutinin from *Urtica dioica* (UDA), and the cyanobacterium-derived lectin (CV-N). These lectins show the dual mode of antiviral action, blockage of virus entry, and triggering the action of immune system by exposing cryptic immunogenic epitopes of the virus surface.

The discovery of these unique anti-HIV properties of lectins that have never been observed in any of the existing chemotherapeutics has spurred intensive study of lectins as therapeutic agents.

While lectins are of great value as possible drug candidates as well as chemical tools in glycobiology research, they have several disadvantages due to their nature as proteins. First of all, large-scale production of proteins is generally both costly and time-consuming, and their relative low chemical stability may be problematic for long storage. From a therapeutic viewpoint, the bioavailability of proteins is likely to be poor, and concerns are also raised about the unfavorable biological properties of lectins, such as immunogenicity, inflammatory activity, cellular toxicity, mitogenic stimulation of human peripheral lymphocyte cells, and hemagglutination of human red blood cells. Under these situations, non-protein small-size molecules having carbohydrate-binding properties have been recently attracting a great deal of attention as alternatives to lectins of protein structure. Since small molecules can be prepared in reliable quantities and are easily "tuned" for structural optimizations, these "lectin mimics" may have potential benefit as research tools and drug candidates.

Lectin mimics of synthetic origin have been developed progressively during recent years.[14–17] On the basis of molecular-design principles, the synthetic lectin mimics may be divided roughly into two groups, boronic acid-dependent and boronic acid-independent lectin mimics. The former group possesses boronic acid motifs, which form tight and reversible covalent bonds with 1, 2- or 1, 3-diol groups of carbohydrates. While boronic acid itself has a low affinity for most common carbohydrates at physiological pH, the affinity can be enhanced by introduction of neighboring functional groups and/or use of two boronic acid motifs of proper alignment. In contrast, boronic acid-independent lectin mimics rely solely on non-covalent interactions for binding to carbohydrates. These types of lectin mimics have been intensively developed in the field of supramolecular chemistry. Representative molecular architectures include tripod- and cage-types, both of which incorporate, respectively, hydrophilic and hydrophobic surfaces for hydrogen bonding and CH/π interaction (a weak hydrogen bond occurring between CH groups and aromatic rings)[18] with carbohydrates. A combination of these non-covalent interactions collectively realizes three-dimensional recognition of carbohydrates.

This article outlines the design concept, molecular architecture, and carbohydrate-binding properties of synthetic lectin mimics by using representative examples, and discusses their therapeutic potential by reviewing recent attempts to develop antiviral and antimicrobial agents using their architectures. We also focus on naturally occurring lectin mimics, pradimicins (PRMs) and benanomicins.[19] They are the only class of non-protein natural products having a C-type lectin-like property of being able to recognize D-mannopyranosides (Man) in the presence of Ca^{2+} ions. Recent investigations have revealed an interesting similarity in molecular architecture between these natural products and tripod-type lectin mimics. Moreover, accumulated evidence suggests that they have unique antimicrobial profiles that differ from those of the major classes of antibiotics, and exhibit anti-HIV activity in a manner similar to lectins. These emerging concepts of the molecular basis of carbohydrate recognition and therapeutic potential of naturally occurring lectin mimics are also described.

II. Synthetic Lectin Mimics

1. Molecular Architecture of Boronic Acid-Dependent Lectin Mimics

Boronic acid-dependent lectin mimics contain more than one boronic acid moiety, which is responsible for their carbohydrate binding. Boronic acids act as Lewis acids because of the empty *p*-orbital on boron, and in the presence of Lewis bases such as hydroxyl groups, interconversion from sp^2 to sp^3 hybridization readily occurs.[20–22] In aqueous media, a water molecule reversibly adds to a neutral trigonal form (I) of boronic acids to produce an anionic tetrahedral form (II) with a release of one proton (Scheme 1). In the alkaline pH range (pH > 10), diols can react with the tetrahedral form (II) to give stable cyclic boronate esters (IV). The general affinity of boronic acids for diols in carbohydrates is as follows: *cis*-1, 2-diol > 1, 3-diol ≫ *trans*-1, 2-diol. On the other hand, boronic acids exist predominately as the trigonal form (I) in the neutral pH range. As a result, reaction with diols produces chemically unstable

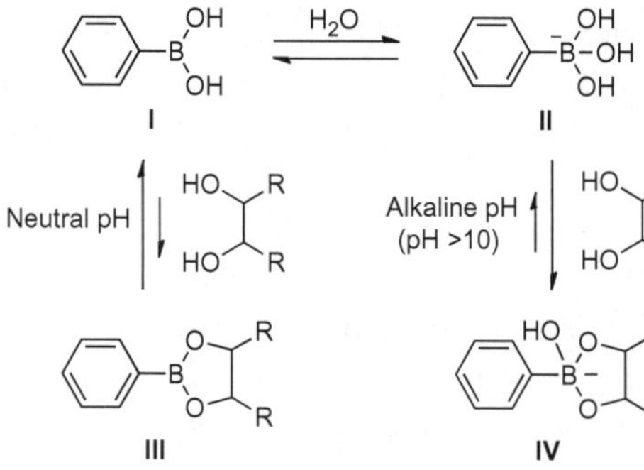

SCHEME 1. Binding of phenylboronic acid with a diol.

adducts (III), which are readily hydrolyzed. Therefore, a high pH is generally required in order to favor the equilibrium toward cyclic boronate esters (IV).

To realize arylboronic acid/diol association at neutral pH, several attempts have been made to stabilize their adducts. One of the most promising strategies is the introduction of dialkylaminomethyl groups to the *ortho* position of arylboronic acids (Scheme 2).[23,24] The covalently appended amino groups facilitate formation of a boronic ester and accelerate arylboronic acid/diol association by direct B–N coordination or "solvent-insertion" mechanism. Recent investigations by X-ray crystallography, ^{11}B-NMR, and computational analysis support that the latter mechanism is predominant in protic solvents.[25–27] By using this molecular architecture as a key structure, a number of boronic acid-dependent lectin mimics have been reported.[14,28] The majority of these works have been directed toward the development of molecular sensors for monitoring blood D-glucose as a key component of insulin-releasing implants for diabetes patients. In early landmark studies, Shinkai and coworkers developed carbohydrate-based photoinduced electron-transfer sensors (**1, 2**, Fig. 1).[29–31] These compounds bound carbohydrates at neutral pH (pH 7.77) and showed increased fluorescence through suppression of the photoinduced electron transfer (PET) from the tertiary amino group to the anthracenyl group upon binding of the carbohydrate. While the binding selectivity of **1** for D-glucose was poor (the order of selectivity for **1** is D-fructose > D-allose ≈ D-galactose > D-glucose > ethylene glycol), compound **2** exhibited binding preference for D-glucose (the order of selectivity for **2** is D-glucose > D-allose > D-fructose ≈ D-galactose > ethylene glycol). This selectivity of **2** for D-glucose was explained by the spatial disposition of the two boronic acid moieties. The two inwardly facing boronic acid moieties are perfectly spaced and aligned for two pairs of diol groups (1,2- and 4,6-hydroxyl groups) of D-glucose.

Using the same fluorescent reporter system, Wang and coworkers developed a series of dimer analogues of **1** (**3–8**, Fig. 1).[32,33] Of three dimers having different linkers (**3–5**), compound **3** showed the highest selectivity for D-glucose over

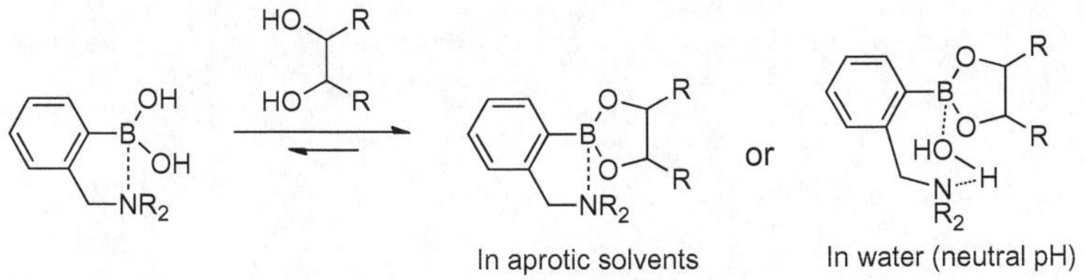

SCHEME 2. Binding of *o*-(*N*,*N*-dialkylaminomethyl)phenylboronic acid with a diol.

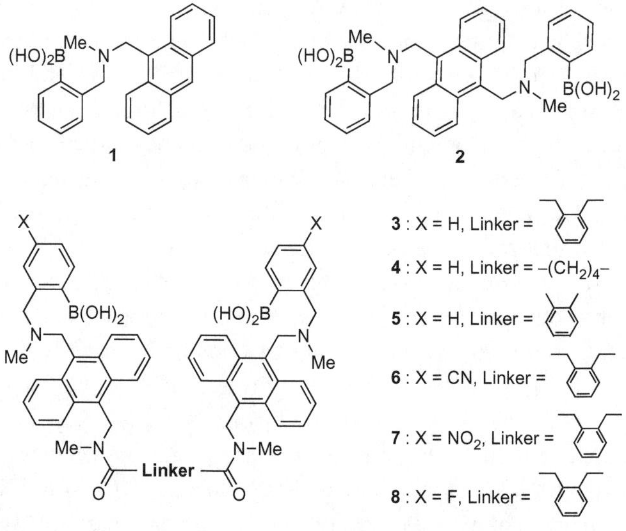

FIG. 1. Boronic acid-dependent lectin mimics (**1–8**).

D-fructose and D-galactose (Table I), suggesting that the linker of **3** offers the appropriate orientation and distance of two boronic acid moieties for binding with D-glucose. Although the affinity of **3** for D-glucose was about three times lower than that of **2**, compound **3** represents about 3-fold enhancement in selectivity for D-glucose over D-fructose. Based on these results, an additional three derivatives with electron-withdrawing groups, which generally enhance the carbohydrate affinity of

TABLE I
Binding Constants of 2–8 for Carbohydrates in Phosphate Buffer (pH 7.4)

Compound	K_a (M^{-1})		
	D-Glucose	D-Fructose	D-Galactose
2[a]	3981	316	158
3[a]	1472	34	30
4[a]	638	77	105
5[a]	178	283	33
6[b]	2540	968	271
7[b]	1808	198	132
8[b]	630	42	46

[a] Data cited from Ref. 32; [b] Data cited from Ref. 33.

arylboronic acids, were tested for their affinity and selectivity for carbohydrates. Although the introduction of cyano and nitro groups enhanced the affinity for D-glucose, the effects were more significant for binding to D-fructose and D-galactose. As a result, the selectivity of **6** and **7** for D-glucose was lower than that of the parent compound **3**. The fluoro-substituted derivative (**8**) was inferior to **3** in both affinity and selectivity. Although the substituent effect on carbohydrate selectivity of arylboronic acids is hard to predict, it is interesting that a simple modification of the aryl group changes the binding preference for carbohydrates.

Introduction of chirality into the PET sensor was successfully performed by Shinkai and coworkers.[34] The sensor **9** having the (R)-1,1′-binaphthyl moiety as the chiral and fluorophore building block discriminated D-glucose from L-glucose (Fig. 2). A maximum of a 4-fold increase in fluorescence intensity was observed upon binding of **9** to D-glucose, whereas L-glucose induced only a 2-fold increase. Competition experiments showed that **9** had the ability to recognize D-glucose selectively in the presence of L-glucose. Very recent examples of PET sensors using chiral units are the R and S enantiomers of **10**, which contain (1R)- and (1S)-1-(1-naphthyl)ethanamines, respectively.[35] The binding of D-glucose to R-**10** caused a significant increase in fluorescence intensity, whereas smaller changes in fluorescence were observed in the case of S-**10**. These studies clearly demonstrate that selectivity of boronic acid-dependent lectin mimics can be finely tuned toward one optical isomer of carbohydrates.

In addition to these glucose sensors, chemical probes for carbohydrate chains have been also reported. One example is a fluorescent probe (**11**) for sialyl Lewis X (sLe^x) developed by Wang and coworkers (Fig. 3).[36] The tetrasaccharide sLe^x is a well-known cell-surface carbohydrate antigen associated with the development and

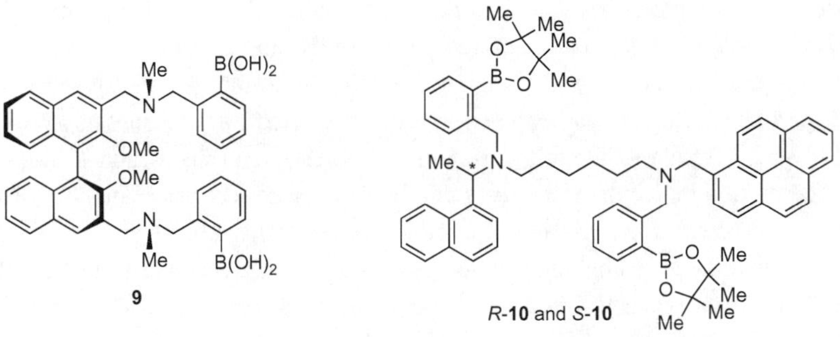

FIG. 2. Boronic acid-dependent lectin mimics (**9**, **10**).

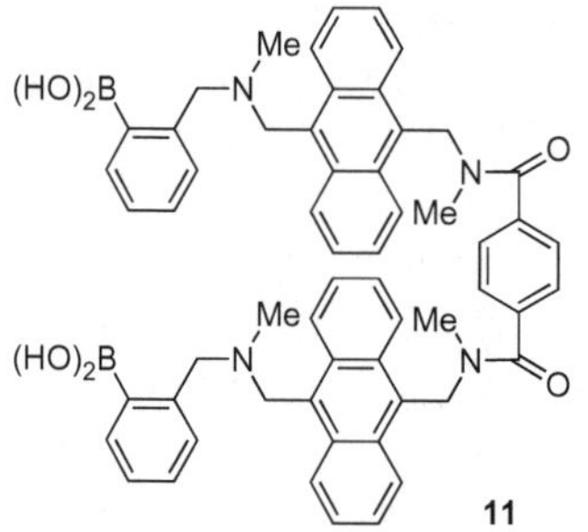

FIG. 3. Probes for sialyl Lewis X (**11**, **12**).

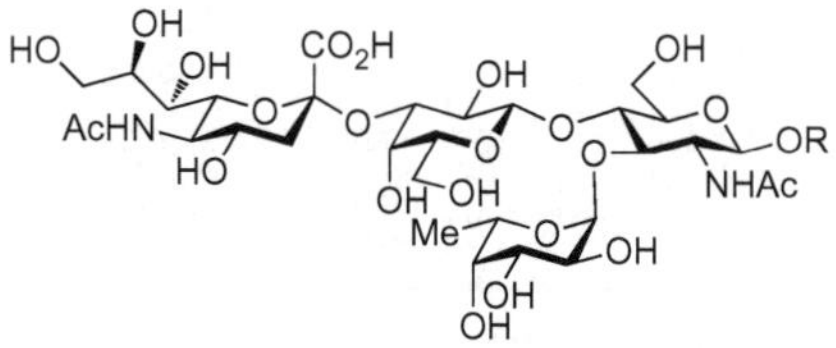

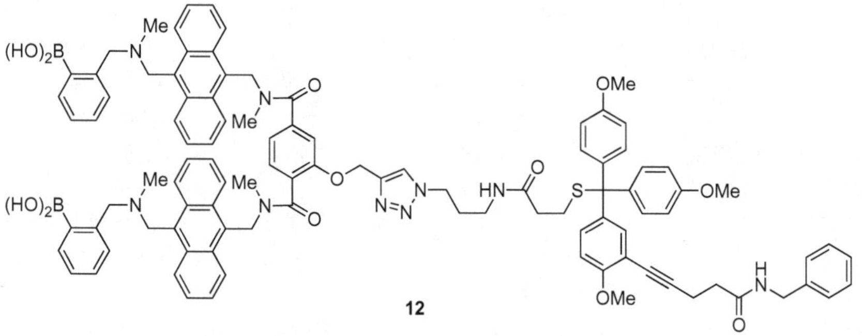

progression of many types of cancer.[37–40] Probes for sLe^x are therefore promising as the diagnostic agents for cancer. Compound **11** showed strong fluorescence enhancement upon binding to sLe^x in a 1:1 mixture of methanol and phosphate buffer (pH 7.4). Moreover, **11** was found to fluorescently label HEPG2 cells expressing high levels of sLe^x, while **11** stained neither HEP3B cells expressing Lewis Y nor COS7 cells lacking fucosylated antigens. Using this unique fluorescent sLe^x probe, matrix-assisted laser desorption/ionization mass spectrometry (MALDI-MS)-based imaging of cancer tissue has been performed.[41] Targeted imaging mass spectrometry (TIMS) is a powerful method of imaging for histological analysis, which allows spatial visualization of a molecule of interest directly from tissue sections by the use of laser-reactive photo-cleavable molecular tags attached to affinity molecules.[42,43] Wang and coworkers synthesized a conjugate molecule (**12**) of **11** and a trityl-based tag, and TIMS analysis of cancer tissues having a high level of sLe^x was performed. The tumor region expressing sLe^x was successfully detected by the imaging, suggesting that MS-based imaging of cancer tissues using lectin mimic-based probes is feasible.

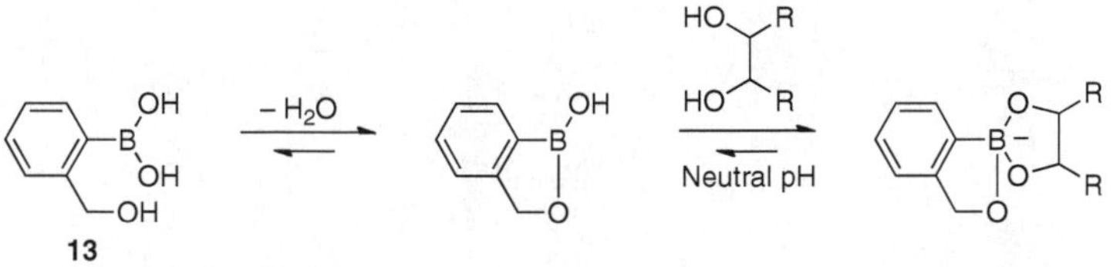

SCHEME 3. Binding of benzoboroxole (**13**) with a diol.

Although *o*-dialkylaminomethyl arylboronic acids have long stood as the established standard for the design of boronic acid-dependent lectin mimics, Hall and coworkers have subsequently introduced *o*-hydroxymethyl phenylboronic acid (benzoboroxole) as a promising alternative.[44,45] Benzoboroxole (**13**) is considered to exist in its cyclic dehydrated boronophthalide form (Scheme 3). The unusually small C–B–O dihedral angle of **13** facilitates production of an anionic boronate ester with a diol even at neutral pH by opening up the cone angle in the resulting tetrahedral structure. The boronate ester formation of **13** is also entropically more favorable than that of the simple phenylboronic acid, which requires the external hydroxyl ligand to form the anionic tetrahedral complex (Scheme 1). The most important and interesting feature of **13** is its capability to bind the pyranose form of carbohydrates in neutral water. Several lines of evidence suggest that carbohydrates bind to boronic acids, including *o*-dialkylaminomethyl arylboronic acids in their weakly populated furanose forms, and not in their pyranose form.[46,47] In contrast, **13** was shown to bind hexopyranosides of D-glucose, D-galactose, D-mannose, and D-fucose with weak but encouraging affinities (Table II). Although the exact mode of complexation of **13** with them is not fully understood, the relatively high Lewis

TABLE II
Binding Constants of 13 for Hexopyranosides in Phosphate Buffer (pH 7.4)

Hexopyranoside	K_a (M^{-1})[a]
D-Glucose	31
Methyl α-D-glucopyranoside	22
Methyl β-D-glucopyranoside	9
Methyl α-D-galactopyranoside	29
Methyl β-D-galactopyranoside	23
Methyl α-D-mannopyranoside	24
Methyl α-D-fucopyranoside	25

[a] Data cited from Ref. 45.

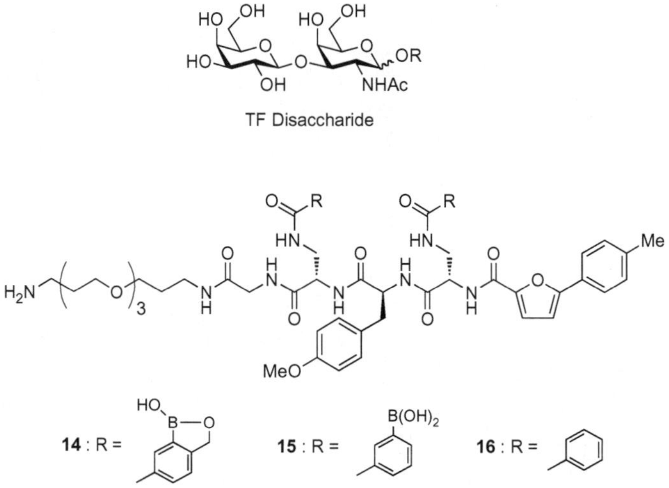

FIG. 4. Receptors for TF disaccharide (**14–16**).

acidity and strained cyclic structure of **13** are thought to be possible contributing factors for its exceptional pyranoside-binding behavior.

By taking advantage of the pyranoside-binding capability, Hall and coworkers have developed a water-soluble receptor (**14–16**, Fig. 4)[48] for the Thomsen–Friedenreich (TF) disaccharide, an important tumor-associated carbohydrate antigen that is present in over 90% of cancers.[49] The design of the receptor was based on the assumption that two units of benzoboroxoles would bind two diol units of the disaccharide, and the peptide backbone could be involved in hydrogen bonding and CH/π interaction with the disaccharide. The tri(ethylene glycol) component was introduced to increase the water solubility. A library of 400 examples having different peptide backbones was prepared by solid-phase peptide synthesis and was screened for binding to the TF antigen disaccharide by use of a competitive enzyme-linked immunosorbent assay (ELISA). The most potent receptor, showing an IC_{50} of 20 μM, was compound **14**, having an electron-rich *p*-methoxyphenylalanine component and a furan, both of which are favorable for CH/π interaction with hydrogen atoms on the carbohydrate rings. The importance of the benzoboroxole moieties was confirmed by the weak affinities of control compounds (**15**: $IC_{50} = 54$ μM, **16**: $IC_{50} = 100$ μM) for the TF disaccharide. These studies demonstrate that benzoboroxole (**13**) holds a great promise for the design of lectin mimics for carbohydrate biomarkers of biological importance.

2. Antiviral Potential of Boronic Acid-Dependent Lectin Mimics

The use of lectins has been recently proposed as a novel and promising approach for antiviral chemotherapy.[11–13] Lectins exhibit antiviral activity against a range of viruses such as HIV,[50,51] HCV,[52,53] coronavirus,[54,55] and influenza virus.[56,57] The antiviral effects of these lectins are ascribed to their specific binding to carbohydrate chains (glycans) on the viral envelope. It is well known that several proteins of the viral envelope are densely glycosylated. For example, gp120, the HIV-1 envelope glycoprotein, possesses 18–30 (an average of 24) N-linked glycans, which constitute about half of the molecular weight of the gp120 molecule.[58] These glycans on the HIV-1 envelope play crucial roles in enabling an efficient entry of HIV into its susceptible target cells.[59,60] Binding of lectins to these glycans thus inhibits the efficient functioning of the envelope molecules during viral entry.[61–63] Another important role of the envelope glycans is to hide the highly immunogenic epitopes from the immune system of the host.[64,65] Because of this glycan shield, the neutralizing antibody response is not elicited, and as a result, the immune system is not able to suppress HIV. However, these glycans are progressively deleted by long-term exposure of HIV to lectins through mutation of HIV.[66–69] Weakening of the glycan shield triggers the production of neutralizing antibodies against the immunogenic epitopes of gp120 previously hidden.

Despite the promising dual mode of anti-HIV action of lectins, concerns regarding their therapeutic use have been raised on account of the unacceptably high cost of large-scale production, low chemical stability, and unfavorable immunogenic response. Investigations are therefore ongoing to identify lectin mimics having high synthetic accessibility and high chemical stability. Boronic acid-dependent lectin mimics are undoubtedly potential candidates that exhibit lectin-like anti-HIV activities. The first attempt to evaluate antiviral activities of small-size boronic acid derivatives has been made by McGuigan and coworkers.[70] In an initial screening, a diverse range of *ortho*-, *meta*-, and *para*-substituted monophenylboronic acids were tested for their antiviral activities against a broad range of viruses, including HIV-1, HIV-2, varicella zoster virus, and influenza virus (Fig. 5). Unfortunately, none of the compounds showed antiviral activity at subtoxic concentrations or bound to gp120 of HIV-1. The same research group therefore performed the second screening using a library of bisphenylboronic acids having a variety of linkers (Fig. 6).[71] Although these compounds have a potential to bind two sets of diol groups in such glycans as **2–8**, surface plasmon resonance (SPR) experiments revealed that none of the bisphenylboronic acids bound to gp120 of HIV-1, even at 50 μM. Obviously, all of the compounds lacked the ability to inhibit virus replication. Although the results were

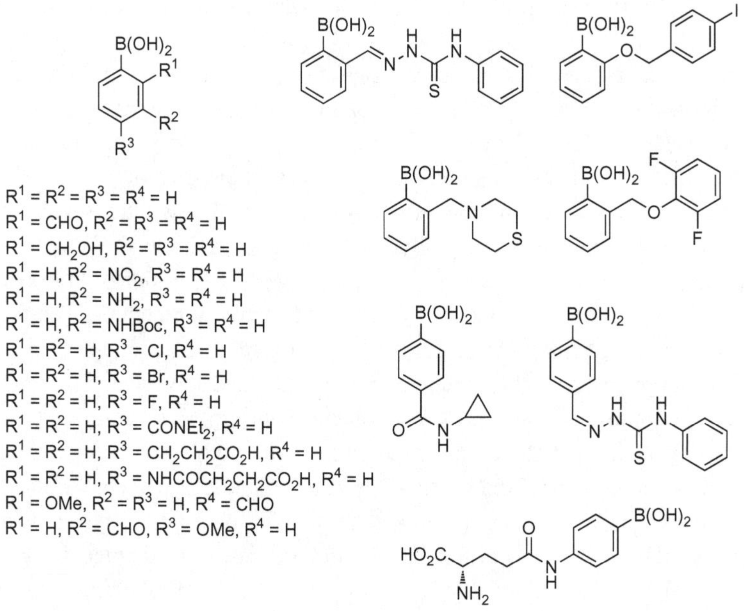

FIG. 5. Library of monophenyl boronic acid compounds.

disappointing, these data are valuable for future research aimed at designing small-size phenylboronic acid derivatives having antiviral activity.

On the other hand, Kiser and coworkers have successfully developed water-soluble polymers of phenylboronic acids as promising anti-HIV agents (Fig. 7).[72] The advantage of using the polymer structure is that multivalent interaction can be realized between the multiply linked boronic acid groups in the polymer and the abundant glycans on the viral envelope. Multivalent interactions can compensate for weak binding affinity of ligands because of entropic advantage and steric stabilization, and are often found in natural systems, including the carbohydrate–lectin interaction.[73,74] Moreover, they utilized benzoboroxole (**13**) for the boronic acid components, which can bind to such non-reducing carbohydrates as methyl α-D-mannopyranoside and methyl α-D-galactopyranoside at physiological pH, as already described.[44,45] The use of benzoboroxole is quite rational, because 30–50% of glycans in HIV-1 gp120 are high mannose-type oligosaccharides containing three terminal non-reducing mannose residues, and 50–63% are complex-type oligosaccharides terminating to some extent with non-reducing galactose groups (Fig. 8).[58] Based on these molecular design

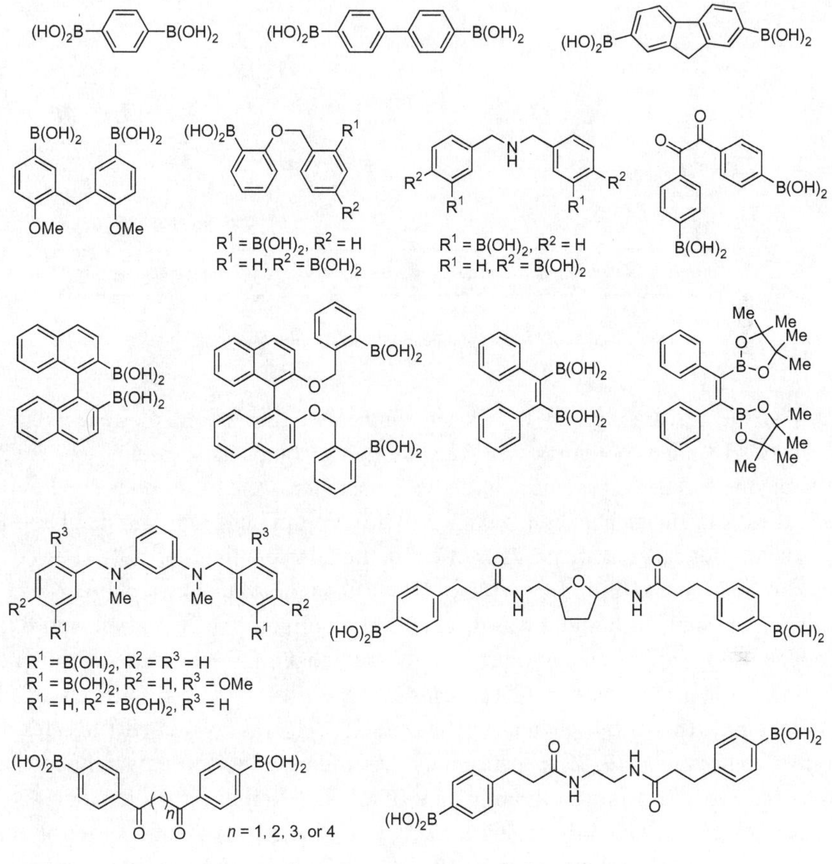

MOLECULAR ARCHITECTURE AND THERAPEUTIC POTENTIAL OF LECTIN MIMICS 13

FIG. 6. Library of bisphenylboronic acid compounds.

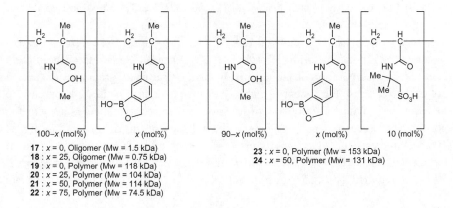

17 : $x = 0$, Oligomer (Mw = 1.5 kDa)
18 : $x = 25$, Oligomer (Mw = 0.75 kDa)
19 : $x = 0$, Polymer (Mw = 118 kDa)
20 : $x = 25$, Polymer (Mw = 104 kDa)
21 : $x = 50$, Polymer (Mw = 114 kDa)
22 : $x = 75$, Polymer (Mw = 74.5 kDa)

23 : $x = 0$, Polymer (Mw = 153 kDa)
24 : $x = 50$, Polymer (Mw = 131 kDa)

FIG. 7. Benzoboroxole-functionalized polymers (**17–24**).

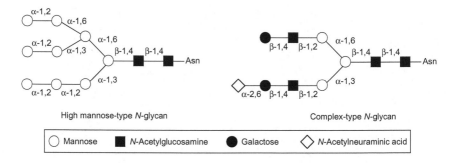

FIG. 8. Structures of high mannose-type and complex-type N-glycans.

concepts, the research group of Kisher synthesized benzoboroxole-functionalized oligomers (**18**) and polymers (**20–22**). In these compounds, 5-methacrylamido-2-hydroxymethylphenyl boronic acid-derived monomers were incorporated at different feed ratios into linear backbones with 2-hydroxypropylmethacrylamide. The SPR experiments revealed that the 25 mol% functionalized oligomer (**18**) showed an about 8-fold higher affinity for gp120 of HIV-1 than the simple benzoboroxole (**13**). On the other hand, no binding was detected for an oligomer lacking benzoboroxole components (**17**) at any concentrations, indicating that multivalent interaction of the benzoboroxole groups with gp120 glycans is responsible for the significant affinity of compound **18**. Antiviral activity against HIV-1 was evaluated only for the polymer compounds (**19–22**) because polymers of larger molecular weight decrease cytotoxicity by preventing cellular uptake. While the 25 mol% functionalized polymer **20** showed only slightly higher activity ($EC_{50}=15$ μM) than the control polymer without benzoboroxole groups (**19**, $EC_{50} \geq 50$ μM), increasing the degree of functionalization markedly decreased the EC_{50} values. The 75 mol% functionalized polymer (**22**) exhibited a strong activity ($EC_{50}=0.015$ μM) comparable to that of the natural lectin CV-N.

Kisher's group has also developed a second generation of the benzoboroxole-functionalized polymer, with 10 mol% of 2-acrylamido-2-methylpropanesulfonic acid (**24**).[75] Incorporation of the negatively charged sulfonate groups was expected not only to increase the aqueous solubility of the polymer at neutral pH but also to increase the binding affinity to gp120 through electrostatic interaction with the positively charged peptide fragments in the V3 loop of gp120. In fact, an almost 100-fold increase in aqueous solubility was observed with the incorporation of the methylpropanesulfonic acid moieties, and compound **24** showed enhanced anti-HIV-1 activity ($EC_{50}=4$ nM). A physical mixture of **21** and **23** did not demonstrate any significant increase in activity, suggesting that simultaneous ionic and covalent interactions possibly contribute to the observed synergistic improvement in antiviral activity of **24**. Moreover, biocompatibility evaluations on VEC-100-reconstructed human vaginal tissue suggest that **24** is likely to be nontoxic. These studies provide the first example of boronic acid-dependent lectin mimics having anti-HIV activity and demonstrate that these polymers are promising candidates for further development as microbicides.

3. Molecular Architecture of Boronic Acid-Independent Lectin Mimics

In contrast to the boronic acid-dependent lectin mimics that utilize non-natural covalent interactions for carbohydrate binding, boronic acid-independent lectin mimics bind carbohydrates through natural non-covalent interactions. These receptors generally realize the carbohydrate recognition by a combination of hydrogen bonding, CH/π interaction, and metal coordination, in a manner similar to that of natural lectins.[15–17] From this characteristic, boronic acid-independent lectin mimics are often referred to as "biomimetic carbohydrate receptors." By imitating the binding motifs observed in the crystal structures of lectin–carbohydrate complexes, a number of boronic acid-independent lectin mimics have been designed. Among them, this section focuses on tripod-type and cage-type architectures and reviews the design concept and carbohydrate-binding properties of representative compounds.

The tripod-type receptors, in principle, incorporate three hydrophobic units interconnected by an arene spacer (Fig. 9). Whereas the apolar bottom face of this

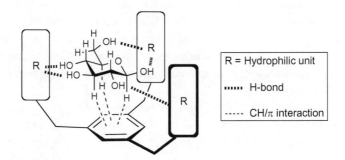

FIG. 9. Typical architecture of tripod-type boronic acid-independent lectin mimics.

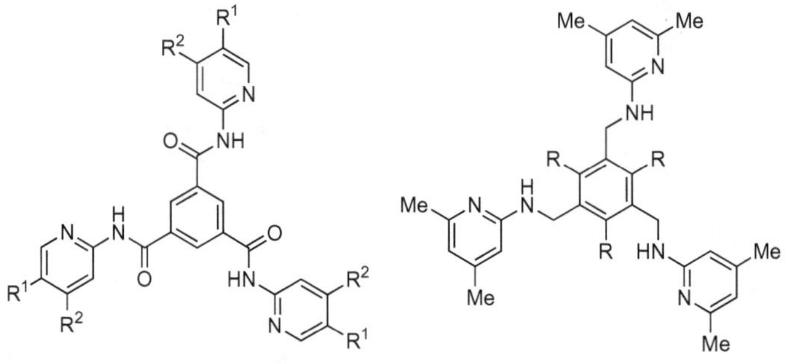

FIG. 10. Tripod-type boronic acid-independent lectin mimics (**25–28**).

architecture interacts with CH groups of carbohydrates through CH/π interaction, the polar lateral face forms hydrogen bonds with the hydroxyl groups of carbohydrates. These non-covalent interactions collectively realize three-dimensional recognition of carbohydrates by the tripod-type receptors. Early examples are compounds **25** and **26** developed by Mazik and coworkers.[76–79] These receptors contain a benzene platform as the π-donor for CH/π interaction and three amidopyridine moieties as hydrophilic units (Fig. 10). ^{1}H-NMR spectroscopic titrations in chloroform (Table III) revealed that both receptors formed very strong 2:1 receptor–carbohydrate complexes with *n*-octyl β-D-glucopyranoside (Oct-β-Glc), *n*-octyl α-D-glucopyranoside (Oct-α-Glc), and *n*-octyl β-D-galactopyranoside (Oct-β-Gal). Replacement of the amidopyridine moieties by aminopyridine groups and incorporation of methyl or ethyl groups into the central phenyl ring (**27**, **28**, Fig. 10) significantly increased the binding selectivity for Oct-β-Glc. In addition, curve fitting of the titration data for receptors **27** and **28** with Oct-β-Glc suggested the existence of both 1:1 and 1:2 receptor–carbohydrate complexes in chloroform, with strong association constants for 1:1 complexes. The results of these initial studies demonstrated that the tripod-type architecture is effective for the carbohydrate recognition, and subtle structural variation can lead to remarkable changes of the receptor properties.

Since these discoveries, Mazik's group has performed systematic studies using the tripodal architecture and found that a variety of functional groups, such as heteroaromatic, guanidinium, carboxylate, crown ether, hydroxy, amide, and oxime-based groups, were suitable for the hydrophilic units.[87] The tripod-type receptors reported by Mazik's group are also shown in Fig. 11 (**29–40**). Receptors **29**, **30**, and **31** contain,

TABLE III
Binding Constants of 25–40 for n-Octyl D-Glycosides in Chloroform

Receptor	β-Glc		α-Glc		β-Gal	
	$K_{11}{}^a$	$K_{21}{}^b$ or $K_{12}{}^c$	K_{11}	K_{21} or K_{12}	K_{11}	K_{21} or K_{12}
25^d	660	24,200	3640	82,450	420	50,770
26^d	440	22,600	2100	47,600	300	29,350
27^d	20,950	790	800	–	1360	211
28^d	48,630	1320	1310	–	3070	470
29^e	144,520	4330	24,880	1750	NT^f	NT
30^e	39,800	1610	2280	–	NT	NT
31^e	18,900	2850	1840	–	NT	NT
32^g	191,730	8560	3160	1540	3320	300
33^g	156,100	10,360	2820	350	7470	1100
34^h	>100,000	10,000	>100,000	ND^i	9550	1030
35^j	>100,000	ND	>100,000	ND	13,360	800
36^k	>100,000	ND	7450	1150	>100,000	ND
37^k	45,900	730	1280	250	38,000	1100
38^l	28,800	530	4360	210	44,540	1680
39^l	12,600	450	1660	280	19,400	940
40^m	69,500	1060	6810	100	148,700	1580

^{a} 1:1 receptor–carbohydrate association constant; ^{b} 2:1 receptor–carbohydrate association constant;
^{c} 1:2 receptor–carbohydrate association constant; ^{d} Data cited from Ref. 79; ^{e} Data cited from Ref. 80;
^{f} Not tested; ^{g} Data cited from Ref. 81; ^{h} Data cited from Ref. 82; ^{i} Not detected; ^{j} Data cited from Ref. 83;
^{k} Data cited from Ref. 84; ^{l} Data cited from Ref. 85; ^{m} Data cited from Ref. 86.

respectively, the primary amide, amino, and hydroxyl groups instead of one aminopyridine unit of **28**.[80] These functional groups were incorporated as analogues of the side chains of Asn, Gln, Lys, and Ser, which are often observed in the carbohydrate-binding motifs of natural lectins. While the binding profiles of **30** and **31** with the amino and hydroxyl groups, respectively, were almost similar to that of the parent compound **28**, significant enhancement in affinity for Oct-β-Glc was observed in **29** with two primary amide groups (Table III). Receptors **32** and **33** containing imidazole and indole rings, which mimic the side chains of His and Trp, respectively, were also tested for their binding to Oct-β-Glc, Oct-α-Glc, and Oct-β-Gal.[81] Marked enhancements in binding to all glycosides were observed in both compounds. ^{1}H-NMR analysis suggested that CH/π interaction might contribute significantly to the complexation of **32** and **33** with Oct-β-Glc. On the other hand, introduction of an unnatural phenanthroline ring to the hydrophilic unit (**34**) was shown to be particularly effective in the binding to Oct-β- and Oct-α-Glc.[82] Molecular-modeling

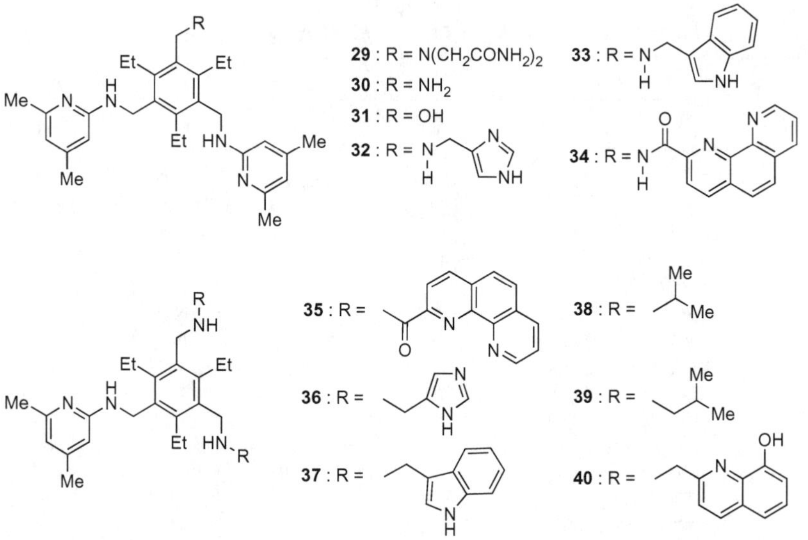

FIG. 11. Tripod-type boronic acid-independent lectin mimics (**29–40**).

calculations suggested that the binding pocket of **34** has the correct shape and size for encapsulation of these glucopyranosides.

Based on the observation that the phenanthroline, imidazole, and indole rings were valuable building blocks for hydrophilic units, a new series of receptors (**35–37**) containing two units of these building blocks were designed.[83,84] Although the affinities of both **34** and **35**, having, respectively, one and two phenanthroline units, for Oct-β-Glc and Oct-α-Glc were too large to be calculated in chloroform, titration experiments in 5% Me$_2$SO–chloroform revealed that **35** was a superior β-Glc binder to **34** (K_{11} for **34** = 36,530 M^{-1}, K_{11} for **35** = 78,400 M^{-1}). These results indicate that the incorporation of the second phenanthroline unit causes further enhancement of the binding affinity of the receptor. On the other hand, significant changes in binding preference were observed in **36** and **37**, containing, respectively, two imidazole and indole units. Whereas the affinities of these receptors for Oct-β-Glc were markedly decreased as compared to those of **32** and **33** with, respectively, one imidazole and indole unit, they displayed a high level of affinity toward Oct-β-Gal. The binding selectivity for Oct-β-Gal was further increased in **38** and **39**, which have the alkyl side chains of Val and Leu in the hydrophilic units, respectively.[85] Their binding affinities for Oct-β-Gal were slightly higher than those for Oct-β-Glc. Van der Waals contacts between the branched alkyl groups and glycosides probably contribute to the binding

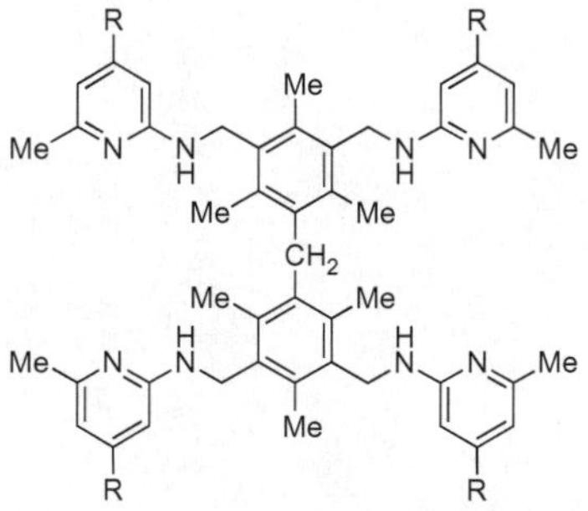

FIG. 12. Tetrapod-type boronic acid-independent lectin mimics (**41, 42**).

capacity of **38** and **39**. The most effective receptor for Oct-β-Gal reported to date is **40**, a compound with two 8-hydroxyquinoline units.[86] Receptor **40** displayed more than 2-fold selectivity for Oct-β-Gal over Oct-β-Glc and Oct-α-Glc, and its affinity for Oct-β-Gal was extremely potent. In addition to these tripod-type receptors for monosaccharides, Mazik's group has developed the tetrapod-type receptors **41** and **42** (Fig. 12).[88] Combination of the dimesitylmethane scaffold and four aminopyridine units was expected to provide a cavity suitable for disaccharides. Indeed, these receptors exhibited a strong binding preference for n-dodecyl β-maltoside ($K_{21} \geq 100{,}000$ M^{-1} for **41** and **42**) over Oct-β-Glc ($K_{12} = 630$ M^{-1} for **41**, $K_{12} = 560$ M^{-1} for **42**) in chloroform, indicating that these relatively large architectures fit well to the disaccharide structures.

Roelens and coworkers have developed a series of pyrrolic tripodal receptors using the triethylbenzene scaffold, representatives of which are shown in Fig. 13 (**43–47**). These receptors were evaluated against a set of n-octyl glycosides of biologically relevant monosaccharides, including Glc, Gal, Man, and GlcNAc. Roelens's group has assessed the binding affinities of the receptors using the original parameter, BC_{50}^{0} (intrinsic median binding concentration) descriptor, which is defined as the total concentration of receptor necessary for binding 50% of the ligand when the fraction of bound receptor is zero.[89] The BC_{50}^{0} value can be viewed as a global dissociation constant in chemical systems involving any number of complex species. Table IV summarizes BC_{50}^{0} values of **43–47** for the set of n-octyl glycosides in chloroform or acetonitrile. The progenitors **43** and **44**, having imino and amino groups, respectively, displayed modest to high affinity for all n-octyl glycosides in chloroform.[89] Both receptors had a preference for Oct-β-Glc and Oct-β-GlcNAc, indicating that the pyrrolic tripodal architecture is well suited for binding to carbohydrates having all-equatorial arrays of polar functionality. However, the binding selectivity was

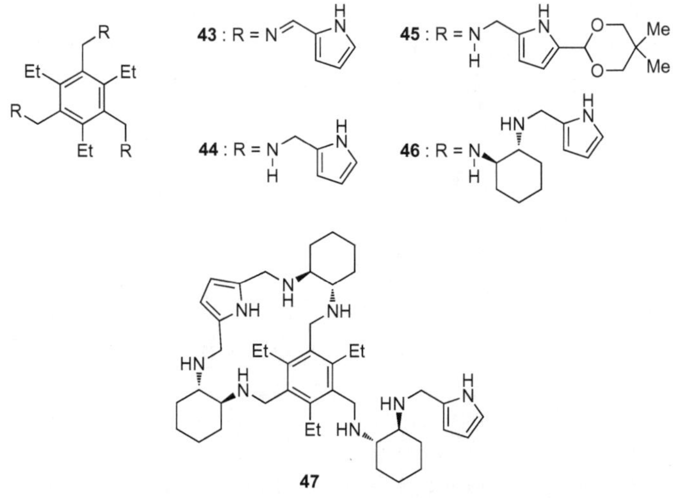

FIG. 13. Tripod-type boronic acid-independent lectin mimics (**43–47**).

TABLE IV
Intrinsic Median Binding Concentrations of **43–47** for *n*-Octyl D-Glycosides in Chloroform or Acetonitrile

Receptor	BC_{50}^{0} (μM)							
	α-Glc	β-Glc	α-Gal	β-Gal	α-Man	β-Man	α-GlcNAc	β-GlcNAc
43[a,b]	268	4.8	368	120	262	660	1179	30
44[a,b]	570	24	790	70	43	37	72	18
45[a,c]	570	39	2250	185	2.8	<1	6.4	6.9
45[c,d]	25,600	7940	28,300	10,290	5850	680	5880	6990
46[d,e]	1251	929	1245	2450	127	873	1070	905
47[d,e]	2493	1248	3183	1532	286	83	1573	1298

[a] Measured in CDCl$_3$; [b] Data cited from Ref. 89; [c] Data cited from Ref. 90; [d] Measured in CD$_3$CN; [e] Data cited from Ref. 91.

completely different in **45**, which has an acetal substituent in the α position of the pyrrole.[90] Receptor **45** showed a striking affinity for Oct-β-Man, which was too strong to be evaluated in chloroform. Titration experiments in acetonitrile revealed that the affinity of **45** for Oct-β-Man was almost one order of magnitude greater than those of the other glycosides. NMR experimental data combined with molecular-modeling calculations suggested that this notable selectivity of **45** for Oct-β-Man was derived from a conformational restriction induced by the cyclic acetal substituents on the pyrrole unit.[92] The resulting narrower binding-cleft of **45** fits preferentially to Oct-β-Man. Based on these promising results, Roelens's group developed a new generation of Man-selective receptors incorporating chiral *trans*-1,2-diaminocyclohexane units (**46**, **47**).[91,93,94] Receptors **46** and **47** displayed substantial affinity and selectivity for Oct-α- and Oct-β-Man, respectively. Replacement of the diamines of **46** and **47** with the corresponding enantiomers decreased both affinity and selectivity, indicating that the chirality of the receptor plays a key role in determining the recognition properties of the receptors. In regard to affinity and selectivity, **46** and **47** are the most effective Man receptors reported to date.

Miller and coworkers have generated a novel class of tripod-type receptors having a *cis*-1,3,5-trisubstituted cyclohexane scaffold (Fig. 14).[95] To achieve the conformational restriction of the receptors, three pyridine (**48**) or quinoline (**49**) rings were directly attached to the cyclohexane ring. ^{1}H NMR, UV, and fluorescence titration experiments demonstrated that both compounds strongly bound to *n*-octyl glycosides in chloroform and even in a polar solvent, methanol (Table V). It is noteworthy that **48** displayed the highest binding affinity reported to date for Oct-α-Glc in chloroform and retained its micromolar range of affinities in methanol. These results indicate that cyclohexane ring can also serve as an effective scaffold for designing tripod-type receptors.

Another representative architecture of boronic acid-independent lectin mimics is the cage type, which consists of two parallel aromatic surfaces held apart by hydrophilic groups (Fig. 15).[16,96] Carbohydrates can be fully enclosed by this architecture, in which the roof and floor of the aromatic rings provide apolar surfaces for CH/π interaction, and polar pillars of hydrophilic units are capable of hydrogen bonding with carbohydrates. Because of this structural feature, the cage-type receptors

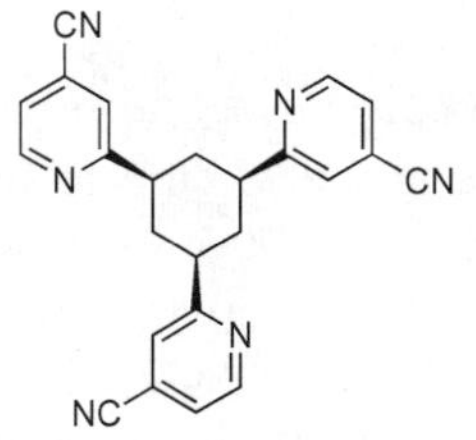

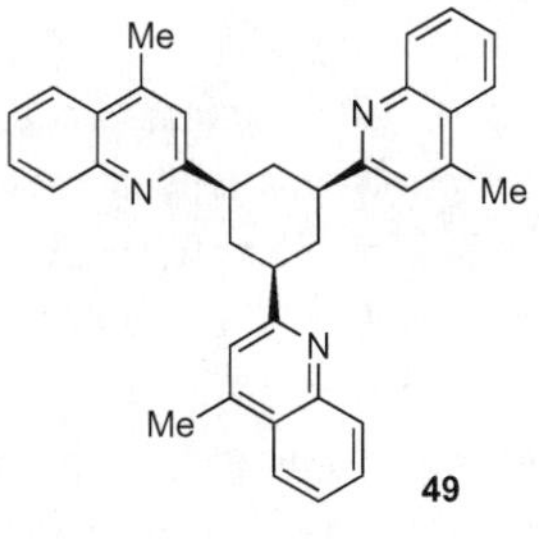

FIG. 14. Tripod-type boronic acid-independent lectin mimics (**48**, **49**).

TABLE V
Binding Constants of 48 and 49 for n-Octyl D-Glycosides in Chloroform or Methanol

Receptor	K_a (M^{-1})^a					
	α-Glc	β-Glc	α-Gal	β-Gal	α-Man	β-Man
48^b	212,000	29,000	17,700	10,900	6500	ND^c
48^d	52,000	9700	9600	5200	>1300	ND
49^d	15,300	9200	6600	3200	2600	>400

^{a} Data cited from Ref. 95; ^{b} Measured in CHCl$_3$; ^{c} Not determined; ^{d} Measured in CH$_3$OH.

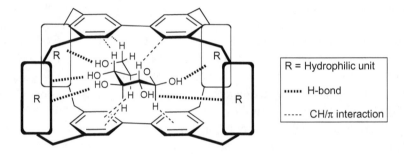

FIG. 15. Typical architecture of cage-type boronic acid-independent lectin mimics.

specifically recognize carbohydrates such as β-Glc and β-GlcNAc that bear all-equatorial arrays of polar functionality and all-axial arrays of hydrophobic CH groups.

In a pioneering study, Davis and coworkers constructed a prototype **50** having two biphenyl and eight isophthalamide groups (Fig. 16).[97] Molecular modeling suggested that the cavity created by this architecture was sufficiently large to accept a β-Glc molecule, making up to six intermolecular hydrogen bonds and several CH/π interactions. This prediction was confirmed by binding studies employing fluorescence-titration experiments in chloroform. Receptor **50** bound Oct-β-Glc with substantial affinity (K_a=300,000 M^{-1}) and reasonable selectivity over Oct-α-Glc (K_a=13,000 M^{-1}) and Oct-β-Gal (K_a=100,000 M^{-1}). Moreover, NMR experiments revealed that **50** was capable of binding Oct-β-Glc (K_a=980 M^{-1}), even in the presence of methanol (CDCl$_3$:CD$_3$OH=92:8). Davis's group has also developed an extended variant of **50** (compound **51**) as a selective receptor for disaccharides having

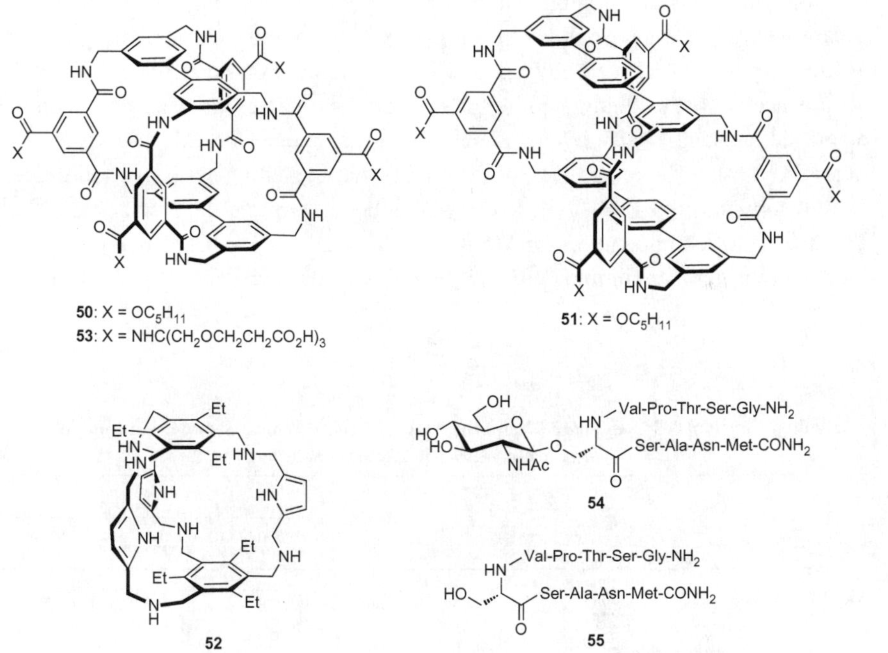

50: X = OC$_5$H$_{11}$
53: X = NHC(CH$_2$OCH$_2$CH$_2$CO$_2$H)$_3$

51: X = OC$_5$H$_{11}$

FIG. 16. Cage-type boronic acid-independent lectin mimics (**50–53**) and (glyco)peptides (**54, 55**).

all-equatorial arrays of polar functionality.[98] While no binding was detected with Oct-β-Glc, n-octyl β-lactoside [β-Gal-(1→4)-Glc], n-octyl β-maltoside [α-Glc-(1→4)-Glc], and n-dodecyl α-maltoside in CDCl$_3$:CD$_3$OH (92:8), compound **51** significantly bound n-octyl β-cellobiose [β-Glc-(1→4)-Glc] with K_a of 7000 M^{-1}.

Roelens and coworkers also reported a cage-type receptor incorporating aminopyrrolic groups (**52**, Fig. 16), which was active in an organic solvent.[99] The roof and floor of **52** were constructed from 1,3,5-triethylbenzene, which has been extensively employed for the core scaffold in tripod-type receptors, as already discussed. Although the binding cavity of **52** is smaller relative to that of Davis's receptor (**50**), compound **52** was capable of enclosing Oct-β-Glc with a K_a of 48,300 M^{-1} in chloroform. Moreover, compound **52** was found not to bind either Oct-α-Glc or both isomers of Oct-Gal and Oct-Man, indicating that the architecture of **52** is also effective for the recognition of carbohydrates having an all-equatorial array of polar functionality.

Based on the promising results of this early work, the Davis group investigated the effectiveness of the cage architecture in aqueous media. After the successful development of a water-soluble variant of compound **50** (**53**, Fig. 16),[100] in which a tricarboxylate group is installed in each hydrophilic component, its binding property against a panel of 22 carbohydrates was thoroughly examined (Table VI).[101] In accordance with the binding preference of **50** in organic solvents, compound **53** displayed substantial affinity for methyl β-D-glucopyranoside (Me-β-Glc) in water ($K_a = 28$ M^{-1}). However, it was found that methyl 2-acetamido-2-deoxy β-D-glucopyranoside (Me-β-GlcNAc) was a better target for **53** ($K_a = 630$ M^{-1}). The affinity of compound **53** for Me-β-GlcNAc was comparable to that of a natural lectin, wheat-germ agglutinin (WGA, $K_a = 730$ M^{-1}), which has classically been used

TABLE VI
Binding Constants of Compound 53 and the Natural Lectin, Wheat-Germ Agglutinin (WGA), for Carbohydrates in Aqueous Solution

Carbohydrate	K_a (M^{-1})^a	
	53	WGA
Me-β-GlcNAc	630^b	730^c
N-Acetyl-D-glucosamine (GlcNAc)	56^b	410^d
Methyl β-D-glucopyranoside	28^b	
Me-α-GlcNAc	24^e	480^c
Cellobiose	17^b	
D-Glucose	9^b	
2-Deoxy-D-*arabino*-hexose	7^b	
Methyl α-D-glucopyranoside	7^b	
D-Xylose	5^b	
D-Ribose	3^b	
D-Galactose	2^b	
L-Fucose	2^b	
N-Acetyl-D-galactosamine (GalNAc)	2^b	60^c
N-Acetyl-D-mannosamine (ManNAc)	2^b	60^c
D-Arabinose	2^b	
D-Lyxose	$\leq 2^b$	
D-Mannose	$\leq 2^b$	
L-Rhamnose	$\leq 2^b$	
Maltose	$\leq 2^b$	
Lactose	$\leq 2^b$	
N-Acetylmuramic acid (MurAc)	0^b	
N-Acetylneuraminic acid (Neu5Ac)	0^b	560^b
N,N'-Diacetylchitobiose	0^b	5,300^d

^{a} Data cited from Ref. 101; ^{b} Measured by ^{1}H NMR titration in D$_2$O; ^{c} Estimated from inhibitory activity of WGA precipitation induced by N-acetyl-D-glucosamine; ^{d} Measured by isothermal titration calorimetry in H$_2$O; ^{e} Measured by induced circular dichroism in H$_2$O.

to bind GlcNAc residues. It is also noteworthy that **53** was far more discriminatory than WGA. Whereas WGA exhibits significant binding to several 2-acetamido sugars, including GlcNAc, Me-α-GlcNAc, *N*-acetylneuraminic acid, and *N,N'*-diacetyl-D-chitobiose, the affinities of **53** for other 21 carbohydrates were more than one order of magnitude weaker than that for Me-β-GlcNAc. Since β-GlcNAc attached to the hydroxyl group of serine or threonine is a common posttranslational modification of proteins, binding experiments of **53** with a glycopeptide containing a GlcNAc residue were also conducted using **54** (Fig. 16). This was based on the sequence from GlcNAc-linked casein kinase II.[102] ^{1}H NMR titration experiments in D_2O demonstrated that **53** bound to **54** with significant affinity ($K_a = 1040$ M^{-1}). On the other hand, binding to the aglycosyl peptide **55** was negligible, supporting the conclusion that **53** binds **54** through the GlcNAc residue.

Elegant work by Davis and coworkers has also developed a water-soluble cage-type receptor for disaccharides.[103] Although the extended receptor **51** was successful in discriminating *n*-octyl β-cellobiose from monosaccharides and other disaccharides in chloroform, the cavity of this relatively flexible architecture had a tendency to collapse in aqueous solution by the hydrophobic contacts of the aromatic rings of the roof and floor. To avoid this phenomenon, Davis's group employed a *meta*-terphenyl structure in place of the *para*-terphenyl group of **51** and incorporated the fifth isophthalamide spacer. The resulting "rigid" variant of **51** (**56**, Fig. 17) bearing carboxylate groups was shown to bind cellobiose ($K_a = 560$–600 M^{-1}) and methyl β-cellobioside ($K_a = 850$–910 M^{-1}) with good affinities in aqueous solutions (Table VII). The binding affinities of compound **56** for carbohydrates were evaluated by ^{1}H NMR, induced circular dichroism (ICD), and fluorescence titrations. To obtain

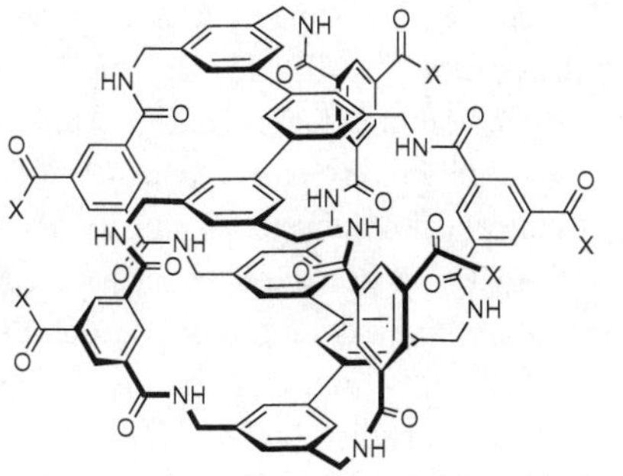

56: X = NHC(CH$_2$OCH$_2$CH$_2$CO$_2$H)$_3$ 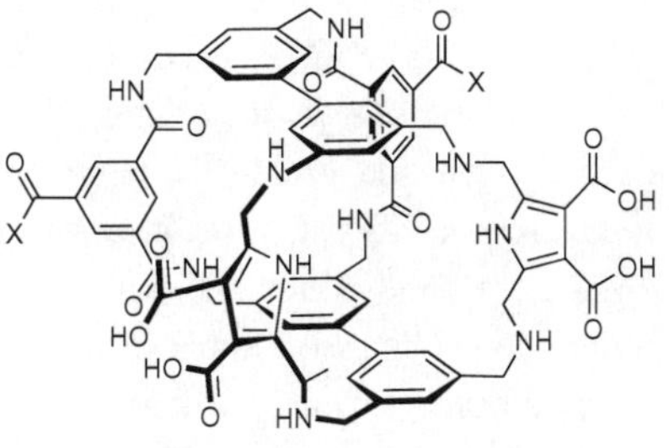   **57**: X = NHC(CH$_2$OCH$_2$CH$_2$CO$_2$H)$_3$

FIG. 17. Cage-type boronic acid-independent lectin mimics (**56, 57**).

TABLE VII
Binding Constants of Compound 56 for Carbohydrates in Aqueous Solution, as Measured by ^{1}H NMR, ICD, and Fluorescence Titrations

Carbohydrate	K_a (M^{-1})^a		
	^{1}H NMR	ICD	Fluorescence
Cellobiose	600	580	560
Methyl β-cellobioside		910	850
Xylobiose		250	270
N,N′-Diacetylchitobiose	120		120
Lactose		11	14
Mannobiose		13	9
Maltose		15	11
Gentiobiose		12	5
Trehalose	ND^b	ND	
Sucrose		ND	ND
D-Glucose	11	12	ND
D-Ribose		ND	ND
N-Acetyl-D-glucosamine	24		19

^{a} Data cited from Ref. 103; ^{b} Not detected.

reliable K_a values, at least two methods were used for each carbohydrate. While **56** showed moderate affinities for xylobiose ($K_a = 250$–270 M^{-1}) and N,N′-diacetylchitobiose ($K_a = 120$ M^{-1}), both of which are also all-equatorial disaccharides, only weak binding was detected for monosaccharides and other disaccharides ($K_a \leq 25$ M^{-1}). Interestingly, compound **56** can distinguish between cellobiose and lactose, the structures of which differ from each other only by one asymmetric center.

In subsequent work, an interesting hybrid molecule of **52** and **53** has been generated.[104] In the new water-soluble receptor **57** (Fig. 17), two isophthalamide spacers of **53** are replaced by two 2,5-bis(aminomethyl)pyrrole (BAMP) spacers of **52**. The isophthalamide and BAMP spacers are similar to each other in length, but very different in hydrogen-bonding characteristics. In the isophthalamide spacer, two sets of the amide hydrogen and amide carbonyl groups can act as hydrogen-bond donors and acceptors, respectively. On the other hand, the BAMP spacer contains one pyrrole ring as a hydrogen-bond donor and two amino groups, which can act either as donors or acceptors, depending on the pH of the solution. ^{1}H NMR titration experiments under basic conditions (pD = 13) demonstrated that the affinity of **57** for D-glucose ($K_a = 18$ M^{-1}) was twice as high as that of **53** ($K_a = 9$ M^{-1}). However, **57** bound weakly to Me-β-Glc ($K_a = 4$ M^{-1}) and GlcNAc ($K_a = 7$ M^{-1}), in contrast to **53** ($K_a = 28$ M^{-1} for Me-β-Glc, $K_a = 56$ M^{-1} for GlcNAc). Although the binding test under neutral conditions has not been performed because of solubility problems with **57**, this study suggests that selectivity among all-equatorial carbohydrates can be tuned though modification of the hydrogen-bonding property of the cage-type receptors.

As already discussed, the affinity and selectivity of some cage-type receptors for carbohydrates are almost comparable with those of natural lectins. In this regard, the cage-type receptors are often called as "synthetic lectins." Since synthetic lectins are superior to natural lectins in terms of chemical stability, synthetic accessibility, and ease of structural modification, they have great potential for applications in glycobiology research.

4. Antiviral and Antimicrobial Potential of Boronic Acid-Independent Lectin Mimics

Coupled with the successful development of tripod-type and cage-type lectin mimics, their biomedical applications have begun to be explored. When considering their application to anti-HIV and antimicrobial therapies, it would be desirable to design receptors capable of binding to Man, which is a main component of high mannose-type oligosaccharides on the HIV envelope, and cell-wall mannans of pathogenic yeasts and bacteria.[105,106] In this respect, the cage-type receptors are unlikely to be useful in applications as anti-HIV and antimicrobial agents because of their weak preference for Man. On the other hand, several tripod-type receptors have been shown to bind Man in organic solvents. Although none of tripod-type receptors thus far reported can bind carbohydrates in aqueous media, the tripod-type architecture has an advantage with regard to its simplicity in structure, which allows for easy preparation of libraries of compounds for systematic studies. By exploiting this advantage, attempts to develop tripod-type receptors with anti-HIV and antimicrobial activities have been recently made.

Pérez-Pérez and coworkers have synthesized a library of tripod-type compounds and have evaluated their anti-HIV activities.[107] The library is classified into three groups, the monomers (**58–66**, Fig. 18), dimers (**67–75**, Fig. 19), and trimers (**76–81**, Fig. 20) of the 1,3,5-triazine motif, incorporating two aromatic amino acid components (Phe, Tyr, or Trp). Each triazine motif is considered to have the tripod-type architecture, in which the 1,3,5-triazine ring is a core scaffold and the amino acid and amine groups can act as hydrophilic components. The carboxylate groups of the amino acid residues are also crucial for increasing the water solubility of the compounds. The dimers and trimers were designed based on the assumption that multivalent interaction can be realized between the two or three triazine motifs and glycans

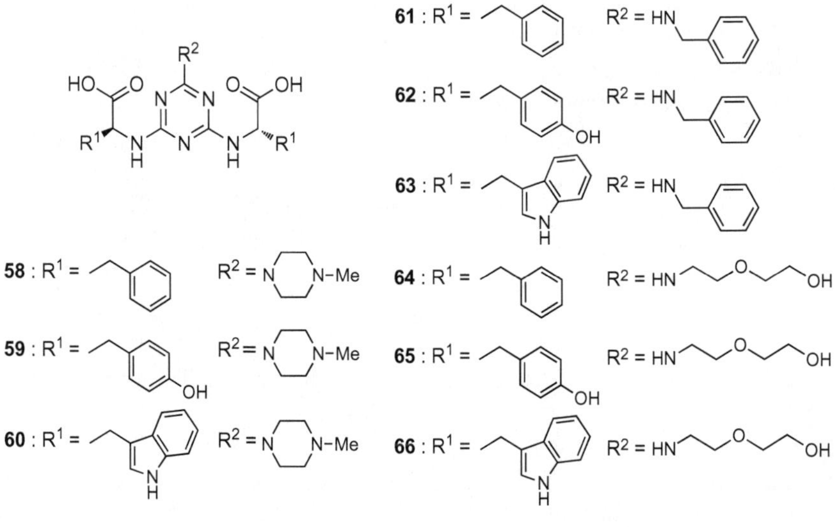

FIG. 18. Monomers (**58–66**) of 1,3,5-triazine compounds.

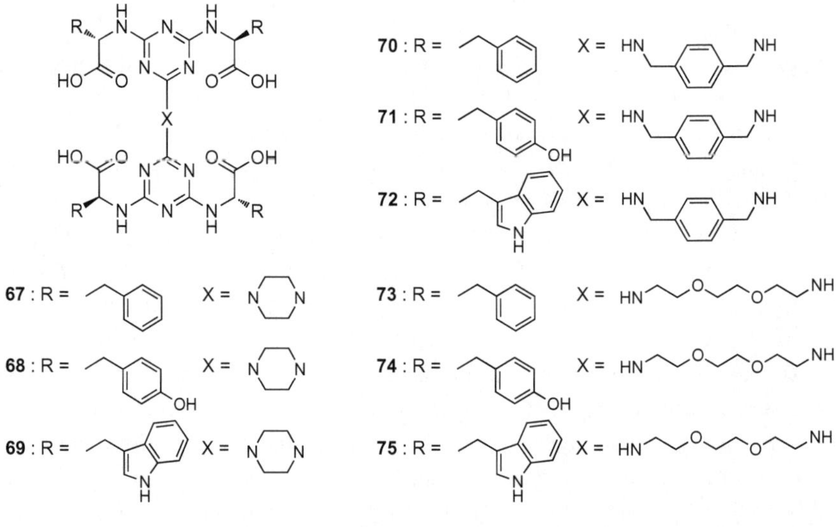

FIG. 19. Dimers (**67–75**) of 1,3,5-triazine compounds.

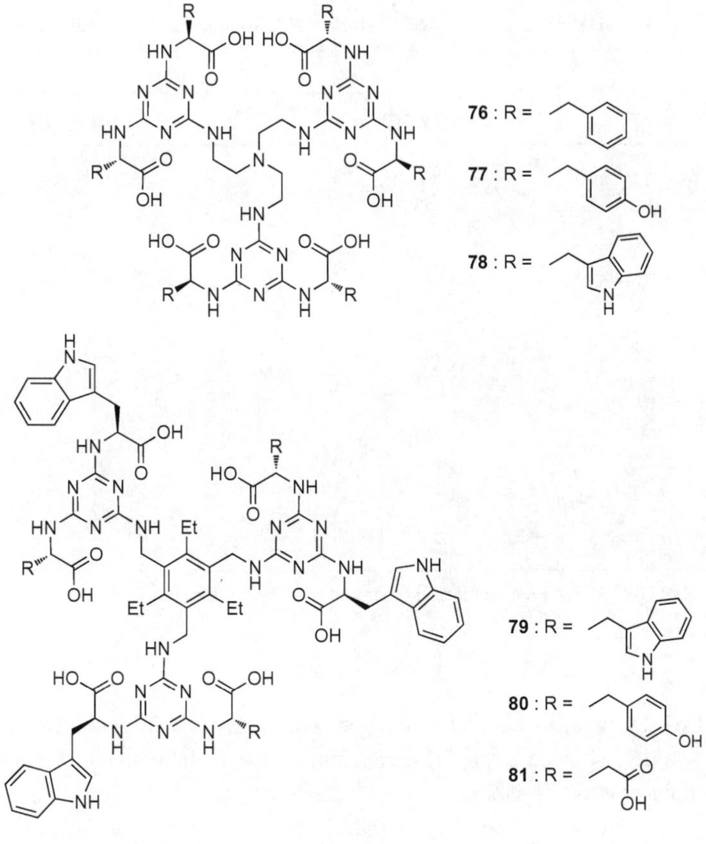

FIG. 20. Trimers (**76–81**) of 1,3,5-triazine compounds.

on the viral envelope. The anti-HIV activity of these compounds was evaluated as the inhibitory activity against HIV-1(III$_B$) and HIV-2(ROD) replication in the CEM T-cell culture (Table VIII). While the monomers (**58–66**) were devoid of anti-HIV activity even at 250 μM, significant activity was observed in several dimers and trimers. Among the dimers, the Trp-containing **69** and **72** showed moderate anti-HIV activity, with EC$_{50}$ values of 65 and 56 μM for HIV-1(III$_B$) and 63 and 81 μM for HIV-2(ROD), respectively. Enhanced activity was observed in the trimers, especially in those having triethylbenzene as the central unit. Thus, **79–81** afforded EC$_{50}$ values of 16–33 μM for both HIV-1(III$_B$) and HIV-2(ROD). The same order in anti-HIV activity (trimer > dimer ≫ monomer) was also observed in binding studies to HIV-1 gp120. SPR analysis revealed that the trimer **79** bound strongly to gp120, with K_d

TABLE VIII
Anti-HIV Activities of 67–81 in Human T-Lymphocyte Cell Cultures

Compound	EC_{50} (μM)[a]	
	HIV-1(III$_B$)	HIV-2(ROD)
67	106	121
68	≥250	≥250
69	65	63
70	>10[b]	>10
71	148	80
72	56	81
73	≥250	≥250
74	>250	>250
75	112	120
76	65	216
77	≥250	≥250
78	20	65
79	16	22
80	26	24
81	18	33

[a] Data cited from Ref. 107; [b] Precipitation was detected at higher concentration.

value of 1.61 μM. On the other hand, the corresponding dimer **72**, which exhibited low antiviral activity, showed a much lower amplitude of binding, and binding of the corresponding inactive monomer **63** was barely detected. The strong correlation for the mono-, di-, and trimer between antiviral activity and binding affinity to HIV-1 gp120 possibly indicates that the anti-HIV activity of these tripod-type compounds is ascribed to the binding to HIV-1 gp120, and that a certain level of multivalent interactions is required to show both antiviral and binding activities. One remaining question is whether the observed affinity to HIV-1 gp120 is due to binding to the glycans or do the polyanionic compounds interact with protein areas of gp120, such as the positively charged V3 loop. Additionally, the anti-HIV activities of these compounds seem still modest compared to the activities of natural lectins. For example, the reported EC_{50} values of CV-N for HIV-1(III$_B$) and HIV-2(ROD) are 0.003 and 0.002 μM, respectively,[68,108] and thus both of these are more than three orders of magnitude lower than those of **79**. Nevertheless, this study provides strong evidence that anti-HIV drug discovery based on the tripod-type architecture is a viable approach.

In other work, Roelens and coworkers have explored the potential of tripod-type receptors as the antimicrobial agents.[109] As already described, Roelens's group has developed a family of aminopyrrolic tripod-type compounds, some of which proved to be effective receptors for Man in polar organic solvents. The pradimicins and benanomicins (see the later section) are small-size Man receptors that occur in nature. They are reported to show potent antimicrobial activity against a wide variety of fungi and yeasts, including clinically important pathogens, by binding to mannans on the cell wall. Focusing on the similarity of Man-binding properties of the aminopyrrolic tripod-type compounds and pradimicin-related natural products, Roelens's group performed the first investigation on the antimicrobial activity of aminopyrrolic tripod-type receptors. In preliminary screening, the progenitor of the family, compound **44**, and the most effective Man receptors, **46** and **47**, were tested for their antimicrobial activities against *Candida tropicalis*, *Pichia norvegensis*, *Prototheca wickerhamii*, and *Prototheca zopfii* (Table IX). All compounds were shown to inhibit markedly the growth of the four microorganisms. In particular, the minimum inhibitory concentration (MIC) values of **44** were comparable for those of the well-known antibiotics, amphotericin B and ketoconazole. Structural segments (**82–87**, Fig. 21) of compound **44** were less active or inactive, indicating that all structural components are essential for antimicrobial activity. Analogues of **44** were also prepared, and representative examples (**88–92**) are shown in Fig. 22. Antimicrobial assay using a variety of yeast and yeast-like microorganisms revealed that none of the modifications, including introduction of functional groups into the pyrrole rings (**88–90**) and variation of substitution pattern on the benzene platform (**91, 92**), enhanced the activity of the parent compound **44** (Table X). The binding affinity of these compounds for Oct-α-Man was also evaluated in acetonitrile. Although this affinity may not reflect the binding for mannans in water, a broad correlation was detected between the antimicrobial activity and Man-binding affinity; compounds without antimicrobial activity lacked Man-binding ability, whereas compounds having antimicrobial

TABLE IX
Antibiotic Activities of Compounds 44, 46, 47, and 82–87 Against Yeast and Yeast-Like Microorganisms

Species	MIC (μg/mL)[a]								
	44	46	47	82	83	84	85	86	87
Candida tropicalis	4	8	16	512	128	64	256	ND[b]	ND
Pichia norvegensis	2	4	16	512	64	16	64	ND	ND
Prototheca wickerhamii	2	4	16	ND	16	64	256	ND	ND
Prototheca zopfii	8	4	16	ND	32	128	256	ND	ND

[a] Data cited from Ref. 109; [b] Not determined (>1024 µg/mL).

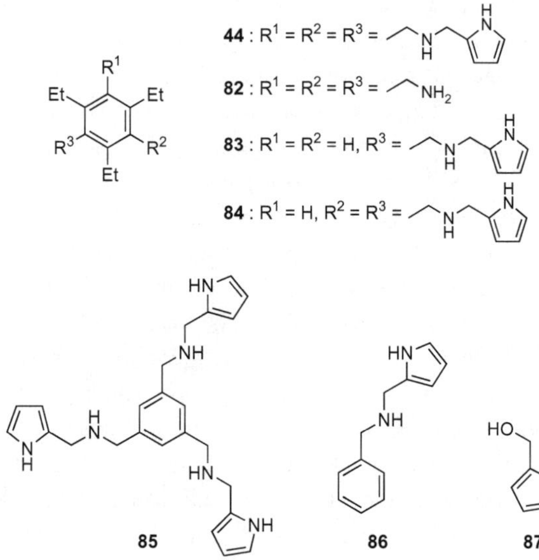

FIG. 21. Aminopyrrolic tripod-type receptor (**44**) and its structural segments (**82–87**).

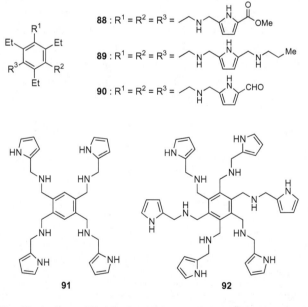

FIG. 22. Analogues (**88–92**) of the aminopyrrolic tripod-type receptor (**44**).

TABLE X
Antibiotic Activities of 44 and 88–92 Against Yeast and Yeast-Like Microorganisms

Species	MIC (μg/mL)[a]					
	44	88	89	90	91	92
Candida albicans	16	256	64	512	1024	256
Candida glabrata	8	64	128	ND	1024	256
Meyerozyma guilliermondii	8	64	64	512	1024	64
Candida parapsilosis	8	64	ND[b]	ND	ND	512
Candida tropicalis	4	32	8	128	512	32
Pichia norvegensis	2	8	8	32	64	8
Clavispora lusitaniae	32	128	128	512	ND	256
Yarrowia lipolytica	8	128	32	256	512	64
Oichia kudriavzevii	16	64	128	128	512	64
Kluyveromyces marxianus	8	8	8	128	256	32
Prototheca wickerhamii	2	128	256	256	256	32
Prototheca zopfii	8	128	256	1024	256	64

[a] Data cited from Ref. 109; [b] Not determined (>1024 µg/mL).

activity showed detectable binding for Man. However, perfect correlation was not obtained. On the basis of these results, Roelens and coworkers proposed the interesting hypothesis that Man binding of the tripod-type compounds would facilitate the approach to the microbial cells. This would trigger a process that would subsequently involve penetration across the membrane into the cytoplasm, where the compounds might exert antimicrobial activity by interacting with a specific target. Although further research is necessary to validate this hypothesis, this pioneering study emphasizes the promising potential of tripod-type carbohydrate receptors as antimicrobial agents having a novel mode of action.

III. Naturally Occurring Lectin Mimics

1. Antimicrobial and Carbohydrate-Binding Profiles of Pradimicins and Benanoimicins

While there are a number of publications describing synthetic lectin mimics, only a few natural products possessing carbohydrate-binding ability have been reported. Pradimicins (PRMs) and benanomicins are the only family of naturally occurring lectin mimics with non-peptidic skeletons. In 1988, three PRMs (PRM-A, B, C) and two benanomicins (benanomicin A, B) were first isolated from actinomycetes as potential antibiotics, by Oki's and Kondo's groups, respectively.[110,111] Interestingly, these compounds were found to share a common aglycon, namely benzo[*a*]naphthacenequinone, and differ only in the glycon components (Fig. 23). Since then, benanomicin B was shown to be the same as PRM-C,[112,113] and 12 additional members of the PRM class demonstrating antimicrobial activity have been identified.[114–124] These structurally unique natural products show potent *in vitro* antimicrobial activity against a wide variety of yeasts and fungi, and high *in vivo* therapeutic efficacy in murine

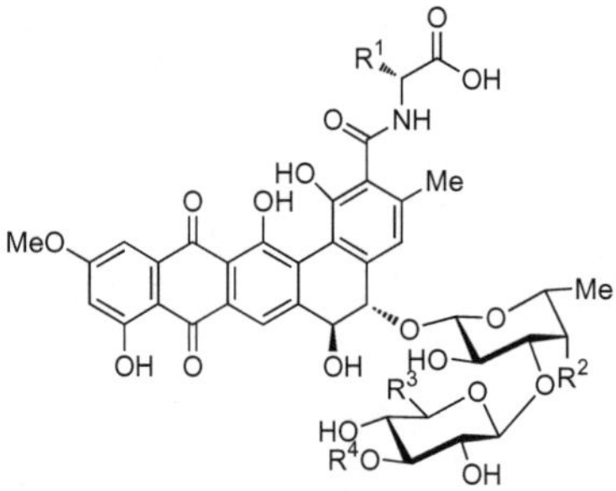

PRM-A : R^1 = Me, R^2 = NHMe, R^3 = R^4 = H
Benanomicin A : R^1 = Me, R^2 = OH, R^3 = R^4 = H
PRM-C (Bemanomicin B) : R^1 = Me, R^2 = NH_2, R^3 = R^4 = H
PRM-D : R^1 = H, R^2 = NHMe, R^3 = R^4 = H
PRM-E : R^1 = H, R^2 = NH_2, R^3 = R^4 = H
PRM-FA1 : R^1 = CH_2OH, R^2 = NHMe, R^3 = R^4 = H
PRM-FA2 : R^1 = CH_2OH, R^2 = NH_2, R^3 = R^4 = H
PRM-L : R^1 = Me, R^2 = NHMe, R^3 = OH, R^4 = H
PRM-FL : R^1 = CH_2OH, R^2 = NHMe, R^3 = CH_2OH, R^4 = H
PRM-FS : R^1 = CH_2OH, R^2 = NHMe, R^3 = CH_2OH, R^4 = SO_3H
PRM-S : R^1 = Me, R^2 = NHMe, R^3 = CH_2OH, R^4 = SO_3H

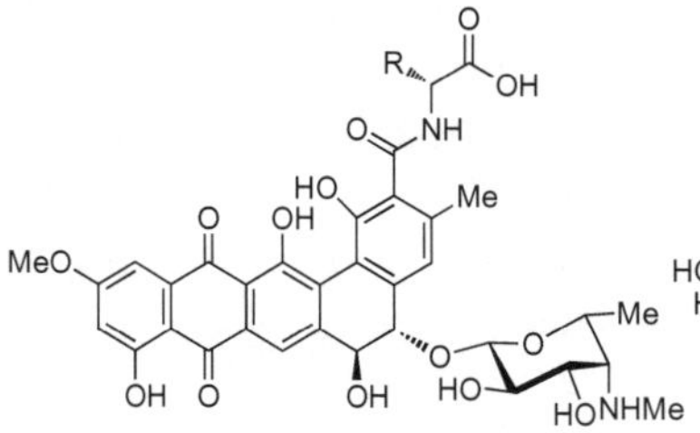

PRM-B : R = Me
PRM-FB : R = CH_2OH

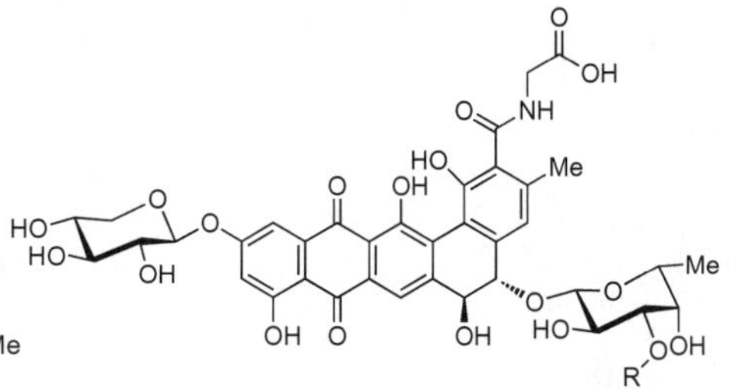

PRM-T1 : R = H
PRM-T2 : R = D-Xyl

FIG. 23. Pradimicins (PRMs) and benanomicins.

models against *Candida albicans*, *Cryptococcus neoformans*, and *Aspergillus fumigatus*. Tables XI and XII list the *in vitro* and *in vivo* antimicrobial profiles of PRM-A. Since there is no cross-resistance to other antimicrobial agents, such as amphotericin B, 5-fluorocytosine, and ketoconazole, this family of compounds is chemically and functionally different from the other major classes of antimicrobial agents. The initial biochemical study using *C. albicans* found that PRM-A causes K^+ leakage from the cells and binds to the outer surface of *C. albicans* cells in a Ca^{2+}-dependent manner.[126] The absorption of PRM-A on *C. albicans* cells was found to

TABLE XI
In Vitro Antibiotic Activities of PRM-A and BMY-28864 (93)

Species	MIC (μg/mL)[a]	
	PRM-A	BMY-28864 (93)
Saccharomyces cerevisiae		
ATCC 9763	12.5	3.1
Candida albicans		
IAM 4888	50	6.3
A9540	50	6.3
ATCC 38247	3.1	0.8
ATCC 32354	12.5	6.3
Candida tropicalis		
IFO 10241	>100	50
CS-07	25	6.3
Candida parapsilosis		
CS-08	>100	3.1
Candida krusei		
A15052	25	3.1
Cryptococcus neoformans		
D49	1.6	1.6
IAM 4514	0.8	1.6
CS-01	1.6	1.6
Aspergillus fumigatus		
IAM 2034	3.1	3.1
IAM 2530	1.6	3.1
Aspergillus flavus		
FA 21436	6.3	6.3
CS-18	25	100
Sporothrix schenckii		
IFO 8158	1.6	3.1
Trichophyton mentagrophytes		
No. 4329	6.3	12.5
D155	3.1	12.5

[a] Data cited from Ref. 125.

TABLE XII
In Vivo Activities of PRM-A and BMY-28864 (93) Against *Candida*, *Cryptococcus*, and *Aspergillus* Systematic Infection in Mice ($n = 5$)

Species	PD_{50} (mg/kg)[a,b]	
	PRM-A	BMY-28864 (93)
Candida albicans A9540	8.9	9
Cryptococcus neoformans IAM 4514	11	11
Aspergillus fumigatus IAM 2034	16	36

[a] 50% protective dose; [b] Data cited from Ref. 125.

originate from binding to mannan in the yeast cell wall. These results collectively suggest that the antimicrobial action of PRM-A is associated with alteration of the membrane permeability by binding to cell-wall mannan. This putative mechanism explains well the fact that PRM-A does not induce K^+ leakage from human erythrocytes or other mammalian cells in the presence of Ca^{2+} ion.

The structure–microbial activity relationship of PRM-A has been also thoroughly examined by using naturally occurring congeners and artificial derivatives (Fig. 24). Replacement of the D-Ala component with L-amino acids, including L-Ala, abolished antimicrobial activity.[127,128] On the other hand, D-α-aminobutanoic acid, D-Phe, D-Asp, D-2,3-diaminopropanoic acid, D-Lys, Gly, and D-Ser derivatives retained the activity, indicating that stereochemistry at position 17 is crucial for the antimicrobial

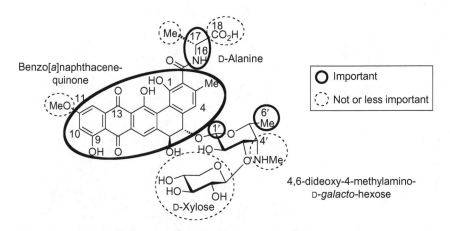

FIG. 24. Structural requirements for antifungal activity of PRM-A.

action of PRM-A, while the side chain of D-Ala plays only a marginal role. The amide hydrogen at position 16 seems to be important, because a complete loss of activity was observed in sarcosine (*N*-methylglycine) and D-Pro derivatives. The role of the carboxyl group at position 18 was also examined, using the ester and amide derivatives.[129] Whereas the esters, butyl amide, and dipeptidyl derivatives showed much diminished activity, the amide, *N*-methylamide, and *N*,*N*-dimethylamide derivatives were as effective as PRM-A. These observations suggest that the carboxyl group at position 18 is not essential, but there are some electronic and steric requirements around this group. On the other hand, the antimicrobial activity of PRM-A is quite sensitive to chemical modifications of the benzo[*a*]naphthacenequinone group.[130] Methylation of the 1- and 9-hydroxyl groups, substitution at positions 4 and 10, and reduction of 13-ketone group significantly decreased or abolished the activity. The position 11 is the sole site to be modified without losing activity. 11-Demethoxy-, 11-*O*-demethyl-, and 11-*O*-ethyl derivatives were as active as PRM-A. These results indicate that the benzo[*a*]naphthacenequinone group of PRM-A, except for position 11, plays a pivotal role in the antimicrobial action. Regarding the disaccharide group of PRM-A, the D-xylose component is apparently not essential, because PRM-B, which lacks the D-xylose group, is indistinguishable from PRM-A in its antimicrobial activity.[115] Similarly, comparison of PRM-A with benanomicins A and B (PRM-C) suggests that the methylamino group at position 4′ of the 4,6-dideoxy-4-methylamino-D-galactose substituent is of little importance,[110] which is further supported by the observations that the activity is retained after introduction of various alkyl and acyl groups onto the methylamino group.[131] In contrast, antimicrobial evaluations of synthetic derivatives having D-fucose, L-arabinose, and D-galactose in place of the 4,6-dideoxy-4-methylamino-D-galactose group showed that the stereochemistry at position 1′ and the existence of the methyl group at position 6′ are fairly important for the biological activity of PRM-A.[132]

During the foregoing studies on structure–activity relationships of PRM-A, Oki's group has developed a semisynthetic derivative showing augmented biological and physicochemical properties. BMY-28864 (**93**, Fig. 25), which was prepared by reductive *N*-methylation of PRM-FA2, displays an *in vitro* and *in vivo* antimicrobial profile (Tables XI and XII) similar to that of PRM-A, but has greater water solubility than PRM-A (0.02 mg/mL for PRM-A, >20.0 mg/mL for **93** in phosphate-buffered saline containing 0.9 mM $CaCl_2$ and 0.5 mM $MgCl_2$, pH 7.2, 25 °C).[125,133,134] The availability of sample and handling convenience enabled further studies on the mode of antimicrobial action of PRMs by using compound **93**. As the first and foremost step, the carbohydrate binding of **93** was thoroughly examined. In the early studies, PRM-A was found to produce an insoluble precipitate with yeast mannan in the

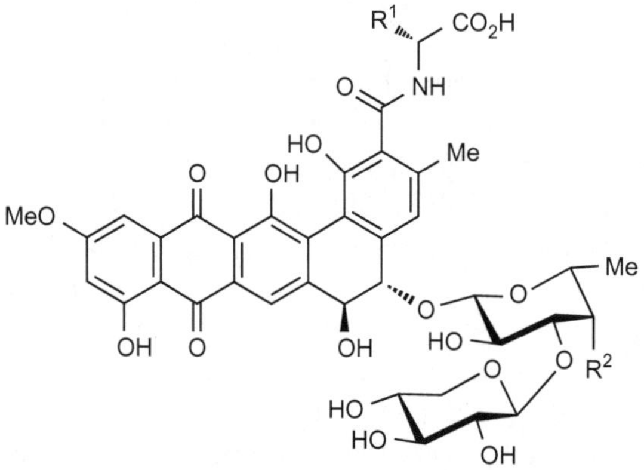

PRM-A : R^1 = Me, R^2 = NHMe
BMY-28864 (**93**) : R^1 = CH_2OH, R^2 = NMe_2

FIG. 25. PRM-A and BMY-28864 (**93**).

presence of Ca^{2+} ion.[126] Thus, quantitative analysis of the precipitate of **93** with Ca^{2+} ion and methyl α-D-mannopyranoside (Me-α-Man) was initially performed.[135] The molar quantity of **93**, Ca^{2+} ion, and Me-α-Man in the precipitate was determined by UV/visible spectrophotometry, atomic absorption spectrometry, and the phenol–sulfuric acid methods, respectively. The molar component ratio of **93**:Ca^{2+}:Me-α-Man was found to be 2:1:4. The same ratio was obtained when the concentration of **93** was varied, or D-mannose was used instead of Me-α-Man for forming the precipitate, indicating that the precipitate of **93** with Ca^{2+} ion and Me-α-Man is a true chemical complex and not a simple mixture. By taking advantage of this complex-forming property, the carbohydrate specificity of **93** was subsequently evaluated.[136] As shown in Table XIII, compound **93** produced precipitates with Me-β-Man, the α and β anomers of p-nitrophenyl D-mannopyranoside, p-aminophenyl D-mannopyranoside, D-fructose, D-arabinose, and D-lyxose, as well as D-mannose and Me-α-Man, while other carbohydrates did not form precipitates at all. The complex formation with these carbohydrates was also confirmed by UV–visible spectrophotometric analysis. It is noteworthy that the carbohydrate specificity of **93** is quite high; the compound has the ability to discriminate D-mannose from L-mannose, and the 2-, 3-, and 4-epimers of D-mannose (D-glucose, D-altrose, and D-talose, respectively), D-mannosamine, and N-acetyl-D-mannosamine. These results indicate that **93** recognizes the 2-, 3-, and 4-hydroxyl groups of D-mannose, and the configuration and substituent at position 1 are not essential. However, it remains unclear as to whether

TABLE XIII
Precipitation of Compound 93 with Carbohydrates in the Presence of Ca^{2+} Ion

Carbohydrate	Precipitation (%) in Water (pH 7.0)
Hexoses	
D-Allose[a]	0
D-Altrose[a]	0
D-Galactose[a]	0
L-Galactose[a]	0
D-Glucose[a]	0
Methyl α-D-glucopyranoside[a]	0
D-Gulose[a]	0
D-Idose[a]	0
D-Talose[a]	0
D-Mannose[a]	100
L-Mannose[b]	0
D-Mannose 6-phosphate[a]	0
Methyl α-D-mannopyranoside[a]	100
Methyl β-D-mannopyranoside[b]	100
p-Nitrophenyl α-D-mannopyranoside[b]	100
p-Nitrophenyl β-D-mannopyranoside[b]	100
p-Aminophenyl α-D-mannopyranoside[b]	100
D-Fructose[a]	100
D-Fucose[b]	0
D-Rhamnose[b]	0
D-Sorbose[a]	0
N-Acetyl-D-glucosamine[a]	0
N-Acetyl-D-mannosamine[a]	0
D-Glucosamine[a]	0
D-Mannosamine[a]	0
Pentoses	
D-Arabinose[b]	100
L-Arabinose[b]	0
D-Ribose[b]	0
D-Lyxose[b]	100
D-Xylose[b]	0
Nonose and disaccharides	
N-Acetylneuraminic acid[a]	0
Lactose[b]	0
Maltose[b]	0
Sucrose[b]	0

[a] Data cited from Ref. 135; [b] Data cited from Ref. 136.

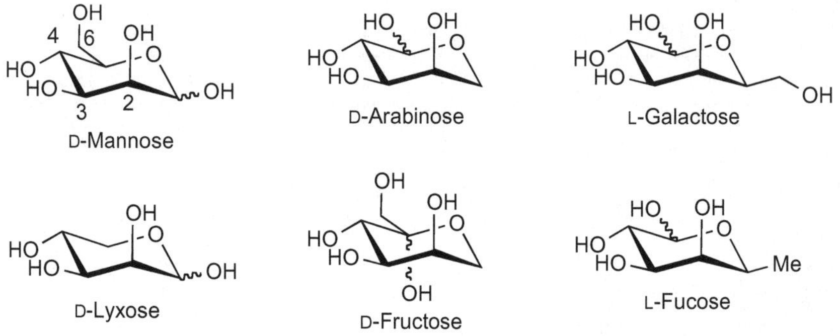

FIG. 26. Comparison of D-mannose with D-arabinose, D-lyxose, D-fructose, L-galactose, and L-fucose.

the 6-hydroxyl group of D-mannose is necessary for binding to **93**. In addition, there is no plausible explanation available concerning the reason why D-fructose, D-arabinose, and D-lyxose formed precipitates with **93**. Oki's group proposed that these carbohydrates have an arrangement similar to 2-, 3-, and 4-hydroxyl groups of D-mannose (Fig. 26), and thus bind to **93** in the same manner as D-mannose. However, L-galactose and L-fucose, both of which do not form precipitates with **93**, can also mimic the array of 2-, 3-, and 4-hydroxyl groups of D-mannose. Further studies are necessary to clarify these issues. In addition, however, the metal specificity of **93** was examined by UV–visible spectrophotometric analysis and two biological assays, yeast-cell absorption, and potassium leakage induction tests.[136] The results collectively suggest that Sr^{2+} and Cd^{2+} ions, which have ionic radii similar to that of the Ca^{2+} ion (1.05 Å for Ca^{2+}, 1.18 Å for Sr^{2+}, 0.99 Å for Cd^{2+}), can also form the ternary complexes with **93** and mannan.

2. Molecular Basis of Carbohydrate Recognition by PRMs

Since the discovery that pradimicins have a C-type lectin-like property, interaction analysis of compound **93** with Ca^{2+} ion and D-mannose has been attempted in an effort to understand the molecular basis of carbohydrate recognition by pradimicins. Oki's group examined the Ca^{2+} salt-forming ability of PRM-A, compound **93**, and their derivatives (**94–100**, Fig. 27) and found that the ester derivatives (**94, 95, 98**) formed no Ca^{2+} salts, whereas PRM-A, **93**, and the other derivatives (**96, 97, 99**, and **100**), having the carboxyl group free, produced the Ca^{2+} salts with a PRM:Ca^{2+} ratio of 2:1.[137] Although the hydroxyl groups on the benzo[*a*]naphthacenequinone group

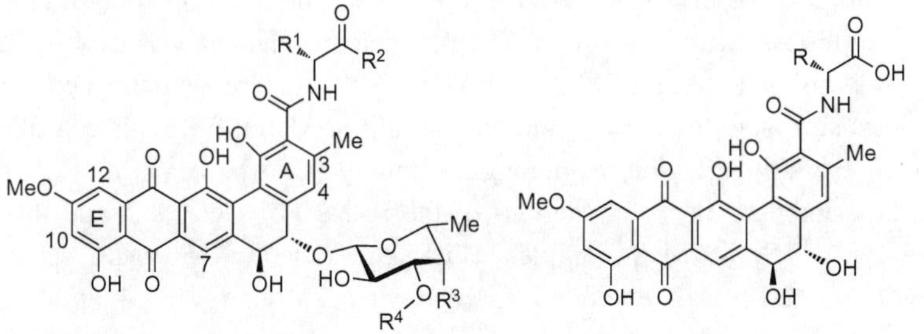

93 : $R^1 = CH_2OH, R^2 = OH, R^3 = NMe_2, R^4 = $ D-Xyl
94 : $R^1 = CH_2OH, R^2 = OMe, R^3 = NMe_2, R^4 = $ D-Xyl
95 : $R^1 = Me, R^2 = OEt, R^3 = NHMe, R^4 = $ D-Xyl
96 : $R^1 = CH_2OH, R^2 = OH, R^3 = NMe_2, R^4 = H$
97 : $R^1 = Me, R^2 = OH, R^3 = NHMe, R^4 = H$
98 : $R^1 = Me, R^2 = OMe, R^3 = NHMe, R^4 = H$

99 : R = OH
100 : R = Me

FIG. 27. Derivatives (**94–100**) of BMY-28864 (**93**).

also have the potential to bind Ca^{2+} ion, these observations suggest that the carboxyl group of PRMs is a possible binding site for Ca^{2+} ion. This hypothesis was partially supported by solution NMR analysis of the formation of the Ca^{2+} salt of **93**.[138–140] Significant change in chemical shifts upon addition of $CaCl_2$ was observed for the proton signals at positions 3-Me, 4, and 7 of **93**. On the other hand, the proton signals at positions 10 and 12 showed little change, indicating that the Ca^{2+}-binding site is near the A ring but not around the E ring of PRMs. The Man-binding process of PRMs was investigated using **93** by Lee and coworkers.[141] UV–visible spectrophotometric analysis demonstrated that two molecules of PRM bind four molecules of Man in two separate steps. As shown in Scheme 4, the [PRM_2/Ca^{2+}] salt initially binds two molecules of Man to form the ternary [$PRM_2/Ca^{2+}/Man_2$] complex, which then incorporates another two molecules of Man to form the ultimate ternary [$PRM_2/Ca^{2+}/Man_4$] complex. This complicated three-component equilibrium, along with the aggregation-forming property of PRMs, has collectively made it difficult to analyze the molecular interaction of PRMs and Man by conventional methods of solution NMR.

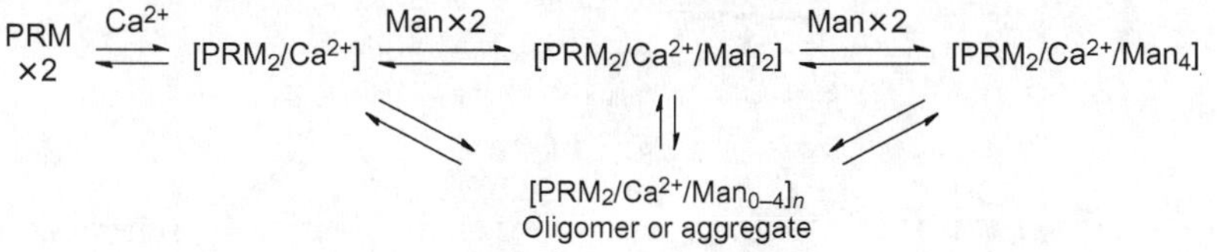

$PRM \times 2 \xrightleftharpoons{Ca^{2+}} [PRM_2/Ca^{2+}] \xrightleftharpoons{Man \times 2} [PRM_2/Ca^{2+}/Man_2] \xrightleftharpoons{Man \times 2} [PRM_2/Ca^{2+}/Man_4]$

$[PRM_2/Ca^{2+}/Man_{0-4}]_n$
Oligomer or aggregate

SCHEME 4. Complex-forming equilibrium of PRM with Ca^{2+} ion and Man.

Our group has performed solid-state NMR experiments using solid aggregates of PRM-A complexed with Me-α-Man. The first stage of the analysis explored the role of Ca^{2+} ion on the binding of Man with PRM-A.[142] As already described, Cd^{2+} ion can act as a surrogate for Ca^{2+} ion in the binding of Man to PRMs. Accordingly, we conducted cross-polarization/magic angle spinning (CP/MAS) ^{113}Cd-NMR spectroscopic experiments with solid samples of the PRM-A/^{113}Cd^{2+} salt and the ternary PRM-A/^{113}Cd^{2+}/Me-α-Man complex (Fig. 28). The ^{113}Cd NMR spectrum of the PRM-A/^{113}Cd^{2+} salt exhibited a broad signal around $\delta = -50$ ppm, which is similar to those reported for solid cadmium compounds containing two carboxyl groups, such as $Cd(OAc)_2 \cdot H_2O$ ($\delta = -46$ ppm) and $Cd(O_2CCH_2CH_2CO_2) \cdot 2H_2O$ ($\delta = -52$ ppm).[143,144] This result indicates that the ^{113}Cd^{2+} ion binds to the carboxyl group of PRM-A and also supports the hypothesis that Ca^{2+} ion bridges the carboxylate groups of two PRM molecules to produce the Ca^{2+} salts having a PRM:Ca^{2+} ratio of 2:1. In contrast, the ternary PRM-A/^{113}Cd^{2+}/Me-α-Man complex exhibited a sharp signal at $\delta = -135$ ppm, which was markedly shifted upfield (>80 ppm) in comparison with that of the PRM-A/^{113}Cd^{2+} salt ($\delta = -50$ ppm), suggesting the occurrence of a change in ^{113}Cd^{2+} coordination upon binding of Me-α-Man to the PRM-A/^{113}Cd^{2+} salt. Since ^{113}Cd signals upfield of $\delta = -100$ ppm are observed only for ^{113}Cd^{2+} coordinated with more than six oxygen ligands,[143] it is reasonable to assume that the hydroxyl group(s) of Me-α-Man coordinates to ^{113}Cd^{2+} ion in

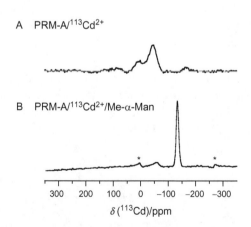

FIG. 28. Solid-state CP/MAS ^{113}Cd-NMR spectra of (A) PRM-A/^{113}Cd^{2+} and (B) PRM-A/^{113}Cd^{2+}/Me-α-Man complexes. The signals with an asterisk are the spinning side bands of the ^{113}Cd signal at $\delta = -135$ ppm.

the PRM-A/^{113}Cd^{2+} salt. The realistic implication of these results is that PRM-A binds at least one molecule of Man in a Ca^{2+}-mediated manner through its carboxylate group.

The solid-state ^{113}Cd-NMR spectroscopic analysis suggested the possibility that one of the binding sites for Man is located in the proximity of the carboxyl group of PRM-A. On the basis of this assumption, our subsequent analysis was directed at identifying the Man-binding site in PRM-A.[145] Several lines of evidence have suggested that PRM-A possesses two binding sites exhibiting different affinities for Man.[135,141] Although it is undoubtedly difficult to search simultaneously two Man-binding sites of PRM-A, we successfully prepared an aggregate composed solely of the [PRM-A$_2$/Ca^{2+}/Me-α-Man$_2$] complex, in which only the Man-binding site of stronger affinity was occupied by Me-α-Man. Me-α-Man was found to be released from the weaker binding site during the process of washing the aggregate composed of the [PRM-A$_2$/Ca^{2+}/Me-α-Man$_4$] complex. The simple 1:1 complexes of biosynthetically ^{13}C-enriched PRM-As and Me-α-[^{13}C$_6$]Man facilitated the analysis, by two-dimensional dipolar-assisted rotational resonance (2D-DARR), of the stronger binding site for Man.[146–148] In 2D-DARR spectra, dipolar interactions between ^{13}C nuclei that are located within a 6 Å distance can be detected as cross peaks, permitting evaluation of intermolecular close contacts between PRM-A and Me-α-Man. The 2D-DARR analyses detected close interactions between the D-alanine residue (C-17, 17-Me, C-18) and the 3-Me, C-4, C-5, and C-14 of PRM-A with Me-α-Man (Fig. 29). These close contacts are simultaneously compatible when Man is located on the same face of the A ring as the D-alanine group and the C-14 carbon atom, suggesting that the cavity consisting of the D-alanine group and the ABC rings is the stronger binding site for Man (Fig. 30). Similar CP/MAS ^{113}Cd-NMR and DARR experiments using compound **93** confirmed that the mode of binding of Ca^{2+} ion and Man is nearly identical between PRM-A and **93**.[149]

Based on the results of the solid-state NMR analyses and the previous structure–activity relationship studies, we proposed a model for the Man binding of PRM-A (Fig. 31).[145] The coordination of Ca^{2+}, hydrogen bonding, and CH/π interaction might all be involved in the interaction of PRM-A with Man. It is particularly interesting that the architecture of PRM-A, the naturally occurring lectin mimic, seems conceptually similar to that of tripod-type lectin mimics (Fig. 9); the A ring provides the hydrophobic surface for CH/π interaction, and the D-alanine, anthraquinone, and disaccharide components serve as hydrophilic units. Although more investigations are necessary to validate this model, these studies provide an important step toward the full elucidation of the molecular basis of the recognition of Man by PRM-A.

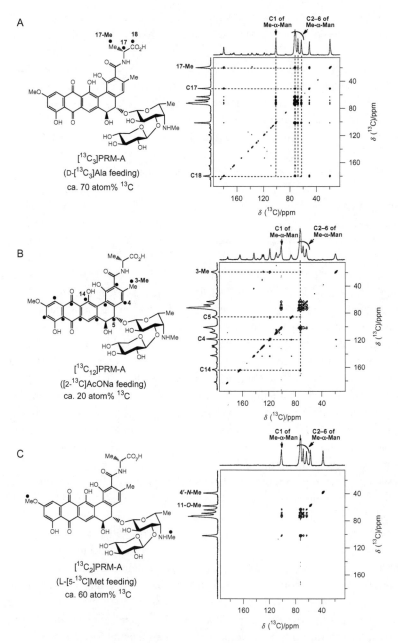

FIG. 29. 2D-DARR spectra of $[^{13}\text{C-labeled PRM-A}_2/\text{Ca}^{2+}/\text{Me-}\alpha\text{-}[^{13}\text{C}_6]\text{Man}_2]$ complexes.

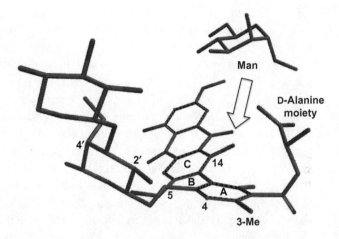

FIG. 30. Possible Man-binding conformation of PRM-A. (For color version of this figure, the reader is referred to the online version of this article.)

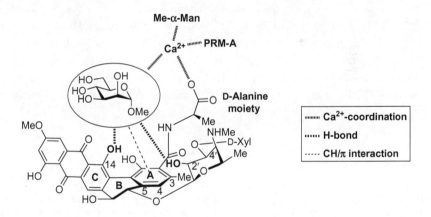

FIG. 31. Man-binding model of PRM-A.

3. Antiviral Profile and Mode of Action of Pradimicins

In the late 1980s, PRM-A and benanomicins A and B were discovered to block effectively HIV infection in cell-free as well as in cell-to-cell infection systems *in vitro*.[150,151] The subsequent study by Oki's group provided evidence that PRM-A inhibits an early step in HIV infection.[152] Preincubation of the cells with HIV at 0 °C, followed by incubation with PRM-A at 37 °C, resulted in complete inhibition of HIV infection, whereas PRM-A did not inhibit the infection after preincubation at 37 °C. No inhibitory effect was observed when the viruses treated with PRM-A were washed prior to infection. Moreover, addition of mannan or ethylene glycol-bis(2-aminoethyl ether)-N,N,N',N'-tetraacetic acid (EGTA) prevented the inhibitory effect of PRM-A. Thus, it has been assumed that PRM-A interferes with the viral fusion process through binding to the glycan on the HIV envelope in the presence of Ca^{2+} ion.

The antiviral properties of PRM-A were later studied in more detail by Balzarini and coworkers.[153] PRM-A was shown to inhibit the cytopathic effects induced by HIV-1 and HIV-2, although its EC_{50} values are in the lower micromolar range, and thus quite higher than those of natural lectins (Table XIV). Time-of-drug-addition studies demonstrated that PRM-A as well as the mannose-specific plant lectin HHA needed to be present from the beginning of the virus infection to fully suppress the infection. This observation supports the assumption that an early event in the infection cycle of HIV is the target of therapeutic intervention by PRM-A. The inhibition of the HIV entry process by PRM-A through interaction with gp120 was further confirmed by the complete loss of the inhibitory effect of PRM-A against VSV-G-pseudotyped HIV-1, which lacks gp120 in the envelope. The SPR analysis revealed that PRM-A significantly bound to gp120 in a Ca^{2+}-dependent manner, with a K_d value of $\sim 2.7\ \mu M$, suggesting the possibility that PRM-A inhibits entry of the virus by binding to high mannose-type glycans of gp120 in a manner similar to those of natural lectins. In accordance with this hypothesis, HIV-1 strains selected under increasing pressure of PRM-A were found to contain up to eight different glycan deletions in

TABLE XIV
Anti-HIV Activities of Natural Lectins, PRM-A, and PRM-S in MT-4 Cell Cultures

Carbohydrate Receptor	EC_{50} (μM)	
	HIV-1(III$_B$)	HIV-2(ROD)
GNA[a]	0.018	0.011
HHA[a]	0.006	0.016
NPA[a]	0.009	0.013
CA[a]	0.031	0.018
LOA[a]	0.004	0.003
EHA[a]	0.004	0.001
CV-N[a]	0.003	0.002
UDA[a]	0.140	0.330
PRM-A[b]	5.2	5.2
PRM-S[b]	5.0	4.7

[a] Data cited from Ref. 12; [b] Data cited from Ref. 154.

gp120, most of which were high mannose-type glycans. Such deletion at similar glycosylation sites was also observed for such mannose-specific plant lectins as HHA, GNA, and CN-V. Thus, PRM-A can act as a lectin in terms of glycan recognition, antiviral activity, and pattern of drug resistance.

Balzarini's group has also examined the antiviral action of PRM-S, a water-soluble derivative of PRM-A having a D-glucose 3-sulfate group instead of the terminal D-xylose group (Fig. 23).[154] PRM-S also inhibited the cytopathic effect of HIV-1 and HIV-2, showing EC_{50} values similar to those of PRM-A (Table XIV), and the binding capacity of PRM-S for gp120 was comparable to that of PRM-A. Drug resistance selection experiments using PRM-S demonstrated that PRM-S, like PRM-A, selected for drug-resistant mutant virus strains contained deletions at various N-glycosylation sites of HIV-1 gp120, showing that PRM-S as well as PRM-A has a strong preference for binding to high mannose-type glycans. Subsequently, Bewley and coworkers have made a direct observation of the binding of PRM-S with the trisaccharide, α-Man-$(1 \rightarrow 3)$-[α-Man-$(1 \rightarrow 6)$]-Man, using saturation-transfer difference (STD) NMR.[155] Moreover, PRM-S and PRM-A were found to inhibit, efficiently and dose dependently, the binding of HIV-1 to the DC-SIGN-expressing cells, indicating that these compounds prevent virus capture by DC-SIGN-expressing cells and subsequent transmission to T lymphocytes. PRM-S and PRM-A were also investigated for their potential to induce chemokines and cytokines. In contrast to the prokaryotic mannose-specific lectin CV-N, which dramatically stimulates the production of a wide range of cytokines and chemokines,[156] neither compound had any stimulatory effect on the chemo/cytokine production levels in mononuclear cells of peripheral blood.

On the basis of the unique antiviral properties of PRM-A and PRM-S, as well as lectins, Balzarini has proposed a novel therapeutic concept for HIV treatment that is entirely different from all currently existing therapeutic methods.[11–13] Exposure of HIV to carbohydrate receptors puts the virus in the dilemma of either becoming eliminated from its host by being kept suppressed by the carbohydrate receptors or escaping the pressure from the carbohydrate receptors by mutating glycosylation sites in gp120, and thereby becoming prone to neutralization and elimination by the immune system. In contrast to the conventional strategies of designing drugs that have as high a genetic barrier as possible to delay the development of drug resistance, the proposed approach makes use of viral variability and the inherent viral replication to generate mutant virus strains having deletions of glycosylation sites on gp120. Moreover, this therapeutic concept is unique in regard to its use of concerned action of drug chemotherapy—triggering the immune system in the host ("self-vaccination") by the administration of one single drug. It is also noteworthy that carbohydrate receptors can interact with a target gp120 molecule in a non-stoichiometric manner.

All existing anti-HIV drugs, such as reverse-transcriptase inhibitors and protease inhibitors, bind to only one target molecule. As a result, one or a few mutations that appear in a target usually elicit a marked degree of drug resistance. On the other hand, several carbohydrate receptor molecules can bind to one gp120 molecule bearing 20–29 glycans at the same time, resulting in a high genetic barrier for HIV to overcome. Obviously, this effect becomes more significant for the smaller carbohydrate receptors, PRM-A and PRM-S, because a larger number of molecules can simultaneously bind to gp120 without steric hindrance. They also have additional advantages over protein lectins in regard to chemical stability and productivity. Although their antiviral activity is less pronounced than that of lectins, these lectin mimics possessing beneficial characteristics for drug development may be very suitable candidates as conceptually novel anti-HIV drugs.

In addition to the anti-HIV activity, PRM-A has been shown to possess antiviral activity against HCV,[52] coronaviruses,[55] and simian immunodeficiency virus.[157] PRM-A also inhibits the entry process of these viruses by binding to the viral envelope-associated glycans. Therefore, PRM-A and its derivatives may qualify as potential antiviral drug candidates for infectious diseases derived from these viruses.

IV. CONCLUSION AND FUTURE PROSPECTS

The emerging biological and pathological significance of carbohydrates has increased the importance of carbohydrate receptors as diagnostic and therapeutic agents as well as research tools in glycobiology. A number of efforts have been made to utilize lectins for glycobiology research and have yielded fruitful results. However, their application in medicinal research has been hampered by the potential disadvantages presented by their protein nature. Consequently, there is a rapidly growing interest in small molecules having carbohydrate-binding properties (lectin mimics). Progress in the study of synthetic lectin mimics has laid an excellent foundation for the future development of diagnostic and therapeutic agents based on carbohydrate recognition. The boronic acid-dependent lectin mimics are designed to form covalent bonds with carbohydrates in aqueous media, and several examples have been successfully shown to act as possible sensors for such biologically important biomarkers as sialyl Lewis X and the TF antigen. An especially encouraging result was obtained for boronic acid-dependent lectin mimics having peptide backbones, in which such non-covalent interactions as hydrogen bonding and CH/π interaction also contribute to carbohydrate recognition. The combination of covalent interactions with boronic acid components and non-covalent interactions with other functional groups may be a promising strategy for designing a new generation of boronic acid-dependent lectin mimics that can act as diagnostic and therapeutic agents.

In parallel development, intensive work in the area of supramolecular chemistry has generated the boronic acid-independent lectin mimics, which rely solely on non-covalent interactions with carbohydrates. A number of the tripod-type lectin mimics have been shown to bind effectively carbohydrates in organic media. However, exact prediction of the binding preference is still difficult, and carbohydrate recognition in aqueous media by means of the tripod-type architecture remains an important goal for future studies. Meanwhile, the cage-type lectin mimics have successfully demonstrated recognition in aqueous solution of those carbohydrates having the all-equatorial substitution mode. However, at present, no strategy exists for designing cage-type lectin mimics that recognize other carbohydrates. Rapidly evolving computational methods may overcome these problems.

While only initial attempts have been made up to the present to develop synthetic lectin mimics having antiviral and antimicrobial activities, the naturally occurring lectin mimics, pradimicins and benanomicins, have become established as promising candidates as therapeutic agents having a novel mode of action. They express antiviral and antimicrobial activities by binding to Man residues of glycans on the viral envelope and the microbial cell wall. It is particularly interesting that these natural products appear to constitute highly sophisticated tripod-type lectin mimics, and they possibly provide a unique guide for designing a new therapeutic leads having the tripod-type architecture. With further advancements in these fields of research, novel therapeutic drugs based on lectin mimics are likely to emerge in the near future.

Acknowledgments

The research described in this article was conducted in collaboration with Professor Yasuhiro Igarashi of Toyama Prefectural University, and Professor Kiyonori Takegoshi, Dr. Yuichi Masuda, Mr. Takashi Doi, and Mr. Keita Yamada of Kyoto University. The work was partly supported by a Fund for Seeds of Collaborative Research and an Incentive Research Grant from RIKEN, and a MEXT Grant-in-Aid for Young Scientists (B) and Scientific Research on Innovative Areas "Chemical Biology of Natural Products."

References

1. A. Larkin and B. Imperiali, The expanding horizons of asparagine-linked glycosylation, *Biochemistry*, 50 (2011) 4411–4426.
2. Y. C. Lee, Warfare between pathogens and hosts: The trickery of sugars, *Trends Glycosci. Glycotechnol.*, 22 (2010) 95–106.

3. H. E. Myrrey and L. C. Hsieh-Wilson, The chemical neurobiology of carbohydrates, *Chem. Rev.*, 108 (2008) 1708–1731.
4. K. S. Lau and J. W. Dennis, N-Glycans in cancer progression, *Glycobiology*, 18 (2008) 750–760.
5. R. A. Dwek, Glycobiology: Toward understanding the function of sugars, *Chem. Rev.*, 96 (1996) 683–720.
6. I. J. Goldstein, R. C. Hughes, M. Monsigny, T. Osawa, and N. Sharon, What should be called a lectin? *Nature*, 285 (1980) 66.
7. N. Sharon and H. Lis, History of lectins: From hemagglutinins to biological recognition molecules, *Glycobiology*, 14 (2004) 53R–62R.
8. J. Hirabayashi, Concept, strategy and realization of lectin-based glycan profiling, *J. Biochem.*, 144 (2008) 139–147.
9. Z. Dai, J. Zhou, S. Qui, Y. Liu, and J. Fan, Lectin-based glycoproteomics to explore and analyze hepatocellular carcinoma-related glycoprotein markers, *Electrophoresis*, 30 (2009) 2957–2966.
10. G. Gupta, A. Surolia, and S. Sampathkumar, Lectin microarrays for glycomic analysis, *OMICS*, 14 (2010) 419–436.
11. J. Balzarini, Targeting the glycans of glycoproteins: A novel paradigm for antiviral therapy, *Nat. Rev. Microbiol.*, 5 (2007) 583–597.
12. J. Balzarini, Carbohydrate-binding agents: A potential future cornerstone for the chemotherapy of enveloped viruses? *Antivir. Chem. Chemother.*, 18 (2007) 1–11.
13. K. O. François and J. Balzarini, Potential of carbohydrate-binding agents as therapeutics against enveloped viruses, *Med. Res. Rev.*, 32 (2012) 349–387.
14. S. Jin, Y. Cheng, S. Reid, M. Li, and B. Wang, Carbohydrate recognition by boronolectins, small molecules, and lectins, *Med. Res. Rev.*, 30 (2010) 171–257.
15. M. Mazik, Recent developments in the molecular recognition of carbohydrates by artificial receptors, *RSC Adv.*, 2 (2012) 2630–2642.
16. D. B. Walker, G. Joshi, and A. P. Davis, Progress in biomimetic carbohydrate recognition, *Cell. Mol. Life Sci.*, 66 (2009) 3177–3191.
17. Y. Nakagawa and Y. Ito, Carbohydrate-binding molecules with non-peptidic skeletons, *Trends GlycoSci. Glycotechnol.*, 24 (2012) 1–12.
18. M. Nishio, The CH/π hydrogen bond in chemistry. Conformation, supramolecules, optical resolution and interactions involving carbohydrates, *Phys. Chem. Chem. Phys.*, 13 (2011) 13873–13900.
19. Y. Fukagawa, T. Ueki, K. Numata, and T. Oki, Pradimicins and benanomicins, sugar-recognizing antibiotics: Their novel mode of antifungal action and conceptual significance, *Actinomycetologia*, 7 (1993) 1–22.
20. G. Springsteen and B. Wang, A detailed examination of boronic acid–diol complexation, *Tetrahedron*, 58 (2002) 5291–5300.
21. J. P. Lorand and J. O. Edwards, Polyol complexes and structure of the benzeneboronate ion, *J. Org. Chem.*, 24 (1959) 769–774.
22. J. Yan, G. Springsteen, S. Deeter, and B. Wang, The relationship among pK_a, pH, and binding constants in the interactions between boronic acids and diols—It is not as simple as it appears, *Tetrahedron*, 60 (2004) 11205–11209.
23. G. Wulff, Selective binding to polymers *via* covalent bonds: The construction of chiral cavities as specific receptor sites, *Pure Appl. Chem.*, 54 (1982) 2093–2102.
24. G. Wulff, M. Lauer, and H. Böhnke, Rapid proton transfer as cause of an unusually large neighboring group effect, *Angew. Chem. Int. Ed.*, 23 (1984) 741–742.
25. S. Franzen, W. Ni, and B. Wang, Study of the mechanism of electron-transfer quenching by boron–nitrogen adducts in fluorescent sensors, *J. Phys. Chem. B*, 107 (2003) 12942–12948.

26. W. Ni, G. Kaur, G. Springsteen, B. Wang, and S. Franzen, Regulating the fluorescence intensity of an anthracene boronic acid system: A B–N bond or a hydrolysis mechanism?*Bioorg. Chem.*, 32 (2004) 571–581.
27. L. Zhu, S. H. Shabbir, M. Gray, V. M. Lynch, S. Sorey, and E. V. Anslyn, A structural investigation of the N–B interaction in an o-(*N*, *N*-dialkylaminomethyl)arylboronate system, *J. Am. Chem. Soc.*, 128 (2006) 1222–1232.
28. J. Yan, H. Fang, and B. Wang, Boronolectins and fluorescent boronolectins: An examination of the detailed chemistry issues important for the drug design, *Med. Res. Rev.*, 25 (2005) 490–520.
29. T. D. James, K. R. A. S. Sandanayake, and S. Shinkai, Novel photoinduced electron-transfer sensor for saccharides based on the interaction of boronic acid and amine, *J. Chem. Soc. Chem. Commun.* (1994) 477–478.
30. T. D. James, K. R. A. S. Sandanayake, and S. Shinkai, A glucose-selective molecular fluorescence sensor, *Angew. Chem. Int. Ed.*, 33 (1994) 2207–2209.
31. T. D. James, K. R. A. S. Sandanayake, R. Iguchi, and S. Shinkai, Novel saccharide-photoinduced electron transfer sensors based on the interaction of boronic acid and amine, *J. Am. Chem. Soc.*, 117 (1995) 8982–8987.
32. V. V. Karnati, X. Gao, S. Gao, W. Yang, W. Ni, S. Sankar, and B. Wang, A glucose-selective fluorescence sensor based on boronic acid-diol recognition, *Bioorg. Med. Chem. Lett.*, 12 (2002) 3373–3377.
33. G. Kaur, H. Fang, X. Gao, H. Li, and B. Wang, Substituent effect on anthracene-based bisboronic acid glucose sensors, *Tetrahedron*, 62 (2006) 2583–2589.
34. T. D. James, K. R. A. S. Sandanayake, and S. Shinkai, Chiral discrimination of monosaccharides using a fluorescent molecular sensor, *Nature*, 374 (1995) 345–347.
35. Z. Xing, H. Wang, Y. Cheng, C. Zhu, T. D. James, and J. Zhao, Selective saccharide recognition using modular diboronic acid fluorescent sensors, *Eur. J. Org. Chem.* (2012) 1223–1229.
36. W. Yang, S. Gao, X. Gao, V. V. R. Karnati, W. Ni, B. Wang, W. B. Hooks, J. Carson, and B. Weston, Diboronic acids as fluorescent probes for cells expressing sialyl Lewis X, *Bioorg. Med. Chem. Lett.*, 12 (2002) 2175–2177.
37. A. Takada, K. Ohmori, T. Yoneda, K. Tsuyuoka, A. Hasegawa, M. Kiso, and R. Kannagi, Contribution of carbohydrate antigens sialyl Lewis A and sialyl Lewis X to adhesion of human cancer cells to vascular endothelium, *Cancer Res.*, 53 (1993) 354–361.
38. R. Kannagi, Molecular mechanism for cancer-associated induction of sialyl Lewis X and sialyl Lewis A expression—The Warburg effect revisited, *Glycoconj. J.*, 20 (2004) 353–354.
39. T. Jorgensen, A. Berner, O. Kaalhus, K. J. Tveter, H. E. Danielsen, and M. Bryne, Up-regulation of the oligosaccharide sialyl Lewis X: A new prognostic parameter in metastatic prostate cancer, *Cancer Res.*, 55 (1995) 1817–1819.
40. H. A. Idikio, Sialyl Lewis X, gleason grade and stage in nonmetastatic human prostate cancer, *Glycoconj. J.*, 14 (1997) 875–877.
41. C. Dai, L. H. Cazares, L. Wang, Y. Chu, S. L. Wang, D. A. Troyer, O. J. Semmes, R. R. Drake, and B. Wang, Using boronolectin in MALDI-MS imaging for the histological analysis of cancer tissue expressing the sialyl Lewis X antigen, *Chem. Commun.*, 47 (2011) 10338–10340.
42. G. Thiery, E. Anselmi, A. Audebourg, E. Darii, M. Abarbri, B. Terris, J. C. Tabet, and I. G. Gut, Improvements of targeted multiplex mass spectroscopy imaging, *Proteomics*, 8 (2008) 3725–3734.
43. R. Lemaire, J. Stauber, M. Wisztorski, C. Van Camp, A. Desmons, M. Deschamps, G. Proess, I. Rudlof, A. S. Woods, R. Day, M. Salzet, and I. Fournier, Tag-mass: Specific molecular imaging of transcriptome and proteome by mass spectrometry based on photocleavable tag, *J. Proteome Res.*, 6 (2007) 2057–2067.
44. M. Dowlut and D. G. Hall, An improved class of sugar-binding boronic acids, soluble and capable of complexing glycosides in neutral water, *J. Am. Chem. Soc.*, 128 (2006) 4226–4227.

45. M. Bérubé, M. Dowlut, and D. G. Hall, Benzoboroxoles as efficient glycopyranoside-binding agents in physiological conditions: Structure and selectivity of complex formation, *J. Org. Chem.*, 73 (2008) 6471–6479.
46. J. C. Norrild and H. Eggert, Evidence for mono- and bisdentate boronate complexes of glucose in the furanose form. application of $^1J_{C-C}$ coupling constants as a structural probe, *J. Am. Chem. Soc.*, 117 (1995) 1479–1484.
47. M. Bielecki, H. Eggert, and J. C. Norrild, A fluorescent glucose sensor binding covalently to all five hydroxyl groups of α-D-glucofuranose. A reinvestigation, *J. Chem. Soc. Perkin Trans.*, 2 (1999) 449–455.
48. A. Pal, M. Bérubé, and D. G. Hall, Design, synthesis, and screening of a library of peptidyl bis (boroxoles) as oligosaccharide receptors in water: Identification of a receptor for the tumor marker TF-antigen disaccharide, *Angew. Chem. Int. Ed.*, 49 (2010) 1492–1495.
49. L. Yu, The oncofetal Thomsen-Friedenreich carbohydrate antigen in cancer progression, *Glycoconj. J.*, 24 (2007) 411–420.
50. J. Balzarini, Inhibition of HIV entry by carbohydrate-binding proteins, *Antiviral Res.*, 71 (2006) 237–247.
51. I. Botos and A. Wlodawer, Proteins that bind high-mannose sugars of the HIV envelope, *Prog. Biophys. Mol. Biol.*, 88 (2005) 233–282.
52. C. Bertaux, D. Daelemans, L. Meertens, E. G. Cormier, J. F. Reinus, W. J. Peumans, E. J. Van Damme, Y. Igarashi, T. Oki, D. Schols, T. Dragic, and J. Balzarini, Entry of hepatitis C virus and human immunodeficiency virus is selectively inhibited by carbohydrate-binding agents but not by polyanions, *Virology*, 366 (2007) 40–50.
53. F. Helle, C. Wychowski, N. Vu-Dac, K. R. Gustafson, C. Voisset, and J. Dubuisson, Cyanovirin-N inhibits hepatitis C virus entry by binding to envelope protein glycans, *J. Biol. Chem.*, 281 (2006) 25177–25183.
54. Y. Kumaki, M. K. Wandersee, A. J. Smith, Y. Zhou, G. Simmons, N. M. Nelson, K. W. Bailey, Z. G. Vest, J. K.-K. Li, P. K. Chan, D. F. Smee, and D. L. Barnard, Inhibition of severe acute respiratory syndrome coronavirus replication in a lethal SARS-CoV BALB/c mouse model by stinging nettle lectin, *Urtica dioica agglutinin*, *Antiviral Res.*, 90 (2011) 22–32.
55. F. J. U. M. van der Meer, C. A. M. de Haan, N. M. P. Schuurman, B. J. Haijema, M. H. Verheije, B. J. Bosch, J. Balzarini, and H. F. Egberink, The carbohydrate-binding plant lectins and the non-peptidic antibiotic pradimicin A target the glycans of the coronavirus envelope glycoproteins, *J. Antimicrob. Chemother.*, 60 (2007) 741–749.
56. B. R. O'Keefe, D. F. Smee, J. A. Turpin, C. J. Saucedo, K. R. Gustafson, T. Mori, D. Blakeslee, R. Buckheit, and M. R. Boyd, Potent anti-influenza activity of cyanovirin-N and interactions with viral hemagglutinin, *Antimicrob. Agents Chemother.*, 47 (2003) 2518–2525.
57. Y. Sato, M. Hirayama, K. Morimoto, N. Yamamoto, S. Okuyama, and K. Hori, High mannose-binding lectin with preference for the cluster of α1-2-mannose from the green alga *Boodlea coacta* is a potent entry inhibitor of HIV-1 and influenza viruses, *J. Biol. Chem.*, 286 (2011) 19446–19458.
58. C. K. Leonard, M. W. Spellman, L. Riddle, R. J. Harris, J. N. Thomas, and T. J. Gregory, Assignment of intrachain disulfide bonds and characterization of potential glycosylation sites of the type 1 recombinant human immunodeficiency virus envelope glycoprotein (gp120) expressed in Chinese hamster ovary cells, *J. Biol. Chem.*, 265 (1990) 10373–10382.
59. P. M. Rudd, T. Elliott, P. Cresswell, I. A. Wilson, and R. A. Dwek, Glycosylation and the immune system, *Science*, 291 (2001) 2370–2376.
60. A. E. Smith and A. Helenius, How viruses enter animal cells, *Science*, 304 (2004) 237–242.
61. J. Balzarini, D. Schols, J. Neyts, E. Van Damme, W. Peumans, and E. De Clercq, Alpha-(1-3)- and alpha-(1-6)-D-mannose-specific plant lectins are markedly inhibitory to human immunodeficiency virus and cytomegalovirus infections *in vitro*, *Antimicrob. Agents Chemother.*, 35 (1991) 410–416.

62. J. Balzarini, Targeting the glycans of gp120: A novel approach aimed at the Achilles heel of HIV, *Lancet Infect. Dis.*, 5 (2005) 726–731.
63. J. Balzarini, K. Van Laethem, S. Hatse, K. Vermeire, E. De Clercq, W. Peumans, E. Van Damme, A. M. Vandamme, A. Bölmstedt, and D. Schols, Profile of resistance of human immunodeficiency virus to mannose-specific plant lectins, *J. Virol.*, 78 (2004) 10617–10627.
64. B. Losman, A. Bolmstedt, K. Schønning, A. Björndal, C. Westin, E. M. Fenyö, and S. Oloffson, Protection of neutralization epitopes in the V3 loop of oligomeric human immunodeficiency virus type 1 glycoprotein 120 by N-linked oligosaccharides in the V1 region, *AIDS Res. Hum. Retroviruses*, 17 (2001) 1067–1076.
65. X. Wei, J. M. Decker, S. Wang, H. Hui, J. C. Kappes, X. Wu, J. F. Salazar-Gonzalez, M. G. Salazar, J. M. Kilby, M. S. Saag, N. L. Komarova, M. A. Nowak, B. H. Hahn, P. D. Kwong, and G. M. Shaw, Antibody neutralization and escape by HIV-1, *Nature*, 422 (2003) 307–312.
66. J. Balzarini, K. Van Laethem, S. Hatse, M. Froeyen, W. Peumans, E. Van Damme, and D. Schols, Carbohydrate-binding agents cause deletions of highly conserved glycosylation sites in HIV GP120: A new therapeutic concept to hit the achilles heel of HIV, *J. Biol. Chem.*, 280 (2005) 41005–41014.
67. J. Balzarini, K. Van Laethem, S. Hatse, M. Froeyen, E. Van Damme, A. Bolmstedt, W. Peumans, E. De Clercq, and D. Schols, Marked depletion of glycosylation sites in HIV-1 gp120 under selection pressure by the mannose-specific plant lectins of *Hippeastrum* hybrid and *Galanthus nivalis*, *Mol. Pharmacol.*, 67 (2005) 1556–1565.
68. J. Balzarini, K. Van Laethem, W. Peumans, E. Van Damme, A. Bolmstedt, F. Gago, and D. Schols, Mutational pathways, resistance profile, and side effects of cyanovirin relative to human immunodeficiency virus type 1 strains with N-glycan deletions in their gp120 envelopes, *J. Virol.*, 80 (2006) 8411–8421.
69. M. Witvrouw, V. Fikkert, A. Hantson, C. Pannecouque, B. R. O'keefe, J. McMahon, L. Stamatatos, E. de Clercq, and A. Bolmstedt, Resistance of human immunodeficiency virus type 1 to the high-mannose binding agents cyanovirin N and concanavalin A, *J. Virol.*, 79 (2005) 7777–7784.
70. P. C. Trippier, C. McGuigan, and J. Balzarini, Phenylboronic-acid-based carbohydrate binders as antiviral therapeutics: Monophenylboronic acids, *Antivir. Chem. Chemother.*, 20 (2010) 249–257.
71. P. C. Trippier, J. Balzarini, and C. McGuigan, Phenylboronic-acid-based carbohydrate binders as antiviral therapeutics: Bisphenylboronic acids, *Antivir. Chem. Chemother.*, 21 (2011) 129–142.
72. J. I. Jay, B. E. Lai, D. G. Myszka, A. Mahalingam, K. Langheinrich, D. F. Katz, and P. F. Kiser, Multivalent benzoboroxole functionalized polymers as gp120 glycan targeted microbicide entry inhibitors, *Mol. Pharm.*, 7 (2010) 116–129.
73. M. Mammem, S. K. Choi, and G. M. Whitesides, Polyvalent interactions in biological systems: Implication for design and use of multivalent ligands and inhibitors, *Angew. Chem. Int. Ed.*, 37 (1998) 2754–2794.
74. C. F. Brewer, M. C. Miceli, and L. G. Baum, Clusters, bundles, arrays and lattices: Novel mechanisms for lectin-saccharide-mediated cellular interactions, *Curr. Opin. Struct. Biol.*, 12 (2002) 616–623.
75. A. Mahalingam, A. R. Geonnotti, J. Balzarini, and P. F. Kiser, Activity and safety of synthetic lectins based on benzoboroxole-functionalized polymers for inhibition of HIV entry, *Mol. Pharm.*, 8 (2011) 2465–2475.
76. M. Mazik, H. Bandmann, and W. Sicking, Molecular recognition of carbohydrates by artificial polypyridine and polypyrimidine receptors, *Angew. Chem. Int. Ed.*, 39 (2000) 551–554.
77. M. Mazik and W. Sicking, Molecular recognition of carbohydrates by artificial receptors: Systematic studies towards recognition motifs for carbohydrates, *Chem. Eur. J.*, 7 (2001) 664–670.
78. M. Mazik, W. Radunz, and W. Sicking, High α/β-anomer selectivity in molecular recognition of carbohydrates by artificial receptors, *Org. Lett.*, 4 (2002) 4579–4582.
79. M. Mazik, W. Radunz, and R. Boese, Molecular recognition of carbohydrates with acyclic pyridine-based receptors, *J. Org. Chem.*, 69 (2004) 7448–7462.

80. M. Mazik and M. Kuschel, Amide, amino, hydroxyl and aminopyridine groups as building blocks for carbohydrate receptors, *Eur. J. Org. Chem.* (2008) 1517–1526.
81. M. Mazik and M. Kuschel, Highly effective acyclic carbohydrate receptors consisting of aminopyridine, imidazole, and indole recognition units, *Chem. Eur. J.*, 14 (2008) 2405–2419.
82. M. Mazik and A. Hartmann, Phenanthroline unit as a building block for carbohydrate receptors, *J. Org. Chem.*, 73 (2008) 7444–7450.
83. M. Mazik, A. Hartmann, and P. G. Jones, Highly effective recognition of carbohydrates by phenanthroline-based receptors: α- versus β-anomer binding preference, *Chem. Eur. J.*, 15 (2009) 9147–9159.
84. M. Mazik and A. Hartmann, Recognition properties of receptors consisting of imidazole and indole recognition units towards carbohydrates, *Beilstein J. Org. Chem.*, 6 (2010) 9.
85. M. Mazik and C. Sonnenberg, Isopropylamino and isobutylamino groups as recognition sites for carbohydrates: Acyclic receptors with enhanced binding affinity toward β-galactosides, *J. Org. Chem.*, 75 (2010) 6416–6423.
86. M. Mazik and C. Geffert, 8-Hydroxylquinoline as a building block for artificial receptors: Binding preferences in the recognition of glycopyranosides, *Org. Biomol. Chem.*, 9 (2011) 2319–2326.
87. M. Mazik, Molecular recognition of carbohydrates by acyclic receptors employing noncovalent interactions, *Chem. Soc. Rev.*, 38 (2009) 935–956.
88. M. Mazik and A. C. Buthe, Recognition properties of receptors based on dimesitylmethane-derived core: Di- vs. monosaccharide preference, *Org. Biomol. Chem.*, 7 (2009) 2063–2071.
89. C. Nativi, M. Cacciarini, O. Francesconi, A. Vacca, G. Moneti, A. Ienco, and S. Roelens, Pyrrolic tripodal receptors efficiently recognizing monosaccharides. Affinity assessment through a generalized binding descriptor, *J. Am. Chem. Soc.*, 129 (2007) 4377–4385.
90. C. Nativi, M. Cacciarini, O. Francesconi, G. Moneti, and S. Roelens, A β-mannoside-selective pyrrolic tripodal receptor, *Org. Lett.*, 9 (2007) 4685–4688.
91. C. Nativi, O. Francesconi, G. Gabrielli, A. Vacca, and S. Roelens, Chiral diaminopyrrolic receptors for selective recognition of mannosides, part 1: Design, synthesis, and affinities of second-generation tripodal receptors, *Chem. Eur. J.*, 17 (2011) 4814–4820.
92. A. Ardá, C. Venturi, C. Nativi, O. Francesconi, F. J. Cañada, J. Jiménez-Barbero, and S. Roelens, Selective recognition of β-mannosides by synthetic tripodal receptors: A 3D view of the recognition mode by NMR, *Eur. J. Org. Chem.* (2010) 64–71.
93. A. Ardá, C. Venturi, C. Nativi, O. Francesconi, G. Gabrielli, F. J. Cañada, J. Jiménez-Barbero, and S. Roelens, A chiral pyrrolic tripodal receptor enantioselectively recognizes β-mannose and β-mannosides, *Chem. Eur. J.*, 16 (2010) 414–418.
94. A. Ardá, F. J. Cañada, C. Nativi, O. Francesconi, G. Gabrielli, A. Ienco, J. Jiménez-Barbero, and S. Roelens, Chiral diaminopyrrolic receptors for selective recognition of mannosides, part 2: A 3D view of the recognition modes by X-ray, NMR spectroscopy, and molecular modeling, *Chem. Eur. J.*, 17 (2011) 4821–4829.
95. P. B. Palde, P. C. Gareiss, and B. L. Miller, Selective recognition of alkyl pyranosides in protic and aprotic solvents, *J. Am. Chem. Soc.*, 130 (2008) 9566–9573.
96. A. P. Davis, Synthetic lectin, *Org. Biomol. Chem.*, 7 (2009) 3629–3638.
97. A. P. Davis and R. S. Wareham, A tricyclic polyamide receptor for carbohydrates in organic media, *Angew. Chem. Int. Ed.*, 37 (1998) 2270–2273.
98. G. Lecollinet, A. P. Dominey, T. Velasco, and A. P. Davis, Highly selective disaccharide recognition by a tricyclic octaamide cage, *Angew. Chem. Int. Ed.*, 41 (2002) 4093–4096.
99. O. Francesconi, A. Ienco, G. Moneti, C. Nativi, and S. Roelens, A self-assembled pyrrolic cage receptor specifically recognizes β-glucopyranosides, *Angew. Chem. Int. Ed.*, 45 (2006) 6693–6696.

100. E. Klein, M. P. Crump, and A. P. Davis, Carbohydrate recognition in water by a tricyclic polyamide receptor, *Angew. Chem. Int. Ed.*, 44 (2005) 298–302.
101. Y. Ferrand, E. Klein, N. P. Barwell, M. P. Crump, J. Jiménez-Barbero, C. Vicent, G. J. Boons, S. Ingale, and A. P. Davis, A synthetic lectin for O-linked β-*N*-acetylglucosamine, *Angew. Chem. Int. Ed.*, 48 (2009) 1775–1779.
102. L. K. Kreppel and G. W. Hart, Regulation of a cytosolic and nuclear *O*-GlcNAc transferase, *J. Biol. Chem.*, 274 (1999) 32015–32022.
103. Y. Ferrand, M. P. Crump, and A. P. Davis, A synthetic lectin analog for biomimetic disaccharide recognition, *Science*, 318 (2007) 619–622.
104. G. Joshi and A. P. Davis, New H-bonding patterns in biphenyl-based synthetic lectins; pyrrolediamine bridges enhance glucose-selectivity, *Org. Biomol. Chem.*, 10 (2012) 5760–5763.
105. D. Gozalbo, P. Roig, E. Villamon, and M. L. Gil, Candida and candidiasis: The cell wall as a potential molecular target for antifungal therapy, *Curr. Drug Targets Infect. Disord.*, 4 (2004) 117–135.
106. H. M. Mora-Montes, P. Ponce-Noyola, J. C. Villagómez-Castro, N. A. R. Gow, A. Flores-Carreón, and E. López-Romero, Protein glycosylation in Candida, *Future Microbiol.*, 4 (2009) 1167–1183.
107. V. Lozano, L. Aguado, B. Hoorelbeke, M. Renders, M. J. Camarasa, D. Schols, J. Balzarini, A. San-Félix, and M. J. Pérez-Pérez, Targeting HIV entry through interaction with envelope glycoprotein 120 (gp120): Synthesis and antiviral evaluation of 1,3,5-triazines with aromatic amino acids, *J. Med. Chem.*, 54 (2011) 5335–5348.
108. M. R. Boyd, K. R. Gustafson, J. B. McMahon, R. H. Shoemaker, B. R. O'Keefe, T. Mori, R. J. Gulakowski, L. Wu, M. I. Rivera, C. M. Laurencot, M. J. Currens, J. H. Cardellina, II,, R. W. Buckheit, Jr.,, P. L. Nara, L. K. Pannell, R. C. Sowder, II,, and L. E. Henderson, Discovery of cyanovirin-N, a novel human immunodeficiency virus-inactivating protein that binds viral surface envelope glycoprotein gp120: Potential applications to microbicide development, *Antimicrob. Agents Chemother.*, 41 (1997) 1521–1530.
109. C. Nativi, O. Francesconi, G. Gabrielli, I. D. Simone, B. Turchetti, T. Mello, L. D. C. Mannelli, C. Ghelardini, P. Buzzini, and S. Roelens, Aminopyrrolic synthetic receptors for monosaccharides: A class of carbohydrate-binding agents endowed with antibiotic activity versus pathogenic yeasts, *Chem. Eur. J.*, 18 (2012) 5064–5072.
110. T. Takeuchi, T. Hara, H. Naganawa, M. Okada, M. Hamada, H. Umezawa, S. Gomi, M. Sezaki, and S. Kondo, New antifungal antibiotics, benanomicins A and B from an *actinomycete*, *J. Antibiot.*, 41 (1988) 807–811.
111. T. Oki, M. Konishi, K. Tomatsu, K. Tomita, K. Saitoh, M. Tsunakawa, M. Nishio, T. Miyaki, and H. Kawaguchi, Pradimicin, a novel class of potent antifungal antibiotics, *J. Antibiot.*, 41 (1988) 1701–1704.
112. S. Gomi, M. Sezaki, and S. Kondo, The structures of new antifungal antibiotics, benanomicins A and B, *J. Antibiot.*, 41 (1988) 1019–1028.
113. M. Tsunakawa, M. Nishio, H. Ohkuma, T. Tsuno, M. Konishi, T. Naito, T. Oki, and H. Kawaguchi, The structures of pradimicins A, B, and C: A novel family of antifungal antibiotics, *J. Org. Chem.*, 54 (1989) 2532–2536.
114. K. Tomita, M. Nishio, K. Saitoh, H. Yamamoto, Y. Hoshino, H. Ohkuma, M. Konishi, T. Miyaki, and T. Oki, Pradimicins A, B, and C: New antifungal antibiotics. I. Taxonomy, production, isolation and physico-chemical properties, *J. Antibiot.*, 43 (1990) 755–762.
115. T. Oki, O. Tenmyo, M. Hirano, K. Tomatsu, and H. Kamei, Pradimicins A, B, and C: New antifungal antibiotics. II. In vitro and in vivo biological activities, *J. Antibiot.*, 43 (1990) 763–770.
116. S. Kondo, S. Gomi, D. Ikeda, M. Hamada, and T. Takeuchi, Antifungal and antiviral activities of benanomicins and their analogues, *J. Antibiot.*, 44 (1991) 1228–1236.

117. Y. Sawada, M. Nishio, H. Yamamoto, M. Hatori, T. Miyaki, M. Konishi, and T. Oki, New antifungal antibiotics, pradimicins D and E glycine analogues of pradimicins A and C, *J. Antibiot.*, 43 (1990) 771–777.
118. Y. Sawada, M. Hatori, H. Yamamoto, M. Nishio, T. Miyaki, and T. Oki, New antifungal antibiotics pradimicins FA-1 and FA-2: D-serine analogs of pradimicins A and C, *J. Antibiot.*, 43 (1990) 1223–1229.
119. K. Saitoh, K. Suzuki, M. Hirano, T. Furumai, and T. Oki, Pradimicins FS and FB, new pradimicin analogs: Directed production, structures and biological activities, *J. Antibiot.*, 46 (1993) 398–405.
120. K. Saitoh, Y. Sawada, K. Tomita, T. Tsuno, M. Hatori, and T. Oki, Pradimicins L and FL: New pradimicin congeners from *Actinomadura verrucosospora* subsp. *neohibisca*, *J. Antibiot.*, 46 (1993) 387–397.
121. T. Furumai, T. Hasegawa, M. Kakushima, K. Suzuki, H. Yamamoto, S. Yamamoto, M. Hirano, and T. Oki, Pradimicins T1 and T2, new antifungal antibiotics produced by an actinomycete. I. Taxonomy, production, isolation, physico-chemical and biological properties, *J. Antibiot.*, 46 (1993) 589–597.
122. T. Hasegawa, M. Kakushima, M. Hatori, S. Aburaki, S. Kakinuma, T. Furumai, and T. Oki, Pradimicins T1 and T2, new antifungal antibiotics produced by an actinomycete. II. Structures and biosynthesis, *J. Antibiot.*, 46 (1993) 598–605.
123. K. Saitoh, O. Tenmyo, S. Yamamoto, T. Furumai, and T. Oki, Pradimicin S, a new pradimicin analog. I. Taxonomy, fermentation and biological activities, *J. Antibiot.*, 46 (1993) 580–588.
124. K. Saitoh, T. Tsuno, M. Kakushima, M. Hatori, T. Furumai, and T. Oki, Pradimicin S, a new pradimicin analog. II. Isolation and structure elucidation, *J. Antibiot.*, 46 (1993) 406–411.
125. T. Oki, M. Kakushima, M. Nishio, H. Kamei, M. Hirano, Y. Sawada, and M. Konishi, Water-soluble pradimicin derivatives, synthesis and antifungal evaluation of N, N-dimethyl pradimicins, *J. Antibiot.*, 43 (1990) 1230–1235.
126. Y. Sawada, K. Numata, T. Murakami, H. Tanimichi, S. Yamamoto, and T. Oki, Calcium-dependent anticandidal action of pradimicin A, *J. Antibiot.*, 43 (1990) 715–721.
127. M. Kakushima, M. Nishio, K. Numata, M. Konishi, and T. Oki, Effect of stereochemistry at the C-17 position on the antifungal activity of pradimicin A, *J. Antibiot.*, 43 (1990) 1028–1030.
128. S. Okuyama, M. Kakushima, H. Kamachi, M. Konishi, and T. Oki, Synthesis and antifungal activities of alanine-exchanged analogs of pradimicin A, *J. Antibiot.*, 46 (1993) 500–506.
129. M. Nishio, H. Ohkuma, M. Kakushima, S. Ohta, S. Iimura, M. Hirano, M. Konishi, and T. Oki, Synthesis and antifungal activities of pradimicin A derivatives modification of the alanine moiety, *J. Antibiot.*, 46 (1993) 494–499.
130. S. Aburaki, S. Okuyama, H. Hoshi, H. Kamachi, M. Nishio, T. Hasegawa, S. Masuyoshi, S. Iimura, M. Konishi, and T. Oki, Synthesis and antifungal activity of pradimicin derivatives modifications on the aglycon part, *J. Antibiot.*, 46 (1993) 1447–1457.
131. H. Kamachi, S. Iimura, S. Okuyama, H. Hoshi, S. Tamura, M. Shinoda, K. Saitoh, M. Konishi, and T. Oki, Synthesis and antifungal activities of pradimicin derivatives, modification at C4'-position, *J. Antibiot.*, 45 (1992) 1518–1525.
132. S. Aburaki, H. Yamashita, T. Ohnuma, H. Kamachi, T. Moriyama, S. Masuyoshi, H. Kamei, M. Konishi, and T. Oki, Synthesis and antifungal activity of pradimicin derivatives modifications of the sugar part, *J. Antibiot.*, 46 (1993) 631–640.
133. Y. Sawada, T. Murakami, T. Ueki, Y. Fukagawa, T. Oki, and Y. Nozawa, Mannan-mediated anticandidal activity of BMY-28864, a new water-soluble pradimicin derivative, *J. Antibiot.*, 44 (1991) 119–121.
134. M. Kakushima, S. Masuyoshi, M. Hirano, M. Shinoda, A. Ohta, H. Kamei, and T. Oki, In vitro and in vivo antifungal activities of BMY-28864, a water-soluble pradimicin derivative, *Antimicrob. Agents Chemother.*, 35 (1991) 2185–2190.

135. T. Ueki, K. Numata, Y. Sawada, T. Nakajima, Y. Fukagawa, and T. Oki, Studies on the mode of antifungal action of pradimicin antibiotics I. Lectin-mimic binding of BMY-28864 to yeast mannan in the presence of calcium, *J. Antibiot.*, 46 (1993) 149–161.
136. T. Ueki, M. Oka, Y. Fukagawa, and T. Oki, Studies on the mode of antifungal action of pradimicin antibiotics III. Spectrophotometric sequence analysis of the ternary complex formation of BMY-28864 with D-mannopyranoside and calcium, *J. Antibiot.*, 46 (1993) 465–477.
137. T. Ueki, K. Numata, Y. Sawada, M. Nishio, H. Ohkuma, S. Toda, H. Kamachi, Y. Fukagawa, and T. Oki, Studies on the mode of antifungal action of pradimicin antibiotics II. D-Mannopyranoside-binding site and calcium-binding site, *J. Antibiot.*, 46 (1993) 455–464.
138. M. Hu, Y. Ishizuka, Y. Igarashi, T. Oki, and H. Nakanishi, NMR study of pradimicin derivative BMY-28864 and its interaction with calcium ions in D_2O, *Spectrochim. Acta A Mol. Biomol. Spectrosc.*, 55 (1999) 2547–2558.
139. M. Hu, Y. Ishizuka, Y. Igarashi, T. Oki, and H. Nakanishi, NMR, UV–Vis and CD study on the interaction of pradimicin BMY-28864 with divalent cations of alkaline earth metal, *Spectrochim. Acta A Mol. Biomol. Spectrosc.*, 56 (1999) 181–191.
140. M. Hu, Y. Ishizuka, Y. Igarashi, T. Oki, and H. Nakanishi, Interaction of three pradimicin derivatives with divalent cations in aqueous solution, *Spectrochim. Acta A Mol. Biomol. Spectrosc.*, 56 (2000) 1233–1243.
141. K. Fujikawa, Y. Tsukamoto, T. Oki, and Y. C. Lee, Spectroscopic studies on the interaction of pradimicin BMY-28864 with mannose derivatives, *Glycobiology*, 8 (1998) 407–414.
142. Y. Nakagawa, Y. Masuda, K. Yamada, T. Doi, K. Takegoshi, Y. Igarashi, and Y. Ito, Solid-state NMR spectroscopic analysis of the Ca^{2+}-dependent mannose binding of pradimicin A, *Angew. Chem. Int. Ed.*, 50 (2011) 6084–6088.
143. M. F. Summers, ^{113}Cd NMR spectroscopy of coordination compounds and proteins, *Coord. Chem. Rev.*, 86 (1988) 43–134.
144. H. Xia and G. D. Rayson, ^{113}Cd-NMR spectrometry of Cd^{2+} binding sites on algae and higher plant tissues, *Adv. Environ. Res.*, 7 (2002) 157–167.
145. Y. Nakagawa, T. Doi, Y. Masuda, K. Takegoshi, Y. Igarashi, and Y. Ito, Mapping of the primary mannose binding site of PRM-A, *J. Am. Chem. Soc.*, 133 (2011) 17485–17493.
146. K. Takegoshi, S. Nakamura, and T. Terao, ^{13}C–^{1}H dipolar-assisted rotational resonance in magic-angle spinning NMR, *Chem. Phys. Lett.*, 344 (2001) 631–637.
147. K. Takegoshi, S. Nakamura, and T. Terao, ^{13}C–^{1}H dipolar-driven ^{13}C–^{13}C recoupling without ^{13}C rf irradiation in nuclear magnetic resonance of rotating solids, *J. Chem. Phys.*, 118 (2003) 2325–2341.
148. C. R. Morcombe, V. Gaponenko, R. A. Byrd, and K. W. Zilm, Diluting abundant spins by isotope edited radio frequency field assisted diffusion, *J. Am. Chem. Soc.*, 126 (2004) 7196–7197.
149. Y. Nakagawa, T. Doi, K. Takegoshi, Y. Igarashi, and Y. Ito, Solid-state NMR analysis of calcium and D-mannose binding of BMY-28864, a water-soluble analogue of pradimicin A, *Bioorg. Med. Chem. Lett.*, 22 (2012) 1040–1043.
150. A. Tanabe, H. Nakashima, O. Yoshida, N. Yamamoto, O. Tenmyo, and T. Oki, Inhibitory effect of new antibiotic, pradimicin A on infectivity, cytopathic effect and replication of human immunodeficiency virus in vitro, *J. Antibiot. (Tokyo)*, 41 (1988) 1708–1710.
151. H. Hoshino, J. Seki, and T. Takeuchi, New antifungal antibiotics, benanomicins A and B inhibit infection of T-cell with human immunodeficiency virus (HIV) and syncytium formation by HIV, *J. Antibiot.*, 42 (1989) 344–346.
152. A. Tanabe-Tochikura, T. S. Tochikura, O. Yoshida, T. Oki, and N. Yamamoto, Pradimicin A inhibition of human immunodeficiency virus: Attenuation by mannan, *Virology*, 176 (1990) 467–473.
153. J. Balzarini, K. Van Laethem, D. Daelemans, S. Hatse, A. Bugatti, M. Rusnati, Y. Igarashi, T. Oki, and D. Schols, Pradimicin A, a carbohydrate-binding nonpeptidic lead compound for treatment of

infections with viruses with highly glycosylated envelopes, such as human immunodeficiency virus, *J. Virol.*, 81 (2007) 362–373.
154. J. Balzarini, K. O. François, K. Van Laethem, B. Hoorelbeke, M. Renders, J. Auwerx, S. Liekens, T. Oki, Y. Igarashi, and D. Schols, Pradimicin S, a highly soluble nonpeptidic small-size carbohydrate-binding antibiotic, is an anti-HIV drug lead for both microbicidal and systematic use, *Antimicrob. Agents Chemother.*, 54 (2010) 1425–1435.
155. S. Shahzad-ul-Hussan, R. Ghirlando, C. I. Dogo-Isonagie, Y. Igarashi, J. Balzarini, and C. A. Bewley, Characterization and carbohydrate specificity of pradimicin S, *J. Am. Chem. Soc.*, 134 (2012) 12346–12349.
156. D. Huskens, K. Vermeire, E. Vandemeulebroucke, J. Balzarini, and D. Schols, Safety concerns for the potential use of cyanovirin-N as a microbicidal anti-HIV agent, *Int. J. Biochem. Cell Biol.*, 40 (2008) 2802–2814.
157. K. O. François, J. Auwerx, D. Schols, and J. Balzarini, Simian immunodeficiency virus is susceptible to inhibition by carbohydrate-binding agents in a manner similar to that of HIV: Implications for further preclinical drug development, *Mol. Pharmacol.*, 74 (2008) 330–337.

ENZYMATIC CONVERSIONS OF STARCH

Piotr Tomasik[a] and Derek Horton[b]

[a]Krakow College of Health Promotion, Krakow, Poland
[b]Chemistry Department, Ohio State University, Columbus, Ohio, USA

I. Introduction	61
1. Introduction and General Remarks	61
2. Historical Background	62
3. Former Reviews	64
II. Enzymes and Microorganisms for Conversion of Starch	65
1. Introduction	65
2. Alpha Amylases (EC 3.2.1.1)	80
3. Beta Amylases (EC 3.2.1.2)	98
4. Glucoamylase (EC 3.2.1.3)	103
5. Other Amylases	111
6. α-Glucosidase (EC 3.2.1.20)	111
7. Pullulanase (EC 3.2.1.41)	114
8. Neopullulanase (EC 3.2.1.135)	117
9. Isoamylase (EC 3.2.1.68)	117
10. Other Hydrolases	119
11. Enzymatic Cocktails	120
12. Glycosyltransferases (EC 2.4.1)	123
13. Microorganisms	124
III. Hydrolysis Pathways and Mechanisms	127
1. Role of Adsorption	127
2. Mechanism of Inhibition	130
3. Mathematical Models of Enzymatic Hydrolysis	132
4. Effect of Light, Microwaves, and External Electric Field	135
5. Kinetics	136
IV. Amylolytic Starch Conversions	144
1. Introduction	144
2. Pulping	145
3. Malting	146
4. Mashing	146
5. Liquefaction	148
6. Saccharification	151
7. Effect of the Botanical Origin of Starch	159
8. Role of Starch Pretreatment	174

ISBN: 978-0-12-396523-3
http://dx.doi.org/10.1016/B978-0-12-396523-3.00001-4

9. Role of Temperature	180
10. Role of the Substrate Concentration	181
11. Role of Water	181
12. Role of Elevated Pressure	182
13. Role of pH	182
14. Role of Admixed Inorganic Salts	183
15. Role of Inhibitors	186
16. Stimulators of Hydrolysis	188
17. Engineering Problems	189
18. Applications of the Enzymatic Processes	191
V. Starch as a Feedstock for Fermentations	208
1. General Remarks on Fermentation	208
2. Alcohol and Alcohol–Acetone Fermentations	209
3. Carboxylic Acid Fermentations	232
VI. Nonamylolytic Starch Conversions	238
1. Glycosylation	238
2. Esterification and Hydrolysis	241
3. Methanogenic and Biosulfidogenic Conversions	243
4. Isomerization	243
5. Hydrogen Production	245
6. Trehalose	246
7. Bacterial Polyester Formation	247
8. Branching of Starch	247
9. Oxidation	247
10. Polymerization	247
11. Cyclodextrins	249
VII. Starch Metabolism in Human and Animal Organisms	258
1. Digestible Starch	258
2. Resistant Starch	261
VIII. Starch Analytics Involving Enzymes	262
1. Starch Evaluation and Analysis	262
2. Enzyme Evaluation	266
References	268

ABBREVIATIONS AND DEFINITIONS

AA, amylase active; AGU, anhydroglucose unit; α-CD, α-cyclodextrin, cyclomaltohexaose; β-CD, β-cyclodextrin; cyclomaltoheptaose; γ-CD, γ-cyclodextrin, cyclomaltooctaose; CGTase, cyclodextrin glycosyltransferase; CMC, O-(carboxymethyl)cellulose; DA, dextrinase-active; DE, dextrose equivalent; DEAE, diethylaminoethyl; DMA, degree of multiple attack; DP, degree of polymerization; EC, Enzyme Classification; FTIR, Fourier-transform infrared; Glucose ("dextrose"), D-glucopyranose;

Isomaltose, α-D-Glc-(1→6)-α-D-Glc; MALDI-TOF, matrix-assisted, laser-desorption, time-of-flight mass spectrometry; Pancreratin, digestive enzymes of pancreatic juice; Panose, α-D-Glc-(1→6)-α-D-Glc-(1→4)-D-Glc; Pepsin, proteolytic enzyme of gastric juice; PEG, poly(ethylene glycol); PEO, poly(ethylene oxide); PPO, poly(propylene oxide); Ptyalin, salivary alpha amylase; Pullulan, an α-(1→6)-linked polymer of maltotriose residues; RS, resistant starch; SP, saccharifying power; TEMPO, (2,2,6,6-tetramethylpiperidin-1-yl)oxyl.

I. INTRODUCTION

1. Introduction and General Remarks

Next to cellulose and hemicelluloses, starch is the most abundant biopolymer on our planet. It is renewable and biodegradable. It is widely available at low cost. During the period from the 17th century to the present time, several tens of thousands of papers have been published on applications of native and modified starches from various botanical origins in numerous areas of technology and nutrition.[1]

Contemporary trends in the manufacture of biodegradable and environmentally benign products from natural, nonfossil resources have targeted polysaccharides as potential source materials. The outlook in the European Union for the chemical and para-chemical industries designates starch as a promising and versatile resource that can limit and even eliminate the utilization of fossil source materials.[1–3]

In former articles published in this series, the author presented modifications of starch involving physical,[4,5] physicochemical,[6,7] and chemical[8] methods. The latter article has been supplemented with other, newer surveys.[9] More and more attention is now being paid, not only to the exploitation of renewable resources but also to "green" processes. A wide range of possibilities for the development of such processes is offered by enzymatic transformations of starch. The advantage of enzymatic over acid- and base-catalyzed transformations results from their selectivity. For instance,

---

[1]Editor's note. This article covers an enormous body of work on starch and its enzymology. It relates not only to fundamental aspects but also to many applied areas of food technology, nutrition, industrial applications, and other fields. In addition to the cited publications in peer-reviewed journals, there are many reports in obscure or discontinued periodicals and other sources that may not have had a rigorous peer-review policy. The numerous patents cited document claims that have not been independently reviewed or verified and may not have been the subject of actual experimental work. Professor Tomasik's article provides a very extensive overview of the literature on starch enzymology, but caution is advised in assessing the validity of information from sources that have not been peer-reviewed.

the hydrolysis of starch by amylases and glucoamylases provides only dextrose (D-glucose)-containing products, and no bitter-tasting components are formed.[10]

A wide range of available enzymes offers a variety of products substantially different from these produced by acid-catalyzed hydrolysis (see, for instance, Refs. 11, 12). Aside from the huge number of reports dealing with the hydrolytic scission of the amylose and amylopectin chains of starch, articles on the enzymatic oxidation, glycosylation, cyclization, and production of various materials are also available. Frequently, starch is used as a nutrient for technologically useful microorganisms, for instance, yeast.[13]

Enzymatic transformations of starch may utilize isolated single enzymes or blends (cocktails) thereof, or producers of enzymes, such as microorganisms, or plant tissue fragments, animal body fluids, and animal organs.

Raw starch in plant waste (biomass), as well as so-called green plastics, namely, synthetic plastics formed with such biopolymers as starch may also be enzymatically degraded. This question merits a separate article.

2. Historical Background

The earliest notes documenting the isolation and application of starches from various origins are at least 5500 years old. Perhaps the first scientific paper on the enzymatic conversion of starch appeared in 1833.[14] Initially, aside from human consumption, starch was used in gelatinized form as a size for papyrus and paper. This application was replaced with starch "dextrinized" (partially depolymerized) by acid hydrolysis with vinegar.[15] The report by Kirchhoff [16] describing acid-catalyzed "saccharification" of starch (depolymerization to a sweet-tasting product) opened an era of systematic studies on the acid-catalyzed degradation of starches.

Starch is an unstable system. On storage, particularly when moist, and when dried at elevated temperature, the granules of starch undergo autolysis by enzymes naturally residing inside the granules.[17,18] The enzymatic autolysis of granules observed when stored starch rotted induced interest in this as a method for transformation of starch in the second decade of the 20th century. One of the milestones at this stage was American patent for the enzymatic clarification of apple juice.[19] Soon it became apparent that enzymes hydrolyzing starch are common in the environment and in living organisms.[20]

At the outset, several sources of enzymes were used arbitrarily for modification of starches. These sources were, for example, human and animal saliva,[21–23] dried saliva (provided it had not been sterilized),[24–27] dried urine,[24] extracts of such tissues as calf muscle,[28] dried thyroid glands,[29] such body fluids as gastric juice (pepsin),[29] pancreatic juice,[30,31] plants and parts thereof, such as nonmalted starch,[32] malt[33–36] and its extracts,[37] wort,[38] soil bacteria,[39] other microorganisms, particularly from the groups of *Bacilli*,[40–43] *Sarcini*,[43] also *Cocci*,[39] *Corynebacteria*,[44–46] and fungi and yeasts such as *Eurotium oryzae*,[47] *Glycini*,[28] *Saccharomyces*,[48,49] and *Mucor*.[50,51] Subsequently, isolated impure enzyme preparations termed "diastase" (later characterized as a blend of various amylases, and pepsin) were introduced. These were then supplemented by other complex enzyme preparations. They included "albumose," a preparation derived from albumin and formed by the enzymatic breakdown of proteins during digestion, and containing peptone and neutral salts.[52–55] Other incompletely identified preparations included pancreatin (a mixture of amylase, lipase, and protease),[56] a mixture of enzymes from sweet and bitter almonds,[57] and "biolase," an enzyme preparation extracted from the lupin plant.[58–61] Some of these starch-hydrolyzing preparations are still in use. The name "diastase" was later abandoned in favor of the term "amylase." Differentiation between various amylases and other hydrolytic enzymes was realized in the middle of the 20th century.[62–64] Subsequently, stable, dried, enzyme-containing mixtures were introduced for use in molecular biology.[65]

Either isolated starch or starchy material can be subjected to enzymolysis. The results, namely, the products formed and their rate of formation, depend on the substrate and enzyme.[66–68] All flora and fauna contain various endo- and exoenzymes, among them hydrolases that offer potential for conversion of starch into useful products. However, several limitations result from economic and technical problems attendant on their isolation and purification. Most of the reported conversions of starch involve hydrolysis by various enzymes in the hydrolase group, to afford products ranging from dextrins through oligosaccharides, and further to maltose and glucose, and these are then utilized either in a dry crystalline state or as syrups. Aside from hydrolases, other enzymes, such as phosphorylases, synthases, oxidases, invertases, cyclodextrin transferases, and others, are used for specific processes. They utilize either starch or such products of its conversion as dextrins, oligosaccharides, maltose, and glucose.

Enzymes immobilized on support media are in common use. They may be used in single or complex preparations. Their application may present problems arising from possible synergistic or antagonistic interactions of the components. Such antagonism leads to decrease in activity of the enzyme preparation[69] and may change its mode of action and narrow the pH range tolerated. At the same time, the thermostability and control of the profile of the products obtained may be enhanced.[69–72] Early applications utilized a clay support as a carrier for bacterial amylase, and such a catalyst could be used even in fluidized-bed reactors.[73]

Progress in genetics and enzymology has offered efficient mutants for various types of starch conversion. The motivation for these manipulations included increase in the enzyme's thermostability, its stability at a specifically selected pH, and either increase or decrease in selectivity. Such enzymes can be isolated from genetically modified plants and microorganisms. For instance, *Saccharomyces* cells were transformed to secrete both yeast glucoamylase and mouse alpha amylase.[74] Several mutants of *Bacillus licheniformis* alpha amylase were prepared.[75] Modifications of the electrostatic environment at the active sites of the enzyme can be the determining factor for the changes in the action pattern of the enzyme. The structure of natural enzymes may be modified by genetic manipulations: see, for instance, Refs. 76–104 or by using ionizing radiation.[105] Chemical modifications of enzymes involve mainly acylation of their amino groups[106,107] and sulfhydryl bonds.[108]

3. Former Reviews

The structure, action pattern, and mechanisms of hydrolases have been reviewed several times in various monographs[109–112] and articles.[113–118] Hollo and Laszlo[119] compared the acidic, alkaline, and enzymatic hydrolysis of starch. Other articles have dealt with individual hydrolases, such as acid alpha amylase,[120] microbial alpha amylase,[121,122] taka amylase (amylase of *Aspergillus oryzae*),[123,124] microbial acid- and thermostable alpha amylase,[120,125] isoamylases,[126,127] glucoamylases,[128–134] and pullulanases,[126,135] α-1,4-glucan glucohydrolase,[136] beta amylase,[137,138] amylases for production of maltose from starch,[139–141] enzymes for production of anhydrofructose,[142] thermophilic bacteria for production of α-glucosidase,[143] and CGTase, the enzyme isolated from *B. macerans*[144] and anaerobic bacteria.[145] Engineering of enzymes for the starch industry has also been reviewed.[146]

Hydrolysis of starch involving a variety of enzymes has been surveyed in numerous papers,[13,118,121,122,129,136,139,146–226] some of them emphasizing the mechanisms of the process [160,214,215] and its technological importance.[216] Other articles have focused on particular steps of starch conversions, including malting and mashing,[227] liquefaction,[228,229] glucose reversion,[230] saccharification,[152,229] and simultaneous saccharification and fermentation.[231]

Starch as a substrate has been the subject of a number of reviews,[151,156,160,163,164,184,232–239] and the bioavailability of starches and reasons for their resistance to amylolysis in the context of starch metabolism were reviewed in 1993.[240]

Many articles have focused on the products of enzymolysis: these include glucose, maltose, coupling sugars, pullulans, hydrolyzates,[162,241–252] trehalose,[174,253,254] anhydrofructose,[255] conversion of glucose syrups into fructose syrups,[229,256] cyclic tetrasaccharides,[257] and various organic compounds.[258,259] Current "hot topics" include bioconversion in general,[128,157,158,260] alcohol fermentation,[261–265] and "bioethanol" production.[266–269] Other applications include production of such beverages as traditional Chinese spirits,[270] ethanol from sweet potatoes[271] and grain,[272] continuous ethanol production with immobilized cells,[273] and processes yielding acetone and butanol,[274,275] citric acid,[276,277] and vinegar.[278] Cyclodextrins have attracted much attention and have been the subject of several reviews.[279–290]

Reviews are also available on a range of areas, including the inhibitory role of pectins toward hydrolases,[291] processes in such specific fields as starch for paper industry,[292,293] desizing in the textile industry,[294] fruit-juice processing,[295] the function of starches in diet,[296] membrane reactors for starch conversion,[297] methods for characterizing the activity of maltooligosaccharide-forming amylases and diastatic decomposition,[298,299] and enzymatic reactions of value in discovering details of the starch structure.[300]

II. ENZYMES AND MICROORGANISMS FOR CONVERSION OF STARCH

1. Introduction

Animals, plants, bacteria, yeasts, and fungi are the commonest sources of hydrolases. Naturally occurring hydrolases (EC 3.2.1) capable of converting starch into specific products are mainly amylases.

The alpha amylase family (glycoside hydrolase family 13, GH 13) (1,4-α-D-glucan glucanohydrolases) contains about 30 enzymes, distinguished from one another in their specific properties.[165,178,182,301,302] The selectivity in their action[10] upon starch is, perhaps, their most significant property discussed in this section. There are 21 different reaction and product specificities found in this family.[182]

The alpha amylase evolutionary tree is presented in Fig. 1.

The amylase family also includes beta amylase (1,4-α-D-glucan maltohydrolase) (EC 3.2.1.2), gamma amylase (1,4-α-D-glucohydrolase) (EC 3.2.1.3) known as glucoamylase or amyloglucosidase, α-glucosidase (α-D-glycoside glucohydrolase) (EC 3.2.1.20), oligo-1,6-glucosidase (dextrin-6-α-D-glucanohydrolase) (EC 3.2.1.10), isoamylase (glycogen 6-glucanohydrolase) (EC 3.2.1.68), pullulanases type I (α-dextrin 6-glucanohydrolase) (EC 3.2.1.41) and pullulanase type II (alpha amylase–pullulanase) (EC 3.2.1.1.41), and limit dextrinase (dextrin α-1,6-glucanohydrolase) (EC 3.2.1.142).

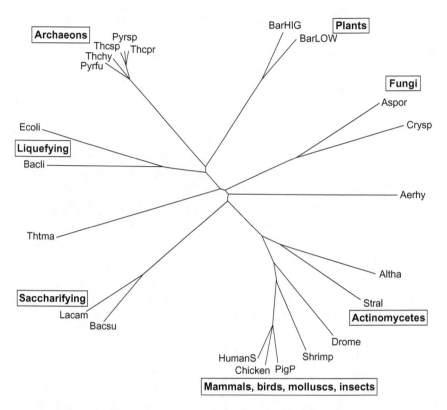

Fig. 1. Alpha amylase evolutionary tree, including only eubacterial, eukaryal, and archeal enzymes, constructed on the basis of a sequence alignment, starting at strand β2 and ending at strand β8 of the $(\alpha/\beta)_8$ barrel, and including the entire B domain. The branch lengths are proportional to the divergence of the sequences of the individual alpha amylases. The sum of the lengths of the branches linking any two alpha amylases is a measure of the evolutionary distance between them. Aerhy=*Aeromonas hydrophila*, Altha=*Alreromonas haloplanctis*, Bach=*Bacillus licheniformis*, Bacsu=*Bacillus subtilis*, Ecoli=*Escherichia coli*, Lacam=*Lactobacillus amylovorus*, Stral=*Streptomyces albidoflavus*, Thtma=*Thermotoga maritime*, Pyrfu=*Pyrococcus furiosus*, Pyrsp=*Pyrococcus* sp. Rt-3, Thchy=*Thermococcus hydrothermalis*, Thcpr=*Thermococcus profundus*, Aspor=*Aspergillus oryzae*, Crysp=*Cryptococcus* sp. BarHIG=barley high pI isoenzyme, BarLOW=barley low pI isoenzyme, Drome=*Drosophila melanogaster*, Chicke=chicken, HumanS=human saliva, PigP=pig pancreas, Shrimp=shrimp. (reproduced from Ref. 177 with permission)

Generally, fungi secrete alpha amylase, but some fungi are known that secrete alpha and beta amylases depending on the composition of the medium in which they are produced.[303,304] Around 100 yeast species produce hydrolases, usually alpha amylase and glucoamylase.[167,305,306]

For technical and economic reasons there is a great interest in thermophilic amylolytic enzymes. Several such thermophilic and hyperthermophilic Archaea have been found in hot springs, hydrothermal vents, and deep sea marine habitats.[177] They are sources of alpha amylase, pullulanase, and α-glucosidase. Depending on the source, optimum conditions for their operation are between 75 and 115 °C and, with two exceptions, function at pH values below 7.0. The alpha amylase from *Pyrococcus furiosus* and the α-glucosidase from *Thermococcus zilligii* have their optimal pH at 7.5 and 7.0, respectively.

Xylanases (EC 3.2.1.8) do not hydrolyze starch but are utilized in the conversion of some starch substrates. These enzymes are secreted by microorganisms that thrive on plant sources and fungi, and degrade plant matter into usable nutrients. They break down hemicelluloses, one of the major components of plant cell walls. Thus, xylanases are used in processing wheat starch and wheat material, converting β-(1 → 4)-xylan into xylose.[307] Microorganisms of the rumen,[308] bacteria, and fungi also produce fiber-degrading enzymes[309] and amylases.[309–312] The action pattern of the hydrolases upon starch is illustrated in Fig. 2.

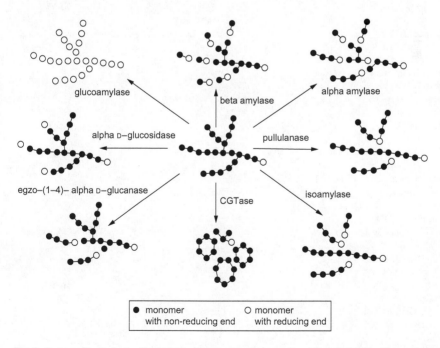

FIG. 2. Pattern of hydrolysis of hydrolases digesting starch. (reproduced from Ref. 260 with permission)

There is an essential difference in hydrolytic behavior between gelatinized, soluble, and raw starch. Not every enzyme or microorganism is capable of degrading raw starch.[312,314] Table I lists microorganisms producing enzymes that digest raw starch. Selected properties of those enzymes are included in Tables II–VI.

TABLE I
Microorganisms Producing Enzymes Decomposing Raw Starch[243,313,315]

Microorganisms	Source	Enzyme type[a]
Fungi		
Acremonium sp.	Forest tree	GA
Aspergillus awamori		GA
A. awamori var. kawachi		GA
A. awamori var. kawachi F-2035		
A. awamori KT-11	Air	alpha
A. carbonarius	Rotten cassava	beta
A. cinnamomeus		GA
A. ficum		alpha
A. fumigatus K27		alpha
A. niger		GA
A. niger	Cassava waste	
A. niger AM07	Soil	alpha, GA
A. niger NIAB 280		GA
A. niger sl. 1	Soil	
Aspergillus sp. GP-21	Soil	GA
A. oryzae		GA
Aspergillus sp. N-2	Cassava chips	GA
Aspergillus sp.	Garden soil	GA
Aspergillus sp. K-27	Soil	alpha
A. terreus	Cassava waste	
Candida antarctica CBS 6678		GA, alpha
Chalara paradoxa	Pith of sago palm	GA, alpha
Cladosporium gossypiicola ATCC 38016		GA
Corticum rolfsii	Tomato stem	GA
Endomycopsis fibuligera		GA
Fusidium sp. BX-1		GA
Giberella pulicaris	Tree	GA, alpha
Lentinus edodes Sing.		GA
Mucor rouxianus		GA
Nodilusporium sp.		GA
Penicillium brunneum No. 24		alpha
P. oxalicum		
Peniclllium sp. S-22	Soil	
Penicillium sp. X-1		GA
Rhizoctonia solani		GA

TABLE I (*Continued*)

Microorganisms	Source	Enzyme type[a]
Rhizomucor pusillus		GA
Rhizopus niveus		GA
Rhizopus sp. A-11	Ragi	GA
Rhizopus sp MB46		GA
Rhizopus sp. MKU 40		GA
Rhizopus sp. W-08	Mildewed corn	GA
Rhizopus stolonifer	Cassava waste	
Schizophyllum commune		GA
Synnematous sp.		alpha
Streptomyces bovis		alpha
S. hygroscopicus		MA
S. limosus		alpha
S. precox NA-273		MA, alpha
Streptomyces sp. E-2248	Mud	
Streptomyces sp. No. 4	Soil	alpha
S. thermocyaneoviolaxeus		
Thermomucor indicae-seudaticae	Soil	GA
Thermomyces lanuginosus F1	Municipal compost	alpha
Yeast		
Aureobasidium pullulans N13d	Deep sea	GA
Candida antartica		
Cryptococcus sp. S-2		alpha
Endomycopsis fibuligera		GA
Lipomyces starkeyi		
Saccharomycopsis fibuligera		GA
Bacteria		
Anoxybacillus contaminans		alpha
Bacillus alvei		
B. amyloliquefaciens	Soil	alpha
B. cereus		beta
B. circulans F-2	Potato starch	M6
B. firmus	Soil	CGT
B. firmus lentus		
B. licheniformis (Termamyl)		alpha
B. macerans BE101		CGT
B. polymyxa TB1012		beta
Bacillus sp. B1018		alpha
Bacillus sp. 2718	Soil	beta
Bacillus sp. I-3		alpha
Bacillus sp. IMD-370		alpha
Bacillus sp. IMD 434		alpha
Bacillus sp. IMD 435	Mushroom compost	alpha
Bacillus sp. TS-23		alpha
Bacillus sp. WN11	Hot spring	alpha

Continued

TABLE I (Continued)

Microorganisms	Source	Enzyme type[a]
Bacillus sp. YX-1	Soil	alpha
B. stearothermophilus		MA
B. steraothermophilus NCA 26		alpha
B. subtilis 65		alpha
B. subtilis IFO 3108		alpha
Clostridium butyricum T-7	Mesophilic methane sludge	alpha
C. thermosulfurogenes		beta
Cryptococus sp. S-2		alpha
Cytophaga sp.	Soil	alpha
Geobacillus thermodenitificans HRO10		alpha
Klebsiella pneumonia		CGT
Lactobacillus amylophilus GV6	Starchy waste	AMP
L. amylovorus		alpha
L. plantarum		
Rhodopseudomnas gelatinosa T-20, T-14		alpha
Streptococcus bovis 148		alpha
Thermoanaerobacter sp.		CGT
Plants		
Barley		alpha
Barley evolved		alpha
Canadian poplar		alpha

[a] alpha = alpha amylase; beta = beta amylase; AMP = amylopullulanase; CGT = cyclomaltodextrin glucotransferase, GA = glucoamylase, MA = maltogenic amylase, M6 = maltohexahydrolase.

TABLE II
Comparison of the Degrading Ability of Porcine Pancreatin and Alpha Amylase from *Bacillus* sp. and *A. fumigates*.[617]

	Hydrolysis (percent) with alpha amylases (reaction time)			
			A. fumigatus	
Starch	Porcine pancreatin (29 h)	*Bacillus* sp.(29 h)	(29 h)	(70 h)
Smooth pea	91	78	100	100
Wheat	90	71	94	100
Waxy maize	88	54	94	100
Normal maize	79	46	94	100
Wrinkled pea	72	66	77	82
Potato	5	3	12	28
Hylon maize	2	1	47	

TABLE III
Sources of Alpha Amylases, Their Molecular Weights, and Optimum Temperature and pH

Source	Molecular weight (kDa)	Optimum Temperature (°C)	pH	Ref.
Human and animal				
Human saliva			5.6–9.0	367
Human pancreas		30–50	7.0–7.2	376
Cow saliva		45	6.5–6.6	369
Horse saliva		50	6.2	369
Euphausia superba	17	35	6.4	397
Honey			5.3–5.5	398
Psammechinus miliaris			7	399
Bombycx mori				400
Kyphosidae				401
Sulculus diversicolor	55.7	45	6	402
aquatitis	65	50		
Tenebrio molitor			5.8	637
Periplaneta americana				638
Plant				
Amygdalin				417,418
Cholam (Sorghum vulgare)				413
Barley				408
Broad bean				408
Almonds		52.5	5.5	57
Malt		50–55	4.4–5.4	408,428,431
Sweet potato	45	71.5	5.8–6.4	442
Safflower (Carthamus tinctorius L.)	35	55	6	639
Oriza sativa rice:				
Y	44	70	4.35	410
Z	44	70	4.45	
A+B	44	70	4.6–4.7	
A	44	70	4.6	
E	42	26	5	
F	44	37	5.1	
G	44		5.4	
H	44	37	5.7	
I	44	37	6.3	
J	44		6.6	
Soy bean				408,409
Sprouted potatoes				440
Wheat flour				432

Continued

TABLE III (*Continued*)

Source	Molecular weight (kDa)	Optimum Temperature (°C)	pH	Ref.
Bacterial				
Aeromonas caviae				640
Alicyclobacillus acidocaldonus	160	75	3	641
Alteromonas haloplanetis	49.34			642,643
Anaerobic bacterium				644
Archeobacterium pyrococcus woesei				645
Bacillus acidocaldarius				646
B. amyloliquefaciens	49–52	70	6	180,647–651
B. brevis				652–654
B. brevis HPD 31		45–55	6	655
B. circulans				656,657
B. coagulans				658
B. flavothermus		60	5.5–6.0	659,660
B. globisporus				661
B. lentus	42	70	6.1	662
B. licheniformis NCIB 6346	62–65	70–90	7.9	180,663
B. licheniformis	22.5–28.0	76–90	9	451,480,651, 664–668
B. licheniformis (hyperactive)	64	50	6.0–8.0	669
B. licheniformis M72	56	85–90	6.5–7.0	670
B. megaterium	59	80	5,5	671
Bacillus sp. WN11	53–76	70–80	5.5	476,672,673
Bacillus sp. IMD434	69.2	66	6	477
Bacillus sp. IMD435	63	65	6.0–6.5	674
Bacillus sp. TS-23	42	70	9	673
Bacillus sp. YX-1	56	40–50	5	675
Bacillus sp. I-3		70	7	676
Bacillus sp. US 100		82	5.6	677
Bacillus sp. XAL 601		70	9	525
B. stearothermophilus	47 exo 58 endo	55–70	4.6–5.1	508,678–683
B. stearothermophilus ATCC 12980	59	70–80	5.0–6.0	684
B. stearothermophilus MFF4		70–75	5.5–6.0	685
B. subtilis	48.0–57	50–80	5.4–6.5	686–689
B. subtilis 65	68	60	6.5	690

TABLE III (Continued)

Source	Molecular weight (kDa)	Optimum Temperature (°C)	pH	Ref.
Bacillus subtilis var. amylosaccharificus	41			180
Bifidobacterium adolescentis	66	50	5.5	691
Chloroflexus aurantiacus	210	71	7,5	692
Clostridium acetobutylicum	84	45	5.6	693–695
C. butricum				696
C. perfringens	76	30	6.5	697
C. thermosulfuricum				696
C. thermosulfurogenes				698,699
Cytophaga sp.	59	50–60	4.5–9.5	700
Escherichia coli	48	50	6.5	686
Eubacterium sp.				701
Halobacterium halobium				702
H. salinarium				668
Halomonas meridiana DSM 5425		37	7	703
Humicola insolens				704
H. lanuginosa				704
H. stellata				704
Lactobacillus amylovorus	140	60–65	5.5	705
L. brevis	75.9	55	6.5	706,707
L. cellobiosus				708
L. plantarum A6	50	65	5.5	709
Micrococcus luteus	56	30	6	706,707
M. varians	14–56	45	7	710
Micromonospora melanosporea	45	55	7	711
M. vulgaris				712
Myxococcus coralloides	22.5	45	8	713
Nocardia asteroids	56–65	50	6.9	714
Pseudomonas stutzeri	12.5	47	8	447,715
Streptococcus bovis JB1	77		5.0–6.0	716
Streptomyces sp. IMD 2679	47.8	60	5.5	717
Streptomyces sp. No. 4	56	50	5.5	718
Thermoactinomyces sp.				719
T. vulgaris	53	62.5	4.8–6.0	720–722
Thermonospora curvata	60.9	80	6	723–725
T. fusca XY		60	6	726

Continued

TABLE III (*Continued*)

Source	Molecular weight (kDa)	Optimum Temperature (°C)	pH	Ref.
T. profundus DT5432	42	80	5.5–6.0	727
T. viridis				728
Thermotoga maritima	61	85–90	7	729
Thermus sp.	59	70	5.5–6.5	557,720
T. filformis	60	95	5.5–6.0	730
Yeast				
Candida famata		50	5	731
C. fennica		50	5	731
Candida tsukubaensis				732
C. tsukubaensis CBS 6389				733
Filobasidium capsuligenum				734
Fusarium vasinfectum			4.4–5.0	735,736
Lipomyces kononenkoe CBS 5608	76	70	4.5–5.0	689,737
L. starkei				738,739
Lipomyces sp.				740
Saccharomyces cerevisae	54.1	50	6	741
Saccharomycopsis bispora				742
S. capsularis				743
S. fibuligera				744–746
Schwanniomyces alluvius	61.9	40	6.3	561
S. alluvius UCD 5483	61.9	40	6.3	747,748
S. castelli				749
Trichosporon pullulans				750
Fungal				
Aspergillus awamori		40	5	751
A. awamori ATCC 22342	54	50	4.8–5.0	752
A. carbonarius	32		6.0–7.0	753
A. chevalieri NSPRI 105	68	40	5.5	754
A. flavus	52–75	30–55	5.25–7.0	751,752,755
A flavus LINK	52.5	50	6.06	756
A. foetidus ATCC 10254	41.5	40–45	5.0–5.5	757,758
A. fumigatus	65	50	6	759,760
A. hennebergi Blochweitz	50	50	5.5	761
A. kawachi				762
A. niger	58–61	40–60	4.0–6.0	611–614, 763–766

TABLE III (*Continued*)

Source	Molecular weight (kDa)	Optimum Temperature (°C)	pH	Ref.
A. niger ATCC 13469		50	5.0–6.0	767
A. niger van Tieghem CFTRI 1105	56.23	60	5.0–6.0	768–770
A. oryzae	50–65	35–50	4.0–6.6	771–780
A. oryzae M13	52	50	5.4	521
A. oryzae ATCC 9376		30–40	5.0–6.0	781,782
A. usamii	54	60–70	3.0–5.5	783
Cladosporium resinae				586,587
Cryptococcus sp. S-2	66	50–60	6	784
Malbrachea pulchella var. *sulfurea*				704
Myceliophtora thermophila				785
Neocallimastix frontalis				786
Paecilomyces ATCC 46889	69	45	4	787
Penicillium brunneum				788
Pycnoporus sanguineus				789
Rhizopus sp.	64	60–65	4.0–5.6	789,790
Schizophyllum commune				791
Scytalidium sp.	87	50	6.5	780,792
Talaromyces thermophilus				704
Trichoderma viride			5.0–5.5	793
Thermomyces lanuginosus IISc 91	42	60–65	4.5–5.6	794–796
Archeal				
Desulfurococcus mucosus		100	5.5	797
Pyrococcus furiosus	66	100	6.5–7.5	797,798
Pyrococcus woesei	70	100–130	5.5	799
Staphylothermus marinus		100	5	797
Sulfolobus solfataricus				800
Thermococcus aggregans		100	5.5	797
T. celer		90	5.5	797,801
T. guaymagensis		100	6.5	797
T. hydrothermalis		75–85	5.0–5.5	177
T. profundus	42	80	5.5–6.0	802

TABLE IV
Sources of Beta Amylases, Their Molecular Weights, and Optimum Temperature and pH

Source	Molecular weight (kDa)	Optimum Temperature (°C)	pH	Ref.
Human and animal				
Sitophilus zeamais larvae mitgut			4.75	868
Sitophilus granaries larvae mitgut			5	868
Plant				
Eleusine coracana	59.1 kDa	50	5	923
Panicum miliaceum L.	58 kDa	55	5.5–6.0	924
	122 kDa	55	5.1–5.5	925
Barley	64			926
Potato	122	55	5.1–5.5	927
Sweet potato	206	53	5.3–5.8	442,828,876,927
Soybean	Two forms			928
Malt			5.65–5.85	356,840,878
Broad bean Vicia faba	107			929
Alfalfa Medicago sative L	61		7	930
Synapsis alba	58			931
Bacterial				
Bacillus cereus var. mycoids	35	50		180
B. circulans	53–64	50–60	7	180,932
B. megaterium NCIB 9323	35	50		180,720,933
B. megaterium NCIB 9376	32	50		180
B. polymyxa	44–59	37–45	5.5	180,934
B. stearothermophilus	39 extra 67 intra			508
Bacillus sp. IMD 198	58	55		180
Bacillus sp. BQ 10	160	45–55		180
Clostridium thermocellum				935
C. thermosulfurogenes	180	70	6	936–938
Pseudomonas sp. BQ 6	37	45–55		180
Rhizopus japonicus				939
Streptomyces sp.				939
Yeast				
Hendersonula toruloidea	60	60 60	6 5	940 941
Fungal				
Aspergillus carbonarius	32	40	6.0–7.0	753
A. niger			3.5–4.0	763

TABLE V
Sources of Glucoamylases, Their Molecular Weights, and Optimum Temperature and pH

Source	Molecular weight (kDa)	Optimum Temperature (°C)	pH	Ref.
Human and animal				
Intestine	21			944
Plant				
Sugar beet cells	83	65	4.4	945
Bacterial				
Bacillus firmus/lentus				947
B. stearothermophilus				508
Clostridium sp.G0005	77	65	4.5	1036
C. acetobutylicum				556,694,695
C. thermohydrosulfuricum		75	5–6	948,1037
C. thermosaccharolyticum	75	70	5.0	949
C. thermosulfurogenes				936–938
Flavobacterium sp.				952
Halobacter sodamense		65	7.5	1038
Lactobacillus brevis				950
Monascus kaoliang	48	50	4.5	951,953,1008
	68		4.7	
Yeast				
Arxula adeninivorans				1039
Candida famata		60	6.0	731
C. fennica		60	6.0	731
C. tsukubaensis				696
C. tsukubaensis CBS 6389				697
Filobasiium capsuligenum				698
Humicola lanuginosa		65–70	4.9–6.6	951
Lipomyces kon	76	70	4.5–5.0	689,741
L. starkei				738,739
Lipomyces sp.				740
Saccharomyces diastaticus		50	5.3	1040,1041
Saccharomycopsis bispora				742
S. capsularis				743
S. fibuligera				744–746
S. alluvius	117–155	50	4.5–5.0	561,748
S. alluvius UCD 5483	61.9	40	6.3	747,748
S. castelli				749
S. occidentalis				1042
Trichosporon pullulans				750

Continued

TABLE V (Continued)

Source	Molecular weight (kDa)	Optimum Temperature (°C)	pH	Ref.
Fungal				
Acremonium zonatum				1043
Amylomyces rouxii		60	4.5	1044
Aspergillus sp.				638,1045
Aspergillus awamori	83.7–110	60	4.5	180,1045–1050
A. awamori var. kawachi HF-15	57–250		3.8–4.5	977,1051
A. candidus	78I, 60II			1052
A. foetidus	75I, 60II			1052
A. niger I	58–112			180,638,1048,1049, 1051, 1053–1067
A. niger II	112			180
A. niger C-IV-4				1068
A. oryzae I	76	60		180
A. oryzae II	38	50		180
A. oryzae III	38	40		180
A. oryzae	38–76	50–65	4.5	979,1067,1069,1070
A. phoenicus		60	4.5	1071–1074
A. saitoi	90		4.5	180,1073
A. terreus	70	60	5.0	103,1073–1076
Aspergillus sp. GP-21		65	5.5	1077
Cephalosporium eichhormonie	26.85	45–62	5.4	180,1078
C. charticola	69	60	5.4	1078
Chalara paradoxa	82			996,997
Cladosporium resinae				585,586
Collectotrichum gloesporoides				1079
Coniophora cerebella			4.0–4.5	131,1080
Corticium rolfsii	78	40–60	4.5	172,1012,1081,1082
Endomyces sp.	55			131,1083
Endomycopsis capsularis	53	40–50	4.5	131,1084,1085
E. fibuligera				1086,1087
Fusidum sp.				1088
Lipomyces kononenkoae	811.5	50		180
Mucor javanicus	61	50	5.0	1089
M. rouxianus I	59	55	4.7	131,180,1049
M. rouxianus II	49	55	4.7	180
Neurospora crassa				1090
N. sitophila				1091
Paecilomyces globosus				1014

TABLE V (Continued)

Source	Molecular weight (kDa)	Optimum Temperature (°C)	pH	Ref.
P. varioti	69		5.0	131,1092
Penicillium italicum				1093
P. oxalicum I	84	55–60	4.5	180,1049
P. oxalicum II	86	60	4.6	180,1049
Penicillium sp. X-1	65		6.5	1094
Pricularia oryzae	94	50–55	4.5	1095
Rhizoctania solani				1061
Rhizopus sp.	58.6–74		4.5–5.0	990,1086,1096,1097
R. delemar	100	40	4.5	180,1098–1100
R. javanicus	48			1069,1101
R. niveus		60	4.5–6.0	647,1069,1102
R. oligospora				1103
R. oryzae				1067,1099
Rhizopus sp. A-11	72.4			1104
Schizophyllum commune	66	40	5.0	131,1105
Thermomucor indicae-seudaticae	42	60	7.0	527,1106
Thermomyces lanuginosa	57	65–70	4.9–6.6	951,1107
Torula thermophila				1108
Trichoderma resei				1109
T. viride				1027,1110

TABLE VI
Sources of Other Amylases, Their Molecular Weights, and Optimum Temperature and pH

Source	Molecular weight (kDa)	Optimum Temperature (°C)	pH	Ref.
Bacterial				
Bacillus clausii BT-21		55	9.5	96
Cytophaga sp.	59	60	6.5–9.5	700
Pseudomonas saccharophila				96
Fungal				
Aspergillus niger			6.0	763
Streptomyces sp. E-2248		50	6.0	1153
Gibberella pulicaris		40	5.5	1154

2. Alpha Amylases (EC 3.2.1.1)

Alpha amylase, also called dextrinogenic amylase, causes nonselective, random endohydrolysis of α-(1→4) bonds in amylose and amylopectin. That amylase produces maltose, maltotriose, and higher oligosaccharides from amylose as well as maltose, glucose, and, additionally, limit dextrins from amylopectin.[67,316–335] Based on the result of the reaction with KI_5, limit dextrins are differentiated into amyloamyloses (blue color), erythrodextrins (red color), and achroedextrins (yellow color).

Alpha amylases may be divided into (i) maltogenic amylase (1,4-α-D-glucan α-maltohydrolase) (EC 3.2.1.133), which produces maltose by splitting the α-(1→4) bonds from the nonreducing ends of the amylose molecules, (ii) maltotriohydrolase (1,4-α-D-glucan maltotriohydrolase) (EC 3.2.1.116) splitting α-(1→4) bonds from nonreducing ends to produce maltotriose, (iii) maltotetrahydrolase (1,4-α-maltotetrahydrolase) (EC 3.2.1.60) splitting the same bonds to produce maltotetraose, (iv) maltohexahydrolase (1,4-α-D-glucan maltohexahydrolase) producing maltohexaose, and so on. These variants differ from one another in their type of attack, which can be either random multichain or one-chain multiple attack[336,337] (see Section III).

Alpha amylases may be divided into liquefying and saccharifying categories. Amylases of the first category hydrolyze starch to a 30–40% extent; the latter to a 50–60% extent.[338]

Alpha amylases usually perform best in the pH range of 6.7–7.0 and tolerate alkaline media better than acidic solutions. Their activity decays over time, with increased temperature, and with increased concentration of such lower saccharides as glucose and maltose, which inhibit the hydrolytic action.[339]

There are some differences in the action pattern and specificity of alpha amylases from various sources.[340] Potato starch hydrolyzed with alpha amylase from *A. oryzae* produces chiefly maltose and maltotriose, but a considerable proportion of glucose is also produced. Small amounts of maltotetraose–maltoheptaose were also observed. The highest levels of maltotetraose and maltopentaose resulted[341] at 20 and 70 °C, respectively. Pancreatin, an enzyme mixture isolated from pancreas, produced from the same substrate chiefly maltose and maltotriose at 20–50 °C and maltose at 70 °C, whereas malt amylase gave mainly maltose, accompanied by glucose at 20 °C, and small amounts of maltotetraose and maltopentaose at 70 °C.[342,343] However, the behavior of alpha amylases as being dependent on their origin should not be considered as a rule. Alpha amylases from dog and pig pancreas performed almost identically.[344] Thoroughly purified alpha amylases from various sources exhibited the same digestibility of a given substrate.[51]

There are considerable differences in the results of hydrolysis when using purified and nonpurified enzymes. This is illustrated by the saccharification of starch with purified and nonpurified taka diastase. In the initial step the differences are negligible, but after 2 h of hydrolysis there are noticeable differences in the course of the reaction. In the first 2 h, a normal amylolysis with alpha amylase takes place. Later, the accompanying contaminating enzymes initiate digestion of the limit dextrins that are formed.[345–347] The differences can also result from the enzyme's source. For instance, it has been suggested[348] that amylases isolated from sorghum are, in fact, associated with α-glucosidases. Commercial alpha amylases from various manufacturers show different patterns of hydrolysis. This can result from subtle differences in the structure and purity of enzyme preparations, and these are frequently mixtures of both alpha and beta amylases.[349–353]

The state of the substrate is also a factor, as demonstrated in the case of rice. For liquefaction of rice starch, the optimal pH for the crude enzyme extracted from nonhulled rice was between 5.2 and 5.4, but for unpolished rice bran and embryo the optimum pH was 4.8, and for polished rice that pH was[354] between 6.1 and 6.2. For liquefaction, the purified alpha amylase from nonhulled, unpolished rice and rice bran had[355] an optimum pH 6.3 at 25 °C, and in germinating rice its optimum pH was[356,357] 4.5 at 50 °C. Differences in the action of alpha amylases can also result from the properties of the digested substrate.[358–360]

The different hydrolyzing behavior of pure enzymes can be defined as a degree of multiple attack (DMA), expressed as the number of bonds broken during the lifetime of an enzyme–substrate complex minus one, in case of potato amylose taken as the polysaccharide standard.[361] In terms of this measure, porcine pancreatin and alpha amylase from *Bacillus stearothermophilus* display high DMA values, alpha amylase from *Aspergillus oryzae* displays a low DMA, and alpha amylases of *B. licheniformis*, *Thermoactinomyces vulgaris*, *B. amyloliquefaciens*, and *B. subtilis* display borderline DMA values. The level of multiple attack (LMA), defined as the relation between decrease in KI_5 binding and the increase in total reducing value, increases with temperature for endoamylases to an extent dependent on the individual amylases.[361]

a. Human and Animal Alpha Amylases.—Human and animal alpha amylase is isolated mainly from saliva (ptyalin) and pancreas (pancreatin). The liver or muscle of animals is sometimes used as a source of that amylase.[28,360–364] Serum shows positive maltase and no amylase activity.[365] These isolates consist of several enzymes, and they act maximally at various stages of hydrolysis. Thus, only maltose is formed in the earlier stages, whereas hydrolytic activity appearing in the subsequent stages results in decomposition of maltose to glucose.[365]

(i) Ptyalin. Ptyalin with the AMY1A, AMY1B, and AMY1C genes has several genetic variations associated with the gene copy number, which in particular individuals may result from a long-established traditional composition of the human or animal diet.[366] Ptyalin performs best at pH 5.6–5.9, and it works until approximately 70% of the starch is converted.[367] Dextrins produced from potato starch by ptyalin are slowly split by action of the same enzyme to give glucose and maltose. With the pure enzyme preparation, splitting begins from the reducing ends of the maltotetraose to maltooctaose components that constitute the dextrin.[368]

Ptyalins of cow and horse are alkaline. Their pH ranges are 8.55–8.90 and 8.5–8.6, respectively. During rumination, the pH of ptyalins decreases slightly, remaining between 7.9 and 8.0. Optimum activity of bovine ptyalin is pH 6.5–6.6 and 45 °C, and for horse ptyalin it is pH 6.2 and 50 °C. Adsorption onto starch and oligosaccharides protects ptyalin,[369] even at pH 3. The activity of those ptyalins increases when either yeast or blood serum is added.[370]

Crude ptyalin is free of beta amylase.[371] Even glycosylated ptyalin is capable of hydrolyzing starch and maltotriose. Glucose is formed from oligosaccharides[372] and maltotetraose–maltoheptaose fragments containing one isomaltose linkage are the main components of the limit dextrins produced by that enzyme.[373,374] Ptyalin is inhibited by the alpha amylase II inhibitor from *Triticum aestivum*.[375]

(ii) Pancreatin. Pancreatin with AMY2A and AMY2B genes is accompanied by trypsin, which digests proteins, and lipase which digests lipids. Many commercial enzyme preparations are so constituted. It performs best[376] at pH 7.0–7.2 and between 30 and 50 °C. Pancreatin exhibits some selectivity in its hydrolysis. Thus, barley and bean flours are digested, whereas wheat and corn starches are not. Purified hog pancreatin digests potato starch without splitting phosphate groups from amylopectin.[377]

Studies on the susceptibility of numerous legume starches to hydrolysis by porcine pancreatin gave the following order of susceptibility: black bean > lentil > smooth pea > pinto bean > wrinkled pea. Insight into the structure of granules of these starches clearly showed that the extent of hydrolysis depends on the organization of the amylose and amylopectin components in the native granules.[378]

Amyloamyloses are completely digested by pancreatin and also by barley and malt amylases.[378] Erythrodextrins undergo hydrolysis by pancreatin, and also by several alpha amylases.[379] Maltose is the predominating end-product, accompanied by maltotriose, maltotetraose, and maltopentaose.[3,325,380–383] Pancreatin slowly splits maltotriose into maltose and glucose.[382,384]

The susceptibility of cooked starches to pancreatin decreases in the order: barley > corn > bean > wheat. Flours are digested more readily than starches.[385] The extent of hydrolysis depends on the enzyme concentration.[386,387]

The results of digestion of starch with pancreatin obviously depend on the enzyme's purity and, to a much lesser extent, on the starch variety treated.[388] Potato starch treated with an isolated enzyme affords fermentable and nonfermentable saccharide fractions. The fermentable fraction contains neither maltose nor glucose, but 33% isomaltose, 31% maltotriose, and 36% dextrins.[342,389] However, the formation of up to 90% maltose has also been reported.[390] Purified pancreatin produced maltose and glucose from corn starch and corn amylose; glucose appeared in the later period of digestion.

The nonbranched amylose component was hydrolyzed more readily than whole starch and, therefore, more glucose was formed.[388] Hydrolysis of waxy maize starch afforded similar results.[387] Porcine, bovine, and ovine alpha amylases showed significant differences from one another in their action on various starch substrates.[391] Porcine pancreatin hydrolyzed soluble starch to the extent of 54%, and human ptyalin hydrolyzed it to an extent of 80%.[392] Porcine pancreatin inhibits retrogradation of gelatinized legume starch, but not cereal starches. The limit after 4 h of hydrolysis was 92.2% for potato-starch gel, while that of wrinkled pea was only 70.5%. The maltose/maltotriose ratio for cereal starches was 1:0.90, and for legume starches that ratio varied from 1:0.84 to 1:0.60. These differences are related to the amylose-amylopectin ratio in the substrates.[393]

Human and animal alpha amylases can cooperate in their stabilization, as shown in the example of hog pancreatin alone and in admixture with human saliva (Fig. 3). Dog and human serum can similarly cooperate with hog pancreatin (Fig. 4).[394]

(iii) Other Animal Alpha Amylases. Alpha amylases of human body fluids were investigated by Brock.[395] Enzymes were isolated from ox small intestine and examined for hydrolysis of alpha-limit dextrins, as well as both amylose and amylopectin of potato.[396] The alpha amylase isolated from *Euphausia superba* krill is a mixture of 4 enzyme forms. Optimum parameters for the most active component, whose molecular weight is 17 kDa, are pH 6.4 and 35 °C. Its activity is stimulated by Co^{2+} ions, L-cysteine, glutathione, and thioglycolic acid. The thiol groups are particularly important for the activity of that enzyme.[397] The amylase of honey is thermostable and operates[398] at pH 5.3–5.5. The amylase of the sea urchin, *Psammechinus miliaris*, has an optimum pH of 7.0.[399]

The silkworm *Bombycx mori* has a three-component alpha amylase, one component having a high activity.[400] Carbohydrases capable of degrading starch have been

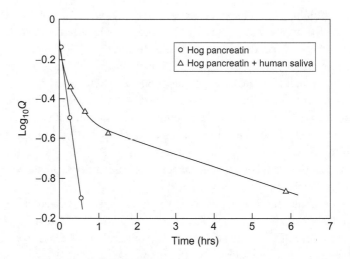

FIG. 3. Spontaneous loss of amylolytic potence of hog pancreatin and of a mixture of that pancreatin with human saliva. (reproduced from Ref. 394 with permission)

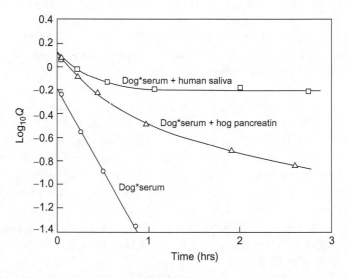

FIG. 4. Spontaneous loss of amylolytic potence of dog serum and its mixture with human saliva and hog pancreatin. (reproduced from Ref. 394 with permission)

isolated from posterior gut sections of various herbivorous marine fish *Kyphosidae* from New Zealand. They were used in an impure state.[401] Two amylases have been isolated from the small abalone shellfish *Sulculus diversicolor aquatilis*. Their molecular weights were 55.7 and 65 kDa, respectively, and their respective optimum temperatures were 45 and 50 °C. Both had the same optimum pH value of 6.0. Both cleave α-(1→4) and α-(1→6) glucosidic bonds. The first one appears to be an endo/exo-amylase and the second is exclusively an exoenzyme.[402]

b. Plant Alpha Amylases.—Alpha amylase from plant sources can be extracted from seeds, fruits, leaves, stems as well as from roots and, for instance, wheaten flour.[403] Soy bean, barley, malted barley, and broad bean are most commonly used. Although practically all plants contain enzymes capable of hydrolyzing starch, only certain of them, for instance, those from soy beans, barley, rice and wheat, and flours made from them, possess saccharifying ability.[404–406]

The composition and hydrolyzing ability of plant amylases depend on the stage of maturation of the plant substrate.[407] Isolated alpha amylases have been purified to remove accompanying beta amylase and complexed with glycogen.[408,409] These complexes exhibited 150–700 fold higher amylolytic activity.

(i) Alpha Amylases from Cereals. Ten isoforms of this enzyme were isolated from rice (*Oryza sativa* L.) cells.[410] They differ subtly from one another in their optimum pH. Rice alpha amylase converted only 50% of starch during 3 h at 35 °C.[411] Alpha amylase isolated from cholam (*Sorghum vulgare*) contained a small proportion of beta amylase,[412] which could be deactivated by heating the enzyme blend for 10–15 min at 62–63 °C. Cholam diastase is less active than barley diastase,[413] and it leaves intact some dextrins produced from potato starch.[414]

Some plants can secrete thermostable enzymes, as shown with a waxy, self-liquefying starch from hull-less barley. This enzyme performed better at 60 °C, producing predominantly maltose and maltohexaose.[415,416] Optimal operational conditions for this enzyme were different from those for alpha amylase isolated from amygdalin.[417,418] Barley is a good source of alpha amylase. The germinated substrate is better than the nongerminated one, and the enzyme isolated from it permits more-complete hydrolysis of starch. Alpha amylase is present in barley before its conversion into malt, but beta amylase also appears on malting.[419] Both enzyme preparations generate no more than an 80% yield of maltose, regardless of the applied enzyme concentration,[420,421] Independently of the cereal starch hydrolyzed, barley alpha amylase produces maltose and maltohexaose as the main products.

Four enzymes, namely alpha and beta amylases, α-glucosidase, and debranching enzyme were isolated from the barley seeds.[422] Incubation of potato starch with malt alpha amylase gave 3% of glucose and 13% of maltose after 1 h of treatment, and after 48 h of processing there resulted 7.8% of glucose and 48.4% of maltose. Glucose was formed from the limit dextrins, but ptyalin could not release glucose from these dextrins.[423]

A mathematical model, the so-called subsite model, based on kinetics[424] predicts the result of hydrolysis. It was evaluated on thirty distillery barley malts checked for reaction progress and the limiting degree of starch hydrolysis. These malts showed some differences in behavior. All provided a limiting degree of hydrolysis[425] in the range of 51.7–56.2%. In contrast to this report, 95.5% of hydrolysis was reported[426] when the concentration of malt amylase was increased. Enzymatic digestion at 50 °C proceeds better than at lower temperature,[427] and the optimal temperature range was settled[428] as being between 50 and 55 °C, although for the hydrolysis of cooked starch, the optimum temperature for ptyalin and for barley amylase was established[429] as 70 °C. With potato and malt starches, malt amylase performs best at 71 °C, providing a 10% yield of maltodextrins.[430] The optimum pH for purified malt alpha amylase ranged between 4.4 and 5.4.[431]

Amylase isolated from wheat flour is specific and differs from amylase isolated from other plants.[432]

Alpha amylase isolated from the cereal materials is usually complex. Three alpha amylases were isolated from ragi (*Eleusine coracana*). They hydrolyzed germinated ragi starches, but the yield did not exceed 70%.[433,434] Two alpha amylases were isolated from seeds of winter triticale. Their hydrolyzing ability was tested on triticale starch granules. After 22 h of treatment, both amylases produced practically the same amount of reducing sugars, but the rates of hydrolysis were different, as shown in Fig. 5.

Three forms of alpha amylase isolated from maize have been distinguished.[436] They differed from one another in their isoelectric point (pI). They all showed substantially different action patterns, particularly when hydrolysis of gelatinized and granular starch was compared.

Amylases isolated from various wheat cultivars exhibited some differences in their activity and susceptibility towards various substrates.[437,438] Alpha amylase isoenzymes of germinating winter wheat were also distinguished from one another in their pI values. Those of pI 6.0–6.5, constituting 84% of total isolated isoenzymes, adsorbed readily on undamaged wheat starch granules and digested them. The remainder of these isolated enzymes did not adsorb on these granules.[439]

(ii) Alpha Amylases from Tubers. The alpha amylase extracted from sprouted potatoes differs from the enzyme extracted from nongerminated wheat. The first

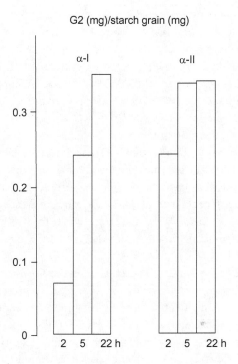

FIG. 5. Amount of reducing sugars (mg maltose/mg starch granules) produced on incubation of triticale starch granules with both alpha amylases isolated from triticale.[435]

contains a phosphorylase capable of digesting potato-starch granules, whereas the second leaves the granules intact.[440] An alpha amylase present in potato is responsible for the instability of potato pastes.[441] This enzyme, termed biolase, appeared to be a combination of various enzymes having proteolytic and oxidative ability, in which alpha amylase prevailed.[59] It rapidly hydrolyzed thick-boiled starches to sugars. Sweet potatoes secrete alpha and beta amylases. Optimal activity of alpha amylase of M_w of 45 kDa occurred at 71.5 °C and pH between 5.8 and 6.4, provided it was stabilized by Ca^{2+} ions. Otherwise, it was deactivated[442] already at 63 °C.

(iii) Alpha Amylases from Other Sources. An amylase isolated from almonds, called salicinase or emulsin at the outset,[57,358,414] performed best at 52.5 °C during the 2 h of action and at 42.5 °C on action for 15 h. A slight acidification of the digest was beneficial. The optimum pH for this enzyme is[57] 5.5.

A starch-hydrolyzing enzyme has been extracted from apple tissues.[443]

c. Bacterial Alpha Amylases.—Bacterial amylases decomposed amylopectin more readily than did malt amylase.[444] They are capable of forming (from amylopectin) a range of maltooligosaccharides from maltotriose (*B. subtilis*,[445] *Bacillus* sp. MG-4[446]), through maltotetraose (*Pseudomonas stutzeri* NRRL B-3389,[447] *Bacillus* sp. GM 8901,[448] *Pseudomonas* sp. IMD 353[449,450]), and maltopentaose (*B. licheniformis*[584,451]) to maltohexaose (*B. caldovelox*,[452] *B. circulans* F-2,[453–455] *Bacillus* sp. H 167,[456,457] and *Aerobacter aerogenes*[458,459]).

Several meso- and thermo-philic anaerobic nonspore-forming bacteria can be sources of enzymes for degrading starch. These enzymes have lower molecular weights (about 25 kDa) than enzymes produced by aerobic bacteria.[460,461] Bacteria delivering alpha amylase can be found in legumes, seeds, soil, sea water, river sewage water, air, and animal feces. Bacterial amylases can retain their activity up to even 95 °C at pH values close to 7.0,[462,463] and up to 105 °C for 2 h.[464] Intercellular alpha amylase from the thermophilic *Bacillus* sp. AK-2 is thermostable, with optimum pH 6.5 at 68 °C.[465]

Perhaps, the most active amylase is produced by the *Bacillus mesentericus* group belonging to *B. amyloliquefaciens* nov. species.[466,467] That amylase produces a considerable amount of maltotetraose.[468–471] Optimum operational conditions for the amylase from these bacteria are pH 6.0 and 55–60 °C, and its thermal stability is unusual.[472] As compared to other amylases, this amylase readily liquefies starch but saccharifies it fairly slowly. The liquefaction/saccharification ratio for that amylase, counted as grams converted starch/gram amylase, is 833.3, whereas for nonsprouted barley amylase, malt amylase, spleen amylase, and takaamylase this ratio is 0.0088, 1.7211, 0.1253, and 5.65, respectively.[473]

Optimization of the conditions for hydrolysis of starch with *B. amyloliquefaciens* was conducted[474] in the pH range of 5.5–7.0 and temperature 30–37 °C. The extracellular alpha amylase from *B. amyloliquefaciens* was stable at 50 °C and pH 5.5, but already at 60 °C it lost 85% of its activity within half an hour.[475] Similar optimum conditions were found for the alpha amylase isolated from the thermophilic *Bacillus* WN-11. Actually, two amylases, of 76 and 53 kDa, were isolated. Their peak activity[476] was 75–80 °C and pH 5.5. The alpha amylase isolated from *Bacillus* IMD 434 had M_w 69.2 kDa, and acted[477,478] preferably at pH 6.0 and 65 °C.

Bacillus licheniformis produced a thermostable alpha amylase performing best at 85 °C at pH between 6 and 7. From corn starch, it produced mainly maltotriose, maltopentaose, and maltohexaose.[479] After immobilization on a silica support that enzyme retained about 60% of its original activity.[480] Some strains of that bacteria (CUMC 305) as well as *B. coagulans* CUMC 512 provided a thermostable alpha amylase working best at 91 °C and pH 9.5. That enzyme retained[481] up to 50% its activity at 110 °C. The possibility of using that enzyme at 100 °C was also reported. That enzyme is inhibited by substrate at high concentration and by glucose.[482]

These properties motivated wider studies on novel sources of that bacterium[483,484] and on its engineering.[485–490] They are typical liquefying amylases, as is the acidic alpha amylase from *B. acidocaldarius* KSTM-2037. Optimum condition for that enzyme, of M_w 55–60 kDa, was pH 4.0 at 80–90 °C. This enzyme did not require Ca^{2+} ions for its activity and stability.[491] A *B. licheniformis* mutant produced two alpha amylases. One of them exhibited hydrolytic activity, whereas the second was capable of transglycosylation.[492]

Another thermostable alpha amylase, actually a combination of two forms, was isolated from *B. subtilis*. That blend operated best at 60 °C, but its thermostability decreased after purification. The enzyme, after reconstitution from a lyophilized preparation, had lower activity.[493] That bacterium, strain 65, produced an alpha amylase of M_w 68 kDa that readily digested maltotriose and γ-CD, an alpha amylase of M_w 48.2 kDa liquefying raw corn starch but not raw potato starch, and an alpha amylase of M_w 45 kDa. The latter slowly digests raw starches and low-molecular-weight substrates.[494] The strain DM-03 isolated from traditional fermented Indian food worked best[495] at pH 9.0 and 52–55 °C. The *B. subtilis* strain A-32 produced an alpha amylase[496] working at pH 5.0–7.0 at up to 80 °C, and the enzyme from *B. subtilis* strain JS-2004 performed best[497] at pH 8.0 and 70 °C. A *B. subtilis* strain isolated from sheep milk secreted alpha amylase, working best at 135 °C and pH 6.5, and its thermostability was further enhanced by the presence of Ca^{2+} ions. Potato starch, when digested by that enzyme, provided a higher level of reducing sugars, but soluble and rice starches were less susceptible to digestion by that enzyme.[498] Alpha amylase from *B. subtilis* splits maltose into glucose at 60 °C.[499] The resulting β-limit dextrins can contain branched oligosaccharides up to maltodecaose.[500] The α-[501] and β-[502] limit dextrins can be purified by alpha amylase. As might be anticipated, (1 → 6) linkages were left intact. *B. thermoamyloliquefaciens* KP 1071 produces alpha amylase together with glucan 1,4-α-maltohydrolase. It hydrolyzes (1 → 4) bonds of glycogen, limit dextrins, amylase, and amylopectin in the endo manner. It was suggested to be the most thermostable endoamylase known at that time.[503] *Bacillus* CNL-90 isolated from grains also produces two enzymes. One of them is a protease and the other hydrolyzes starch.[504] A novel alpha amylase (alpha amylase K) isolated from *B. subtilis*, produced, from amylopectin and number of starches,[505] mainly maltohexaose, accompanied by maltoheptaose, maltopentaose, and maltotriose.

Eighty-eight amylolytic *Bacilli* strains of 18 species have been checked for secreting amylases capable of hydrolyzing granular starches. Only strains of *B. stearothermophilus* and *B. amylolyticus* exhibited a high activity against corn starch, and alpha amylase from *B. stearothermophilus* NCA26 digested granules of corn and wheat starches. The activity of the enzyme could be enhanced by increasing the temperature to 60 °C, when the enzyme could digest potato-starch granules with up to 45% conversion.[506] The activity of that alpha amylase depends on the concentration of maltohexaose in the early stage of the process, and that is dependent on the starch variety.[507] There are two *B. stearothermophilus* alpha amylases: one, an exoenzyme, has M_w 47 kDa and the second, an endoenzyme, has M_2 58 kDa.[508]

Alpha amylase of enhanced thermostability, working at 100 °C during 70 min, was obtained through mutation of *B. stearothermophilus*. This amylase (EC 3.2.1.98) known as maltohexaose-forming alpha amylase, produced maltohexaose.[509] The amylase from *B. stearothermophilus* ATCC is heat- and acid-stable.[510] *Bacillus macerans*, which is known for producing cyclodextrins, also converts starch into dextrins of high molecular weight.[511] The other strains of the *B. stearothermophilus* bacterium produced both alpha and beta amylases. The first, of M_w 48 kDa, performed best at 80 °C and pH 6.9. Glutathione and cysteine inhibited the enzyme activity, and K^+ as well as cations of the second principal group stimulated it.[512] Amylase originating from that soil bacterium, working at 70 °C, degrades wheat starch better than potato starch.[513] It has M_w 52 kDa and works at pH 7.0.[514]

B. circulans F-2 grown on potato starch delivered an exo-alpha amylase that effectively degraded raw starch granules, producing mainly maltohexaose.[515] The activity of that enzyme was higher when *B. circulans* was cultivated on corn starch. Epichlorohydrin-crosslinked starches were even better sources for cultivating that bacterium.[516–518] A highly acid-stable, extracellular alpha amylase is produced by *B. acidocaldarius* A-2. Its M_w is 66 kDa and the optimum pH is 3.5 at 70 °C, but it can survive for 30 min at the same temperature at pH 2.0, retaining up to 90% its original activity.[519] *B. megaterium* delivers a maltose alpha amylase that converts starch into maltose.[520] Glucose, maltose, and oligosaccharides are produced by the action of an alpha amylase isolated from *Bacillus* sp. B-1018[521] and I-3.[522] The latter is superior in digesting potato starch. The majority of alpha amylases isolated from *Bacillus* strains are active at pH values well below 7.0. An alkaline alpha amylase was isolated from *Bacillus* IMD 370. This amylase digests raw starch at pH 8.0–10.0 producing glucose, maltose, maltotriose, and maltotetraose.[523] Another alkaline alpha amylase active at pH 8.0–12.0 and below 70 °C was isolated from *Halobacterium* species H-371.[524] Thermophilic and, simultaneously, alkaliphilic *Bacillus* sp XAL601 provided an alpha amylase working best at pH 9 at 70 °C. It produced maltose, maltotriose, and maltotetraose from raw corn starch, and only maltotriose from pullulan.[525] A *Bacillus* strain isolated from soda lakes secreted two alpha amylases. One of them, of M_w 51 kDa, was deactivated above 55 °C, but hydrolysis could be performed at pH values up to 11 [Ref. 526].

An alpha amylase isolated from *Geobacillus thermoleovorans* NP33 and NP54 saccharified starch best at pH 7.0 and 100 °C. It is a high-maltose-forming enzyme acting independently of Ca^{2+} ions.[527] *B. thuringiensis* produces an alpha amylase capable of degrading raw starch. This property is useful in degradation of sewage sludge.[528] A fungal amylase of *Corticium rolfsii* likewise exhibits such properties.[529] A new alpha amylase has been isolated from *Anoxybacillus contaminans*. It hydrolyzes granules of raw starch very efficiently below the gelatinization temperature. In combination with *A. niger* glucoamylase it achieves 95% liquefaction.[530]

Zooglea ramigera, isolated from soil, produced an extracellular alpha amylase of M_w 63 kDa, which first produces maltopentaose exclusively, and then gradually converts it into maltose and maltotriose. It worked preferably at pH 6.0–6.5 and below 50 °C.[531] An enzyme isolated from *Pseudomonas* sp. KO-8940 provided similar results, working at pH 6.5 and 55 °C, and lower oligosaccharides were not formed when the process was conducted at pH 8.4.[532] A membrane-associated amylase of M_w of 91.3 kDa isolated from *Ruminobacter amylophilus* hydrolyzes amylose to maltose and maltotriose, whereas hydrolysis of pullulan produces panose.[533]

A thermostable alpha amylase of M_w 135 kDa isolated from the thermophilic bacterium *Thermus* produced maltose, maltotriose, and higher sugars from starch, splitting specifically α-(1→4) linkages.[534] The optimum pH for this enzyme is 5.5–6.0. An alpha amylase from *Thermobifida fusca* produced only maltotriose, showing high activity particularly against raw sago starch.[535] Its optimum parameters are pH 7.0 and 60 °C. Assay of alpha amylases from ordinary *B. subtilis*, thermostable *B. subtilis*, and *B. stearothermophilus* showed that their thermal stability increased in this order.[536] The saccharifying/liquefying activity ratio, and the amount of glucose formed, also decreased in the same order.

Testing 30 strains of *Streptococci* revealed that *Streptococcus bovis* PCSIR-7B provided the most active alpha amylase against cereal starches; it was less active against tuber starches.[537] The alpha amylase isolated from *S. bovis* could split α-(1→6) linkages.[538] *Thermococcus profundus* produced an alpha amylase that splits (1→4) linkages of starch and (1→6) linkages of pullulan. It performed best[539] at pH 5.5 between 80 and 90 °C. An extremely thermostable alpha amylase was designed by cloning the *Pyrococcus furiosus* alpha amylase gene and expressing it in *Escherichia coli* and *Bacillus subtilis*. It rapidly decomposes starch to glucose[540,541] at pH 7.0 and 105 °C. Thermostable alpha amylases isolated from *Thermomonospora viridis* TF-35[542] and *Thermomyces lanuginosus* IISc 91[543] could not compete with it, as their optimum temperatures were 55 and 65 °C, respectively. Enzymes from *T. viridis* and *T. lanuginosus* produced maltose and glucose, respectively. Various strains of *Serratia* also produce amylase for hydrolysis of starch.[544] Along with alpha amylase,

Thermomyces lanuginosus IISc 91 also secretes glucoamylase. The former enzyme, of 42 kDa, is a dimeric protein acting best at pH 5.0–6.0. Its thermal stability increases to 73 °C after inclusion of Ca^{2+} ions.[543]

A nonsulfur, purple photosynthetic bacterium (a protobacterium), produced two enzymes that digest raw potato, corn, and cassava starches. Optimum conditions for one of them, probably an alpha amylase, were pH 6.0 at ≤ 45 °C, but it could function at pH up to ≤ 12.0.[545] *Bacillus* H-167 produces amylase H, which manufactures mainly maltohexaose among the maltooligosaccharides.[546]

Chimeric alpha amylases were constructed from the DNA of *B. amyloliquefaciens* and *B. licheniformis*. These enzymes can be used for the net conversion of starch into highly saccharified syrups.[547,548] Behaving similarly, the thermostable alpha amylase isolated from *B. apiarius* CBML 152 bypassed $(1 \rightarrow 6)$ linkages to produce glucose, maltose, and maltotriose at a high rate. No maltotetraose was formed.[549] Another approach involved isolated amino acid sequences of *Athelia rolfsii* glucoamylase, *Pachykytospora papayrycea* glucoamylase, *Valsaria rubicosa* alpha amylase, and *Meripilus giganteus* alpha amylase. These polypeptides revealed superior hydrolytic activity, allowing hydrolysis of nongelatinized starches.[550]

Lysobacter brunescens produced an alpha amylase of M_w 47–49 kDa, and its activity did not require Mn^{2+}, Ca^{2+}, Mn^{2+}, and Zn^{2+} ions. It was active over the pH range 5.0–7.5.[551] A thermostable alpha amylase, a pullulanase, was isolated from *Clostridium thermohydrosulfuricum* DSM 567.[552,553] It was useful for producing lactate and/or acetate. The presence of Fe^{2+} ion was helpful. *Clostridium acetobutylicum* ATCC 824, grown on starch, provided an extracellular alpha amylase of M_w 61 kDa, functioning best at pH 4.5–5.0 at 65 °C. It cleaved $(1 \rightarrow 4)$ bonds in the endo fashion.[554] Action of this amylase is retarded by sodium acetate, phosphate, and chloride.[555]

Characteristic for the enzyme of this origin is that it converts starch completely into maltose without other persistent by-products.[556] A novel extracellular alpha amylase has been isolated from an extreme thermophile *Thermus* sp.[557] The optimum parameters for this amylase of M_w 59 kDa were pH 5.5–6.5 at 70 °C. It is not selective and produces maltose to maltoheptaose from amylose and can hydrolyze maltoheptaose and maltohexaose mainly to maltose and maltopentaose.

d. Yeast Alpha Amylases.—The yeast *Lipomyces starkeyi* HN-606 produces a novel alpha amylase of M_w 56 kDa which converts raw starch into glucose, maltose, and maltotriose. Its specificity is attributable to the lower temperature requirement for its culturing than for most processes for culturing microorganisms.[558] *Lipomyces kononenkoae* yeast produces an alpha amylase and glucoamylase acting on starch in a specific process involving an additional growth-linked regulatory mechanism,[559] and *Lipomyces starkeyi* KSM-22 produces an enzyme exhibiting both amylase and dextrinase activity. It is capable of hydrolyzing rice starch to maltodextrin.[560] Similarly, *Schwanniomyces alluvius* produces an alpha amylase and glucoamylase. The alpha amylase of M_w 61.9 kDa released glucose from starch, but not from pullulan. The optimum conditions for that enzyme is pH 4.5–7.5 and 40 °C. Above that temperature the enzyme deactivates rapidly.[561–563] Alpha amylase from the *S. castelli* 1402 and 1436 mutants was more active, and maltose was the stronger inducer of activity than starch.[564] *Thermoactinomyces vulgaris* cultivated on corn starch produced an alpha amylase having amylolytic as well as proteolytic activity. Amylolytic activity appeared after 4 h and reached a maximum in the 16th hour.[565] *Saccharomycopsis capsularis*, isolated from an Indian fermented food based on cereals, produced an extracellular alpha amylase that liquefied and saccharified starch completely within 24 h at pH 4.5–5.0 and 50 °C.[566] There is a Japanese patent using S*accharomyces* for the manufacture of maltotriose.[567]

e. Fungal Alpha Amylases.—Fungal amylases are, usually, more efficient than malt amylases in the hydrolysis of starch, although glucose is formed in an earlier stage of hydrolysis with malt enzyme.[568,569]

Taking the liquefaction stage as a criterion, pancreatic amylase is the most efficient, followed by plant (cereal), fungal, and bacterial amylases.[570] It is also documented that fungal amylases are needed at higher concentration than malt plant amylases followed by bacterial amylases, to achieve the same degree of conversion of starch.[571]

Eighteen genera of molds tested showed amylolytic activity,[572] but only those of the genera *Syncephalastrum*, *Catenularia*, *Aspergillus*, and *Thamnidium* exhibited high activity. Mold amylases are combinations of labile and nonlabile forms of alpha and beta amylases and, additionally, α-glucosidase (maltase).[573,574] Through examination of about 300 strains of *Aspergilli*, *Rhizopus*, *Mucor*, *Penicilli*[573] and *Streptococci*,[574] five groups of amylases have been distinguished[573] based on their composition.

Fungal alpha amylases are distinguished from human and bacterial alpha amylases in their optimum pH.[575] Takadiastase performs[576] best at pH 3.0–4.0, although it may operate at pH between 4.0 and 7.0 at 50 °C within less than 4 h.[577] Comparative hydrolysis of starch with amylases originating from *Aspergillus niger* and *Aspergillus oryzae* revealed that the first amylase produced mainly glucose, whereas the second enzyme gave maltotriose and maltotetraose as dominating products, accompanied by glucose and maltose.[578]

Streptomyces strains produce extracellular amylases. Among 97 strains tested,[579] *Streptomyces limosus*, acting as an endoenzyme, appeared the most active against granular starch, gelatinized starch, maltose, and malt. In contrast to that amylase, the amylase from *S. griseus* NA-468 is an exo-acting enzyme producing maltotriose.[580] It was unstable above 45 °C but at 25 °C it was more active than the amylase from *B. amyloliquefaciens*, producing mainly maltose. In other experiments, 26 thermophilic strains of *Actinomycetes* were investigated.[581] Among them, *Streptomyces thermocyaneoviolaceus* IFO 14271 produced the most active alpha amylase, having M_w 49 kDa and working best at pH 6.5 and 40 °C. It was efficient against most starches, either waxy or normal, along with tuber, cereal, or sago starch. *Streptomyces* IMD 2679 secreted an alpha amylase producing 79% maltose from starch.[582] *Rhizopus* sp. produce an alpha amylase that splits from amylose only glucose through maltotriose; from amylopectin it also produces maltopentaose and maltohexaose.[583] Thermostable amylases having alpha amylase and glucoamylase activity were isolated from thermotolerant *R. microsporus* var. *rhizopodiformis*.[584] That enzyme hydrolyzes starch mainly to glucose, and its optimum conditions are 65 °C and pH 5.0. *Cladosporium resinae* produces an alpha amylase, an exopullulanase, and two glucoamylases.[585,586]

In *Pichia burtoni* Boldin, two alpha amylases were found.[587] One of them was localized within cells and was not secreted into the culture medium, whereas the second was tightly bound to the membranes. The cell-bound enzyme hydrolyzed starch, amylose, dextran, and glycogen. The amounts of saccharides produced from starch by *Aspergilli* enzymes depended logarithmically on the concentration of the enzymes applied. Basically, all *Aspergilli* can be divided into *A. oryzae* and *A. niger* groups.[588] Alpha amylases (takadiastases) from various *Aspergilli* differ from one another in their thermostability. Thus, in contrast to the enzyme from *A. oryzae*, the one from *A. niger* cannot be regenerated after being subjected to 75 °C, even in the presence of the substrate. Maintaining takadiastase at 60 °C for 30 min is tolerated.[589] Taka amylase hydrolyzes α-(1 → 4) bonds and does not affect either α-(1 → 6) linkages or maltose.[590] The hydrolysis of starch to low-molecular-weight dextrins (limit dextrins) is fast in the initial period, and then it slows down.[591,592] The optimum temperature for the amylase from *A. oryzae* to produce maltose is 60 °C, when applying 4% of enzyme into a 30% starch mash.[593] In the course of the first 3 h at 50 °C, maltose is accompanied by maltotriose, which within the next 3 h is converted into glucose.[594]

That takadiastase is thermally stabilized by starch. In a 0.6% starch solution at pH 5.6, the inactivation takes place at a temperature 3.5 °C above that in the absence of starch, and this stabilization increases[595] with starch concentrations up to 2.0%, and this effect is pH dependent.[596] At that optimum temperature and at pH 5.0–6.0, 7.6% of glucose is formed, along with maltose and limit dextrin.[597] The alpha amylase isolated from *A. awamori* has a pH optimum dependent on the substrate being digested. Thus, for converting starch into dextrin pH 6.2–7.2 is recommended, whereas for the action of tryptase on gelatin and on fibrin, pH ranges of 7.7–8.3 and 5.2–6.7, respectively, are the best.[598,599] Other sources[600,601] report a complete hydrolysis of starch to glucose at pH 4.5–5.0 and 40 °C. Inclusion of a buffer plays a role in determining the final yield.[601,602] The action of takadiastase upon kaoliang starch has been widely studied.[603–606] The alpha amylase isolated, along with glucoamylase, from *A. awamori* NRRL 3112 was the most active at pH 4.4–5.0 at 65 °C. It could be thermally stabilized with D-glucitol (sorbitol). That enzyme satisfactorily saccharified potato starch in a one-step process.[607]

Alpha amylases from *Aspergilli* also differ from one another in their activity and the products formed from starch.[608] *A. awamori* KT-11 digested raw and soluble corn starch to give maltose and maltotriose, accompanied by small amount of glucose. That fungus also produced a glucoamylase.[609] *A. niger* provides an extracellular thermophilic alpha amylase, the molecular weight of which was 56 230 Da. As compared with other fungal amylases, this one is characterized by a lower activation energy, tolerance to low pH, and enhanced affinity to starch.[610] That fungus actually secretes two alpha amylases. The preponderant one is acid-stable and the minor one is acid-unstable.[611–614] *A. niger* TSA-1 produces an acid alpha amylase operating at pH 3.2–3.5.[615] An acid alpha amylase accompanying *A. niger* alpha glucosidase appeared suitable for saccharification of starch at pH 3.5–4.0 and 65–70 °C. The use of that enzyme in a cocktail with amyloglucosidase, and even better with pullulanase, was beneficial.[616]

Aspergillus fumigatus (K-27) produced only glucose, regardless of the botanical origin of the granular starch substrate. Corn, waxy corn, smooth pea, and wheat starches were digested more readily than hylon corn, wrinkled pea, and potato starches.[617] The overall yield of glucose after hydrolysis with that amylase was higher than with porcine pancreatin and commercial alpha amylase from *Bacillus* sp. (see Table II).

The alpha amylase from *A. fumigatus* efficiently hydrolyzed hylon maize starch.

Admixture of methyl α-D-glucopyranoside to a culture of *Aspergillus fumigatus* K-27 doubled the production of alpha amylase and glucoamylase during 5 days of incubation.[618] Alpha amylases isolated from *A. niger* AM07 in soil were active against all tuber starches.[619] An alpha amylase isolated from *Penicillium expansum* produced maltose at the exceptionally high level of 74%, thus 14% more maltose and 17% less maltotriose than other enzymes previously used. It is an extracellular enzyme of M_w 69 kDa and pI 3.9, and the pH for its optimum activity is 4.5, although it is stable in the range of 3.6–6.0.[620]

Alpha amylases isolated from *A. terreus*, *A. carneus*, *Fusarium moniliformis*, and *Phoma sorghina* were tested in the hydrolysis of starch, ground millet, and acha.[621]

Aspergillus foetidus ATCC 10254 when cultivated on rice starch produced highly active extracellular alpha amylase for which the optimum parameters were pH 5.0 and 45 °C.[622] An alpha amylase produced by Trichoderma harzianum has very high affinity towards starch.[623]

Whole cells, and especially microsomal fractions, of Fusarium oxysporum, F. semitectum, F. sporotrichiella, F. gibossum, and F. moniliforme contain amylase and inulinase.[624] F. oxysporum and F. scirpi grown on starch produced an alpha amylase performing best at pH 6.9 and between 30 and 40 °C when it was used for saccharifying starch, but dextrinization proceeded more effectively at somewhat lower pH.[625]

Corticium rolfsii is presented as one of the most promising species for producing alpha amylase to saccharify raw starch,[626] with optimum activity at pH 4.0 and 65 °C. Starch is rapidly hydrolyzed to glucose, with only weak inhibition by starch at high concentration. The production of alpha amylase by Blastobotrys proliferans has been patented.[627] The higher fungus Ganoderma lucidum produced an extracellular alpha amylase working preferably at pH 5.5 and 50 °C. It was activated by the Mn^{2+}, Ca^{2+}, and Cu^{2+} ions. It readily hydrolyzed boiled amylaceous polysaccharides, but raw starch was hydrolyzed only slowly. Maltose inhibited that enzyme.[628] Candida antartica CBS 6678 produces extracellular alpha amylase (M_w 48.5 kDa) and glucoamylase selectively hydrolyzing some starches and CDs. Optimum conditions[629] for the alpha amylase are pH 4.2 at 62 °C. An alpha amylase of M_w 52 kDa was isolated from Rhizomucor pusillus.[630] The fungus Thermitomyces clypeatus produces an amylase demonstrating endohydrolytic activity against, among others, amylose and amylopectin. Its optimum conditions [631] were pH 5.5 at 55 °C. Some imperfect mucoral fungi and Ascomycetes also produce alpha amylase, as well as α-glucosidase.[632]

Comparative studies on the activity of yeast and fungal alpha amylase[633] from Schwanniomyces castelli, Endomycopsis fibuligera, and Aspergillus oryzae showed that the activity of the enzyme from the third source was 18 and 26 times higher than that from either E. fibuligera or S. castelli. The enzymes from these three sources had similar low activity against maltose, were passive against isomaltose and pullulan, and very active against starch. Alpha amylase of S. castelli is specific for α-(1 → 4) bonds and produces short-chain oligomers.[634,635] Heat- and acid-stable alpha amylases were also secreted by Mucor pusillus.[636]

Sources of alpha amylases and their selected properties are listed in Table III.

f. Immobilized Alpha Amylases.—Alpha amylases are often used after immobilization on various supports.

Silica supports for B. licheniformis alpha amylase provides 60% retention of the enzyme activity.[485] Oxidized bagasse cellulose,[803] coconut fiber,[804] highly porous cross-linked cellulose,[805] and a composite temperature-sensitive polyester membrane[806] are other supports used for immobilization of that enzyme. Alpha amylase immobilized onto corn grits and porous silica produced less polymerized products, perhaps because the number of transglycosylation reactions is limited.[807] The enzyme of *B. subtilis* immobilized onto a resin hydrolyzed starch into dextrins of high molecular weight, although the yield of conversion was almost 93%.[808] Immobilization of that enzyme within capsules of calcium alginate, with tailoring the characteristics of the capsule, has been described.[809] Covalent immobilization of the enzyme on diazotized silica glass decreased the enzyme activity by 45%. Concurrently, the optimum conditions[810] for use of the enzyme change from pH 7.3 and 50 °C to pH 6.4 and 55 °C. Alpha and beta amylases immobilized onto an epoxypropylsilanized support produced a limit dextrin from a 2% aqueous slurry of potato starch.[811] Covalent immobilization of alpha amylase on Eudargit [a poly(meth)acrylate support] makes that enzyme tolerant to pH changes without decrease in activity.[812] Eudargit and poly(ethyleneimine) as soluble carriers for alpha amylase covalently bound by carbodiimide linking retained 96% of the original enzyme activity in first case and further activating the enzyme in the second case.[813] Epichlorohydrin-cross-linked potato starch appeared to be unsuitable as an adsorbent for immobilizing bacterial alpha amylase. The degree of cross-linking was a crucial factor. The higher the degree of cross-linking, the worse were the results of digestion. This was related to the affinity of the enzyme to adsorption, which decreased with increase in the extent of cross-linking.[814] Immobilization of alpha amylase on poly(*N*-vinylformamide) led to an increase in the enzyme's activity.[815] A methacrylic acid–*N*-isopropylacrylamide copolymer served as a temperature-dependent, reversibly soluble–insoluble support for alpha amylase.[786] Amylase immobilized together with pullulanase on a chitosan-derived support provided a 7–10% higher yield of hydrolysis products; however, immobilization of both enzymes on separate supports was more effective.[816] Chitin was also applied as a support for immobilization of the alpha amylase.[817] Covalent immobilization of amylolytic enzymes on soluble polysaccharides increases the thermal stability of those enzymes,[818] and immobilization of alpha amylase on acrylic acid-grafted cellulose, using carbodiimide linking, provided a preparation displaying high activity; this activity, however, reached only 65% of the activity of the free enzyme.[819] Alpha amylase immobilized on polystyrene beads converted corn starch into glucose at 45 °C and pH 4.5.[820] Immobilization of pancreatin onto collagen membranes decreased the optimum pH from 8.0 to 7.0, increasing simultaneously the thermal stability of that enzyme.[821]

Immobilization of alpha amylase by encapsulation in the starch matrix[822] and by a microencapsulation has also been reported.[823] Immobilization of alpha amylase of

B. subtilis on hydrolyzed $TiCl_4$-coated Fe_3O_4 increased the optimum pH by 0.3 units to 6.1, and addition of $CaCl_2$ decreased that optimum to pH 5.4. The optimum temperature after immobilization was increased from 63 to 73 °C, and adding $CaCl_2$ further elevated[824] that temperature to 86 °C.

A novel result was demonstrated by Inoue-Japax Research, Inc.[825] who immobilized diastase by attaching it to either powdered magnetite or powdered ferrite (40–200 Å) and then performed the enzymatic digestion of starch under control by a magnetic field. The yield of the saccharide products reached 85%.

3. Beta Amylases (EC 3.2.1.2)

As compared to alpha amylase, beta amylase—an exoenzyme earlier termed saccharogenic amylase—has a relatively simple mechanism of hydrolysis, and therefore there have been fewer relevant studies on its behavior in the hydrolysis of starch. Frequently, saccharification of starch with beta amylase constitutes the second step in the production of maltose, and it is preceded by liquefaction of starch with alpha amylase.[826,827]

Beta amylase is specific for amylose chains of six glucose units,[828,829] although there are examples of its attacking maltotetraose.[830] Maltotetraose is the shortest normal saccharide attacked.[829] That enzyme produces the disaccharide β-maltose, cleaving α-(1→4) bonds successively from the nonreducing end of the amylose chains and converting it completely into maltose,[831,832] provided that the chain contains an even number of glucose units.[380,833]

Beta amylase was employed jointly with glucoamylase for making high-maltose syrup.[834] The chains having an odd number of glucose units are split by beta amylase to release glucose.[835] There is a report on a bacterial beta amylase producing 30% of glucose and 40% of maltose from starch.[836] In contrast to alpha amylase, that enzyme cannot split maltotriose into glucose and maltose.[383,829] However, a slight thermal shock, namely heating to 40 °C, results in the formation of glucose and dextrin.[381] The hydrolysis could be enhanced by pretreatment of granules with warm water.[837,838]

With soluble starch, the enzyme releases approximately 64% of the theoretical amount of maltose.[838–842] Amylopectin and related glucans are also digested, but since the enzyme is unable to bypass branches, the hydrolysis is incomplete and a macromolecular "limit dextrin" remains.[840,843–847]

Controversies concerning the results of digestion starch and its separate polysaccharide components can be rationalized in terms of the purity of the enzyme. The crude enzyme from various sources is accompanied by β-glucosidase (Z-enzyme), as well as alpha amylase and α-glucosidase.[848] The latter enzyme is able to split maltose into glucose. At pH 4.6 and 35 °C, beta amylase jointly with β-glucosidase splits amylose by a single-chain mechanism (see Section III), producing mainly maltose; the residual "amylose" had a high DP, indicating the intervention of reversion.[849]

In general, beta amylase is less efficient than alpha amylase in degrading starch granules.

Beta amylase splits (1 → 4) linkages less readily than alpha amylase. The affinity constant for the latter was established as 200 and that for beta amylase was only slightly lower.[850,851] For Lintner starch it reached 170 and did not change until two thirds of that starch had been turned into maltose. The affinity constant was thus independent of the chain length. The activity of both alpha and beta amylases is dependent on their concentration, but only in the initial stage of hydrolysis. Then, when the dextrins (actually phosphodextrins) are formed, the hydrolysis is controlled by the level of bound phosphorus. At this stage a phosphatase is involved in the hydrolysis.[852,853]

Beta amylase can digest starch inside coacervates of starch with gelatin, forming a three-component coacervate.[854] However, when the enzyme is within the coacervate, starch cannot penetrate it.[855]

The susceptibility of starch sources to beta amylase obviously depends on their botanical origin[856] and the provenance of the enzyme. Wheat starch is not very well digested by beta amylase, and digestion depends on the variety of wheat. Sweet-potato starch is only partially hydrolyzed by beta amylase, although a 76–79% yield of maltose can be achieved.[857] Amylopectin is more readily digested than amylose up to approximately 40% of hydrolysis.[858] Corn starch is poorly digested by beta amylase unless it is lintnerized.[859] Once it is hydrolyzed, it becomes resistant to retrogradation. Gelatinized corn starch is more readily hydrolyzed by beta amylase, but it more readily retrogrades. In order to overcome this inconvenience, gelatinization should be performed in the presence of beta amylase.[860]

Optimum operational conditions are[861] 50 °C and pH 5.6. Because of its specificity, beta amylase can hydrolyze even strongly oxidized starches.[862]

The enzyme is inhibited by several compounds, such as ascorbic acid, which forms a complex with the substrate.[442,863-865] Indole-3-acetic acid inhibits beta amylase in the ripening banana.[866] Cyanides do not inhibit the enzyme,[867] but maltose does.[868] Metal ions usually inhibit the enzyme; however, Ca^{2+}, Sr^{2+}, and Mn^{2+} activate it.[465,514]

Beta amylase can be of animal, plant, bacterial, yeast, or fungal origin (see also Table IV).

a. Animal Beta Amylases.—Animal endo-beta amylases were found in the midguts of the weevil *Sitophilus zeamais* and *Sitophilus granarius* larvae. They effect hydrolysis of starch and amylopectin.[869] They were activated by Cl^- anions and inhibited by $NaHCO_3$.

b. Plant Beta Amylases.—Soybean, barley, rice, wheat, koji, sweet potato, and broad bean are the commonest sources of plant beta amylase. They are found primarily in the seeds of higher plants and in sweet potatoes, and it resides in an inactive form until germination and/or ripening.[223,357,405,839,870–873] Extracts of sugar cane contain beta amylase, which can be used in the production of maltose syrup. That enzyme is readily isolated in high yield and is thermostable.[874]

In the digestion of corn-starch granules, soybean beta amylase was 60% less active than bacterial beta amylase.[875]

The isolation and purification of potato beta amylase has been described. Its M_w is 122 kDa and optimum conditions for its activity are pH 5.1–5.5 at 55 °C. It was strongly inhibited by sodium dodecylsulfate and 4-chloromercuribenzoate.[873] The optimum conditions for the beta amylase of sweet potato are similar,[442,828,876] namely pH 5.3–5.8 and 53 °C, and the molecular weight of this beta amylase is[877] 206 ± 1 kDa.

The optimum pH for beta amylase from malt is[356,840,878] between 5.65 and 5.85. The action of beta amylase is not blocked by phosphate or fatty acyl groups.[879,880] The enzyme from malt is totally deactivated at 70 °C within 15 min.[881] It appears[882] that barley beta amylase isolated from germinated and that from matured grains are different in their thermostability. Germinated barley produces the more-thermostable enzyme.

Genetic manipulations showed that exhaustive removal of the enzyme's 4-C-terminal glycine-rich repeating units provides an "engineered" enzyme of higher stability and higher affinity towards starch. Barley beta amylase exposed to hydrostatic pressure as high as 200 MPa becomes significantly more stable to heat. However, this effect is accompanied by a decrease in the conversion rate. Optimum conditions for maltose production were established[883] as 106 MPa, 63 °C, and pH 5.6.

Wheat beta amylase digests starch in proportion to its concentration and demonstrates a peak of activity in the presence of 1% NaCl.[884] In contrast to alpha amylase, beta amylase does not adsorb on starch, and Ca^{2+} ions have no influence on it. However, Voss[885] postulated the formation of aggregates of starch with beta amylase through mutual entanglement. Wheat bran can also be a source of that enzyme.[886]

Nongerminated rice contains only beta amylase, or actually a mixture of two beta amylases, one soluble and the other insoluble in salt solutions. They are similar to such enzymes from grains of other cereals.[887] Tap-root alfalfa provides two beta amylases of M_w of 41.7 and 65.7 kDa, respectively. They are accompanied by two and one isoenzymes, respectively. These amylases hydrolyze amylopectin to maltose, but do not hydrolyze pullulan or β-limit dextrins. Both amylases operate best at pH 6.0 and they both follow Michaelis–Menten kinetics. The amylase of low molecular weight is inhibited by maltose.[888]

In flours hydrolyzed with their own beta amylase, small granules are digested before large granules, but in soluble starch the large granules are digested first.[866] It should be noted that there are two beta amylases, one free and the other latent, in wheat.[889]

c. Bacterial Beta Amylases.—In contrast to plant beta amylases which hardly attack nongelatinized granules, bacterial beta amylases from *Bacilli* will adsorb on raw starch. The enzymes of *B. megatermium* and *B. polymyxa* adsorb on starches, most efficiently on rice starch. Ammonium sulfate enhanced the adsorption. This factor can be useful for isolating that enzyme from culture filtrates.[890]

The ratio of saccharifying/liquefying activity for bacterial beta amylase is twice as high as that for bacterial alpha amylase. Inhibition of beta amylase with maltose and glucose, competitive and noncompetitive, respectively, is stronger than the inhibition of alpha amylase.[891,892] The action of that enzyme is hindered by retrogradation, which makes amylose inaccessible to the enzyme. This effect can be avoided[892] by dropwise addition of alkali-treated amylose to the enzyme buffered to pH 4.8. When the enzyme is used in the form of a nonpurified wheat bran extract, the hydrolysis proceeds at its natural pH between 6.0 and 7.0, but the thermal stability of that enzyme is lower. It is decomposed completely at 60 °C within 1 h.[863,893]

B. cereus, *B. megaterium*, *B. polymyxa*, and *B. circulans* are common sources of bacterial beta amylase.[894] They can also be genetically modified to make them thermostable.[894] The 46 kDa beta amylase from *B. polymyxa* worked best at pH 6.8 and 45 °C, producing mainly maltose.[895] It digested raw starch at a high rate,[896,897] and its activity was stimulated by pullulanase.[898] The same result could be demonstrated[899] with the enzyme isolated from *B. megaterium*, under slightly different optimum conditions, namely, pH 6.9 and 60 °C. The enzyme from *B. cereus* seems to be especially effective, and therefore it is suitable for the conversion of biomass. Toward starches is efficacy decreases in the order: wheat > corn > potato > sweet potato.[900]

Adsorption of the beta amylase from *B. cereus mycoides* onto corn starch reached a yield of 95%, and the activity of adsorbed enzyme was 85% of that of the original enzyme. The enzyme could be eluted with 10% maltose solution.[901] The beta amylase of *B. carotarum* B6, isolated from soil, was also readily adsorbed onto native raw starches, that of arrowroot being the superior adsorbent, and adsorption was complete within 30 min of interaction. The mobility of beta amylase adsorbed on the surface of starch gel was studied with the fluorescently labeled enzyme. The enzyme migrated across the surface through lateral diffusion and exchange between free and anchored enzyme molecules in the solution covering the gel.[902]

The beta amylase from *B. stearothermophilus* consists of an exoenzyme of M_w 39 kDa and an endoenzyme of M_w 67 kDa.[509]

d. Fungal Beta Amylases.—*Aspergillus oryzae* is a good source of beta amylase.[903–908] The enzyme, after purification, produced glucose during the initial stage of starch hydrolysis.[908] Beta amylase secreted by *A. awamori* also hydrolyzes potato starch in a similar manner, with yields reaching 90%. Phosphate groups are not cleaved.[909]

A thermostable beta amylase is produced by *Clostridium thermosulfurogenes*.[910] In the presence of 5% soluble starch, it is stable[911] even above 80 °C at pH 5.5–6.0. With raw starch, the optimum pH decreases to 4.5–5.5 at 75 °C. Pullulanase also stimulates that amylase,[912] but another report[913] contradicts this statement. The same report claims a stimulating effect of isoamylase upon beta amylase.

The fungus *Syncephalastrum racemosum*, when cultivated on starch as the sole source of carbon, produces a beta amylase stable in the presence of heavy metal ions, thiols, and 4-chloromercuribenzoate inhibitors. It performs best at pH 5 and 60 °C, and it is suitable for hydrolysis of waste starch.[914]

e. Immobilized Beta Amylases.—As with alpha amylase, beta amylase can be used immobilized on a support. In this form it loses some of its activity, but its stability is increased.[915] Beta amylase from barley was covalently attached to aminated derivatives of epichlorohydrin-cross-linked Sepharose. The enzyme thus immobilized retained up to 35% its of original activity. The pH and ionic strength optima remained the same, but the stability of the enzyme was enhanced.[916] Beta amylase in conjunction with pullulanase covalently coimmobilized on an acrylamide–acrylic acid copolymer displayed higher operational stability in a packed-bed column.[917] Immobilization of beta amylase on chitosan beads brought an increase by 10 °C in its thermal stability and the optimum temperature by 20 °C. Coimmobilization of pullulanase appeared also beneficial.[918] Beta amylase and pullulanase can be used together coimmobilized on a ceramic support.[919] The use of beta amylase and isoamylase, both nonimmobilized[920] and coimmobilized on such a ceramic support has also been reported.[921] Sweet-potato beta amylase was activated 1.3–3.0 fold after immobilization on polycations, but polyanions inhibited it.[922]

Table IV lists sources of beta amylases and selected properties of those enzymes.

4. Glucoamylase (EC 3.2.1.3)

Glucoamylase splits α-(1 → 4) bonds of amylose, amylopectin, and related glucans and α-(1 → 6) bonds of amylopectin yielding glucose. In digesting granular starch, glucoamylases are more efficient than ptyalin.[942] However, the origin of the particular glucoamylase is a key factor in its performance. They all degrade starch components from the nonreducing end of the chains, and their action upon potato starch stops when a glucose residue bears a 6-phosphate group.[943]

Glucoamylases may originate from various fungal and yeast sources, although they are also present in human and animal intestines.[944] The cells of the sugar beet plant are a rich source of glucoamylase.[945] The optimum temperature for the enzyme is between 40 and 60 °C, and this is higher than for fungal alpha amylase, which operates preferably between 30 and 50 °C. The optimum pH for glucoamylases at these temperature intervals is between 3.6 and 6.5, and between 5.0 and 6.5, respectively.[633] Frequently, organisms secreting glucoamylase also secrete alpha amylase, and the secretion of only one form of glucoamylase is rather uncommon.

Glucoamylases may be divided into those converting starch and β-limit dextrins completely into glucose, and those which convert these substrates only incompletely into glucose.[946]

a. Bacterial Glucoamylases.—Glucoamylases have been isolated from *B. firmus*[947] and *B. stearothermophilus*,[131] various *Clostridium* species,[131,948,949] *Flavobacterium* species,[131] *Halobacter sodamense*,[131] *Lactobacillus brevis*,[950] and *Monascus kaoliang*.[948,951] Glucoamylase from *B. firmus* digested potato, maize, and wheat starches yielding at 37 °C principally glucose, but small amounts of maltose, maltotriose, and maltotetraose are also formed.[947] Glucoamylase isolated from a *Flavobacterium* species degrades CDs.[952] *M. kaoliang* secretes two glucoamylases that are slightly different in their optimum operational pH, and differ considerably in M_w, being 48 kDa and 68 kDa for the endo and exo forms, respectively.[953]

b. Fungal Glucoamylases.—Two forms of glucoamylase were isolated from *A. niger*. One of them was readily adsorbed onto corn starch in a manner dependent on pH, ionic strength, and temperature, whereas the second form adsorbed weakly in a manner independent of the pH and ionic strength. It was shown that the second form does not utilize its binding sites in the adsorption.[954] An inhibitory factor accompanying glucoamylase in *A. niger* also adsorbs tightly onto starch. Most probably, glucoamylase and the inhibitor are adsorbed onto a common binding-site for raw starch.[955,956] Among glucoamylase-secreting fungi, *A. niger* strains are the ones most commonly utilized. The M_w of that enzyme is around 63 kDa and the optimum operational conditions are pH 4.5 and 70 °C, indicating a fair measure of thermostability. That glucoamylase is stabilized by starch, glycerol, and alditols. Polyvalent cations stimulate that enzyme, in contrast to monovalent [Ag(I)] and bivalent [Cu(II), Ni(II), and Co(II)] cations which inhibit it.[957] When glucoamylase is incubated with a highly concentrated solution of glucose, a small amount of isomaltose is formed as a consequence of reversion.[957] *Cladosporium resinae* produces alpha amylase, exopullulanase, and two glucoamylases.[586,587]

In the hydrolysis of wheat starch, glucoamylase from *A. niger* had hydrolytic activity twice as high as the activity of that enzyme from *Rhizoups nivea*. Both enzymes digested starch granules uniformly.[958] The higher activity of glucoamylase from *Aspergilli* than that from *Rhizopus* sp. was also confirmed in other studies.[959] The enzymes from various strains differ slightly from one another.[960] Two isoenzymes of *A. niger* glucoamylase have been characterized.[961] They appear in the late stage of production of glucoamylase in submerged culture. Both isoenzymes split α-(1 $\rightarrow$ 6) linkages. Glucoamylase isolated from *A. niger* may be contaminated with an inhibiting factor that competes with that enzyme for adsorption onto the starch substrate. This inhibiting factor (whose nature was not discussed) can be isolated from glucoamylase by heat treatment at pH 7.2 followed by a decrease in pH to 3.4 and centrifugation.[962,963] On the other hand, there is a report that, in contrast to glucoamylase from other *Aspergilli*, *A. niger* glucoamylase does not adsorb onto starch granules. The other glucoamylases adsorb best at pH 3.5. When pullulanase is incorporated, the optimum pH is 5.0.[617,964]

The success in practical application of *A. niger* glucoamylase has stimulated interest in modifying the enzyme by mutations.[965–968]

The glucoamylase from *A. awamori* has M_w of 62 kDa. It adsorbs onto starch,[964] and its optimum pH was 4.6–4.8. Experiments with numerous saccharide substrates showed that the enzyme splits solely the α-(1 $\rightarrow$ 4) and α-(1 $\rightarrow$ 6) bonds and leaves α-(1 $\rightarrow$ 1), β-(1 $\rightarrow$ 4), and β-(1 $\rightarrow$ 6) bonds intact.[969,970] That enzyme cooperates effectively with isoamylase, but does not cooperate with alpha amylase.[969] There is a report[971] that a glucoamylase of M_w 90 kDa from that source is a mixture of two enzymes. One of them adsorbs onto starch but does not digest it, and the other one can digest starch. Glucoamylase produced from *A. awamori* NRRL 3112 worked best at pH 4.4–5.0 and 55 °C. Starch and D-glucitol increase its thermostability.[608,972]

In fact, the glucoamylase of *A. avamori* consists of three species differing from one another by their molecular weight (M_w values of 90, 83, and 62 kDa) and, importantly, in their ability to hydrolyze starch.[973–975] The highest molecular weight component adsorbed at its active domain onto the glycosidic bonds of raw starch and cleaved the bonds. The component of intermediate molecular weight neither adsorbed onto starch nor split it. The component of lowest molecular weight probably operated on the oligosaccharides formed by digestion. In every instance, homogenization of starch with the enzyme results in increased conversion.[976] Mutants of *A. awamori* glucoamylases have also been designed.[977,978]

Currently, more attention is being paid to glucoamylase from *A. oryzae*. The optimum conditions for that enzyme are[634] pH 3.6–6.5 at 40–60 °C. It adsorbs onto starch and its cooperation with pullulanase is beneficial for the hydrolysis of raw starch.[964] *A. oryzae* in a solid-state culture produces an extracellular glucoamylase of M_w 65 kDa and pI 4.2. Its activity against raw starch is low. The same fungus in submerged culture produces glucoamylase of M_w 63–99 kDa and pI 3.9. It is more active in the hydrolysis of starch.[979] From a traditional Korean enzyme preparation for the brewing of rice wine (nuruk), utilizing *A. oryzae* NR 3-6, there was an isolated glucoamylase of 48 kDa whose optimum conditions are pH 4.0 and 55 °C. Raw wheat starch was digested 17.5 times faster than soluble starch.[980] The effect of temperature (T) and time (t) on the hydrolysis of starch with glucoamylase produced by *A. oryzae* 8500 is given[981,982] by Eq. (1):

$$k+2(T,t) = 5200k + 2(55)\exp(-2770/T - k(T)\text{gtorsim}(t)) \qquad (1)$$

where gtorsim(t) expresses initial temperature at a given time.

The glucoamylase of *A. oryzae* was also used after its immobilization in calcium alginate.[983]

Both *A. candidus* and *A. foetidus* each secreted two glucoamylases. They differed from one another in their M_w. Glucoamylases from the first fungus had more acidic amino acids than those from the other.[984] *A. saitoi* produces two glucoamylases of M_w 90 and 70 kDa and their pH optima are 3.5 and 4.0, respectively. Both glucoamylases digest soluble starch almost completely, but at different rates. The rate for the first glucoamylase is 51 times less than the other, as it binds strongly to starch.[985,986] *A. terreus* 4 secretes a glucoamylase working best[987] at pH 5.0. *Aspergillus* K-27 produces an extracellular thermophilic glucoamylase of higher activity against raw starch than that showed by the glucoamylase of *A. niger*.[619] It was tested in the hydrolysis of starch from sweet potatoes.[988,989] The glucoamylase of *A. shirousamii* of M_w 89 kDa was stable below 55 °C at pH 4.5. Calcium ions markedly increased the thermostability of that enzyme and most anions inhibited it.[969]

Glucoamylase isolated from *Rhizopus* sp. is in fact a combination of three hydrolases. The species of M_w 74 kDa with optimum pH 4.5 binds tightly to starch and is the most active of the three components. The species of M_w 58 kDa and optimum pH 5.0, as well as the species of M_w 61.4 kDa of the same optimum pH 5.0, are, respectively, 22 and 25 times less active than the first enzyme.[990] In contrast, only two glucoamylases could be isolated from *Rhizopus oryzae*.[991] They hydrolyze amylose, amylopectin, soluble starch, and oligosaccharides (including maltose), yielding only glucose, but they do not hydrolyze β-CD, raffinose, sucrose, or lactose.[991,992] Glucoamylase isolated from recombinant yeast containing the *Rhizopus* glucoamylase gene, although similar to the natural glucoamylases,[990] hydrolyzed starch with a higher yield.[993] Comparative studies on the hydrolysis of raw starch of sweet potatoes[988] showed that glucoamylases of *Rhizopus*, *Aspergillus* K-27, and *Chalara paradoxa* hydrolyzed that starch at a similar rate. Five forms of glucoamylase were isolated from *Rhizopus niveus*. They adsorbed very well onto raw starch, but the best pH for efficient adsorption differed for the individual forms.[994] Hydrolysis of starch with *Rhizopus niveus* glucoamylase showed that for a substrate DP between 3 and 7 a random attack occurred, whereas when the substrate DP was 17, a multiple-chain attack took place.[995]

The fungus *Chalara paradoxa* was isolated from the pith of sago palm. Its glucoamylase is a combination of 6 enzymes. Its combined M_w is 82 kDa and optimum conditions are pH 5.0 at 45–50 °C. The enzyme is deactivated above 50 °C but Ca^{2+} ions increase its thermostability.[996,997] That glucoamylase is particularly suitable for the hydrolysis of sago starch; wheat and waxy corn starch are hydrolyzed to a lesser extent.[998] That glucoamylase resembles the glucoamylase isolated from *Rhizopus niveus*. It exhibits some ability to digest CDs. On hydrolysis of corn-starch granules, this glucoamylase penetrates solely to the center of the granule, whereas granules of other starches are digested on the surface.[999,1000] Its activity differed according to the variety of starch and was the highest for rice starch followed by waxy corn, wheat, corn, cassava, sweet potato, sago, and potato, the last two being significantly resistant to that enzyme under the conditions employed.[1001] Glucoamylase isolated from the endophytic fungus *Acremonium* sp. was also utilized for the hydrolysis of sago starch.[1002]

A novel glucoamylase was isolated from *Thermomucor indicae-seudaticae*.[527] Its optimum parameters for starch saccharification are pH 7.0 and 100 °C.

c. **Yeast Glucoamylases.**—*Schwanniomyces alluvius* produces an extracellular glucoamylase which under its optimum pH at 5.0 at 50 °C hydrolyzes soluble starch and pullulan exclusively to glucose.[561] Its M_w is 117 kDa.[562] Glucoamylase of M_w 155 kDa isolated from *Schwanniomyces castelli*, although less active than glucoamylases isolated from *Endomycopsis fibuligera* and *A. oryzae*, is the sole glucoamylase known capable of splitting α-(1 → 4) and α-(1 → 6) glucosidic linkages, and its activity toward α-(1 → 4) bonds is higher than to the others bonds. Polymers of glucose are converted into maltose and panose.[633,635,1003,1004] There are several mutants of the enzyme from *Schwanniomyces occidentalis* formerly classified as glucoamylase, but later it was recognized as a typical α-glucosidase.[1005]

The optimum pH for the glucoamylase isolated from *Endomycopsis* species depends on the time of hydrolysis,[969,1006,1007] being[633] between 3.6 and 6.5 at 40–60 °C. *Cladosporium resinae* produces two glucoamylases, an α-glucosidase—an alpha amylase—and an exopullulanase. Both glucoamylases degrade starch and pullulan to glucose and also split α-(1→6) linkages. They are activated by surfactants.[585,586] *Monascus kaoliang* F-1, cultured on wheat bran, produces two glucoamylases, which differ in activity toward starches of various origin.[1008] For hydrolysis of cassava flour, moldy wheat bran was used.[1009] *Monascus rubiginosus* Sato,[1010] and other *Monascus* sp.,[1011] also produce two glucoamylases. They are similar to one another in their M_w, high activity, structure, and rate of hydrolysis of maltose, maltotriose, and amylose. *Corticium rolfsii* IFO 4878 produces an acid-stable glucoamylase in a medium containing sucrose. It was stable in the pH range 2.0–9.0, with maximum activity[1012] at pH 4.5 and 40–50 °C. *C. rolfsii* AHU 9627 and its mutant produced a glucoamylase of similar tolerance to low pH.[1013] *Paecilomyces subglobosum* also secretes an acid-stable amylase working best at pH 4.0 and 55 °C and retaining 75% of its activity even at pH 2.0.[1014] The optimum pH for glucoamylase of *Paecilomyces varioti* AHU 9417 is higher by an 0.5 unit, but this enzyme exhibits a very high activity.[1015] *Talaromyces emersonii* produces a thermostable glucoamylase of M_w 70 kDa and pI below 3.5. It performs best at pH 4.5 and 70 °C, and in the saccharification of starch, it is superior to the glucoamylase from *A. niger*. Its mutants were engineered by expression of the enzyme on various fungi.[1016,1017] Thermostable glucoamylases are also available from *Clostridium thermoamylolyticum*,[1018] *Clostridium thermohydrosulfuricum*,[1019] and *Thermomyces lanuginosus*.[543] Their optimum temperature is between 65 and 70 °C.

The yeast *Candida antarctica* produces alpha amylase and glucoamylase. The latter has M_w 48.5 kDa and its optimum parameters are pH 4.7 and 57 °C. It exhibits debranching activity and it is strongly inhibited by acarbose and trestatins.[629] The extracellular glucoamylase from *Saccharomyces capsularis* yeast has a fairly low optimum temperature, 28–32 °C and its optimum pH is 4.5–5.0.[566] *Saccharomyces fibuligera* IFO 0111 produces a glucoamylase useful in hydrolysis of raw starches without the need for preliminary thermal modification.[1019]

Glucoamylases from other fungi have also been isolated and characterized. That from *Lentinus edodes* (Berk.) Sing., having M_w of 55 kDa, hydrolyzes maltose, maltotriose, phenyl α-maltoside, soluble starch, amylose, and amylopectin to β-glucose. Phenyl α-maltoside is converted into phenyl α-D-glucopyranoside.[1020] The phytopathogenic fungus *Colletotricum gloelosporioides* produces a single extracellular glucoamylase.[1021] A novel glucoamylase for saccharification of barley grain mash was isolated from the creosote fungus *Hormoconis resinae*, produced by the heterologous host *Trichoderma reesei*.[1022] Either *Humicola grisea themoidea* GSHE gene or a similar gene of *A. awamori* was expressed in *Trichoderma resei* to design a novel glucoamylase for making glucose syrup from granular starch.[1023]

Commercial glucoamylase preparations are frequently contaminated with alpha amylase and other enzymes. Usually, they cooperate satisfactorily.[1024,1025] When their activity is high, the production of maltooligosaccharides increases considerably. At high concentrations of substrate, inhibition of the process has to be taken into account because of the high level of glucose formed. The resynthesizing action of glucoamylase, yielding maltose and isomaltose from glucose, is postulated.[1026] The same could be observed when glucoamylase alone was used and the concentration of substrate was high. Reversion products can be avoided when using *Saccharomyces cerevisiae* embedded in calcium alginate.[657] Under normal conditions, only glucose results from the action of glucoamylase.[1027,1028]

Glucoamylase performs better on gelatinized starches.[1029] Sometimes, prior to saccharification of starch with glucoamylase, the substrate is digested with hydrochloric acid. Such pretreatment increases the yield of crystalline glucose.[1030,1031] Instead, pretreatment of starch with alpha amylase can be employed. The degree of saccharification increased with DE of the starch hydrolyzate and was practically independent of starch concentration.[1032] The use of glucoamylase instead of beta amylase in the hydrolysis of starch in membrane reactors prevents formation of limit dextrins, because of a higher permeation rate resulting from a low level of starch gelatinization.[1033] Use of a very active glucoamylase accompanied by small amount of transglucosidase gave almost total hydrolysis of starch.[1034] When starch digested with alpha amylase was then treated with glucoamylase, glucose was the principal product along with small amounts of maltose, isomaltose, and panose. Crude glucoamylase contaminated with transglucosidase converted starch into glucose with a maximum yield after 72 h, after which time reversion took place.[1035]

Table V collects sources of glucoamylase and characteristics of the enzymes isolated from these sources.

d. Immobilized Glucoamylases.—Immobilized glucoamylase has excellent operational and storage stability as well as reproducibility, particularly at high substrate concentration.[1111] Several mineral supports have been utilized for immobilization of glucoamylase, for instance, various ceramics.[1112] After immobilization on porous silica, glucoamylase in pilot plant tests[1113] provided a yield of up to 93.5% of glucose, even after 80 days of continuous operation. Immobilization of glucoamylase on macroporous silica also does not deactivate the enzyme.[1114]

When glucoamylase was immobilized on alumina, a sugar syrup could be produced at 45–70 °C and 1–1000 psi pressure without the reversion that imparts an objectionable taste to the syrup.[1115] Porous alumina was also used as a support for that enzyme, affording stable activity of the enzyme for at least 7 days of continuous action at 45 °C.[1116–1118] The longest half-life, 113 days, was observed for the enzyme immobilized on a 3:1 $SiO_2:Al_2O_3$ support.[1119]

For that enzyme, covalently bound to a glutaraldehyde–silochrome support (silochrome is a siliceous carrier treated with 3-aminopropyltriethoxysilane), an increase in temperature of hydrolysis accelerated the process, but decreased the activity of that enzyme.[1102] Glucoamylase covalently immobilized onto glass fibers and 125 μm beads hydrolyzed corn starch and maltose.[1120] The optimum pH for glucoamylase from *Rhizopus* sp. after immobilization on ZrO_2-coated glass rose to 7.0 and the optimum temperature rose to 40–60 °C.[1121] Molecular sieves appeared to be a poor support, as the enzyme lost 25% of its activity within 30 days of operation to produce a syrup of DE 95.[1121] Immobilization onto montmorillonite involved intercalation of the enzyme between layers of the mineral. The enzyme may also be bound covalently to that support, but in so doing, it loses its activity.[1122] Charcoal supports were very sensitive to the state of the surface. Carriers covered with a layer of catalytic filamentous and pyrolytic charcoal were the most suitable, and the enzyme immobilized on them was the most suitable for hydrolysis of dextrins.[1123] Granulated chicken bones were used as a support for immobilizing glucoamylase together with pullulanase.[1124]

Mineral supports can be coated with inorganic and organic activators, for instance, silica gels activated with BCl_3 and with aliphatic amines,[1125] as well as keratin- and polyamide-coated matrices.[1126] Another procedure involves the covalent binding of glucoamylase to poly(ethyleneamine) and the mycelium of *Penicillium chrysogenum*, and blending the resulting product with silica gel. The process carried out at 50 °C and pH 4.5 provided a 92% yield of glucose.[1127] Decreasing the temperature from 70 to 50 °C drastically lowered the enzyme's activity because of deposition of impurities on the enzyme.[1128]

Supports based on synthetic organic polymers also offer several advantages, although the selection of good support presents a difficult task. For instance, a polyamide support can decrease the activities of glucoamylase, alpha amylase, and glucose isomerase, as shown in the use of a methacrylic acid-*N*-isopropyl(acrylamide) copolymer, which is a temperature-dependent, reversibly soluble–insoluble support for alpha amylase.[767] Beads of diethylaminoethylcellulose (DEAE-cellulose) deactivated the enzyme to 44–62% of its original free-state activity,[1129] and that support required additional activation with cyanuric chloride.[1130] A pH value of 4.0 is advantageous for the efficiency of the process at the outset,.[1131] and it then increases to 4.5

Glucoamylase from *A. niger* immobilized on a poly(acrylamide)-type carrier performed best at pH 3.8,[1132] and it retained 50% of its activity during 110 days.[1133] That enzyme immobilized on an acrylic polymer support dextrinized corn starch, sodium carboxymethyl starch, and starch phosphate.[1134] A poly(N-isopropylacrylamide) gel was suitable for hydrolysis of starch and pullulan, but this gel is temperature sensitive.[1135]

When granular poly(acrylonitrile) resin was used as a support for the enzyme,[1136] its activity was stimulated by Mn^{2+} and Co^{2+} ions.[1137] Glucoamylase immobilized on an acrylonitrile–acrylic acid–divinylbenzene–styrene copolymer exhibited long-lasting activity[1138] (up to 30 days) and that on DEAE-cellulose provided long-lasting production of glucose, with 92% yield.[1139] A copolymer of ethylene glycol, dimethylacrylate, and glycidyl methacrylate was also a useful support for glucoamylase.[1140]

Glucoamylase complexed to resin performed well during a long period of continuous utilization. After 15 days of digestion with corn syrup of DE 25 to yield a syrup of DE 33, 28% of the bound enzyme was lost.[1141] An anionic resin, Amberlite IRA 93, performed better and the enzyme did not lose its activity up to pH 4.0. A lower pH caused deactivation[1142] to the extent of $\sim 50\%$. Glucoamylase immobilized onto poly(styrene) anion-exchange resin produced glucose and maltose from liquefied cassava starch.[1143] Immobilization of glucoamylase on a series of commercial ion exchangers was reported[1144] and patented.[1145]

There is a difference in the course of hydrolysis of those polysaccharides that can enter the gel and those that cannot. In the latter case, the enzyme attacks the polysaccharides on the periphery, whereas there is random enzyme attack on the polysaccharide entering the gel.

Glucoamylase encapsulated in glutaraldehyde-cross-linked gelatin was deposited either on activated glass, on a polyester, or on aluminum foil. It operated at 50 °C during 34 days without any significant loss of its activity.[1146] Glucoamylase was also coimmobilized with pullulanase onto polyurethane foam.[1147] Both preparations also provided higher yields of glucose from insoluble starch. Immobilization of glucoamylase on a semipermeable membrane permitted the continuous hydrolysis of starch.[1148] Polysaccharide supports coated with poly(ethyleneimine) were also used. Glucoamylase was attached to them via ionic adsorption.[1149]

Some biopolymers can serve as suitable supports. Chicken egg-white can be used in its native state as a support for glucoamylase.[1150] Glucoamylase supported on corn stover of granularity below 44 µm allows the processing of 10% soluble starch at pH 3.5 and 40 °C during 3 h.[1151]

Glucoamylase immobilized jointly with pullulanase on beads of calcium alginate hydrolyzed starch more efficiently than the enzymes attached sequentially.[1152] Glucoamylase immobilized in calcium alginate gel is useful for starch liquefaction at pH 5.1 and 40 °C as well as saccharification at pH 4.8 and 50 °C. Raw cassava and potato starches hydrolyzed to the same (82%) extent, whereas hydrolysis of shati starch gave only 70% conversion, and soluble starch was hydrolyzed to almost 97%.[983]

5. Other Amylases

There are a few amylases that thus far cannot be attributed to a particular group of amylases. They are listed and selectively characterized in Table VI.

Nonmaltogenic exoamylases isolated from *B. clausii* BT-21 and *Pseudomonas saccharophila* retarded the detrimental retrogradation of starch, cleaving linear maltooligosaccharides and detaching 4–8 glucopyranosyl units from the nonreducing end.[96] These enzymes are useful in preventing the staling of bread.

6. α-Glucosidase (EC 3.2.1.20)

α-Glucosidase is an exoenzyme acting in a manner similar to that of glucoamylase on di- and oligo-saccharides and aryl glucosides. It yields glucose. This enzyme can be of animal, plant, bacterial, or fungal origin. All plants contain α-glucosidase as an endocellular enzyme, and it resides in germinated and nongerminated cereals.[1155] The neutral α-glucosidase from porcine serum appeared very substrate-specific. It hydrolyzes maltooligosaccharides, phenyl α-maltoside, nigerose, soluble starch, amylose, amylopectin, and β-limit dextrins. Isomaltose and phenyl α-glucoside were hydrolyzed with difficulty. Isomaltooligosaccharides built of 3 and more glucose units were not attacked by that enzyme.[1156]

α-Glucosidase in plants controls the plant polysaccharide composition already at the stage of plant maturation, as shown, for instance, in the case of rice[1157] or potato.[1158]

Corn α-glucosidase splits glucose from starch as the sole product, with no intermediary compounds.[1159,1160] Two glucosidases were isolated from rice.[1157,1161] They were strongly activated by KCl and by monovalent and divalent cations. With maltose as the substrate, such activation did not occur and glucose did not inhibit the enzyme activity.[1162] Their behavior resembled that of porcine serum α-glucosidase.[1163]

Along with alpha amylase, two isoform α-glucosidases were isolated from barley kernels. The rate of hydrolysis of native starch granules with them was comparable to the results of its hydrolysis with alpha amylase isoenzymes. A 10.7-fold increase in the rate of starch hydrolysis was achieved when that enzyme cooperated with alpha amylase. Based on the pattern of hydrolysis of starch granules digested separately and jointly by both enzymes, a suggestion was made that α-glucosidase can split not only α-(1→4) and α-(1→6) bonds.[1164] This suggestion was confirmed in case of α-glucosidase from sugar beet seeds.[1165] Two α-glucosidases isolated from malt digested intact starch granules, but the rate of hydrolysis was lower than the hydrolysis with alpha amylase. Only one of two glucosidases exhibited synergism with alpha amylase.[1166] An α-glucosidase from buckwheat[1165] and Welsh onion[1167] was also isolated and characterized.

α-Glucosidase secreted by bacilli that contaminate sizing starches decreases the quality of paper and paperboard.[1168] α-Glucosidase together with alpha amylase and pullulanase, all secreted by *B. subtilis*, after purification produced chiefly maltotetraose.[1169] *Bacillus* sp. APC-9603 produces a novel thermostable α-glucosidase exhibiting isoamylase and pullulanase activity. Optimum conditions for that enzyme are pH 4.5–6.0 and 60–70 °C. This enzyme is recommended for a use jointly with glucoamylase for making glucose syrup and with beta amylase for making maltose syrup.[1170] The *Thermus thermophilus* bacterium produces thermostable α-glucosidase that works best at pH 6.2 and 85 °C and is suitable for one-step starch processing, being also remarkably active against maltose and maltotriose.[1171] The bacteria *Sulfolobus solfataricus*[1172] and *Thermococcus* strains[1173] also produce α-glucosidase. It has been proposed[1174] to use genes of the organisms for making thermostable α-glucosidase for hydrolysis of maltooligosaccharides and liquefied starch. These genes are available from bacteria present in the environment.

Mycelia of the *Mucor javanicus* fungus produce α-glucosidase having glucosyltransferase activity.[1175] That enzyme, together with glucoamylase, is also produced by *Lentinus edodes* (Berk.) Sing. Its M_w is 51 kDa, and it hydrolyzes maltose, maltotriose, phenyl α-maltoside, amylase, and soluble starch.[1020] *Cladosporium resinae* produces alpha amylase, α-glucosidase, exopullulanase, and two glucoamylases.[585,586] *A. niger* and *Rhizopus* also produce α-glucosidase. Among several enzymes tested, only that enzyme by itself, and more efficiently in combination with pullulanase, could split isomaltose and panose.[1176] Optimum conditions for the *A. niger* α-glucosidase are pH 5.5 and 55 °C.[1177] An α-glucosidase from *Paecillomyces lilacinus*, acting as a transferase, synthesizes α-(1→3)- and α-(1→2)-linked oligosaccharides. That enzyme is most active at pH 5.0 and 65 °C.[1178]

Covalent immobilization of α-glucosidase on porous silica increased its stability and the deactivation energy increased by almost 14%.[1179] As with other enzymes, acylation of α-glucosidase results in its partial deactivation.[1180]

Table VII lists sources of α-glucosidases and selected characteristics of the enzyme.

TABLE VII
Sources of α-Glucosidases, Their Molecular Weights, and Optimum Temperature and pH

Source	Molecular weight (kDa)	Optimum Temperature (°C)	pH	Ref.
Human and animal				
Kidney	315–352			1181
Porcine serum				1156
Periplaneta americana				638
Plant				
Barley				1164
Buckwheat				1182
Corn				1159,1160
Malt				1166
Potato				1158
Rice				1157,1161
Sugar beet seeds				1165
Welsh onion				1167
Bacterial				
Bacillus subtilis P-11	33	45		180
B. amyloliquefaciens	27	38		180
B. brevis	52	48–50		180
B. cereus	12	40		180
B. amylolyticus		40		180
Bacillus sp. KP 1035	43–53	65		180
Bacillus sp. APC-9603		60–70	4.5–6.0	1170
Pseudomonas fluorescens W				180
P. amyloderamosa		50		180
Thermococcus sp.				1173
Thermus thermophilus	67	85	6.2	1171,1183
Yeast				
Saccharomyces cerevisiae				
Maltase	68.5			180
α-Methyl glucosidase	64.7			180
α-Glucosidases I-III		36–42		180
Sacharomyces italicus	85 ± 30			180
S. logos	270	40		180
Schwanniomyces occidentalis				1184
Fungal				
Aspergillus awamori	125–140	55		180
A. flavus	63	35		180
A. fumigatus	63.56			180
A. niger		55	5.5	1177

Continued

TABLE VII (Continued)

Source	Molecular weight (kDa)	Optimum Temperature (°C)	pH	Ref.
A. oryzae		50–56		180
Cladosporium resinae				586
Lentinus edodes	51			1020
Mucor javanicus	124.6	55		180,1177
M. pusillus				1027
M. rouxii		50–60		180
M. racemosus	97–114	50		180
Paecillomyces lilacinus		65	5	1178
Penicillium oxalicum		50		180
P. purpurogenum	120	50		180
Rhizopus sp.				1176
Archeal				
Pyrococcus furiosus		105–115	5.0–6.0	1185
P. woesei		110	5.0–5.5	1186
Sulfolobus shibatae		85	5,5	1187
S. solfataricus		105	4.5	1172,1188, 1189
Thermococcus hydrothermalis		110	5.0–5.5	1190
T. zilligii		75	7	1191

7. Pullulanase (EC 3.2.1.41)

Pullulanase type I cleaves α-(1→6) bonds in amylopectin, dextrins, and pullulan, and pullulanase type II, which is less specific than type I, hydrolyzes α-(1→4) and α-(1→6) bonds in starch and dextrins. The latter type is the debranching enzyme and lacks a link-forming ability. It splits α-(1→6) linkages, but is inactive toward α-(1→4) linkages, and therefore it is suitable for splitting amylopectin. The hydrolysis of starch begins with a rapid decrease in the molecular size of amylopectin.[1192]

Plants, bacteria, and less commonly fungi, are sources for pullulanase. The fungus *Streptomyces* strain NCIB 12235 is a source of pullulanase. The enzyme performs best at pH 4–6 and 55–65 °C.[1193] Pullulanase secreted from *A. oryzae* and *A. usamii* could not be separated from amylases.[1194] Pullulanase was also extracted from brewery yeast.[1195]

Potato and broad bean are, among others, common plant sources of that enzyme.[1196] Pullulanase from these sources produced maltose and maltotetraose, accompanied by traces of maltohexaose, from waxy maize β-dextrin with an overall yield of 12.8%.[1197] In contrast to pullulanase from potato, the one isolated from corn produced a high yield of maltohexaose.[1198] Pullulanase isolated from malted barely had only low activity[1199] until it had been thoroughly purified. After purification, the enzyme of 103 kDa operated under optimum pH 5.0–5.5 and 50 °C. Among other substrates, it could split isopanose.[1200] Rice can also serve as a source of pullulanase.[1201] Plant pullulanase may be more active than bacterial pullulanase, as shown from comparison of the activity of rice and *Arthrobacter aerogenes* pullulanases.[1201] The latter exhibits a strong synergism with beta amylase.[1202] That pullulanase has M_w 51–52 kDa and operates best at pH 5.3–5.8 and 50 °C.[1203]

Bacterial pullulanases are usually thermostable, as shown by those from the genera *Clostridium*, *Thermoanaerobacter*, and *Thermobacteroides*.[1204] In addition to α-(1→6) linkages, they cleave α-(1→4)-linkages as with, for instance, the pullulanase from *Bacillus* sp. 3183.[1205] Pullulanase from *B. subtilis*[1169,1206,1207] is thermostable, but that from *B. acidopullulyticus* (M_w 102 kDa) decomposes at 60 °C within 1 h of working at pH 5.0, the optimum for this enzyme. A slight activation by Ca^{2+} was observed.[1208] This enzyme, along with that from *Klebsiella pneumonia*, does not adsorb onto raw starch and cannot hydrolyze it.[913] The latter enzyme is suitable for production of maltotriose at pH 8.0–9.0.[1209] A highly thermostable, cell-associated pullulanase of M_w 83 kDa was produced by *Thermus aquaticus* YT-1 strain working at its optimum pH 6.4 and tolerating 85 °C for 10 h. The thermostability was increased by Ca^{2+} ions.[1210] Pullulanase of similar thermostability, but of M_w 120 kDa, was isolated from *Thermoanaerobium* Tok6-B1.[1211] Pullulanases produced by *Thermococcus litoralis*, M_w 119 kDa, and *Pyrococcus furiosus*, M_w 110 kDa, retained their activity up to 130–140 °C. They did not hydrolyze maltohexaose and lower oligosaccharides.[1212]

The hyperthermophilic pullulanases from *Thermococcus hydrothermalis*[1213] and *Clostridium thermohydrosulfuricum*[1214] provided glucose and maltose from starches. A syrup comprising maltotetraose (58.1%), maltotriose (14.6%), glucose (11.6%), maltose (7.9%), and maltopentaose (4.1%) was produced by the pullulanase of *Micrococcus* sp NO 207.[1215] *Microbacterium imperiale* produces another thermostable pullulanase working at pH 4–9 at optimum 60 °C; the enzyme is deactivated by Ca^{2+} at 60 °C. This enzyme produces mainly maltotriose when operating at pH 8–9.[1209,1216,1217] *Cladosporium resinae* releases an exopullulanase (exo-1,6-α-glucanase) whose ability to cleave α-(1→6) linkages exceeds that of *A. niger* pullulanase.[585,1217]

Pullulanase can be used when immobilized, for instance, on poly(N-isopropylacrylamide) gel[1135] or γ-alumina beads.[1218] Pullulanase cooperates effectively with glucan 1,4-α-maltohydrolase (EC 3.2.1.133).[1219]
Table VIII presents sources of pullulanases and characteristics of the enzyme.

TABLE VIII
Sources of Pullulanases, Their Molecular Weights, and Optimum Temperature and pH

Source	Molecular weight (kDa)	Optimum Temperature (°C)	pH	Ref.
Plant				
Broad bean				1196
Corn				1198
Malted barley	103	50	5.0–5.5	1199
Potato				1196
Rice				1199
Bacterial				
Aerobacter aerogenes	114.3	50		180
Arthrobacter aerogenes	51–52	50	5.3–5.8	1203
Bacillus acidopullulyticus	102	60	5	1208
B. cereus var. mycoides	112±20	50		180
B. polymyxa	48	45		180
B. subtilis				1169,1206,1207
Bacillus sp. No. 202-1	92	55		180
Bacillus sp. 3183				1205
Cladosporium resinae				586,1218
Clostridium thermohydrosulfuricus				1214
Klebsiella pneumonia				913,1209
Microbacterium imperiale		60		1209,1216
Micrococcus sp. NO207				1215
Thermus aquaticus	83	85	6.4	1210
Yeast				
Saccharomyces cerevisiae				1195
Fungal				
Aspergillus oryzae				1194
A. usamii				1193
Cladosporium resinae				585,1217
Streptomyces sp. No. 280		50		180
Streptomyces sp. NCIB 12235		55–65	6-Apr	1194

TABLE VIII (Continued)

Source	Molecular weight (kDa)	Optimum Temperature (°C)	pH	Ref.
Archeal				
Desulfurococcus mucosus		100	5	792
Pyrococcus furiosus	110	98	5.5	1212,1220
P. woesei		100	6	1221
Thermoanaerobium Tok6-B1	120			1221
Thermococcus aggregans		100	5.5	797
T. celer		90	5.5	797,801
T. guaymagensis		100	6.5	797
T. hydrothermalis		95	5.5	1213,1222
T. litoralis	119	98	5.5	1212,1220
Thermomyces ST 489		80–95		1223

8. Neopullulanase (EC 3.2.1.135)

This extracellular enzyme is produced by *B. stearothermophillus*, *B. subtilis*,[1224,1225] and *B. polymyxa*.[1226] It hydrolyzes pullulan to panose and can induce transglycosylation. Neopullulanase from *B. polymyxa* has M_w 58 kDa and works best at pH 6.0 and 50 °C.

9. Isoamylase (EC 3.2.1.68)

Isoamylase hydrolyzes α-(1→6) bonds of amylopectin and dextrins, forming the corresponding oligosaccharides depending on the size of their side chains. That enzyme was initially identified as an amylosynthetase. The (1→6) branches are indispensable for its hydrolytic action.[1227] Eight debranching isoenzymes have been characterized from human pancreatin. They split starch into fragments from glucose to maltodecaose, and the result depends on the characteristics of the substrate hydrolyzed.[1228]

The organism *Pseudomonas amyloderamosa* is the most generally used source for isoamylase.[913,1229–1231] Isoamylase from that source adsorbed onto starch and caused hydrolysis,[913] but afforded few products of low molecular weight. However, addition of a surfactant increases the number of components in the final product because of formation of a series of maltooligosaccharide.[1229] Isoamylase decomposes at room temperature, but its adsorption onto raw corn starch increases its stability. It is stable for 8 months at 4–6 °C, and after vacuum drying it remains almost intact during 8 months of storage at room temperature.[1232] Isoamylase can be used after immobilization on a polysaccharide matrix composed of agarose, cellulose, and raw corn starch.[1227]

Flavobacterium odoratum is another source of an isoamylase, of M_w 88 kDa, which specifically splits the $(1 \rightarrow 6)$ linkages.[1233] *Flavobacterium odoratum* KU secreted an isoamylase of M_w 78 kDa, which operated best at pH 6.0 at 45 °C, and it was stabilized and activated by Ca^{2+} ions. It hydrolyzed neither pullulan nor CDs.[1234]

Susceptibility of starches to that enzyme depends on their origin. Tuber starches are less susceptible than cereal starches. That enzyme hydrolyzes $(1 \rightarrow 6)$ branch linkages in amylopectin to a limited extent and that extent also depends on starch origin. In the reaction performed on granular starches at 37 °C, the yield increases in the order: cassava < potato < barley < maize < waxy maize < shoti < hylon maize, whereas in gelatinized starches that order is shoti < hylon maize = waxy maize < potato.[1230]

The isoamylase isolated from *Rhodotermus marinus* DSM 4252 was a thermostable enzyme of M_w 80 kDa and optimum conditions for its activity are pH 5.0 and 50 °C.[1235] Its use leads to an increase in the yield of glucose. Ohata et al.[1236] compared the hydrolysis of granular starches of waxy and normal corn, wheat, potato, sweet potato, and cassava with the isoamylase of *Pseudomonas amyloderamosa* and the glucoamylase of *Rhizopus nivea*. Prior to digestion, the starch was preheated for 30 min at 50, 60, and 95 °C. Native and preheated wheat starch was hydrolyzed much more readily with these two enzymes than the other starches.

Table IX presents sources of isoamylases and characteristics of that enzyme.

TABLE IX
Sources of Isoamylases, Their Molecular Weights, and Optimum Temperature and pH

Source	Molecular weight (kDa)	Optimum Temperature (°C)	pH	Ref.
Human and animal				
Human pancreatin				1228
Bacterial				
Cytophaga sp.	120	40		180
Escherichia coli		45–50		180
Flavobacterium odoratum	88			1233
Flavobacterium odoratum KU	78	45	6.0	1234
Pseudomonas amyloderamosa	90			180
Rhodotermus marinus DSM 4252	80	50	5.0	1235
Yeast				
Saccharomyces cerevisiae		25		180

10. Other Hydrolases

Glucan 1,4-α-maltotetraohydrolase (EC 3.2.1.60) is usually isolated from *Pseudomonas stutzeri*[447,1237] and *Bacillus circulans* MG-4.[446,1238] It produces maltotetraose from starch. Its peak activity is at pH 7.5 and 50 °C, and it is stabilized by Ca^{2+} ions. The enzyme isolated from *P. stutzeri* has M_w 12.5 kDa.[447]

Glucan 1,4-α-maltohydrolase (EC 3.2.1.133) also called maltogenic amylase and secreted by *Thermomonospora viridis* THF-35 is an extracellular thermostable hydrolase converting maltotriose, maltotetraose, maltopentaose, amylase, and amylopectin into maltose, along with glucose.[1239] Optimum conditions were pH 6.0 and 60 °C.

β-Glucosidase (EC 3.2.1.21),[179] also called Z-enzyme and supporting the action of beta amylase and phosphorylase, was first isolated from soy bean and sweet almonds.[1240] β-Glucosidase splits anomalous linkages sometimes present in amyloses from certain sources,[848,1241,1242] but it does affect neither α-(1→2)-, α-(1→4)-, or α-(1→6)-glycosidic linkages nor β-linked disaccharides. Preheating to 40–50 °C for 15–60 min prior to use increases the enzyme activity by 30–40% toward wheat starch and by 70–80% in case of whole grains. Preheating above 50 °C decreases activity of that enzyme.[1243]

β-Glucanase (EC 3.2.1.6) is present in malt. It splits fiber. It is beneficial to supplement with this enzyme to support hydrolysis of starchy materials containing cellulose and hemicelluloses.[1244] Glucodextranase (EC 3.2.1.70) of M_w 120 kDa from *Arthrobacter globiformis* T-3044 is stable up to 40 °C at pH 6.0–7.5. It produces β-glucose from dextran and exhibits high activity toward starch.[1245]

β-Glucosiduronase (EC 3.2.1.31) breaks down complex carbohydrates. Human β-glucosiduronase catalyzes hydrolysis β-D-glucuronic acid residues from the nonreducing end of such glycosaminoglycans (mucopolysaccharides) as heparin sulfate. It is also present in the majority of flavobacteria. The enzyme from this source can also hydrolyze starch.[1246] A novel thermostable enzyme termed amylopullulanase was isolated from *Geobacillus thermoleovorans* NP33, which either alone or in combination with alpha amylase saccharifies starch at 100 °C and pH 7.0.[527]

Dextrins and isomaltose produced from starch by alpha amylase can be further converted by detachment of terminal α-(1→6)-linked glucose using oligo-1,6-glucosidase (EC 3.2.1.10).[1247] The enzyme releases the terminal α-(1→6)-linked side chains, splits the dextrins resulting from digestion of starch by alpha amylase, and also splits isomaltose.

Maltotriohydrolase, 1,4-α-D-glucan maltotriohydrolase (EC 3.2.1.116) splits α-(1→4) bonds, releasing maltotriose from the nonreducing ends of starch chains.[1247]

Glucodextrinase (EC 3.2.1.70) from *Arthrobacter globiformis* I42 hydrolyzes α-(1→6) glucosidic linkages of dextrin from the nonreducing end to produce β-glucose.[1248] Likewise, the limit dextrinase (dextrin α-1,6-glucanohydrolase) (EC 3.2.1.142) also splits (1→6) linkages in dextrins.

An enzyme defined as an α-1,4-glucan lyase was isolated from *Glacilaria lemaneiformis* red seaweed which at pH 5.0–5.8 and 50 °C acted on floridean and soluble starch and was only slightly inhibited by the glucose formed.[1249]

An extract of intestinal mucosa contains an α-limit dextrinase (EC 3.2.1.142) that is acid-labile, and the optimum pH for it is around 6. It is much less thermostable than alpha amylase.[1250] Potatoes contain a nonphosphorylating transglucosidase having cross-linking ability, which in the plant together with phosphorylase participates in the synthesis of amylopectin.[1251–1253]

11. Enzymatic Cocktails

Combined acid–enzyme processes or enzymatic cocktails are commonly used in starch hydrolysis.[1254–1258] The hydrolytic agents can be applied in the form of a cocktail or sequentially.[1259,1260] Frequently, there is a synergism between hydrolases, which in the form of enzymatic cocktails digest starch more efficiently. Synergism can be observed between two alpha amylases, as shown with a combination of these thermostable enzymes from *B. licheniformis* and *B. stearothermophilus* used in the liquefaction of corn starch,[1261] with ptyalin and pancreatin, ptyalin and malt amylase, and pancreatin with malt amylase.[1262] Regardless of the combination used, maltose was always formed, with no traces of glucose.

There is a synergism between alpha and beta amylases in their hydrolytic action,[1263] although the synergism between bacterial alpha amylase and barley beta amylase may be weak,[1264] and dependent on the proportion of both enzymes.[1265] Alpha amylase extracted from black koji mold had very low activity, but it was very strongly activated by beta amylase from the same source.[1266] Based on a measure of the reducing power, at pH 4.6 and 21 °C, there was a linear relationship against time, up to 15% hydrolysis of soluble starch with beta amylase, and 20–25% hydrolysis with alpha and beta amylases together. For alpha amylase used alone, this relationship is somewhat uncertain. The action of both enzymes up to 20–25% hydrolysis and a beta-to-alpha ratio between 4/1 and 1/4 was additive, but beyond this range it was variable.[1267] In the saccharification of soluble starch with blends of alpha amylase from *A. oryzae* and beta amylase from *A. usamii*, when the latter was in excess the proportion depended on the degree of polymerization (DP) of the starch. Starches of intermediate DP required an excess of beta amylase, whereas for dextrins of high DP a lesser proportion of beta amylase in the blend was sufficient.[1268] The conversion of starch with both enzymes is accelerated by inclusion of various salts, and this inclusion works only if both enzymes are present. It is suggested that salts interact with the branched starch derivatives produced by beta amylase, thereby facilitating their alpha-amylolysis.[1269] The hydrolyzing action of alpha and beta amylases is effectively enhanced by animal ptyalin and pancreatin. In such a manner the yield of maltose can be increased to as high as 70%.[1270]

Poplar wood contains three amylases and two phosphorylases. The alpha amylase is more active in degrading granular potato starch than the beta amylase. The third amylase does not attack soluble starch, but only the granules. There is no synergism between these three amylases, but one of the two phosphorylases exhibits very strong synergism with alpha amylase. The second phosphorylase attacks only starch granules. When phosphorylases are combined with all three amylases, the formation of glucose 1-phosphate is inhibited and maltose is the principal hydrolysis product. Maltose at the concentration existing in poplar wood during the starch-degradation phase completely inhibits beta amylase.[1271] Another example of synergism between enzymes of the same origin is demonstrated by *Clostridium thermosulfurogenes* EM1. This organism secretes seven different pullulanases and one form of alpha amylase. Acting separately, the pullulanases split α-(1 → 4) and α-(1 → 6) linkages, whereas alpha amylase preferably attacks insoluble starch, producing maltohexaose. In conjunction, all of these enzymes show synergistic action.[1272]

Alpha amylase together with glucoamylase rapidly converts starch into glucose with a yield as high as 99%[1273] The enhanced yield of maltose is the most visible result of cooperation of both enzymes. Maltose is not formed when both enzymes perform separately.[1274] Frequently, alpha amylases cooperate with glucoamylases in the hydrolysis of various starches,[609,1275–1280] but the extent of synergism depends on the molecular weight of the substrate. For instance, with starches having $M_w < 5$ kDa, no synergism could be observed.[1281] The amylase/glucoamylase ratio is also a factor.[1282] As the proportion of fungal amylase increased, the DE of the product increased and its viscosity declined.[1282] For *A. oryzae* alpha amylase (4 U/mg) and *A. niger* glucoamylase (70U/mg), both immobilized on porous glass, the 1:4 ratio provided the highest yield of glucose.[72]

A thermophilic amylase secreted by *Anoxybacillus amylolyticum* also utilized galactose, trehalose, and maltose. The enzyme performs best at 61 °C and pH 5.6.[1283] A four-domain alpha amylase isolated from contaminants of *Anoxybacillus*, together with a glucoamylase from *A. niger*, is able to liquefy starch to 99% at low temperature.[530]

The hydrolysis with pancreatic amylase significantly accelerates after admixture with enteric amylases.[1284,1285] Satisfactory results are achieved when amylases are combined with a yeast.[1286] Hog pancreatin and *A. oryzae* enzyme readily hydrolyze raw starch, provided that either a mineral component such as $CaCl_2$ or ash is present.[1287] α-Glucosidase jointly with alpha amylase produces glucose-transferred products from starch hydrolyzates. These hydrolyzates were available from liquefied starch treated with beta amylase and pullulanase.[1288]

The cocktail of bacterial alpha amylase, a maltogenic amylase, and glucoamylase converts starch into glucose syrup in a single step.[1289] For the manufacture of high-conversion starch syrup containing isomaltose, a cocktail of alpha amylase (0.02%), glucoamylase (0.01%), and transglucosidase (0.03%) was applied at 55 °C and pH 5.0, yielding a product containing 43.2% glucose, 4.1% maltose, 20.6% isomaltose, 31.7% isomaltooligosaccharides, and 0.4% maltooligosaccharides.[1290]

α-Glucosidase from *A. oryzae* combined with pancreatin rapidly hydrolyzes raw starch.[1291] Synergism of alpha amylase and α-glucosidase, both isolated from the cockroach *Periplaneta americana*, hydrolyzed starch to maltose and maltodextrins.[638] The combined action of soy bean beta amylase and isoamylase from *Flavobacterium*, incubated first for 46 h at pH 6.0 and 40 °C, produced maltose and maltotriose when the DE of the hydrolyzate was between 1.6 and 4.0%. At a DE between 6.6 and 9.1%, maltose was the principal product apart from moderate amount of maltotriose and traces of glucose.[1292] Synergism was found for glucoamylase and pullulanase in the saccharification of liquefied starch.[1293] Hydrolysis of starch with a combination of barley alpha amylase and *A. oryzae* beta amylase was studied by an artificial neural network numerically modeled by the authors.[1294]

Cooperative hydrolysis by means of glucoamylase and pullulanase was advantageous with potato-starch slurry; however, both enzymes had to be applied sequentially.[1295] Applied in the same sequential manner, glucoamylase and mutarotase provided a higher yield of glucose from corn starch,[1296] and glucoamylase treatment followed by the addition of alpha amylase of *B. subtilis* was of value in the saccharification of wheat flour.[1297]

Studies performed on the cocktail of three hydrolases, namely taka amylase, beta amylase, and maltose-oligosaccharide transglucosidase from *A. usamii*, showed that the correct proportion of the components in the cocktail is essential. The ratio of two first components is crucial for maximizing the amount of fermentable sugars formed, and it increases with increase in that ratio. Admixture of transglucosidase increases the yield of fermentable sugars.[1298,1299] Another cocktail of three hydrolases contained alpha and beta amylases with pullulanase. It was suitable for production of maltose, but no synergism between these hydrolases was observed[1300–1305] in barley.

The cooperative action of alpha and beta amylases, α-glucosidase, and a debranching enzyme, all present in barley seeds, was studied to control retrogradation in barley.[422] (For the ability of various common cereals to form RS starch, see paper[1306]).

In an isoamylase cocktail with a saccharifying enzyme, the presence of isoamylase decreased the required amount of beta amylase, increased the saccharifying time, and increased the yield of glucose. The optimum conditions[1307] for the cocktail were 60–61 °C and pH 4.2–4.5.

Enzymatic cocktails can be prepared employing coimmobilized enzymes, as shown for various amylases and glucoamylase attached to a DEAE-cellulose support and hydrolyzing starch to glucose.[1308]

12. Glycosyltransferases (EC 2.4.1)

Although these enzymes belong to the class of transferases, they are unable to transfer glucose but they effectively participate in the hydrolysis of starch. Cyclodextrin glycosyltransferase (EC 2.4.1.19) (CGTase) liquefies and saccharifies starch. The thermostable enzymes from *Thermoanaerobacter* sp. ATCC 63.627,[1309] *Nocardiopsis dassonvillei*, and *B. pabuli*[1310] operate best at pH 3.5 and 95 °C. The CGTase from *Klebsiella pneumonia* M 5 modified at the His residue yielded branched saccharides from maltotetraose to maltononaose.[1311] The use of CGTase covalently immobilized on porous silica, manifesting no change of the activity, has been described.[1312] A 4-α-glucanotransferases (EC 2.4.1.25) isolated from *Pyrobaculum aerophilum* IM2[1313] transfers α-(1→4)-linked glucans to other acceptors, such as glucans and even glucose. It produces thermoreversible gels of gelatin-like properties. A glucosyltransferase resembling 4-α-glucanotransferase was produced by *B. megaterium*. On saccharification of starch at pH 4.5 and 60 °C, it provided a product containing of 97.1% glucose.[1314] Another glucosyltransferase-like enzyme was secreted by *Streptomyces sanguis*. It converted amylopectin into amyloglucan.[1315] Other transferases useful in starch conversion are a glucan-branching enzyme (EC 2.4.1.18) originating from *Streptococcus mutans*,[1316] and amylomaltase (EC 2.4.1.15), which catalyzes the formation of cyclomaltosaccharides.

In contrast to the well-studied CGTases, which synthesize cyclomaltosaccharides having a ring size (degree of polymerization or DP) of 6–8, the amylomaltase from *Thermus aquaticus* produces cyclomaltosaccharides having a DP of 22 and higher.[1317] Some CGTases digest raw starch. They include the one secreted by *B. firmus*[1318] having M_w of 82 kDa, and optimum pH and temperature 5.5–8.8 and 65 °C, respectively, and another secreted by *Klebsiella pneumoniae* AS-22[1319] with M_w of 75 kDa and optimum pH and temperature of 7.0–7.5–8.0 and 65 °C, respectively.

A thermostable α-glucanotransferase from *Thermus scotoductus* used with 5% rice-starch paste at 70 °C significantly liquefied it and increased its freeze–thaw stability.[1320]

13. Microorganisms

Before isolated hydrolytic enzymes became available on the market, microorganisms—producers of hydrolases—were used for conversion of starch. Bacteria[1321] and yeasts[1322] were commonly used. Among bacteria, the thermophilic ones performed better than mesophilic ones.[1323] Several anaerobic strains have been isolated.[460,1324] All of them were strictly anaerobic, producing soluble starch-degrading enzymes. These enzymes had molecular weights around 25 kDa and were thus of molecular weight lower than that of enzymes produced by aerobic bacteria. Anaerobic thermophiles, such as those from *Clostridium*, *Thermoanaerobacter*, and *Thermobacteroides* sp. secrete amylolytic and pullulytic enzymes capable of cleaving α-$(1\rightarrow 4)$ and α-$(1\rightarrow 6)$ bonds in starch polysaccharides by a random attack.[1325] A *Paenibacillus granivorans* species has been isolated from native potato starch. This species belonging to the *B. firmus/lentus* group, degrades native potato-starch granules.[939,1326]

Some bacteria convert starch directly into CO_2 and H_2. *Bacilli* display pullulanase and amylolytic activity, and split starch into maltose and glucose.[1327] *B. cereus* provided maltopentaose from starch.[1328] Specific bacteria are required for particular situations. For instance, in starch-based drilling mud, the hydrolyzing bacteria had to withstand 12 h of contact with 0.1% formaldehyde.[1329] Diphtheria *bacilli* decompose starch rapidly, but they are highly virulent.[45] *Pectinobacter amylophilum*, isolated from soil, digested starch but in an erratic manner. It effectively digested wheat and rice starches directly to CO_2 and hydrogen.[1330] Also *Clostridium butyricum*, a mesophilic bacterium isolated from sludge, utilized cereal and tuber starches, and in comparison with the formerly mentioned bacteria, it utilized potato starch more effectively, generating CO_2 and hydrogen.[1331] However, there is a report[1332] that this bacterium and also *Streptococcus bovis* rapidly digest corn starch but not potato starch. Alpha amylase isolated from both bacteria behaved similarly against these starches, forming maltotriose and other products, but both bacteria and their alpha amylases were distinguished from one another in their behavior in the conversion of maltotriose. *C. butyricum* and its amylase did not digest maltotriose.

Mixed bacteria from sheep rumen digested corn and potato starch at comparable rates, but amylase isolated from them did not adsorb onto starch granules and did not hydrolyze them. Rumen contains several bacteria of different starch-digesting ability, which produce glucose, maltose, and maltosaccharides through pentaose and above. The bacteria *S. bovis* JN1, *Butyrivibrio fibrisolvens* 49, and *Bacterioides ruminicola* D31d rapidly hydrolyzed starch, along with the maltooligosaccharides formed. *S. bovis* produced the range glucose–maltotetraose, *B. fibrisolvens* produced additionally maltopentaose, and *B. ruminicola* gave glucose–maltotetraose, maltohexaose, and maltoheptaose, but only a limited amount of maltopentaose. *Selenomonas ruminatum* HD4, also present among these bacteria, grew poorly on starch.[1333] Rumen oligotrich enzymes of the genus *Entodinium* digested starch in plant cells, yielding CO_2 and other gases as well as lower fatty acids, namely butanoic, acetic, propanoic, and formic acids in the molar proportion 51:35:10:4. Additionally, lactic acid was also formed.[1334]

Strains of *Rhizobium* species have been isolated from sweet clover, the succulent clover berseem, and pea.[1335] These strains are known as producers of beta amylase. They caused appreciable hydrolysis of potato starch, but reducing sugars were not formed when using the species isolated from sweet clover because it appeared that these strains had produced solely alpha amylase. A viscous polysaccharide syrup was produced from starch by *Klebsiella pneumonia*.[1336] *Pseudomonas* species digested soluble starch to give reducing sugars[1337] and *Xanthomonas campestris* liquefied starch in the presence of glucoamylase.[1338] Hydrolysis of starch with *Bacterium casavanum* generates an insoluble gel containing uronic acid, glucosamine, and nucleic acids from the microorganism used.[1339] Amylose alone was also hydrolyzed by these last noted bacteria, but amylopectin was resistant to them without prehydrolysis with 1.5% hydrochloric acid.[1340]

The fermentation of starch by yeast still evokes interest. Brewery yeast added to saccharified cereal starch decreases the requirement for alpha amylase and glucoamylase.[1341] Corn amylose, Zulkowsky soluble starch, and corn dextrins were hydrolyzed[1342] by yeast juice at pH 6.9. A hybrid *Saccharomyces* strain secreting yeast glucoamylase and mouse alpha amylase converts 92.8% of starch into reducing sugars within 2 days.[74,1343] Starch of rice crumb was liquefied and saccharified by yeast, followed by treatment with transglucosidase to give an isomaltooligosaccharide syrup. The latter was converted by *Saccharomyces carlsbergensis* or *S. cerevisiae* into glucose.[1344] A 2-deoxyglucose-resistant *Schwanniomyces castelli* R68 mutant produced three to four times more alpha amylase and glucoamylase, but it still could not hydrolyze cooked wheat starch completely.[1345] Yeast expressing glucosylase on the cell surface was patented for degradation of starch.[1346] A recombinant strain of *S. cerevisiae* producing glucoamylase and isoamylase has been shown to cause over 95% conversion of starch.[1347]

Rhizopus species have also been evaluated for hydrolysis of starch, particularly raw starch in the solid-state fermentation of carbohydrate waste.[1348] As compared to *Aspergilli*, *Rhizopus javanicus* exhibited only low activity, but it produced mainly glucose.[1349,1350] *Rhizopus oligosporus* brought about[1351,1352] 92.8–99.4% hydrolysis of cassava starch, with a 91.6% yield of glucose in the processing of a 2.6–5% slurry of that starch at pH 3.8–4.0 and 45 °C. *Rhizopus achlamydosporus* immobilized on nonwoven fabrics has been used.[1353] *Rhizopus oryzae* 28627 appeared suitable for the solid-state fermentation of cassava bagasse.[1354]

Earlier studies revealed that such fungi as *Aspergillus oryzae* can saccharify starch, and that process reaches its maximum rate at the time of conidia formation. The conidia provide a glucoamylase-type enzyme. The rate of fungus growth decreases in proportion to the amount of starch. At a low concentration of starch, the fungus consumes it rapidly. With conidia of *A. wentii* NRRL 2001, the maximal yield of glucose was achieved within 3 days.[1355] Diastase isolated from *A. oryzae* can hydrolyze starch to 95% extent, provided that a large amount of enzyme is used.[1356]

Various strains of *Aspergilli* have been used for the hydrolysis of starch.[1357–1359] In the formation of soluble starch in the presence of various mono- and di-saccharides, using either *A. niger* or *A. oryzae*, some differences were observed in the effect of the sugars. In the hydrolysis of starch with *A. niger*, all of the sugars stimulated the process in the order: fructose > sucrose > glucose > maltose > galactose, whereas the hydrolysis with *A. oryzae* was inhibited by fructose.[1360] *A. oryzae* could digest barley bran, wheat flour, and potato and sweet-potato starch.[1361] On digestion of potato starch with various *Aspergilli*, *A. oryzae* produced large amount of dextrins and maltose. *A. awamori* gave less maltose and more glucose. The results with *A. niger* and *A. usamii* were similar; they produced less maltose and more glucose.[593,1349] *A. awamori* converted sago starch into a glucose syrup[1362] and was also useful in the conversion of palm starch into glucose.[1363] There is a patent[1364] for hydrolysis of starch with that fungus where the starch was prehydrolyzed with alpha amylase from *B. licheniformis*.

The beta amylolytic activity of *Aspergilli* toward potato starch decreased in the order: *A. niger* > *A. usamii* > *A. awamori*.[1350] *A. niger* activity against glutinous starches was higher than that against nonglutinous starches, and malt starch was digested more readily than barley starch.[1365] In the digestion of corn, potato, and rye meals, *A. niger* shows a clear advantage over malt.[1366] The efficiency of *A. niger* in the hydrolysis of starch is associated with adsorption of its amylase onto the substrate.[1367] *A. awamori* liquefies starch well, but saccharifies it less readily. *A. usamii* is a strain that is poorly liquefying; it acts better as a means for debranching.[1368] *A. niger* produced from starch chiefly glucose, accompanied by small amounts of maltose, isomaltose, panose, and maltotriose. Under the same reaction conditions *A. oryzae* produces mainly maltotriose and considerably less maltose and glucose.[1369]

Some complex bacterial preparations for hydrolysis of starch have been patented. One of them is composed of yeast (*Saccharomyces* and brewery yeast), bacteria (*B. subtilis* and *B. licheniformis*), *Dialister bacterium*, *Actinomyces israelii*, *Lactobacillus*, and *A. niger*.[1370] Another one is simply a complex of *A. awamori* with the product of fermentation of whole wheat flour with that enzyme.[1371] An *Acetobacter* has been said to produce cellobiose from starch.[1372]

Fungi can be immobilized. Thus, *A. awamori* and *A. oryzae* were immobilized on beech chips and performed satisfactorily at pH 5.5 at 45–55 °C to bring about saccharification, even with repeated use.[1373] *A. niger* was used in a form immobilized on polypropylene tiles.[1374]

III. Hydrolysis Pathways and Mechanisms

1. Role of Adsorption

The adsorption of enzymes onto starch is a key condition for starch hydrolysis;[1375,1376] if the enzyme is not adsorbed, it cannot hydrolyze that starch. Adsorption is generally favored by low temperature and high pH, and it is proportional to the area of the starch granule, following the Freundlich isotherm.[1375,1377] Hence, the temperature and pH are the key factors. The length of the polysaccharide chain, forming a template for the enzyme, is also a factor.[1378] Furthermore, adsorption is also controlled by the state of the enzyme in the reaction mixture: the enzyme should stay in solution and in colloidal form.[1379] Dissolution of the enzyme in boiling water appeared to be the best approach.[1380] There is a proportional relationship between the activity of enzymes and their adsorption onto raw starch, and the adsorption curves are inversely related to the digestibility of starch.[1381–1383]

The complexation of the enzyme onto the starch substrate[1384] and onto such modified starches as "starch dialdehyde" stabilizes the enzyme thermally.[1385]

The enzymatic digestion causes liquefaction of starch along with the formation of reducing sugars. The susceptibility of the starch to liquefaction depends on the botanical origin of the starch, its state, either granular or gelatinized, and the enzyme applied.[340,1385–1395] The composition of the final products thus depends on the botanical origin of the starch,[1396] the enzyme provenance, the temperature, the pH,[1397] and also the amount of enzyme.[1398,1399] A mathematical model presented for the course of liquefaction of starch involves a role for each of these factors.[1400]

Amylase decay is thus given by Eq. (2)

$$C_1 = C_o \exp(-bt) \qquad (2)$$

where C_o, b, and t denote initial concentration, decay constant equal to $\ln 2/T_{1/2}$, and time in min, respectively.

DE is given by the differential equation Eq. (3)

$$d(DE)/dt = aC_t f(DE) \qquad (3)$$

where a is a factor depending on pH, temperature, and the amount of enzyme, and f (DE) is a correction term for nonlinearity. For DE < 12, this term is equal to 1.

Earlier studies on the hydrolytic action of diastase on starch[1401–1405] already proposed a mechanism in which not only fixation of the enzyme to starch but also presence of certain inorganic salts in the fermenting broth, were indispensable for the hydrolysis.[1406]

The fate of the starch–enzyme complex is essential for defining the mechanistic course of hydrolysis. Thus, when the enzyme forms a complex with starch that is then completely degraded before the enzyme diffuses away to attack another polysaccharide chain, the single-chain degradation mechanism is in effect. The number of chain termini steadily decreases, and since the reaction rate is proportional to the number of the terminals, the process is kinetically of the first order. Multichain degradation takes place when the enzyme diffuses away from its complex with starch after splitting a single bond. The number of terminals remains constant along the hydrolysis and, therefore, the process follows zero-order kinetics. The action of beta amylase action does not follow the single-chain pattern.[1407] Most probably, starch undergoes multiple attack by that enzyme.

Three mathematical models suitable for different mechanisms for the splitting of polysaccharide chains by hydrolases have been presented.[992] Thus, for the random splitting of one glycosidic bond by the attack and desorption of the enzyme, the rate of the liberation of glucose from a substrate comprising of glucose$_n$ mers (Glc$_n$) is given by Eqs. (4) and (5)

$$-d[\text{Glc}_n]/dt = d(\text{Glc})/dt \tag{4}$$

and

$$d[\text{Glc}_{n-1}]/dt = d[\text{Glc}]/dt \tag{5}$$

For the single chain attack, desorption of the enzyme takes place after splitting of the substrate chain is completed. In such a case, Eqs. (6) and (7) are followed

$$-d[\text{Glc}_n]/dt = (1/n)d[\text{Glc}]/dt \tag{6}$$

and

$$d[\text{Glc}_{n-1}]/dt = 0 \tag{7}$$

and in the multiple-chain attack desorption occurs after a maximum of eight successive glycosidic bonds have been cleaved. For such a case, Eq. (8) is applicable.

$$-d[\text{Glc}_n]/dt = 1(f)d[\text{Glc}]/dt \tag{8}$$

where Glc is glucose and $1 < f < n$.

A new model for amylolysis of starch has been published.[1408]

The extent of adsorption of enzyme on starches correlates with the surface area of the adsorbent, and the adsorption follows the Langmuir mechanism.[1409]

The active-site domains of the enzymes play a key role. Such domains have been recognized in several alpha amylases,[215,1410–1417] beta amylases,[1410,1411,1413,1414] glucoamylases,[130,132,971,1410–1412,1418–1426] CGTase,[1413,1427–1430] pullulanase,[1431] and maltoporin.[1432]

Histidine residues in hydrolase enzymes play a key role in the active domains.[1410,1411,1433] The carboxylic group of this residue acts to protonate the glycosidic bonds of starch to generate a transition stage that distinguishes enzymatic hydrolysis from proton-catalyzed hydrolysis, where the glycosidic bond is simply protonated from the medium. The enzymatic hydrolysis has a much lower energy of activation and causes splitting of the bond between the oxygen atom of the glycosidic bond and the anomeric carbon atom. Since that cleavage is not accompanied by Walden inversion, the reaction proceeds by a double-displacement mechanism.[1434–1436]

In fact, this picture is a simplification, as enzymes isolated from various bacterial or fungal substrates are complexes of two or more species, which may play different role on starch digestion.[954,973,974] Moreover, it has been suggested[1437] that amylose losses its helical structure and adopts a linear chain structure during enzymatic hydrolysis.

Generally, the rate of amylolyis can be predicted from the Michaelis–Menten relationship.[1438] The viscosity of the reaction medium can be a factor,[1439] and the viscosity of a starch solution digested with diastase changes according to Eq. (9).[1440]

$$\tau_i - \tau_w = (v_t - v_w)/\{1 - [(v_o - v_f)/(v_o - v_w)]e^{-rt}\} \tag{9}$$

where v_t is the flow of the mixture, τ_w is the time of flow of water, v_f is the time of flow of the mixture when the hydrolysis is completed, τ_{i_t} is the initial time, t is the time of observation, and r is a constant.

The activity of amylase increases on hydrolysis[1441] but an increase in the concentration of amylase may inhibit the reaction.[1442] On hydrolysis of starch, the cleavage of each glycosidic bond consumes one water molecule, which is used to terminate the cleaved bonds. In concentrated solutions, that reaction eliminates water to such an extent that the rate of hydrolysis decreases, and it finally stops after reaching the decomposition limit of starch, irrespective of the temperature applied (it is the so-called hydrolytic gain[1443]). Limit dextrins are formed, and regardless of whether the substrate is tuber or cereal starch, the dextrins consist of maltotriose up to maltohexaose.[37]

The rate of adsorption is inversely dependent on temperature, and the amount of adsorbed enzyme is likewise dependent on the temperature over the pH range of 4.7–7.7 (see also Ref. 1404). A pH dependence of the optical activity of the enzyme can be observed in the pH range between 4.7 and 5.2.[1444] Inclusion of salts in the hydrolyzing broth is favorable, as metal ions may control the adsorption. Thus, the adsorption is controlled by the concentration of the Na^+ cation, but not by that of the Ca^{2+} cation.[1445,1446] Instead of NaCl, KCl can be used.[1447] A report[1400] describes the stabilization of the alpha amylase of *B. licheniformis* in the liquefaction of starch at 102 °C at pH 5.0–6.5 by adding $CaCl_2$. The adsorption of alpha amylase is not immediate, and the rate of adsorption depends on the temperature and, to a limited extent, also on the pH. Adsorption from 15 to 40% ammonium sulfate is advantageous and reaches the level of 80% with fungal amylase. The 75–100% desorption of the enzyme can be performed with 0.1% aq. $NaHCO_3$ and 0.5% aq. calcium acetate at 50 °C for 15–20 min.[1448,1449]

Adsorption is also controlled by the provenance of the enzyme. Adsorption of crude pancreatin, which is an indispensable condition for the enzyme action (see Section III), is fairly selective, and this process can be used for purification of the enzyme. It was found[1450] that pancreatic amylase adsorbs more readily onto starch than bacterial amylase from *Bacillus subtilis*, with malt amylase adsorbing less so. The fungal amylase from *Aspergillus oryzae* did not adsorb at all. Pancreatic alpha amylase adsorbs specifically onto starch crystallites specifically and reversibly, with equilibrium achieved very slowly.[1451]

The alpha amylase of *B. subtilis* adsorbs specifically and reversibly onto starch crystallites of the B-type polymorph spherulites, forming a monolayer. The binding energy of that enzyme to spherulites was -20.7 kJ/mol and is, therefore, higher than the energy of binding to soluble amylose chains (<-30 kJ/mol).[1452,1453]

Beta amylase isolated from a mutant strain of *Aspergillus nidulans* adsorbed strongly onto raw starch at pH 5.0, but there was no correlation between capacity for digestion of raw starch and adsorption of the enzyme.[1454]

Glucoamylases isolated from *A. niger* adsorbed almost completely onto wheat starch at the isoelectric point, pH 3.4. The adsorption was inhibited by an amylase inhibitor from *Streptococcus* species, but only when the concentration was 7-fold higher than that normally encountered. Comparison of the adsorption of that enzyme onto raw wheat and high-amylose corn starch suggested that in both processes the enzyme employed different active sites.

Adsorbates can be eluted with either borax–boric acid buffer or sodium borate.[1455]

2. Mechanism of Inhibition

Several components residing in the substrates, as well as others added on purpose, inhibit the hydrolysis. Such nonstarchy polysaccharides as pectins and gums may obscure the enzymatic process, influencing swelling and then adsorption of the hydrolase onto starch.[1456] The mechanism of inhibition may follow one of the four patterns (Schemes 1–4),[1457] where E, S, I, P, and Q stand for enzyme, substrate, inhibitor, and product, respectively:

Random-type noncompetitive inhibition

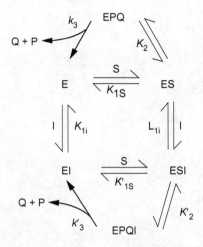

SCHEME 1. Random-type noncompetitive inhibition.

Noncompetitive inhibition

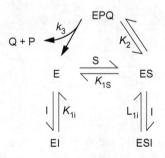

SCHEME 2. Noncompetitive inhibition.

Competitive inhibition

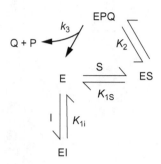

SCHEME 3. Competitive inhibition.

Uncompetitve inhibition

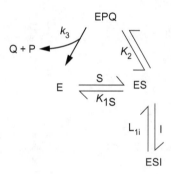

SCHEME 4. Uncompetitve inhibition.

3. Mathematical Models of Enzymatic Hydrolysis

Enzymatic depolymerization of starch has been described by an iteration model based on the Monte Carlo method.[1458,1459] There is a distinction between productive and nonproductive collisions of enzyme and substrate. The effect of productive collisions is iterated until the glycosidic bonds capable of enzymatic scission are cleaved. This approach is suitable for interpretation of single and multiple attacks, for hydrolysis with enzyme cocktails, and for predicting the result of the reactions. Liquefaction of starch with alpha amylase from *B. globigii* was mathematically modeled assuming the action of two isoforms of that enzyme.[1460]

TABLE X
Coefficients for the Taylor Equation[1461]

β_o		Center point
β_1	Time	Linear effects
β_2	Dry substance	
β_3	Glucoamylase dose	
β_4	Pullulanase dose	
β_{11}	(Time)^2	Squared effects
β_{22}	(Dry substance)^2	
β_{33}	(Glucoamylase dose)^2	
β_{44}	(Pullulanase dose)^2	
β_{12}	Time × dry substance	Interactions
β_{13}	Time × glucoamylase dose	
β_{14}	Time × pullulanase dose	
β_{23}	Dry substance × glucoamylase dose	
β_{24}	Dry substance × pullulanase dose	
β_{34}	Glucoamylase dose × pullulanase dose	

A mathematical model for glucoamylase/pullulanase saccharification of liquefied starch was constructed using the Response Surface Methodology, based on the Taylor expansion equation, which involves four variables and 15 terms, as detailed in Table X.[1461]

Figure 6 presents the effect of use of the enzyme cocktail, showing that the effect of admixture of pullulanase ceased after about 60 h of the process.

The validity of this approach was verified with laboratory and plant-scale experiments. The three-dimensional response surface is presented in Fig. 7.

Apar and Oezbek[1462] developed a mathematical model for predicting the effect of various reaction conditions upon hydrolysis of corn starch with alpha amylase. They provided the following equations for the particular effects upon the activity of alpha amylase. First the temperature effect [Eq. (10)]:

$$S_1 = S_o - a \quad \exp\ (b \times [T]) \tag{10}$$

where S_1 and S_o are the residual and initial starch concentrations at time, $t=0$, and a and b are constants estimated for 0.0291 g starch/L and 0.0844 °C^{-1}, respectively.

For the pH effect [Eq. (11)]:

$$S_1 = a - b[\text{pH}] + c[\text{pH}]^2 - d[\text{pH}]^3 \tag{11}$$

where values of the a, b, c, and d constants were estimated for 73.2892, 28.8903, 3.9249, and 0.1670 g/L, respectively.

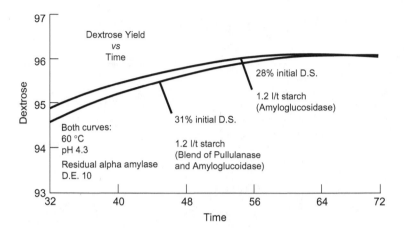

Fig. 6. Increase in the dry substance yield of dextrose (glucose) in the process conducted with a glucoamylase/pullulanase cocktail as compared to the yield of dextrose from the processing of liquefied starch with glucoamylase alone.[1461]

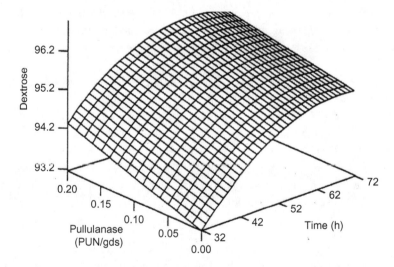

Fig. 7. Three-dimensional response surface.[1461]

For the effect of impeller speed [Eq. (12)]:

$$S_1 = a - b[N] + 10^{-5}c[N]^2 - 10^{-8}d[N]^3 \tag{12}$$

where the a, b, c, and d constants are estimated for 9.9667 g starch/L, 0.0284 g starch/1 rpm, 6.2028 g starch/1 rpm^2, and 3.4308 g starch/1 rpm^3, respectively.

The effect of processing time [Eq. (13)]:

$$S_1 = a\exp(-b[t]) + c \tag{13}$$

where a, b, and c were estimated for 4.4243 g starch/L, 0.1828 min^{-1}, and 5.5657 g starch/L, respectively.

The effect of enzyme concentration [Eq (14)] (see also Komolprasert and Ofoli[1463]):

$$S_1 = a\exp(-b[E]) + c \tag{14}$$

where a, b, and c were estimated for 4.6074 g starch/L, 2.3713 (g enzyme/L)$^{-1}$ and 5.4542 g enzyme/L, respectively.

The effect of viscosity [Eq. (15)] (see also Hill et al.[339]):

$$S_1 = S_o - a\exp(-b[\mu]) \tag{15}$$

where a and b are estimated for 72.7864 g starch/L and 2.7631 cp^{-1}, respectively.

The effect of hydrolyzate [Eq. (16)]:

$$S_1 = S_o - S_{H_o}\exp(-a[H]) \tag{16}$$

where H is the percent of hydrolyzate added (v/v), S_{H_o} is the starch concentration when $H = 0$ estimated for 0.0030 (v/v)$^{-1}$, and a is estimated for 4.1998 g starch/L.

The effect of maltose [Eq. (17)]:

$$S_1 = S_o - S_{M_o}\exp(-a[M]) \tag{17}$$

where M is the maltose concentration [g/L], S_{M_o} is the starch concentration at $M = 0$ estimated for 0.0192 (g maltose/L)$^{-1}$, and a is 4.3179 g starch/L.

The effect of glucose [Eq. (18)]:

$$S_1 = S_o - S_{G_o}\exp(a[G]) \tag{18}$$

where S_{G_o} is the starch concentration at glucose $= 0$ estimated for 0.0150 (g glucose/L)$^{-1}$ and a is 4.0781 g glucose/L.

For hydrolysis of starch, the saturation of alpha amylase with Ca^{2+} was estimated to be 0.15 g/L.[1464]

The Response Surface Methodology was applied for laboratory evaluation of the removal of starch from juice produced in sugarcane mills with the thermostable alpha amylase.[1465]

4. Effect of Light, Microwaves, and External Electric Field

Studies on the effect of illumination of enzymes with light gave only equivocal results. The illumination effect depends slightly on temperature.[1466] It was realized back in the 1920 s[1467–1469] that polarized light activates enzymes that hydrolyze starch. Fiedorowicz et al.[1470–1474] showed that alpha amylase and glucoamylase, when activated by linearly polarized white light, accelerated the depolymerization of starch. The kinetics depends to a certain extent on the botanical origin of the starch. Prolonged illumination results in the repolymerization of depolymerized fragments of starch. Polarized infrared light also causes hydrolysis of starch, probably through activation of enzymes in the granules.[1475]

It has been shown[1476] that several enzymatic reactions can be accelerated by microwave irradiation. The use of microwaves for enzymatic digestion of starch leads to formation of RS.[1477] Control of the enzyme activity with weak rotating, alternating and static magnetic fields of 10 µT to 10 mT has been patented.[1478]

An exterior electric field up to 110 V/cm showed no effect on starch hydrolysis and the stability of alpha amylase was slightly enhanced.[1479] A dose of 0.5 kGy γ-radiation upon alpha amylase and glucoamylase increased the activity of those enzymes by 5 and 10%, respectively. The enthalpy of denaturation of those enzymes decreased inversely with the radiation dose.[1480]

5. Kinetics

Michaelis–Menten kinetics [Eq. (19)] generally reflects correctly the effect of enzyme concentration upon the reaction rate, provided that the reactions are irreversible and not perturbed by such side effects as inhibition of reaction participants with the hydrolysis products:

$$v_o = \frac{v_{max}[S]}{K_M + [S]} \quad (19)$$

where v_o is the current reaction rate, which can be expressed as the rate of conversion of bound substrate to product (k_2) multiplied by the concentration of enzyme currently binding the substrate, [ES], that is, as $k_2[ES]$. The term v_{max} is the maximum k_2 multiplied by the concentration of enzyme binding substrate when the enzyme binds the substrate $[E_o]$, that is, $k_2[E_o]$. The K_M term is the reciprocal of the enzyme affinity expressed as $[k-1+k_2]/k_1$, that is, as the rate of debinding of enzyme by substrate. The term k_1 is the rate at which enzyme binds substrate.

Several factors can be responsible for deviation of the reaction course from this kinetics. They may arise from the intervention of more than one reaction path toward product formation and/or regeneration of the enzyme. The enzyme concentration, participation of the enzyme in formation of the intermediate complex, and whether or not the enzyme was in the equilibrium along the pathway for formation of the major product seem to be the principal reasons for such deviation.[1481,1482] The effect of the average chain length of hydrolyzed starch was taken into account in the model derived by Adachi et al.[1483] The specificity of the enzyme, the substrate, and other factors are reflected by the order of the reaction. Inhibition with the reaction products should also be taken into account.[632]

The rate constant for hydrolysis of potato starch is zero-order until about 20% of the substrate is hydrolyzed. For the next 30% of hydrolysis the reaction became second order.[1484] The initial stage of saccharification of soluble starch also proceeds in the zero-order manner, provided that the starch concentration exceeded 0.7%.[1485] It was suggested[1486] that hydrolysis of granular starch is zero-order, whereas the hydrolysis of gelatinized starch is a first-order process. The absolute rate of the reaction is dependent on size of the granules and the amylose/amylopectin ratio.

In the hydrolysis of starch with malt alpha amylase, the formation of each product and the enzyme deactivation accord with the first-order process.[1487]

The elucidation of kinetic parameters for simultaneous hydrolysis of commercial amylose and amylopectin with the alpha amylase of *B. licheniformis* and isoamylase of *Pseudomonas amyloderamose* accords with Eq. (20).[1488,1489]

$$V'_{max}/K'_{max} = \left[(V_{max,B}/K_{m,B}) - (V_{max,L}/K_{m,L})\right]W_B + V_{max,L}/K_{m,L} \quad (20)$$

where V'_{max}, K'_{max}, and W_B are the apparent maximum reaction rate, apparent Michaelis constant, and the initial mass fraction of amylopectin, respectively.

The apparent kinetic parameters of alpha amylase-catalyzed depolymerization of the amylose (L) and amylopectin (B) blend are presented in Table XI.[1489]

It may be seen that the degradation slows down as the contribution from amylopectin in the reaction mixture increases. This model was applied in consideration of four types of reactions in solution, assuming inhibition by reaction products of low molecular weight.

The dependence on temperature of the initial rate of starch hydrolysis with pancreatin is expressed by Eq. (21).[1490]

TABLE XI
The Apparent Kinetic Parameters of Alpha Amylase-Catalyzed Mixed Amylose and Amylopectin Depolymerization[1489]

W_B	V'_{max} [(mg/mL)]/min	K'_{max} (mg/mL)	V'_{max}/K'_m (min^{-1}) × 10^{-2}	$1/K'_m$ (mL/mg)
0	0.51	5.72	8.91	0.18
0.2	0.43	4.38	9.92	0.23
0.4	0.39	2.72	10.41	0.27
0.6	0.32	2.61	12.13	0.38
0.8	0.29	2.03	14.17	0.49
1	0.25	1.75	14.66	0.58

$$v = v_{max}S^n/(K_s + S^n) \quad (21)$$

with $n=3$ and 2 at 27 and 17 °C, respectively. The K_s term depends on the enzyme concentration, indicating that the reaction is heterogenous. This formula is valid up to 65% hydrolysis.

In the case of waxy corn starch digested with porcine pancreatin, the temperature-dependent rate of hydrolysis deviated from that predicted by an Arrhenius plot over the range of 15–40 °C. The authors of this work also introduced a reproducible and comparable enzyme-affinity constant,[1491] and its applicability was then evaluated in the hydrolysis of waxy maize starch with porcine pancreatin.[1492] Liquefaction of starch with pancreatin is unimolecular until at least 75% conversion, and it is linear with respect to the enzyme concentration.[1493] The course of the saccharification varies with the amount of enzyme applied, and at a high enzyme concentration the reaction is initially also unimolecular. The rate of saccharification is also dependent on the concentration of enzyme, but the sensitivity of this reaction to that concentration is different from that of liquefaction. The rate of saccharification is inversely proportional to the concentration of substrate.[1493] The limit of the unimolecular course in the overall hydrolysis of starch with pancreatin was at 50% conversion.[1494]

In the case of hydrolysis with soybean amylase, the course of liquefaction followed Eq. (22)

$$k_1 = x/c\sqrt{Et} \quad (22)$$

and the course of saccharification, being unimolecular, proceeded according to Eq. (23)

$$k_s = (c/Et)\log[75/(75-x)] \quad (23)$$

where x, E, and c are concentrations of the product formed in time t, of the enzyme and substrate, respectively. The liquefaction was proportional to the square root of the temperature.[1495]

The kinetics of hydrolysis by pancreatin of native and initially heat-treated starches of various botanical origins was studied by Slaughter et al.[1496] Hydrolyses of buckwheat starch[1497] and barley starch[1498] with alpha amylase followed Michaelis–Menten kinetics and were unimolecular, although for the barley starch hydrolyzed at 50 °C, the unimolecular character of the process was maintained for only the first 7–8 min. Hydrolysis with takadiastase is unimolecular only at very low concentration of that enzyme.[1499]

Hydrolysis of starch with the alpha amylase of *B. stearothermophilus* was a first-order process following Michaelis–Menten kinetics, taking into account the temperature, the pH, and Ca^{2+} ions added.[1500] Behaving similarly were the alpha amylase of *B. amyloliquefaciens* taking into account elevated pressure,[1501] the alpha amylase of

B. subtilis comparing hydrolysis of native, partly solubilized, and Zulkowski potato starch,[1502] and diastatic hydrolyses of corn[1503] and potato[1504] starches.

The same kinetics is followed for hydrolysis by alpha amylase in batch reactors and extruders.[1463] Hydrolysis of 35 and 55% slurries of wheat starch with *B. licheniformis* alpha amylase at 80 °C required a nonlinear model,[1505] and that reaction performed with 30 and 50% slurries at 90 and 120 °C appeared to be an irreversible first-order process.[1506] The same kinetics was observed when using that enzyme in the hydrolysis of cassava starch[1507] and when the hydrolysis of a 40% stock in a batch reactor was preceded by hydrolysis in a twin-screw extruder.[1463]

Since beta amylase attacks the terminal groups of starch, it might be supposed that the rate of hydrolysis of starch with that enzyme should remain constant until either the hydrolysis is complete or a branching point in the chain is encountered.[1508] However, at high concentration of substrate, the rate reaches a limit because of formation of a complex between beta amylase and starch. Moreover, the maltose formed inhibits the reaction.[1509] When a 20% starch paste was hydrolyzed with beta amylase, the formation of maltose up to 18% conversion proceeded through the zero-order process.[1510] The two beta amylases isolated from alfalfa, which differed in their molecular weights, hydrolyzed amylose, amylopectin, and soluble starch according to the normal Michaelis–Menten kinetics.[888] The hydrolyses of two starch fractions with alpha amylase from *B. subtilis* and with beta amylase from sweet potatoes were considered as two consecutive first-order reactions with single and multiple attack on several chains, respectively.[828]

The kinetics of starch hydrolysis with glucoamylase has been widely studied. Comparison of the hydrolysis of maltose and starch with that enzyme showed that in both instances two different active centers of the enzyme are involved.[1511] The processes with that enzyme involve not only such phenomena as concentration effects and inhibition from the reaction products but also by the ability of that enzyme to link two glucose molecules to give maltose and isomaltose.[1512]

In the temperature range of 20–35 °C, the process is unimolecular, but above this range the reaction order increases with temperature. Denaturation of enzyme is involved, but the temperature of denaturation depends on the enzyme provenance.[1513–1515] For glucoamylase, the initial rate of hydrolysis increases with concentration of the enzyme, but then a saturation point is reached that is proportional to the concentration of substrate and, in the case of granular starch, to the surface area of the substrate. These observations are also valid for hydrolysis in concentrated solutions.[1516,1517] In concentrated solutions, the hydrolysis of viscous starch and amylopectin is controlled by the mass-transfer rate.[1518] According to some authors,[1519–1521] the hydrolysis corresponds to typical Michaelis–Menten behavior. Other authors refined that equation by introducing the factors of product inhibition,[1522,1523] two-step saccharification of amylopectin,[1524] the degradation pattern of the branch point,[1525] and structural features of glucoamylase.[1526] Equations were developed to express equilibrium constants for the hydrolysis of α-(1→4) (δ'_n) and α-(1→6) (δ''_n) linkages with glucoamylase, and these need not be equal[1527] [Eqs. (24) and (25)]:

$$\delta'_n = [DP'_n]/[Glc_o] = (\gamma)^{n+1}(\alpha)^n [K_2/(K_1+K_2)] \quad (24)$$

and

$$\delta''_n = [DP''_n]/[Glc_o] = (\gamma)^{n+1}(\alpha)^n [K_1/(K_1+K_2)] \quad (25)$$

where $[DP_n]$ is the concentration of glucose oligomer of degree of polymerization expressed by the term n; [Glc] is the concentration of free glucose units; $[Glc_o]$ is the total concentration of glucose subunits, polymerized or free in the solution (mol/dm^3) $\gamma = [Glc]/[Glc_o]$ where [Glc] is the concentration of free glucose at thermodynamic equilibrium; and:

$$\alpha = ([Glc_o]/[H_2O])[(K_1+K_2)/K_1K_2]$$

where K_1 and K_2 are equilibrium constants for hydrolysis of α-(1→4) and α-(1→6) linkages, respectively. Comparison of the kinetics of the hydrolysis with soluble glucoamylase and covalently immobilized enzyme on CMC showed that that the rate constants in both instances are slightly different up to DP 8, probably due to diffusion. Above DP 8, the reaction with immobilized glucoamylase slowed down slightly.[1528]

The model given by Eq. (26) was applied[1529] for the hydrolysis of a 33% cassava starch slurry performed with alpha amylase of *B. licheniformis* in conjunction with commercial glucoamylase at 85 and 100 °C at pH 5.2 and 7.0.

$$dC/dt = -KC^n \quad (26)$$

Kinetic parameters showed that etherification of potato starch with ethylene oxide increased the values of K_M and v_{max} in Eq. (21), and that increase was higher when the degree of etherification was higher.[1530]

Soluble and liquefied cassava starch was hydrolyzed with glucan-1,4-α-maltohydrolase (maltogenase) (EC 3.2.1.133), and Michaelis–Menten kinetics was followed at the optimum conditions of pH 5.0 and 60 °C.[1531] The rate, v, of production of maltose could be determined with Eq. (27):

$$v(t) = (AB)/(B+t)^2 \quad (27)$$

and the initial rate of maltose formation, v_o, is given by Eq. (28)

$$v_0 = A/B \quad (28)$$

The Michaelis A and B constants depend on the initial substrate and enzyme concentrations, S_o and E_o, respectively, in the manner expressed by Eqs. (29)–(31):

$$A = 0.6378 S_o^{0.997} \tag{29}$$

$$B = (3.73 \times 10^{-4}) S_o^{1.72} \tag{30}$$

$$1/B = (2/69 \times 10^{-3}) E_o^{1.03} \tag{31}$$

The constant A was independent of the enzyme concentration.

The kinetics of hydrolysis by systems of two enzymes has been studied for a few examples. For hydrolysis with a combination of alpha amylase and a debranching enzyme, the ratio of both enzymes was an essential factor in determining the rate in the second period of the process. The debranching enzyme could cleave only peripheral α-(1→6) bonds and, therefore, high-molecular dextrins were formed.[1532] Because beta amylase has a more-uniform action, the kinetics for hydrolysis of starch with a system of that enzyme plus a debranching enzyme (such as isoamylase or pullulanase) could be limited to the derivation of an expression for the production of maltose by beta amylase, utilizing products of the debranching action of the other enzyme.[1533]

The hydrolysis kinetics of potato starch with two amylases isolated from *A. awamori* and *A. batatae*, respectively, did not follow a constant order. A kinetic model for such a situation was constructed.[1534] When potato and barley starches were hydrolyzed with alpha and beta amylases, the zero-order reaction proceeded for only the first 35 min, and it then turned into a first-order process.[1535]

Simultaneous hydrolysis of starch with the endoenzyme alpha amylase plus the exoenzyme glucoamylase has elicited considerable interest. Hydrolysis of soluble starch of M_w 8.9 kDa and of M_w 14.5 kDa utilized the synergism between both enzymes. The rate of formation of glucose can be expressed with Eq. (32):

$$d\,Glc/dt = V_{max2} S_o e^{kt} / [K_{m2} + S_o e^{kt}] \tag{32}$$

where $k = V_{max1}/K_{m1}$, S_o is the initial concentration of polysaccharide and indices 1 and 2 correspond to endoenzyme and exoenzyme, respectively. On reaction with the endoenzyme, the concentration of substrate for the exoenzyme increased because of synergism. Finally, the concentration of that substrate reaches the level allowing utilization of all molecules of exoenzyme in formation of the complex, and the expression for that portion of the overall reaction becomes Eq. (33):

$$d\,Glc/dt = V_{max2} \tag{33}$$

as synergism disappears.[1536] On the other hand, since alpha amylase splits the chains into shorter fragments, the activity of glucoamylase on the substrate was decreased.[1537] In the pair of enzymes, glucoamylase preponderated, although its action was inhibited by glucose.[1538] The Chinese group[1539] proposed the following expressions for the synergizing stage [Eqs. (34) and (35)]:

$$dS/dt = V_{m1}S_o/K_{m1} + S_o \qquad (34)$$

and

$$d\,Glc/dt = V_{m2}S_o/K_{m2}S_o(1+Glc/K_i) \qquad (35)$$

and for the postsynergizing stage [Eqs. (36) and (37)]:

$$ds/dt = V_{m2}S_{1/1}1/K_{m2}(1+Glc/K_i) + S_1 \qquad (36)$$

and

$$d\,Glc/dt = V_{m2}S/K_{m2}(1+Glc/K_i) + S \qquad (37)$$

Studies on raw potato starch[1540] suggest that only its hydrolysis with glucoamylase follows the Michaelis–Menten kinetics. Glucoamylase hydrolyzed first the nonreducing terminals on the granule surfaces.

Retrogradation of starch increases the K_m and V_m parameters in the Michaelis–Menten equation, but the equation remains valid.[1541] Pretreatment of starch by microwave heating increases the concentration of potentially enzyme-inhibiting components and influences these parameters, but not the applicability of the equation.[1542]

Saccharification of corn starch with glucoamylase and pullulanase has been represented in the form of a mathematical model based on factorial analysis and called Response Surface Methodology.[1543]

Hydrolysis of starch with glucoamylase and inulase (EC 3.2.1.7), both immobilized on ion exchangers, increased the activation energy for the hydrolysis of starch and changed the enthalpy of activation as compared to nonimmobilized enzymes. The Michaelis–Menten constant increased and the rate of hydrolysis decreased. The kinetics of that process does not follow the Michaelis–Menten equation.[1544]

Hydrolysis was also performed with three-enzyme mixtures. For hydrolysis with alpha amylase, beta amylase, plus α-glucosidase, the rate, defined by the production of glucose, can be described by a classical Michaelis–Menten mechanism, but the equation should be modified to account for inactive enzymes and enzyme–substrate encounters in which no chemical bonds are broken.[1545]

For treating complicated mechanisms of the enzymatic hydrolysis, some rate laws have been derived that are valid regardless of a number of enzymes involved.[1546] The overall rate of hydrolysis, determined by fitting experimental data, $x - f(t)$, to cubic spline functions (polynomial functions of the third order) yielded after differentiation, dx/dt.[1547] Equations are available from fitting dx/dt vs. x. The procedure assumes that all E_o, S_o, pH and temperature values remain constant.

Kinetics of starch hydrolysis has been subjected to several numerical simulations involving the Monte Carlo approach (see, for instance, Ref. 1548)

For the hydrolysis of potato starch with *Bacillus* sp. IIA, performed at pH 4.5 at 20 °C, the integral model with the numerical Monte Carlo simulation provided Eq. (38).[1549]

$$t = (kc_g)^{-1}[(c_{50} - c_s) + K_m \ln(c_{50}/c_s)] \tag{38}$$

Hydrolysis of potato starch pretreated with alpha amylase and then with *Aerobacter aerogenes* pullulanase was described with the polynomial Eq. (39).[1550]

$$f(x) = x^5 + ax^4 + bx^3 + cx^2 + dx + e \tag{39}$$

A multisubstrate Michaelis–Menten model[1551] designed principally for hydrolysis of polymers appeared to be too complex to be applicable to hydrolysis of starch, assuming single and multiple attack of hydrolases on α-(1→4) and α-(1→6) glycosidic bonds.[1549]

Accordingly, kinetic models for the hydrolysis of starch with endo- and exoenzymes were also developed[1549] assuming that only α-(1→4) and α-(1→6) glycosidic bonds are attacked by enzymes on productive and nonproductive collisions. This iterative model can be adjusted for various types of attack, for inhibition with substrate and reaction products (competitive, acompetitive, and noncompetitive), for inactivation of the enzyme, and mutual antagonistic and synergetic enzyme–substrate interactions, as well as the steric shape of the enzymes applied. The model is applicable for reactions with more than a single enzyme. Thus, for the overall reaction:

$$A + D \rightarrow 2G$$

separated into a sequence of the following reactions composing it:

$$A + D + E \rightarrow B + G + E \quad (k_1)$$

$$B + D + E \rightarrow D + G + E \quad (k_2)$$

$$C + D + E \rightarrow 2G + E \quad (k_3)$$

$$2G + E \rightarrow C + D + E \quad (k_4)$$

$$2G + E \rightarrow D + F + E \quad (k_5)$$

$$D + F + E \rightarrow 2Glc + E \quad (k_6)$$

where A denotes the α-(1→4) glycosidic bond at C-1–C-4 and the α-(1→6) bond at C-6, B is the α-(1→4) glycosidic bond at C-1 and C-4, C is the α-(1→4) glycosidic bond at C-4, and D denotes the α-(1→4) glycosidic bond at C-1. E denotes the molecule of enzyme, F is the α-(1→6) glycosidic bond at C-6, and Glc is the glucose molecule liberated on the hydrolysis. Thus, these reactions involve also incorporate the reversion process.

These reactions are characterized subsequently by the following Eqs. (40)–(45).

$$d\,A/dt = -k_1 ADE \qquad (40)$$

$$d\,B/dt = -k_1 ADE - k_2 BDE \qquad (41)$$

$$d\,C/dt = -k_3 CDE + k_4 \text{maltose } E \qquad (42)$$

$$d\,D/dt = -k_1 ADE - k_2 CDE + k_5 \text{maltose} E - k_6 DFE \qquad (43)$$

$$dE/dt = 0 \qquad (44)$$

$$d\,F/dt = k_5 \text{maltose } E - k_6 DFE \qquad (45)$$

The overall reaction is described by Eq. (46).

$$d\,\text{Glc}/dt = k_1 ADE + k_2 BDE + 2k_3 CDE - 2k_4 \text{maltose } E - 2k_5 \text{maltose } E + 2k_6 DFE \qquad (46)$$

The validity of this model has been verified experimentally, and the experimental data provide an excellent fit to the model in over 95% of examples.

Kinetic models for the reactions proceeding in bioreactors fed with varying amounts of enzymes have also been elaborated.[1552,1553]

IV. AMYLOLYTIC STARCH CONVERSIONS

1. Introduction

Amylolytic hydrolysis of starch is most frequently the objective in the practical conversion of starch. The processes used are generally simple and relatively clean. The activity of amylases depends on the mode of dissolution of the enzyme preparation.[1380] Glucoamylase provides the possibility of using higher temperatures and, at the same time, causing less reversion. Thus, the reactions can proceed faster, fewer by-products are formed, and processing with more-concentrated solutions is possible.

Enzymatic hydrolysis, known for centuries, has received many modifications, not only in the area of enzyme selection, but by using combinations of two or more enzymes, either jointly or sequentially, and by immobilization and genetic modifications. Obviously, selection of such physical parameters as temperature, pH, time, concentration, and so on has received considerable attention. However, some less-common factors have also been taken into account. Thus, hydrolysis in degassed water has been patented as a means for improving the process efficiency.[1554] Enriching the wort in oxygen has also been patented.[1555] Shaking the wort with 3-mm glass beads has been shown useful.[1556]

The hydrolysis is completed by inactivating the enzyme, either by lowering the pH, by raising the temperature, or both.[1557] Copper(II) sulfate can be used as the enzyme deactivator.[1558] The deactivated enzymes cannot be recovered and reused, but when an organic solvent comprising lower alcohols (C_1–C_4), acetone or butanone, methyl to butyl acetate, or diethyl ether is used, the hydrolyzing enzyme can be recovered and recirculated.[1559]

Admixture of such activators as ethylene chlorohydrin, thiourea, or KCNS at low concentration is advantageous. The effect is approximately proportional to the concentration of activator, but above a certain concentration level, the rate of hydrolysis declines.[1560] Sometimes, starch is hydrolyzed in a mixture containing other additives, such as powdered milk, soybean meal, malt flour, and inorganic salts, preferably NaCl.[1561] This process can be operated in a continuous manner on gelatinizing starch.[1562]

Hydrolysis usually proceeds in two stages, namely dextrinization (liquefaction) and saccharification.[852,862,902,1563–1566] Liquefaction is an enzymatic process performed on a starch slurry to afford a solution of maltodextrins (oligosaccharides and dextrins). Saccharification follows liquefaction to convert maltodextrins completely into glucose along with smaller amounts of maltose, isomaltose, and a number of other lower saccharides.[1567] Both the liquefaction and saccharification stages can be conducted directly, but they are frequently multistage processes: liquefaction can be preceded by pulping, malting, and mashing. The products of liquefaction serve either as final products of technological or commercial value, or they are saccharified to glucose and maltose prior to conversion in subsequent steps to afford ethanol or such fine chemicals as amino acids, vitamins, and other products. Fermentation products are used for the manufacture of enzymes.[258,1568]

2. Pulping

The pulping of starch is usually a mechanical process performed to secure maximum cell-wall breakdown without degrading the starch. Vacuum pulping has been proposed for potatoes.[1569] Starch is mainly degraded upon mashing.[1570]

3. Malting

Malting generally utilizes cereal grains, particularly barley[1571] which on soaking in water germinate and in so doing develop the enzyme capable of hydrolyzing starch in the grain to give glucose and maltose. During the germination, barley changes only slightly its heterodispersity. There is generally a correlation between the quality of malting and the protein content in the malted substrate, as demonstrated in the germination of wheat. The germinating ability, as well as the beta amylase and protease activity in the substrate, correlates positively with the protein content.[1572,1573]

Germination is arrested at a certain stage by kiln drying, preferably in hot air. Sprouting in barley can be accelerated by the use of gibberillic acid.[1574] An extended time of malting is not beneficial because of the loss of starch. With a short malting time, enzymes consume accompanying proteins prior to consuming starch.[1575] The use of high-amylose barley does not improve the process.[1576]

Diastase is suitable for the production of malt.[1577] When acid-treated taka diastase was employed, glucose but not maltose was formed as that diastase has a high maltase activity.[1578] Dextrins of degree of polymerization (DP) approximately 12 were also formed. Their yield correlated with the fermentation efficiency.[1579] Generally, pretreatment of starch with 1 molar aqueous solutions of KCl, $MgCl_2$, $CaCl_2$, or $BaCl_2$ increases the activity of the enzyme, whereas treatment with solutions of $NaBr$, NaI, $NaCl$, or Na_2SO_4 decreased it.[1572,1580]

4. Mashing

Milled grain blended with other cereal grains is heated in water in order to allow malt enzymes to penetrate and break down starch to maltose.[1581] This process is time, temperature, and pH dependent. Up to 75 °C, the rate of the process increases with temperature and concentration of substrate and enzyme. The effects of the substrate and enzyme concentration are not proportional to one another. An extended time of mashing is not beneficial.[1582] Above 75 °C and pH 7.6, the action of amylase ceases completely.[1583] The origin of the malt has a minor effect on the result of mashing.[1584] Unmalted millet was preliminarily mashed with ground malt sprouts and ground malt.[1585]

Mashing should be performed on gelatinized starch to permit its total degradation.[1586] Gelatinization is supported by added amylotic enzymes.[1570] However, the processing of starch without gelatinization is possible when milled starch is saccharified with a glucoamylase from *Athelia rolfsii* and with an added alpha amylase.[1587]

Mashing can be carried out by either an infusion or a decoction manner. In the first procedure, there is a single vessel heating the blend, whereas in the second one a portion of the grains in water is first boiled and then returned to the mash to maintain the desired temperature. The operations employed depend on the origin of the starch.[1586] The increase in temperature should not be progressive. At some temperatures (rest temperatures), time is given for the enzymes to act effectively in the mash. The rest temperatures are specific for a given enzyme. Typically, β-glucanase, degrading β-glucan of cell walls, has its optimum activity around 40 °C. At 50 °C, protease decomposes protein (if any).[1570] According to some sources,[1588] the use of proteases prior to heating is beneficial, as the thermal pretreatment lowers the rate of protein hydrolysis. At 62 and 72 °C, beta and alpha amylases, respectively, decompose starch most efficiently.

Alpha amylase in malt is responsible for the production of glucose and some maltotriose. Beta amylase contributes to the production of glucose and a small amount of maltotriose, but maltose is the principal product of its action. Alpha amylase also produces maltotetraose from maltopentaose and, jointly with beta amylase, gives additionally maltohexaose.[1589,1590] A kinetic model was elaborated and verified experimentally in industrial-scale reactors.[1487] α-Glucosidase significantly increases the efficiency of mashing. Since that enzyme is present in cells, it may be supposed that increased temperature of mashing facilitates the excretion of that enzyme from the cells.[1591,1592]

The joint use of amylases isolated from *Rhizopus*, *A. oryzae*, and A. *niger* glucoamylase has been patented.[1593] Acidification of the mash is not beneficial for the production of hydrolases because it inhibits the propagation of fungal mycelia. Hence, the composition of the culture medium for the production of fungal amylase should be thoroughly adjusted. Corn meal is not recommended as it acidifies the culture medium, but either wheat or defatted rice bran was a good medium for this purpose.[1594]

Studies of the effect of added amylase and adjustment of pH upon the rheology of mashes[1595] showed correlations that contribute to understanding the mechanism of the process. Thus, the correlation of rheological parameters against the level of amylase demonstrates the importance of the primary swelling of starch and its subsequent digestibility. The viscosity of gelatinized starch, which increased with the content of small starch granules, correlated with the level of amylase in the mash. The viscosity of the mash can be decreased by the application of thermostable β-glucanase. That enzyme was isolated from *Thermoascus aurantiacus*.[1596]

5. Liquefaction

Liquefaction, extending up to 44%[902] or 30%[851] of conversion depending on the concentration of the processed solution, is characterized by an affinity constant of 200 and is independent of the length of the carbohydrate chain. This stage leads to soluble dextrins, the so-called amylodextrins, with accompanying maltose, but sometimes soluble starch is the target of that operation. For that, a 3% starch paste is treated with 0.4% of enzyme and kept for 15 min at 60 °C, followed by boiling for 3 min.[1597] Another method, patented in China,[1598] is based on hydrolysis of starch pulped in basic solution together with the selected enzyme.

The proportion of enzyme to substrate is an essential factor, and it should take into account the desired amount of conversion of the cereal added and malt applied.[1599] An excess of enzyme does not increase the maltose yield.[1583] The pH optimum for liquefaction depends on the temperature and rises from 4.0 at 20–40 °C to 6.2 at 65 °C.[1583] The liquefied product is sometimes bitter, and to avoid this effect, the initial pH of the substrate should be between 0.75 and 2.0. The pH should then be gradually increased.[1600]

The malting and mashing stages can be omitted and starch, usually in form of paste, is treated with enzymes of various provenance. Liquefaction with enzyme can be preceded by treating the substrate with hydrochloric acid,[1601–1604] but the use of acid hydrolysis may be eliminated.

A two-step liquefaction is commonly recommended,[1605] according to the properties of the substrate. For instance, rice starch consists of normal and vitreous starch. The former can be liquefied to a 70% extent within 2 h at 30 °C, whereas only 20% of the latter is liquefied under such conditions.[1606] Two steps of liquefaction of a given starch may differ from one another solely in the temperature and time of the process. For instance, the first step requires heating to 50–80 °C for 10–180 min, and the second step involves heating to a higher temperature for 1–60 min.[1607]

On liquefaction a certain amount of insoluble components is formed. Their formation can be limited by the addition of soluble calcium compounds, for instance, $CaCl_2$.[1608] The starch:water:$CaCl_2$:amylase ratio should be 1:1–1.5:0.0025–0.0035:0.0025–0.0035. The addition of diatomite is recommended.[1609] Once the insoluble products have appeared in the product, they can be liquefied with glucoamylase,[1610] but even then some insoluble matter remains. Corn starch produces significantly more such matter than sweet-potato starch.[1611] To solve the problem of an insoluble component of liquefied starch, the use of alpha amylase followed by fungal glucoamylase[1612–1614] or both enzymes jointly[1615,1616] may be employed. A continuous two-step liquefaction of starch was developed by subsequently using either alpha amylase and glucoamylase[1614] or alpha amylase alone in two portions.[1617,1618] When the liquefaction is performed on wheat flour, such a product can be used in bread making.[1619] Instead of alpha amylase, beta amylase can be used in combination with glucoamylase.[1620] In each step of the two-step process, alpha amylase alone may be used. The first step is, in fact, the production of a soluble starch that can be blue-stained with KI_5, and in the second step, a deeper hydrolysis takes place.[1621,1622] The second step can be performed under elevated pressure.[1623,1624]

A three-step liquefaction of starch with alpha amylase has also been proposed.[1625] These three steps differ from one another in the temperature and the pH applied. The first step proceeded at 93–102 °C and pH 5.7–7.5. The initial reaction mixture is then supplemented with fresh starch, NaCl, and $CaCl_2$, the pH is adjusted to 6.2, and a subsequent portion of the enzyme is added. Finally, the temperature of the reaction mixture is lowered to 85 °C and a third portion of starch is added. This process was improved by using NaOH instead of NaCl and applying different temperatures in particular steps.[1626]

Because the process employs elevated temperature, heat-stable enzymes are preferred, for instance, those of *Bacillus subtilis* or *Bacillus mesentericus*, but useful enzymes can also be selected from those of fungal origin. Blending with NaCl, $CaSO_4$, $CaCl_2$, NaH_2PO_4, and Na_2HPO_4 activates and stabilizes these enzymes.[1606,1625] Thus, a starch slurry of adjusted pH and blended with a suitable enzyme (alpha amylase for instance) is injected into a jet cooker, where at about 105 °C gelatinization of starch takes place. Steaming or hydrothermal treatment at 120 °C can be applied.[1627] Subsequently, the temperature is lowered to 90–100 °C, and the gel is digested with the enzyme until the desired DE of a low-sweetness syrup is reached. A secondary liquefaction step utilizing either a thermostable acid alpha amylase or a thermostable maltogenic acid alpha amylase has been patented.[1628] However, cooking at low temperature has some advantage over high temperature cooking, as it increases starch conversion by 8–10%.[1629,1630]

The rate of amylolytic liquefaction can be expressed with the linear Eq. (47)[1631]:

$$1/Z = a + btE \tag{47}$$

where Z is the ratio flow of the paste/flow of the standard (aqueous glycerol), t is the time in min from the moment of combining the paste with enzyme, E is the enzyme concentration, and a and b are constants.

There are several patented procedures for hydrolysis with alpha amylase of soluble starch,[1632–1634] potato,[1635–1638] corn,[1639–1648] wheat[1639,1641,1642,1649,1650] rice,[1651] cassava,[1638,1652–1654] sorghum,[1638] and triticale[1655] starches.

Amylolysis of potato starch to give maltodextrins is performed after the gradual gelatinization of starch at 95–100 °C, followed by centrifugation to remove coagulated protein, and spray-drying the final product.[1656] Rice starch in a 10–20% paste can be degraded into maltodextrins at 95 °C within 10–15 min to provide a product of DE 0.75–5.72, depending on the enzyme concentration.[1657] The hydrolysis of wheat starch was optimized.[1658] At 72 °C, the smallest amount of solids remained at pH 4.8 and 9.0, but the deepest extent of hydrolysis (measured as I_5^- uptake through complexation to dextrins) was observed at pH 7.8. At pH 6.8, the smallest amount of insoluble dextrins was found at 85 °C and the most extensive degradation occurred between 72 and 65 °C.

Corn starch was liquefied with an enzyme extract of *A. oryzae*[1603] or *A. awamori*.[1659] In the latter case, a product containing 75–80% of reducing compounds was obtained after 3–4 h of processing at 45 °C. The enzyme needed to be added in one portion at the outset.[1659] Bacterial alpha amylase can be used in soluble[1611,1660–1663] or immobilized form.[1664,1665] A short period of contact of processed starch with high humidity is advised to inhibit the formation of sparingly soluble starch that is resistant to the enzyme.[1666,1667] The same behavior is observed on treating starch with the glucoamylase of *Endomyces fibuligera*. The use of glucoamylase from the outset of the liquefaction process usually produces a solid product of low DE. Glucoamylase is frequently contaminated with transglucosidase. Elimination of this contamination slightly facilitates liquefaction with glucoamylase.[1668] Neither thermal pretreatment[1667,1668] nor agitation of the liquefied wort[1669] is recommended. The temperature effect depends on the concentration of the wort. As its concentration increases, the increase in temperature becomes less significant, and that increase from 75 to 95 °C does not decrease the hydrolytic yield.[1670]

Liquefaction of starch with beta amylase,[1671,1672] neopullulanase,[1673] glucoamylase,[1671,1674 1677] and pullulanase[1674] has been patented. Enzymatic cocktails have also been used. Alpha amylase has been used jointly with glucoamylase,[1678–1681] beta amylase,[1682] pullulanase,[1683,1684] CGTase,[1685] glucan 1,4-α-maltohydrolase,[1686] the phytase of *Buttiauxella* sp.,[1687] beta amylase and isoamylase,[1233] glucoamylase, and α-glucosidase.[1688] Glucoamylase was used jointly with pullulanase[1689] and beta amylase.[1690] Addition of glucose and maltose prior to the process suppresses the liquefaction and saccharification.[1691] Some novel enzymes have been examined. For liquefaction of sorghum, the alpha amylase isolated from a *Bacillus* species originating in hot-spring waters operated at 60 °C and pH 6.5.[1692] In other examples, liquefaction has been performed with a strong mineral acid, with subsequent enzymatic saccharification.[1693,1694] The use of glucoamylase isolated from one of the groups *A. niger*, *A. awamori*, *Rhizopus* or *Endomycopsis fibuligera* obviates the need for cooking.[1695]

Thinned, nonretrograding hydrolyzates can be prepared from starch slurry with alpha amylase by heating at 120 °C and pH 3.6–6.5 to DE < 2.0. Further hydrolysis can be performed at 90–100 °C and finally at 85–90 °C to DE > 5.[1696] A product containing over 80 wt% of maltopentaose–maltododeacose was obtained by treating granular starch, swollen in a slurry by heating for 10 min at 65 °C, and then hydrolyzed during 24 h with *B. subtilis* alpha amylase at pH 5.5. The resulting product has DE ~ 30.[1697] Alpha amylase from *B. circulans* provides maltohexaose[1698] as well as maltotetraose with maltopentaose,[1699,1700] maltotetraose,[1701] and maltotriose.[1649] A high yield of maltopentaose results from the reaction with alpha amylase conducted in such hydrophobic solvents as dodecane[1702] or hexane.[1703] The manufacture of branched-chain starch hydrolyzates has also been patented.[1704] Maltodextrins of DE 9–15 result from the hydrolysis of starch in slurry by the alpha amylase of *B. stearothermophilus*.[1705] Some hylon starches can be liquefied with alpha amylase at pH ~ 6.5.[1706]

Branched dextrins and oligosaccharides can be manufactured from starch and alpha amylase at > 120 °C for 3–10 min for the primary liquefaction followed by 6–12 h of hydrolysis at 80–100 °C.[1707] Optimum conditions for liquefaction of corn and potato starch were presented by Grabovska et al.[1708]

6. Saccharification

Saccharification intervenes when chains of fewer than 12 glucose residues are present. This process is slower than liquefaction, and the affinity constant decreases to 12.5 as compared to 200 for the liquefaction step.[851,902] Limit dextrins generated by alpha amylase generally have much lower molecular weights than those generated with beta amylase.[1709] The saccharification period can be shortened by over 50% provided that fresh enzymes and substrates are used; liquefaction is completed and the reaction mixture is continuously agitated.[1710] The manufacture of sugar syrups of various DE, along with maltose, glucose, and eventually such other disaccharides as isomaltose and maltulose, is the target of this operation. The results of saccharification depend on the enzymes used[1711] as well as the reaction time and enzyme concentration.[1712]

a. Saccharification to Syrups.—Figure 8 presents the relationship between the DE of the resulting syrups, the enzyme used, and its amount, as it applies to the saccharification of potato starch liquefied with *B. subtilis* alpha amylase.

It may be seen that *A. niger* glucoamylase alone exhibited adequate hydrolyzing ability at the level of 0.01% added, and the DE increased fairly linearly with increase in the amount of enzyme. The production of glucose–fructose and maltose syrups using alpha and beta amylases has been patented.[1713,1714] The alpha amylase of *A. oryzae*, a mixture of *A. niger* glucoamylase and *B. acidopullulyticus* pullulanase,

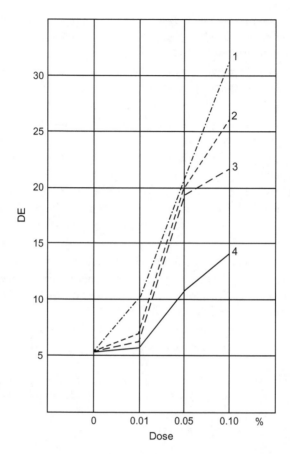

FIG. 8. Relationship between the DE of resulting syrups, the enzyme used, and its concentration, applied to the saccharification of potato starch initially liquefied with *B. subtilis* alpha amylase.[1711] A—*A. niger* glucoamylase, B—*A. oryzae* alpha amylase, C—cocktail of *A. niger* glucoamylase and *B. acidopullulyticus* pullulanase, D—*B. stearothermophilus* maltogenic amylase.

and *B. stearothermophilus* maltogenic amylase, used at a level up to 0.01%, performed poorly. Alpha amylase was more efficient than maltogenic amylase. The efficiency of a cocktail of both of these enzymes did not increase, and even declined, when its concentration was above the 0.25%. Up to this point, the DE increased fairly linearly with the enzyme level applied.[1711]

Gelatinization of starch prior to saccharification is not essential, and such processes were patented already in the early 1950s.[1715] Inclusion of such cationic surfactants as benzyldodecyldimethylammonium chloride or *tert*-butyl-(2-ethoxyethyl) phenoxydimethylammonium chloride activates enzymes.[1716] Saccharification can be accelerated when starch blended with the enzyme is subjected to malting.

Saccharification usually involves alpha and beta amylases and glucoamylase of various origin.[1561] Saccharification with beta amylase was more beneficial in case of substrate of higher molecular weight.[1717] Recently, also pullulanase was added to the list of saccharifying enzymes.

One- and two-step saccharifications have been described and patented. One-step processes are designed for saccharification of separately liquefied starch, and two-step processes include liquefaction and saccharification of the substrate.[1718] Fat and proteins can be removed between both stages, and this operation lowers the consumption of enzymes.[1719] In two-stage processes, the second stage requires more-precise temperature adjustment, depending on the concentration of the hydrolyzed substrate. Hydrolysis of more-concentrated substrate requires a slightly lower temperature.[1720] Moreover, these processes can be divided into these conducted with only one enzyme or with several enzymes, used either as a cocktail or sequentially.

There is also the possibility for conversion of starch with microorganisms. For instance, sago starch was hydrolyzed with *A. awamori*, and it is noteworthy that the hydrolyses proceeded at a temperature lower by 30 °C than enzymatic hydrolysis, which usually requires heating to 60–90 °C.[1721] Wheat starch could be hydrolyzed to 95% extent by using *A. fumigatus* K27.[1722]

Increase in the efficiency of the enzymatic conversion is possible by employing such mechanochemical effects as shear stress, tossing, high mechanical frequency, or irradiation, for instance by microwaves and sonication,[1723,1724] and recirculation with supplementation by fresh enzyme.[1506] Employment of a jet cooker requires the use of a thermostable alpha amylase.[1725] When vegetable material is subjected to saccharification, pretreatment with pectinase is advantageous.[1726] The use of xylanase for hydrolysis of pentosans is advised in the hydrolysis of cereals and their starches to glucose syrups.[1727] Efficient hydrolysis of wheat B-starch requires the use of alpha amylase in combination with pentosanase and lipase.[1728] Sliced cassava roots should be first liquefied by acid, then filtered, and then saccharified with alpha amylase.[1729,1730]

At one time, the saccharification of pasted starch involved either malt for liquefaction at 74 °C followed by treatment with diluted hydrochloric acid[1731] or use of malt diastase first for mashing and then for saccharification.[1732,1732] Regardless of whether either malt or malt diastase was used, the conditions of the process depended on whether raw or cooked starch was processed. The botanical origin of starches also had some impact on the selection of optimum conditions. Diastatic activity reached a maximum within 15 min and then ceased.[1733]

The idea of combining enzymatic and mineral acid for degradation of starch was then developed into a process whereby liquefaction of starch was performed with mineral acid followed by saccharification with either glucoamylase[1734,1735] or alpha amylase.[1736] This idea was based on the fact that the glucoamylase of *Rhizopus* used for liquefaction produced mainly isomaltose, and when the concentration of glucose exceeded 40%, the enzyme utilized it to form maltotriose.[1734] However, better results in the production of glucose were afforded when alpha amylase was used instead of mineral acid.[1737] A contradictory conclusion comes from another study[1738] showing that preliminary mild acid-catalyzed hydrolysis followed by treatment with alpha amylase is beneficial. Another modification of that idea involved the use of oxalic acid for liquefaction, followed by saccharification with glucoamylase.[1739,1740] The acid-catalyzed liquefaction can be performed under pressure.[1741,1742] A 1939 French patent described the saccharification of starch with a combination of enzymes, superclastases, usually used for treating sewage and feces.[1743] A comparative analysis of starch conversion with mineral acid, acid–enzyme, and dual enzyme catalysts, together with evaluation of the costs of operations, is available.[1744]

Alpha amylase alone can convert starch into syrups, but for practical application of the process an elevated temperature is required. Activation of the enzyme by preheating to 50 °C is advisable.[1745–1747] Bacterial amylases are valuable for the saccharification of starch, thus a 40% corn-starch paste is hydrolyzed with bacterial amylase at pH 5.5 for 10 min at 90 °C, followed by 96 h of storage at 60 °C. The resulting syrup contains 9.15% glucose, 27.7% maltose, and 33.2% maltotriose.[1748] In another study, corn-starch slurries, even over 50%, were digested by bacterial alpha amylase at pH 7.5–8.0, first at 90 °C and then heated to 100–150 °C, filtered and the filtrate digested with a fresh portion of enzyme at 80–85 °C. Low-dextrose syrups of up to 65% solids are available in such manner.[1749] The temperature of hydrolysis can be lowered to 45–55 °C after inclusion of Mg, K, and Mn salts.[1750] Alpha amylases from *B. licheniformis* and *B. stearothermophilus* were used in hydrolysis of starch by jet cooking of a 35% slurry. An initial 5 min treatment was performed at 105 °C, and then after flash-cooling to 95 °C digestion was conducted for 1 h at pH 5.8. The resulting maltodextrins were then hydrolyzed for further 48 h.[1751] With a continuous process conducted at pH 6.55 and 80 °C, a 26.6% syrup can be obtained from corn starch.[1752] *Bacillus subtilis* TU produces a pullulanase-like alpha amylase that gives on saccharification a syrup having a high content of maltose and maltotriose.[1753]

Fungal alpha amylases have also been used. Saccharification with the enzyme from *A. oryzae* required a lower temperature than that for the conversion with the *B. subtilis* enzyme. Both amylases can be used in combination.[1754] Comparative studies[1755] performed with alpha amylases from *A. usamii* and *A. oryzae* showed that the efficiency of both microorganisms depended on the DP of the substrate. A highly polymerized substrate usually promoted saccharification in both cases. However, the amylase of *A. usamii* performed better with highly polymerized material, whereas amylase from *A. oryzae* was more efficient in the saccharification of material of lower molecular weight.

Amylases can be used in both soluble and immobilized forms. A collagen membrane can be used as a support.[1756] Alpha amylase isolated from *A. niger* and covalently immobilized provided high-conversion syrups when operated for 140–160 h at pH 3.2–3.5 and 34 °C.[1757] For the liquefaction and saccharification of dry-grind corn, the alpha amylase from *A. kawachi* was used. It performed below 48 °C.[1758] Alpha amylase from *Rhizopus delemar*, immobilized on CMC–azide, saccharified liquefied starch in a continuous process conducted at 40 °C. The enzyme rapidly lost its initial activity at 50 °C.[1759]

Hydrolysis of liquefied starch as well as chemically modified starch with beta amylase has been patented.[1760]

Glucoamylase may act as a single hydrolyzing enzyme.[1761–1763] Thoroughly purified enzyme from one of the groups *A. phoenicis*, *A. diastaticus*, *A, usamii*, and *A. niger* produced syrups of a higher dextrose content (DE 94–98) than nonpurified enzyme (DE 92).[1764] Continuous hydrolysis of liquefied starch with glucoamylase attached to active carbon produced a syrup of DE 97 in the process conducted at pH 5.0 and 50 °C.[1765] Starch liquefied enzymatically or by acid treatment can be used.[1766] Glucoamylase isolated from the *Humicola grisea thermoidea* fungus appeared very efficient in the hydrolysis of raw starch to a syrup containing 97.5% glucose.[1767] An increase in the thermostability of glucoamylase is crucial, as the saccharification at higher temperature and higher dry-solid levels offers shorter reaction times and enhanced economic viability.[1261]

Starch has frequently been saccharified with a combination of hydrolyzing enzymes. Among them, the combined action of alpha amylase and glucoamylase has been the most exploited. Those instances where a single-batch liquefaction/-saccharification process uses alpha amylase for liquefaction and, sequentially, glucoamylase for saccharification[1768–1775] are simple examples of the dual enzyme system. The order of the application of alpha amylase and glucoamylase may be reversed. Glucoamylase-saccharified starch and alpha amylase applied subsequently digested the saccharification product at 50–80 °C and pH 5.5–6.0, making filtration and purification by ion exchange unnecessary and facilitating crystallization.[1776] The enzymes of either barley malt or *B. subtilis*, and *A. niger* or *A. awamori*, respectively, were used as an enzyme cocktail for the saccharification of ground whole corn, wheat, and sorghum.[1777,1778] A mixture of fungal alpha amylase and glucoamylase hydrolyzed corn starch to a syrup of DE 18 with 50% solids.[1779] The DE of the resulting syrups depended on the proportion of both cooperating enzymes.[1780] Cassava starch hydrolyzed with such a cocktail gave a syrup of DE 80–85.[1781] A similar cocktail was also used for saccharification of grain mash,[1782] corn and potato starch,[1556] and rye starch.[1783] Saccharification of potato starch with *Actinomyces* alpha amylase in conjunction with glucose oxidase provided a syrup containing 85.5% maltose and 10.8% maltotriose.[1784]

In technical applications, liquefied starch is converted into maltose syrups with either fungal alpha amylase or beta amylase, jointly with either pullulanase or isoamylase. The combination of amylase with pullulanase provides a maltotetraose syrup. The liquefied starch treated with beta amylase jointly with α-glucosidase yields an isomaltooligosaccharide syrup. Highly saccharified syrups result from liquefied starch processed with alpha amylase and glucoamylase.[232]

When amylases and maltase are used, the true saccharification power (SP) can be expressed by means of formula (48):

$$SP = (0.947M + 0.900G)/S \qquad (48)$$

where M, G, and S are the amounts of maltose and glucose produced, and starch employed, respectively.[1785]

Alpha amylase can cooperate with pullulanase. A sequential liquefaction with the first enzyme at 70–90 °C and saccharification with the second one at 40–60 °C has been used.[1786] There are some examples of cooperative use of liquefying alpha amylase with saccharifying beta amylase in the production of high-maltose[1787,1788] and glucose syrups.[1789] For the latter, it has been suggested that the use of beta amylase is unnecessary. For the production of maltose syrup, initial liquefaction with alpha amylase, saccharification with beta amylase, and debranching with an enzyme secreted by *Escherichia intermedia* ATCC 21,073 was employed. This syrup contained chiefly not only maltose but also up to 5.2% maltotriose. Inclusion of wheat germ decreases the maltotriose content to below 1%.[1790] Glucoamylase can be used instead. Thus, alpha amylase is used first, in a 40% aq. starch suspension and with a rapid increase of temperature to 95 °C and efficient homogenization; the liquefaction takes 15 min. After cooling to 55 °C, beta amylase and glucoamylase are added for saccharification, which takes 1 h.[1791] A Japanese patent[1792] claims that a 30% starch suspension should be first saccharified with alpha amylase at 90–95 °C and pH 6.0 to DE 5, followed by addition of beta amylase and glucoamylase at pH 5.0 and 55 °C for 50 h. These procedures have been subjected to several modifications. Some strains of *A. batatae* produced glucoamylase of high activity.[1793]

The cooperating alpha and beta amylase system can be supplemented with α-1,6-glucosidase to hydrolyze potato starch.[1794,1795] Saccharification of starch liquefied with alpha amylase utilized a cocktail comprising beta amylase and pullulanase,[1796] and was further saccharified with glucoamylase.[1797] Liquefied starch was saccharified with a combination of glucoamylase and isoamylase to give a glucose syrup of DE 11.[1798] Use of pullulanase increased the yield of maltose slightly.[1799] In the cocktail, pullulanase can be replaced by isoamylase.[1800] The production of high-maltose syrup from corn starch, using the alpha amylase of *B. polymyxa* and pullulanase of *Acetobacter* at pH 5.5 and 91 °C, has been patented.[1801]

b. Saccharification to Maltose.—Procedures for manufacture of maltose are substantially the same as those described for the production of syrups, although there are some modifications leading to high-maltose syrups, possibly free of dextrins, and allowing the isolation and purification of maltose. Processes permitting the economically feasible production of maltose begin with gelatinization and/or liquefaction of starch. Properly adjusted gelatinization conditions can obviate the need for liquefaction. There are also methods based on hydrolyzates prepared nonenzymatically, in either acid or alkaline media.[1802–1804] In the conversion of sweet-potato starch into maltose, a starch slurry was digested directly by a combination of beta amylase present in that starch and pullulanase, operating initially at 77 °C and then overnight at 60 °C, to yield approximately 80% of maltose.[1805]

When the enzymatic liquefaction stage is to be avoided, the gelatinization should be performed above, for example, 100 °C, and under elevated pressure.[1806–1813] The gelatinized starch is then cooled down, its pH is adjusted, if necessary, and subjected to saccharification. If the liquefaction step is selected, alpha amylase is preferentially used.[1814–1820] Liquefaction of starch was also performed by the action of barley malt amylase for 48 h at 52–55 °C and pH 5.0.[1821–1823] Saccharification is the next step, and it can be a continuation of the liquefaction with either alpha amylase or α-glucosidase. The resulting syrup is passed through a column packed with active carbon and Celite to separate maltose (97%) from glucose (0.1%) and maltotriose (2.9%).[1816]

Most frequently, saccharification is performed with beta amylase,[1804,1811,1815,1817,1819,1824] beta amylase with alpha amylase,[1814,1815,1825] with α-1,6-glucosidase,[1807] with pullulanase,[1808,1810,1820,1826] with isoamylase,[1820] and with pullulanase and dextrinase.[1813] Other saccharifying enzymes used in production of maltose are maltogenic amylase,[1827] glucoamylase,[1818] α-1,6-glucosidase with beta amylase,[1809,1812] maltose amylase,[1828] maltose amylase with maltotetraose amylase,[1829] maltotetraose amylase with beta amylase,[1830] and maltose amylase with pullulanase.[1831]

In later work, extractive bioconversion of starch with amylases in two-phase water–PEO–PPO–MgSO$_4$ systems has been recommended, as it increases the hydrolysis yield.[1832] Once the saccharification is complete, maltose is isolated by filtration,[1808] ultrafiltration,[1811,1815,1819] ion-exchange chromatography,[1814,1817] passing through a column packed with glucoamylase immobilized on alumina,[1818,1819] or Amberlite[1825] columns packed with cysteine and/or glutathione,[1824] or by precipitation with acetone.[1804]

c. Saccharification to Glucose.—Procedures for production of glucose resemble, to a certain extent, those employed for the manufacture of maltose (see, for instance, patents).[1833–1836] Earlier, in 1924, the use of diastase in paste was patented[1837] but later either glucoamylase or α-1,6-glucosidase was introduced.[1838] An essential difference between the older and the current processes concerns the use of a saccharifying enzyme. This may be a glucoamylase of varied provenance, but sometimes microorganisms excreting that enzyme, such as *A. phoenicis*[1839] and *B. cereus*,[1840] are used. Because of this difference, a random gelatinization of the substrate is used as an alternative to liquefaction. Corn and wheat starch were prepared for saccharification by applying high-pressure gelatinization. This approach is more efficient with wheat starch.[1841] Cassava starch was gelatinized by extrusion.[1842] Starches from such oily plants as maize (corn) should be defatted by extraction prior to hydrolysis.[1843] Liquefaction and saccharification in supercritical CO_2 has been reported,[1844,1845] and both enzymes retain their activity well under such conditions. An aqueous two-phase system composed of PEG, crude dextran, and solid starch has also been studied.[1846] That system operated successfully in mixer–settler reactors for during 8 days.

There are several fungi producing glucoamylase suitable for saccharification of starch to glucose. Thus, starch was saccharified with *Endomycopsi*,[1847] *A. phoenici*,[1848] *Humicola* fungus,[1849] and *A. niger*.[1850–1852] The *A. niger* ATCC 13497 strain produced highly active alpha amylase and glucoamylase. It offered 90% conversion of starch into glucose within 48 h at 60 °C.[1853] The optimal pH is said to be 4.0,[1852] but pH 4.0–4.2,[1854] 4.2,[1735,1855] and 4.5–5.0,[1855–1859] have also been claimed. Bacterial glucoamylase was also used. *Bacillus* species isolated from soil readily produced a thermostable amylase. For saccharification to glucose, it required pH 7.1 at 57.5 °C,[1860] but the process operating at pH 4.5–5.0 has also been reported.[1855]

Immobilized glucoamylase is also in use, especially in the continuous processes.[1861,1862] That enzyme was immobilized by adsorbing it on DEAE-cellulose to form a complex,[1863] or adsorbed onto silica,[1864] glutaraldehyde–1,3-phenylenediamine copolymer developed on granulated pumice,[1865] polysulfone hollow fiber,[1866] bone ash,[1865] arylamino porous glass beads,[1867] SiO_2,[1868] porous alumina,[1869] and covalently bound to silica.[1870]

For saccharification of liquefied potato starch, glucoamylase was used in cooperation with cellulase, providing 97% conversion into glucose.[1871] Processes employing glucoamylase jointly with either pullulanase,[1872] isoamylase[1873] or a novel pullulanase-like enzyme isolated from *B. subtilis* TU[1874] are also described. The last-noted enzyme, exhibiting alpha amylase-like activity when acting independently, produces mainly maltose and maltotriose. Used in combination with glucoamylase, it produces over 97% glucose, whereas when normal pullulanase is used instead the yield of glucose is below 96%. The use of glucoamylase results in the formation of (1 → 6)-linked oligosaccharides. The cooperative action of glucoamylase with α-1,6-glucosidase eliminates (1 → 6)-linked oligosaccharides.[1875,1876] Pullulanase can also be added to act cooperatively with both former enzymes.[1877] On saccharification with glucoamylase, maltulose is also formed. Alpha amylase, glucoamylase, or pullulanase cannot hydrolyze this ketose. Consequently, the use of maltulose was patented.[1878]

Glucose can be isolated from the reaction mixture by filtering it and passing the filtrate through ion exchangers,[1825,1854,1856] by crystallization in the presence of NaCl to form complex monohydrate,[1741] by ultrafiltration,[1815] by passing through membranes,[1842,1861,1879] and by centrifugation.[1880]

d. Saccharification to Other Sugars.—Saccharification of a 1% solution of soluble starch with alpha amylase from *B. circulans* for 44 h at 55 °C produced maltotetraose in 40.6–72.6% yield. The yield depended on the enzyme concentration.[1881] Maltopentaose is available via hydrolysis of various starches with alpha amylase in a biphasic system of aqueous slurry–immiscible organic solvent (such as dodecane).[1882] Amylase isolated from *Streptomyces griseus* hydrolyzed potato starch to a syrup containing mainly maltotriose.[1883] That enzyme could produce maltotriose syrup from soluble starch when used in cooperation with α-1,6-glucosidase.[1884] Cooperation of that enzyme with either pullulanase,[1884] beta amylase,[1885] or debranching pullulanase with beta amylase[1885] also resulted in the formation of maltotriose syrup.[1885] Maltohexaose is produced from liquefied starch with the alpha amylase of *B. circulans* used jointly with α-1,6-glucosidase.[1886] Syrups formed after saccharification with alpha amylase, particularly those saccharified with *A. niger* alpha amylase could, after separation of glucose, be fermented further with yeast to give isomaltose.[1850]

7. Effect of the Botanical Origin of Starch

The optimum conditions of hydrolysis depend on the origin of the starch (see Table XII).[1887]

Starches from different sources differ in their susceptibilities to liquefaction and saccharification[1888,1889] and also on the digesting enzyme. For instance, using the alpha amylase of *B. subtilis*, the digestibility of various starches decreases in the

TABLE XII
Selection of Suitable Methods[a] for Hydrolysis of Starch Depending on the Botanical Origin of the Substrate[1887]

Starch substrate	Process			
	Acid-catalyzed	One-step enzymatic	Combined acid-catalyzed and enzymatic	Two-step enzymatic
Maize	+	−	+	++
Wheat	+	−	++	+
Rice	+	−	+	++
Potato	+	++	+	+
Cassava	+	++	+	+
Corn flour and grits	−	++	−	−

[a] +, suitable; ++, particularly recommended; −, unsuitable.

order: barley > rice > potato > corn > wheat, and the differences between the Michaelis constants for particular starches varied differently with increase in temperature.[1445] Corn, wheat, rice, and taro starches are readily digested by pancreatic amylase, whereas potato, banana, lily, gingko, high-amylose corn, lotus, chestnut, kudzu, and sweet-potato starches are fairly resistant to that enzyme.[450,1890] Another report[1891] states that, after boiling, among the list: potato, wheat, rice, corn, barley, rye, and oatmeal starches, the first one is the most susceptible to breakdown by pancreatin. However, notwithstanding these differences, all of these starches ultimately suffer breakdown to the same extent.[1890]

Beta amylase digests potato starch somewhat more readily than corn starch.[1892] Starches from tropical sources are less susceptible to hydrolysis by porcine pancreatin and *B. subtilis* alpha amylase.[1893] Studies with amylases isolated from *A oryzae*, wheat malt, and *B. subtilis* indicated that waxy starches are more susceptible to hydrolysis than nonwaxy starches. The susceptibility was higher than that of pregelatinized potato starch, a substrate known for its superior susceptibility to amylolysis.[1894] The solubilization of corn starch was more efficient than the solubilization of potato starch. Alpha amylases, regardless of their origin (human, plant, bacterial), performed better than glucoamylase followed by beta amylase.[1895]

The solubilization of starches was also dependent on their origin and decreased in the order: waxy maize > cassava = waxy sorghum > sorghum = corn > wheat > rice > sago > arrowroot > potato > heat–moisture-treated potato > hylon corn.

Some differences in the course of digestion of various components of the substrates of the same botanical origin with the same enzyme may be noted, as in case of Merck soluble starch,[1896] granular wheat,[1897] and wheat starch.[1898] These differences may be rationalized in terms of differences in the structure of the outer part of the granules, which depend on the genetics and conditions of growth of the plant.[1897] The cooperative action of two or more enzymes that accompany one another in some of the enzyme preparations used might also be taken into account.[1899] These controversies probably result from different criteria used to judge the breakdown of starch. The rate of hydrolysis of wheat starch depends on whether it is the soft or hard variety of that cereal. The hard variety hydrolyzes more readily.[1900]

Different distributions of the sugars produced by amylolysis of the substrates can result from different levels of native amylases present in the starches from different botanical origins.[1901] The kinetics of hydrolysis of three different gelatinized potato starches with *B. subtilis* alpha amylase suggested that the DP and the number of branches in amylopectin (impeding access of the enzyme) could be a key factor.[1502] It implies that the amylase/amylopectin ratio in starches must also be taken into account.

The effect of bacterial alpha amylase, glucoamylase, and pullulanase in relation to the variety of starch is summarized in Table XIII.[1902]

a. Tuber and Root Starches

(i) Potato Starch. The first systematic studies on the suitability of potato starch, and its preliminary modification for efficient enzymatic degradation, date back to the beginning of the past century.[1903] Detailed studies on the degradation of potato starch with malt amylase revealed limit dextrins up to maltooctaose, along with maltose, glucose, and isomaltose among the degradation products.[1904] For digestion of waxy corn starch with that enzyme, an empirical relationship was established between the viscosity of the conversion liquor, the concentration of the enzyme, and the time of hydrolysis.[1905]

Potato starch was digested with bacterial alpha amylase to the extent expressed as dextrose equivalent (DE) at 8 °C and pH 6.5. The procedure was varied to study changes on formation of the products of DE varying from 3.4 to 20.6. As the liquefaction progressed there was an initial decrease in the content of dextrins and maltoheptaose. Up to DE 3.4, a rapid decrease in molecular weight and the number of branches was observed. The rate of these changes slowed down after exceeding this DE.[1906] The hydrolysis catalyzed by *B. subtilis* and malt alpha amylases proceeded in a similar manner.[1907] When, after liquefaction with bacterial alpha amylase, saccharification was performed with a beta amylase–pullulanase combination, the resulting hydrolyzate contained 60–85% of maltose, no more than 17.5% of maltotriose, and less than 1% of glucose.

Beta amylase without cooperation by pullulanase was unable to debranch amylopectin.[1908] In another study, potato starch liquefied to DE 20 was saccharified jointly

TABLE XIII
Enzymatic Hydrolysis of Starch from Various Origins with Selected Hydrolases

Enzyme and its activity (U/g)	Time (h)[a]	DE	η_{20}[b] (mPa s)	G1	G2	i-G2	G3	i-G3	G4	G5	G6	G7	Gn[d]
Potato starch													
Bacterial alpha amylase 0.18 KNU	0.33	13.4	1.60	0.9	5.2	0	6.7	0	4.8	5.9	10.8	7.2	58.3
Glucoamylase 0.24 AG	1	45.7	1.38	19.0	10.3	5.2	10.4	2.0	9.1	6.5	5.2	4.0	28.5
	2	57.8	1.30	35.1	17.5	3.5	9.3	0.9	9.0	5.0	4.0	3.1	12.5
Glucoamylase 0.24 AG + pullulanase 0.18 PUN	1	46.9	1.36	24.0	13.5	0	12.9	0	10.5	5.1	4.7	3.7	25.5
	4	70.9	1.19	43.1	28.0	0	5.1	0	5.5	2.5	3.5	1.8	10.5
Pullulanase 0.40 PUN	24	16.7	1.52	0.9	6.9	0	7.3	0	5.2	6.5	10.0	5.9	54.5
	96	17.9	1.49[c]	0.9	7.9	0	8.1	0	5.5	6.2	9.5	8.4	54.5
Maize (corn) starch													
Bacterial alpha amylase 0.18 KNU	0.33	13.5	1.54	1.0	4.9	0	6.5	0	3.0	6.2	11.4	6.5	58.5
Glucoamylase 0.24 AG	1	47.3	1.26	21.2	10.5	5.3	10.5	2.1	8.7	5.0	4.8	3.8	28
	2	58.8	1.23	36.0	18.1	3.2	9.0	0.8	9.0	5.1	3.7	3.0	12
Glucoamylase 0.24 AG + pullulanase 0.18 PUN	1	48.2	1.25	25.1	13.9	0	13.0	0	11.1	4.8	4.7	3.2	24
	4	71.4	1.20	44.0	29.5	0	5.0	0	5.5	2.4	3.3	1.7	7.5
Pullulanase 0.40 PUN	24	16.8	1.50	1.0	7.0	0	7.0	0	5.1	7.0	10.5	7.5	55
	96	18.0	1.47[c]	1.0	7.8	0	8.2	0	6.0	6.6	9.9	6.9	53.5
Wheat starch													
Bacterial alpha amylase 0.18 KNU	0.42	13.5	1.74	0.7	4.7	0	6.0	0	4.5	6.3	11.5	7.3	59
Glucoamylase 0.24 AG	1	50.9	1.51	23.5	11.5	3.0	9.0	0.8	8.6	6.8	5.7	4.8	26.5
	2	59.0	1.19	34.9	17.5	3.3	7.9	0.6	8.7	5.7	4.5	4.4	12.5
Glucoamylase 0.24 AG + pullulanase 0.18 PUN	1	51.8	1.50	26.9	14.2	0	13.5	0	11.8	4.7	4.5	3.5	21
	4	68.1	1.40	40.0	25.2	0	4.9	0	5.2	2.8	3.8	2.8	15.5
Pullulanase 0.40 PUN	24	16.2	1.68	0.7	5.9	0	5.8	0	5.0	7.5	11.0	9.0	55
	96	17.5	1.66[c]	0.7	6.2	0	7.1	0	5.5	7.0	10.5	8.5	54.5

[a] The time of heating to 85 °C. [b] Viscosity at 20 °C. [c] G1 = glucose; G2 = maltose; i-G2 = isomaltose; G3 = maltotriose; i-G3 = isomaltotriose; G4 = maltotetraose; G5 = maltopentaose; G6 = maltohexaose; G7 = maltoheptaose. [d] Taken after 72 h. (Based on Data From Ref. 1902)

with glucoamylase and mycolase (a novel fungal amylase from *A. oryzae*) to a product of DE 64.9–72.3, having glucose and maltose content of 37.3–49.8 and 54.9–34.2%, respectively.[1909] However, the sequential application of those enzymes seems to be more efficient.[1910] When using the alpha amylase of *B. licheniformis*, the inclusion of Ca^{2+} is beneficial, and the optimum conditions, based on the rate of hydrolysis as the criterion, are 90 °C at pH 6. The concentration of enzyme is also an essential factor. Neither substrate nor products were inhibitory. The composition of the products was dependent to a certain extent on the temperature of hydrolysis. The level of maltopentaose (15–24%) was controlled chiefly by temperature.[1911] The conditions for liquefaction of potato starch with thermostable alpha amylase were optimized by using orthogonal tests. The optimized process provided a maltose syrup of DE 9.8% within 10 min at 94 °C.[1912,1913] In such a manner, parameters for liquefaction and saccharification of potato starch with either alpha amylase, beta amylase, or pullulanase were also optimized. A 17% (or less) slurry of the starch of red potatoes was saccharified with a blend of alpha amylase and α-glucosidase at 65–78 °C.[1914]

Phosphate ester groups in starch usually protect it from enzymatic attack.[1915–1918] However, changes of total phosphorus, and phosphate ester groups on the glucose residues of amylopectin, in the digestion of granular potato starch revealed that the alpha amylase of *B. subtilis* preferentially attacks the phosphate-bound sites. Pure pancreatin can digest phosphated starch without splitting the ester moiety.[378,1919] The phosphatase activity of pancreatic amylases in liver could be observed in some diseases,[1919] but amylases usually fail to liberate phosphorus also from limit dextrins.[390,1917,1918,1920–1922] During the saccharification of starch on sonication in the presence of either *Endomycopsis bispora* or *A. batatae*, phosphodextrins were formed. They inhibited hydrolysis by alpha amylase. The addition of *A. awamori* phosphatase split the phosphate groups from these phosphodextrins.[1923]

(ii) Sweet-Potato Starch. Sweet-potato starch is known for its low susceptibility to amylolysis. This starch has much higher gelatinization temperature (72–74 °C) and lower heat of adsorption.[1924] A higher pH is required for digestion of this starch.[1925] Preheating the starch prior to hydrolysis is advisable, and the optimum temperature is about 70 °C.[1926] Preliminary removal of outer and starchy inner tissues, and heating them separately, followed by addition of amylase isolated from the outer tissues of the substrate roots, and incubation are recommended.[1927] A complex mixture of the final products was obtained when alpha amylase from *B. subtilis* was employed. That complexity results specifically from the hydrolysis of amylopectin. The process is

nonrandom (see Section III).[1928] Complex enzyme mixtures for decomposing starch have been proposed.[1929]

(iii) Cassava Starch. Cassava starch is fairly resistant to alpha amylase. It is less susceptible to enzymatic action as compared to sweet-potato starch.[1925] On the other hand, there is a report[1930] that cassava starch undergoes hydrolysis by alpha amylase at a higher rate than corn starch. The maltodextrins produced from the two substrates differed from one another. Cassava starch can be saccharified with a 30% yield by alpha and beta amylases.[1931] Malt, mold bran, and bacterial amylases produce fermentable sugars from it with over 95% yield.[1932]

(iv) Alfalfa Tap Root, Ginseng, Mango, and Canna Starches. In contrast to beta amylase, alpha amylase released water-soluble carbohydrates from starch granules of alfalfa tap root, and as the granules are more soluble the enzyme hydrolyzed them more readily. Beta amylase produced reducing sugars preferably from the more readily gelatinized amylopectin than from sparingly soluble carbohydrates.[1933] Ginseng starch in a 10% slurry can be hydrolyzed with alpha amylase at 85 °C.[1934] Starch of mango kernel was saccharified by glucoamylase from *A. niger* at 55 °C, and an increase in the enzyme concentration increased the sugar yield.[1935]

The resistance of edible canna starch to alpha amylase is similar to that of potato starch.[1936]

b. Cereal Starches.—Generally, cereal starches are readily hydrolyzed by alpha amylase from either *B. subtilis* or *B. licheniformis* to give a mixture of maltotriose and maltose, with the components in 0.35–0.40 proportion. The process is initially conducted at 50–56 °C, and then the temperature should be elevated to 64–70 °C, with the pH maintained between 4 and 7.[1937] The course of hydrolysis of corn and wheat starches was found practically independent of pH over a wide range.[1938] There is no link between the diastatic activity of flours and their susceptibility to alpha amylase. However, flours having highly susceptible starch produced more maltose than flours with poorly susceptible starch.[1939]

(i) Corn Starch. Even after defatting, corn starch contains a residual amount of lipid[1940] which impedes enzymatic hydrolysis of that starch. Therefore, the use of amylases jointly with lipases is advantageous. A combination of lysophospholipase with glucoamylase is especially successful.

Differences were observed in the susceptibility to amylolysis between German and American varieties of corn. The latter was less suitable for that process, as demonstrated in the two-stage hydrolysis of 15 German and 2 American cultivars.[1941]

For hydrolysis of corn starch, whether normal or waxy, saccharification with pullulanase provides a higher level of glucose.[1942]

A synergism is observed between added alpha amylase and the intrinsic enzymes residing in the starch granules. Such synergism was not observed with wheat or rice starch.[1943] Hydrolysis within corn-starch granules can be performed with endogenous amylases. Germination reaches a maximum on the 4th day. Inactivation of the corn amylase produced by *Bacillus* sp. was observed as a result of changes in several reaction parameters and presented in the form of mathematical models.[1462]

Malt amylases performed best at pH 4.0 and temperatures close to 30 °C.[1944] An improved procedure for liquefaction to maltodextrin and glucose includes a two-stage hydrolysis. In the first stage, the slurry plus alpha amylase was heated with agitation from 30 °C to 120–130 °C with the temperature increasing at the rate of 1.1–1.6 °C/min. After reaching that temperature, the mixture was cooled to 70–80 °C and maintained for 5 min of hydrolysis without stirring, and then the enzyme was inactivated by boiling.[1945,1946]

A. fumigatus alpha amylase is suitable for hydrolysis of corn starch. In mutant corn starch, pancreatin attacked A-type crystallites more readily than B-type crystallites. This behavior might also be associated with distribution of the crystallites inside granules, as amorphous regions were likewise poorly digested by that enzyme.[1947]

A novel thermostable alpha amylase from *Pseudomonas fluorescens* was evaluated in the amylolysis of waxy corn starch.[337] Hydrolysis of corn hylon starch with pancreatin showed a higher extent of digestion of the amylose component and concurrently a higher content of RS.[1948] There is a synergism of that enzyme with bacterial alpha amylases, the pullulanase of *Klebsiella pneumonia* and *B. pullulyticus*, as well as the oligo-1,6-glucosidase from alkalophilic *Bacillus*.[1949] *Aureobasidium pullulans* secretes an extracellular alpha amylase together with two glucoamylases. The alpha amylase operating at 55 °C and pH 4.5 provided a maltodextrin syrup of DE 10. The composition of that syrup depended on the concentration of substrate in the slurry and not on the amount of enzyme used. A syrup containing 93% of glucose was available when the temperature of hydrolysis rose to 65 °C.[1950] Glucose of high purity is available when the hydrolysis is initially performed with bacterial alpha amylase, with admixture of glucoamylase in subsequent steps.[1259,1950,1951] *Lactobacillus amylovorus* secretes an alpha amylase of the unusually high molecular weight of 150 kDa. It rapidly solubilizes corn starch.[1952] Thermostable alpha amylase in cooperation with pullulanase from *Pseudomonas* KO 8940 at 50 °C produced a syrup containing 55.4% maltopentaose together with higher oligosaccharides. Without that amylase, more higher oligosaccharides were formed.[1953] Nonreducing linear and branched oligosaccharides were produced when, after hydrolysis with thermostable alpha amylase at 105–160 °C, the product was autoclaved with Raney nickel at 130 °C under hydrogen.[1954] For the RS produced from corn starch by either retrogradation or cross-linking, the advantage of enzymatic hydrolysis over acid-catalyzed hydrolysis is not clear.[1955]

Genetic modifications of maize resulted in mutants that differ from one another in the macromolecular composition and crystallinity of their starch, including the proportions of A-, B-, and V-type allomorphs. Hydrolysis of these starches with porcine pancreatin showed that those starches with the B-crystalline type predominating were more resistant than the other ones. The resistance of those starches to amylolysis was related to the distribution of B-type crystallites within the granules rather than to the proportion of the crystallites.[1947]

Corn flour was hydrolyzed with a glucoamylase–alpha amylase cocktail with a 1:8 ratio of the two components. Dry-milled corn flour was hydrolyzed more readily than wet-milled flour.[1956]

(ii) Wheat Starch. Several alpha amylases effect the hydrolysis of wheat starch. Low-glucose hydrolyzates are thermoreversive.[1957] The use of alpha amylase from *Thermoactinomyces vulgaris*, either at 54 °C for 90 min, at 58 °C for 3 h, or at 64 °C provided a syrup with a maximum 66.5% yield. It contained 54.5% maltose, 20.5% maltotriose, and 4.3% glucose.[1958] Wheat starch can be efficiently hydrolyzed in one step provided that high temperatures, even up to 140 °C, are employed.[1959] The *B. licheniformis* alpha amylase used for digestion of that starch has its optimum at 110 °C and pH 5.5. It cooperates with α-glucosidase in converting maltose into glucose and a mixture of oligosaccharides ranging from maltotriose to maltoheptaose.[1960] Hydrolysis at a temperature below that for gelatinization is also possible. Barley amylase performs best working at 45 °C and pH 4.5. Under these conditions, 98% of the granules were hydrolyzed within 3 h.[1961] Glucose is the principal product when a cocktail of bacterial alpha amylase, glucoamylase, lysophospholipase, proteinase, and a cellulolytic enzyme is used.[1962,1963] Combinations of alpha amylases with pullulanase, whether or not pullulanase was used jointly or in pretreatment, offered products with decreased amounts of glucose, maltose, and maltopentaose. This approach provides an increased yield of maltotriose and maltohexaose.[1964]

Thermal pretreatment of wheat prior to isolation of starch has some influence upon further amylolysis of the material, as demonstrated in studies with porcine pancreatin. Preincubation with pepsin is also helpful.[1965] Alpha amylase retards the firming and retrogradation of wheat-starch gel.[1966]

Two alpha amylase isoenzymes have been isolated from germinated wheat. There was no difference between them in the hydrolysis of β-limit dextrins, amylose, and amylopectin but only one of them adsorbed onto starch granules. Large starch granules degraded less readily.[1967] The complexation of wheat amylose with phospholipids does not provide any preference for the site of complexation to be digested.[1968]

Wheat flours from various cultivars, treated with different diastases, showed some significant differences of behavior.[1969,1970] When *A. oryzae* alpha amylase at pH 5–6 and 50 °C was used, the hydrolysis rate increased with substrate concentration and with the enzyme/substrate ratio.[1971]

Wheat grains exhibit some proteolytic and alpha amylase activity. After milling, these activities change, and the particular flow fractions differ from one another in these activities. Fractions from the central kernel have lower activities than those of fractions from the peripheral parts of the grains. The middle fraction contains most of the enzymes and, therefore, it is the most enzymatically active.[1972]

(iii) Barley Starch. Gelatinized barley and waxy barley starches are partially hydrolyzed by porcine pancreatin[1973] and by *B. licheniformis*[1974] and *B. acidopullulyticus* pullulanase.[1975] The process proceeds in two stages, with a rapid initial depolymerization. The enzymatic attack occurs mainly between the clusters without significant hydrolysis of external chains.[1976] Branched and linear dextrins are formed.[1977] The liquefaction can be effected at 95 °C and pH 6.5, either on starch slurry or on an extrudate prepared in a twin-screw machine. The second step involves digestion with immobilized *A. niger* glucoamylase at pH 4.5 and 60 °C,[1978] providing a glucose syrup of DE $\leq$ 96. Hydrolysis of amylopectin of that waxy starch occurred in a nonrandom manner (see Section VI).

Liquefaction and saccharification of naked barley starch was performed with glucoamylase immobilized on chitin and, optionally, encapsulated in calcium alginate. The nonencapsulated enzyme performed at a much higher rate.[1979] Insoluble barley starch was amylolyzed at pH 6.0 and 60 °C, and during 5 h the concentration of reducing sugars formed increased with concentration of the substrate.[1980]

Waxy, normal, and hylon starches of naked barley were digested with porcine pancreatin, an alpha amylase from *Bacillus* sp. and *A. niger* glucoamylase. The results depended on the starch type and also on the digesting enzyme, and were manifested by different yields of glucose–maltotriose and their ratio, as well as the pattern of erosion of the granules.[1981,1982]

(iv) Rice Starch. Earlier procedures for the liquefaction of rice starch involved a two-step process. First, starch was heated in water at 50–80 °C followed by the addition of malt extract, and digesting the entire mixture for 2 h at 30 °C. Hard vitreous rice required an elevated temperature.[1983]

It has been shown[1984] that varieties of rice starch can be characterized by their different behavior with respect to pancreatin, ptyalin, trypsin, and pepsin. Also different varieties of pea starch are hydrolyzed by amylase with different rates.[1985]

Optimized hydrolysis of rice starch by the alpha amylase of *Bacillus* sp., involving additives, proportion of components, pH, temperature, hydrolysis time, and agitation has been presented.[1986] Beta amylase digests glutinous (waxy) rice into branched dextrins.[1987] Cooperation of both enzymes and, additionally, pullulanase was advantageous.[1988] The starch can be protected from retrogradation by a decrease in its amylose content. This can be effected by incubation of that starch with CGTase (EC 3.2.1.54) from alkaliphilic *Bacillus cyclomaltodextrinase* I-5, which selectively digests amylase, leaving amylopectin intact.[1989]

High-protein rice flour was converted into maltose syrup and protein-enriched rice flour by treatment with thermostable alpha amylase. That enzyme liquefies the substrate to a higher extent than does beta amylase. Thermostable alpha amylase converted normal rice flour in one step into maltodextrins. At 80 °C, there were more low-molecular products, and at 70 °C there were more reducing sugars formed.[1990]

(v) Sorghum, Amaranth, and Triticale Starches. Sorghum starch was allowed to germinate for 4 days prior to amylolysis. The later stage was controlled by the rate of gelatinization, which was dependent on the starch cultivar and the pH.[1991]

Native amaranthus (kiwicha) starch, which is a natural waxy starch, was hydrolyzed with bacterial alpha amylase to give a product of low viscosity. The process was optimized by a technique of Response Surface Methodology.[1992]

Raw amaranth[1993] and sorghum[1994] flours were enriched in proteins by digesting the raw substrate with thermostable alpha amylase.

Acid-catalyzed, acid–enzyme-catalyzed, and enzyme-catalyzed direct hydrolyses of ground triricale grains were studied.[1995] Biopentosanase X (EC 3.2.1.8) and alpha amylase were used sequentially for the liquefaction. A yellow hydrolyzate of DE 1.8 resulted from the fully enzymatic process, and this appeared to be the most productive method.

(vi) Oat Starch. Hydrolysis of milled oat with a variety of hydrolases was conducted at pH 6.5 with alpha amylase, and a temperature that rose from 60 to 95 °C.[1996] The DE of the resulting product increased at a given concentration of the slurry from about 20 to 40. The reaction progress was reciprocally dependent on the slurry concentration. At a given temperature and the least concentrated (10%) slurry, the DE maximum of 40 was attained already after 1 h, whereas with the 30% slurry it took about 2 h (Fig. 9).

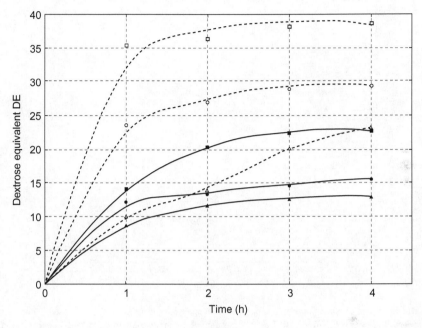

FIG. 9. Effect of the concentration of oat starch slurry on the degree of hydrolysis carried out at 60 and 95 °C with a 0.1% concentration of Thermamyl 120L at pH 6.5.1711 Solid and open points correspond to the processes carried out at 60 and 90 °C, respectively. Concentration of the slurry in both experiments increased from 10 (upper) through 20 to 30% dry substances (lower).

The reaction progress was favored by an increase in the amount of the enzyme (Fig. 10).

Similar tendencies were observed when a combination of alpha amylase, glucoamylase, and pullulanase was used, and the DE rose by approximately 100% with respect to that achieved with alpha amylase. Figures 11 and 12 present the difference in the course of the hydrolysis of a 30% slurry with the combination of these enzymes applied jointly and sequentially. It may be seen that initially the sequential mode of addition of the enzymes is less efficient, but the overall result (expressed as DE) is approximately 30% better than the result observed when the enzymes were used jointly. The joint use of alpha amylase and 1,4-α-glucan-α-maltohydrolase was less beneficial in terms of DE of the final product. A hydrolysis employing alpha amylase in conjunction with glucoamylase was patented[1997] for the hydrolysis of oat cereals.

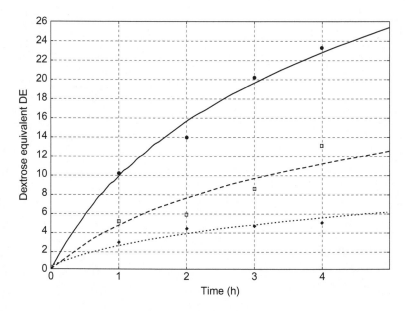

FIG. 10. Effect of enzyme concentration upon the degree of hydrolysis of 30% oat starch slurry, using Termamyl 120L at pH 6.5 and 95 °C. The concentration of the enzyme declined from 0.1% (upper) through 0.05% to 0.01% (lower).[1996]

c. Sago Starch.—Sago starch is a poor substrate for enzymatic hydrolysis. Alpha amylase and glucoamylase act after incubation of the substrate in acetate buffer at pH 3.5 and 60 °C.[1998] Liquefaction followed by saccharification provides glucose.[1999] Alpha amylase isolated from *Penicillium brunneum* from the sago palm tree can be used. The optimum conditions are 60 °C and pH 2.0.[2000] Hydrolysis with a thermostable alpha amylase from *B. licheniformis* at 90 °C and pH 6.0, with Ca^{2+} added, follows Michaelis kinetics in which both constants and rate are temperature dependent.[2001] When *B. subtilis* β-glucanase and pullulanase were used, the C-type crystals disappeared at 100 °C.[2002] The rate of hydrolysis increases with concentration, but the yield of glucose simultaneously decreases. An excessively high concentration of alpha amylase negatively influences the rate of glucose release.[2003] Pullulanase produced linear, long-chain dextrins.[2004] The effect of pH and substrate concentration was studied by Bujang et al.[2005]

d. Chestnut Starch.—Optimum conditions for liquefying chestnut starch with thermostable alpha amylase are 20% slurry, 90 °C and pH 6.0. Saccharification proceeds best at 60 °C and pH 4.5.[2006] When thermostable alpha amylase was used

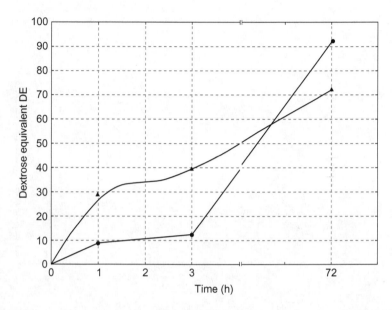

FIG. 11. Separate and simultaneous action of 0.1% concentration of Termamyl 120L and 0.1% of dextrozyme E upon 30% oat starch slurry at pH 6.4 and 60 °C. Triangles and circles are related to joint and sequential action, respectively, of the enzyme.[1997]

in combination with glucoamylase, the total conversion of that starch into glucose was achieved within 15 min, provided that a high concentration of the enzymes was used. Because of the thermal deactivation of glucoamylase, the processing temperature had to be lowered. The amylase/glucoamylase ratio should be 0.35/0.65.[2007,2008] Total starch conversion in the solid state requires 70 °C.[2009] A chestnut beverage was prepared by liquefying 30–40% starch slurry with alpha amylase at 95 °C and pH 6.0 in the presence of $CaCl_2$ followed by saccharification with glucoamylase at 65 °C and pH 5.0.[2010]

e. **Legume Starches.**—Digestion of legume starches is generally more difficult than digestion of cereal starches. Therefore, for maximum efficiency, the enzymatic hydrolysis of legume starches should be preceded by autoclaving, pressure cooking, extrusion, or sprouting.[163] Pea and bean starches are digested by alpha amylase in a similar manner, forming low-molecular saccharides. The results of amylolysis with beta amylase show that pea starch has more branched amylopectin.[2011] On hydrolysis with alpha amylase from *B. amyloliquefaciens*, starch of smooth (less soluble) and wrinkled (more soluble) starch initially produced dextrins of DP 2–100, whereas in

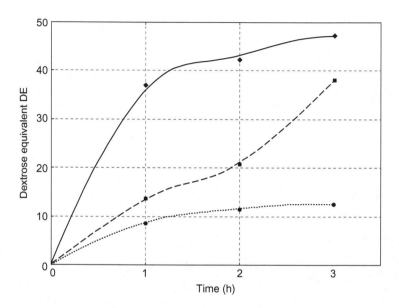

FIG. 12. Simultaneous action of Termamyl 120L and Maltogenase 4000L on 30% oat starch slurry at pH 5.5 and 60 °C. Circles—0.1% Termamyl 120L, squares—0.1% Termamyl 120L with 0.1% Maltogenase 4000L, rhombs—0.15% Termamyl 120L with 0.15% Maltogenase 4000L.[1997]

further stages the dextrins reached DP 3–6. The majority of them were linear. The enzyme seems to attack preferentially the interior of the granules.[2012]

f. Lichen and Millet Starches.—Lichen starch ("isolichenin")[2013] is hydrolyzed by diastase to give maltose.[2014] Millet starch was hydrolyzed with alpha and beta amylases of *A. oryzae*. The saccharification began immediately, but it was incomplete because of inhibition from bound phosphorus.[1922]

g. Amylose and Amylopectin.—Amylose is hydrolyzed more readily than amylopectin, and the ratio of the rate of hydrolysis of amylose to amylopectin may reach 2:1.[1488,2015–2019] For the hydrolysis of amylopectin, the alpha amylase from *B. stearothermophilus* is recommended as the enzyme providing nonrandom (see Section III) cleavage of the molecule. The resulting fragments have 20–50 kDa and all contain α-(1 $\rightarrow$ 6) side chains.[2020] Models of predictive value were developed[2021,2022] for results of saccharification of starches based on their structure and, especially, on the pattern of their branching. These models are, however, not uniformly valid, as no relationship between starch constitution and the results of hydrolysis of starch with takadiastase and maltogenic amylase, performed under a variety of conditions could be found.[2023]

h. Synthetic Starches.—Rendleman[2024] produced high-amylose starches of DP 61–71 by digesting α-CD with CGTase. These starches were hydrolyzed by ptyalin at 37 °C and pH 7.0. They were poorly digested as compared to natural starches, among which hylon VII hybrid corn was also digested with difficulty. Natural starches produced mainly maltose, maltotriose, and glucose along with minor amounts of maltohexaose and maltoheptaose, but "synthetic" starches produced mainly maltose, followed by decreasing amounts of glucose and maltotriose.

i. Chemically Modified Starches.—Chemically modified starches have also been subjected to enzymatic hydrolysis. Deuterated starch is hydrolyzed more readily by amylase than native starch.[2025] Sweet-potato starch is sometimes bleached with hypochlorites; this bleaching involves oxidation, and the rate of hydrolysis of such oxidized starch is similar to that of intact starch.[2026] A haze-free maltodextrin is obtained.[2027]

Etherification of corn starch with ethylene oxide in the starch:ethylene oxide ratio of 7:1, 7:2, and 7:3 increased the rate of hydrolysis with glucoamylase by the factor of 1.5, and further increase in the degree of etherification did not further influence the rate.[2028]

Digestion of hydroxyethylated starch resulted in a decrease in the viscosity of the digested samples, but that effect was not accompanied by the formation of reducing centers. That observation was interpreted in terms of anhydro moieties formed instead.[2029] Hydroxypropyl cassava starch and a phosphate derivative were hydrolyzed by hog pancreatin to glucose–maltotetraose mixtures. The molecular weights of oligosaccharide products were lower than those of products from nonmodified starch.[2030] Hydroxypropylated maize, waxy maize, and hylon maize starches were digested also by porcine pancreatin. The product profile depended on the starch and its degree of polymerization.[2031]

Acetylation of starch resulted in lower production of glucose and/or maltose when alpha and beta amylases were used, and this decrease is fairly proportional to degree of substitution.[2032] Acetylated starch of DS 1.5 blended with native starch was hydrolyzed with takaamylase from *A. oryzae*, and only the nonmodified component underwent hydrolysis. The acetylated component was degraded with the acetylesterase from *A. niger* interacting synergistically with alpha amylase.[2033,2034] *Bacillus liquefaciens* digests acetylated starch to give glucose and maltose. The net result depends on the degree of acetylation. At high degrees of acetylation, promoters of the acetylesterase activity have to be added.[2035,2036] Because of their higher hydrophobicity, starch esterified with such higher fatty acids as dodecanoic (lauric) and hexadecanoic (palmitic) was digested by alpha amylase with more difficulty than unmodified starch.[2037]

Epichlorohydrin-treated starch microspheres were hydrolyzed by porcine alpha amylase, and the process was surface controlled.[2038] Digestion of a cationic potato-starch derivative with alpha amylase, pullulanase, and isoamylase showed more intense fragmentation as the DS of the starch derivatives increased.[2039,2040] The modification of the starches to a higher DS probably degraded the substrates prior to their enzymatic treatment. Starch cross-linked by $POCl_3$, Na_3PO_4, or epichlorohydrin was digested by the alpha amylase from *Rhizopus* after heating or treating with alkali to decrease its crystallinity.[2041] Cross-linked starch microspheres prepared by emulsion polymerization were readily digested by pancreatin.[2042]

8. Role of Starch Pretreatment

The susceptibilities of granular and gelatinized starches to digestion are not necessarily the same. For corn granules digested with *Streptomyces hygroscopicus* or *A. oryzae*, glucose and maltose were the main products, whereas from gelatinized corn starch there was less glucose and more maltotriose.[2043] The starch origin-dependent order of susceptibility to hydrolysis may also be different for granular and gelatinized starches. Thus, granular waxy maize starch was definitely more susceptible to digestion with *Rhizopus nivei* than granular barley, maize, and cassava starches. The susceptibility of these three latter starches to such hydrolysis is similar. Granular hylon maize, shoti, and potato starches are fairly resistant to glucoamylase[2044] (Fig. 13).

It is noteworthy that raw starches are usually hydrolyzed by malt amylase more effectively than soluble starch.[2045] This phenomenon may result from the fact that raw starches contain amylases that are partly deactivated and removed during the solubilizing processing.

Studies with fungal and bacterial alpha amylases and glucoamylases on granular wheat starch point to preferences for the digestion of small granules.[2046,2047] This observation contradicts the other findings that document the higher susceptibility of large granules to hydrolysis.[2048] In barley malt alpha amylase, two components are distinguished from one another by their pI. The component of low pI hydrolyzed large granules more readily, but the original alpha amylase not separated into its components digested small granules more efficiently. Thus, that original enzyme can be

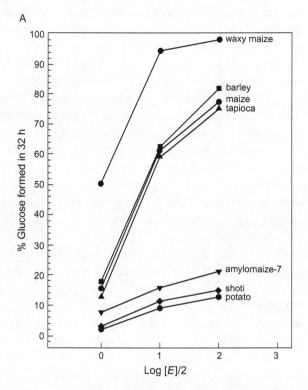

FIG. 13. Effect of glucoamylase on granular starches.[2044]

more suitable for hydrolysis of either small or large granules, depending on the proportion of each component.[2049]

The size of the granules exerts a certain influence. Among several properties of native granular starches, it is perhaps a significant factor controlling the enzymatic digestibility. Other factors, such as the shape of granules; their amylase, lipid, and phosphorus content; and their crystallinity, architecture, and other factors, are frequently interrelated and complicate the drawing of any firm conclusions.[164] There is more maltose produced by malt amylases from small granules than from large ones.[2050] The degradation of granules of woody starches with alpha amylase,[2051] and of a series of granular cereal and tuber starches with glucoamylase,[2052] was inversely proportional to the size of the granules. This suggests that not the size but the state of the granule surface and the crystallinity of the granules control the digestion rate.[2052,2053] The higher affinity of algal starches to amylolysis as compared to plant starches was explained in such manner. Such a relation exists between eight different algal starches.[2054] However, experiments conducted with large and small granules of normal and waxy barley, digested with two alpha amylases isolated from barley malt, suggest that properties of enzymes can also be involved[2055] (see also Ref. 238).

As shown in the hydrolysis by porcine pancreatic alpha amylase of native granular starch from potato, corn, and rice, the rate of the process is proportional to the size of the granules.[2056] The observed effect can probably be associated with the mineral content of granules, and this depends on the size of granules.[2057] This series of apparently contradictory effects may also result from contamination of the hydrolases. In 30 varieties of sweet-potato starches digested with *Rhizopus niveus* glucoamylase, a negative relationship was found[2058] between granule size and their digestibility.

With barley starch, the higher surface area and lower crystallinity of small granules facilitated their amylolysis over the large granules.[2059] In cassava and corn starches, there is a relationship between granule size and their susceptibility to hydrolysis with the alpha amylase–glucoamylase cocktail. Large granules are hydrolyzed more readily.[2060]

Granules of different botanical provenance show different patterns of enzymatic digestion that are specific for each given starch and the hydrolase employed,[2061–2064] and this variation can be associated with the structure of the granule envelopes.[2065] Alpha amylases showed both centrifugal and centripetal erosion of corn, rice, and wheat granules, but only centrifugal erosion of potato granules.[875]

Detailed studies have been performed on the mechanism of digestion of granules of various starches, such as barley,[2055,2066–2069] waxy barley,[2055,2069] corn,[875,1895,2043,2070–2076] waxy corn,[1895,2043] rice,[875,1895,2043,2076,2077] legume seeds,[163,2078,2079] cassava,[1895,2080] potato,[516,875,1895,2043,2075,2081,2082] sweet potato,[875,2058,2074,2083] wheat,[1961,2076,2084–2088] sorghum and waxy sorghum,[1895] sago,[1895,1998,2005] arrowroot,[1895] hylon corn,[2076] mung bean,[2076] several legume starches,[2089,2090] banana, lily, gingko, Chinese yam,[2090] finger millet,[434] buckwheat starch,[1363] and soluble starch.[436]

Ultramicroscopic studies of granules of potato and arrowroot starches digested with pancreatin[2091] revealed peripheral damage of the granules. In other studies,[2092] starch granules remained intact on treatment by the pancreatic enzyme,[2092] but they were attacked by takadiastase (alpha amylase from *A. oryzae*),[2093] malt, and to a lesser extent by the bacterial amylase form *B. subtilis*.[2094] Granules of Chinese yam (*Diascorea batata*) and maize starch, which suffered attack by the alpha amylase and glucoamylase from *Rhizopus amagasakiens*, showed in scanning electron micrographs numerous pin holes on the surface of the granules, with pores penetrating into the inner layers. Inner layers in contact with enzymes were digested more readily than peripheral regions of the granules. Small-size starch granules of black pepper appeared resistant to fungal glucoamylase and human salivary alpha amylase. Starch granules of black gram were fully resistant to these enzymes, regardless of the granule size.[2063] Granules of black gram and ragi, after treatment with either urea or periodate, lost their resistance to attack by the enzymes.[2095]

Extensive erosion of granules was observed in waxy starch granules, native corn, and sorghum.[1895] Crystallites of the A-type are more susceptible to digestion with *Bacillus* sp. alpha amylase than crystallites of the B-type.[2096]

Limited enzymatic digestion of granular starch leads to porous starches useful in several fields.[2074] Digestion of starch granules can be performed to effect retention of glucose inside the granules. Such a possibility was demonstrated[2097] in studies with starch granules of waxy maize, hylon maize, and normal maize, digested with 1 IU of the glucoamylase enzyme per 50 mg of starch in either a slurry, a sealed vessel, or an open vessel. In the two last instances, granules containing 50% w/w water were used. As may be seen in Fig. 14, waxy maize starch granules retained the most glucose, followed by normal maize and hylon maize. The reaction is time dependent, and the peak content of glucose is afforded after 28 h of processing, regardless of the starch variety.

Retrogradation decreases the susceptibility of starches to amylolysis.[2098] Retrograded corn and potato starches on hydrolysis with alpha amylase provide glucose–maltononaose and branched dextrins.[2098] It appears that a certain amount of low-molecular (DP 61–71) hylon starches are formed on production of cyclodextrins with *Bacillus macerans* enzyme. Their hydrolysis with ptyalin was attempted before and after cooking. They showed fairly good resistance to hydrolysis.[2024]

Any mode of damaging starch granules affects the digesting of starch by enzymes.[2099,2100] However, the result depends on how the granules are damaged. Initially, swelling of the starch for several hours was used.[1894,2101] A 20-min pretreatment of starch granules with steam increases its ability to adsorb alpha amylase.[2102] Autoclaving starch at 120 °C for 1 h, or mixing starch with 10% diatomite and water also facilitates adsorption of the enzyme.[2103] Preheating starch at 80–85 °C for 30 min to 2 h followed by drying at 30–35 °C is another way of activating starch for amylase adsorption.

Mechanical damaging (kneading) starch in the presence of such polyols as alditols is more efficient than kneading in water.[2104] Generally, such polyols as glycerol and alditols decrease the inactivating effect of temperature.[2105] Aliphatic alcohols facilitate the digestion of granular starches, and that effect increases as the alcohol chain extends from methanol to 1-butanol.[2044] The milling of starch in the presence of the enzyme is another solution.[2106]

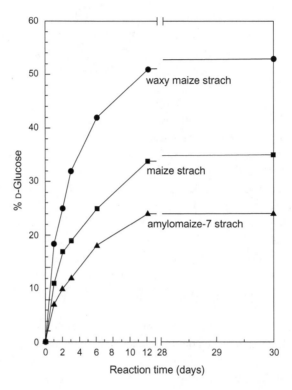

FIG. 14. Time-course of the formation of glucose in the reaction of glucoamylase with three varieties of maize starch performed in a sealed vessel at 37 °C.[2097]

Crushing granular starch followed by digestion with glucoamylase[2107] or amylase[2108] provided a higher yield of glucose. With corn starch, defatting with ethanol and removal of zein causes sufficient damage to the granules that the resulting granular starch is readily hydrolyzed by alpha amylase and glucoamylase to give glucose in over 97% yield.[1752,2108]

Preheating granular starch containing alpha amylase is also an efficient way of activating the enzyme.[1375,2109,2110] Use of UV radiation favors the enzymatic process, but the treatment of starch with calcium hypochlorite (bleaching powder) inhibits enzymatic action, probably because of the effect of adsorbed chlorine upon alpha amylase.[2048] Although ionizing radiation damages starch granules, the damage is associated with a liberation of such compounds as formaldehyde,[5] which can deactivate enzymes. Inhibition of enzymatic hydrolysis of such irradiated starch was also interpreted in terms of structurally modified glucose residues formed in the starch structure upon irradiation.[2111] These factors can be eliminated by decreasing the radiation dose to ≤ 20 megarads of γ-radiation.[2112] Coagulation of starch by irradiation can also be involved.[2113] A dose of X-ray radiation of 10^8 rad increased the formation of reducing sugars by approximately 15%.[2114]

The use of elevated pressure is another approach.[2115] Saccharification of starch that has been extruded prior to digestion also augments the rate of amylolysis.[2116] Preextrusion of the substrate is more effective than preautoclaving.[2117]

Starches altered by phosphation to starch monophosphate,[2118] or by hydroxypropylation,[2119] were much more susceptible to digestion with diastase than native, unprocessed starch.[2118] A special conditioning of starch prior to hydrolysis was patented.[2120] Starch is blended with monocalcium phosphate, with up to 5 ppm of Cu^{2+} added, and the blend is then cooked. Another US patent[2121] describes the production of a bleached starch composition of improved convertibility with alpha amylase. The pretreatment of starch with dilute hydrochloric acid is probably the most efficient.[2122,2123] Activation of starch by rapid heating to 200 °C, followed by rapid cooling to 20 °C was proposed.[2124] The time of the heating stage should be as short as one second, and the moisture content of treated starch should not exceed 20%.

It should be noted that storage influences, to a certain extent, the properties of starch granules. The swelling capacity of the granules increases with time of storage, and thus their susceptibility to liquefaction and saccharification increases.[2125] Microwave irradiation also makes starch more susceptible to amylolysis.[2126] Starch from germinated sources, for instance barley, shows higher susceptibility.[2127]

The freezing of moist starch also aids its digestibility, and slow freezing is more efficient than rapid freezing. Because of retrogradation, the digestibility of gelatinized starch decreases with time.[2099]

Starch can be physically modified by hydrothermal treatment. With such changes in the structure of native granules, changes in the affinity of granules to enzymatic attack are to be anticipated. The heat–moisture treatment causes gelatinization of the amorphous regions of the granules, leaving the crystalline residue component unmodified. The amorphous, gelatinized portion of the starch thus becomes more susceptible to amylolysis, while the susceptibility of the crystalline residue to enzymatic attack is not changed.[2128,2129] The procedure is, among others, a mode for the production of resistant starch. The heat–moisture treatment of the gels does not affect the crystalline portion up to approximately 12.5% of hydrolysis, and the amorphous fraction of the gel declines from 75 to 62.5%.[2130] The outcome of the heat–moisture treatment depends on the level of moisture employed. The crystalline fraction of normal and waxy corn starches treated at the 27% moisture level was more susceptible to enzymatic attack than that from the treatment at 18% moisture, as degradation of the starch was more extensive at the higher moisture level. Such effects are particularly pronounced in hylon starches.[2128,2131,2132] Hydrothermal treatment of starch in the presence of alpha amylase increases the breakdown of starch.[2133,2134] Such treatment of wheat, rye, and potato starches in the presence of alpha amylase from *Thermoactinomyces vulgaris* in a heterogenous phase provided a hydrolyzate containing 50% maltose and 15–20% maltotriose. The conversion yield was 85%. For wheat and rye starch, 64 °C is the optimum temperature, whereas for potato starch 70 °C is required.[2135] That improvement in adsorption ability is further enhanced by the addition of up to 30% v/v of ethanol. In contrast, addition of ethanol to starch that had been prehydrolyzed with alpha amylase resulted in a decrease in the ability of that starch to adsorb the enzyme.[2136]

9. Role of Temperature

The temperature may significantly alter the susceptibility of starch to the enzymatic hydrolysis.[1365] The temperature effect depends on the botanical origin of the starch. At 100 °C, the rate of hydrolysis of waxy rice starch accelerated from that at ambient temperature by 13-fold, whereas that for potato starch was accelerated by 239-fold.

The temperature is a crucial factor in the enzyme activity. Too high a temperature can deactivate the enzyme completely, although incompletely deactivated enzymes can have their activity restored.[2137] Below this inactivation temperature, the hydrolysis product profile changes with temperature. On the other hand, mild preheating of an enzyme prior to its use makes it more tolerant of an increase in temperature on the hydrolytic process.[2138]

Hydrolysis at a higher temperature provides a narrower molecular weight range in the products, and variation of the pH varying between 5.1 and 7.6 in hydrolysis by the alpha amylase of *B. licheniformis* had a little effect on that range.[2139] Changes in temperature of hydrolysis altered in an irregular manner the susceptibility of corn, wheat, rice, and potato starches to hydrolysis with *B. subtilis* alpha amylase at 50 and 60 °C at pH 6.0.[2140] The order of susceptibility of those starches, based on the extent of hydrolysis after 1 h at 50 °C, changed from rice > potato = wheat >> corn to potato > wheat >> rice > corn after 1 h of hydrolysis at 60 °C.

Different starches digested with pig pancreatin reacted differently to the temperature. Between 17 and 27 °C, potato starch was hydrolyzed to a greater extent than rice and corn starch, while both of the latter behaved identically, but between 27 and 37 °C corn starch was hydrolyzed more readily than potato starch, followed by rice starch.[2141]

The temperature influences the interaction of the enzyme with a coenzyme as well as the normal reaction of the activated enzyme.[2142] The optimum pH for maltatic action of the enzyme is less sensitive to temperature changes than that for amylolytic action.[2143] At a given pH and temperature, amylolytic activity remains constant with time, whereas the maltatic activity decreases slightly.[2144] The role of temperature during starch hydrolysis in a batch reactor was studied for alpha amylases of *Bacillus* species and *A. oryzae* in relation to the time of hydrolysis. Inactivation of the enzymes was noted in the range of 50 and 60 °C. In the hydrolysis with malt amylase below 50 °C, maltose and dextrin of reducing power 5–10% lower than that of maltose were formed. Above 50 °C, a limit maltodextrin of reducing power 30% of that of maltose was formed, and its yield increased with temperature.[2145] Some mathematical models for the inactivation were proposed and their validity was confirmed.[2146]

10. Role of the Substrate Concentration

The concentration of the starch slurry is also important. Bacterial amylases perform well with 2–30% slurries, above which concentration the hydrolysis was inhibited. Decomposition of limit dextrins is not perturbed by a high slurry concentration.[1375,2147] It is noteworthy that an increase in the substrate concentration diminishes the thermostability of the enzyme.[2148] Studies on the mode of action of amylases on starch[2149] confirmed earlier findings that the rate of hydrolysis is a function of the starch type, and the affinity of starches to enzymes is the reciprocal of their concentration.[2150] The amount of enzyme used increases the reaction rate up to a certain maximum, and further increase in the amount has little effect.[2151]

11. Role of Water

Water is crucial for the hydrolysis, mainly because it enables swelling and gelatinization of starch.[2152] The quality and quantity of the products of starch hydrolysis with amylases depends on the hydrating ability of the medium. The available water is distributed between solvent and nonsolvent water. Only a small amount of solvent water is required to initiate amylolysis; this amount satisfies the requirements of both alpha and beta amylase, and it is independent of temperature.

A certain minimum amount of water is needed to initiate amylolysis, and the minimum amount of water required at particular stages of the hydrolysis is different. These amounts can be predicted from the adsorption isotherms.[2153,2154] The rate of starch hydrolysis increases with the water content over the range of 14–42% and then remains constant. The enzymes retain their activity at 5.5% content of water.[2155] An increase in humidity decreases the thermal stability of amylase. In the air-dried state, the amylase does not suffer any inactivation up to 90 °C, but an increase in the humidity from 1.1 to 9.2% decreased the activity of amylase at 90 °C by over 63%. The interaction of amylase with the starch substrate increased the thermal stability of the enzyme in the air-dried state.[2156–2158] On the other hand, amylolysis of granular starch under a deficiency of water provides starch granules that retain the products of hydrolysis inside the granule.[2159]

Under low-water conditions, depending on the level of the water deficiency, a two- and even three-stage hydrolysis may take place.[2160] At a low water content, the shear stress necessary to melt starch is so high that it can deactivate enzymes. Therefore, melting and liquefying should be spread into two independent steps. With a severe deficiency of water, more isomaltose and isomaltotriose are formed.

12. Role of Elevated Pressure

Elevated pressure up to 1000 kg/cm^2 applied at pH ≥ 6 accelerated hydrolysis of starch with human saliva, but higher pressure does not further enhance amylolysis.[2161,2162] Applied pressure of 0.01–600 MPa at 40 °C on starch hydrolyzed with *B. subtilis* alpha amylase had a major effect on the product profile;[2163,2164] maltopentaose was the principal product and there was a decrease in the amount of glucose, maltose, and maltotriose. It was suggested[2165] that this effect is associated with the gelatinization of starch. When the pressure was elevated to 800 MPa, alpha and beta amylases showed a loss of activity because of association of the enzymes by linkage through the sulfhydryl bonds. Beta amylase is more susceptible to pressure than alpha amylase. In experiments with sorghum and corn grains and the glucoamylase enzyme, the susceptibility of the substrates to enzymatic attack increased[2166] with increase in applied pressure from 1.6 to 6.0 kg/cm^2. As shown by FTIR spectroscopy, the alpha amylases of *B. licheniformis*, *B. amyloliquefaciens*, and *B. subtilis* changed their conformations at 6.5, 7.5, and 11 kbar, respectively. Their thermal and high pressure stabilities correlated with one another. Under combined heat–pressure action, the alpha amylase of *B. licheniformis* was considerably more stable than the two others, which exhibited almost equal stability under the conditions applied.[2167]

13. Role of pH

Alpha amylases enable processes operating at lower pH, as their optimum performance is usually around pH 4.5.[1380] The pH optimum is specific for each given enzyme. With the alpha amylases, a lower pH suppresses the formation of maltotetraose and higher oligosaccharide products. As the pH increases the yield of maltotetraose and higher oligosaccharides increases.[1397]

14. Role of Admixed Inorganic Salts

Calcium ion is indispensable for the functioning of alpha amylase, specifically for one of its components. That ion stabilizes the enzyme thermally,[2148,2168,2169] but Mn^{2+} cations inhibit amylolysis.[2170] The alpha amylases of B. subtilis, A. oryzae, human ptyalin, and hog pancreatin contain in their basic constitution at least 1 gram-atom of very firmly bound Ca^{2+} ion per mole of enzyme.[2171] In the alpha amylolysis of granular oat starch with the enzyme from B. subtilis, the Ca^{2+} ion influences the reaction by accelerating solubilization of the granules.[2172]

In alkaline media cations play an essential role, whereas in acidic media the effect of anions is important.[1379,2173] Detailed studies performed on the effect of sodium halides in the amylolysis of starch provided a rationalization of that effect: chemical or adsorption equilibria between halide anions and amylase is established very rapidly, and when starch diffuses towards that complex and binds to it, the hydrolysis begins.[2174] Sodium sulfate behaves neutrally[2175] unless it is employed at high concentration.[2176] Low concentrations of 2-chloroethanol (ethylene chlorohydrin), potassium thiocyanate, and urea also activate amylase.[1561] Aluminum salts, particularly aluminum lactate, also stabilize alpha amylase in solution.[2177]

Pancreatin is stimulated by pyrophosphates of the Group I and II metals[2178] and the hydrolysis of starch by ptyalin and pancreatin is stimulated by halides of the Group I and II metals, and also ammonium halides. The effect of the halide ions decreases in the order chlorides > bromides > iodides > fluorides. The effect of these salts is strongly and nonlinearly dependent on the concentration, as shown in Figs. 15–18 for fluorides, chlorides, bromides, and iodides, respectively.[2179]

Some comparative studies have shown the following orders of increasing activation of potato amylase by cations and anions: $Na^+ < K^+ < NH_4^+$ and $SO_4 = < Cl^- < Br^- < F^- NO_3^- < PO_4^{3-}$.[2180,2181] In the autolysis of potato starch, these orders change to $K^+ < Na^+ < NH_4^+$ and $F^- < Cl^- < SO_4 = < NO_3^- < Br^-$.[2180] A 1.0–1.75 molar aqueous solution of NaF increases the activity of potato amylase.[2182] These orders are dependent on the particular enzyme and, for instance, for ptyalin the order of anions is $CNS^- < NO_3^- < Cl^- = F^-$.[2183] The effect of salts also depends on the pH and substrate concentration, even if these salts are neutral. A slight inactivation of ptyalin

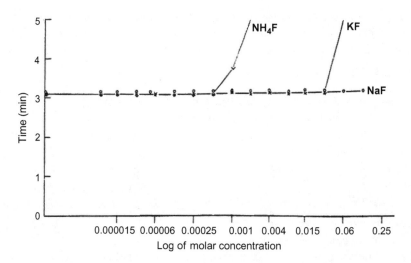

FIG. 15. Stimulation of ptyalin and pancreatin by fluorides. Na: ○; K: x; NH$_4$: v. Higher concentrations of KF hydrolyze soluble starch within 10–40 min, and a higher concentration of NH$_4$F did not hydrolyze starch within 4 h.[2179]

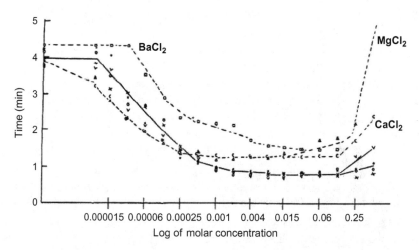

FIG. 16. Stimulation of ptyalin and pancreatin with chlorides of Li: ●; Na: ○; K: x; NH$_4$: v; Mg: Δ; C: c; Ba: □. An 0.5 M aq. solution of MgCl$_2$ at 37 °C takes 9 min 38 s for hydrolysis.[2179]

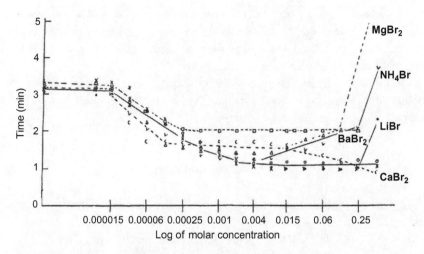

FIG. 17. Stimulation of ptyalin and pancreatin by bromides. Notation of the points is the same as in Fig. 16. An 0.25 M aq. solution of $MgBr_2$ hydrolyzes soluble starch within 5 min 23 s.[2179]

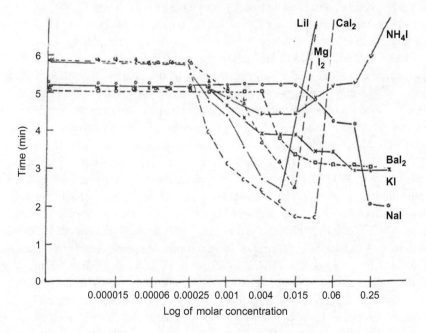

FIG. 18. Stimulation of ptyalin and pancreatin by iodides. Notation of the points is the same as in Fig. 16. An 0.06–0.5 M aq. solution of MgI_2 hydrolyzes soluble starch within 13 min to 2 h, respectively.[2179]

by Cd salts has been observed.[2184] Concentrations of Fe(III) up to 850 ppm had a little effect on malt amylase,[2185] but Hg(II) ions strongly inhibit hydrolases (see, for instance, Ref. 2186). Microorganisms have different demands for metal ions, for instance *A. niger* is tolerant to heavy metal ions.[2187]

Inclusion of larger quantities of K_2HPO_4 activates amylase,[2188–2190] but smaller amounts of inorganic phosphates stimulate the growth of *S. cerevisiae*, which under such conditions secretes more glucoamylase.[2191]

15. Role of Inhibitors

The simplest way to inhibit amylolyis is mechanical; shearing partially deactivates enzymes.[2192]

Amylolysis is inhibited by several factors, such as inhibitors residing in the substrate,[2193] the aforementioned concentration of substrates and products, numerous metal ions, particularly transition-metal ions, and organic additives. Considering that amylolysis is a two-stage process of liquefaction and saccharification, each stage can be perturbed differently by a given factor. Thus, $CuSO_4$ suppresses both stages in parallel, whereas aniline affects only the liquefaction component.[2194] Phenols[2195] and certain alcohols[2196] suppress amylolysis completely. Lipids selectively inhibit digestion of starch with alpha amylase, according to whether or not the lipid forms a membrane that impedes contact of the enzyme with the starch substrate.[2197] The inhibitory action of lipids upon the hydrolysis of starch is interpreted in terms of formation of a helical complex of the lipid with starch, which thereby becomes less susceptible to undergoing conformational changes to form a complex with the hydrolyzing enzyme.[2198] Dodecanoic, tetradecanoic, hexadecanoic, and (9Z)- octadec-9-enoic (oleic) acids all inhibit this process, whereas octadecanoic acid does not. Phytic acid present in the plant material decreases the activity of alpha amylase, particularly on preincubation and in the presence of Mn^{2+} and Ca^{2+} ions.[2199] Lysolecithin also inhibits amylolysis, whereas cholesterol has no effect.[2200,2201] Salts of higher fatty acids inhibit hydrolysis, and their action is attributable to inhibition of adsorption of enzymes on the substrate.[2202,2203] The effect of lower fatty acids upon amylolysis depends to a certain extent on the origin of the enzyme used. Usually, fatty acids below decanoic acid do not impede the hydrolysis.[2201] However, hydrolysis by ptyalin is strongly inhibited by these acids. Very probably, the latter enzyme is unable to desorb acids from the substrate.[2204,2205]

Heparin acts to inhibit malt alpha amylase,[2206] and tea polyphenols, catechins, have been shown to inhibit ptyalin.[2207] The alpha amylases of *B. licheniformis* in the presence of such polyols as glycerol, D-glucitol, D-mannitol, or sucrose are thermally stabilized, even when immobilized.[2107] It appears that certain buffers used to maintain an optimum pH can deactivate some amylases.[2208] However, it has been shown[2204] that buffering with sodium acetate in the presence of a proteinaceous material such as cereal flour is beneficial.

Generally, an increase in the total sugar in the hydrolyzing medium inhibits hydrolysis.[2209] The inhibiting properties of lower saccharides can be selective with respect to various enzymes. The alpha and beta amylases from malt are both adsorbed onto starch, however, when maltose is added, the alpha amylase component alone is selectively adsorbed.[2210] The inhibiting properties can be also selective with respect to the particular process involved. Glucose inhibits liquefaction with alpha amylase and glucoamylase more than maltose does.[2211,2212] Added glucose inhibits saccharification by decreasing the amylase activity,[2213–2215] and maltose acts similarly.[2216] However, there is a report that these saccharides stimulate the saccharification stage.[2211] Acarbose (1), used clinically for diabetes, inhibits the hydrolysis of amylase by porcine pancreatin in noncompetitive manner.[2212]

1 Acarbose

Acarbose is bound to the secondary active site of amylase, and its primary active site is engaged in formation of the amylose–enzyme complex. The α-, β-, and γ-cyclodextrins also inhibit hydrolysis of amylose in a competitive manner, although in contrast to acarbose they do not bind to the complex.[1447] Gelatin inhibits the action of amylase on starch by impeding diffusion of the enzyme to the substrate.[2217] Citrate does not inhibit the enzymatic action at the optimum pH, but below or above that pH range it retards hydrolysis.[2218] Some alkaloids of the quinine series also retard the reaction.[2219] The amylolytic activity of enzymes can be decreased by proteins present in plants.[2220] The action of ptyalin in the mouth in digesting some semi-solid meals is an undesired factor, as it diminishes the taste of the meals. In such a situation, such ptyalin inhibitors as acarbose may be added to such foodstuffs.[2221]

16. Stimulators of Hydrolysis

Several organic compounds stimulate enzymatic hydrolysis. Their effect can result from binding to specific or nonspecific active sites of the enzyme, promoting a favorable mutual orientation of enzyme and substrate. The additive–substrate interactions may induce conformational changes. Organic solvents added to the reaction mixture may promote the dissociation of inhibitors from the active sites of the enzyme.[2222]

Polyethylene glycol (PEG) and various detergents,[2223,2224] together with polyvinyl alcohol, chitosan, and other hydrophilic polymers,[2225] as well as bile salts,[2226] activate alpha amylase and other hydrolases and prevent its denaturation. There is a report that surfactants strongly inhibit the adsorption.[2227] On the other hand, sodium dodecylsulfate, which is used to remove proteases from starch, stimulates the hydrolysis of granular sorghum starch by alpha amylase.[2228] Aliphatic amines inhibit amylolysis, whereas their hydrochlorides stimulate the process. The effect is concentration dependent, and inactivation by primary amines is more pronounced than with tertiary amines.[2229–2232] *N*-Ethylmaleinimide had no effect upon the liquefaction activity of the enzyme from soy flour.[2233]

Inhibitors present in starch can be removed or deactivated by treating starch at low pH with phytase or acid phosphatase prior to, or simultaneously with, liquefaction.[2234] The inactivation effects on enzymes attributable to temperature changes and/or some components of the reaction mixture can be diminished and even eliminated by inclusion of either peptone, albumin, gelatin, glutathione, or casein. Papain and pepsin act similarly, but to a lesser extent.[2174,2235–2239] The stimulating effect of these additives can be perturbed by electrolytes. In the hydrolysis of sago starch, cellulase aids digestion by the alpha amylase from *Penicillium brunneum*.[2240] Furthermore, sodium chloride augmented the effect on amylolysis of proteases, whereas calcium chloride retarded it.[2241] The stimulation from added amino acids differs from one enzyme to another and depends also on the amino acid. Among seven amino acids tested, only alanine supported amylolysis, and the other amino acids initially accelerated the reaction, but later inhibited it. Another variable is the toxic effects of heavy metal ions.[2242–2245] Urea initially accelerated amylolysis, but this effect ceased with time.[2246] Autolyzed yeast readily liquefies starch because, together with amylolytic enzymes, it contains protecting proteases.[2247–2249]

In a coacervate of starch with protamine sulfate, gelatin showed a weak protective effect on the hydrolysis of starch by alpha amylase,[2250,2251] but questions of solution viscosity make this effect concentration dependent.[2217]

17. Engineering Problems

The suitability of starches for specific industrial processes has been evaluated.[2252] Factors influencing susceptibility to amylolysis include flow characteristics of slurries of nongelatinized starches, and changes in the viscosity and microscopic appearance of slurries on heating. A link between the type of alpha amylase and the molecular weights of the resulting dextrins has been noted. It enabled the optimization of the alpha amylolytic conversion of starch into glucose.[2253]

Several improvements in the apparatus used for the starch conversion have been proposed.[2254] Already in early 1970s, comparative studies[2255–2257] confirmed the advantage of membrane reactors over solid-walled reactors. Semipermeable membranes allowed separation of enzymes (glucoamylase, alpha and beta amylases) based on osmotic considerations, enabling their reuse and permitting the products of hydrolysis to pass through.

Fixed-bed reactors (see, for instance, Refs. 2258, 2259) were once in common use, and the critical problem of controlling the process rate by mass-transfer limitations was attempted by the introduction of pulsed-flow[2260] and expanded-bed techniques.[2261] In the latter type, two-phase liquid expanded-bed, and three-phase air expanded-bed, solutions were examined. Two-phase reactors performed better, offering 20 days of continuous operation with barely a 6% decrease in the amount of conversion, which reached 96%. An aqueous two-phase reactor employing PEG-dextran was also patented.[2262] The rotational speed of agitation was crucial for maximizing the hydrolytic yield. It need to be controlled, as under more-vigorous rotation the production of glucose declined. Reactors employing rotating inner containers improved the process.[2263–2265] Agitation facilitates swelling of the granules, and this type of swelling differs from that caused by cooking.[2266,2267]

Further progress in processes for starch hydrolysis was achieved when packed-bed reactors were replaced with fluidized-bed reactors.[2268] In such reactors, the enzymes were used either free or immobilized. Immobilized enzymes usually had higher thermal stability, but the Arrhenius function for the reaction with either immobilized or free enzyme was identical. Based on the criteria of external mass-transfer limitation, dispersion effects, enzyme activity, and operational stability, the fluidized-bed reactor appears to be superior.[2268–2274] Models proposed by the authors[2272] are valid for cassava and other liquefied starches at various concentrations. In some extensions, immobilized microorganisms, for instance, yeast for hydrolysis of maize starch,[2275] *Escherichia coli* BL 21 for production of glucose from maltodextrins,[2276] and *Lactococcus lactis* IO-1 for production of lactic acid from sago starch,[2277] were also employed.

Membrane reactors have been widely explored.[2278–2291] These reactors were designed to work with various substrates, operated with either free enzymes[2278,2280,2281,2283,2287,2289,2291,2292] or immobilized ones,[2279,2282,2283] including glucoamylase,[2278,2281–2284] alpha amylase,[2279,2289,2291–2293] beta amylase,[2285] beta amylase with pullulanase,[2274] beta amylase with isoamylase,[2293] and glucoamylase and pullulanase.[2294] Reversed osmosis through the membranes enabled the separation of the glucose, maltose, and maltotriose mixture.[2295] Reasons for the loss of enzyme activity during the hydrolysis of starch have been addressed by Paolucci-Jeanjean et al.[2288,2290] That problem is encountered in continuous recycling membrane reactors. The effects of temperature and denaturation by adsorption on the membrane are the most important factors responsible.

An electro-ultrafiltration bioreactor has been proposed.[2296] The electric field caused foaming and denaturation. Multistage bioreactors equipped with ultrathin layers of immobilized glucose oxidase and glucoamylase have also been described.[2297,2298]

For production of maltose[2299] and sucrose,[2300] hollow-fiber reactors were employed. They provided pressure filtration followed by isolation of the products by reversed osmosis. The reactors operated with immobilized amyloglucosidase,[2299] glucoamylase,[2300] invertase,[2299] alpha amylase,[2301–2303] beta amylase and isoamylase,[2304] and alpha amylase plus isoamylase.[2286] Alpha and beta amylases were used for liquefaction and saccharification, respectively.[2305] Such reactors suffer from external and internal diffusion limitations and inhibition by the products of hydrolysis.[2299] Also other types of reactors have been described, for instance, a scraped-surface heat exchanger as an enzyme reactor for the hydrolysis of wheat starch.[2306]

Hydrolysis in extruders has also been investigated. The water content in the extruded material controls the torque and energy requirements of the operation. With a high water content, the so-called wet extrusion, the extruders are conveniently also used as bioreactors. On extrusion of starch, enzymatic liquefaction and saccharification provides syrups of high DE.[2307] At a fixed flow rate and temperature, the extent of conversion depends on the moisture content, residence time, and level of enzyme added.[2308] Continuous hydrolysis of barley, corn, and wheat starches, and also potato peel, was performed with alpha amylase in the presence of calcium ions inside an extruder boiler,[2309–2313] and with alpha and beta amylases with either starch or flours in a single-screw extruder.[2314] Corn starch was hydrolyzed in a twin-screw extruder. In the first two barrels, gelatinization and liquefaction occurred without enzyme. In the third barrel, the hydrolysis then proceeded with alpha amylase.[2315] The process has been optimized with bacterial and fungal amylases in a corotating twin-screw extruder.[2316,2317] Rice bran could be made soluble after two-stage extrusion cooking with alpha amylase.[2318] A countercurrent chromatographic bioreactor was simulated[2319] for continuous saccharification of modified starch and the continuous biosynthesis of dextran from sucrose. This reactor offered simultaneous separation of the products. A cloth-strip reactor operated with glucoamylase chemically bound to poly(ethyleneimine)-coated cotton strips.[2320] This reactor could hydrolyze starch continuously for 21 days.

Direct resistive heating for continuous cooking and liquefaction of starch has been proposed. Alpha amylase of *B. subtilis* was used.[2321]

An immobilized enzyme catalyst for the saccharification of starch has also been designed, and a mathematical model based on spatial characteristics of the immobilized enzyme bed was presented.[2322]

The viscosity of the processed solutions is an important factor which controls diffusion, heat exchange, mixing, and so on. In the course of enzymatic starch conversion, the viscosity of the medium changes. Because of the complexity of the mechanism, this change cannot be used for monitoring the progress of the process. The changes in viscosity are controlled by the enzyme activity and the rate of deactivation, as well as the temperature. A block diagram has been presented correlating these effects upon the changes in viscosity.[2323]

A continuous starch-liquefying ejector is considered a key item of equipment in any manufacturing plant making products by starch hydrolysis and/or fermentation. Saccharification of starch with enzymatic cocktails provides uniform liquefaction, high yields of pure glucose, facilitates filtration, and produces a filtrate that is only slightly colored.[2324] Dielectric heating of starch with amylase and 10–35% water causes hydrolysis.[2325]

18. Applications of the Enzymatic Processes

a. Starch Isolation and Purification.—Swelling of starch granules and their bacterial and/or enzymatic digestion begins already at the stage of isolation.[2326] This factor can influence the quality of the starch product designed for commercial purposes. However, when the starch is destined for further liquefaction and saccharification, this stage can be modernized and employed efficiently for the intended conversion.

An enzymatic cocktail of cellulose, endo-$(1 \to 3),(1 \to 4)$-D-glucanase, and xylanase applied on isolation of starch from naked barley, significantly decreased the viscosity of the slurry and, therefore, less ethanol was needed for precipitation of β-glucan from the reaction mixture. Although the yield of the glucan was decreased, the yield of extracted starch was higher by 10%. The physicochemical characteristics of starch so isolated did not change.[2327]

Starch from dehusked cereal grains can be isolated by using enzymes that digest the nonstarch components of the grains.[2328] The wet-milling process can be modernized by performing it with enzymes. It decreases the time of the operation and eliminates the use of sulfur dioxide.[2329] When alpha amylase is added to a 30–32% slurry of dry-ground corn, the centrifugation and filtration steps are facilitated.[2330] The use of lysophospholipase, β-glucanase, and pentosanase (a xylanase preparation from *Trichoderma*) for this purpose has been proposed.[2331] The simultaneous action of alpha amylase and lysophospholipase increased the filtration rate by only 12%, but a cocktail of lysophospholipase, β-glucanase, and pentosanase enhanced the rate of filtration by the order of 2.5.

Fiber-degrading enzymes facilitate the wet milling of corn and sorghum starch and allow a shorter steeping time.[2332,2333] To isolate potato starch from sliced tubers, aerobic bacteria have been employed. The optimum time for their contact with the slices was only 1 h at 40 °C.[2334] To produce rice starch from the milling process, proteases acting at pH 7–12 and 40–70 °C gave the starch in high yield.[2335] Enzymes have been used in the isolation of rye starch in order to remove mucus and fiber from the substrate.[2336,2337] To isolate the residual native starch remaining in sago pith, the cell-wall degrading pectinase from *Aspergillus aculeatus* was used. Acceptable results were observed after only 30 min. Small granules were preferentially digested.[2338]

Starch from sweet potato was isolated with participation of the cellulase from *Trichoderma viride*, which degraded cellulose, hemicelluloses, and protopectin. The yield of starch was increased by 0.7–2.0%. The isolated starch contained slightly elevated contents of ash, crude protein, and small granules.[2339] Optionally, the enzyme isolated from *T. viride* or *Aspergillus niger* could be used in conjunction with the enzyme isolated from *Rhizopus*.[2340] An enzyme isolated from *Clostridium acetobutylicum* was also suitable. Starch isolated according to this procedure was characterized by a slightly lower blue value than starch isolated without the intervention of that enzyme.[2341] Pepsin employed in the isolation of wheat starch from the flour at pH 1.9–2.2 and 40 °C provided pure starch and a fraction of solubilized protein.[2342] Some additives to the fermentation broth may influence the enzymatic action. For instance, soluble vitamins, particularly vitamin C, activated amylase. Gelatin and casein had no effect on the enzyme, but gluten slightly inhibited the amylase action, as demonstrated by processes employing wheat flour.[2343] Isolation of rice starch from rice flour requires the use of a protease as well as alpha amylase. The milling of rice and other steps should be performed in the pH range of 7–12.[2339]

Enzymatic digestion of starch is a convenient mode of pretreatment prior to use of the starch in various chemical modifications. A suitable approach calls for treating starch in an aqueous emulsion with amylase, first at 55–65 °C and pH 5.0–6.6 for 2–5 h, and then at pH 9.0–11.0 for from 30 min to 2 h.[2344]

Amylolysis of starch in refrigerated food with alpha amylase and glucoamylase improves its texture, increases the glucose content, lowers the freezing point of food thus preserved, and lessens the aging of starch on storage.[2345]

Microporous granular starch, used as an adsorbent, is generally produced by digestion with alpha amylase.[2346–2348] The adsorbability effect depends on the botanical origin of the starch. Comparative studies showed that microporous cassava starch was a better adsorbent than microporous corn starch.[2347] Further studies[2349] on the production of microporous starch showed that use of glucoamylase was superior to alpha amylase and pullulanase. However, a cocktail of glucoamylase with alpha amylase was the most efficient in this respect. Globular amylose of 2–10 μm particle diameter and a polydispersity of 1.4–1.7 was prepared as an adsorbent for the food, pharmaceutical, and cosmetic industries. It was made with CGTase from either CD or starch.[2350] Porous starch powder, a sorbent for liquid fat, was made by treating raw starch with hydrolases.[2351,2352] A porous starch made by cross-linking starch with chloromethyloxirane, $POCl_3$, sodium trimetaphosphate, adipic acid, a mixture of citric acid with acetic anhydride, formaldehyde, or acetaldehyde has been patented.[2353] The cross-linked substrates were digestible by alpha amylase. Adsorbents for removing color and odor can be made from enzymatically produced branched dextrins of average molecular weight of 25 kD.[2354]

Alpha amylase has also been employed for making porous films of biodegradable "green" polymers, such as those made from synthetic polymers blended with starch.[2355–2358] Decreasing the proportion of amylopectin in starch gels by using pullulanase provided films of high tensile strength.[2358]

Because neither alpha amylase nor mycelia exert any pectolytic activity, they could be used for removal of contaminating starch from pectin preparations. Either the alpha amylase from *A. niger*[2359] or mycelia of *A. oryzae*, *Rhizopus oryzae*, *R. japonicas*, or *Mucor rouxianus* provided a successful one-step operation.[2360,2361] For the extraction of biologically active compounds from roots of *Panax ginseng* and *Panax notgoginseng*, the starch therein was digested with alpha amylase.[2362] This enzyme in a cocktail with glucoamylase was also employed for the degradation of starch in flue-cured tobacco.[2363] Biologically active components, present in association with starch, in *Pueraria*, a herb from Thailand, were extracted with water. The extract was freed from that starch by means of alpha and beta amylases.[2364] The fractionation of cereal bran required first a decrease in the starch content by the use of alpha amylase, and subsequent hydrolysis of the crude bran with xylanase, arabinase, β-glucanase, cellulase, hemicellulase, and/or pectinase.[2365] A hypoallergenic wheat flour was prepared by using enzymes.[2366] The process involves enzymatic fragmentation of the allergens.

b. Starch in the Pulp Industry.—The pulp and paper-making industry utilizes large amounts of starch and its derivatives, particularly cationic starches as fillers and binders. Moreover, starch hydrolyzed specifically for that use[2367–2373] has found its application as a paper coating. Hydrolysis can be conducted in either acidic or alkaline media, or in an enzyme-assisted manner. Amylolytic enzymes are pH sensitive, but they can be stabilized by calcium cations.[2367,2374] Conversion performed in the presence of urea allows[2374] processing at pH values up to 8.8. Sizing compositions are usually made from raw starch hydrolyzed with amylolytic enzymes and such salts as copper sulfate, zinc acetate, or titanium(III) chloride. They are used in stoichiometric amounts with respect to the enzymes used in order to inactivate them without increasing the viscosity and lowering the pH of the sizing product.[2375] Dextrins resulting from the hydrolysis can also be combined with a clay.[2376] The tear strength of paper was enhanced after treatment with a hydrolyzate of potato starch.[2377] A US patent[2378] describes a preparation for sizing paper made from (2-hydroxyethyl)starch blended with a styrene copolymer by use of either disodium maleate or maleic monoamide monoammonium salt, and enzymatic digestion of the product as a slurry.

Cationized starches can also be enzymatically degraded with alpha amylase. Paper containing such strengthening components disintegrates readily and is suitable for recycling.[2379,2380] Paper for ink-jet printers usually disintegrates with more difficulty than common office waste paper. To make the paper more readily recyclable, a special approach is needed to disintegrate the fiber, and treatment with amylase at the production stage of the pulp is advisable.[2381] Semi-aqueous amylase poulticing methods were tested to detach "silking" from archival documents weakened by corrosion from the gall ink used.[2382]

Paper is sometimes coated with kaolin. The kaolin is suspended in gelatinized starch, which then is treated with alpha amylase. Amylolysis of gelatinized starch prior to admixture with kaolin is also possible.[2383]

c. Textile Sizing and Desizing.—The textile industry consumes large amounts of starch for use as a size. Enzymatic reactions are used for sizing, and also for desizing, that is, for the solubilization of starch and washing the size out.[2384,2385] Various types of amylases are utilized for desizing. Pancreatic amylase rapidly hydrolyzes starch to maltose, and this process proceeds at 55 °C, which is below the swelling temperature of starch and it functions at pH 6.5 regardless of the type of fiber and dye involved. Malt amylase functions between 60 and 65 °C under slightly acidic conditions. Bacterial enzymes, mostly those of *Bacillus subtilis*, react in the range of 75–90 °C, and so they attack swollen starch. Dextrins result from this reaction.[2385] A high-solid starch adhesive for paper coating was prepared from corn-starch slurry with alpha amylase from that microorganism.[2386,2387] The efficiency of these bacteria to produce the enzyme could be increased by adding about 6% of calcium salts such as chloride, sulfate, or citrate, with chloride being the superior additive.[2388] An adhesive was patented in Japan based on starch hydrolyzed with cyclodextrin glucotransferase.[2389] An aqueous composition of enzymatically hydrolyzed starch produced in the presence of octadecanoic (stearic) acid was also suitable for coating paper to make it more resistant to water and to mechanical stress.[2390] Hydrolyzing corn starch to give a material suitable for tub sizing was also discussed.[2391]

Glucoamylase and beta amylase were also tested for their potential in desizing. In contrast to alpha amylases, which under selected reaction conditions produce dextrins of low molecular weight, beta amylase provides mainly maltose, and amyloglucosidase gives maltose, maltotriose, and other maltodextrins.[2392] A mixture of highly and moderately thermostable amylases has also been used. This procedure permits desizing at a lower temperature and, because of synergism, the use of less enzyme.[2393] A cocktail of alpha amylase and glucoamylase to hydrolyze raw starch slurry was patented[2394] and recombinant alpha amylase mutants, variants of that enzyme from *Bacillus licheniformis*, have also been patented for textile desizing.[2395] Between potato, corn, and rice starch, the hydrolysis of rice starch was the slowest.[2396]

The extent of enzymatic desizing is controlled by the type of wetting agent used, and this agent should preferably be nonionic. Supplementation of the alpha amylase-containing preparation with either hydrogen peroxide or cellulase is sometimes advantageous.[2397] The desizing of cotton fabrics with amylosubtilin G 3X was recommended,[2398] and the process should be conducted between 70 and 80 °C using $CaCl_2$ as the stabilizer. For desizing cellulose-containing fabrics, an aqueous composition comprising the peroxidase from *Basidiomycetes coriolus hirustus*, NaCl, $MgSO_4$ as an enzyme stabilizer, and N-C_{10-13}alkyl-N,N'-bis(polyoxyethylene)-1,3-trimethylenediamine as a wetting agent was patented.[2399] The loss of enzymatic activity on desizing is caused by ionic surfactants. This effect can be neutralized by addition of cyclodextrins.[2400] On the other hand, the reaction can be stopped by addition of phenol and ethylenediamine after the product achieves the desired viscosity. These additives inhibit the adhesive from losing viscosity on storage.[2401] Starch in waste water from offset printing can be removed by using enzymes.[2402]

d. Adhesives.—Attention has been directed to enzymatic conversions of starch to provide adhesives. They are prepared from starch[2403] or grain flour,[2404] with alpha amylase activators (surfactants) added.[2403,2405] They act as antigelling and complexing agents. High-solids, storage-stable adhesives based on maltodextrins can be obtained by blending a maltodextrin syrup with poly(vinyl acetate) and/or ethylene vinyl acetate copolymer. Maltodextrins useful as remoistenable adhesives are obtainable from hydroxypropylated waxy maize starch hydrolyzed with alpha amylase.[2406] The blending of hydrolyzed starch with such vinyl polymers as poly(acrylamide) provides a paste for gummed tapes.[2407]

Proteinases are employed for releasing and/or separating starch from gluten.[2408] In using wheat-flour hemicellulose, the joint action of cellulase and proteinase increases the yield of gluten. The effect of combining these enzymes depends upon the type of flour.[2409,2410] The effect of endoxylanases and arabinoxylan upon coagulation of gluten was studied.[2411] Material for adhesives can be prepared by stepwise, multiple hydrolysis of starch with alpha amylase.[2412]

e. Washing and Cleaning.—Starch-based washing and cleaning aids have been proposed, for instance, a combination of starch (500 g), amylase or sucrase (15 g), proteinase (10 g), and lipase (10 g) in water (1500 mL).[2413] The use of enzymes for removing starch from glass plates was found to be temperature dependent. Between 47 and 57 °C, the enzymes worked satisfactorily, but above that range some residues remained on the glass surface and normal alkaline detergents performed better.[2414] A cleaning agent containing a novel enzyme isolated from *Bacillus agaradherens* DSM 9948, which produces CGTase, was found suitable for cleaning textiles and hard surfaces.[2415] Mutants of *B. licheniformis* were proposed for use in the production of laundry and dishwasher detergents.[2416] Detergents for removal of starch stains in laundering and dishwashing have been designed; they incorporated the alpha amylase from *Bacillus licheniformis* together with such polymers as poly(vinylpyrrolidone), poly(vinylimidazole), and poly(vinylpyrrolidone *N*-oxide) as dye-transfer inhibitors. Such detergents might also incorporate either hydrogen peroxide or peroxidase.[2417] These detergents can also contain alkali metal peroxycarbonates,[2418] such peroxycarboxylic acids as 1,12-diperoxydodecanedioic or phthalimidoperoxyhexanoic acid and their alkali metal salts[2419] as well as ingredients already listed. Instead of peroxycarbonates and peroxy acids, these detergents may contain quaternary ammonium salts,[2420] anionic and nonionic surfactants,[2421] the alpha amylase of *Bacillus amyloliquefaciens*, and a protease from *Bacillus lentus* optionally genetically modified,[2422,2423] and transition-metal complexes that activate bleaching.[2424] Enzymes can also be utilized for removing contaminations from ion exchangers.[2425] Starch treated with pullulanase has good flow properties for use in color coatings.[2426]

f. Pharmaceutical Industry.—The pharmaceutical industry has a major demand for excipients, drug microencapsulating preparations, and drug carriers. Amylose resulting from the debranching of amylopectin was considered suitable as an excipient, as it affords superior compressibility and structural integrity to the tablets. Debranching employed α-1,6-D-glucanohydrolase.[2327–2429] Quick-release compositions were prepared from corn starch, alpha amylase, soybean oil as a plasticizer, lactose for making the material more porous, and the pharmaceutically active agent. The whole mixture was extruded.[2430] Encapsulating agents were prepared by hydrolyzing, with either beta amylase or glucoamylase, starches that had been chemically modified with (1-octenyl)succinic anhydride.[2431,2432] Acetylated starch with alpha amylase incorporated,[2433,2434] and starch cross-linked with epichlorohydrin or phosphates and with alpha amylase added, were also suitable materials.[2435] Amylase could be immobilized by periodate oxidation of blends of the enzyme with starch.[2436]

A hydrophobic starch powder used as adsorbent in cosmetics was prepared by a short-duration treatment of granular starch with glucoamylase.[2437] Corn starch is known for its content of lipids, which form innate complexes with starch and are therefore difficult to remove. Digestion of native corn starch with alpha amylase and glucoamylase, followed by extraction of the lipids liberated, permits the lipid content to be decreased from an initial 619 mg% to 97 mg% after 95% conversion of starch into glucose. The fatty acids extracted from starch were mainly unsaturated.[2438] Hydrolysis of octenylsuccinylated starch with glucoamylase, beta amylase, or pullulanase with beta amylase, provided a material useful for encapsulation of various active agents which, after such encapsulation, exhibit good bioavailability.[2439]

g. Food Industry.—The food industry makes wide use of enzymatic reactions in many fields of food production, employing several different enzymes.[2440] Alpha amylase, hemicellulases, cellulases, lipases, and optionally proteases are used in an increasing number of foodstuffs, bakery products in particular. The stability of products through control of staling, shelf-life, and freshness is maintained with alpha amylases and hemicellulases. For developing the desired texture of foodstuffs, alpha amylases, hemicellulases, and lipases are useful.

The color of bakery products and a bleaching effect are imparted by employing alpha amylases and lipooxygenases, and optionally hemicellulases also. Alpha amylases, lipases, proteases, lipoxygenases, and glucose oxidases aid in developing flavor. The nutritional properties are improved with hemicellulases, and overall quality is enhanced by each of the enzymes just mentioned. Alpha amylase added to cooked rice imparts luster and improved texture.[2441]

(i) Bakery Production. Baking is one of the most important processes where the quality of the products depends strongly on the enzymes employed. The origin of the alpha amylases used is an essential determinant of the rheology of the dough and the manner of its processing.[2442,2443] The use of alpha amylase in conjunction with glucoamylase enhances the bread quality.[2444] The selection of the enzymes used is also important. For instance, in the baking of sponge dough, fungal preparations can be used provided that, in the proteinase to alpha amylase ratio, the first does not prevail. Sometimes, an excessive amount of alpha amylase may give rise to a sticky and gummy bread crumb because of the low affinity of that enzyme to dextrins. That fault can be minimized by decreasing the water content.[2445] The use of high-amylose wheat flour in dough resulted in the formation of resistant starch (RS) on baking.[2446] The selection of an enzymatic cocktail rich in either proteinase or alpha amylase depends on the character of the flour.[2447] Bread dough that is deficient should be supplemented not only with alpha amylase and proteinase but also with pentosanase and such swelling agents as corn starch, guar flour, galactomannan of carob bean flour, or pectin.[2448] Jimenez and Martinez-Anaya,[2449] working with eleven commercial enzyme preparations, paid particular attention to their pentosanase and/or amylase activity. They looked for relationships between their action and the functional properties of the resulting breads. They found that the preparations were mainly of the endo-type. The dextrinogenic properties of the preparations depended on the substrate and the enzyme characteristics. Part of the pentosans were solubilized in the preparations. The amount of dextrins of low molecular weight correlated with some physical properties of fresh bread crumbs, and the level of water-insoluble pentosans correlated with crumb elasticity and hardness during storage. The enzymatic action induced changes of the level of soluble starch and solids. Dextrins of low molecular weight inhibited these changes and permitted retention of the bread texture.

All alpha amylases provided bread with a darker crust and decreased the effect of resting and mixing time upon the loaf volume.[2450] Enzymes digesting raw starch in bread increased the level of glucose, changing the crust color and increasing the loaf volume.[2451] They had no effect on the rate of firming of bread. A high level of enzyme weakens the side walls of the loaf. Enzymes retard the firming of bread crumbs. Because alpha amylases convert starch into water-soluble dextrins comprising maltose–maltohexaose, they probably act by decreasing the level of starch that could retrograde.[2452] Formation of dextrin–starch–protein complexes must also be taken into account.[2453] Enzymatic cocktails containing bacterial and fungal alpha amylases supplemented with lipase (which is also known for retarding the staling of bread) have also been used.[2454] Such additives exert a positive effect upon the rate of crumb firming, crumb springiness, and amylopectin recrystallization.

Table XIV presents characteristics of nine commercial enzymes and their effect upon bread making, based on the fermentation time, the bread volume (Fig. 19), and the sensory properties of the bread (Fig. 20).[2455]

It may be seen that all admixed enzymes decreased the fermentation time, and the combination of pentosanase, hemicellulase, and amylase performed best. Except for the fungal amylase of *A. niger*, all other enzymes tested increased the bread volume, and the blends of fungal amylase with pentosanase, as well as pentosanase with hemicellulase and fungal amylase, were the most efficient.

As a rule, added enzymes improved the bread aroma and overall acceptance, although the effect of those enzymes upon such other sensory properties as taste and its intensity, as well as aroma and its intensity, was diverse. The lowest overall

TABLE XIV
Summary of Commercial Enzyme Characteristics

Sample No.	Enzyme[a]	Principal activity	Source	Actvity	Dose (mg/100 g)
1	None (control)				
2	Pentopan 500	Pentosanase, hemicellulase	*Humicola insolens*	700 FXU/g	6
3	Fungamyl SHX	Pentosanase, fungal amylase		60 FAU/g	16
4	Fungamyl	Fungal amylase	*A. niger*	800 FAU/g	5
5	Novamyl	Maltogenic amylase	*B. subtilis*	1500 MANU/g	30
6	Pentopan 500 + Fungamyl	See each component			6 + 5 resp.
7	Pentopan 500 + Novamyl	See each component			6 + 30 resp.
8	Fermizyme H14001	Pentosanase, amylase, hemicellulase	*A. niger*	150 FAU/g	18
9	Fermizyme HL 200	Pentosanase, amylase, hemicellulase[b]	*A. niger*	150 FAU/g	35
10	Fermizyme FFCP	Pentosanase, amylase, hemicellulase[c]	*A. niger*	150 FAU/g	10
11	Fermizyme 200	Amylase	*A. oryzae*	4500 FAU/g	1
12	Pluszyme IV	Hemicellulase, fungal amylase			18

[a] Enzymes 2–5 are from Novo Nordisk (Denmark), enzymes 8–11 are from Gist Brocades (The Netherlands), and enzyme 12 is from Tecnufar Iberica (Spain). [b] Fermentation of derivatives as a secondary activity. [c] Protease as a secondary activity.
(From Ref. 2455)

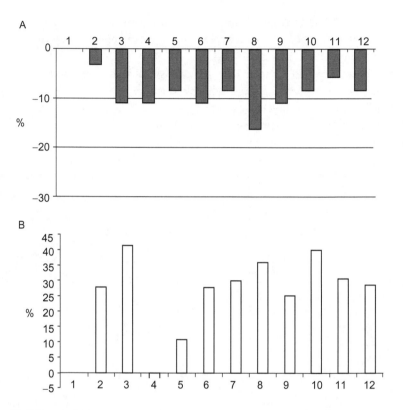

FIG. 19. Effect of added enzymes upon the fermentation time (upper) and bread volume (lower). All data are in percent.[2455] See Table XIV for notation.

appreciation was for breads that have a more open and irregular grain. They were prepared with blends of pentosanase, hemicellulase, and fungal amylase.

A patent[2456] describes that adding dry yeast to the dough adds a specific flavor and aroma to bread that is not achieved by using fresh yeast. It has also been found[2457] that enzymes of amylolytic activity between 100 s < FFN < 200 s (where FFN denotes the Fungal Falling No.) provides bread loaves having the largest volume and superior sensory properties. Among fungal glucoamylases, the one isolated from *Saccharomyces fibuligera* IFO 0111 (glucoamylase Glm) is the only glucoamylase that can cleave native starch, as demonstrated in studies conducted on wheat and rye flour under conditions simulating kneading and ripening at 28 °C for 55 min.[2458] When rice flour is added to wheat flour dough, more alpha amylase should be used, and a fermented flavor mixture made of sour milk, sugars, and other ingredients is recommended as an additive to mask

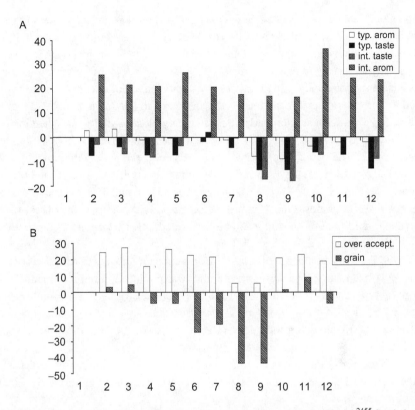

FIG. 20. Influence of the addition of enzyme on the sensory characteristics of bread.[2455] Symbols: (A) Bars denote sequentially the type of aroma, type of taste, taste intensity, and aroma intensity. (B) Open and solid bars denote overall appreciation and remaining grains, respectively. See Table XIV for notations.

the rice taste and color.[2459] Other compositions which, optionally, can enable lactic acid fermentation during raising of the dough, are proposed.[2460] For microwave baking, a preparation made of a gum and an enzyme has been suggested.[2461]

During dough making, the digestion of starch by amylase is negligible. However, the endosperm rich in alpha amylase furnishes sufficient sugar for the yeast to digest that fraction of starch solubilized by grinding.[2462,2463] Denaturation of alpha amylase takes place between 65 and 90 °C, but the conversion of starch went forward between 57 and 83 °C.[2464] Thermal inactivation of beta amylase occurs between 57 and 72 °C together with the conversion of starch. The joint action of both enzymes provides a higher level of conversion than that anticipated by summation of the effects of single enzymes. The action of both enzymes decreases the amount of dextrins formed. There is no mutual interaction between these enzymes, as their thermal stabilities and times of denaturation are the same, regardless of whether they act jointly or separately.

Gelatinization is an important factor for the fermentation in dough by yeast. Damaged granules gelatinize more readily. Obviously, the rate of hydrolysis of starch depends on its botanical origin. Starches low in branched polysaccharides are hydrolyzed faster. Starch in bread hydrolyzes more readily than in dough, but bread obtained by acid fermentation undergoes slow amylolysis initially, and then it slows down further to the level typical for bread designed with conventional yeast fermentation.[2465,2466] The handling of dough, its consistency, the temperature, acidity, and the intensity of kneading have no influence upon the activity of amylase and the rate of starch decomposition. However, on baking for 30 min at 240 °C, a direct relationship between starch degradation and the enzyme activity and pH was observed.[2466] Enzymatic modification of starch can be performed either before its use for making dough, but also in the dough itself.[2467]

A hybrid enzyme comprising *B. licheniformis* endoamylase fused with *B. flavothermus* amylase and having better thermostability was produced. It improves batter-cake dough and the quality of bread.[2468] Maltogenic amylase (glucan 1,4-α-maltohydrolase, EC 3.2.1.133) from *B. stearothermophilus* has an unusual impact upon the molecular and rheological properties of starch by its effect on the rheology of starch slurries. It is opposite to the effect of other endo-amylases. This effect results from specific digestion of the amylase component, which is responsible for monodispersity of the slurries. This factor can be essential in controlling rheology of the dough and bread making.[2469]

The activity of amylases depends on the available moisture. In air-dried flour (5.1% moisture) as well as in fresh bread this level of moisture is sufficient for enzyme activity.[2470] Alpha amylolysis of grains of low moisture content produces chiefly glucose and maltose. At moisture levels insufficient for the growth of microorganisms, the activity of enzymes decreased, and abnormally large amounts of sugars directly fermentable by yeast are produced.[2471]

For sweet bakery products, for instance biscuits (cookies), the starch should be partially saccharified in conjunction with sucrose added. Then the process is performed using alpha amylase with amyloglucosidase.[2472] Thermostable alpha amylase is suitable for modifying starch *in situ* during the making of biscuits, crackers, and wafers on hot surfaces.[2473]

A Japanese patent[2474] describes an emulsifier-free composition containing hydrogenated and nonhydrogenated rapeseed oils, palm oil, "cross-linked α-starch," and alpha amylase suitable for the making of bread of good moisturized texture.

The staling and retrogradation of bread is a consistent problem. Generally, retrogradation inhibits amylolysis,[2475,2476] and alpha amylase decreases the rate of staling and firming.[2477,2478] It appears[2479] that the rate of staling is controlled by the kind of alpha amylase added. Amylases decrease the crystallinity of starch in bread following the order: bacterial > cereal > fungal. The order of the efficiency of these amylases upon firming is the opposite. Similar studies performed on concentrated wheat-starch gels provided the opposite order of the effect of these amylases on firming.[2480,2481] The use of mixtures of amylases from different sources is recommended for inhibiting staling.[2482] This effect seems to be related to the ability of amylases to convert amylopectin. A later source[2483] claims that the effect of amylases upon staling and retrogradation does not depend on whether the amylase is of bacterial or fungal origin. The alpha amylases of *B. subtilis* and *A. oryzae* had limited impact on the conversion of amylopectin, whereas there was a notable effect with porcine pancreatic and *B. stearothermophilus* maltogenic amylase.

When alpha amylase was supplemented with lipase, the volume of loaves increased significantly, together with the rate of staling. On the other hand, lipase alone retards retrogradation.[2484] The effect of amylases on staling has been discussed from a mechanistic viewpoint.[2485]

Emulsifiers decreased the firmness of bread, but had no effect upon the initial susceptibility of starch to the enzyme within first hour after baking or the rate of staling.[2486] Staling decreases the susceptibility of bread crumbs to beta amylase, and this susceptibility in terms of amount of the hydrolysis products and rate of their formation decreases with staleness.[2487]

Workability and storability of bread can be enhanced when whole wheat flour is enzymatically degraded in order to hydrolyze exclusively the carbohydrates, leaving other flour components intact. Thus, the flour is first digested in the presence of Ca^{2+} with thermostable bacterial alpha amylase at 25 °C and then 95 °C at pH 6.0. At this stage, the total carbohydrates comprise 2% glucose, 42% maltose, and 22% maltotriose. Subsequently, this product is exposed to hydrolysis with immobilized beta amylase and glucoamylase at 55 °C and pH 5.4 to furnish a product for making bread containing 72% glucose and 22% maltose.[2488]

Amylases form complexes with CO_2. Their performance with respect to starch in wheat is not uniform. Some samples of wheat produce chiefly sugars of low molecular weight and little gas, while from other samples mainly gas is evolved.[2489] This property can have practical impact on controlling bread properties.

Amylase-containing flour preparations are used as improvers of starch liquefaction on baking. Their efficiency can be enhanced by adding salt.[2490]

(ii) Fruit and Vegetable Juices and Pomace Processing. The diminution of starch content in crude cane juice[2491–2494] was performed with the enzymes of *Bacillus licheniformis, Aspergillus oryzae*,[2494] as well as *Bacillus subtilis*.[2495] Fungal microorganisms performed better.[2496] Thermostable amylases from *Bacillus subtilis* permit 85% degradation of that starch within 15 min.[2495] Other alpha amylases can also be used in immobilized form.[2497,2498] Amylolytic removal of starch from fruit juices can be performed with amylase, pectinase, and naringinase,[2498] but glucoamylases seem to have been more frequently used.[2499–2501]

Pectins usually inhibit action of hydrolases, and so if pectins are present, increased amounts of hydrolases are recommended for complete hydrolysis of starch.[2502]

Treating fruit juices with CGTase increases their oligosaccharide content. Oligosaccharides are also formed from starch by that enzyme.[2503]

In the removal of starch from pomaces, for instance fresh or dried apple pomace, a problem arises from the presence of pectin and pectic acids, which inhibit the digestion of starch by diastase. The recommended optimum conditions for the hydrolysis are 30 °C at pH 3.2–3.8, depending on the pectic acids present. At higher temperature, the hydrolysis of starch accelerates, but it is accompanied by decomposition of the pectin and a consequent decrease in viscosity of the pomace.[2504]

(iii) Sweeteners. Starch for the production of sweeteners does not need to be fully saccharified.[2422] It is beneficial to terminate the enzymatic hydrolysis at the oligosaccharide stage. For this purpose, starch can be treated with the alpha amylase of malt and then the product is debranched with pullulanase to provide a syrup composed of 4.2% glucose, 13.1% maltose, 7.9% maltotriose, 7.2% maltotetraose, 7.7% maltopentaose, 40% maltohexaose and maltoheptaose, and 19.9% higher maltosaccharides. Such a syrup is suitable for sweetening beverages and food products.[2505] A patented process[2506] to produce a sweetener for food and beverages involves treating starch with alpha amylase, pullulanase, and/or isoamylase to yield a mixture of maltohexaose and maltoheptaose, which is reduced by hydrogen over Raney nickel to give the corresponding alditols. A strategy based on a 5 min liquefaction of potato starch at 130 °C with alpha amylase, followed by storage at ambient temperature for 24 h, and then saccharification for 24 h with takaamylase was also patented.[2507] Sweetening can be effected enriching foodstuffs with maltotetraose made available from starch by using alpha amylase.[2508] When starch was first digested with a bacterial liquefying amylase at pH 5.0–7.0 followed by *Streptomyces amylase* at pH 4.5–5.0, a blend of oligosaccharides composed of chiefly maltose and maltotriose, a minor amount of glucose, and oligosaccharides higher than maltotetraose, plus a dextrin and 75% of solids was formed. Such a blend was designed for a candy

production.[2509] *Streptomyces hygroscopicus* was also used.[2509–2511] Syrups resulting from amylolysis of starch can be supplemented with minerals, amino acids, and other ingredients to provide better color and taste.[2512] The use of alpha amylase for converting amylopectin in either sweet or waxy corn starch is patented.[2513] Thus, the enzyme is blended with substrate, $CaCl_2$ is added, and the whole is boiled or steamed, and then the enzyme is inactivated.[2514] There is a patent for production of sweeteners from corn using a combination of various hydrolases. In such a manner glucose, maltose, maltotriose, and fructose can be made available.[2515]

A hydrolyzate of hydroxypropyl corn starch, prepared by combined acid and alpha amylase hydrolysis, was proposed as a low-calorie sweetener.[2516] Oligosaccharides from starch hydrolyzed by 4-α-glucanotransferase and terminated on their reducing ends with *N*-acetylglucosamine attached by applying an *Escherichia coli* homogenate were patented as mild sweeteners.[2517] Glycosylation of starch by use of glycyrrhizin and CGTase yielded another low-calorie sweetener.[2518] The combined use of both amylases and glucoamylase, together with acid, converted corn starch into a non-crystallizing corn syrup of high sweetness.[2519] The process can be performed by initiating the hydrolysis with 1,6-α-glucosidase, followed by either glucoamylase[2520] or pullulanase,[2521] alpha amylase followed by either pullulanase,[2522] glucoamylase[2523,2524] or beta amylase,[2525] beta amylase followed by isoamylase,[2526] or alpha amylase and/or pullulanase, all of these being thermostable.[2527] Use of glucoamylase in conjunction with glucose isomerase affords the widely used glucose–fructose syrup.[2528]

Maltotetraose, proposed as a low-sweetness, retrogradation-resistant syrup, has evoked interest;[2529] it is used in food technology to improve properties. It also controls the growth of putrefactive bacteria in the intestine. It is produced in a continuous process utilizing an immobilized enzyme from a *Pseudomonas stutzeri* NRRL B 3389 mutant identified as maltotetrahydrolase (1,4-α-maltotetrahydrolase) (EC 3.2.1.60). The production of maltotetraose is supported by the pullulanase from *Klebsiella pneumoniae*. The enzymes are immobilized on porous chitosan beads.

(iv) Resistant Starch and Dietary Fiber. Resistant starch (RS) has received considerable attention as a substitute or rather a supplement of dietary fiber. It is by definition, resistant to enzymatic digestion, and it passes into the colon in either unchanged or slightly changed form. Apart from its therapeutic and prophylactic role in the diet, resistant starch has found a number of applications for improving the functional properties of foods, in particular cakes and other bakery products, where it reduces the tempering step.

Resistant starch accompanies normal starch in various proportions in the starch of various plants. Among cereal sources, native high-amylose corn starch is the most important source. Because of the considerable demand for RS, caused mainly by the fashion for low-calorie food, several processes have been developed to manufacture it on an industrial scale. The nonenzymatic ones involve several approaches, for instance, gelatinized starch is subjected to heat treatment under pressure. Commonly, RS is enzymatically developed in foodstuffs. Thus, starch is first digested with alpha amylase, gelatinized, and then debranched with pullulanase or isoamylase.[2530–2535] Optionally, high-amylose starch containing over 30% of amylose is treated with CGTase.[2536]

Fiber was preferably removed from potato starch by initial digestion with hydrochloric acid, followed by incubation with alpha amylase and glucoamylase,[2537,2538] and subsequent digestion with pectolytic and cellulolytic enzymes, followed by amylolysis.[2539] Potato dregs can also serve as a substrate, which are digested with alpha amylase and proteinase.[2540] Dietary fiber was prepared from wheat bran by multienzyme hydrolysis employing phytase, alpha amylase, beta amylase, neutral protease, and lipase.[2541] Wheat flour rich in resistant starch can be prepared from brown wheat starch blended with sucrose, on treatment with a microorganism, for instance, *Leuconostoc mesenteroides*.[2542] Controlling the time for hydrolysis of starch suspensions by pullulanase can produce meals either readily digested or resistant to digestion.[2543] An alternative approach involves treating starch with alpha amylase and pullulanase, followed by heat–moisture treatment.[2544]

Resistant maltodextrin is obtainable by acid-catalyzed hydrolysis of starch to dextrin, thermal conversion into pyrodextrin, followed by enzymatic hydrolysis.[2545]

(v) Starch Processing and Use of Enzymes in Food. It should be noted that the behavior of starch when incorporated into foodstuffs is different from that exhibited by starch in the isolated state. Starch in foodstuffs is accompanied by fiber, natural enzymes, and inhibitors of hydrolases. These substances, together with factors resulting from the method of processing starchy components (for instance, production of flour), influence the rate of starch hydrolysis therein.[2546]

Food of higher quality is frequently supplemented with modified starch additives. These are incorporated as taste improvers, fillers, stabilizers, texture modifiers, edible foils, gels, agents decreasing calorific value, preventing spoilage, and so on (see, for instance, Refs. 1657, 2443, 2547–2561). Oligosaccharides for promoting the growth of *bifidus* bacteria are available by treating a mixture of starch and lactose with α-glucosidase. The products are used as prebiotics in supplementing health foods.[2562]

Particular attention has been paid to improving rice-based food. The taste and flavor of cooked rice was improved by adding a mixture of amylase and cellulase which breaks down the cell walls of rice grains.[2563–2565] Rice starch having α-maltosyl- and/ or α-maltotetraosyl branches introduced at C-6 by glycosylation with maltose saccharides served as an antiaging agent that inhibits food hardening. However, the elasticity of cakes decreases.[2566,2567]

In the production of food, enzymes are frequently allowed to act on raw substrates. Thermostable α-glucanotransferase added to food reconstructs the starch structure, enhances storage stability, and improves taste.[2568,2569] Food thus supplemented has to be heated to 75–80 °C under pressure of 150–200 U. Isoamylase was used for debranching amylopectin to improve the quality of edible foil derived from starch.[2570,2571] Processing of grains and oil seeds with protease, alpha amylase, and a third enzyme optionally added depending on the processed substrate, generates very soluble food products for making beverages.[2572] B-Starch from wheat liquefied with saccharogenic amylase gave material for the production of "ammonia caramel."[2573] Wholemeal cereal flour is produced by treating that flour with alpha amylase, in optional conjunction with glucoamylase.[2574,2575] Bean jam had better taste and storage stability after hydrolysis of the starch therein with amylase.[2576] A hydrolyzate of waxy corn starch, after saccharification to DE 5–20, is a suitable coating for dried frozen foods, as it does not contain the amylase responsible for staling on frying or freezing.[2577] Starch partially hydrolyzed by alpha amylase is proposed as a fat replacement in food technology[2578] and a starch hydrolyzate after extrusion exhibited a high absorption capacity for lipids[2579] Treatment of a slurry of oats with glucoamylase and alpha amylase provided a nondairy base enriched in glucose.[1999]

Using pullulanase to debranch amylopectin, an edible instant kudzu powder was prepared.[2580] The reaction of a solution of saccharified starch with hydrogen and the repeated action of glucoamylase and/or debranching enzyme and yeast produced maltitol of high purity.[2581] Potato starch treated with thermostable alpha amylase produced maltodextrins potentially useful as fat substitutes.[2582] A rice protein concentrate of high purity was obtained by enzymatic hydrolysis and solubilization of the starch component of a rice substrate with alpha amylase and glucoamylase.[2583] Starch modified by glucan 1,4-α-maltohydrolase is suitable for incorporation into edible emulsions.[2584] The best syrup for manufacture of caramels should contain as much maltohexaose as possible. Such a syrup is obtainable from starch via saccharification with alpha amylase at pH 6.2. The resulting caramel is only slightly hygroscopic.[2585]

Lactic acid fermentation[2586,2587] improves the digestibility of wheat grains and increases their nutritive value. Amylopectin from potato starch digested with isoamylase produces edible starch films of excellent properties.[2588]

More emphasis in food manufacture is now being put on new-generation enzymes, namely genetically modified ones.[2589,2590] These enzymes are useful in many starch conversions: in bakery, fruit and vegetable processing, the dairy, meat, and fish industries, in protein modifications, brewing, and lipid manufacturing.

(vi) Animal Feed. Some animals, for instance dogs, do not synthesize endogenic amylase and, therefore they cannot digest starch or starchy feedstocks. Animal feed containing saccharides and polysaccharides should therefore be supplemented with alpha amylase[1764,2591,2592] or malt amylase.[2593] Supplementation of feed with enzymes is used in the nutrition of broiler chickens[2594,2595] and pigs[2596–2598] to facilitate the metabolizing of starch and increase the growth rate of those animals. For better storage stability such enzymes can be provided in a granular form.[2561,2599] Together with alpha amylase, thermostable glycosidases from *Thermococcus, Staphylothermus*, and *Pyrococcus* are also proposed as adjuvants.[2600] Starchy feed pretreated with β-glucan appeared more acceptable for animals than feed from the same sources that had been pretreated thermally or not pretreated at all.[2601] Other carbohydrases and their combinations have also been utilized for the degradation of cell walls of polysaccharides and increase nutritive value of chicken feed.[2602]

Complexes of hydrolases with starch may have a commercial value as starch-hydrolyzing preparations.[2603]

V. Starch as a Feedstock for Fermentations

1. General Remarks on Fermentation

Fermentation may be defined as the generation of energy involving an endogenous electron acceptor from the bacterial (enzymatic) oxidation of any organic material. The results of fermentation depend on the organic substrate, most frequently carbohydrate or protein, the applied catalyst in the form of either isolated enzyme or its microorganism producer, as well as the process conditions. The character of the process may be mainly aerobic or anaerobic. The different fermentation modes of carbohydrates include ethanol, lactic acid, butanoic acid, citric acid, and acetone fermentation, and aerobic respiration producing CO_2 should also be included in the scope of fermentations starting from glucose. Other types of fermentation, such as acetic acid fermentation, propanoic acid fermentation, and mixed acid fermentations, including butanoic and decanoic acids, butanol, and glyoxalate fermentations, do not utilize glucose directly, but utilize products of its transformation.[2604] In addition to these products, fermentation of starch can deliver a variety of other compounds such as amino acids, antibiotics, vitamins, growth factors, and enzymes.[259]

The use of a sequence of fermenting microorganisms and selected crop products enables the production of a functional food exhibiting certain therapeutic and hygienic properties in human and veterinary practice. A patented approach involves a preliminary fermentation of a cereal crop with *Aspergillus kawachi*. The resulting product is then inoculated with *Saccharomyces cerevisiae* and *Lactobacillus acidophilus*. That fermentation yields a product that may serve as an improver of digestion and a deodorant.[2605]

The well-known glycolysis pathway is the common initial step for all types of fermentation starting from glucose.[2606] The net reaction involves adenosine diphosphate (ADP), inorganic phosphate (P_i), and nicotinamide adenine dinucleotide (NAD^+) and converts glucose into pyruvic acid, adenosine triphosphate (ATP), and the reduced form of NAD^+, namely NADH.

$$C_6H_{12}O_6 + 2ADP + 2P_i + 2NAD^+ \quad 2CH_3COCO_2^- + 2ATP + 2NADH + 2H_2O + 2H^+$$

Pyruvic acid resulting from glycolysis is then converted enzymatically into the final products of various types of fermentation. Some of the processes are anaerobic and others are aerobic, for instance, aerobic respiration. The anaerobic/aerobic character of such processes depends not only on the type of fermentation but also on the microorganisms involved.

2. Alcohol and Alcohol–Acetone Fermentations

a. Ethanol Fermentation

(i) Introduction. The availability of appropriate enzymes and their ability to hydrolyze native starch and granules is a key factor in the energy-efficient production of ethanol.

Ethanol fermentation, usually carried out with yeast, is anaerobic, whereas with some other microorganisms and enzymes, the process may proceed aerobically. In aerobic fermentation, limitation in the availability of oxygen can change the type of fermentation. This familiarly occurs in mammalian muscles, where physical fatigue resulting from a limitation in the availability of oxygen turns the common aerobic respiration into a lactic acid fermentation. Fermentation of either starch-containing plant sources, such as cereal grains, tubers or other parts of plants (fruits, roots, stems) or isolated starch, should be understood as a sequence of processes hydrolyzing starch to glucose and then, via the glycolysis stage, producing the desired fermentation product. These technical processes are pulping, malting, mashing, liquefaction, and saccharification.

Malting, as in the preparation of malted grain, serves for the production of malted beverages, malt vinegar, and malt bakery goods. Usually, a barley malt is used,[1571] but triticale malts[2607,2608] have also been employed. In the production of ethanol from potatoes, malt from a deep culture of *Aspergillus niger*[1571,2609] or *Rhizopus*[1577,1578] was used instead.

Mashing utilizes milled grain, usually barley malt blended with other cereal grains and, sometimes, also triticale preparations.[2610] The mashing of tuber and cereal starches in the presence of thermostable bacterial and fungal amylase has been patented.[1983] Such a mash is suitable for making beer. Limit dextrinase (α-dextrin endo-1,6-α-glucosidase) performs better than barley malt and provides a higher yield of ethanol.[2611] In the presence of barley malt, that dextrinase becomes largely ineffective, as it is inhibited by proteins present in malt.[2612] At 62.8 °C, such malt can convert over 70% of cooked grain into maltose within 1 min.[34] If the mashes are prepared in the form of a not too fine, gritty meal, saccharification requires a longer period.[2613] Using a stator–rotor dispersion machine, the grinding of grains on mashing allows the processing of pulp in a pressureless manner, increasing the yield of ethanol in the final step of fermentation.[2614]

Equation (49) provides a simple description, followed with 3% precision, of the activity (A) of amylases in mashing in brewing, and prediction of fermentability upon joint action of alpha and beta amylases.[2615]

$$A = a^{bt} - c^{dt} \tag{49}$$

where t is temperature of the mash and b and d are constants.

The following scheme of starch hydrolysis during mashing for production of beer has been presented[2616] (Fig. 21).

Thus, products of the hydrolysis, namely fructose and glucose, are available either from gelatinized starch or from sucrose. Sucrose employs invertase for its conversion into glucose and fructose. Naturally, starch cannot be a direct source for the production of fructose. Gelatinized starch, when hydrolyzed with alpha amylase gives rise to glucose, maltotriose, and dextrins. For the preparation of maltose, alpha and beta amylases are used cooperatively. The same blend is used for conversion of dextrins into maltose, but for conversion of dextrins into glucose and maltotriose, alpha amylase alone is sufficient.[2616]

Liquefaction is performed to produce fermentable products. Among other procedures, starch and starch-containing material can be liquefied by digestion for

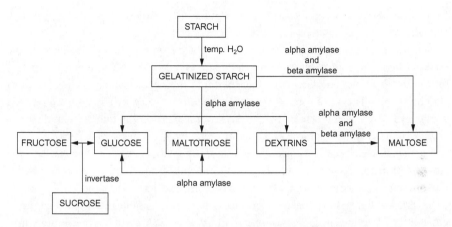

FIG. 21. Hydrolysis of starch during mashing.[2616]

1–2 h with bacterial alpha amylase at 65–70 °C.[2617] Enhanced amounts of that enzyme augment the conversion of dextrins and decrease the use of enzyme during saccharification, thus increasing the yield of ethanol.[2618]

Saccharification follows liquefaction to convert maltodextrins into glucose containing only a minor amount of maltose and isomaltose. The joint use of *A. batatae* and *Rhizopus* to produce glucoamylase also increases the fermentation yield.[2619,2620]

Dextrins of wort and beer are poorly digested by amylases, but the limit dextrinase of malt splits them readily, as it is capable of cleaving $(1 \rightarrow 6)$ linkages. Such an enzyme, glucoamylase, is secreted by *Saccharomyces diastaticus*. Hybrids of *S. diastaticus*, single-spore yeasts, offer selective splitting of either dextrins or maltose.[2621]

The efficiency of saccharification controls the efficiency of fermentation. That process determines the composition of the fermentation products;[2622] see, for instance, production of the traditional Chinese spirit—Xiaoqu—following ancient recipes. Supporting the traditional fermentation with saccharifying enzymes enhanced the flavor and yield of that liquor.[2623]

Saccharification of cereal starches can also be performed with the enzyme of rice koji added.[2624] The use of glucoamylase together with thermostable yeast permits simultaneous saccharification and fermentation.[2625] The enzymes from *A. niger* and *A. awamori* were also both used.[2626] The production of ethanol with joint use of amylase and pullulanase isolated from *Fervidobacterium pennavorans*[2627] and *Staphylothermus* amylase[2628] has been patented. A mixture of glucoamylase from *A. oryzae* and *A. awamori*, as well as a mixture of natural barley and glucoamylase from *A. awamori*, also accelerated fermentation.[2629] The joint use of glucoamylase and acid proteinase was reported to be effective.[2630]

The use of combined alpha amylase and glucoamylase allows omission of the cooking step, and so granular starch can be fermented.[2631,2632] There is also a patent on the use of alpha amylase alone, and without the necessity for gelatinization of starch.[2633,2634] The organism *Zymomonas mobilis* is very efficient in converting starch hydrolyzates into ethanol.[2635,2636] An organism originating from Indochina, *Mucor* Boulard No. 5, can grow and act in septic, well-aerated media.[50,2637,2638]

Saccharomyces cerevisiae accelerates the fermentation step.[2639] Saccharification can be performed in the presence of cytolytic enzymes from *Trichotecium roseum*.[2640] Performing saccharification in a fluidized bed also speeds up the process.[2641]

Several methods have been elaborated for direct, simultaneous production of ethanol.[2642] For instance, *Schwanniomyces aluvius* was aerobically cultured in a medium of 2% potato starch. A 5% proportion of starch was then added and the whole was anaerobically incubated at 25 °C. A 99.3% yield of ethanol was achieved within 72 h.[2643] Production of ethanol with a *S. diastaticus* culture grown in a starch–salt–yeast extract takes the same amount of time. Fermentation of 12.5% starch slurry proceeded at 37 °C and pH 6.0–6.5.[2644] An approximately 80% yield of ethanol was obtained with a crude amylase preparation immobilized on a reversible soluble–insoluble copolymer of methacrylic acid–methyl methacrylate. This conjugated copolymer saccharified starch, and the amylase culture fermented glucose. Amylase could be recovered after its precipitation by decreasing the pH from 5 to 3.5.[2645] The joint use of *Schizosaccharomyces pombe* and an enzyme isolated from *Corticium rolfsii* provided ethanol within 48 h on incubation of 30% raw starch at 27 °C and pH 3.5.[2646] Direct fermentation with *Schwanniomyces castelli* provided ethanol from 15% soluble starch, dextrin, or glucose.[2647] A *Saccharomyces* expressing amylase genes from *Schwanniomyces* were proposed for saccharification of starch for brewing.[2648] *Streptococcus bovis* expresses rice glucoamylase on the cell surface and secretes alpha amylase.[2649] These properties are superior for the direct production of ethanol from starch as boiling is omitted. In the processing of granular starch, contact of yeast and starch granules is important.[2650] For this reason, starch and yeast should be combined together. The size of the granules should possibly be uniform. In the production of strong aromatic Chinese spirits, the initial concentration of the starch substrate is critical.[271]

Coimmobilized microorganisms are coming into common use, and they also offer a method for direct production of ethanol. Coimmobilized *A. awamori* and *Z. mobilis*, combined with an immobilized culture of the anaerobic *S. cerevisiae*, was employed in the production of ethanol. Calcium alginate beads,[2651,2652] a cellulose carrier,[2653] or a pectin gel[2654] was used for the immobilization. The latter catalyst could be reused three times. Cells of *A. niger* were coimmobilized with yeast, and this catalyst could be reused several times. The cycle for direct production was shortened to 36–48 h, and 88% of the starch was consumed.[2655,2656] Coimmobilized *R. japonicus* and *Z. mobilis* considerably enhanced the production of ethanol.[2657] Coimmobilized *Z. mobilis* and glucoamylase,[2658,2659] as well as *S. cerevisiae* with glucoamylase, performed successfully with raw starch,[2660] and a starch hydrolyzate.[2661]

(ii) Fermentation. Fermentation of the saccharified material may involve various microorganisms, for instance, *Lactobacillus plantarum*, *Streptococcus themophilus*, and/or *Lactobacillus bulgaricus*; fermentation conditions are 25–45 °C for 3–20 h.[2662,2663] A strain of *S. cerevisiae* named amylolytic nuclear petite could increase the ethanol yield by over 54%.[2664] Enzymes isolated from strains of *Rhizopus* and *Aspergillus* increased the starch consumption up to 90.5%, along with a shortened fermentation period and decrease of microbial contamination.[2665] A cocultivated *Schwanniomyces occidentalis* mutant and *Saccharomyces cerevisae* fermented sugars liberated from sorghum starch very efficiently. The concentration of slurry could reach 28% without any decrease in the yield of ethanol, although the time required for total fermentation had to be extended with increasing concentration of the slurry.[2666]

Solid-phase fermentation is also possible.[2667] The use of *Schwanniomyces castellii* provided a solid-state method for production of ethanol in an aerobic–anaerobic process.[2668] In an aerobic cycle, growth of the microorganism and hydrolysis of starch took place, and fermentation in a subsequent anaerobic cycle produced ethanol. Foaming needed to be controlled.[2669] When corn mashes are fermented, the distillation is facilitated when the mash contains alpha amylase and glucoamylase.[2670] After cloning and expressing on either *A. niger* or *Trichoderma resei*, an *A. kawachi* acid-stable alpha amylase in conjunction with glucoamylase, acting at pH 4.5, readily converted granular starch into ethanol. The amount of ethanol produced is higher and the amount of residual starch is considerably decreased.[2671]

A common problem with fermentation is to assure sterility of the process and avoiding contamination of the fermenting wort with undesired, deleterious microorganisms. In the production of ethanol from amylaceus material by use of yeast, such contamination effects could be eliminated by the addition of 0.5% boric acid.

However, the yeast had to be acclimatized to that acid.[2672] When NaF is used for inhibiting the fermentation with yeast, hydrolysis begins to predominate over fermentation.[2673] Fluoride anions can protect *Mucor mucedo* from developing secondary strains.[51] The fertility of pure cultures of microorganisms can be controlled by pH, as shown with *Rhizopus japonicus* as used for fermenting corn starch.[2674]

Some *Vibrio* strains can be inhibited by glucose and sucrose.[2675] In some instances, heating and vigorous agitation,[2676] and overnight steeping of the substrate in 0.15% sulfuric acid[2677] can be helpful. Apart from the common bactericides used in the food industry, such as sorbic, benzoic, dehydroacetic (3-acetyl-2-hydroxy-6-methyl-1*H*-pyran-4-one), and 4-hydroxybenzoic acids, various microbiocides such as poly(hexamethylene biguanidine) and others can be employed.[2678,2679] The latter method was used in the ethanol fermentation of raw starch with coimmobilized *Aspergillus awamori*, *Rhizopus japonicas*, and *Zymomonas mobilis*. Papain and cysteine hydrochloride were also used.[2680] However, in a fluidized-bed reactor filled with immobilized *Z. mobilis*, hydrolyzed B-starch could be treated without sterilization.[2681]

In the production of high-quality oriental wine, amylase is used to decompose starchy material to make it more readily fermentable and, in general, to increase the fermentation efficiency.[2682] Degradation of starch with pullulanase and beta amylase has been used in the brewing of beer.[2683]

(iii) Production of Ethanol. Enzymes. Production of ethanol can proceed either batch-wise or in a continuous manner. Various such versions have been used. Thus, cassava starch liquefied in an acid-catalyzed step was neutralized with ammonia, and then simultaneously saccharified and fermented by glucoamylase in a series of separate vessels.[1693] In modern processes, the acid-catalysis step can be omitted.[2684] In another adaptation, an aqueous two-phase system with ultrafiltration of the upper phase is employed. The filtered phase with amylolytic enzymes is recycled continuously.[2685] Recirculation of the enzyme is possible, even when the process is performed in a slurry. After the process is complete, the liquid is taken off and the slops fraction (suspension remaining after removal of ethanol by distillation) is recycled.[2686,2687] Alternatively, the enzyme fraction can be recycled from the sedimented layer of immobilized amylase.[2688] Immobilized glucoamylase, together with two different strains of *S. cerevisiae*, immobilized in beads of either calcium[2689] or aluminum[2690] alginate, serves as a continuously working bioreactor. The process is anaerobic.

In another version, steam-cooked starch paste is saccharified with an enzyme complex containing alpha amylase, protease, and other enzymes. The saccharified material is inoculated with *S. cerevisiae* and *Schizosaccharomyces pombe* on aeration.[2691] Since *Schizosaccharomyces pombe* yeast is highly flocculent, it can be used in the flock form.[2692] The use of flocculable *Saccharomyces* hybrids has also been patented.[2693] Some Russian patents[2694,2695] describe the continuous production of ethanol from raw starch material by using a mixture of alpha amylase, glucoamylase, cellulase, and hemicellulase. A combination of alpha amylase with glucoamylase and cytase (a plant-seed extract capable of solubilizing plant cell walls) increased the yield of alcohol by 8.5%.[2696] Coimmobilized enzymes and yeast also provide a method for continuous fermentation.[2697–2701] When glucoamylase is covalently immobilized and its action is supported by *S. cerevisiae* immobilized in calcium alginate, the continuous process could be operated in the solid phase.[2702]

A novel alpha amylase for liquefaction of corn-starch slurries and mashes at pH 4.5 and with the absence of Ca^{2+} ions was developed based on the enzyme isolated from *Thermococcales* and then expressed on *Pseudomonas fluorescens*. This amylase is especially promising for the production of "bioethanol."[2703] Cocktails of alpha amylases from *B. licheniformis*, *A. oryzae*, glucoamylase from *Rhizopus*, *A. niger*, and dry barley malt were tested in the production of "bioethanol" (anhydrous ethanol intended for use as a motor fuel).[2704]

Corn hulls, the outer peel covering the corn grain, were used for preparation of koji, a material used for alcoholic fermentation of raw starchy materials without cooking. Corn-hull koji has a lower saccharifying power than alpha amylase, CMCase, xylanase, and wheat-bran koji, but has higher protease and pectinase activities, Its use in alcoholic fermentation of cassava starch and sweet potato appeared superior to fermentation employing wheat-bran koji. Corn-hull koji gave 10.3% (v/v) of alcohol in 93% yield from 20 g of cassava starch, whereas wheat-bran koji gave 9.4% (v/v) alcohol and a 90.4% yield. Sweet potatoes processed with corn-hull koji gave 9.1% (v/v) alcohol and a 92.6% yield from 50 g of sweet potato, whereas wheat-bran koji gave only 8.1 (v/v) of alcohol with 88.6% yield.[2705]

Two types of process are commonly utilized in the fementative production of ethanol. These are the classical processes performed with various modifications, and the amylo process usually employed for the fermentation of grain mashes and utilizing mold fungi, as they exhibit more beneficial amylase activity (AA) and dextrinase activity (DA).[2706–2708] *Aspergillus oryzae* appears more useful than *Aspergillus niger*, as the AA of the latter is considerably lower, but, on the other hand, *A. niger* exhibits a considerably higher DA.

It should be noted that molds are more sensitive to both temperature and pH than are enzymatic preparations. *Aspergillus batatae* is used effectively in fermentation of starch, but although it has a high AA, its glucoamylase activity is low, and therefore, a mixture with a culture of *Endomycopsis bispora* of high glucoamylolytic activity is recommended. The *E. bispora* organism does not secrete the transglucosidases which would inhibit formation of ethanol.[2709,2710]

The use of *Aspergilli* molds derived from the Japanese tradition, whereas in the Chinese culture *Amyloses rouxii* and other organisms from the group of *Mucor and Rhizopus delemar* were used. The fungi traditionally used in China are distinct from *Aspergilli* with their lower production of acids and higher tolerance to alcohol. Studies on the use of coimmobilized systems of *A. awamori* and *Z. mobilis*, *Rhizopus japonicus* with *Z. mobilis*, and *A. awamori*, *R. japonicas*, and *Z. mobilis* revealed[2711] that the triple yeast system provided 96% yield of ethanol within 18 h of anaerobic fermentation at pH 4.8. *Clostridium thermohydrosulfuricum* was employed in a bioreactor equipped with a pervaporation membrane that enabled continuous operation.[2712] *Clostridium thermosulfurogenes* secretes a thermostable exocellular beta amylase that ferments starch to ethanol at 62 °C. Its activity was the highest at pH 5.5–6.0 and its stability range was between pH 3.5 and 6.5.[1936] That microorganism also secretes a thermostable glucoamylase and a pullulanase.[948] *Bacillus polymyxa* also secretes a beta amylase suitable for fermentation of starch to ethanol.[2713]

The amylo process involves the conversion of grains by the sequence of pressure heating, thinning the gelatinized starch with mineral acid, followed by saccharification with a mold producing diastase, and finally fermentation of the resulting glucose with a yeast. Thus, in the amylo process the step of malting is omitted. The amylo process offers a reduction in the cost of grain, an increase in the yield of alcohol, and the saving of energy, water, and labor.

The classical fermentation process utilizes enzymes isolated from various microorganisms. The mode whereby the enzymes are applied depends on the substrate and the proposed process. The latter involves, among others, concentration of the mash. In highly concentrated mashes, the use of *Saccharomyces cerevisiae* powder increases the alcohol content by 1–2 vol.% and accelerates the process, which takes 6–15 h. These conditions were evaluated for corn, sorghum, rice, wheat, and other amyloceus substrates.[2714.] The problem with the limited tolerance of *S. cerevisae* to ethanol can be solved by addition of pulverized soybean flour to >40 mesh into the fermenting liquid[2715] as well as by the application of genetically modified enzyme strains.[2716,2717] If the fermentation proceeds in a sequence of fermentors, there is a progressive decrease in concentration of the saccharide and a corresponding increase in the concentration of ethanol. In the initial fermentors, the concentration of sugars is high and that of ethanol is low. In the first reactors, *Saccharomyces bayanus* performs better, whereas in later fermentors having an enhanced concentration of ethanol and low concentration of saccharide, *S. cerevisiae* works better.[2718] Separation of stillage (distillery backset) into four value-added product streams can enhance the production of ethanol.[2719] The addition of distillery backset to mashes had an adverse effect on the enzymes. That effect upon alpha amylase depended on the composition of the mash, but beta amylase was always deactivated and the limit dextrinase activated. These results can be induced by pH changes brought about by the backset added.[2719] The use of glucoamylase can make cooking unnecessary.[2720,2721]

The use of enzymes in saccharification of potato, sweet potato, cassava, corn, and other cereal starches has been the subject of several publications[224,1629,2622,2663, 2697,2716,2722–2725] and patents.[1694,1791,1792,2622,2726–2738] Several factors, such as the activity of the enzymes, and their tolerance to pH and alcohol, have to be taken into account. *Schwanniomyces castelli* appeared to be inhibited by alcohol, and this inhibition was reversible.[2739] Similarly, *Arxula adeninivorans* produced ethanol at 30 °C, but as this yeast has low tolerance of alcohol, the yield was low.[2740]

The commercial importance of alcohol fermentation was a driving force for use of genetic engineering in the quest for novel enzyme mutants. Cocultivation of microorganisms in the course of which a transfer of genes takes place is a common approach. *Clostridium thermosulfurogenes* producing beta amylase, cocultivated with and *C. thermohydrosulfuricum* producing pullulanase and glucoamylase, markedly enhanced the rate and yield of ethanol production under aerobic conditions at 60 °C.[2741] Cocultivated *S. cerevisia* and *S. diastaticus* showed enhanced production[2742,2743] of glucoamylase and permitted direct fermentation to ethanol.[2744,2745]

Several genetically engineered microorganisms for improved ethanol production have been evaluated.[2746] Thus, a respiratory deficient strain of *Schwanniomyces castelli* had higher resistance to ethanol.[2739] A novel catabolite repression-resistant *Clostridium* mutant produced a beta amylase that was eight times more thermostable and it therefore accelerated fermentation.[2747] *Klebsiella oxytoca* P2 was modified by adding genes from *Z. mobilis*[2748] and genes of alpha amylase and pullulanase.[2749] Genetic modifications of *S. cerevisiae* to improve the production of ethanol have received particular attention.[2750–2758]

Substrates. Alcohol fermentation utilizes a variety of substrates.[265,2759] With a wide available range of microorganisms and/or enzymes, as well as variations in reaction conditions and technological processes, it is impossible to design any valid general order of substrates arranged from the point of view of process efficiency, duration, yield, and generally, economics. Some comparative studies[2721] conducted on potato, wheat, and other cereals, fermented with several strains of *Aspergillus oryzae, A. awamori, A. usamii, A. niger, A. batatae, Bacillus mesentericus*, and malt revealed that potato starch was the most susceptible to splitting. Use of malt, as well as of *A. oryzae*, required only 1 h to decrease the molecular weight of 268 kDa to 1353–1556 Da, whereas such result with *A. awamori* and *B. mesentericus* required 18–24 h to reach a similar point. Granular starch can be fermented to ethanol in a blend with whey lactose, as the latter also provides ethanol.[2760,2761] Starch has also been fermented together with sugar juice.[2762] Ethanol is likewise available from several starchy raw materials.[2763–2767]

Procedures specific for particular substrates have been elaborated.

Potatoes   Potato starch mash, following treatment with mold malt, was saccharified in a deep culture of *A. niger* with 150 units of alpha amylase per 100 mL. For 100 g of saccharified starch, 100 amylase units was used.[2609] Potato starch was also saccharified with alpha amylase (≥ 100 units of amylolytic activity/1 g starch) and glucoamylase isolated from *Bacillus subtilis*. Simultaneous admixture of both enzymes allowed the amount of glucoamylase to be decreased by 30–35%. However, a higher content of glucoamylase decreased the fermentation time by one third. The same applies to the fermentation of barley and maize starches.[2768] *Schwanniomyces alluvius* cultured aerobically and then incubated anaerobically in a 5% potato-starch slurry converted the starch directly into ethanol.[2643] Also, the glucoamylase from *Rhizoctonia solani*, used jointly with *S. cerevisiae*, simultaneously converted either raw or cooked potato starch into ethanol at 35 °C within 4 days.[2769] *Aspergillus niger* was also utilized simultaneously with *S. cerevisae* in a semicontinuous two-stage process.[2770] This process produces a large amount of biomass. Satisfactory results in production of ethanol from potato starch were achieved when alpha amylase and *S. cerevisiae* in 1:2 proportion were applied jointly to potato flour.[2771] In a Chinese patent,[2772] bran koji composed of *A. oryzae* and *Rhizopus* (0.02–0.08 g/g) saccharified pulverized potatoes at 60–85 °C, and *S. cerevisiae* at 28–42 °C during 48–60 h fermented the product to glucose. Powdered or homogenized starch can be gelatinized with $>1\%$ alkali and saccharified.[2736] Potatoes for production of ethanol may be pretreated by microwave heating.[2773] Reactors working with glucoamylase immobilized on cellulose,[2774] and with *A. niger hyphae* and *S. cerevisiae*,[2775] were designed for production of ethanol.

Potato waste, after removal of starch, was considered as a substrate for fermentation. In a first step, cellulolytic enzymes decomposed cellulose, and in the next step that enzyme was supplemented with an alpha amylase and glucoamylase cocktail.[2776]

Micronized potato-starch waste from ball milling could likewise be degraded with symbiotic cocultures of microorganisms.[2777] High-temperature wastewater from the potato-starch industry was fermented by mixed cultures of thermophilic aerobic bacteria.[2778]

For production of the Japanese liquor shochu, potatoes coarsely ground with bacterial amylase were processed without cooking. Filamentous glucoamylase and, optionally, microbial protease were introduced. Shochu fermentation proceeded at pH 3–5. Similarly rice, wheat, and buckwheat,[2779] as well as sweet potatoes,[2780] could be processed. In the last example, alpha amylase was supported by cellulases from *A. niger* and *Trichoderma viride*. Uncooked cereal grain mash could be saccharified and then fermented with *Rhizopus* glucoamylase and *Aspergillus kawachii*.[2781]

Cassava Extended studies[2782] on saccharification of cassava starch revealed that the use of a double-enzyme combination has advantages over that with the use of acids. The latter approach produces maltooligosaccharides of DP 10. The double-enzyme system can consist of coimmobilized *S. diastaticus* and *Zymomonas mobilis*,[2783] *Z. mobilis* and glucoamylase,[2784] or *S. cerevisiae* and glucoamylase.[2785] Such systems allow the continuous production of ethanol. The same outcome was provided by strains[2786] of *Endomycopsis fibuligera* and mixed cultures of *E. fibuligera* and *Z. mobilis*.[2787] A continuous process has been described that utilizes a dry meal from which *Bacillus subtilis* amylase and *A. awamori* amyloglucosidase produced a syrup that was fermented by *S. cerevisiae*.[2788] When the syrup was fermented with *Z. mobilis*, a considerable amount of CO_2 was produced.[2789] The technology for producing fuel alcohol by fermentation has been elaborated.[2790]

An 112% increase in the yield of ethanol as compared with substrate without molasses was possible after combining the residue from decanting cassava juice with sugarcane molasses and fermenting the resultant blend.[2791] A blend of starch with lactose was also fermented by using coimmobilized[2792] β-galactosidase and *S. cerevisiae*. Higher yields of ethanol could be realized. Using only wheat-bran koji from the *Rhizopus* strain, raw cassava starch and cassava pellets were converted at 35 °C and pH 4.5–5.0 reasonably well into ethanol without cooking, and an 85.5% conversion yield was obtained. The addition of yeast cells and a glucoamylase preparation is not essential.[2793,2794] *Rhizopus oryzae* and *Rhizpopus delemar* produced a glucoamylase capable of digesting granular starch during solid-state fermentation. The optimum conditions were pH 4.5 at 32 °C and 72 h of fermentation, and the first of the two strains was more productive.[2795]

In contrast to corn starch, liquefaction of cassava starch may be carried out at low temperature.[1629,2796] Sliced material can also be fermented.[2797] As with corn starch, A. niger cultures appeared suitable for the production of ethanol from cassava starch, with a 90% reported yield.[2798] *Aspergillus niger* and *A. awamori* cultivated on wheat bran could serve as a koji extract, which converted starch into a syrup that was fermented by a yeast.[2799] Glucoamylase from *A niger*[2799] and *Rhizopus*[1092] is suitable for production of ethanol from a 30% starch slurry, employing 3 days of fermentation with 5600 units of that enzyme. A cooking step is dispensable.[2720,2677] In such processes, shaking the broth significantly enhances the yield of ethanol.[2800] After pretreatment of starch with alpha amylase and amyloglucosidase, the 30% broth was treated with *Z. mobilis*, providing a good yield of ethanol.[2801] Slurry that had been enzymatically liquefied and then saccharified could be converted into ethanol by *Saccharomyces ellipsoideus*[2802] and *S. uvarum* Wuxi 58.[2803] The use of a glucoamylase from *Aspergilli* and a yeast at pH 3.6 with shaking allowed operation during 5 days with slurry up to 21% concentration.[2804] In some solutions, starch was hydrolyzed with dilute hydrochloric acid then fermented with *S. cerevisiae*.[2805] A 25% concentration of the slurry gave the best results.[2806] Digesting the slurry with alpha amylase and crude glucoamylase from *Aspergillus* sp NA21 provides the most efficient liquefaction and requires 45 min when steam under pressure is used.

Sweet Potatoes Production of ethanol from sweet potatoes requires *A. niger* glucoamylase, as the source is rich in the pectin depolymerase necessary for decomposition of plant tissues. Thus, the uncooked plant is macerated with the enzyme at 40 °C and pH 4.5. Prior to that, the plant had been steeped overnight in 0.15% H_2SO_4 for sterilization. The glucoamylase employed had 5000 units of pectin depolymerase and 3000 units of glucoamylase. The mash prepared during 2 h was cooled to 30 °C and the pH was brought to 5.2–5.4, and the mash was then treated with alpha amylase.[2677] The presence of a cellulase-producing *Paecilomyces* species increased the yield of alcohol.[2807] Mashing was first performed with pectin depolymerase and 5 min of heating at 75–85 °C, then saccharification was effected with 0.05% glucoamylase.[2808] An *A. niger* mutant strain 145 produces an amylase capable of rapidly saccharifiying raw sweet potato to glucose.[2809] The procedure omitting the use of pectin depolymerase, but with heating in 3% aq. NaOH for 2–5 h, was also patented.[2726] In another variant, sweet potatoes are first steeped for 4 h at pH 1.8 and crushed. The resulting slurry is then liquefied with a complex mixture of cellulase, arabinogalactanase, arabinoxylanase, pectinase, acidophilic alpha amylase, and glucoamylase, and the fermentation is performed with a yeast.[2810] Starch from raw sweet potato can be fermented, without cooking, with the glucoamylase of *Endomycopsis flbuligera* IFO 0111[2811] or *A. niger*.[2812]

Dried sweet-potato powder was treated with alpha amylase at 80 °C for about 1 h then fermented with *S. cerevisiae*. A disadvantage of this method is the production of a higher level of methanol.[2813] A continuous process with immobilized yeast cells has been described.[2814] Fermentation of the syrup can be performed with *S. uvarum* Wuxi 58, and the process requires 48 h at 40–42 °C.[2803] *Saccharomyces cerevisiae* effects direct fermentation of liquefied milled sweet potato at 30–37 °C.[2815] A continuous fermentation procedure was patented in Brazil.[2816]

Yam   Tubers of yam (*Dioscorea*) for production of ethanol are ground, a thick, 30–35% slurry is prepared, and its pH is brought to 6.0–7.0. Alpha amylase is added, and after gelatinization and liquefaction at 90–93 °C, filtration and vacuum concentration to 30–35%, the pH is brought to 1.2–1.8 with hydrochloric acid. Saccharification proceeds under 0.3 MPa to DE 90–95.[2817] In another procedure, pulp of *Diascorea sativa* was gelatinized at 90 °C with alpha amylase, saccharified with α-amyloglucosidase, and then fermented with yeast.[2818]

Buffalo Gourd   Roots of carrot-sized buffalo gourd (*Cucurbita doetidissima*) also merit attention as a possible feedstock for production of ethanol. Slurried roots were dextrinized and saccharified with thermostable alpha amylase and glucoamylase, followed by fermentative action of *S. cerevisiae*, producing ethanol with 82–86.5% of the theoretical yield.[2819]

Corn   The conversion of corn has usually involved the use of mineral acids to decrease the content of nonreducing dextrins. Such an approach appeared unnecessary with corn starch, where acidification of the substrate with mineral acids had a little effect.[2820]

In the production of ethanol from corn, the amylase from *A. niger* koji was convenient.[2821] Alternatively, *Rhizopus* bran koji can be used.[2822] The use of *A. niger* koji enabled the production of 7 vol.% of ethanol from a 30% suspension of raw corn starch during a 6-day process. Precooking was not necessary.[2800] In another version, *A. niger* cultivated on rice koji was inoculated with a yeast, and this preparation was allowed to digest a 40% mash at 30 °C.[2823] Corn starch that had been heated (liquefied) for 20 min at 120 °C could be readily saccharified with alpha amylase (1500 units/100 mg) at pH 6.0. Such pretreatment was advantageous, as raw nonprocessed starch was saccharified only very slowly with that enzyme. Defatting of that starch, and/or addition of the Ca^{2+} cations, promoted digestion.[2824] Periodic replenishment of the substrate and water, addition of isoamylase, and stirring the fermented broth increase the yield of ethanol.[2825] Glucoamylase from a mutant of *A. niger*, with the equivalent of 63–79 glucoamylase units/g substrate, operated at pH 4.5 on a mixture of 8–10% corn powder, 2% corn steep liquor, and 2% soybean powder. Saccharification proceeded first at 55–60 °C, followed by fermentation at 30 °C for 72 h.[2826] Other strains of *A. niger* are also suitable sources of glucoamylase.[2677,2720,2827] The joint use of either alpha amylase and glucoamylase,[2722,2727,2828] or glucoamylase and *Saccharomyces* sp. HO,[2829] for instance the *S. cerevisiae* fusant with *S. diastaticus*,[2830] is recommended. Genetically modified *S. cerevisiae* expressing *A. awamori* glucoamylase utilized approximately 95% of the substrate and produced significantly less glycerol.[2831] *Saccharomyces cerevisiae* GA7458, which includes genes of alpha amylase of *Bacillus stearothermophilus* and glucoamylase of *S. diastaticus*, performed well at pH 5.5.[2832] A flocculable *Saccharomyces* hybrid (FERM-P 7794), engineered by protoplast fusion, provided an increase in the ethanol production and facilitated its recovery.[2693] Glucoamylase of *A. niger* and yeast cells, coimmobilized in calcium alginate, provides continuous production of ethanol.[2833,2834] The simultaneous fermentation of raw corn starch is possible by using *Schizosaccharomyces pombe* jointly with the saccharifying enzyme from *Corticium rolfsii*. An 18.5% concentration of ethanol was obtained within 48 h from a 30% slurry at pH 3.5 and 27 °C.[2646] An almost 19% concentration of ethanol was available when cooked and simultaneously liquefied corn starch was digested with thermophilic alpha amylase at 95–105 °C, saccharified at 60 °C with glucoamylase, and fermented with *Saccharomyces* sp. W4 with ammonium sulfate added.[2835] The triple system composed of *S. cerevisiae*, glucoamylase, and β-glucanase improved the fermentation of corn mash.[2836] In another adaptation, the liquefaction is performed with alpha amylase, followed by acid-catalyzed saccharification using hydrochloric acid at pH 1.2–1.8.[2837]

Fermentation of corn starch without cooking was also proposed.[2838] A wasteless, environmentally benign technology was described using dry-milled corn, which was liquefied at 80–85 °C with simultaneous saccharification and fermentation with *S. cerevisiae* to effect 90% conversion of substrate and yielding 15% ethanol.[2815] Dry-milled corn starch was also processed in a fluidized-bed bioreactor filled with glucoamylase and *Z. mobilis*, coimmobilized in beads of κ-carrageenan. Prior to fermentation, the milled substrate was dextrinized.[2839] Corn-starch hydrolyzates were processed by liquefaction and dextrinization with heat-stable alpha amylase, saccharification with glucoamylase and pullulanase, and fermentation with *S. cerevisiae*,[2840] to produce "bioethanol."[2841] The yield of bioethanol from corn can be enhanced by utilization of the kernel-fiber fraction. This fraction is first digested with diluted mineral acid, followed by fermentation of the resulting syrup with either *S. cerevisiae* or ethanologenic *Escherichia coli*, with the addition in both instances of cellulase, β-glucosidase, and glucoamylase.[2842–2844] Alternatively, a combination of pectase, cellulase, acid proteinase, hemicellulase, xylanase, phytase, bacterial proteinase, and fungal proteinase can be used.[2845]

When corn (maize) grits are liquefied, the temperature of the slurry with enzyme is initially elevated to 105 °C to release starch and is then decreased to 85 °C.[1630]

A new alcoholic beverage resulted[2846] when cooked grits were treated with amylase, protease, and lipase, and then with lactic acid and rice koji extract, followed by fermentation with sake yeast. Prior to the fermentation, degermed corn grits should be extruded, and then *A. awamori* on bran should be used together with acid protease and cellulolytic enzymes.[2847] Raw corn flour could be saccharified directly and fermented with enzymes of *Rhizopus* sp. W-08 and *S. cerevisiae* Z-06. A concentration of 21% (v/v) of ethanol was obtained after 48 h. The conversion efficiency of raw corn flour to ethanol reached 94.5% of the theoretical ethanol yield.[2848]

Solid-state fermentation of the grits was performed with success in a stream of CO_2, using a moderately thermophilic strain of *S. cerevisiae*.[2849] Sake, which is traditionally produced by fermentation of rice, can also be manufactured from corn. It had a lower content of amino acids than traditional rice sake.[2850] Waste from corn wet milling was used for continuous production of ethanol, with *S. cerevisiae* immobilized in calcium alginate.[2851] Glucoamylase from *Chalara paradoxa* was very efficient for the fermentation of corn starch without cooking. It exhibited higher activity in digesting raw starch than the conventionally used glucoamylases. The optimum conditions for that process were pH 5.0 and 30 °C for five days. The yields of ethanol were between 63.5 and 86.8% of the theoretical value on using baker's yeast (*S. cerevisiae*) and between 81.1 and 92.1% of the theoretical value using sake yeast (*S. sake*).[2852]

Wheat Wheat used for production of ethanol can be processed either directly or after extraction of gluten.[2853,2854] The following approaches have been applied for production of ethanol from wheat. Preliminary pressure heating in water at 180 °C eliminated gluten from wheat flour. Next were performed either short-time liquefaction involving heat-stable amylase at ≥ 70 °C and/or saccharification with heat-unstable amylase from *B. subtilis* at 50–65 °C. Finally, the liquor was fermented with yeast at pH 5.6.[2855] The *Aspergillus* K27 organism isolated from soil and used as the source of alpha amylase enabled fermentation of raw starch with omission of the steaming process.[2856] Amylase can be used jointly with glucoamylase.[2827] Raw wheat starch can be presaccharified with glucoamylase, followed by simultaneous saccharification and fermentation with *S. cerevisiae*.[2857] The use of a complex of amylolytic and cellulolytic enzymes from *Penicillium janthinellum*[2858] and *S. distaticus*[2859] has also been described. A continuous process involved coimmobilized *S. cerevsiae* and *Z. mobilis*.[2860] Damaged grains of lower quality can be digested with a crude amylase from *B. subtilis* and *S. cerevisiae* for 4 days at 35 °C and pH 5.8.[2861] Grain can be fermented jointly with whey.[2760]

Wheat flour, usually freed from gluten, germ, and bran, is frequently used in fermentation with *Schwanniomyces castellii*.[2862] A two-step biphasic conversion of the flour into ethanol was also proposed.[2863] In the first stage, a wort prepared from the flour is incubated with *S. cerevisiae* at 35 °C for 36 h. In the second stage, at 33 °C, the resulting broth is incubated with *Schizosaccharomyces pombe*. This process produces a relatively low amount of biomass. Wheat flour can be completely fermented with *Z. mobilis* when ammonium sulfate is added as a source of nitrogen. Within 70 h, a 99.5% conversion can be achieved.[2864] Employing glucoamylase isolated from *Rhizoctonia solani*, mashing can be performed on a 30% substrate. The latter is simultaneously saccharified and fermented with *S. cerevisiae*.[2770] A combination of glucoamylase with *S. cerevisiae* is also useful. First, the substrate is presaccharified with glucoamylase, followed by simultaneous saccharification with the yeast.[2865] Wheat bran could be processed in the solid state with bacterial alpha amylase from *Bacilli* and *Aspergilli*, and fungal glucoamylase. The same combination efficiently hydrolyzed wheat mash.[2866]

Rye A procedure for the production of ethanol from rye was patented in Germany.[2867] Mash was fermented with distillery yeast, such as *S. cerevisiae*, and brewery yeast. Thoroughly milled rye grain, mixed with alpha amylase, was brought to 52 °C and the pH raised to 6.1 with $Ca(OH)_2$. The mixture was warmed to 90 °C and then cooled to 55 °C, the pH was lowered to 5.4, and then SUN 150 L enzyme was added. Fermentation with *S. cerevisiae* at 34 °C took 3 days.[2868] A complex of amylolytic and cellulolytic enzymes from *Penicillium janthinellum* was also effective.[2858] Mashes containing 10–14% rye starch are also hydrolyzed by *Z. mobilis* or *S. cerevisiae*. The second yeast performed better.[2869]

Triticale Varieties of triticale were converted into ethanol with commercial alpha amylase and glucoamylase preparations. The study confirmed that this cereal can serve as a substrate for the production of ethanol.[2870] Triticale was also mashed together with rye in 7:3 proportion at 50 °C with alpha amylase (120×10^6 units/ton starch) and then at 58 °C with glucoamylase (9×10^6 units/ton starch). After 30 min, the mash was cooled to 35 °C, diluted with water to 3.5 L/kg raw starch, and fermented for 3 days.[2871]

Rice Ethanol can be produced from whole-grain rice[2872] or rice flour.[2873] When grain is the substrate, either glucoamylase or alpha amylase is used, and the process proceeds without precooking. Saccharification of rice flour was performed with alpha amylase on autoclaving. The process is promoted by Ca^{2+} cations.[2873,2874]

Rice is a traditional substrate for production of such Oriental beverages as sake and alcoholized spices such as mirim. Rice flour is usually fermented with *A. oryzae*.[2875] In China, beer is produced from rice using bifidobacteria.[2876] There is also patent for making sake involving alpha amylase for saccharification.[2877] Waxy rice starch was fermented to ethanol in one-step process involving strains of *Endomycopsis fibuligera*.[2786]

Various microorganisms such as *E. coli*, *Streptomyces cansus*, *S. prunicolor*, and others produce a fertilizer from rice bran.[2878]

Sorghum The substrate for production of ethanol from sorghum employs either whole grains[2666] or sorghum starch.[2879] With whole grains, amylases from cocultivated *Schwanniomyces occidentalis* and *S. cerevisiae* were employed in a stationary-phase system. The concentration limit of the slurry was 28%. Higher concentrations led to a decrease in the yield of ethanol. Alpha amylase isolated from *B. subtilis* and *S. cerevisiae* also could be used for granular sorghum in a 25% slurry at 35 °C.[2861] A 5% higher yield of ethanol could be obtained when the sorghum was prepared by supercritical-fluid extrusion prior to liquefaction.[2880] Sorghum starch was processed with alpha amylase isolated from a thermophilic *Bacillus* strain isolated from hot-spring waters. The process was promoted by Ca^{2+} cations.[2880]

Sorghum bran containing 30% starch, 18% hemicellulose, 11% cellulose, 11% protein, and 3% ash can serve as a source of "bioethanol." The bran is pretreated with water at 130 °C for 20 min to enable access of the enzymes to cellulose and hemicelluloses. The starch was then hydrolyzed with H_2SO_4, and the cellulose and hemicelluloses were subsequently digested for 60 h with cellulases and hemicellulases. The total sugar yield reached[2874] 75%. Sorghum has been used without cooking as a raw material for alcoholic fermentation. Two Thai varieties of sorghum, containing, respectively, 80.0 and 75.8% of total sugar, were fermented in 30 and 35% (w/v) broths to give 84 and 91.9% yields, respectively, based on the theoretical value of the starch content.[2881]

Nonpasteurized juice of sweet sorghum was fermented with *S. cerevisae* at pH 3.0–4.1 for 56–98 h.[2882] Alternatively, sorghum stalks could be simultaneously saccharified and fermented by using cellulase from *Penicillium decumbens* together with *S. cerevisiae*.[2883]

Barley Barley can be fermented with *S. cerevisiae*.[2884] Barley grain was macerated with galacturonase from *A. niger* and the glucoamylase of *Rhizopus*. The mash was pretreated with sulfuric acid at 55 °C for 2 h and then fermented at pH 4.5 and 30 °C for 96 h.[2885] Glucoamylase and *S. cerevisiae* coimmobilized in calcium alginate were also used.[2785] Barley is commonly utilized for the production of beer, and making wort is the initial step. For making the wort, a combination of alpha amylase, malt diastase, glucoamylase, and protease Amano A (EC 3.4.24.28) in the proportion 1:1:1:2:0.7 w% with respect to barley is recommended.[2886]

Amaranthus Amaranth grain can be fermented in a pressureless process. This substrate produced nine times more methanol and four times more 1-butanol than rye mash.

Ground grain can be fermented as an additive to rye mash. Admixture of 10% of amaranthus provides an increase in the ethanol yield of up to 30%.[2887]

Pearl Millet Pearl millet can be fermented to ethanol with *S. cerevisiae*. As compared to corn and sorghum, this substrate has more protein, but the yield of ethanol is comparable. Millet can replace other crops as a feedstock for ethanol in areas too dry for cultivation of corn and sorghum.[2888]

Sago Sago starch has been fermented to ethanol with glucoamylase and *Z. mobilis* in nonimmobilized, immobilized, and coimmobilized forms.[2679,2802,2889–2895] Granular sago starch could be hydrolyzed with amylase isolated from *Penicillium brunnenum* on heating below the temperature of gelatinization, namely 60 °C, at pH 2.0. *Saccharomyces cerevisiae* was then employed for saccharification and fermentation.[2896] A recombinant *S. cerevisiae* strain was also used.[2897] Sago starch was first processed with alpha amylase followed by a mixture of glucoamylase and pullulanase, and the resulting syrup was fermented with *Z. mobilis*.[2898]

Other Substrates Field pea (*Pisum sativum*), optionally dehulled, was separated into starch-enriched, protein, and fibrous fractions. The starch-enriched fraction contained between 73% and 78% of starch. This fraction was liquefied and saccharified by simultaneous fermentation with alpha amylase and glucoamylase during 48–52 h. When the liquefaction involved autoclaving, 97% of the starch could be converted.[2899]

An instant beverage can be produced enzymatically from mung bean.[2900] The process is performed by using a thermostable alpha amylase and a saccharifying maltogenic enzyme.[2901]

A chestnut beverage has been produced from chestnut starch liquefied in a 30–40% slurry by using alpha amylase at 95 °C and pH 6.0 in the presence of 0.01 M CaCl$_2$. After 70 min glucoamylase was introduced, and the process was continued at 65 °C and pH 5 for a further 120 min.[2902]

Waste and Biomass Various forms of starchy waste have been utilized in the production of ethanol (termed "bioethanol" and utilized chiefly as a motor fuel). Thus, natural-rubber waste was saccharified with enzymes commonly used for starch saccharification, followed by fermentation with *Z. mobils*.[2895] Raw refuse containing 50% starch was liquefied in acidified water and then saccharified with glucoamylase at pH 5.0 and 37 °C.[2903] *Zea mobilis* in conjunction with *S. cerevisiae* was used for fermentation of residual starch from flour milling, supplemented with crushed wheat grain.[2904] Composite microorganism preparations have also been applied for degradation of waste material. Compositions containing *Bacillus subtilis*, *B. licheniformis*, and/or *B. thermodenitrificans*, together with other *Bacilli*, are capable of degrading nonsaccharide components of waste.[2905] A mixture of alpha amylase, glucoamylase, xylanase, and cellulase is suitable for degrading vegetable and food waste.[2906] Genetically engineered yeasts have also been used to prepare commercially useful products from starch waste. Starch plays a role of both substrate and nutrient for the yeast.[2907] Banana peel was first saccharified with *Aspergillus sojae* and then fermented with *S. cerevisiae*.[2908] Cassava waste could be converted into fuel ethanol in one step provided that *S. diastaticus* was used.[2909]

Ethanol has been frequently prepared from renewable arable crops,[2910-2913] grass,[2914] and starch of wild plants.[2915] To produce ethanol from such marine plants as algae, saccharification can be performed with alpha amylase, but salt-resistant yeast is required for fermentation.[2916] Microalgae in a concentrated slurry, in the presence of a nitrogen source such as NO or NH_4^+, were fermented anaerobically in the dark to ethanol.[2917] Utilizing bamboo as the substrate for production of ethanol, the starch fraction was hydrolyzed with alpha amylase at 80–100 °C for 20–30 min and with amyloglucosidase at 50–60 °C for 30–40 min. The pentosan and cellulose fractions were digested either by xylanase and cellulase at 50–60 °C for 24–48 °C[2918] or hydrolyzed with sulfuric acid.[2919]

Biomass containing soluble starch and/or dextrins could be fermented in either batch or in immobilized-cell systems. In the batch process, *S. diastaticus* performed better than *Schwanniomyces castelli*. An immobilized combination of *Endomycopsis fibuligera* with *S. diastaticus* can also be used.[2920] Biomass having a high content of dry matter could be processed with a combination of common hydrolytic enzymes, provided that a specific system of agitation of the reactor content was employed.[2921]

Corn silage was digested with ruminal microorganisms, augmented by supplementary fibrolytic enzymes.[2914]

Fermentation Stimulators. When sucrose, salts, particularly ammonium salts, protein, and carbohydrates are present in the wort, the fermentation is rapid and the amount of by-products depends only slightly on the amount of yeast used. Without protein and other carbohydrates, fermentation of the sucrose present in the wort was slow, and larger amounts of by-products were formed. When protein was present and ammonium salts and carbohydrates other than sucrose were absent, the fermentation accelerated and smaller amounts of by-products were formed. Without salts other than ammonium salts, and without protein, fewer by-products were formed and the process was rapid.[2922]

Among the by-products formed on fermentation of starch are glycerol, some esters, and aldehydes. Admixture of these by-products to the consecutive fermentation batch leads to a higher yield of ethanol and decreases the output of glycerol and other by-products.[2923] Inclusion of ammonium salts or vitamins had little effect on the fermentation yield.[2924]

Specific Ethanol Fermentations. Alcoholic fermentation may produce various other compounds besides ethanol. Fermentation of raw starchy substrates with *Saccharomyces cerevisiae* under certain conditions leads to aldehydes and esters. Kinetic studies[2925] showed that formation of aldehydes, mainly acetaldehyde and acrolein, showed an irregular dependence on the substrate and is, to a certain extent, controlled by the pH. Esters present in the product are mainly methyl and ethyl acetates. The formation of by-products, as well as the entire fermentation process, can be controlled by electrostimulation of the microorganisms used. Pulsating, electromagnetically induced currents, simple alternating currents (ac), and even direct currents (dc) have been applied to cells, tissues, and entire organisms in order to stimulate membrane permeability and some metabolic pathways. It is probable that polarization effects, ion displacement, and dipole induction, leading to conformational changes of pore proteins, enzymes, and phase-transition phenomena of membranes may be involved.

Electrostimulated production of CO_2 by yeast is one of the examples of the application of that technique in fermentation.[2926–2928] When either a 10 mA dc current or a 100 mA ac current was applied to the culture broth of *Saccharomyces cerevisiae*, the cell growth and production rates of alcohol increased considerably. These conditions also influenced the content of higher alcohols, esters, and organic acids produced. However, several compounds, such as acetaldehyde and acetic acid, were formed from ethanol as the a result of an electrode reaction.[2929] As shown in Fig. 22, both dc and ac current stimulated *S. cerevisiae*. The yield of ethanol from glucose

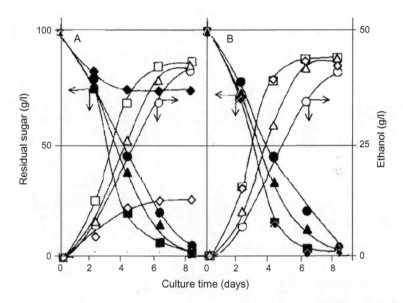

FIG. 22. Effect of dc (A) and ac (B) electric current on alcohol fermentation by yeast. Open and closed symbols correspond to concentrations of ethanol and residual sugar, respectively. Circles: control culture; triangles: 2 mA dc and 30 mA ac; squares: 10 mA dc and 100 mA ac; rhombs: 30 mA dc and 200 mA ac.[2929]

within 8 days increased with an increase in the current up to 10 mA. An increase in the yield by about 5% was achieved by the 6th day. A dc current of 30 mA decreased the yield of ethanol considerably. The application of an ac current up to 30 mA resulted in a similar (~5%) increase in the ethanol yield by the 6th day, but further increase to 30 mA ac was unnecessary because the results were practically identical to those observed at 10 mA (Fig. 22).

The differences between the use of dc and ac currents are more clearly illustrated in terms of the yield of by-products, as seen in Table XV.[2929]

Except for ethyl acetate and 2-butanol, whose yields were slightly lower, use of 100 mA ac produced the by-products listed in Table XV with higher yields than were obtained with 10 mA dc. 1-Propanol was produced in the same low yield, regardless of whether ac or dc was used. Aeration increased the yield of 2-butanol and 2-pentanol slightly, and lowered the yield of other by-products. Acetic and pyruvic acids were major by-products.

Starch–cellulose substrates can be fermented after saccharification of the material with amylase and cellulase from *Trichoderma reesei*.[2930]

TABLE XV
Formation of Organic Acids and Aromas by Yeast Under Different Culture Conditions

Components	Concentration (mg/dm)			
	Control	dc 10 mA	ac 100 mA	Aeration
Citric acid	11	17	19	13
Pyruvic acid	280	356	379	298
Malic acid	120	125	121	112
Succinic acid	109	136	123	120
Lactic acid	96	248	232	102
Acetic acid	241	1047	1143	351
Acetaldehyde	23	109	174	26
Ethyl acetate	5	22	20	9
1-Propanol	22	24	24	23
2-Butanol	24	39	16	34
2-Pentanol	36	27	29	34

(By Permission From Ref. 2929)

The hydrolysis of starch in alcoholic fermentation has been the subject of several mathematical models.[2931–2939] It follows unimolecular kinetics. The average rate constants in continuous sugar fermentation are 0.043–0.056 when malt was used and 0.048–0.061 when the fungal sponge *A batatae*-61 was applied.[2933] One of the most credible models for the hydrolysis of starch by amylase takes into account the loss of amylolytic activity resulting from interaction of the products of hydrolysis with the enzyme.[2931]

The continuous production of ethanol from wheat mashes[2940] involving externally introduced enzymes turning starchy materials into maltohexaose and then turning this into ethanol by *S. cerevisiae*, the kinetics may be presented as follows:.

The time-dependent unimolecular process is explained by Eq. (50).

$$P = 0.622a\{t - \exp(-k_1 t)\} \qquad (50)$$

where a is the initial concentration of the fermenting material in the medium, t is the time, P is the concentration of ethanol produced, and k_1 is the reaction rate constant. Figure 23 presents kinetic curves of ethanol formation in the continuous plant-scale production. There is a relationship between the alcohol yield and the specific growth rate of yeast.[2940]

b. Acetone–Butanol Fermentation.—This process was discovered by Weizmann.[2941] He found that a microorganism then called *Clostridium acetobutylicum* produces acetone and butyl alcohol from saccharide substrates. This fermentation

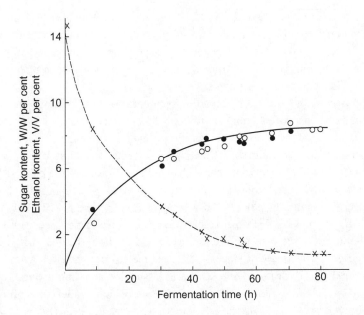

Fig. 23. Kinetic curves of the formation of ethanol in a continuous fermentation plant. Symbols: ○—based on data calculated from Eq. (1), ●—based on experimental data, x—maltohexaose content.[2940]

first produces butanoic acid, accompanied by minor amounts of propanoic and acetic acids, and oxygen. Then gradually there is evolution of CO_2 and hydrogen and butanol forms. It appeared that butanoic acid underwent oxidation to acetoacetic acid with the liberation of hydrogen. Reduction of the acid produces butanol, together with ethanol.[2942–2944] Such fermentation can be performed not only with glucose but also with mashes.[2945] In order to produce acetone, fermentation with *Clostridium* should be conducted[2946] between 28 and 32 °C, maintaining the pH between 5.8 and 6.1.

Generally speaking, the amylase system providing the acetone–butanol fermentation contains amylolytic, dextrinolytic, and saccharifying components. Inclusion of beta amylase accelerates the fermentation.[2947] Saccharification of starch is complete within 12 h, and the sugars formed are totally fermented within the next 36 h. Pentosans remain intact.[2948]

In later work, continuous fermentations have been described. They utilize *C. acetobutylicum*[2949,2950] and *C. beijerinckii* BA101.[2951] Propanol and butanol can be produced from wheat flour after extracting the gluten component.[2852]

c. Isopropyl Alcohol Fermentation.—This process is associated with the acetone–butanol fermentation. For the fermentation of corn, several enzymes can be used, among them *Clostridium* sp. 172 CY-02. The strain is cultivated at 35 °C and pH 5.5 for 72 h in liquefied starch containing 6% of glucose. Corn starch was then treated for 1 h at 80 °C, adding 0.01% of that enzyme, plus polypeptone and ammonium sulfate as the source of nitrogen.[2952] Fermentation with engineered strains of *Escherichia coli*[2953,2954] has attracted later interest.

d. 1,3-Propanediol and 2,3-Butanediol Fermentations.—2,3-Butanediol is available through fermentation with aeration of 15% wheat mash with *Aerobacillus polymyxa*. The aeration inhibited the parallel formation of ethanol. Generally, under aerobic conditions, the ratio of 2,3-butanediol to ethanol is 3:1. Under anaerobic conditions, the ratio of the two alcohols changed to 1.3:1.0, and the process accelerated. Under aerobic conditions, the process required 96 h for completion, whereas under anaerobic conditions, only 48 h was necessary.[2955] The pH had only a minor effect upon the process.[2956] Proper solubilization of the fermenting species appeared essential and positively affected the rate and yield of the fermentation.[2957] *Enterobacter aerogens* selectively produces[2958] 2,3-butanediol from a starch hydrolyzate at 39 °C and pH 6.0. Hydrolyzed and saccharified starch-containing material could be converted by *Candida krusei* into glycerol from which, under the influence of *Klebsiella pneumoniae*, *Clostridium butyricum*, or *Clostridium pasteuranum*, 1,3-propanediol or 2,3-butanediol could be obtained under anaerobic followed by aerobic conditions.[2959] *Bacillus licheniformis* B-05571 converts glycerol into 1,3-dihydroxypropanone.[2960] Under certain conditions, *E. fibuligera* can produce extracellular amylases, comprising mainly alpha amylase accompanied by a low level of beta amylase. This product is useful in fermentation of starch to 2,3-butandiol, ethanol, and acetylmethylcarbinol (acetoin, butan-3-one-2-ol).[2961]

3. Carboxylic Acid Fermentations

Starch is a suitable substrate for the fermentative production of a number of carboxylic acids. Usually, these anaerobic fermentations utilize glucose from the hydrolysis of starch, and then various enzymes convert glucose into carboxylic acids. Thus, in the production of carboxylic acids from corn meal, a coculture of *Lactobacillus lactis* and *Clostridium formicoaceticum* converted glucose into acetic acid, *Propionibacterium acidipropionici* produced propanoic acid from that source, and *Clostridium tyrobutyricum* provided butanoic acid.[2962] In order to produce itaconic acid from a starch hydrolyzate, *Aspergillus terreus* was employed.[2963] Poly(3-hydroxybutanoate) is produced from starch by using *Raistonia euthropha*.[2964,2965] Starch waste can also be a suitable substrate.[2966]

Some organisms isolated from soil[2967,2968] are capable of fermenting hemicelluloses and starch into butanoic acid and the latter into butanol. Some of them were thermophilic. The production of butanol can be suppressed by decreasing the acidity of the mash with sodium and/or calcium carbonates[2969] Polyunsaturated fatty acids could be produced from starch-processing wastewater by using a fatty acid acidophilus.[2970]

a. Lactic Acid Fermentation.—The pyruvate resulting from glycolysis is further oxidized completely, generating additional ATP and NADH in the citric acid cycle and by oxidative phosphorylation. However, this process can occur only in the presence of oxygen. Oxygen is toxic to organisms that are obligate anaerobes and is not required by facultative anaerobic organisms. Lactic acid fermentation is one of the processes for regenerating NAD^+ in the anaerobic processes, that is, in the absence of oxygen.[2971]

Once glucose is generated from starch, it is split through glycolysis into pyruvic acid, and lactic acid fermentation may start. There are two pathways, homolactic and heterolactic (phosphoketolase). In the first of them, one glucose molecule produces two molecules of lactic acid:

$$C_6H_{12}O_6 \to 2CH_3COCO_2 \to 2CH_3CH(OH)CO_2H$$

whereas in the second process, from one glucose one molecule of lactic acid is produced, together with one molecule of ethanol and one molecule of CO_2:

$$C_6H_{12}O_6 \to 2CH_3COCO_2H \to CH_3CH(OH)CO_2H + C_2H_5OH + CO_2$$

Several fungi and bacteria cause lactic acid fermentation. *Lactobacillus* is the most common among them. *Leuconostoc mesenteroides*, *Pediococcus cerevisiae*, *Streptococcus lactis*, and *Bifidobacterium bifidus* are also quite common.[2972] There are also other, less-common, microorganisms that are capable of lactic acid fermentation, for instance, *Diphtheria bacilli*.[46] The capacity of microorganisms for lactic acid fermentation can be ranked on the basis of their ability to ferment 0.5% raw potato starch in agar.[2973] When starch is fermented, not only does it play a role as the source of lactic acid via glucose and pyruvic acid but it is also an indispensable nutrient for fermenting microorganisms.[2974]

Various fermenting organisms are used for production of fermented food. Yogurt is, perhaps, the most common such food. Its production involves *Lactobacillus bulgaricus* and *Streptococcus thermophilus*. For making probiotic yogurt, *Lactobacillus acidophilus* is used. In Central Europe kefir, sauermilk, and buttermilk are common milk-derived drinks.

Sauerkraut is usually produced by fermentation of cabbage with *Leuconostoc*. From the common Korean fermented food—kimchi—*Lactobacillus kimchii* was isolated.[2975,2976] Different parts of the world have local fermentation products. Thus, in Poland, cucumber pickles and fermented rye-based soup—zhur—are common in the national cuisine. Fermented olives are typical for countries of the Mediterranean region. Fermented vegetables feature widely in Japanese cuisine. A fermented maize porridge called magou is produced in South Africa. Thai cuisine offers fermented fresh pork called bham, and in the Philippines, a fermented rice dish with shrimps, termed balao balao, is served. Throughout the world, a wide variety of fermented cheeses are common. Cottage cheese and cheese of ewe's milk (bryndza, feta, ricotta, pecorino, romano, and roquefort) are produced by thermal coagulation of fermented milk. Fermented hard cheeses are produced by various bacteria and technologies; all of them basically involve lactic acid fermentation. In the Orient, fermented soy protein is offered as tofu; when it is fermented with various additives such as herbs, shrimps, and other ingredients, it is presented as "stinky tofu" (chou tofu).

The enzymatic synthesis of lactic acid in the muscles plays an important physiological role as a means of signaling the physical exhaustion of the organism. The lactic acid enzymes of the muscles can also produce lactic acid *in vitro* from starch and glycogen.[2977,2978]

Lactic acid fermentation is employed for the industrial-scale production of lactic acid (see, for instance, Refs. 2979, 2980). Simple lactic acid fermentation of cereal or tuber starches with *L. thermophilus* at 50 °C proceeded to 50% within 5 days and to a 60% fermentation ratio after 10 days. The yield of lactic acid reached 80%. Neither bran added as a nutrient nor an increase in the concentration had any effect on that fermentation ratio. Inclusion of *L. delbrueckii* increased that ratio by 10–15%, and the addition of 1% aq. NaCl positively influenced the fermentation ratio.

The lactic acid molecule is chiral, and both the R and S enantiomers are available. Some methods provide the pure enantiomers. S-Lactic acid [L(+)-lactic acid] results from fermentation of corn starch with *Rhizopus* strains,[2981] mutant strain *Rhizopus* sp. MK-96-1196,[2982] *Rhizopus arrhizus*,[2983] *Rhizopus oryzae*,[2984,2985] *L. delbrueckii*,[2986] *Sporolactobacillus inulinus*,[2986] *L. amylophilus* GV6,[2987–2989] and *L. manihotiovorans* LMG 18010T.[2990] *Sporolactobacillus inulinus* was demonstrated to be suitable for production of R-lactic acid [D-(−)-lactic acid] in high optical purity and with up to 82.5% yield.[2991]

Streptococcus bovis 148 was found to produce D-(−)-lactic acid directly from soluble and raw starch substrates at pH 6.0. Productivity was highest at 37 °C, with 14.7 g/L lactic acid produced from 20 g/L raw starch. *Leuconostoc mesenteroides* provided D-lactic acid and D-mannitol from a starch hydrolyzate converted first into a glucose/fructose blend. D-Glucitol was a by-product.[2992]

Polysaccharides, among them starch and even starchy plants, have frequently been subjected to lactic acid fermentation. The highest (97%) yield of lactic acid was provided by adding barley bouillon paste added to 1% aq. NaCl and 5 days of culturing. Corn starch appeared more resistant to that fermentation.[2993] *Leuconostoc delbrueckii* ferments not only potatoes (up to 69% yield)[2994] but also molasses and starch hydrolyzates after 50–100 h at 45–50 °C.[2995] The organism provides simultaneous saccharification and fermentation of potato and pearl cassava starch. The rate of the two steps operated simultaneously was always higher than the rate of the separate processes operated consecutively.[2996] The same microorganism was applied for production of lactic acid from corn starch[2997] and rice starch.[2991,2998] Corn starch could serve as a substrate for the production of lactic acid in a simultaneous process when glucoamylase was added to the fermented broth.[2999] When the fermenting broth was supplemented with 5% of reducing sugars, *L. plantarum* performed well with cassava starch, cassava flour, cassava hydrolyzate,[3000,3001] with potato starch, preferably of damaged granules,[3002] and raw sweet-potato starch.[3003] *Leuconostoc plantarum* can convert the lactic acid formed into acetic acid. When glucose, maltose, or cellobiose are added to the reaction mixture, lactic acid remains left intact. The presence of oxygen favors stoichiometric conversion of lactic acid into acetic acid.[3004] Starch of sweet potato was also fermented with *Rhizopus oryzae*.[2984] Enrichment of cassava starch with yeast protein was beneficial.[3005] The same microorganism was used in the production of pickles from sweet potatoes.[3006] Among the microflora in fermented, sour cassava starch, *Streptococcus*, *Bacillus*, *Lactobacillus*, and *Saccharomyces* could be identified, but lactic acid bacteria predominated and the contribution from the molds was insignificant.[3007] Waste from potato processing can also be readily fermented by *Lactobacilli* at pH 5.5. Because that waste is rich in proteins, the resulting fermentation product wherein lactic acid was converted into its ammonium salt appeared to be an excellent source of nitrogen for ruminants.[3008] In addition, *L. amylovorus* is suitable for the fermentation of raw-potato starch.[3009] That microorganism can be used in an immobilized form.[3010] *Lactobacillus amylophilus* GV6 produced L-(+) lactic acid from wheat bran in an anaerobic, solid-state process.[2988]

Potato hydrolyzates resulting from treatment of the substrate with alpha amylase and glucoamylase, containing 8–13% starch, are suitable sources for the production of lactic acid by the action of *Lactobacilli* at 50 °C for 4 days. A minor amount of acetic acid is a by-product.

Pretreatment, namely the liquefaction of starch in a solution of lactic acid has been proposed. Such pretreatment led to a more extensive hydrolysis of the starch substrate.[3011] *Lactobacillus* L2G-15, isolated from germinated corn, converted saccharified mashes of potato and sweet potato, utilizing almost 95% of the starch.[3012] Corn in swine-feed waste was fermented with *L. amylophilus*.[3013]

Fifteen isolates from 81 *Streptococci*, screened for fermenting raw rice, were evaluated for their ability to ferment sago starch and were found useful.[3014] Comparative studies on fermentation of sago and cassava starches using *Lactococcus lactis* IO-1[3015] showed that cassava starch was saccharified more readily, in times of 48 and 24 h, respectively. This strain permitted continuous fermentation by using a pH-dependent feed system.[3016,3017] Sago palm was saccharified with *Pycnoporum coccineus* PH1h, a fungus of wood. The hydrolyzate was then fermented with *Streptococcus bovis* JCM 5802. Within a week of culturing, a 40% yield of lactic acid could be achieved.[3018]

The use of *Lactococcus lactis* enabled simultaneous saccharification and fermentation of wheat starch to lactic acid.[3019] *Lactobacillus fermentum* Ogi E1, isolated from Benin maize sour dough, is an acid-tolerant microorganism that performed well at pH 4, producing lactic acid from sour dough. It is a dominant strain in such African food.[3020] A traditional Mexican fermented maize beverage—pozol—requires weakly amylolytic but fast-growing lactic acid bacteria, and *Streptococcus bovis* predominated among the other microorganisms involved; these were *Streptococcus macedonis*, *Lactococcus lactis, and Enterococcuws sulfurous*.[3021]

Bacteria can be used either in a free or immobilized form, in continuous, semi-continuous, or batch-wise processes.[3022] Continuous operation is possible with raw starch by using *L. casei* immobilized on κ-carrageenan.[3023]

Starch and food waste appeared to be a good source of L-(+)-lactic acid when *Lactobacillus manihotivorans* LMG 18011 was employed.[3024] In the fermentation of barley[3025] and maize[3026] silage, *Lactobacillus buchneri* was used. Various enzymes were used as supplements to ensilaged whole-plant barley, and depending on their nature and their concentration, lactic, acetic, and propanoic acids were produced, together with some ethanol. Wheat forage, as well as varieties of pea, has been fermented with heterolactic and homolactic bacteria.[3027]

b. Citric Acid Fermentation.—This commercially important product comes, among others, from enzymatic processes in which *Aspergillus niger* and its mutants are frequently employed. Glucose or sucrose usually constitute the substrate, but there have also been attempts to produce citric acid directly from starch, generally in solid-state fermentations. Yields of citric acid may reach as high as 90%, but vary according to the strain of *A. niger* applied, the type of fermented substrate, the initial moisture content, the concentration of methyl alcohol present, the temperature, and the duration of the process. Agitation is also an essential factor. Among starches, maize starch and its hydrolyzates have been the most frequently selected as substrates.[3028–3035] There is also one report describing the formation of citric acid from cereals, using the *Lactobacillus fermentum* Ogi E1 strain. The yield of this process exceeded[3036,3037] 88%. Two Russian patents describe the use of bacterial alpha amylase.[3046,3047] Fermentation of gelatinized starch present in cassava bagasse in a horizontal drum bioreactor has been reported.[3037]

c. Acetic Acid (Vinegar) Fermentation.—This fermentation converts ethanol into acetic acid by the involvement of acetic acid bacteria. These genera include aerobic bacteria that are usually gram-negative. Some of them, such as *Acetobacter*, produce acetic acid, which is subsequently oxidized to CO_2, while some microorganisms, for instance, *Gluconobacter*, complete their action at the stage of producing acetic acid. These genera also include such known anaerobic bacteria as *Acetobacter woodii*, which reduce CO_2 to acetic acid.[3038–3042]

Because ethanol is the common substrate in this fermentation, starch should be considered simply as the source of glucose and ethanol. Modification of the pathway from starch to acetic acid by the addition of *Aspergillus niger* and water in the acetic acid fermentation stage was suggested.[3043] It elevates the rate of fermentation and increases the level of conversion of starch.

d. Pyruvic Acid Fermentation.—Pyruvic acid is formed from carbohydrates on glycolysis, as described in Section III. Production of that acid on the technical scale can be performed by using various saccharides and polysaccharides, including starch, and involving *Trichosporon cutaneum*. Saccharides and/or polysaccharides are the carbon source and corn slurry, beef extract, yeast extract, peptone, soybean, ammonium salts, or urea are the nitrogen sources.[3044]

e. Gluconic Acid Fermentation.—*Aureobasidium pullulans* produces D-gluconic acid and D-glucono-1,4-lactone from saccharides, preferably from glucose, and both discontinuous and continuous processes have been designed for this purpose.[3045,3046] Both of these have been produced from starch hydrolyzates by employing *Pullularia pullulans*. The process is performed with aeration and agitation at 26–28 °C and pH 7.8–8.0 and afforded sodium gluconate in 97% yield. The yield of the lactone was 85%.[3047] Starch hydrolyzates were further converted into gluconic acid in 74% yield when *A. niger* ORS-4 was used. Molasses is also a good substrate for production of gluconic acid, provided that this substrate is pretreated with hexacyanoferrate.[3048] Preliminary treatment with glucoamylase at 60 °C is followed by cooling to 30 °C and the addition A. niger.[3049]

f. Glutamic Acid Fermentation.—Glutamic acid can be obtained by fermentation of carbohydrates with one of the *Brevibacterium*, *Arthrobacter*, *Microbacterium*, or *Corynebacterium* species. Prior to the step of fermentation to glutamic acid, corn starch is digested with alpha amylase and a saccharifying enzyme, the optimum for which was 320 units/g starch.[3050]

g. Kojic Acid Fermentation.—*Aspergillus flavus* Link grows in cooked starch, producing kojic acid.[3051] Among sago, potato, and corn starches, the last appeared to be the superior substrate. The optimum concentration of starch slurry was 7.5 w/v%, and inclusion of 10% of glucose as the carbon source increased the yield of kojic acid. Gelatinized sago starch can also be converted directly into kojic acid. Controlling the pH during the process is critical.[3052]

VI. Nonamylolytic Starch Conversions

1. Glycosylation

The use of microorganisms and enzymes is not limited to the degradation of starch and its fermentation. There are several reactions on functional groups of the D-glucose residues in starch. α-Glucosylation is one such reaction. Thus, alpha amylase from *A. niger* can produce methyl glycopyranosides with 90% relative conversion when it acts on starch in 20% aqueous methanol.[3053] Ethyl α-D-glucopyranoside can be prepared in 17% yield from starch when it is digested for 24 h at 30 °C in an ethanol slurry with ethanol-resistant *Aspergillus kawachi* N-3. In this instance, α-glucosidase is the functioning enzyme.[3054] Ethyl α-D-glucopyranoside is responsible for the specific aroma of sake and mirin. Switching to other alcohols in the slurry, and using the alpha amylase of *Aspergillus oryzae*, a series of alkyl glycosides could be prepared.[3055,3056] Alcohols of more complex structure could be used for glycosylation, but hexanol and octanol appeared unsuitable for this purpose.[3055] However, benzyl α-D-glucopyranoside and -maltoside were prepared from soluble starch by transglycosylation with alpha amylase.[3057] α-Glucosidase rapidly hydrolyzes these glycosides. Maltooligosaccharide derivatives incorporating genipin [methyl (1*R*,2*R*,6*S*)-2-hydroxy-9-(hydroxymethyl)-3-oxabicyclo[4.3.0]nona-4,8-diene-5-carboxylate, **2**], used for analysis of alpha amylases, have been prepared from geniposide (**3**) and a starch hydrolyzate in an aqueous hydrophilic solvent with *Pseudomonas stutzeri* amylase added.[3058]

2 Genipin

3 Geniposide

Maltotetraose, which has several applications in the food and pharmaceutical industries, is accessibly by use of maltotetraose-forming (maltotetraose) amylases (EC 3.2.1.60),[447,1238,3059–3061] as, for instance, those from *Pseudomonas stutzeri*.

Several alpha amylases, but not all of them, are capable of glycosylating starch. The liquefying enzyme isolated from *B. licheniformis* and *B. stearothermofilus* does not work.[3053]

Transferases can be effectively used in glycosylation. A cyclodextrin glycosyltransferase (CGTase) was successfully used for the preparation of benzyl and phenyl α-D-glucopyranoside derivatives.[3062] The production of glycosides with the CGTase of *Brevibacterium* species has been patented.[3063] Using an alkalophilic CGTase in a slurry containing 13% sucrose and 13% potato starch at 85 °C, the starch is first liquefied and then at 55 °C both components are stated to undergo coupling.[3064] The 3-*O*-β-D-glucopyranosyl derivative **4** of quercetin [2-(3,4-dihydroxyphenyl)-3,5,7-trihydroxy-

4H-chromen-4-one] could be converted into 3-O-glycosides by using corn starch in the presence of CGTase.[3065] Using *B. macerans* CGTase [EC 2.4.1.19] moranoline (1-deoxynojirimycin, 5-amino-1,5-dideoxy-D-glucopyranose, **5**) could glycosylate soluble starch.[3066] With the same transferase, α-CD was glycosylated by compound **5**.[3067] Using CGTase and glucoamylase together, 6-O-α-D-glucosyl-α-cyclodextrin,[3068,3069] and 3- as well as 4-glycosylated catecholamines[3070] could be prepared from hydrolyzed potato starch.

4 3-O-β-D-Glucopyranosylquercetin

5 1-Deoxynojirimycin

Starch can be bound to xylitol using CGTase.[3071] With the same enzyme, transglycosylation can be effected between an acyl monosaccharide or acyl oligosaccharides and starch.[3072] Several oligosaccharides can be coupled with starch, and CGTase is not the only useful enzyme.[3073,3074] Alpha amylolysis of starch in the presence of alditols leads to heterooligosaccharide derivatives.[3075] It has been found that a branching enzyme (RBE1) from rice, under laboratory conditions can introduce branches on rice, potato, sweet potato, wheat, corn, sago, and cassava starches. The reaction is controlled by branches existing initially in these starches.[3076]

Candida transglucosyl amylase was isolated from *Candida tropicalis* yeast.[3077] It transferred α-D-glucopyranosyl groups from starch onto such alcohols as glycerol and alditols. Amylomaltase (EC 2.4.1.25) transfers one α-(1→4) glucan chain either onto another such glucan or to glucose. Amylomaltase resides in numerous plants and microorganisms where it acts as a disproportionating enzyme. The one isolated from *Thermus thermophilus* performs best at pH 6.2 and 68 °C, yielding a white gel-like thermoreversible product.[3078,3079] *Pyrobaculum aerophilum* IM2 is another microorganism secreting amylomaltase.[1313]

Glycosylations of glycyrrhizin,[3080] capsaicin,[2518] and ascorbic acid[3081,3082] have been described. Glycosylation of glycyrrhizin to prepare a low-calorie sweetener was performed by incubation of glycyrrhizin with starch and CGTase. L-Ascorbic acid was transglycosylated with a maltooligosaccharide or starch involving either cyclomaltodextrin glucanotransferase or α-glucosidase. The α-glucosyl-L-ascorbic acid that is formed is a stable source of vitamin C.[3081,3082] A compound termed "lactoneotrehalose" [β-D-galactopyranosyl-(1→4)-β-D-glucopyranosyl-(1↔1)-α-D-glucopyranose] was also prepared[3083] from starch and lactose in the presence of CGTase. These products serve as a sweeteners for chewing gum and chocolate, and as additives in pharmaceuticals and cosmetics.

2. Esterification and Hydrolysis

Acylation of starch was performed with (*R*)-3-hydroxybutanoic acid in the presence of Novozyme 435 lipase B. All of the hydroxyl groups in the glucose residues of starch participate in esterification of the (*R*)-3-hydroxybutanoic acid. D-Glucose by itself as the acceptor produced tri- and tetra-substituted products in 30% yield.[3084] 2-*O*-α-D-Glucopyranosyl-L-ascorbic acid was claimed to be made from L-ascorbic acid and an α-glucosyl donor saccharide and an α-isomaltosylglucosaccharide-producing enzyme.[3085] The cross-linking of maltodextrins by enzyme-facilitated esterification has been patented.[3086] Phosphorus oxychloride, sodium trimetaphosphate, or sodium polymetaphosphate was used in conjunction with the esterifying agent, octenyl, succinic, or dodecyl succinic anhydride, and alpha amylase was the enzyme. The sodium salt of octenyl succinylated starch was hydrolyzed in a continuous process in a membrane reactor, using either bacterial or fungal alpha amylase. Bacterial amylase performed better.[3087] Amylases normally hydrolyze nonmodified starches, but acetylated starches may require acetylesterases.[3088]

"Nanostarch" (a starch composed of particles of size 10–100 nm produced by milling and/or either multiple freezing and thawing or controlled enzymatic digestion) in emulsions could be esterified with vinyl octadecanoate, 2-oxepanone (ε-caprolactone), and maleic anhydride, using *Candida antarctica* lipase B, either in free or immobilized form. The esterification was regioselective, taking place at C-6 of the D-glucose residues.[3089,3090] Maize and cassava starches could be esterified with coconut oil in the presence of microbial lipase. The use of microwave radiation facilitated the process. The degree of esterification of both starches was greater than unity.[3091]

Phosphatase of potato readily hydrolyzes natural starch phosphates (potato amylopectin)[3092] and synthetically prepared starch phosphates,[3093] leaving glycosidic bonds intact. On the other hand, corn starch can be phosphorylated with Na_2HPO_4 in the presence of isoamylase.[3094]

Phosphorylases (EC 2.4 and 2.7.7), as found in potato juice split reversibly the $(1 \rightarrow 4)$ but not the $(1 \rightarrow 6)$ bonds in starch. They differ from the phosphorylase of autolyzed yeast[3095] which is capable of phosphorylating glucose at the anomeric position.[3096] Potato phosphorylase can, however, utilize glucose 1-phosphate in the synthesis of amylopectin[1251,3097,3098] and transfer the phosphate group from glucose 1-phosphate onto amylose, amylopectin, and starch.[3099] Phosphorylating phosphorylase is present in potatoes, turnips, and pumpkins.[3098] In the presence of sucrose phosphorylase, glucose 1-phosphate and fructose combine to give sucrose.[3100] When starch phosphorylase is used, sucrose is produced directly from starch and fructose.[3101] The starch phosphoylase generates glucose 1-phosphate, which then combines with fructose to give sucrose. Potato and rabbit-muscle phosphorylases are inhibited by hydrosulfite anion at pH 6.0, but not by sulfate, azide, and cyanide anions. The hydrogencarbonate anion exhibited a weak inhibition at pH 8.0.[3102] Phlorizin (**6**) also inhibits potato phosphorylase slightly.[3097] Pea cotyledon starch phosphorylase phosphorylates glucose into glucose 1-phosphate and the product does not inhibit the enzyme.[3103]

6 Phlorizin

Cooperation of phosphorylase with pullulanase in phosphate buffer at pH 6.8 provided α-D-glucose 1-phosphate within 72 h at 25 °C.[3104] Waste potato juice from the manufacture of starch is a good source for production of glucose 1-phosphate, using alpha amylase and phosphorylase.[3105] Seeds of wrinkled pea possess the enzyme system capable of synthesizing glucose 1-phosphate from starch.[3106] The enzymes can be used in immobilized form, for example, on a barium alginate or DEAE-cellulose matrix. Starch phosphorylase immobilized on a barium alginate and DEAE-cellulose matrix showed a higher specific activity than that immobilized solely on DEAE-cellulose. Immobilization on that matrix raised the optimum pH, whereas immobilization on alginate shifted it down.[3107]

Amylase hydrolyzed starch in the presence of inorganic phosphates to produce glucose 1-phosphate, which is then converted into α-glucose 6-phosphate with the phosphoglucomutase isolated from potatoes and activated by Mn^{2+} ions.[3108,3109] The use of alkaline phosphomonoesterase from dog intestinal mucosa has been described.[3110] Aqueous extracts and plasma pressed from rabbit skeletal muscles caused rapid phosphorylation of starch by inorganic phosphates. Extracts from brain, kidney, and liver appeared to be inactive.[3111] An extract of rabbit muscle, autolyzed and purified by dialysis, phosphorylated starch to produce glucose 1-phosphate.[3112] Formation of starch from glucose 1-phosphate is reversible when catalyzed by potato phosphatase.[249] Brain phosphatase can detach the phosphate group in potato starch.[3113] Amylophosphorylases can be isolated from potato, sweet potato, pumpkin, and turnip. They are also available synthetically, preferably when a 1% starch solution is blended with Michaelis phosphate buffer at pH 7.0 and conditioned at 37 °C for 1 h.[3908]

In the pollen of *Typha latifolia*, two phosphorylases were found. One of them has M_w 112 kDa. Both worked best at pH 5.5–5.8 to form inorganic phosphates and at pH 7.3–8.0 to produce glucose 1-phosphate.[3114]

Starch, especially cassava starch, underwent esterification by coconut oil upon microwave heating the mixture with added lipase.[3091] Lipase-catalyzed acylation of starch can be carried out without microwave assistance.[3115]

3. Methanogenic and Biosulfidogenic Conversions

Glucosidases are sensitive to sulfides. Sulfides inhibit α-glucosidase and stimulate β-glucosidase.[3116] This behavior suggests an approach to degradation and utilization of starch and cellulose substrates in wastewater sludge under biosulfidogenic (biological sulfate reduction) conditions. Under anaerobic conditions, potato starch waste is converted by methanogenic bacteria into methane and CO_2.[3117]

4. Isomerization

Glucose isomerase (EC 5.3.1.5) can be isolated from several sources,[3118–3120] and a genetically engineered glucose isomerase is also available[3121,3122] and can be used following immobilization on silica beads.[3123,3124]

Glucose isomerase converts the products of hydrolysis of granular starch with alpha amylase and glucoamylase into D-fructose (levulose).[3125–3140] The conversion may proceed only to 45–46%,[3123,3141] but if the temperature of the process is close to 110 °C, some of the glucose is converted into fructose to yield a syrup containing 55% fructose.[3142] The yield of isomerized product can be increased if enzymes are added directly to an aqueous starch slurry prior to gelatinization and, preferably, if the process is conducted below the gelatinization temperature.[3143] Optionally, the partially isomerized mixture of fructose and glucose, also containing oligosaccharides, can be hydrolyzed with glucoamylase to afford a purely glucose–fructose product.[3144,3145] A glucoamylase from an *A. niger* mutant could directly saccharify 5% slurries of potato or wheat flour at the optimum 60 °C temperature to a glucose syrup in 86 and 87% yields, respectively.[1164] The resulting syrup was then isomerized at 60 °C with immobilized glucose isomerase from *Streptomyces murinus* providing a product composed of 50% glucose and 50% fructose. When raw cereals served as the substrate, the removal of proteins prior to the glucose–fructose isomerization is advantageous.[3146,3147] In the production of fructose from cassava starch, isomerization of glucose was performed with immobilized glucose isomerase, yielding a blend of 36.9% fructose, 60.7% glucose, and 2.4% of maltose together with higher saccharides.[3148] A higher (42–45%) yield of fructose was afforded when thinned starch was treated with immobilized glucoamylase under saccharifying conditions, followed by treatment with immobilized glucose isomerase.[3149] Isomerization proceeds with higher yields in concentrated glucose solutions.[3150] Glucose isomerase is also suitable for the processing of starch hydrolyzates.[3145,3151] A high-fructose syrup can be produced from bagasse by using immobilized *Lactobacillus* cells having glucose isomerase activity.[3152]

Glucoamylase and glucose isomerase can be immobilized separately on porous silica gel beads and then the beads with both enzymes are mixed together.[2528] Glucoamylase immobilized together with mycelia-bound glucose isomerase facilitates the continuous production of fructose syrup directly from liquefied starch.[3153] Cationic and anionic Sephadexes are suitable supports. Conferring an electric charge on the enzyme opposite to that of support improves the binding capacity of the support and also the thermal stability of the enzyme.[3154]

Separation of fructose from invert sugar or a starch hydrolyzate can be performed by converting the glucose component into maltose with CGTase.[3155]

Glucose 6-phosphate isomerase, derived from *Thermus* species, isomerized glucose 6-phosphate to fructose 6-phosphate. The substrate was available from starch by digestion with pullulanase of the same origin as the isomerase.[3156] Such a conversion of a starch hydrolyzate into fructose is possible on anion-exchange resins.[3157]

Progress of the glucose-to-fructose isomerization is controlled by the concentration of both components, as is the reverse process. In order to modulate that process, the phosphorylation of glucose and fructose into the respective 1- and 6-phosphates is effected by using inorganic phosphate. In a transaldolase-catalyzed reaction in the presence of glyceraldehyde, the fructose 6-phosphate then forms glyceraldehyde 3-phosphate and fructose.[3158]

1,5-Anhydro-D-fructose can be prepared from starch by using an α-1,4-glucan lyase (EC 4.2.2.13) from the marine alga *Gracilariales*.[142]

Syrups containing maltulose are available by isomerization of maltose syrups. The latter are digested with glucose isomerase, an enzyme that does not act on maltulose, but the glucose component of maltose is converted into fructose.[3159] Glucose present in maltose syrups can be isomerized at pH >7 with glucose isomerase.[3160] Maltose and other oligosaccharides are readily attacked by isomerase when the solution contains only a minor amount of glucose, or glucose is absent.[3151,3161,3162]

5. Hydrogen Production

A number of strategies have been employed for the fermentative production of hydrogen, but the selection of economically suitable enzymes and substrates is difficult. Chinese scientists have proposed the utilization of a mixture of paper-mill wastewater sludge for fermentation of its starch content at 35 °C and pH 5–7. They noted a significant increase in the rate of hydrogen production.[3163] Anaerobes present in the sludge can be activated by supplementing the hydrolyzed wort with compounds of nitrogen, phosphorus, and iron to maintain the correct N/C, P/C, and Fe(II)/C ratios. When using hydrolyzed wheat starch as the substrate, that ratio should be[3164] N/C < 0.02, P/C < 0.01, and Fe(II)/C < 0.01. Anaerobic *Enterobacteriaceae* SO5B isolated from soil produced hydrogen efficiently from starch rather than from glucose. However, not all starches were good substrates. Potato and corn starches were more suitable than wheat starch.[3165] *Enterobacter asburiae* SNU-1, also isolated from the soil of landfills, enabled the production of hydrogen, mainly from the solid matter in the reactor. The optimum pH was 7.0 and the optimum concentration of glucose was 25 g/L.[3166] The optimum conditions for *Enterobacter aerogenes* are pH 6.1–6.6 at 40 °C.[3167]

Hydrogen can be produced from starch and other saccharides such as sucrose by using enzymes immobilized on a colloidal solution of poly(vinyl alcohol) with bentonite as the adsorption carrier and with inclusion of cysteine or cystine, ammonium molybdate, oxalate, phenolate, or sulfate, and sulfates of Fe(II), Ni(II), Fe(III), Mg, K, or Mn(II).[3168] In another application, starch was first subjected to digestion in the dark with *Cladimonas taiwanensis* On-1 bacterium to produce reducing sugars at pH 7.0 and 55 °C. In a subsequent step, *Clostridium butyricum* CGS2 at pH 5.8–6.0 and at 37 °C was employed. The effluent from this stage was converted into hydrogen by using *Rhodopseudomonas palustris* WP3-5 under illumination with 100 W/m^2 at pH 7.0 and 35 °C.[3169] *Clostridium butyricum*, either followed by *Rhodobacter* or jointly with *Enterobacter aerogenes*, could also produce hydrogen from sweet-potato residue.[3170] A rapid mode for generation of hydrogen from waste has also been presented.[3171] The continuous production of hydrogen was made possible when the hyperthermophilic archaeon, *Thermococcus kodakarensis* KOD1, was allowed to operate on a substrate supplemented with either starch or pyruvate.[3172] All members of the *Thermatogales* family are capable of generating hydrogen from combined carbohydrate and protein substrates. *Thermatoga neapolitana* seems to be particularly useful, as it provides a high production of hydrogen, with CO_2 as the sole contaminant.[3173]

6. Trehalose

Starch is a good source for the production of trehalose. Among potato, sweet potato, corn, wheat, and cassava starches, the last appeared to be superior for this purpose.[3174] Enzymes causing limited hydrolysis of starch, but producing nonreducing saccharides should be used.[3174,3175] With either maltooligosyl trehalose synthase (EC 5.4.99.15) or maltooligosyl trehalose trehalohydrolase (EC 3.2.1.141) from *Arthrobacter* sp Q36, and isoamylase at 35–40 °C and pH 5.6–6.4, a 0.3% starch slurry yielded 85.3% of trehalose.[172,3174] A thermostable enzyme from *Sulfolobus acidocaldarius* is also suitable,[172,3175] but the isoamylase from *Pseudomonas amyloderamosa* was better for this purpose. It works preferably at pH 5.5 and 55–57 °C.[3176,3177] Maltose phosphorylase, trehalose phosphorylase, beta amylase, and such starch-debranching enzymes as pullulanase were evaluated. Their application as cocktails appeared more suitable than using these enzymes separately. No by-products were formed.[3178] Good results were achieved when liquefied starch was incubated with maltose synthetase, thermostable maltose phosphorylase, and thermostable trehalose phosphorylase.[3179] *Sulfolobus solfataricus*, when immobilized in calcium alginate beads, produces at 70 °C glucose and trehalose, and maltose, maltotriose, and maltotetraose are absent.[3180] The manufacture of trehalose from soluble starch employing *Saccharomycopsis fibuligera* sdu A 11 strain has been described.[3181]

Neotrehalose (α-D-glucopyranosyl β-D-glucopyranoside) and centose [α-D-glucopyranosyl-(1 → 4)-[α-D-glucopyranosyl-(1 → 2)]-β-D-glucopyranose] can be manufactured by using β-cyclodextrin synthase isolated from soil. The process carried out for 2 days at 40 °C and pH 6.0 provided 45% yield of a blend of both products.[3182]

7. Bacterial Polyester Formation

Sago starch hydrolyzed with alpha amylase and either glucoamylase or pullulanase produced a poly(3-hydroxybutanoate) when digested with a photosynthetic bacterium *Rhodobacter spheroids*.[3183] The same polymer is available from either starch, molasses, or glucose fermented with *Bacillus aureus*.[3184]

8. Branching of Starch

A glycogen-branching enzyme from *Neurospora crassa* N2-44 produced an increased number of branches in waxy corn starch.[3185]

9. Oxidation

Laccase, a group of multicopper enzymes of fungal origin, is capable of oxidizing granular starch. Laccase transfers a single electron to an electron acceptor, namely, an oxygen molecule. These processes usually require mediators,[3186] for example, TEMPO [2,2,6,6-tetramethylpiperidin-1-yl)oxyl].[3187] The oxidation takes place at C-6 of the glucose residues, generating aldehyde and carboxylic groups. The process proceeds at pH 5, and the degree of oxidation to aldehyde ranges from 0.16 to 16.4 groups/100 AGU and to carboxylic acid from 0.01 to 3.71 groups/100AGU.

Starch treated with yeast produces CO_2. The yield and rate of that process can be enhanced by preliminary treatment of the starch with lipases, which decompose residual lipids in the starch.[3188]

10. Polymerization

A synthetic starch can be prepared from glucose 1-phosphate and purified potato phosphorylase,[3189] and dextrins of low molecular weight can readily be polymerized by isoamylase (amylosynthase),[1227] regardless of their provenance.[3190] That enzyme, isolated from autolyzed yeast, acting at pH 6.0–6.2 and 20 °C converted glutinous rice starch into an amylopectin fraction of average M_w of 264 kDa and an amylose fraction of average M_w 16 kDa.[3191–3193] Isoamylase is also capable of converting amylopectin into amylase.[3194]

Potato phosphorylase is capable of forming starch from glucose 1-phosphate in a reaction of zero-order rate at the early stage of reaction, until the phosphorus content ratio P_{inorg}/P_{total} in the product reaches 0.2. The optimum pH is 5.9,[3195–3397] but the working pH extends up to 8.3.[3198] The phosphorylase from lima bean can also behave similarly.[3199]

The enzyme that catalyzes transfer of glucose from UDP-glucose (uridine 5′-α-D-glucopyranosyl diphosphate) to starch[3200] is 10 times less efficient than that transferring glucose from ADP-glucose (adenosine 5′-α-D-glucopyranosyl diphosphate) to starch.[3201] Starch synthesis in coacervates involves a phosphorylase isolated from sliced potatoes.[3202] Glucose 1-phosphate enters the droplets and maltose simultaneously migrates out of the droplets.

Production of starch gels in which oligosaccharide chains are terminated with fructose has been patented.[3203] For this purpose, starch gel mixed with fructose is treated with an alpha amylase and/or an enzymatic cocktail of enzymes. The product is sweeter and less prone to crystallization.

When starch is sequentially treated with α-1,6-glucosidase from *B. acidopullulyticus* for 3 h at pH 5.2 and then incubated for 15 min at pH 5.5 and 50 °C with alpha amylase, a short-chain amylose could be isolated.[3204] Maltotriose in the presence of 4-α-glucanotransferase (D-enzyme, EC 2.4.1.25) produces pure maltodextrins with only α-(1 → 4) glucosyl bonds linking the glucose residues.[3205,3206]

Cellulases, for example β-glucanases and β-glucosidase from *Fusarium moniliforme*, were used in the production of starch granules of red bean by selectively digesting nonstarchy components of the beans.[3207]

Q-enzyme, which is a cross-linking enzyme that produces amylopectin from amylose and, moreover, has some liquefying properties; it also shows some phosphatase activity.[1251] It works through a nonphosphorylating mechanism and is thus a transglucosylase-type enzyme. It is different from the isophosphorylase that is also present in starch.[1252] A minimum chain length of amylose to undergo transglucosylation by Q-enzyme has 40 ± 10 glucose residues. With shorter chains, branching still occurs, but is very slow.[1253]

Dextrin dextranase (EC 2.4.1.2) converts dextrins into dextrans, but it is unable to form dextrans from nonhydrolyzed starch. A 55–60% yield of dextrans can be obtained even from slightly hydrolyzed starch in the presence of pullulanase.[3208] A thermostable dextranase of M_w 140 kDa, working best at pH 5.5 and 80 °C was isolated from an anaerobic thermophilic bacterium termed Rt364, found in a New Zealand hot spring. This enzyme hydrolyzes dextrans as well as amylose, amylopectin, and starch but does not hydrolyze pullulan.[3209]

Debranching high-amylose rice starch with pullulanase produced an enzyme-resistant starch (RS III).[3210] A starch hydrolyzate fermented with *Aureobasidium pullulans* produces pullulan.[3211]

11. Cyclodextrins

In 1891 Villiers[3212] discovered that *Bacillus amylobacter* (*Clostridium butyricum*) grown on starch produced cyclic oligosaccharides known currently as cyclomaltoses, Schardinger dextrins, or cyclodextrins (CDs). In 1903 Schardinger[3213–3215] discovered that *Bacillus macerans* produces such cyclic oligosaccharides more efficiently. At present this microorganism is the most common source of the enzyme responsible for producing these important compounds, which have a wide range of practical uses.

Initially, CDs were prepared from starch from various botanical origins, digested either with *B. macerans*[3216–3220] or with that bacterium supplemented with CGTase, the enzyme isolated from it.[3221]

The results of digestion depend on the botanical origin of the starch. The susceptibility of soluble starches for digestion with CGTase decreases in the order: potato > cassava > sweet potato > corn.[3222] Most probably, this order depends on the origin of the CGTase. In the production of CD from granular, raw starch, the ability of starch to swell could be a relevant factor.

In earlier work with isolated, but crude CGTase from that microorganism, gelatinized potato starch was utilized and the optimum procedure[3223] required 50 °C at pH 5.6–6.4. An optimum pH of 6 and a temperature of 40 °C was also reported.[144] With the purified enzyme, a maximum total CD yield of 55% could be achieved. It was observed that at 20 °C, CGTase selectively digested α-CD, leaving β-CD intact.[3224] Consequently, inactivation of the enzyme is sometimes required. Under identical processing conditions, corn starch was an inferior source for CDs than potato starch, yielding,[3225] respectively, 25 and 30.6% of CD, and the β-CD/α-CD ratios for the two sources were 26 and 28. Other strains of *B. macerans* provided a CGTase capable of a 50% yield of CD from ground potato and a higher yield from 12% potato-starch paste.[3226,3227] Granular potato starch can also be digested.[3228] At pH 6 and 45 °C, a soluble fraction of corn amylose provided up to a 70% yield of CD.[3229] First formed was α-CD, which is later converted by CGTase into β-CD, and subsequently γ-CD.[3230] Evidently, the results depend on biospecific adsorption of CGTase onto starch and the CD. Using optimized conditions for the sorption of the enzyme on physically modified starch, together with the use of purified enzyme, provided a 70% yield of CD.[3231]

Although the production of CDs has moved away from the earlier use of bacteria in favor of CGTase, in later times a *Bacillus firmus* strain, immobilized on inorganic matrices (preferably SiO_2/TiO_2 and SiO_2/MnO_2) has been proposed[3232] for the production of CDs.

Bacilli strains belong to the group of alkali bacteria and their growth therefore takes place preferably at pH > 7, but the optimum pH of starch digestion for some strains lies not only between 9 and 10 but also at pH 4^{3218} or pH 6.[3221] Usually, regardless of the reaction conditions, the three homologous CDs, α-, β-, and γ-CD, namely, cyclomalto, hexa -, hepta-, and octa-oses, respectively, are formed simultaneously, and their proportions depend on such reaction conditions as pH, temperature, concentration, and the composition of the reaction broth.[3233–3235] The latter can be modulated appropriately. The α-CD usually preponderates among the products, but the procedure with an immobilized *B. firmus* strain provided solely β-CD.[3232] The choice of the strain is also essential for determining the resultant α-CD:β-CD:γ-CD ratio. Depending on the selected enzyme, the substrate can be not only starch that has been liquefied and hydrolyzed to various extents[3236] but also gelatinized or raw starch.[3237] Beta amylase secreted by *B. macerans* produces CDs which, on extended action of the enzyme, are converted into maltose.[3238,3239]

Because starch can retrograde, particularly in gels, the concentration of the substrate is also an essential factor, and a concentration above 5% is not recommended.[144] However, the concentration of prehydrolyzed substrate can be enhanced[3238] to an extent dependent on the degree of hydrolysis. The decreasing yield of CDs from solutions of high viscosity or higher concentration of substrate can be overcome not only by pretreatment with enzymes, but also with vigorous stirring, heating under pressure, and sonication.[3238]

Bacilli and the GCTases isolated from them also require starch in the broth for their growth (see, for instance, Szejtli[270] and references therein). From *Bacillus firmus* there was a purified novel CGTase of M_w 78 kDa, for which the optimum parameters for CD production were pH 5.5–5.8 and 65 °C. Cassava starch was the superior substrate when gelatinized, whereas wheat starch performs better in the raw state.[3029] The reducing sugars formed impede the formation of CDs.[3240]

Numerous current procedures for preparing CDs involves CGTases isolated from various strains of *B. macerans*. That enzyme can digest intact granules of raw potato starch.[3241]

CGTases have also been isolated from other *Bacillus* species and used for CD production. The CGTase from *Bacillus amyloliquefaciens* appeared to be a thermophilic enzyme, it was effective at 60 °C and pH 9.0, and the total CD yield of up to

95% consisted mostly of α-CD.[3242] The extracellular CGTase from *Bacillus circulans* has received more attention in the production of CDs. At pH 4.5–4.7, this enzyme converted 73% of starch, 65% of amylopectin, and only 45% of glycogen, with a clear preference for the formation of α-CD.[3243] CGTases from another *B. circulans* species worked at 50 °C and pH 10, providing CDs in yield slightly below 50%,[3244] and at 48 °C and pH 6.9 providing a high yield of β-CD.[3245] At 56 °C and pH 6.4, it yielded 66% of CDs consisting of α-CD:β-CD:γ-CD in the ratio of 1.00:0.70:0.16, respectively.[3246] The optimum conditions for that enzyme are 55 °C and pH 7.[3247] The last-mentioned CGTase was used under such conditions for production of CDs from cassava starch. Because CGTase isolated from *B. circulans alkalophilus* had two optimum pH vales, 5.5–6.0 and 9.0, it was suggested that it either consists of two forms or includes isoenzymes.[3248] When the production of CD was carried out in ethanol, the yield of α-CD decreased in favor of the formation of β- and γ-CD. The effect was interpreted in terms of a decrease in the activity of water.[3249] A CGTase isolated from some strains of *B. circulans* appeared particularly suitable for the production of β-CD.[3250,3251] Some CGTase preparations isolated from various *B. circulans* strains are particularly suitable for the production of γ-CD, but the nature of the substrate is also a factor.[3252] A CGTase from *B. circulans alkalophilus*, immobilized on Sepharose or Serdolit, produced principally β-CD, with four times less α- and γ-CD, the last two in practically equal proportion.[3253]

Bacillus coagulans produces a CGTase that is a moderate thermophile,[3254] being active at pH 6.5 and 74 °C. From a 13% solution of potato starch, it produces α-, β-, and γ-CD in 2:2.6:1 proportion, respectively, with about 40% overall yield.[3255] This process was performed in the presence of $CaCl_2$ and at 50 °C, and the proportions of the three CDs were 1.0:0.9:0.3, respectively.[3256]

CGTase from *Bacillus firmus* appeared suitable for production of β- and γ-CD.[3257] When the process was conducted in 30% aqueous ethanol, the yield of γ-CD reached 35%.[3258] Corn starch was the best substrate for making γ-CD, and the optimum conditions were 50 °C and pH 8.[3259] This CGTase is able to digest raw starch. It performs best with cassava starch, and the results are successively inferior for potato and then corn starches. This process gives mainly β-CD (40%), along with 8% of γ-CD (8%), and this ratio depends on the initial concentration of substrate and the reaction time. A small proportion of α-CD is formed on prolonged incubation.[3260] This enzyme performs best at pH 7.5–8.5 and 65 °C.[3261]

CGTase isolated from *Bacillus megaterium* is suitable for production of β-CD in high yield, but also di- through penta-saccharides accompany the CD.[3262–3264] The process performed on a liquefied substrate with inclusion of isoamylase increases the yield of CD, but liquefaction carried out in the presence of CGTase decreased the yield of CD.[3265] There is also a report that such a CGTase is suitable for the production of γ-CD at pH 10 and 40–45 °C.[3266] This enzyme performs best with corn starch.[3267] It has been observed that this CGTase is significantly inhibited by CD. To overcome this problem, the raw, unliquefied corn starch was processed in a membrane bioreactor.[3268,3269] CGTase isolated from *Bacillus mesenterius* offered an efficient mode for production of γ-CD from 5% potato starch solution at 55 °C.[3270]

Branched CDs could be identified among the products resulting from the use of a CGTase from *Bacillus ohbensis*.[3271] This enzyme, when acting in the presence of glycyrrhizin or stevioside (penta- and hexa-terpenoids, respectively) produces from potato starch a considerable amount of γ-CD. The optimum conditions were 50 °C and pH 7.0.[3272]

Bacillus stearothermophilus provides a thermophilic CGTase, whose thermostability can be augmented by inclusion of Ca^{2+} ions. It produced a net 44% yield of α-, β-, and γ-CD in 4:2:5.9:1 proportion, respectively, when a 13% solution of corn starch was processed at 80 °C and pH 6.0.[3273] At pH 7.0 and 60 °C in the presence of sodium naphthalene-2-sulfonate, this enzyme produces mainly β-CD.[3274] Thermophilic CGTases are strongly inhibited by the CD formed, and therefore the processes should be performed with the minimum amount of enzyme added, a shorter reaction time and with the maximum concentration of starch substrate.[3275]

A CGTase from *Bacillus subtilis* offers a 78% yield of CD, performing best at pH 8.5 and 65 °C.[3276] It produces mainly β-CD, along with a minor amount of a mixture of cyclic and acyclic dextrins,[3277,3278] but there is also another report[3279] that a different strain of that bacterium offers mainly γ-CD.

CGTase isolated from *Bacillus 1011* allowed the production of CDs from starch cross-linked with epichlorohydrin. It performed at 50 °C and pH 7, and the yield of CD after 3 h reached 34.2%, decreasing in time to 18% after 24 h of processing.[3280]

Other microorganisms produce CGTases useful for the manufacture of CDs. Thus, strains of *Brevibacterium* delivered an enzyme offering an enhanced yield of γ-CD from starch, amylose, and amylopectin. It worked preferably at 45 °C and a pH between 8 and 9.[3281] The addition of ethanol is advantageous.[3282]

Clostridium thermoamylolyticum provided a CGTase suitable for the production of a mixture of α-, β-, and γ-CD in the ratio of 1.09:2.00:1.10 from liquefied starch at pH 5.5 and 90 °C. The total yield was 16.8%.[3283] The enzyme isolated from *Klebsiella oxytoca* produced CD preferably from amylopectin and soluble starch, and neither amylose nor simple saccharides induced that CGTase. The optimum conditions for production of α-CD were pH 6.0 and 40 °C. That CD was accompanied by a negligible proportion of β- and γ-CD.[3284,3285] Similarly, the CGTase isolated from

Klebsiella pneumonia performed poorly on amylose, although in the initial stage a rapid shortening of the amylose chain was observed.[3286,3287] At pH 6–8 and 40–50 °C, it produced chiefly α-CD, although there is a patent[3288] claiming production of pure γ-CD from potato starch using this enzyme at 40 °C and at a higher pH. Addition of decanol to the reaction mixture was advantageous.[3289] Bender[3290] has studied the mechanism of action of CGTase from *K. pneumonia* and *B. circulans* on amylose. The action of both enzymes was dependent on the chain length of the substrate, but the CGTase from *B. circulans* was less sensitive to that factor. Based on the rate of CD formation, it could be concluded that the active site of both enzymes spans 9 glucose residues and the catalytic sites are located between subsites 3 and 4 in the CGTase of *K. pneumonia* and subsites 2 and 3 in the CGTase from *B. circulans*.

A CGTase isolated from *Micrococcus varians* produced CDs from starch in the presence of $CaCl_2$ at 50–60 °C and pH 6.0 during 2 days with trichloroethylene added to the medium, and the yield reached 60%.[3291] Cyclomaltononaose (δ-CD) could be obtained by using a CGTase isolated from *Paenibacillus* species F8; the yield was 5–11%, and it was accompanied by α-, β-, and γ-CD.[3292] The process involving a CGTase isolated from *Paenibacillus graminis* did not report the formation of δ-CD.[3293]

A thermostable CGTase could be isolated from *Thermoanaerobacter* grown under anaerobic conditions.[3294,3295] It acted on a 15% solution of lintnerized starch at 90 °C and pH 5.0 to produce CDs in 36.1% yield. From 35% solutions of corn starch, glucose was formed.[3294] At 65 °C and from 7.5% solutions of raw corn starch, a 47% yield of CD could be achieved, accompanied by maltodextrin (31.4%).[3295] The efficiency of the process can be increased by use of a bioreactor with an ultrafiltration membrane.[3296] Super thermophilic bacteria of the *Thermotogales* genus, such as *Thermatoga maritime* MSB8 secrete enzymes suitable for production of CDs, and likewise bacteria of the *Aquificales* genus, such as *Aquifex aeolicus* VF5 and *Aquifex pyrophilus* DSM6858, provide enzymes suitable for the same purpose.[3297]

Branched CD can be produced enzymatically. For this purpose, the glycosyl fluorides from either glucose or maltooligosaccharides are treated with either *Pseudomonas amylodramosa* isoamylase[3298] or *Aerobacter aerogenes* pullulanase.[3299]

Both monobranched- and dibranched-CDs are formed, with the first product preponderant.

Developments in genetic engineering have also led to modifications of CGTase. By such techniques, a novel modified enzyme could be isolated from alkalophilic *Bacillus* that gave a higher yield of β-CD and a decrease in the yield of α-CD, while maintaining the total yield of the process on the same level.[3300] The gene for γ-CGTase of *Bacillus* sp was cloned and expressed in *Escherichia coli*, yielding an enzyme of M_w 75 kDa working best at pH 6.0–8.0 and 60 °C. The enzyme is useful for synthesis of γ-CD.[3301] Replacement of histidine by asparagine at the 140, 233, and 327 positions of CGTase from alkalophilic *Bacillus* gave an enzyme producing an enhanced amount of γ-CD from starch.[3302] CGTase from a mutant strain of *Bacillus stearothermophilus* converted highly concentrated solutions of potato starch into CDs in high yield. This enzyme operated with simultaneous liquefaction of the substrate at 80 °C and pH 6.0, yielding 29% of CD.[3303] In the CGTase from *Bacillus ohbensis*, which normally produces mainly β-CD, Tyr188 was replaced by Ala, Ser, or Trp. This led to enhanced production of γ-CD, up to 15%, was obtained when Trp was inserted, whereas insertion of Ala or Ser gave a mixture of α- and β-CD in equal proportion.[3304] Such changes in the CGTase from *B. circulans*, *B. ohbensis*, *Paenibacillus macerans*, and *Thermanaerobacter* have been carried out to furnish enzymes capable of producing enhanced amounts of γ-CD.[3305] A CGTase was modified by inserting a yeast expression-vector pYD1 into it in order to enhance the production of β-CD.[3306] The yield of CDs can be augmented without necessarily resorting to genetically engineered enzymes. The use of an insoluble starch prepared by extrusion facilitates formation of the complex with CGTase.[3307]

CGTase can also be modified chemically, as shown by Nakamura and Horikoshi,[3308] who succinylated CGTase from an alkalophilic *Bacillus*. This modification allowed the operation temperature to be increased from 50 to 55 °C at pH 8, and the enzyme was stable for a longer period of time. The reaction yield of predominantly β-CD was 46%, but joint action of pullulanase increased the yield to 52%.

Either the joint or consecutive use of various enzymes can also improve the process for manufacturing CDs. For example, the reaction mixture producing CDs from potato starch using *B. macerans* CGTase was digested with crude glucoamylase, which selectively digested unprocessed starch and dextrins, leaving the CD intact.[3309,3310] Such enzymes can be used in an immobilized state.[3311] Alpha amylase could be used instead of glucoamylase,[3312–3315] along with membrane ultrafiltration.[3316] Pullulanase[3317] and pullulanase jointly with beta amylase[3318–3320] as well as isoamylase[3321,3322] were applied simultaneously with CGTase as debranching enzymes to produce branched CDs. Wang and coworkers[3323] obtained CDs from starch by first using debranching amylopectase, a pullulanase-type enzyme, followed by amylomaltase.

Based on earlier observations that ethanol in the reaction broth stabilized the formation of β-CD, a process was proposed wherein cyclization proceeded simultaneously with fermentation.[3324] Thus, the process with CGTase proceeded in the presence of *Saccharomyces cerevisiae*.

Immobilized CGTase has also been in use.[3325,3326] A Japanese patent[3327] proposed use of the immobilized enzyme in the presence of mannuronic acid and such mineral salts as $CaCl_2$ to facilitate gelatinization of starch and hence, its liquefaction. An US patent advocates the addition of an C_1–C_6 aliphatic alcohols, for instance, *tert*-butanol. The use of immobilized CGTase in this process is also useful.[3328] Another patent[3329] suggests the passing of saccharified starch through a column filled with CGTase immobilized on a weakly basic anion-exchanger. Immobilization of the enzyme on an amino group-containing, polysulfone-type, anisotropic ultrafiltration membrane facilitates the isolation of CD with a yield exceeding 65%.[3330] Other types of membrane can also be used.[3331] Filtration through diatomite extended the life of the immobilized enzyme and allowed an increase in the processing temperature.[3332]

Reducing sugars are the main by-products in the formation of CDs with CGTase. They not only inhibit formation of CDs but also combine with CGTase to decompose the CDs formed.[3333] Limit dextrins usually accompany CDs, and their formation at the expense of CDs can be suppressed by separating the starch into amylose and amylopectin fractions. Amylose provides a higher yield of CD and less of the limit dextrins.[3334] A Japanese patent[3335] proposes that potato starch should be jointly liquefied with CGTase at 60–70 °C, providing not only α-, β-, and γ-CD but also δ-CD. About 42, 36, 12, and 4% of these CDs, respectively, constitute the total CD yield of 62%. Various methods for pretreating starch prior to its digestion have been tried. For wheat starch, compression at high temperature prior to digestion provided a 66.3% yield of α-CD within 100 h of processing, although only 30.9% CD when the process lasted only 4 h.[3336] Alternatively, the starch may be autoclaved at 70 °C before preparing a 3% aqueous solution, whereupon the yield of α-CD reaches 60%.[3337] The initial liquefaction of starch prior to its digestion by CGTase can be replaced by crushing, steaming, and other procedures, but the total yield of the three CDs does not exceed 24%.[3338] A liquefaction procedure for starch in the presence of CGTase is stated to enable a high-yielding continuous production of CDs.[3339] Corn starch can be pretreated for CD production by moderate heat treatment.[3340]

Admixture of such complexing agents as trichloroethylene,[3341,3342] bromobenzene,[3342,3343] toluene, acetone, and cyclohexane[3342] is useful as they stabilize the CDs formed, facilitate their separation from the reaction broth, and favor the formation of β-CD. Interphase catalysis in which β-CD produced by CGTase in a lower aqueous phase rich in dextran is transferred into a polyethylene glycol upper phase has also been described.[3344] The partition coefficients for β-CD and CGTase were 1.5 and 0.25, respectively.

Inclusion of C_8–C_{16} aliphatic alcohols in the reaction mixture[3345] favors the production of α-CD, and the addition of a surfactant[3346] into the mixture increases the yield of γ-CD. A German team[3347] proposed the inclusion of cyclohexadec-8-en-1-one, whose dimensions fit the cavity of that CD. Cyclic C_{12} compounds, such as cyclododecanone, cyclododecanol, cyclododecylacetamide, and others enhance the yield of γ-CD, whereas the corresponding C_{11} complexants provide a higher yield of β-CD.[3348] Addition of ethanol increases the yield of α-CD by a factor of two, but it has no influence upon the yield of β- and γ-CD.[3349] A Japanese patent[3350] claims that inclusion of C_1–C_9 aliphatic alcohols, C_2–C_4 aliphatic ethers, esters, or ketones increases the yield of CDs. In the example cited, the outcome of a 22-h reaction at 50 °C was the production of α-, β-, and γ-CD in 44.7, 13.0, and 12.4% respective yields. The yield of CDs was also increased by inclusion of hydrocarbons up to C_{10} having an unsaturated 6-membered ring, for instance limonene.[3351] The use of a water-soluble organic solvent permits production of CDs from unliquefied starch.[3352] Polyethylene glycol and polypropylene glycol added to the mixture increased the yield of α- and β-CD, but the amounts of the additives used need to be controlled.[3353] Unliquefied starch can also be successfully processed without any organic solvent in an agitated-bead reaction system.[3354]

There is evidence[3355] that a pair of organic solvents should be used to allow the formation of ternary complexes. Such complexes form a type of clathrate structure, as described in utilizing bromobenzene and chloroform to generate such clathrates.[3356] The simultaneous use of either an aliphatic ketone or aliphatic alcohol together with a C_{12} cyclic compound favors the formation of γ-CD.[3357]

Generally, when CDs are produced from starch pastes, a hydrolysis step prior to digestion with CGTase is recommended, to decrease the viscosity of the reacting medium and limit retrogradation. Such initial hydrolysis can be performed with either alpha amylase[3358] or alpha amylase plus glucoamylase.[3359] Liquefaction of starch can be performed in 1 M aqueous NaOH with overnight refrigeration.[3360] Gelatinization is not needed.[3361]

Fiedorowicz and coworkers[3362] have demonstrated the stimulation of CGTase by the action of linearly polarized white light. The enzyme needs to be illuminated with such light for 1–2 h in a separate flask, and then the enzyme is transferred into a bioreactor, where it operates without illumination. The production of CD is accelerated by up to 30% and the α-:β-:γ-CD ratio changes. Figure 24 shows that the increase reaches a peak after illumination of CGTase for 2 h; prolonged illumination decreased the yield of all three CDs.

The kinetics of formation of the CDs is illustrated in Fig. 25. It may be seen that the concentration of α- and γ-CDs depends only slightly on the reaction time ($k = -4.4 \times 10^{-7}$ and 3.2×10^{-6}, respectively), whereas the concentration of β-CD increases in time ($k = 2.2 \times 10^{-5}$).

Optimization of the reaction conditions for efficient production of γ-CD has been the subject of simulation studies.[3363,3364]

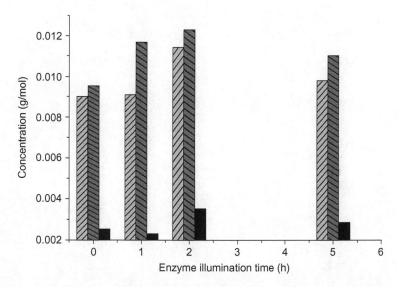

FIG. 24. Impact of illumination time with white linearly polarized light on the concentration of α-, β-, and γ-CDs formed by CGTase acting on sago starch (c_s and c_e denote concentration of starch and enzyme, respectively. They were 0.02 g/cm^3 and 0.64 U/cm^3, respectively. The process was carried out at 37 °C). The bars from left to right correspond to α-, β-, and γ-CD.[3362]

In some processes, the production of pure CD is not a particular target, as in the preparation of a starch–sugar powder containing CDs as a product of low sweetness and hygroscopicity for food and pharmaceutical uses.[3365] Patents have described gels rich in CD,[3366] starch and maltose syrups enriched in CDs,[3367] and starch containing CDs.[3368]

CDs are well known for their propensity to form inclusion complexes with a wide range of compounds having biological activity, and they can also form complexes with various fragrances, colorants, and products that prevent drying when exposed to air.[280,283,284] Other uses include the ability to form microcapsules with enzymes and to modulate enzyme-catalyzed processes. Thus, the alpha amylase of rice-grain is inhibited by β-CD,[3369] whereas CD seems to induce the production of CGTase in *B. macerans*.[3370] Generally, CDs inhibit enzymes, including amylolytic enzymes. This ability is utilized in the purification of these enzymes by affinity chromatography.[3371]

An *Arthrobacter globiformis* bacterium isolated from soil[3372] secretes a glucosyltransferase of M_w 71.7 kDa that converts starch into a novel cyclic tetrasaccharide, cyclic maltosyl-(1→6)-maltose {cyclo[→6)-α-D-Glc$p$-(1→4)-α-D-Glc$p$-(1→6)-α-D-Glc$p$-(1→4)-α-D-Glc$p$-(1→]}. Using that enzyme and α-isomaltosyltransferase, a cyclic

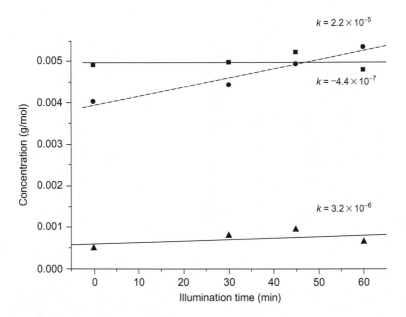

Fig. 25. Changes in the time-dependent concentration of α-, β-, and γ-cyclodextrins resulting from the conversion of sago starch with CGTase, under illumination for 2 h with white linearly polarized light. Starch and enzyme concentrations are 0.01 g/cm^3 and 0.32 U/cm^3, respectively. The process was performed at 37 °C. Squares, circles, and triamgles denote α-, β-, and γ-CD, respectively.[3362]

pentasaccharide, cyclo[→6)-α-D-glucopyranosyl-(1 → 3)-α-D-glucopyranosyl-(1 → 6)-α-D-glucopyranosyl-(1 → 3)-α-D-glucopyranosyl-(1 → 4)-α-D-glucopyranosyl-(1→], and glycosyl derivatives thereof, can be prepared from a starch hydrolyzate.[3373]

Taka amylase hydrolyzes α- and β-CDs to glucose and maltose.[590]

VII. Starch Metabolism in Human and Animal Organisms

1. Digestible Starch

Adsorption of amylase onto starch is the first, key step of starch metabolism.[3374] In humans, starch is then hydrolyzed by ptyalin and pancreatin secreted in the fluids in the initial parts of the gastric tract. The hydrolysis produces oligosaccharides and, finally, glucose, which is cotransported by the Na^+-glucose mechanism across membranes employing the electrochemical gradient of the Na^+ ions.[3375,3376] Resistant starch (RS) alone passes into the intestine, where colonic bacteria digest it, together with nonstarchy polysaccharides and oligosaccharides (prebiotics).[3377,3378]

Human body fluids contain alpha amylases that are synthesized from AMY-1 (nonpancreatic amylases) and AMY-2 (pancreatic amylases) genes located on chromosome No. 1 in man. Human alpha amylases should therefore be considered as a combination of two isoenzymes, with multiple forms dependent on posttransitional modifications.[3379] The two isoenzymes show different kinetics. The nonpancreatic enzyme shows higher activity and lower affinity to soluble starch than the other one.[395] Human males and females digest starch at the same rate.[3380] Studies on the metabolism of raw starch by colonic bacteria of adults, toddlers, and infants revealed that young children digest raw starch more rapidly than adults. Lactic acid fermentation was observed only in the case of children.[3381]

Particular organs secrete enzymes whose role depends on the needs of the organism and in response to the composition of the feed. A starch-containing diet evokes changes in the human organism which can be evaluated, among others, as the plasma (insulin) response and gastric emptying. In healthy humans, these factors depend on the susceptibility of starch to alpha amylase rather than on the viscosity of test meals. The insulin response is independent of the gastric emptying rate.[3382] Pepsin and bile enhance the hydrolysis of slowly digestible starch.[3383] However, dietary fiber from oat can affect binding to bile acids and fermentation, as shown by *in vitro* studies.[3384]

Digestion of intestinal carbohydrates is inhibited by some transition-metal ions.[3385] On the other hand, transition metals may stimulate the metabolism of saccharides as, for instance, the influence of dietary copper in the liver.[3486] There are some drugs and preparations that inhibit amylase and glucosidase, and thereby reduce the absorption of starch in humans. This effect is determined by the rate of uptake of digestible starch, DS, in the intestine.[3387] These agents include (trifluoromethylphenylthio)-2-ethylaminopropane (known as BAY-e 4609 and used as an anorectic drug), and acarbose,[3388,3389] miglitol [($2R,3R,4R,5S$)-1-(2-hydroxyethyl)-2-(hydroxymethyl) piperidine-3,4,5-triol], a compound used as antidiabetic drug,[3389] green and black tea,[3390] and fiber (maltodextrin fiber, oat fiber), alpha trim (an alpha amylase blocker of wheat origin), and epigallocatechin gallate.[3391]

A diet composed of bread and cereals is the most common mode of starch uptake. The fate of starch in bread was being studied already at the beginning of the 20th century, when the role of saliva in the digestion of bread starch was investigated. In the bread crumb, starch granules remain nondisrupted, and only a minor proportion of the granules show exudates. Experiments *in vitro* showed that this exudate portion of the granules is digested by saliva within a few minutes. Digestion of the polysaccharide of the granule envelope under such circumstances took over 24 h. Under physiological conditions in the mouth, amylose is hydrolyzed to dextrins which, together with amylopectin, are digested in the stomach, whereas the granule envelopes are hydrolyzed only in the intestine. The origin of the flour from which the bread was made is important, and differences in the rate of digestion by saliva result from the level of gluten in the flour. Stale bread undergoes slow digestion as long as the gluten remains intact. Saliva decomposes starch in three steps. Proteins are digested first, followed by dextrins of lower and higher molecular weight.[3392] Saliva decomposes starch in bread within the first hour of digestion, and then this process is stopped by the hydrochloric acid of gastric juice. The digestion is impeded by phosphates present in bread. The higher the level of phosphate in flour, or as baking time is extended, the slower is the rate of digestion. Salts of other acids tend to neutralize gastric hydrochloric acid.[3393] Malt amylase, when supplemented to the dough, cooperates with saliva in accelerating digestion of starch.[3394]

The botanical origin of starch plays a part. The digestion of amylase of starches isolated from various tubers, cereals, and fruit plants was compared in *in vitro* experiments.[3395] Hydrolysis of corn starch was the fastest, followed by starches of cassava, sweet potatoes, and breadfruit. Starches of potatoes, plantain, macabo, (*Xanhosoma*) taro, and yam were poorly digested. Other comparative *in vitro* and *in vivo* studies on the digestibility of various starches by saliva[3396,3397] showed that cereal starches are more readily digested than potato starch and legume starches.[3398] High-amylopectin starch is digested more readily, and starch complexes with proteins are equally well digested.[3399] The digestibility of starch can be enhanced by its enzymatic pretreatment, and starch so treated is more readily digested than cooked starch.[3400]

Some mammals digest starch and some do not [See Section IV.18.g (vi)]. In those animals that can digest starch, there is also observed an adaptation of the organism to changes in the dietary starch. For instance, α-glucosidases secreted from mammalian pancreas commonly convert linear oligosaccharides produced by alpha amylase into glucose, but when it is necessary, they also digest α-limit dextrins.[3401] Sometimes, "irritation" of the organs can cause secretion of uncommon enzymes. Thus, the liver of rabbits, after receiving an injection of sucrose, degraded regularly soluble starch and, additionally, produced lactose.[3402] Among rats fed with diet either rich or poor in starch, there was observed a certain enzymatic adaptation to the diet.[3403]

Particular mammals differ from one another as to which are the predominant bacteria responsible for digesting starch.[3404–3406] In pigs, *Clostridium butyricum* is the dominant bacterium,[3404] and the rumen of ruminants is colonized mainly by various protozoa.[3407] In sheep, the amylolytic activity of microflora of their cecum increases with increase in the level of saccharides for digestion, but that increase is not paralleled by an increase in the microorganism count therein.[3408]

Felines (bobcats, pumas, and domestic cats) digest frozen corn starch more readily than nonfrozen starch, whereas no appreciable effect of freezing was noted with potato starch.[3409] The site of starch breakdown in mammals depends on the botanical origin of starches and the animal species. In nonruminants other than hamster, untreated cereal starches are degraded by digestive secretions in the small intestine. Tuber starches, except cassava starch, are hydrolyzed by bacteria in the cecum. In hamsters, digestion of tuber starches involves bacteria in the gastric diverticulum and in the cecum. Ruminants digest with bacterial involvement all starches in the rumen.

The digestion of starch in selected insects has also been studied. The adaptation of their enzymatic system for uptake and digestion of starch also depends not only on the diet but also on their state of maturation. For instance, it has been observed with the fruit fly *Drosophila melanogaster*[3410] that, in feeding either its larvae or imago with starch, alpha amylase was activated to the same extent and such activation was stronger in flies fed with starch. α-Glucosidase activity increased during adult development. The activity of both enzymes was higher in starving organisms than in organisms that had been fed. The style of feeding influenced the functions of alpha amylase and α-glucosidase. Studies on the digestion of starch in the midgut of the *Lygus disponsi* bug, using salivary amylase, revealed that anions, particularly Cl^- and NO_3^-, may play an important role in starch digestion.[3411–3413] In the presence of these anions, insoluble starch might also be digested.[3414] These anions support digestion of other polysaccharides to the same extent.[3413]

2. Resistant Starch

Resistant starch (RS) is fermented only in the colon, where bacteria produce short-chain fatty acids.[296,3415] Resistant starch from corn and pea fed to rats passed through the small intestine and was found in the ileum. There was twice as much pea RS therein as compared to corn RS, that is, 50 and 25% of the RS administered, respectively.[3416] Feeding rats with corn RS showed fecal bulk and excretion of starch higher than normal. The cecal size and content was higher, and the pH was lower.[3417] There was no difference in serum cholesterol, glucagon, and enteoglucagon levels, probably, because RS was partially degraded in the alimentary tract of the animals. However, the level of esterified and liver cholesterol decreased by 24 and 22%.[3417,3418] Total excretion of fecal bile acid and bile acid composition remained unchanged after feeding rats with RS. Total cholesterol concentrations in the plasma showed a negative correlation with excretion of fecal coprostanol in hypercholesterolemic rats.[3419] Studies on the energy balance in rats after administration of either corn RS, pea RS, or sucrose revealed that feeding with these RS increased the intake of digestible energy and led to weight gain in the animals to nearly the same extent as did feeding with sucrose.[3420] However, retrograded starches are fermented differently by *Clostridium butyricum*, and hence, the acetate/butanoate ratio in this instance is different than for fermentation of normal starches.[3421]

It has been reported[3422] that RS forms complexes with enzymes that induce an oxidative stress that is not correlated with increased cell proliferation. Thus, RS may have a tumor-enhancing effect rather than a tumor-protective effect.

VIII. STARCH ANALYTICS INVOLVING ENZYMES

1. Starch Evaluation and Analysis

Demands from starch manufacturers and consumers have led to new methods for the determination of starch in plants and food, controlling the quality of starch and its conversion products, as well as the efficiency of starch processing and storage. Several of the methods elaborated involve enzymatic processes. First of all, such processes require the selection of suitable enzymes (see, for instance, Refs. 3423–3427). Selection of the appropriate enzyme may depend on the botanical origin of the starch and the particular starch sample being analyzed.[3428–3432]

It has been shown[3433] that the degradability of starch can be predicted from its behavior on gelatinization, that is, from the gelatinization enthalpy as measured by differential scanning calorimetry. There is a good correlation between the enthalpy of gelatinization so determined and the degradation of the starchgranules by alpha amylase.

Methods have been described for the determination of starch in food in general[3434–3441] and in particular such food products as meat,[3442–3444] crude and dried potatoes,[3445–3447] cereals,[3445,3465,3448–3456] bran,[3453,3457] corn silage,[3456] fresh, lyophilized, and cured tobacco,[3458] plant tissues,[3459–3461] dietary fiber,[3462] and raw sugar.[3454] All of these methods are based on hydrolysis of starches with suitable enzymes, preferably glucoamylase,[3425,3434–3436,3440–3442,3446,3448,3449,3459,3463–3465] amylases,[3427,3436,3442,3443,3448–3450,3460,3466,3467] takadiastase,[3459] and malt diastase[3468] to glucose, the latter being estimated enzymatically. Sometimes, starch is either gelatinized prior to its enzymatic digestion[3425,3445] or dispersed by sonication in acid buffer.[3446,3449] Drying the source starch, preferably in a microwave oven, also facilitated enzymatic digestion, as shown in case of potato starch.[3447] Solubilization of starch in dimethyl sulfoxide has also been proposed.[3464] Two-step digestions with two types of amylase,[3469] amylase followed by glucoamylase,[3436,3452,3454,3462] glucoamylase followed by maltase,[3458] and amylase followed by a cocktail of glucoamylase and pullulanase[3454] were also described.[3452] The specificity of some microorganisms toward certain saccharides and oligosaccharides was employed for analyzing the hydrolysis products from starch. Thus, *Candida monosa* digested only glucose, whereas *Saccharomyces chodali* readily attacks glucose and maltose, but isomaltose is attacked much more slowly. Other oligosaccharides are not attacked at all.[3470]

Alpha amylolysis also made it possible to determine the damage caused in starch granules by germination and in flours by milling.[3471–3473] A rapid enzymatic determination of starch damage in wheat flours is based on the digestion of starch with *Aspergillus oryzae* alpha amylase. In a digestion lasting 5 min, that enzyme did not attack undamaged granules, and the reducing sugars determined originated solely from damaged starch granules.[3474,3475] An automatic method for determining glucose, maltose, and starch, based on digestion with glucoamylase, was published by Richter.[3464]

A special procedure for the determination of resistant starch has been presented.[3476] In general, use of the foregoing enzymes and the digestion conditions employed do not allow for differentiation between resistant and nonresistant starches. Prior to the determination, protein should be removed by the action of a 2% solution of pepsin for 70 min. The enzymatic conversion of starch can be monitored by means of Mathews' formula[3477] [Eq. (51)].

$$\text{Degree of conversion} = PT/D \tag{51}$$

where P is the polarimetric rotation angle of the solution, T is the result of the Lane–Enyon titration (titrimetric determination of reducing sugars with aqueous $CuSO_4$, in mL), and D is the percent dilution of the titrated solution. This approach, originally developed to determine the fructose/total reducing sugar ratio, has been applied for monitoring the enzymatic conversion of starch with alpha amylase and glucoamylase. It is suitable for determining the extent of conversion employing enzyme cocktails.[1035] The degree of liquefaction depends on the starch quality, and the latter can be monitored nephelometrically in the liquefaction product.[3478] In another method of determination of resistant (retrograded) starch, the fact that *Bacillus subtilis* alpha amylase cannot digest nongelatinized starch is utilized. Thus, a foodstuff containing retrograded starch is subjected to gelatinization. The gelatinized starch is then determined colorimetrically after blue staining with KI_5.[3479]

A Chinese group[3480] has monitored progress of the hydrolysis microcalorimetrically, and a Japanese group analyzed progress of the reaction progress by measuring loss of dissolved oxygen.[3481] The hydrolysis has also been monitored dilatometrically.[3482,3483] In most of the analytical procedures, the glucose resulting from digestion is determined with hexokinase and glucose 6-phosphate dehydrogenase,[3434,3435,3443,3448,3452] glucose oxidase–peroxidase,[3446,3449,3455,3463,3466,3484,3485] coupling with antipyrilquinone imine,[3450] as reducing sugars,[3450,3486] with the latter used for determining maltose,[3487] and by fermentation with yeast.[3488] Sabalitscka and Weidlich[3489] proposed the colorimetric determination of the rate of dextrinization and saccharification based on the changes in blue staining with KI_5, Broeze[3490] monitored the reaction progress viscosimetrically, and Di Paola et al.[3441] monitored the increase in glucose concentration.

A variety of instrumental techniques have been used, including polarimetry,[3443,3486,3491,3492] refractometry,[3447] near infrared spectrometry,[3451,3454,3493] polarography,[3488,3494] oscillopolarography,[3491] gel electrophoresis,[3495] and also after labeling with a suitable dye, preferably 8-aminonaphthalene-1,3,6-trisulfonic acid.[3496,3497] Chromatographic techniques employed include paper chromatography,[3491,3495,3498–3501] thin-layer chromatography (Kieselgel 60 Merck) which enables separation of isomeric CDs, maltose, and glucose, all possibly formed on enzymatic hydrolysis,[3502] high-performance liquid chromatography,[3469,3503] gel-permeation chromatography,[3504–3508] rotating angular chromatography,[3509] and ion-exchange chromatography.[3510] Emmeus and Gorton[3511,3512] proposed a sequential hydrolysis of starch inside two reactors filled with immobilized amylase and glucoamylase, respectively. The glucose that was formed was oxidized inside a third reactor packed with coimmobilized mutarotase and glucose oxidase. The concentration of the resultant hydrogen peroxide was then determined electrochemically (see also the fast method of starch determination according to Cuber[3513]). The latter was determined electrochemically. A similar approach was applied by Karkalas,[3514] who passed the sample hydrolyzed with alpha amylase into a flow-injection analyzer, where it was subsequently hydrolyzed with glucoamylase. Glucose was measured spectrophotometrically at 505 nm through continuous addition of a reagent containing glucose oxidase–peroxidase to generate a pink-colored chromogen. Umoh and Schuegerl[3515] determined glucose amperometrically. Later, enzyme electrodes for determining maltose, glucose, and starch were constructed. These can be O_2- or hydrogen peroxide-selective electrodes employing both immobilized glucose oxidase and glucoamylase.[3516–3524] Raghavan et al.[3525] proposed an aerobic biometric analysis suitable for determination of glucose as well as biodegradation products from starch. This approach is based on determination of the total CO_2 evolved during the course of a microbial process. This method is suitable in environmental studies.

Following acid-catalyzed hydrolysis of starch with hydrochloric acid, an enzyme, glucose oxidase–peroxidase, can be used for colorimetric determination of the resultant glucose.[3444,3457] The course of the enzymatic hydrolysis of starch may be monitored either colorimetrically in terms of changes of the color of the starch complex with KI_5,[3442,3476,3458,3526,3527] microscopically,[3528] with involvement of ^1H NMR,[3529] or thermokinetically.[3530] The method involving determination of glucose with glucose oxidase–peroxidase at pH 7 permitted determination of native normal or waxy starches and distarch phosphate.[3466]

Transglucosylation of soluble starch with buckwheat α-glucosidase produced several disaccharides, including kojibiose, nigerose, and maltose, and they were determined quantitatively by a combination of paper and gas–liquid chromatography.[3531] The products from hydrolysis of starch derivatives with either beta amylase or glucoamylase could be pyrolyzed and the pyrolyzate analyzed by gas–liquid chromatography[3532] and by mass spectrometry.[3533] Glucose in starch hydrolyzates can be quantified by MALDI-TOF mass spectrometry.[3534]

Assays of enzymatic hydrolyzates of starch can be erroneous if mono- and disaccharides are present in the samples, and such contaminants should be removed either by oxidation in alkaline medium[3535] or by borohydride reduction.[3536] The purity of starch could be determined by digesting it with amyloglucosidase from *Aspergillus niger*.[3537]

Enzymatic hydrolysis has been employed in studying the gelatinization of starch. Its progress may be monitored by observing changes in birefringence of granules and affinity to beta amylolysis,[3538] glucoamylolysis,[3539,3540] color reaction with either iodine or anthrone,[3541] color reaction with glucoamylase and *o*-toluidine,[3542] digestion by diastase, and β-amylolysis assisted by pullulanase.[3540] In the last instance, the retrogradation of starch can also be monitored (see also Ref. 3426).

Enzymes are widely used to study the macrostructure of starch granules, as revealed through enzymatic digestion.[3543,3544] Starch isoamylase (EC 3.2.1.68) has a specific pattern of hydrolysis and has been used for the end-group assay of starch and its hydrolysis products.[3545]

The different affinities of starches to enzymes, and the selectivity of enzymes with respect to various sites of the amylose and amylopectin chains, allow controlled splitting into fragments that can then be characterized by various instrumental techniques.[3546–3550] This approach is also applicable to derivatized starches.[3551] The average unit-chain lengths in amylopectin can be determined based on the fact that beta amylase splits odd-numbered chains of glucose units into glucose. If that enzyme from sweet potatoes is supplemented with yeast glucosidase transferase, the results of estimations are not rendered ambiguous through the possible influence of contaminating alpha amylase.[3552] The extent of cross-linking in derivatized starch can be established by digestion with alpha amylase and fermentation with *Lactobacillus bulgaricus*.[3553] Enzymatic degradation of starch has been used to evaluate enzymatic desizing agents.[3554,3555] The profiles of the products of enzymatic conversions of starch are commonly characterized by chromatographic techniques.[3556,3557]

Japanese workers[3558] have proposed an enzyme and starch–iodine complex as a component in packaging materials for food. As the enzymatic reaction on the blue complex progresses, the color changes from blue to purple to yellow, signaling loss of freshness of the packaged product.

2. Enzyme Evaluation

Enzymatic reactions of starch have been employed to determine the hydrolyzing ability of enzymes from various sources.[1247,3559] Estimation of the enzyme activity is of vital importance in the processing of crude starch-containing materials,[3560] production of compost[3561] and various other circumstances. The malting process requires simultaneous estimation of the activity of alpha and beta amylases, as discussed by Podgórska and Achremowicz,[3562] and used in evaluation of enzymes in triticale grains.[3563] The diastatic power of an enzyme is defined as the amount of starch liquefied by the diastase present in 1 g of enzyme preparation.[3564] The thermostability of the enzymes is also a factor of key significance. Based on the correlations of various factors influencing the fermentability of many malt varieties, Evans and coworkers[3565] have revised the conventional method for estimating the diastatic power, which assesses of the diastatic power of the whole medium, in favor of determining the diastatic power for all enzymes separately. This approach provides more realistic data. Other approaches have been proposed[3562] for determining the diastatic power of cereal flours[3566] and grains.[3567]

Related procedures are based on combining a given enzyme with starch under specified conditions, and monitoring changes termed the enzyme-liquefying power in the digested solution. Determination of total reducing sugars is the most common approach. Comparing methods based on acid-catalyzed hydrolysis and enzyme-catalyzed hydrolysis indicates that the latter is the more precise.[3568]

Clearing of the opacity of the solution was observed. Addition of Neutral Red facilitates the precise reading of the end point.[3569] This method was proposed to test the efficiency of removing starch sizes from textiles. Saccharogenic methods are based on the reduction of inorganic salts by products of the enzymatic hydrolysis of starch. Suitable salts include ferricyanide(III),[3570,3571] Cu(II) salts, and either arsenomolybdate[3572,3573] or 4,4′-dicarboxy-2,2′biquinoline.[3574] An assay based on the reduction of 3,5-dinitrosalcylic acid is also in use.[3575] Observing the disappearance of the blue color of the starch–iodine complex is commonly used.[3576,3577] Another method is based on the coagulating effect of iodine on starch pastes.[3578] That approach, which has a rapid effect, was proposed for the study of bacterial cultures with saliva and urine. Changes in viscosity,[3579] onset of the floating of stiff gels,[3580,3581] and dilatometry[3582] could also be used. The progress of hydrolysis may also be monitored photometrically,[3583–3587] with a digital oscillator densimeter,[3588] an amperometric glucose sensor,[1516] and by means of paper chromatography,[3583] high-performance thin-layer chromatography,[3589] high-performance liquid chromatography and scanning electron microscopy,[3589,3590] absorbance of the enzyme in a starch–agar gel in a microplate well,[3591] and electron spin-resonance spectroscopy.[3586] An enzyme assay employing an oxygen electrode with immobilized glucose oxidase has also been described.[3592,3593]

Back in 1925, it was observed that the rate of liquefaction of starch was proportional to the enzyme concentration over a wide range of values. From this observation, it is possible to determine the concentration of the enzyme preparations from the rate of starch liquefaction.[3594] The thermal stability of alpha amylase could be determined by estimating the half-life of that enzyme, based on its reducing power.[3595] The origin of the starch is an important determinant in the final result.[3592–3594] Soluble starch appeared to be the most suitable for determination of amylase activity. However, the choice of starch should not be arbitrary. According to Goryacheva,[3596] the use of potato starch provides the lowest ($\pm 5\%$) error in colorimetric measurement at 565 nm. Seidemann,[3597] who used absorbance as the measured value, concluded that the recorded results depend on the moisture content of the starch used, its ash and mineral content, its solubility, the wavelength of the color of the iodine complex, the granularity, and other factors. Baks et al.[3598] determined the activity of alpha amylase by the Ceralpha method, which eliminates the use of starch. In this method, either cereal flour extract or fermentation broth is used. Carbohydrates, if present, impede the determination.

A radiolabeling assay with soluble starch and ^{33}P-labeled ATP was employed for quantitative determination of α-glucan–water dikinase activity in crude extracts of plant tissues.[3599] This enzyme is responsible for phosphorylation of starch.[3600] When more than one enzyme is present in the investigated solution, another, less enzyme-susceptible substrate should be used, as demonstrated in malted barley with the estimation of endo-amylase in the presence of exoamylases. Hydroxypropylated starch of a suitable degree of substitution was taken as a substrate.[3601]

Some kits for detecting enzymes have been designed. Thus, tablets composed of microcrystalline cellulose, talcum powder, albumin, starch, and sodium

8-amino-5-[3-(sulfoethylsulfonyl)aniline]-6-anthraquinonesulfonate, bound covalently to starch, were proposed for colorimetric determination of microbial alpha amylase.[3602] Another kit was designed for electrophoresis on cellulose acetate membrane with a starch–agarose board.[3603] The so-called Cerealpha method[3604–3606] is specific for measuring the level of alpha amylases in cereals. It is based on measuring spectrophotometrically the result of endo-attack, followed by glucoamylase/α-glucosidase action, on 4-nitrophenyl maltoheptaoside. Another procedure is the Betamyl method, based on exo-attack and action of α-glucosidase on 4-nitrophenyl maltopentaoside.[3607] A dyed-starch method is based on splitting by the enzyme of a blue dye from amylose that had been cross-linked with an azo dye.[3608] A special so-called flow-through diffusion-cell method was elaborated[3609] for *in vitro* estimating degradation of the starch used as an excipient for pellets. It models the enzymatic degradation of starch within a starch-based ternary semi-interpenetrating network. That network simulates gastrointestinal drug delivery.

Preharvest sprout activity of cereal grains is estimated as the so-called Falling Number. Alternatively, Chang and coworkers[3610] have proposed a model involving kinetics of gelation, kinetics of enzymatic hydrolysis with alpha amylase, and kinetics of thermal deactivation of the enzyme. They claim this method to be more precise than the Falling Number. An assay for determination of the activity of amylolytic enzymes involved cross-linked "chromolytic" starches. The degree of cross-linking should exceed 0.02.[3611] Residual enzyme activity during the starch hydrolysis can be estimated after its adsorption on clay.[3612] Ledder et al.[3613] presented an *in vitro* evaluation of hydrolytic enzymes as dental plaque agents.

References

1. C. Okkerse and H. van Bekkum, Towards a plant-based economy? In: H. van Doren and N. van Swaaij, (Eds.). *Starch 96 the Book,* Carbohydrate Research Foundation, Noordwijkerhout, The Netherlands, 1996, Chapter 1.
2. P. Tomasik, Polysaccharides and ecology (in Polish), *Chem. Inz. Ekol.*, 6 (1999) 831–838.
3. P. Tomasik and J. Gładkowski, Polysaccharides and economy of the XXI century (in Polish), *Zywn. (Food)*, 8 (2001) 17–27.
4. P. Tomasik, S. Wiejak, and M. Pałasiński, The thermal decomposition of carbohydrates. Part II, *Adv. Carbohydr. Chem. Biochem.*, 47 (1989) 279–344.
5. P. Tomasik and M. F. Zaranyika, Non-conventional modifications of starch, *Adv. Carbohydr. Chem. Biochem.*, 51 (1995) 243–318.
6. P. Tomasik and C. Schilling, Starch complexes. Part I. Complexes with inorganic guests, *Adv. Carbohydr. Chem. Biochem.*, 53 (1998) 263–343.
7. P. Tomasik and C. Schilling, Starch complexes. Part II. Complexes with organic guests, *Adv. Carbohydr. Chem. Biochem.*, 53 (1998) 345–426.

8. P. Tomasik and C. Schilling, Chemical modifications of starch, *Adv. Carbohydr. Chem. Biochem.*, 59 (2004) 176–404.
9. P. Tomasik, M. Fiedorowicz, and A. Para, Novelties in chemical modification of starch, In: P. Tomasik, V. P. Yuryev, and E. Bertoft, (Eds.). *Starch: Progress in Structural Studies, Modifications and Applications,* Polish Society of Food Technologists, Cracow, 2004, Chapter 24.
10. H. Barfoed, Die Verwendung von Enzymen bei der Herstellung von Dextrose und Stärkesirup, *Starch/Die Stärke*, 19 (1967) 2–8.
11. K. Kanenaga, K. Kako, T. Harada, A. Koreeda, A. Harada, and K. Okuyama, Different behaviors between actions of *Bacillus subtilis* α-amylase and 18% sulfuric acid at 32 °C to starch granule, *Chem. Express*, 4 (1989) 519–522.
12. N. A. Zherebtsov and I. D. Ruadze, Specific features of acid and enzymic hydrolysis of starch (in Russian), *Izv. Vyssh. Uchebn. Zaved., Pishch. Tekhnol.* (3–4), (1994) 14–17.
13. M. E. Guerzoni, P. Lambertini, G. Lercker, and R. Marchetti, Technological potential of some starch degrading yeast, *Starch/Stärke*, 37 (1985) 52–57.
14. A. Payen and J. F. Persoz, Memoire sur la diastase, les principaux produits de ses reactions et leurs applications aux arts industrieles, *Ann. Chim. Phys.*, [2] 53 (1833) 73–92.
15. G. Tegge, Stärke and Stärkederivative, (2004) B. Behrs Verlag, Hamburg.
16. S. M. Roberts, N. Tuener, A. J. Willetts, and M. K. Turner, Introduction to Biocatalysis Using Enzymes and Micro-organisms, (1995) Cambridge Univ Press, Cambridge.
17. W. Biedermann, Die Autolyse der Stärke, *Fermentforschung*, 1 (1916) 474–504.
18. G. Wahl, Die autolytische Abbau von Cereallienstärken durch arteigene Amylasen, *Starch/Stärke*, 19 (1967) 322–327.
19. Z. Kertesz, A new method for enzymatic clarification of unfermented apple juice, US Patent 1,932,833 (1930).
20. W. Biedermann, Die organische Komponente der Diastasen und das währe Wesen der "Autolyse" die Stärke, *Fermentforschung*, 4 (1921) 359–396.
21. K. Hattori, Action of saliva on starch, *J. Pharm. Soc. Jpn.*, 516 (1925) 170–184.
22. E. Pozerski, Verdauung von roher Stärke mit Hilfe der Speichel, *Compt. Rend. Soc. Biol.*, 96 (1927) 1394–1396. *Chem. Zentr.*, 1927, II, 592.
23. C. M. Huang, Comparative proteomic analysis of human whole saliva, *Arch. Oral Biol.*, 49 (2004) 951–962.
24. F. N. Schulz, Wirkung de Speichelasche auf Stärkelösung, *Fermentforschung*, 3 (1919) 71–74.
25. W. Biedermann, Bemerkung zu Wohl gemuths und Sallinges Einwänden gegen meine Versuche über Autolyse der Stärke, *Fermentforschung*, 3 (1919) 70–71.
26. W. Biedermann, Fermentstudien, IV, Zur Autolyse der Stärke, *Fermentforschung*, 2 (1919) 458–472.
27. E. Bachrach, Spaltung der Stärke in Gegenwart von Speichelasche, *Compt. Rend. Soc. Biol.*, 83 (1920) 583. *Chem. Zentr.*, 1921, III, 443.
28. M. Samec, Sur l'hydrolyse enzymatique des amylophosphates naturels et synthetiques, *Compt. Rend. Acad. Sci.*, 181 (1925) 532–533.
29. A. Kaelin, Studien zur Prüfung von Pepsin, Pancreatin und getrockneter Schilddrüse, (1931) A.-G. Gebr. Leemann & Co., Zürich.
30. F. Welzmüller, Abbaufähigkeit der Kuhmilchdiastase gegen Stärkearten, *Biochem. Z.*, 125 (1921) 179–186.
31. E. Pozerski, Verdauung von roher Stärke durch der Pankreasaft aus Hundes, *Compt. Rend. Soc. Biol.*, 98 (1928) 1196–1198. *Chem. Zentr.*, 1928, II, 1686.
32. C. B. Duryea, Producing sugar and syrups from crude green starch, US Patent 1,110,755.
33. E. Kneen, O. C. Beckord, and R. M. Sandstedt, The starch-degrading properties of barley malts, *Cereal Chem.*, 18 (1941) 741–754.

34. F. H. Gallagher, H. R. Billford, W. H. Stark, and P. J. Kolachov, Fast conversion of distillery mash for use in a continuous process, *Ind. Eng. Chem.*, 34 (1942) 1395–1397.
35. B. Drews and B. Lampe, Maltspirit Diastasegriess oder Brennereiderrmalz in stärkehaltige Rohstoffe. Verarbeitenden Brennereien, *Brennerei Ztg.*, 57 (1940) 61–65.
36. D. I. Lisitsyn, Role of maltase in the hydrolysis of starch by different varieties of malt (in Russian), *Biokhimiya*, 1 (1936) 351–358.
37. K. Myrbäck, B. Örtenblad, and K. Ahlborg, Über Grenzdextrine, XIV. Über die Phosphorsäure in Stärkedextrinen, *Biochem. Z.*, 315 (1943) 240–249.
38. P. Petit, Worts and saccharification of starch, *Brasserie Malterie*, 17 (1928) 321–324. *Chem. Abstr.*, 22 (1928) 11957.
39. A. S. Siedich, Two new microbes which decompose starch (in Russian), *Russ. Health Resort Service*, 1 (1923) 11–17.
40. A. Boidin and J. Effront, Liquefying amyloceus substances with bacterial enzymes, US Patent 1,227,374 (1917).
41. H. L. A. Tarr, Hydrolysis of certain polysaccharides and proteins by the endospores of aerobic bacilli, *Biochem. J.*, 28 (1934) 391–393.
42. A. Vetsesol, The saccharification of potato starch by means of bacteria and the symbiosis of the latter with yeast (in Russian), *Brodil. Prom.*, 12(1), (1935) 17–22.
43. A. Janke and B. Schaefer, Beitrage zur Kenntnis für Stärkeabbau durch Bakterien, *Zentr. Bakt. Parasitenk.*, II Abt. 102 (1940) 241–250.
44. R. D. Stuart, Starch fermenting strains of *Corynebacterium diphtheriae* in Newcastle on Tyne: Serologica nad clinical investigation, *J. Path. Bact.*, 46 (1938) 173–179.
45. A. A. Margo, The intensity of starch decomposition and the virulence of diphtheria bacilli (in Russian), *Z. Microbiol. Epidemiol. Immunitaetsforsch (USSR)*, 19 (1937) 97–102.
46. G. Tarnovski and I. Ruessbuelt, Zuckerspaltende Fermente bei der Diphtherie- und Pseudodiphtherie Bak Serien. I. Stärkespaltung der verschiedene D-Typen Dextrosespaltung und Säurebildung, *Zentr. Bakt. Parasitenk.*, I Abt. Orig., 150 (1943) 247–260.
47. J. Takamine and J. Takamine, Jr., Enzymic composition, US Patent 1,460,736 (1923).
48. K. Sjoeberg, Spaltung der Stärke mit, *Saccharomyces sake*, *Z. Physiol. Chem.*, 162 (1927) 223–237.
49. A. Gottschalk, Über den Chemism der Vergärung von Glykogen und Stärke durch maltasefreie Hefe, *Z. Physiol. Chem.*, 168 (1927) 267–273.
50. H. Boulard, The saccharification of starch, British Patent 25,406 (1913).
51. H. Joula, Fluidyfying starch and utilizing fluorides in the fermentation with *Micor mucedo*, French Patent 478,972 (1916).
52. H. Hähn, Abbau der Stäerke durch ein System: Neutralsalze + Aminosäuren + Peptone, *Biochem. Z.*, 135 (1923) 587–602.
53. M. Eckhard, Wirkung der Diastase auf Stärke, *Z. Spiritusind.*, 48 (1925) 249.
54. L. de Hoop and J. A. van Laer, Diastatische Stärkeabbau, *Biochem. Z.*, 155 (1925) 235–244.
55. H. Hähn and H. Berentzen, Stärkeabbau durch das System: Neutralsalze + Aminosaeuren + Pepton, 2, *Chem. Zelle Gewebe*, 12 (1925) 286–316.
56. J. H. Walton and H. R. Dittmar, The hydrolysis of corn starch by commercial pancreatin, *J. Biol. Chem.*, 70 (1926) 713–728.
57. R. Kuhn, Verzuckerung der Stärke durch Emulsin, *Z. Physiol. Chem.*, 135 (1924) 12–15.
58. H. Wrede, Verflüssigung und Aufschliessigung von Stärke mitteln Bilase sowie ihre Vervendung in der Papierindustrie, *Papier-Fabr.*, 27 (1929) 197–202.
59. E. Glimm and I. Gizycki, Zur Kenntinis der Biolase, *Biochem. Z.*, 248 (1932) 449–473.
60. R. Weidenhagen and B. Korotkyj, Biolase, *Z. Ver. Deut. Zucker-Ind.*, 83 (1933) 262–272.
61. H. Pringsheim and E. Shapiro, Über den fermentativen Abbau der Stärke durch Biolase, *Berichte*, 59B (1926) 996–1000.

62. D. H. Cook, Temperature coefficients of enzymic activity and the heat destruction of pancreatic and malt amylases, *J. Biol. Chem.*, 65 (1925) 135–146.
63. H. Pringsheim and J. Leibowitz, Über die Spezialität der Amylasen, *Berichte*, 59B (1926) 991–995.
64. H. Pringsheim and A. Steingröver, Zur Kenntnis der Amylobiose, *Berichte*, 59B (1926) 1001–1006.
65. P. M. Franco de Sarabia Rosado, G. Limones Limones, A. Madejon Seiz, and M.D. Marin Alberdi, Stable, dried enzyme-containing mixtures for use in molecular biology, EU Patent WO 20002072002 (2002); *Chem. Abstr.*, 137 (2002) 212868.
66. K. Myrbäck, Über den s.g. Grenzabbau der Stärke, *Compt. Rend. Trav. Lab. Carslsberg, Ser. Chim.*, 22 (1938) 357–365.
67. G. M. Wild, Action patterns of starch enzymes, *Iowa State Coll. J. Sci.*, 28 (1954) 419–420.
68. E. Husemann and E. Lindemann, Neuere Untersuchungen über die Substratspecifität von alpha-, beta-, Macerans-Amylase und Phosphorylase, *Starch/Staerke*, 6 (1954) 141–148.
69. A. M. Vacca-Smith, A. R. Venkitaraman, R. G. Quivey, Jr., and W. H. Bowen, Interactions of streptococcal glucosyltransferases with α-amylase and starch on the surface of saliva-coated hydroxyapatite, *Arch. Oral Biol.*, 41 (1996) 291–298.
70. S. Rugh, T. Nielsen, and P. B. Poulsen, Application possibilities of a novel immobilized glucoamylase, *Starch/Stärke*, 31 (1979) 333–337.
71. J. Rogalski and J. Jamroz, The optimization of enzymatic hydrolysis of starch using immobilized α-amylase and glucoamylase (in Polish), *Post. Tech. Przetw. Spoz.*, 2 (1993) 14–18.
72. V. Ivanova, E. Dobreva, and M. D. Legov, Characteristics of immobilized thermostable amylases from two *Bacillus licheniformis* strains, *Acta Biotechnol.*, 18 (1998) 339–351.
73. F. G. Ewing and H. S. DeGroot, Evaluation of new continuous starch conversion process at Oxford's Rumford mill, *Paper Trade J.*, 150 (1966) 48–51.
74. K. Kim, C. S. Park, and J. R. Mattoon, High-efficiency one-step starch utilization by transformed *Saccharomyces* cells which secrete both yeast glucoamylase and mouse alpha-amylase, *Appl. Environ. Microbiol.*, 54 (1988) 966–971.
75. J. E. Nielsen, L. Beier, D. Otzen, T. V. Borchert, H. B. Frantzen, K. V. Andersen, and A. Svedsen, Electrostatics in the active site of α-amylase, *Eur. J. Biochem.*, 264 (1999) 816–824.
76. W. Callen, T. Richardson, G. Frey, J. M. Short, J. S. Kerovuo, E. J. Mather, M. Slupska, and K. Grey, Identification, screening, design, characterization and use of novel α-amylase, EU Patent WO 2002068589 (2002); *Chem. Abstr.*, 137 (2002) 212862.
77. W. Callen, T. Richardson, G. Frey, C. Miller, M. Kazaoka, E. J. Mathur, and J. M. Short, Identification, characterization, design and use of novel α-amylase, EU Patent WO 2002068597; *Chem. Abstr.*, 137 (2002) 212864.
78. A. Svendsen, C. Andersen, T. Thisted, and C. Van der Osten, *Bacillus* α-amylase variants with altered properties and their use, EU Patent WO 2002092797 (2002); *Chem. Abstr.*, 137 (2002) 381698.
79. K. S. Siddiqui and R. Cavicchioli, Methods for improving amylase thermostability for starch hydrolysis, EU Patent WO 2003056002 (2003); *Chem. Abstr.*, 139 (2003) 81324.
80. J. Liu, L. Jiang, X. Zhao, P. Li, Y. Tian, and J. Zhang, Study of *Rhizopus* species mutant PE-8-starch degradation enzyme. I. Strain isolation and assay, *Shipin Kexue (Beijing)*, 23(11), (2002) 34–37. *Chem. Abstr.*, 139 (2003) 130494.
81. L. Tang, W. Wu, J. Duan, and P. F. Johannesen, Cloning, purification and characterization of thermistable α-amylase from *Rhizomucor pusillus* and use in liquefying starch, production of alcohol, brewing and baking, EU Patent WO 2004055178 (2004); *Chem. Abstr.*, 141 (2004) 67294.
82. W. Callen, T. Richardson, G. Frey, K. Gray, J. S. Kerovuo, M. Slupska, N. Barton, E. O'Donoghue, and C. Miller, Amylases, nucleic acids encoding them and methods for their production and biological and industrial use, EU Patent WO 2004091544 (2004); *Chem. Abstr.*, 141 (2004) 362388.
83. R. Taira, S. Takagi, C. Hjort, A. Vikso-Nielsen, E. Allain, and H. Udagawa, Fungal α-amylase fusion proteins for starch processing, EU Patent WO 2005003311 (2005); *Chem. Abstr.*, 142 (2005) 129779.

84. E. Ferrari, M. Kolkman, and C. E. Pilgrim, Mutant α-amylases with improved properties for stability and/or activity, EU Patent WO 2005111203 (2005); *Chem. Abstr.*, 143 (2005) 474237.
85. J. Vehrmaanpera, P. M. A. Nybergh, R. Tanner, E. Pohjonen, R. Bergelin, and M. Korhola, Industrial production of α-amylase by genetically engineered *Bacillus*, *Enzyme Microb. Technol.*, 9 (1987) 546–549.
86. B. Tatsinkou Fossi, F. Tavea, and R. N. D. Joneukeu, Production and partial characterization of a thermostable amylase from ascomycetes yeast strain isolated from starchy soil, *Afr. J. Biotechnol.*, 4 (2005) 14–18.
87. T. Krishnan and A. K. Chandra, Purification and characterization of α-amylase from *Bacillus licheniformis* CUMC 305, *Appl. Environ. Microbiol.*, 46 (1983) 430–437.
88. M. J. Crispeels, M. Fatima Grossi de Sa, and T. J. V. Higgins, Genetic engineering with α-amylase inhibitors makes seeds resistant to bruchids, *Seed Sci. Res.*, 8 (1998) 257–264.
89. B. Svensson, K. S. Bak-Jensen, H. Mori, J. Sauer, M. T. Jensen, B. Kramhoft, T. E. Gottschalk, T. Christensen, B. W. Sigurskjold, N. Aghajari, R. Haser, N. Payre, S. Cottaz, and H. Driguez, The engineering of specificity and stability in selected starch degrading enzymes, *Roy. Soc. Chem, Spec. Publ.*, 246 (1999) 272–281.
90. B. E. Norman and H. V. Hendriksen, Enzymatic preparation of glucose syrup from starch, EU Patent WO 9946399 (1999); *Chem. Abstr.*, 131 (1999) 198844.
91. M. Bjornvad, S. Pedersen, M. Schulein, and H. Bisgard-Frantzen, Fusion protein comprising α-amylase and cellulose-binding domain for the degradation of starch, EU Patent WO 9816633 (1998); *Chem. Abstr.*, 128 (1998) 318808.
92. J. R. Cherry, A. Svendsen, C. Andersen, L. Beier, and T. P. Frandsen, Maltogenic α-amylase variants with altered properties, EU Patent WO 9943794 (1999); *Chem. Abstr.*, 131 (1999) 181666.
93. B. J. H. Janes and I. S. Pretorius, One-step enzymatic hydrolysis of starch using a recombinant strain of *Saccharomyces cerevisiae* producing α-amylase, glucoamylase and pullulanase, *Appl. Microbiol. Biotechnol.*, 42 (1995) 878–883.
94. S. C. H. J. Turk, N. Gerrits, J. C. M. Smeekens, and P. J. Weisbeek, Sugar-transferring enzyme recombinant expression, polysaccharide modification in vitro and in transgenic plant, and uses in food and non-food industry, EU Patent WO 9729186 (1997); *Chem. Abstr.*, 127 (1997) 186629.
95. A. Amsal, M. Takigami, and H. Ito, Improvement of raw starch digestibility by ion-beam mutation of *Aspergillus awamori* (in Japanese), *Shokuhin Shosha*, 33(1,2), (1998) 37–40.
96. K. M. Kragh, B. Larsen, P. Rasmussen, L. Duedahl-Olesen, and W. Zimmermann, Non-maltogenic exoamylases and their use in retarding retrogradation of starch, EU Patent WO 8850399 (1999); *Chem. Abstr.*, 131 (1999) 268983.
97. A. Viksoe-Nielsen, C. Andersen, S. Pedersen, and C. Hjort, Hydrolysis of starch using hybrid α-amylase, EU Patent WO 2004113551 (2004); *Chem. Abstr.*, 142 (2004) 92332.
98. J. Narita, K. Okano, T. Tateno, T. Tanino, T. Sewaki, M. H. Sung, H. Fukuda, and A. Kondo, Display of active enzymes on the cell surface of *Escherichia coli* using PgsA ancor protein and their application in bioconversion, *Appl. Microbiol. Biotechnol.*, 70 (2006) 564–572.
99. H. Fuwa, D. V. Glover, Y. Sugimoto, and M. Tanaka, Comparative susceptibility to amylases of starch granules of several single endosperm mutatnts representative of floury-opaque, starch-defficient and modified starch types and their double-mutant combinations with opaque-2 in four inbred lines of maize, *J. Nutr. Sci. Vitaminol.*, 24 (1978) 437–448.
100. A. Goldsbrough and S. Colliver, Starch branching enzyme II (SBEII-1 and SBEII-2) isoforms from wheat, cDNA transgenic plants and altering starch properties for food use, EU Patent WO 2000015810 (2000); *Chem. Abstr.*, 132 (2000) 247996.
101. K. A. Clarkson, N. Dunn-Coleman, S. E. Lantz, C. E. Pilgrim, P. Van Solingen, and M. Ward, Cloning, sequences and biotechnological use of *Trichoderma resei* acid proteases NSP24, NSP25 and PepA, EU Patent WO 2006073839 (2006); *Chem. Abstr.*, 145 (2006) 140134.

102. G. Cenci, G. Caldini, and F. Trotta, Bacterial amylolytic activity enhances β-glucuronidase expression of amylase-negative *Escherichia coli* strain in starch-medium, *J. Basic Microbiol.*, 40 (2000) 311–318.
103. R. M. Dhawale and W. Ingledew, Starch hydrolysis by depressed mutants of *Schwanniomyces castellii*, *Biotechnol. Lett.*, 5 (1983) 185–190.
104. Y. Tanaka, T. Ashikari, N. Nakamura, Y. Tanaka, Y. Shibano, T. Amachi, and H. Yoshizume, Glucoamylase produced by *Rhizopus* and by a recombinant yeast containing the *Rhizopus* glucoamylase gene, *Agric. Biol. Chem.*, 50 (1986) 1737–1742.
105. A. Amsai, M. Takigami, and H. Ito, Increased digestibility of raw starches by mutant strains of *Aspergillus awamori*, *Food Sci. Technol. Res.*, 5 (1999) 153–155.
106. J. Horai, Stabilization of *Bacillus subtilis* α-amylase by amino group acylation, *Biochim. Biophys. Acta*, 310 (1973) 264–267.
107. S. Janeček, S. Balaz, M. Rosenberg, and M. Stredansky, Chemical stabilization of *Bacillus subtilis* α-amylase by modification with D-glucono-δ-lactone, *Biotechnol. Tech.*, 6 (1992) 173–176.
108. Y. Li, P. M. Coutinho, and C. Ford, Hyperthermostable mutant of *Bacillus licheniformis* alpha amylase: Thermodynamic studies and structural interpretation, *Protein Eng.*, 10 (1997) 541–549.
109. D. W. S. Wong, Food Enzymes, Structure and Mechanisms, (1995) Chapman & Hall, New York, Chapter 3, pp. 37–84.
110. D. French, In: H. L. Boyer, H. Lardy, and K. Myrbäck, (Eds.). The Enzymes, 2nd edn. Academic Press, New York, 1960.
111. J. Thoma, J. Spradlin, and S. Dygert, In: P. Boyer, (Ed.). The Enzymes, 3rd edn. Academic Press, New York, 1971.
112. C. Bertoldo and G. Antranikian, Natural polysaccharide-degrading enzymes, Enzyme Catalysis in Organic Synthesis, 2nd edn., pp. 653–685.
113. A. Blennow, Starch bioengineering, In: A. C. Eliasson, (Ed.), *Starch in Food*, Woodhouse Publ. Ltd, Cambridge, 2004, pp. 97–127.
114. H. S. Olsen, Enzymes in starch modification, In: R. J. Whitehurst and B. A. Law, (Eds.) *Enzymes in Food Technology*, Sheffield Academic Press, pp. 200–228.
115. B. J. D. Meeuse, The reversible hydrolysis of starch in connection with some biological problems (in Dutch), *Chem. Weekbl.*, 47 (1951) 17–23.
116. P. Bernfeld, Enzymes of starch degradation and synthesis, *Adv. Enzymol.*, 12 (1951) 379–428.
117. I. A. Preece, Carbohydrate changes in malting and mashing, *J. Incorp. Brewers Guild*, 27 (1941) 19–26.
118. K. Myrbäck, Novel facts on starch constitution (in Sweedish), *Tekn. Samfund. Handl.*, 70 (1941) 79–129.
119. J. Hollo and E. Laszlo, Kinetic comparison of acidic, alkaline and enzymic hydrolysis of starch, *Aust. Chem. Eng.*, 10(4), (1969) 17–22.
120. L. Zhang, Y. Xu, and J. Yan, Application and study of acid alpha amylase, *Niangiju*, 29(3), (2002) 19–22. *Chem. Abstr.*, 138 (2002) 186475.
121. R. Gupta, P. Gigras, H. Mohapatra, V. K. Goswami, and B. Chauhan, Microbial α-amylases: A biotechnological perspective, *Process Biochem.*, 38 (2003) 1599–1616.
122. N. S. Reddy, A. Nimmagadda, and K. R. S. S. Rao, An overview of the microbial α-amylase family, *Afr. J. Biotechnol.*, 2 (2003) 645–648.
123. T. Kitano, Takaamylase (in Japanese), *Chem. Rev. Jpn.*, 3 (1937) 433–444.
124. O. Tanabe, *Aspergillus oryzae* and amylolytic enzymes (in Japanses), *Hakko Kyokaishi*, 7 (1949) 203–207.
125. Y. Takasaki, Acid-stable and thermostable α-amylase (in Japanese), *Shokuhin Kogyo*, 37(12), (1994) 44–50.
126. B. Farjon, Z. Targoński, and J. Pielecki, Enzymes hydrolyzing linkages of α-1,6-D-glucosides, their properties ad applications in starch bioconversion (in Polish), *Przem. Ferm. Owoc-Warz.*, 36(3), (1992) 19–20.

127. R. R. Ray, Microbial isoamylases: An overview, *Am. J. Food Technol.*, 6 (2011) 1–18.
128. J. Wang, Fungal glucoamylase in raw-starch digestion and ethanol fermentation, *Zhongguo Niangzao*, 2(1), (1983) 6–10. *Chem. Abstr.*, 99 (1983) 103599a.
129. Y. Tsujisaka, Applications of microbial enzyme in starch production, *Denpun Kagaku*, 35(1), (1988) 61–67.
130. N. Juge, M. F. Le Gal-Coeffet, C. S. M. Furniss, A. P. Gunning, B. Kramhoft, V. J. Morris, G. Williamson, and B. Svensson, The starch binding domain of glucoamylase from *Aspergillus niger*: Overview of its structure, function and role in raw starch hydrolysis, *Biologia (Bratislava)*, 57 (Suppl. 11), (2002) 239–245.
131. A. Pandey, Glucoamylase research: An overview, *Starch/Stärke*, 47 (1995) 439–445.
132. G. Williamson and N. J. Belshaw, Properties of the binding domain of glucoamylase, *Carbohydr. Polym.*, 21 (1993) 147–149.
133. J. J. Marshall, Some aspects of glucoamylase action relevant to the enzymic production of dextrose from starch, *Proc. 2nd Int. Congr. Food Process Eng., 1979*, 2 (1980) 224–236.
134. M. Goto, Digestion of raw starch by glucoamylase I from *Aspergillus awa, ori* var kawachi (in Japanese), *Baiosai. Indas.*, 59(3), (2001) 183–184. *Chem. Abstr.*, 134 (2001) 256129.
135. B. C. Saha and J. G. Zeikus, Novel highly thermostable pullulanase from thermophiles, *Trends Biotechnol.*, 7 (1989) 234–239.
136. T. Sato, Use of food enzymes. Theories and examples (in Japanese), *Shokuhin Kogyo*, 23(12), (1980) 57–63.
137. F. Kaplan, Y. Dong, and C. L. Guy, Roles of β-amylase and starch breakdown during temperature stress, *Physiol. Plant.*, 126 (2006) 120–128.
138. P. Ziegler, Cereal beta-amylases, *J. Cereal Sci.*, 28 (1999) 195–204.
139. L. Frankova, Basic characteristics of enzymes involved in the amylolytic process (in Slovak), *Biol. Listy*, 68 (2003) 81–107.
140. H. Hidaka and T. Kono, Manufacture of maltose and technical problems (in Japanese), *Shokuhin Kogyo*, 24(2), (1981) 33–43.
141. D. P. Butler, M. J. E. C. van der Maarel, and P. A. M. Steeneken, Starch-acting enzymes, In: A. C. Eliasson, (Ed.), *Starch in Food*, pp. 128–155.
142. S. Yu, K. Kragh, and A. Morgan, New enzymes and products of the anhydrofructose pathway of starch, *ACS Symp. Ser.*, 972 (2007) 49–60.
143. J. Synowiecki and A. Zdziebło, Thermophiles as a source of α-glucosidases suitable for production of glucose syrups (in Polish), *Biotechnologia* (2), (2002) 165–174.
144. F. Cramer and D. Steinle, Die wirkungsweise der Amylose aus *Bacillus macerans*, *Ann.*, 595 (1955) 81–100.
145. K. L. Anderson, Degradation of cellulose and starch by anaerobic bacteria, In: R. J. Doyle, (Ed.), *Glycomicrobiology*, Kluwer Academic/Plenum Publ, New York, 2000, pp. 359–386.
146. J. Hollo, E. Laszlo, and A. Hoeschke, Enzyme-engineering in the starch industry, *Acta Aliment. Pol.*, 10(1–2), (1984) 33–49.
147. K. Myrbäck, Der enzymatische Abbau der Stärke und der Bau der Stärkemakromoleküle, *J. Prakt. Chem.*, 162 (1943) 29–62.
148. K. Myrbäck, The action of malt amylases on starch and glycogen, *Brewers Digest*, 23(3), (1948) 41–44. 47.
149. C. V. Ramakrishnan, Use of enzymes in starch industry, *J. Sci. Ind. Res.*, 33 (1974) 143–147.
150. D. J. Manners, Some aspects of the enzymic degradation of starch, *Annu. Proc. Phytochem. Soc.*, 10 (1974) 109–125.
151. I. Maeda, Starch decomposition enzymes and decomposition of starch grains, *Denpun Kagaku*, 25 (1980) 11–26.

152. T. Ohya, N. Yokoi, and T. Sakai, Enzymic saccharification of starch (in Japanese), *New Food Ind.*, 26(8), (1984) 16–21.
153. P. Reilly, Enzymic degradation of starch, *Food Sci. Technol.*, 14 (1985) 101–142.
154. S. Wang, Research and applications of raw starch-saccharyfying enzymes, *Shipin Yu Fajiao Gongye* (2), (1988) 72–78. *Chem. Abstr.*, 109 (1988) 127245h.
155. P. Colonna, A. Buleon, and C. Mercier, Starch and enzymes: Innovations in the products, processes and uses, *Chim. Oggi* (10), (1988) 9–14.
156. S. A. Barker, Enzymic routes to cereal starch utilization, *Int. Ind. Biotechnol.*, 9(2), (1989) 12–15.
157. J. Hollo and A. Hoeschke, Bioconversion of starch. I. (in Hungarian), *Elelmezesi Ipar*, 45 (1991) 362–369.
158. J. Hollo and A. Hoescke, Bioconversion of starch, *Pol. J. Food Nutr. Sci.*, 2(1), (1993) 5–37.
159. H. Guzman-Maldonado and O. Paredes-Lopez, Amylolytic enzymes and products derived from starch: A review, *Crit. Revs. Food Sci. Nutr.*, 35 (1995) 373–403.
160. V. Planchot, P. Colonna, A. Buleon, and D. Gallant, Amylolysis of starch granules and α-glucan crystallites, *Spec. Publ. Roy. Soc. Chem.*, 205 (1997) 141–152.
161. J. Bryjak, Enzymic hydrolysis of starch to maltodextrin and starch syrups. I. Enzymes. (in Polish), *Biotechnologia* (1), (1999) 180–200.
162. W. D. Crabb and J. K. Shetty, Commodity scale production of sugars from starches, *Curr. Opin. Microbiol.*, 2 (1999) 252–256.
163. R. Hoover and Y. Zhou, In vitro and in vivo hydrolysis of legume starches by α-amylase and resistant starch. Formation in legumes – A review, *Carbohydr. Polym.*, 54 (2003) 401–417.
164. R. F. Tester, X. Qi, and J. Karkalas, Hydrolysis of native starches with amylase, *Animal Feed Sci. Technol.*, 130(1–2), (2006) 39–54.
165. R. A. J. Warren, Microbial hydrolysis of starch, *Annu. Rev. Microbiol.*, 50 (1996) 183–212.
166. W. Błaszków, Enzymes for hydrolysis of starch (in Polish), *Przem. Spoż.*, 21(12), (1967) 22–24.
167. C. T. Greenwood and E. Milne, Starch degrading and synthesizing enzymes. A discussion of their properties and action pattern, *Adv. Carbohydr. Chem.*, 23 (1968) 281–366.
168. M. D. Lilly, S. P. O'Neill, and P. Dunnill, Bioengineering of immobilized enzymes, *Biochimie*, 55 (1973) 985–990.
169. J. J. Marshall, Starch-degrading enzymes old and new, *Starch/Staerke*, 27 (1975) 377–383.
170. H. C. Barfoed, Enzymes in starch processing, *Cereal Foods World*, 21 (1976) 583–589. 592–593, 604.
171. M. K. Weibel, W. H. McMullan, and C. A. Starace, Microbial enzymes in starch the production of nutritive sweeteners from starch, *Dev. Ind. Microbiol.*, 19 (1977) 103–116.
172. W. M. Fogarty and C. T. Kelly, Starch-degrading enzymes of microbial origin. I. Distribution and characteristics, *Progr. Ind. Microbiol.*, 15 (1979) 87–150.
173. Y. Takasaki, Development of new saccharification enzymes for starch. Maltooligosaccharide-producing enzymes (in Japanese), *Hakko to Kogyo*, 41 (1983) 477–489.
174. N. Takahasi, Novel enzymesm maltooligosyl trehalose synthase and maltooligosyl trehalose trehalohydrolase and their application in the production of trehalose from starch, *Trends Glycosci. Glycotechnol.*, 8 (1996) 369–370.
175. J. Synowiecki and B. Grzybowska, Suitability of thermostable enzymes fro improvement of starch processing (in Polish), *Biotechnologia* (2), (2001) 26–35.
176. F. J. Plou, M. T. Martin, A. Gomez de Seruga, M. Alcalde, and A. Ballesteros, Glucotransferases acting on starch or sucrose for the synthesis of oligosaccharides, *Can. J. Chem.*, 80 (2002) 743–752.
177. E. Leveque, S. Janeček, B. Haye, and A. Belarbi, Thermophilic archeal amylolytic enzymes, *Enzyme Microb. Technol.*, 26 (2000) 3–14.

178. A. Pandey, P. Nigam, C. R. Soccol, V. T. Soccol, D. Singh, and R. Mohan, Advances in microbial amylases, *Biotechnol. Appl. Biochem.*, 31 (2000) 135–152.
179. Y. Bhatia, S. Mishra, and V. S. Bisaria, Microbial beta-glucosidases: Cloning, properties and applications, *Crit. Revs Biotechnol.*, 22 (2002) 375–407.
180. P. Nigam and D. Singh, Enzyme and microbial systems involved in starch processing, *Enzyme Microb. Technol.*, 17 (1995) 770–778.
181. W. D. Crabb and C. Mitchinson, Enzymes involved in the processing of starch to sugars, *TIBTECH*, 15 (1997) 349–356.
182. M. J. E. C. van der Maarel, B. van der Veen, J. C. M. Uitdehaag, H. Leemhuis, and L. Dijkhuizen, Properties and applications of starch-converting enzymes of the α-amylase family, *J. Biotechnol.*, 94 (2002) 137–155.
183. I. S. Pretorius and M. G. Lambrechts, The glucoamylase multigene family in *Saccharomyces cerevisia* var. *diastaticus:* An overview, *Crit. Revs. Biochem. Mol. Biol.*, 26 (1991) 53–76.
184. L. Chang, Starch as a good source of sugars, *Huanan Ligong Daxue Xuebao, Ziran Kexueban*, 22(6), (1995) 105–112. *Chem. Abstr.*, 123 (1995) 196807.
185. J. Hollo, Über einige aktuelle theoretische und praktische Probleme der Stärkeenzymologie, *Starch/Stärke*, 26 (1974) 257–263.
186. H. Fuwa and Y. Sugimoto, The degradation of starch granules by starch degrading enzymes (in Japanese), *Denshi Kenbikyo*, 19 (1985) 200–207.
187. D. Howling, Mechanism of starch enzymolysis, *Int. Biodeterior.*, 25 (1989) 15–19.
188. H. Fuwa, Enzyme degradation of starch granules (in Japanese), *Denpun Kagaku*, 29 (1982) 99–106.
189. R. W. Kerr and H. Gehman, The action of amylases on starch, *Starch/Staerke*, 3 (1951) 271–278.
190. J. Hollo, E. Laszlo, and A. Hoeschke, Mechanism of amylolytic starch degradation, *Zesz. Probl. Post. Nauk Roln.*, 159 (1974) 227–255.
191. W. Z. Hassid, The mechanism of breakdown and formation of starch and glycogen, *Wallerstein Lab. Commun.*, 9 (1946) 135–144.
192. K. Myrbäck, The constitution and the enzymic degradation of starch II, *SuomenKemistilehti*, 16 A (1943) 63. *Chem. Abstr.*, 40 (1945) 7676.
193. H. B. Zhou, H. Zhao, and X. D. Gao, Enzymatic modification of polysaccharides, *Yaoxue Jinzhan*, 31 (2007) 349–352. *Chem. Abstr.*, 148 (2007) 314554.
194. P. Diedering, Colloidal behavior and enzymatic degradation of starch and a contribution in decreasing sprout injury, *Z. Ges. Getreidew.*, 29 (1942) 182–184. *Chem. Abstr.*, 37 (1943) 40030.
195. P. Diedering, Colloidal behaviour and enzymatic degradation of starch and a contribution on decreasing sprout injury, *Z. Ges. Getreidew.*, 30 (1943) 1–17.
196. K. Myrbäck and L. G. Sillen, Branched molecules. IV. Enzymic dextrinization of starch and glycogen, *Svensk. Kem. Tid.*, 56 (1944) 60–72. *Chem. Abstr.*, 40 (1946) 14827.
197. S. Ueda, Raw starch digestion with koji fungal amylase (in Japanese), *Chomi Kagaku*, 13(5), (1966) 1–4.
198. V. Horvathova, S. Janeček, and E. Sturdik, Amylolytic enzymes, their specificity origins and properties, *Biologia (Bratislava)*, 55 (2000) 605–615.
199. J. H. Pazur, Enzymes in synthesis and hydrolysis of starch, *Starch Chem. Technol.*, 1 (1965) 133–175.
200. T. Komaki, Occurrence of insoluble starch particles in enzymically saccharidfied starch solution (in Japanese), *Dempun Kogyo Gakkaishi*, 17 (1969) 131–138.
201. M. A. Gromov, Thermophysical characteristics of starch, *Sakh. Prom.*, 45(4), (1971) 67–70. *Chem Abstr.*, 75 (1971) 7699s.
202. A. D. Woolhouse, Starch as a source of carbohydrate sweeteners, *N. Z. Dep. Sci. Ind. Res., Chem. Div.*, (1976) C.D. 2237.

203. J. Hollo and E. Laszlo, Chemical processes of the biosynthesis and enzymic degradation of starch (in Hungarian), *Kem. Ujabb Eredmenyi*, 11 (1972) 7–80.
204. Y. Takasaki, Manufacture of maltose from starch with enzymes produced by bacteria (in Japanese), *Dempun Kogyo Gakkaishi*, 22 (1977) 9–26.
205. H. Fuwa, Digestion of various starch granules by amylase (in Japanese), *Dempun Kagaku Gakkaishi*, 24(4), (1977) 128–140.
206. S. Ueda, Fungal glucoamylases and raw starch digestion, *Trends Biochem. Sci.*, 6 (1981) 89–90.
207. I. Spencer-Martins, Starch bioconversion by yeast, *Proc. 9th Symp. Iberoam. Catal.*, 1 (1984) 373–380. *Chem. Abstr.*, 102 (1985) 60739g.
208. A. W. MacGregor, α-Amylase, limit dextrinase, and α-glucosidase enzymes in barley and malt, *CRC Crit. Rev. Biotechnol.*, 5 (1987) 117–128.
209. H. M. Li, Liquefaction technology in the production of dextrose by enzymic process, *Shih Pin Yu Fa Hsiao, Kung Yeh* (1), (1980) 53–62. *Chem. Abstr.*, 94 (1980) 101367w.
210. T. Yamamoto, Research on amylases and their effects on advances in enzyme chemistry and industry (in Japanese), *Denpun Kagaku*, 32 (1985) 99–106.
211. M. Richter and F. Schierbaum, Wirkung der Glucoamylase auf Stärkepolysaccharide, *Ernaehrungforsch.*, 13 (1968) 385–397.
212. G. Antranikian, Physiology and enzymology of thermophilic anaerobic bacteria degrading starch, *FEMS Microbiol. Revs.*, 75 (1990) 201–218.
213. J. F. Kennedy, V. M. Cabalda, and C. A. White, Enzymatic starch utilization and genetic engineering, *Trends Biotechnol.*, 6 (1988) 184–189.
214. R. A. Oosterbaan, Active groups in hydrolytic enzymes, *Chem. Weekbl.*, 58 (1962) 453–460. *Chem. Abstr.*, 57 (1962) 477697.
215. M. Ohnishi, Is there a "double-headed eagle"? The secret of taking in saccharides by amylase, *Trends Glycosci. Glycotechnol.*, 2 (1990) 343–352.
216. E. Möller, B. Fülbier, and D. Haberland, Kinetic models of starch hydrolysis by exo- and endo-amylases. Problems and technical importance, *Wiss. Z. Ernst-Moritz-Arnd Univ., Greifswald, Math-Naturwiss. Reihe*, 39(1), (1990) 7–13.
217. K. Sjöberg, Stärke und ihre enzymatische Spaltung, *Fermentforschung*, 9 (1928) 329–335.
218. H. Colin, Les diastases, *Bull. Trimestr. Assoc. Eleves Ecole Sup. Brasserie Univ. Louwain*, 29 (1929) 49–60.
219. K. Myrbäck, The structure and enzymatic cleavage of starch, Svenska Bryggarefoeren. Monadsbl., 57 (1942) 69–77. 105–112; *Chem. Abstr.*, 37 (1943) 24576.
220. W. J. Whelan, Enzymic breakdown of starch, *Biochem. Soc. Symp.*, 11 (1953) 17–26.
221. Y. Hirayama, Enzyme digestion of carbohydrates (in Japanese), *Gendai Kagaku*, 350 (2000) 51–53.
222. K. Meyer, The past and present of starch chemistry, *Experientia*, 8 (1952) 405–420.
223. H. C. Gore, Action of amylases on starch, *Ind. Eng. Chem.*, 28 (1936) 86–88.
224. R. Delecourt, Industrial uses of amylases, *Ind. Chim. Belge*, 37 (1972) 1005–1011. *Chem. Abstr.*, 78 (1973) 39763.
225. H. U. Wölk, Enzyme in der Stärketechnologie, *Z. Ernahrungwiss. Suppl.*, 16 (1978) 136–149.
226. Z. Boskov and J. Jakovlević, Enzymic methods for the hydrolysis of starch in dextrose production (in Serbo-Croatian), *Hrana Ishrana*, 12 (1971) 437–444.
227. A. Hesse, Vervendung von Enzymen in der Industrie. 1. Abbau von Stärke und Eiweiss dem Malzen und Maischen, *Ergebnisse Enzymforsch.*, 3 (1934) 95–134.
228. T. Tamura, Liquefaction of starch, *Denpun Kogyo Gakkaishi*, 10 (1963) 93–97.
229. P. Foolbrook and B. Vaboe, Industrial enzymatic conversion of starch to sugars, *Inst. Chem. Eng. Symp. Ser.*, 51 (1977) 31–44.
230. L. Ludvig and E. Szatmar, Starch glycolysis (in Hungarian), *Szeszipar*, 21 (1973) 7–10.

231. K. Hoshino, S. Morohashi, T. Sasakura, and M. Taniguchi, Development of simultaneous saccharification-fermentation process (in Japanese), *Kemikaru Enjiyaringu*, 41(11), (1996) 8–13.
232. C. Weller, M. P. Steinberg, and E. D. Rodda, Fuel ethanol from raw corn, *Trans. Am. Soc. Agric. Eng.*, 27 (1984) 1911–1916.
233. A. A. Kukharenko, M. N. Dadashev, V. M. Korotkii, and A. Yu. Vinarov, Hydrolysis of starch-containing grain materials by amylolytic enzymes (in Russian), *Oboron. Kompl. Nauch.-Tekhn. Progr. Rossii* (3), (2008) 71–75.
234. R. Olered, Starch and its enzymes – Quality problems, *Rec. Plant Breeding Res.* (1963) 284–291.
235. W. M. Doane, Starch: Renewable raw material for the chemical industry, *J. Coat. Technol.*, 50 (1978) 88–98.
236. L. Lima da Rocha and A. Marcondes de Abreu, Manioc as industrial raw material (in Portuguese), *Rev. Quim. Ind. (Rio de Janeiro)*, 48 (1979) 14–29.
237. R. A. Serebrinskaya, Complex use of raw material in the starch-syrup industry (in Russian), *Pishch. Prom. Ser.*, 5(1), (1980) 16–21.
238. J. R. Stark and A. Lynn, Starch granules large and small, *Biochem. Soc. Trans.*, 20(1), (1992) 7–12.
239. P. N. Boyes, Starch in the manufacture of raw sugar, *Proc. Annu. Congr. S. Afr. Sugar Tech. Assoc.*, 34 (1960) 91–97.
240. K. E. M. Linnecar, G. Norton, R. J. Neale, and J. M. V. Blanshard, Bioavailability of raw starches – Causes of granule resistnce to α-amylase attack, *Ber. Bundesforschungsanst. Ernaehr.*, BFE-R-93-01, Bioavailab. Pt. 1 (1993) 187–190; *Chem. Abstr.*, 122 (1995) 159307.
241. K. G. De Noord, Use of enzymes in the manufacture of starch hydrolysates, *DECHEMA Monogr.*, 70 (1972) 247–263.
242. L. M. Marchal, Partial enzymatic hydrolysis of starch to maltodextrins on the laboratory scale, *Methods Biotechnol.*, 10 (1999) 119–127.
243. G. H. Robertson, D. W. S. Wong, C. C. Lee, K. Wagschal, M. R. Smith, and W. J. Orts, Native or raw starch digestion: A key step in energy efficient biorefining of grain, *J. Agric. Food Chem.*, 54 (2006) 353–365.
244. S. Suzuki, Production of glucose by enzymic hydrolysis of starch, *Dempunto Gijutsu Kenkyu Kaiho*, 23 (1961) 52–62.
245. D. J. Manners, Some industrial aspects of the enzymic degradation of starch, *Biochem. J.*, 123(2), (1971) 1P–2P.
246. Z. Kosicki, M. Remiszewski, and L. Słomińska, New trends in the production of starch hydrolysates (in Polish), *Przem. Spoz.*, 32(2), (1978) 58–60.
247. B. E. Norman, The application of polysaccharide degrading enzymes in the starch industry, *Spec. Publ. Soc. Gen. Microbiol.*, 3 (1979) 339–376.
248. S. J. Luenser, Microbial enzymes for industrial sweetener production, *Dev. Ind. Microbiol.*, 24 (1983) 79–96.
249. V. Donelyan, The use of enzymes for starch derivatives (in Italian), *Tecnol. Chim.*, 4(1), (1984) 38–47.
250. P. B. Rasmussen, Enzyme processes in the production of sweeteners from starch, *Acta Aliment. Pol.*, 10(1–2), (1984) 51–66.
251. Y. Tsujisaka, Starch processing by enzymes (in Japanese), *Sen'i Gakkaishi*, 43(9), (1987) P370–P374.
252. J. Bryjak, Enzymic manufacture of maltodextrins and syrup solids from starch. II. Processes (in Polish), *Biotechnologia* (1), (1999) 201–225.
253. T. Sugimoto, Production of trehalose from starch by novel enzymes (in Japanese), *Kagaku Kogyo*, 50 (1999) 465–473.
254. H. Chaen, Studies on novel enzymes for synthesis of trehalose from starch (in Japanese), *Oyo Toshitsu Kagaku*, 44 (1997) 77–82.

255. S. Yu and J. Marcussen, α-1,4-glucanase, molecular features and its use for production of 1,5-anhydro-D-fructose from starch, *Spec. Publ. Roy. Soc. Chem.*, 246 (1999) 243–250.
256. C. S. Hanes, The reversible formation of starch from glucose-1-phosphate catalyzed by potato phosphorylase, *Proc. Roy. Soc (London)*, B129 (1940) 174–208.
257. T. Nishimoto, The development of a new mass-production method of cyclic tetrasaccharide and its functions (in Japanese), *Nippon Nogei Kagaku Kaishi*, 78 (2004) 866–869.
258. B. L. Dasinger, D. M. Fenton, R. P. Nelson, F. F. Roberts, and S. J. Truesdell, Enzymic and microbial processes in the conversion of carbohydrates derived from starch, *Food Sci. Technol.*, 14 (1985) 237–262.
259. L. A. Underkofler and L. B. Lockwood, The production of chemicals from starch by fermentation process, In: J. A. Radley, (Ed.), *Starch Production and Technology,* Applied Science Barking, England, 1976, pp. 399–422.
260. L. Słomińska, Enzymatic methods of the starch transformations (in Polish), *Przem. Spoz.*, 49 (1995) 472–475. 480.
261. C. C. Tseng, Preparation of ethanol from starch by continuous fermentation method (in Chinese), *Shih Pin Yu Fa Hsiao Kung Yeh* (5), (1980) 30–40.
262. S. Ueda, Ethanol fermentation of starch material without cooking (in Japanese), *Denpun Kagaku*, 34 (1987) 113–118.
263. S. Ueda, Glucoamylase and the digestion of starch. Application to alcohol fermentation (in Japanese), *Kagaku to Seiku*, 19 (1981) 178–780.
264. H. Yoshizumi, Breeding of glucoamylase producing yeast and alcohol fermentation using raw starch (in Japanese), *Hakko to Kogyo*, 45 (1987) 579–585.
265. F. Wang and M. Wang, Starch raw materials in energy saving ethanol fermentation, *Zhongguo Niangao* (6), (1987) 7–12.
266. G. Duan, H. Xu, C. Sun, Y. Qian, Y. Li, H. Zhou, X. Jiang, J. Shetty, and O. Lantero, Progress on ethanol production technology-breakthrough of new enzyme technology for producing ethanol from raw starch, *Shipin Yu Fajiao Gongye*, 32(7), (2006) 65–70. *Chem. Abstr.*, 149 (2008) 405262.
267. C. Sun and G. Duan, New enzyme and application technology for ethanol industry, *Niangiju*, 34 (2007) 73–80. *Chem. Abstr.*, 149 (2008) 405263.
268. J. F. Kennedy, J. M. S. Cabral, I. Sa-Correia, and C. A. White, Starch biomass: A chemical feedstock for enzyme and fermentation process, *Crit. Rep. Appl. Chem.*, 13 (1987) 115–148.
269. T. Senn and H. J. Pieper, Classical methods for ethanol production from biological materials, In: M. Röhr, (Ed.), *Biotechnology of Ethanol,* Wiley-VCH Verlag GmbH, Weinheim, 2001, pp. 7–86.
270. G. Zhang, Initial starch concentration in fermentation of strong aromatic Chinese spirits, *Niangiju* (4), (2000) 48–49. *Chem. Abstr.*, 134 (2001) 70435.
271. Y. Kiba and S. Ueda, Saccharification of uncooked starch and fermentation (in Japanese), *Nippon Jozo Kyokai Zasshi*, 75 (1980) 858–861.
272. A. Rosenberger, Gewinnung von Bioethanol aus Getreide mit Hilfe technischer Enzyme, *Getreidetechnol.*, 58 (2004) 303–305.
273. F. Godia, C. Casas, and C. Sola, A survey of continuous ethanol fermentation systems using immobilized cells, *Process Biochem.*, 22(2), (1987) 43–48.
274. S. C. Beech, Acetone-butanol fermentation of starches, *Appl. Microbiol.*, 1(2), (1953) 85–95.
275. D. T. Jones and D. R. Walsh, Acetone–butanol fermentation revisited, *Microbiol. Revs.*, 50 (1986) 484–524.
276. M. Berović and M. Legisa, Citric acid production, *Biotechnol. Annu. Rev.*, 13 (2007) 303–343.
277. S. Anastassiadis, I. G. Morgumov, S. V. Kamzolov, and T. V. Finogenova, Biosynthesis of lipids and organic acids by Yarrowia lipolytica strains cultivated on glucose, *Rec. Pat. Biotechnol.*, 2 (2008) 107–123.

278. D. Li and R. Wang, The application of enzyme preparation in vinegar production, *Zhongguo Tiaoweipin* (1), (2006) 37–42. *Chem. Abstr.*, 147 (2007) 384402.
279. N. Nakamura and K. Horikoshi, Beta cyclodextrin (in Japanese), *Kagaku Gijutsushi MOL*, 17 (1979) 46–50.
280. J. Szejtli, *Cyclodextrins and Their Inclusion Complexes*, Akademiai Kiado Budapest, 1982.
281. S. Okada and S. Kitahara, Hydrolysis of α-1,6-glucosidic bond by *Streptococcus brevis* α-amylase, *Fragrance J.*, 11(6), (1983) 58–62.
282. T. Takahashi and S. M. Smith, The functions of 4-α-glucanotransferases and their use for the production of cyclic glucans, *Biotechnol. Genetic Eng. Revs.*, 16 (1999) 257–280.
283. H. Dodziuk, Introduction to Supramolecular Chemistry, (2002) Kluwer, Dortrecht.
284. H. Dodziuk, (Ed.), *In* Cyclodextrins and Their Complexes, Wiley-VCH Verlag GmbH & Co., Weinheim, 2006.
285. H. Takata, I. Kojima, N. Taji, Y. Suzuki, and M. Yamamoto, Industrial production of branching enzyme and its application to production of highly branched cyclic dextrin (Cluster Dextrin) (in Japanese), *Seibutsu Kogaku Kaishi*, 84(2), (2006) 61–66.
286. T. Nishimoto, K. Oku, and K. Mukai, Discovery of two cyclic tetrasaccharides synthesizing systems from starch (in Japanese), *Kagaku to Seibutsu*, 44 (2006) 539–550.
287. Z. Li, M. Wang, F. Wang, Z. Gu, G. Du, J. Wu, and J. Chen, γ-cyclodextrin: A review on enzymatic production and applications, *Appl. Microbiol. Biotechnol.*, 77 (2007) 245–255.
288. H. Dodziuk, Introduction to Supramolecular Chemistry (a novel edition in Polish), (2009) Wyd. Uniw, Warszawskiego.
289. J. Szejtli, Past, present and future of cyclodextrin research, *Pure Appl. Chem.*, 76 (2004) 1825–1845.
290. T. Loftsson and D. Duchene, Cyclodextrins and their pharmaceutical applications, *Int. J. Pharm.*, 329 (2007) 1–11.
291. L. N. Ceci and J. E. Lozano, Use of enzymes in non-citrus fruit juice production, In: A. Bayindirli, (Ed.), *Enzymes in Fruit and Vegetable Processing, Chemistry and Engineering*, CRC Press, Boca Raton, 2010, pp. 175–196.
292. D. A. Hughes and W. L. Craig, Enzyme conversion of starch for paper coating, *Tappi*, 33 (1950) 253–256.
293. W. Maurer and R. L. Tearney, Opportunities and challenges for starch in the paper industry, *Starch/Stärke*, 50 (1998) 386–402.
294. K. Romaszeder, Theory and practice of enzymic desizing (modern desizing processes), *Magy Textiltech.*, 15 (1963) 164–172.
295. G. Hasselbeck, Starch degradation enzymes in fruit juice processing, *Flüss. Obst*, 67 (2000) 455–458.
296. A. Garcia-Alonso, N. Martin-Carron, and I. Goni, Functionality of starches in the diet. Digestion and fermentation, *Alimentaria (Madrid)*, 284 (1997) 65–69.
297. Q. Li, J. Ciao, and Y. Liu, Enzymatic membrane reactor and its application in production of starch sugar, *Mo Kexue Yu Jishu*, 24(2), (2004) 58–61.*Chem., Abstr.*, 142 (2004) 281845.
298. L. Duedahl-Olesen, L. H. Pedersen, and K. L. Larsen, Suitability and limitations of methods for characterization of activity of malto-oligosaccharide-forming amylases, *Carbohydr. Res.*, 329 (2000) 109–119.
299. W. Schmeider, Diastatic decomposition of starch and practical methods in its determination (in Polish), *Przegl. Zbozowo Mlyn.*, 11 (1967) 374–377.
300. W. J. Whelan, Enzymic exploration of the structures of starch and glycogen, *Biochem. J.*, 122 (1971) 609–622.
301. S. Janeček, B. Svensson, and E. A. MacGregor, Relation between domain evolution, specificity, and taxonomy of the α-amylase family members containing a C-terminal starch-binding domain, *Eur. J. Biochem.*, 270 (2003) 635–645.

302. B. Achremowicz and W. Wójcik, Amylolytic enzymes and other o-glycosidic hydrolases, (2000) Agric. Univ, Lublin Editorial Office, Lublin, (in Polish).
303. A. K. Kundu, S. Das, and S. Manna, Method fort he assay of amylase of *Aspergillus oryzae*, *Indian J. Pharm.*, 31 (1969) 25–26.
304. A. K. Kundu and S. Das, Production of α-amylase in liquid culture by a strain of *Aspergillus oryzae*, *Appl. Microbiol.*, 19 (1970) 598–603.
305. I. Spencer-Martins and N. Van Uden, Yield of yeast growth on starch, *Eur. J. Appl. Microbiol. Biotechnol.*, 4 (1977) 29–35.
306. H. Ebertova, Amylolytic enzymes of *Endomycopsis capsularis*. I. Formation of the amylolytic system, *Folia Microbiol.*, 11 (1966) 14–20.
307. M. L. Polizeli, A. C. Rizzeti, R. Monti, H. F. Terenti, J. A. Jorge, and D. S. Amorim, Xylanases from fungi: Properties and industrial applications, *Appl. Microb. Biotechnol.*, 67 (2005) 577–591.
308. M. A. Cotta, Amylolytic activity of selected species of ruminal bacteria, *Appl. Environ. Microbiol.*, 54 (1988) 772–776.
309. D. O. Mountfort, Rumen anaerobic fungi, *FEMS Microbiol. Rev.*, 46 (1987) 401–408.
310. P. N. Hobson and M. MacPherson, Amylases of *Clostridium butyricum* and *Streptococcus* isolated from the rumen of the sheep, *Biochem. J.*, 52 (1952) 671–679.
311. G. J. Walker, The cell-bound α-amylase of *Streptococcus bovis*, *Biochem. J.*, 94 (1965) 289–298.
312. D. O. Mountfort and R. A. Asher, Production of α-amylase by the ruminal anaerobic fungus *Neocallimastix frontalis*, *Appl. Environ. Microbiol.*, 59 (1988) 2293–2299.
313. H. Sun, P. Zhao, X. Ge, Y. Xia, Z. Hao, J. Liu, and M. Peng, Recent advances in microbial raw starch digesting enzymes, *Appl. Biochem. Biotechnol.*, 160 (2010) 988–1003.
314. R. Sawicka-Żukowska, K. Zielińska, and B. Jędrychowska, Enzymic degradation of various native starches (in Polish), *Przem. Spoz.*, 53(5), (1999) 33–35.
315. H. Sun, X. Ge, L. Wang, P. Zhao, and M. Peng, Microbial production of raw starch digesting enzymes, *Afr. J. Biotechnol.*, 8 (2009) 1734–1739.
316. T. Kushimoto and K. Marumoto, Acid hydrolysis of polysaccharide. V. Comparison of acid hydrolysis and enzymic hydrolysis of starch from the observation on the decrease in starch and increase of erythrodextrin, *J. Jpn. Biochem Soc.*, 21 (1949) 128–130.
317. K. Myrbäck and K. Ahlborg, Über Grenzdextrine und Stärke. XIII. Spezifität der Amylasen und Produkte ihrer Wirkung, *Biochem. Z.*, 311 (1942) 213–226.
318. K. Ahlborg and K. Myrbäck, Grenzdextrine und Stärke. XII. Darstellung und Konstitutionbestimmmung eines schwer hydrolisierbaren Disaccharides aus Stärke, *Biochem. Z.*, 308 (1941) 187–195.
319. K. Myrbäck, Action of the amylases on starch and glycogen, *Actualites Biochim.*, 10 (1947) 12–14.
320. R. L. Lohmar, Jr., The structure of α-amylase-modified waxy-corn starch, *J. Am. Chem. Soc.*, 76 (1954) 4608–4611.
321. K. Myrbäck, Grenzdextrine und Stärke, XIV. Wirkung der Dextrinogenamylasen, *Biochem. Z.*, 311 (1942) 227–233.
322. K. Myrbäck and G. Nycander, Grenzdextrine und Stärke. XV. Dextrinsäure und ihre Spaltung durch Amylase, *Biochem. Z.*, 311 (1942) 234–241.
323. K. Myrbäck, Über Grenzdextrine und Stärke. XVI. Über die Natur der α-Dextrine, *Biochem. Z.*, 311 (1942) 242–246.
324. K. Myrbäck and B. Lundberg, Enzymatische Dextrinierung der Stärke, *Svensk Kem. Tids.*, 55 (1943) 36–41.
325. C. J. Lintner and M. Kirschner, Zur Kenntinis des beim diastatischen Abbau der Stärke auftretenden Grenzdextri, *Z. Angew. Chem.*, 36 (1923) 119–122.
326. J. Blom, B. Brae, and A. Bak, Staerkeabbauprodukten einer α-Amylase, *Z. Physiol. Chem.*, 252 (1938) 261–270.

327. B. Örtenblad and K. Myrbäck, Über Grenzdextrine und Stärke, *Biochem. Z.*, 303 (1940) 335–341.
328. K. Myrbäck, B. Örtenblad, and K. Ahlborg, Über Grenzdextrine und Stärke. VI. Grenzdextrine aus Kartoffelstärke mittels Pankreasamylase, *Biochem. Z.*, 307 (1940) 49–52.
329. K. Myrbäck, B. Örtenblad, and K. Ahlborg, Über Grenzdextrine und Stärke. VII. Schwerhydrolisierbare Glykosidbildungen in der Stärke, *Biochem. Z.*, 307 (1940) 53–68.
330. K. Myrbäck and K. Ahlborg, The limit dextrins and starches. VIII. Constitution of a starch limit dextrins. Proof of existence of α-glucosidic 1,6-combination in dextrin and starch, *Biochem. Z.*, 307 (1940) 69–78. *Chem. Abstr.*, 35 (1941) 29229.
331. B. Örtenblad and K. Myrbäck, Über Grenzdextrine und Stärke. IX. Primäre Spaltung der Stärke durch die Dextrinogenamylase des Malzes, *Biochem. Z.*, 307 (1940) 123–128.
332. K. Myrbäck, Über Grenzdextrine und Stärke. X. Primärprodukte der Stärkespaltung durch Dextrinogenamylase, *Biochem. Z.*, 307 (1940) 132–139.
333. K. Myrbäck, Über Grenzdextrine und Stärke, XI. Spaltung der Stärke durch die Dextrinogenamylase des Malzes, *Biochem. Z.*, 307 (1940) 140–158.
334. J. L. Baker, Hydrolysis of potato and malt starches by malt amylase. III. The stable dextrin, *J. Inst. Brew.*, 47 (1941) 344–348.
335. W. N. Haworth, H. Kitchen, and S. Peat, Amylolytic degradation of starch, *J. Chem. Soc.* (1943) 619–625.
336. S. Sivuramakrishnan, D. Gaugadharan, K. M. Nampathiri, C. R. Socol, and A. Pandey, α-amylases from microbial sources – An overview on recent developments, *Food Technol. Biotechnol.*, 44 (2006) 173–184.
337. N. Atichokudomchai, J. Jane, and G. Hazlewood, Recent pattern of a novel thermostable α-amylase, *Carbohydr. Polym.*, 64 (2006) 582–588.
338. M. Vihinen and P. Manstala, Microbial amylolytic enzyme, *Crit. Rev. Biochem. Mol. Biol.*, 24 (1989) 329–418.
339. G. A. Hill, D. G. Macdonald, and X. Lang, Amylase inhibition and inactivation in barley malt using cold starch hydrolysis, *Biotechnol. Lett.*, 19 (1997) 1139–1141.
340. J. T. Kung, V. M. Hanrahan, and M. L. Caldwell, A comparison of the action of several alpha amylases upon a linear fraction from corn starch, *J. Am. Chem. Soc.*, 75 (1953) 5548–5554.
341. W. Tur, E. Szczepaniak, W. Krzyżaniak, W. Białas, and W. Grajek, Characterization of maltodextrins obtained from potato starch using amylolytic preparations (in Polish), *Żywnosc, (Food)*, 11(4), (2004) 79–94.
342. J. Blom and C. O. Rosted, The formation of trisaccharides during degradation of potato starch by pancreatic amylase, *Acta Chem. Scand.*, 1 (1947) 233–235.
343. M. Ulmann and J. Seidemann, Zur fermentativen Wirkung von Pilzamylase gegenüber Stärke, *Nahrung*, 2 (1958) 910–924.
344. J. Roos and C. Romijn, Digestibility of some species of starch by pancreatic amylase of the dog and the pig, *Arch. Neerland. Physiol.*, 19 (1934) 392–402. *Chem. Abstr.*, 28 (1934) 53064.
345. T. Kitano, Taka amylase. XIV. Selectivity of various adsorbents (in Japanese), *J. Soc. Chem. Ind. Jpn.*, 40(Suppl.), (1937) 37–38.
346. T. Kitano, Taka amylase. XV. Selective adsorption of taka diastase solutions purified by adsorption (in Japanese), *J. Soc. Chem. Ind. Jpn.*, 40(Suppl.), (1937) 38–41.
347. T. Kitano, Taka amylase. XVI. Purification of taka amylose by adsorption (in Japanese), *J. Soc. Chem. Ind. Jpn.*, 40(Suppl.), (1937) 41–43.
348. C. A. Adams, T. G. Watson, and L. Novellie, Lytic bodies from cereals hydrolyzing maltose and starch, *Phytochemistry*, 14 (1975) 953–956.
349. G. Orestano, The liquefaction and saccharification of starch by the action of amylases from different sources (in Italian), *Arch. Farmacol. Sper.*, 56 (1933) 383–406.
350. E. Ohlsson and O. Rosen, The action of various amylases on starch, *Svensk. Farm. Tids.*, 38 (1934) 497–504. 517–523, 537–538, 648–652.

351. H. Pringsheim and J. Leibowitz, Über α- und β-Amylasen, *Berichte*, 58B (1925) 1262–1265.
352. V. Syniewski, Gehalte verschiedenen Getreidearten an extrahiererbarer und nichtextrahiererbarer α-Diastase, *Biochem. Z.*, 162 (1925) 228–235.
353. E. G. Glock, Rates of digestion of starches and glycogen and their bearing on the chemical constitution. II. Liver amylase, *Biochem. J.*, 30 (1936) 2313–2318.
354. G. Yamagishi, Enzymes of grain. III. The relation between the action of the starch- liquefying enzyme of rice and pH (in Japanese), *J. Agr. Chem. Soc. Jpn.*, 11 (1935) 825–835.
355. G. Yamagishi, Enzymic studies on cereals. VII. Dextrinizing amylase of rice seeds (in Japanese), *J. Agr. Chem. Soc. Jpn.*, 13 (1937) 1268–1274.
356. G. Yamagishi, Enzymic studies on cereals. VII. Amylase in germinated rice (in Japanese), *J. Agr. Chem. Soc. Jpn.*, 13 (1937) 1275–1283.
357. G. Yamagishi, Enzymic studies on cereals. IX. Development of amylase during germination of rice (in Japanese), *J. Agr. Chem. Soc. Jpn.*, 13 (1937) 1284–1295.
358. H. C. Sherman and P. W. Punnett, Products of the action of certain amylases upon soluble starch with special reference to the formation of glucose, *J. Am. Chem. Soc.*, 38 (1916) 1877–1885.
359. H. C. Sherman, F. Walker, and M. L. Caldwell, Action of enzymes upon starch of different origin, *J. Am. Chem. Soc.*, 41 (1919) 1123–1129.
360. R. Lecoq, Diastatische Wirkung des Malzmehle, *Compt. Rend. Soc. Biol.*, 91 (1924) 924. 926; *Chem. Zentr.*, 1922, II, 173.
361. A. Bijttebier, H. Goesaert, and J. A. Delcour, Temperature impacts the multiple attack action of amylases, *Biomacromolecules*, 8 (2007) 765–772.
362. X. Remesar and M. Alemany, Glycogen and amylase in the liver and striated muscle of rats under altered thyroid states, *Horm. Metab. Res.*, 14 (1982) 179–182.
363. P. Fantl and M. N. Rome, The breakdown of commercial starches by muscle extract, *Austr. J. Exp. Biol. Med. Sci.*, 20 (1942) 187–188.
364. D. Luzon Morris, Hydrolysis of starch and glycogen by blood amylase, *J. Biol. Chem.*, 148 (1943) 271–273.
365. G. E. Glock, An investigation of the rates of digestion of starches and glycogen. III. Human liver amylase, *Biochem. J.*, 32 (1938) 235–236.
366. G. H. Perry, N. J. Dominy, K. G. Clow, A. S. Lee, H. Fiegler, R. Redon, J. Werner, F. A. Villanea, J. L. Mountain, R. Misra, N. P. Carter, C. Lee, and A. C. Stone, Diet and the evolution of human amylase gene copy number variation, *Nat. Genet.*, 39 (2007) 1256–1260.
367. H. McGuigan, The action of ptyalin, *J. Biol. Chem.*, 39 (1919) 273–284.
368. K. Myrbaeck and E. Willstaedt, Starch degradation by salivary amylase. 3. Residual dextrins and the glucose formation, *Arkiv Kemi*, 7 (1954) 403–415.
369. T. Chrzaszcz and Z. Schechtlowna, Amylase der Speichels der Rinder und Pferde, *Biochem. Z.*, 219 (1930) 30–50.
370. J. L. Rosenblum, C. L. Irwin, and D. H. Alpers, Starch and glucose oligosaccharides protect salivary-type amylase activity at aid pH, *Am. J. Physiol.*, 254(Pt. 1), (1988) G775–G780.
371. K. H. Meyer, F. Duckert, and E. H. Fischer, Sur la liquefaction de l'empois d'amidon par l'α-amylase humaine. Sur les enzymes amylolytiques. XIII, *Helv. Chim. Acta*, 33 (1950) 207–210.
372. I. Koyama, S. I. Komine, M. Yakushijin, S. Hokari, and T. Komoda, Glycosylated salivary α-amylases are capable of maltotriosehydrolysis and glucose formation, *Comp. Biochem. Physiol.*, 120B (2000) 553–560.
373. K. Myrbäck, Starch degradation by salivary amylses. II, Arkiv Kemi., 7 (1954) 53–59.
374. K. Myrbäck and E. Willstädt, Starch degradation by salivary amylses. III End products of the salivary amylase action, Arkiv Kemi., 7 (1954) 443–454.
375. D. J. Goff and J. F. Kull, The inhibition of human salivary α-amylase by type II α-amylase inhibitoe from *Triticum aestivum* is competitive, slow and tight binding, *J. Enzyme Inhib.*, 9 (1995) 163–170.

376. H. C. Sherman, M. L. Caldwell, and M. Adams, Establishment of the optimal hydrogen-ion activities for the enzyme hydrolysis of starch by pancreatic and malt amylases under varied conditions of time and temperature, *J. Am. Chem. Soc.*, 49 (1927) 2000–2005.
377. Y. Zhou, R. Hoover, and Q. Liu, Relationship between α-amylase degradation and the structure and physicochemical properties of legume starches, *Carbohydr. Polym.*, 57 (2004) 299–317.
378. H. Pringsheim and S. Ginsburg, Amylolysis of phosphoric esters of starch and glycogen, *Bull. Soc. Chim. Biol.*, 17 (1935) 1599–1606. *Chem. Abstr.*, 30 (1936) 13640.
379. M. Samec and E. Waldschmidt-Leitz, Enzymatische Spaltbarkkeit der Amylo- und Erythrokörper aus Stärke, *Z. Physiol. Chem.*, 203 (1931) 16–33.
380. K. Myrbäck and R. Lunden, Zwischenprodukte der enzymatichen Stärkespaltung, *Arkiv Kemi.*, A23 (1946) 1–16.
381. K. Myrbäck, On the amylase-substrate compounds, *Arkiv Kemi*, 2 (1950) 417–422.
382. J. H. Pazur and R. M. Sandstedt, Identification of the reducing sugars in amylolyzates of starch and starch oligosaccharides, *Cereal Chem.*, 31 (1954) 416–422.
383. T. T. Kelly, B. S. Collins, W. M. Fogarty, and E. M. Doyle, Mechanism of action of the α-amylase of *Micromonospora melanospora*, *Appl. Microbiol. Biotechnol.*, 39 (1993) 599–603.
384. M. Somogyi, Analysis of diastatic split-products of starch, *J. Biol. Chem.*, 124 (1938) 179–187.
385. J. Labarre and P. Lapointe, Digestibility in vitro of starch and bean flour with pancreatic amylase, *Ann. L'Acfas*, 8 (1942) 85. *Chem. Abstr.*, 40 (1946) 6854.
386. R. B. Alfin and M. L. Caldwell, Further studies of the action of pancreatic amylase: Extent of hydrolysis of starch, *J. Am. Chem. Soc.*, 70 (1948) 2534–2539.
387. F. M. Mindell, A. L. Agnew, and M. L. Caldwell, Further studies of the action of pancreatic amylase. Hydrolysis of waxy maize starch, *J. Am. Chem. Soc.*, 71 (1949) 1779–1781.
388. R. B. Alfin and M. L. Caldwell, Further studies of the action of pancreatic amylse: Digestion of the products of the hydrolysis of potato starch and of a linear fraction from corn starch, *J. Am. Chem. Soc.*, 71 (1949) 128–131.
389. J. Blom and T. Schmith, The degradation of potato starch by pancreatin, *Acta Chem. Scand.*, 1 (1947) 230–232.
390. H. Pringsheim, H. Borhardt, and R. Levy, Verflüssigung des Stärkekleisters, *Naturwissenschaften*, 21 (1933) 299–300.
391. J. J. Clary, G. E. Mitchell, Jr., and C. O. Little, Action of bovine and ovine α-amylase on starches, *J. Nutr.*, 95 (1968) 469–473.
392. H. J. Vonk and J. P. Braak, The limit of hydrolysis of starch by saliva and pancreas amylase, *Proc. Acad. Sci. Amsterdam*, 37 (1934) 188–198. *Chem. Abstr.*, 28 (1934) 34189.
393. R. M. Faulks and A. L. Bailey, Digestion of cooked starches from different food sources by porcine α-amylase, *Food Chem.*, 36 (1990) 191–203.
394. W. R. Thompson and I. Friedman, On the individual characteristics of animal amylases in relation to enzyme source, *J. Gen. Physiol.*, 19 (1936) 807–828.
395. A. Brock, Reaction kinetics of different forms of human α-amylase isoenzymes, *Adv. Clin. Enzymol.*, 5 (1987) 137–143.
396. S. Makoto, Digestion of starch, *J. Jpn. Soc. Food Nutr.*, 5 (1952–1953) 38–41.
397. P. H. Hsu, Purification and properties of α-amylases from *Euphasia superba*, Chung-kuo Nung Yeh Hua Hsueh Hui Chih, 19(1–2), (1981) 74–85. *Chem. Abstr.*, 95 (1981) 216880w.
398. S. Babacan, L. F. Pivarnik, and A. G. Rand, Honey amylase activity and food starch degradation, *J. Food Sci.*, 67 (2002) 1625–1630.
399. E. Hultin and G. Lundblad, Degradation of starch, hydroxyethyl cellulose ether, and chitosan by enzymes in spermatozoa and sperm fluid from *Psammechinus* and *Modiola*, *Exp. Cell Res.*, 3 (1952) 427–432.
400. F. Mukaiyama, Y. Horie, and T. Ito, Amylase of digestive juice and utilization of dextrin and starch in the silkworm *Bombyx mori* L, *J. Insect Physiol.*, 10 (1964) 247–254.

401. G. L. Skea, D. O. Mounfort, and K. D. Clements, Gut carbohydrases from the New Zealand marine herbivorous fishes *Kyphosus sydneyanus* (Kyphosidae), *Aplodactylus arctidens* (Aplodactylidae) and *Odox pullus* (Labridae), *Comp. Biochem. Physiol.*, B 140 (2005) 259–269.
402. C. Y. Tsao, Y. Z. Pan, S. T. Jiang, and Shann-Tzong, Purification and characterization of amylases from small abalone (*Sulculus diversicolor aquatilis*), *J. Agric. Food Chem.*, 51 (2003) 1064–1070.
403. J. E. Carter and J. B. Hutchinson, Liquefaction of gelatinized starch by wheaten flour, *Chem. Ind. (London)* (29), (1966) 1264–1265.
404. G. L. Teller, Plant diastase in evidence as to the formation and structure of starch granules, *Plant Physiol.*, 13 (1938) 227–240.
405. H. C. Gore and S. Jozsa, Sugar-forming enzymes in certain foods, *Ind. Eng. Chem.*, 24 (1932) 102.
406. H. C. Gore and S. Jozsa, Sugar formation by diastatic enzymes of flour, *Ind. Eng. Chem.*, 24 (1932) 99–102.
407. A. Beleia and E. Varriano-Marston, Pearl millet amylases. II. Activity toward intact and heated starch granules, *Cereal Chem.*, 58 (1981) 437–440.
408. C. T. Greenwood, A. W. MacGregor, and E. A. Milne, α-amylolysis of starch, *Starch/Stärke*, 17 (1965) 219–225.
409. G. Orestano and C. Zummo, The liquefaction and saccharification of starch by soy-bean amylase (in Italian), *Bol. Soc. Ital. Biol. Sper.*, 5 (1930) 246–249.
410. T. Mitsui, J. Yamaguchi, and T. Akasawa, Physicochemical and serological characterization of rice alpha amylase isoforms and identification of new corresponding gene, *Plant Physiol.*, 110 (1996) 1395–1404.
411. W. T. Tao, Biochemical studies on rice starch (2) (in Japanese), *Bull. Soc. Chem. Jpn.*, 5 (1930) 91–92.
412. C. N. Acharya, The hydrolysis of starch by the enzymes in cholam malt, *Sorghum vulgare*, *Indian J. Agr. Sci.*, 4 (1934) 476–536.
413. B. Visvanath and M. Suryanarayaba, Hydrolysis of starch by cholam malt extract, *J. Inst. Brew.*, 31 (1925) 425–429.
414. N. V. Patwardhan, Studies on dextrins, I. Action of amylase from cholam (*Sorghum vulgare*) on potato starch, *J. Indian Inst. Sci.*, 13A (1930) 31–37.
415. B. W. DeHaas, D. W. Chapman, and K. J. Goering, An investigation of the alpha-amylase from self-liquefying barley starch, *Cereal Chem.*, 55 (1978) 127–137.
416. K. J. Goering and R. Eslick, Waxy barley starch with unique selfliquefying properties, US Patent, 4,116,770 (1978); *Chem. Abstr.*, 90 (1978) 20833k.
417. G. Bertrand and A. Compton, Über die Gegenwart eines neuen Enzymes, der Salicinase in den Mandeln, *Compt. Rend. Soc. Biol.*, 157 (1914) 797–799. *Chem. Zentr.*, 1914, I, 39.
418. K. Myrbäck, Untersuchungen über Amylase und Stärkespaltung in Gerste und Malz, *Proc. IVth Congr. Int. Tech. Chim. Ind. Agr., Bruxelles*, 3 (1935) 318–325.
419. F. Polak and A. Tychowski, Chemie der Stärke von diastatische Standpunkt, *Biochem. Z.*, 214 (1929) 216–228.
420. K. Sjoeberg and E. Eriksson, Amylase, *Z. Physiol. Chem.*, 139 (1924) 118–139.
421. A. Tychowski, Products of diastatic hydrolysis of starch (in Polish), *Przem. Chem.*, 15 (1931) 346–354.
422. Z. Sun and C. A. Henson, A quantitative assessment of the importance of α-amylase, β-amylase, debranching enzyme and α-glucosidase in starch degradation, *Arch. Biochem. Biophys.*, 284 (1991) 298–305.
423. K. Svanborg and K. Myrbäck, Starch degradation by the α-amylases, *Arkiv Kemi*, 6 (1953) 113–131.
424. E. A. MacGregor and A. W. MacGregor, The action of cereal α-amylases on solubilized starch and cereal starch granules, *Progr. Biotechnol.*, 1 (1985) 149–160.
425. A. Joszt and A. Kleindienst, Limiting decomposition of starch under action of distillery malts of various origins (in Polish), *Przem. Chem.*, 14 (1930) 537–541.

426. S. Pronin, Zur Frage des Grenzabbau der Stärke durch Malzamylase, *Biochem. Z.*, 249 (1932) 7–10.
427. W. T. Tao, Biochemical studies on rice starch. III. Action of enzymes on rice starch (in Japanese), *Bull. Chem. Soc. Jpn.*, 5 (1930) 87–90.
428. H. Russina, Enzymatishe Wirkungen auf Stärkeloesungen, *Z. Ges. Textil-Ind.*, 35 (1932) 281–283.
429. R. Weitz and R. Lecoq, Ähnlichkeiten zwischen Ptyalin und Amylase aus gekeimter Gerste, *Compt. Rend. Soc. Biol.*, 91 (1924) 926–927. *Chem. Zentr.*, 1925, I, 235.
430. J. L. Baker and H. F. E. Hulton, Hydrolysis of potato and malt starches by malt amylase. II. Maltodextrin, *J. Inst. Brew.*, 44 (1938) 514–519.
431. T. Chrząszcz, Z. Bidziński, and A. Krause, Über den Einfluß der Wasserstoffionkonzentration auf die Dextrinierung der Stärke durch gereinigte Malzamylase, *Biochem. Z.*, 160 (1925) 155–171.
432. M. J. Blish, R. M. Sandstedt, and D. K. Mecham, Action of wheat amylases on raw whaet starch, *Cereal Chem.*, 14 (1937) 605–628.
433. M. Nirmala and G. Muralikrishna, Changes in starch during malting of finger millet (Ragi, *Eleusine coracana*, Indaf-15) and its in vitro digestibility studies using purified ragi amylases, *Eur. Food Res. Technol.*, 215 (2002) 327–343.
434. M. Nirmala and G. Muralikrishna, In vitro digestibility studies of cereal flours and starchews using purified finger millet (*Eleusina coracana,* ragi, Indaf-15) amylases, *Carbohydr. Polym.*, 52 (2003) 275–283.
435. L. Salomonsson, W. K. Heneen, M. Larsson-Razinkiewicz, and R. Larlsson, *In* vitro degradation of starch granules by α-amylase isomers from mature triticale, *Starch/Stärke*, 41 (1989) 340–343.
436. C. A. Knutson, Reaction patterns of maize alpha-amylases with soluble starch granular starch and maltooligosaccharides, *Cereal Chem.*, 70 (1993) 534–538.
437. K. Kaur, A. K. Gupta, and N. Kaur, A comparative study on amylases from seedlings of tall and double dwarf wheat, *J. Plant Biol.*, 31 (2004) 111–116.
438. A. Leahu and M. Avramiuc, Research concerning the action of some enzymatic preparations of polysaccharides from wheat, *J. Agroaliment. Proc. Technol.*, 13 (2007) 425–432.
439. J. G. Sargeant and T. S. Walker, Adsorption of wheat alpha-amylase isoenzymes to wheat starch, *Starch/Stärke*, 30 (1978) 160–163.
440. J. R. Turvey and R. C. Hughes, Enzymic degradation of raw starch granules, *Biochem. J.*, 69 (1958) 4P.
441. M. Sprockhoff, Über stärkeabbauende Fermente (Amylasen) in der trockenen Kartoffelstärke, *Z. Spiritusind.*, 42 (1929) 312–313.
442. V. Hagenimana, L. P. Vezina, and R. E. Simard, Sweet potato α- and β-amylases. Characterization and kinetic studies with endogenous inhibitors, *J. Food Sci.*, 59 (1994) 373–377.
443. C. P. Harley, S. F. Fisher, and M. P. Masure, A starch-splitting enzyme in apple tissues, *Proc. Am. Soc. Hort. Sci.*, 28 (1931) 561–565.
444. E. Honorio de Aranjo and W. Schmidell Natto, Deactivation energy for glucoamylase on soluble starch, *Rev. Microbiol.*, 18 (1987) 46–51. *Chem. Abstr.*, 107 (1987) 194108.
445. Y. Takasaki, An amylase producing maltotriose from *Bacillus subtilis*, *Agric. Biol. Chem.*, 49 (1985) 1091–1098.
446. Y. Takasaki, H. Shinohara, M. Tsuruhisa, S. Hayashi, and K. Imada, Studies on enzymic production of oligosaccharides. VII. Maltotetraose-producing amylase from *Bscillus* sp. MG-4, *Agric. Biol. Chem.*, 55 (1991) 1715–1720.
447. J. F. Robyt and R. J. Ackerman, Isolation, purification and characterization of a maltotetraose producing amylase from *Pseudomonas stutzeri*, *Arch. Biochem. Biophys.*, 145 (1971) 105–114.
448. T. U. Kim, B. U. Gu, J. Y. Jeong, S. M. Byun, and Y. C. Shin, Purification and characterization of maltotetraose-forming alkaline alpha amylase from alkalophilic *Bacillus* strain GM8901, *Appl. Environ. Microbiol.*, 61 (1995) 3105–3112.

449. A. C. M. Bourke, C. T. Kelly, and W. M. Fogarty, A novel maltooligosaccharide producing amylase, *Biochem. Soc. Trans.*, 19 (1991) 16S.
450. M. Fogarty, C. T. Kelly, A. C. Bourke, and E. M. Doyle, Extracellular maltotetraose forming amylase of *Pseudomonas* sp. IMD353, *Biotechnol. Lett.*, 16 (1994) 473–478.
451. N. Saito, A thermophilic extracellular α-amylase from *Bacillus licheniformis*, *Arch. Biochem. Biophys.*, 155 (1973) 290–298.
452. W. M. Fogarty, F. Bealin-Kelly, C. T. Kelly, and E. M. Doyle, A novel maltohexaose-forming α-amylase from *Bacillus caldovelox:* Patterns and mechanism of action, *Appl. Microbiol. Biotechnol.*, 36 (1991) 184–189.
453. H. Taniguchi, C. M. Jae, N. Yoshigi, and Y. Maruyama, Purification of *Bacillus circulans* F-2 amylase and its general properties, *Agric. Biol. Chem.*, 47 (1983) 511–519.
454. H. Taniguchi, Maltohexaose-producing amylase of *Bacillus circulans* F-2, *ACS Symp. Ser.*, 458 (1991) 111–124.
455. Y. Takasaki, Production of maltohexaose by α-amylase from *Bacillus circulans* G-6, *Agric. Biol. Chem.*, 46 (1982) 1539–1547.
456. T. Hayashi, T. Akiba, and K. Horikoshi, Production and properties of new maltohexaose forming amylases from alkalophilic *Bacillus* sp. H-167, *Agric. Biol. Chem.*, 52 (1988) 443–448.
457. T. Hayashi, T. Akiba, and K. Horikoshi, Properties of new alkaline maltohexaose-forming amylases, *Appl. Microbiol. Biotechnol.*, 28 (1988) 281–285.
458. K. Kainuma, S. Kobayashi, T. Ito, and S. Suzuki, Isolation and action pattern of maltohexaose producing amylase from *Acetobacter aerogenes*, *FEBS Lett.*, 26 (1972) 281–285.
459. N. Monma, T. Nakakuki, and K. Kainuma, Formation and hydrolysis of maltohexaose by an extracellular exomaltohexaohydrolase, *Agric. Biol. Chem.*, 47 (1983) 1769–1774.
460. H. Weber, H. J. Schepers, and W. Troesch, Starch degrading enzymes from anaerobic non-clostridial bacteria, *Appl. Microbiol. Biotechnol.*, 33 (1990) 602–606.
461. A. R. Plant, B. K. C. Patel, H. W. Morgan, and R. M. Daniel, Starch degradation by thermophilic anaerobic bacteria, *Syst. Appl. Microbiol.*, 9 (1987) 158–162.
462. T. Kotaka and M. Sasaibe, Heat stability of bacterial amylase in the enzymic liquefaction process of heavy starch slurries (in Japanese), *Dempun Kogyo Gakkaishi*, 16 (1968) 92–95.
463. I. A. Smith, Factors affecting enzymic starch hydrolysis in sugar solutions, *Sugar News*, 46 (1970) 454–458, 460.
464. G. B. Madsen, B. E. Norman, and S. Slott, A new heat stable bacterial amylase and its use in high temperature liquefaction, *Starch/Stärke*, 25 (1973) 304–308.
465. R. A. K. Srivastava, S. N. Mathur, and J. N. Baruah, Partial purification and properties of thermostable intercellular amylases from a thermophilic *Bacillus* sp. AK-2, *Acta Microbiol. Pol.*, 33(1), (1984) 57–66.
466. J. Fukumoto, Studies on bacterial amylase. I. Isolation of bacteria excreting potent amylase and their distribution (in Japanese), *J. Agr. Chem. Soc. Jpn.*, 19 (1943) 487–503.
467. A. S. Schultz, L. Atkin, and C. N. Frey, Enzymatic material suitable for starch liquefaction, US Patent 2,159,678 (1938).
468. S. Akiya, Decomposition product of starch with amylase of *Bacillus mesentericus vulgatus*. Constitution of the tetrasaccharides. I. (in Japanese), *J. Pharm. Soc. Jpn.*, 58 (1938) 71–82.
469. S. Akiya, Decomposition product of starch with amylase of *Bacillus mesentericus vulgatus*. Constitution of the tetrasaccharides. II. (in Japanese), *J. Pharm. Soc. Jpn.*, 58 (1938) 117–120.
470. S. Akiya, Decomposition product of starch with amylase of *Bacillus mesentericus vulgatus*. Constitution of the tetrasaccharides. III (in Japanese), *J. Pharm. Soc. Jpn.*, 58 (1938) 120–132.
471. S. Akiya, Decomposition product of starch with amylase of *Bacillus mesentericus vulgatus*. Constitution of the tetrasaccharides. IV (in Japanese), *J. Pharm. Soc. Jpn.*, 58 (1938) 133–138.

472. J. Fukumoto, Amylases. Optimum pH and temperature and heat resistance, *J. Agr. Chem. Soc. Jpn.*, 19 (1943) 853–861.
473. J. Fukumoto, Amylases. Liquefaction and saccharification, *J. Agr. Chem. Soc. Jpn.*, 20 (1944) 23–26.
474. M. J. Syu and Y. H. Chen, A study of the α-amylase fermentation performed by *Bacillus amyloliquefaciens*, *Chem. Eng. J.*, 65 (1997) 237–247.
475. P. E. Granum, Purification and physicochemical properties of extracellular amylose from a strain of *Bacillus amyloliquefaciens* isolated from dry onion powder, *J. Food Biochem.*, 3 (1979) 1–12.
476. G. Mamo and A. Gessesse, Purification and characterization of two raw-starch-digesting thermostable α-amylase from thermophilic *Bacillus*, *Enzyme Microb. Technol.*, 25 (1999) 433–438.
477. L. M. Hamilton, C. T. Kelly, and W. M. Fogarty, Purification and properties of the raw starch-digesting α-amylase of *Bacillus* sp. IMD 434, *Biotechnol. Lett.*, 21 (1999) 111–115.
478. L. M. Hamilton, C. T. Kelly, and W. M. Fogarty, Raw starch degradation by the non-raw starch adsorbing bacterial alpha-amylase of *Bacillus* sp. IMD 434, *Carbohydr. Res.*, 314 (1998) 251–257.
479. J. P. Chiang, J. E. Alter, and M. Sternberg, Purification and characterization of a thermostable α-amylase from *Bacillus licheniformis*, *Starch/Stärke*, 31 (1979) 86–92.
480. V. Ivanova and E. Dobreva, Catalytic properties of immobilized purified thermostable α-amylase from *Bacillus licheniformis* 44MB32-A, *Process Biochem.*, 29 (1994) 607–612.
481. S. Medda and A. K. Chandra, New strains of *Bacillus* licheniformis and *Bacillus coagulans* producing thermostable α-amylase active at alkaline pH, *J. Appl. Bacteriol.*, 48 (1980) 47–58.
482. D. Yankov, E. Dobreva, V. Beshkov, and E. Emanuilova, Study of optimum conditions and kinetics of starch hydrolysis by means of thermostable α-amylase, *Enzyme Microb. Technol.*, 8 (1986) 665–667.
483. Y. Takasaki, S. Furutani, S. Hayashi, and K. Imada, Acid-stable and thermostable α-amylase from *Bacillus licheniformis*, *J. Ferm. Bioeng.*, 77 (1994) 94–96.
484. R. L. Antrim, B. A. Solheim, L. Solheim, A. L. Auterinen, J. Cunefare, and S. Karppelin, A new *Bacillus licheniformis* α-amylase capable of low pH liquefaction, *Starch/Stärke*, 43 (1991) 355–360.
485. J. Pen, L. Molendijk, W. J. Quax, P. C. Sijmons, A. J. J. Van Ooyen, P. J. M. Van den Elzen, K. Rietved, and A. Hoekema, Production of active *Bacillus licheniformis* alpha amylase in tobacco and its application in starch liquefaction, *Biotechnol.*, 10 (1992) 292–296.
486. J. K. Shetty and W. G. Allen, An acid-stable thermostable alpha-amylase for starch liquefaction, *Cereal Foods World*, 33 (1988) 929–934.
487. I. C. Kim, S. H. Yoo, S. J. Lee, B. H. Oh, J. W. Kim, and K. H. Park, Synthesis of branched oligosaccharides from starch by two amylases cloned from *Bacillus licheniformis*, *Biosci. Biotechnol. Biochem.*, 58 (1994) 416–418.
488. Y. Takasaki, Novel acid- and heat-resistant α-amylase of *Bacillus*, its preparation and use for liquefying starch, EU Patent WO 9413792 (1994); *Chem. Abstr.*, 121 (1994) 103080.
489. A. Day and B. Swanson, Amino acid substituted analogs of *Bacillus licheniformis* α-amylase with altered stability or activity profiles, EU Patent WO 9909183 (1999); *Chem. Abstr.*, 130 (1999) 206696.
490. A. Svendsen, T. V. Borchert, and H. Bisgard-Frantzen, α-amylase mutants with improved thermostability for use as detergent additives and for starch liquefaction, EU Patent WO 9919467 (1999); *Chem. Abstr.*, 130 (1999) 308445.
491. I. Kojima, H. Suzuki, G. R. Koh, and T. Adachi, Novel acidic α-amylase from *Bacillus acidocaldarius* for liquefying starch without the need of calcium, Jpn. Kokai Tokkyo Koho 10136979 (1998); *Chem. Abstr.*, 129 (1998) 38117.
492. M. H. Rivera, A. Lopez-Munguia, X. Soberon, and G. Saab-Rincon, α-amylase from *Bacillus licheniformis* mutants near to the catallytic site: Effects on hydrolytic and transglycosylation activity, *Protein Eng.*, 16 (2003) 505–514.

493. J. F. Kennedy and C. A. White, Characteristics of α-amylase K, a novel amylase from starin of *Bacillus subtilis*, *Starch/Stärke*, 31 (1979) 93–99.
494. Y. Teramoto, I. Kira, and S. Hayashida, Multiplicity and preferential inactivation by proteolysis as to raw starch-digestibility of bacterial α-amylase, *Agric. Biol. Chem.*, 53 (1989) 601–605.
495. K. Das, R. Doley, and A. K. Mukherjee, Purification and biochemical characterization of a thermostable alkaliphilic extracellular α-amylase from *Bacillus subtilis* DM-03, a strain isolated from the traditional fermented food of India, *Biotechnol. Appl. Biochem.*, 40 (2004) 291–298.
496. J. Burianek, J. Zapletal, I. Koziak, and F. Škoda, Starch enzyme hydrolyzates, Czechoslovak Patent 196,987 (1982); *Chem. Abstr.*, 97 (1982) 94366b.
497. M. Asgher, M. J. Asad, S. U. Rahman, and R. L. Legge, A thermostable a-amylase from a moderately thermophilic *Bacillus subtilis* strain for starch processing, *J. Food Eng.*, 79 (2007) 950–955.
498. Z. Konsula and M. Liakopoulou-Kyriakides, Hydrolysis of starch by the action of an α-amylase from *Bacillus subtilis*, *Process Biochem.*, 39 (2004) 1745–1749.
499. G. Okada, T. Nakakuki, Y. Tanaka, and M. Muramatsu, Enzymic manufacture of monosaccharides from polysaccharides, Jpn. Kokai Tokkyo Koho 02031685 (1990); *Chem. Abstr.*, 113 (1990) 4716.
500. K. Umeki and T. Yamamoto, Structures of multibranched dextrins produced by saccharifying α-amylase from starch (in Japanese). *J. Biochem. (Tokyo)*, 78 (1975) 897–903.
501. M. Seiji, Enzymic breakdown of limit dextrin. I. The enzymic breakdown of β-limit dextrin by α-amylase (in Japanese). *J. Biochem. Jpn.*, 40 (1953) 515–518.
502. M. Seiji, Enzymic breakdown of limit dextrin. II, α-Limit dextrin (in Japanese). *J. Biochem. Jpn.*, 40 (1953) 509–514.
503. Y. Suzuki, T. Nagayama, H. Nakano, and K. Onishi, Purification and characterization of a maltotriogenic α-amylase I and a maltogenic α-maltogenic α-amylase II capable of cleaving α-1,6-bonds in amylopectin, *Starch/Stärke*, 39 (1987) 246–252.
504. S. H. Choi, Y. K. Lee, and M. S. Nam, Microorganisms *Bacillus* sp. Cnl90 producing both improved protein decomposing enzyme and starch decomposing enzyme, South Korean Patent 2005010434 (2005); *Chem. Abstr.*, 145 (2006) 291276.
505. J. F. Kennedy, C. A. White, and C. L. Riddiford, Action pattern and substrate specificity of α-amylase K, a novel amylase from a strain of *Bacillus subtilis*, *Starch/Stärke*, 31 (1979) 235–241.
506. B. G. Dettori-Campus, F. G. Priest, and J. R. Stark, Hydrolysis of starch granules by the amylase from *Bacillus stearothermophilus* NCA 26, *Process Biochem.*, 27 (1992) 17–21.
507. G. E. Inglett, Amylo-oligosaccharide formation from *Bacillus stearothermophilus* α-amylase action on starches, *J. Food Sci.*, 55 (1990) 560–561.
508. R. A. K. Srivastava, Studies on extracellular and intracellular purified amylases from thermophilic *Bacillus stearothermophilus*, *Enzyme Microb. Technol.*, 6 (1984) 422–426.
509. M. Ben Ali, B. Khemakhem, X. Xavier, R. Haser, and S. Bejar, Thermostability enhancement and change in starch hydrolysis profile of the maltohexaose-forming amylase of *Bacillus stearothermophilus* US 100 strain, *Biochem. J.*, 394 (2006) 51–56.
510. M. Tamuri, M. Kanno, and Y. Ishii, Heat- and acid-stable α-amylase enzyme, US Patent 4,284,722 (1981); *Chem. Abstr.*, 95 (1981) 167146a.
511. Y. Kato, K. Mikuni, K. Hara, H. Hashimoto, T. Nakajima, S. Kobayashi, and R. Kainuma, The structure of high-molecular weight dextrins obtained from potato starch by treatment with *Bacillus macerans* enzyme, *J. Ferment. Technol.*, 66 (1988) 159–166.
512. R. A. K. Srivastava, Purification and chemical characterization of thermostable amylases produced by *Bacillus stearothermophilus*, *Enzyme Microb. Technol.*, 9 (1987) 749–754.
513. J. Kim, T. Nanmori, and R. Shinke, Thermostable raw-starch-digesting amylase from *Bacillus stearothermophilus*, *Appl. Environ. Microbiol.*, 55 (1989) 1638–1639.
514. R. Shinke and T. Nanmori, Production and properties of thermostable amylase and its use in maltooligosaccharide production, EU Patent EP 350737 (1990); *Chem. Abstr.*, 112 (1990) 213086.

515. H. Taniguchi, M. J. Chung, Y. Maruyama, and M. Nakamura, Enzymic digestion of potato starch granules (in Japanese), *Denpun Kagaku*, 29 (1982) 107–116.
516. H. Taniguchi, Raw starch-degrading amylase from *Bacillus circulans* F-1. Induction of the enzyme by epichlorohydrin-crosslinked starch, *Kagaku to Seibutsu*, 23(1), (1985) 7–9.
517. H. Taniguchi and Y. Maruyama, Raw starch digesting α-amylase from *Bacillus circulans* F-2 (in Japanese), *Denpun Kagaku*, 32 (1985) 142–151.
518. H. Sata, H. Taniguchi, and Y. Maruyama, Induction of *Bacillus circulans* F-2 amylase by crosslinked starches, *Agric. Biol. Chem.*, 50 (1986) 2803–2809.
519. M. Kanno, *Bacillus acidocaldarius* α-amylase that is highly to heat under acidic conditions, *Agric. Biol. Chem.*, 50 (1986) 23–31.
520. Y. Takasaki, Novel amylase from *Bacillus* species, Jpn. Kokai Tokkyo Koho 62,126,973 (1987); *Chem. Abstr.*, 107 (1987) 132656f.
521. Y. Udaka, I. Pisheto, N. Tsukagoshi, and O. Shinoda, Novel starch-hydrolyzing enzyme manufacture with *Bacillus* sp. B1018, Jpn. Kokai Tokky Koho 01218587 (1989); *Chem. Abstr.*, 112 (1990) 97013.
522. N. Goyal, J. K. Gupta, and S. K. Soni, A novel raw starch digesting thermostable α-amylase from *Bacillus* sp. I-3 and its use in the direct hydrolysis of raw potato starch, *Enzyme Microb. Technol.*, 37 (2005) 723–734.
523. C. T. Kelly, M. A. McTigue, E. M. Doyle, and W. M. Fogarty, The raw starch-degrading alkaline amylase of *Bacillus* sp. IMD 370, *J. Ind. Microbiol.*, 15 (1995) 446–448.
524. G. Zhan and Y. Dang, Study on properties of an alkaline amylase from halobacterium sp. H-371, *Xian Jiaotong Daxue Xuebao*, 30(2), (1996) 94–98. *Chem. Abstr.*, 124 (1996) 282753.
525. S. P. Lee, M. Morikawa, M. Takagi, and T. Imanaka, Cloning of the aapT gene andcharacterization of its product, α-amylase-pullulanase (AapT) from thermophilic and alaliphilic Bacillus strain XAL601, *Appl. Environ. Microbiol.*, 60 (1994) 3764–3773.
526. E. C. M. J. Bernhardsdotter, J. D. Ng, O. K. Garriott, and M. L. Pusey, Enzymic properties of an alkaline chelator-resistant α-amylase from alkaliphilic *Bacillus* sp. Isolate L1711, *Process Biochem.*, 40 (2005) 2401–2408.
527. T. Satyanarayana, M. Noorwez, S. Kumar, J. L. U. M. Rao, M. Ezhilvannan, and P. Kaur, Development of an ideal starch saccharification process using amylolytic enzymes from thermophiles, *Biochem. Soc. Trans.*, 32 (2004) 276–278.
528. K. Takaoka, T. Nakaya, and R. Irie, *Bacillus thuringiensis* for degradation of starch in organic waste liquid, Jpn. Kokai Tokkyo Koho 2002125660 (2002); *Chem. Abstr.*, 136 (2002) 345047.
529. T. Uno, K. Tominaga, Y. Shimobayashi, and T. Terai, Utilization of biomass in Hokkaido and application of bioreactor. II. Examination of enzymatic degradation of raw potato starch (in Japanese), *Hokkaidoritsu Kogyo Shikenjo Hokoku* (288), (1989) 105–110.
530. A. Viksoe-Nielsen, C. Andersen, T. Hoff, and S. Pedersen, Development of new α-amylases for raw starch hydrolysis, *Biocatal. Biotransform.*, 24 (2006) 121–127.
531. S. Yamamoto, Raw starch-digesting (maltooligosaccharide-producing type) of *Zoogloea ramigera* (in Japanese), *Oyo Toshitsu Kagaku*, 41 (1994) 283–289.
532. H. Takahashi, H. Okemoto, and H. Hara, Production of maltopentaose. I. Properties of enzyme produced by *Pseudomonas* sp. KO-8940 (in Japanese), *Seito Gijutsu Kenkyu Kaishi* (37), (1989) 105–112.
533. K. L. Anderson, Purification and analysis of a membrane assiciated starch-degrading enzyme from *Ruminobacter amylophilus*, *Anaerobe*, 8 (2002) 269–276.
534. N. Nakamura and K. Horikoshi, Sugar manufacture from starch with α-amylase from thermophilic bacterium *Thermus*, Jpn. Kokai Tokkyo Koho 01034296 (1989); *Chem. Abstr.*, 112 (1990) 19965.
535. C. H. Yang and W. H. Liu, Purification and properties of maltotriose-producing α-amylase from *Thermobifida fusca*, *Enzyme Microb. Technol.*, 35 (2004) 254–260.

536. T. Kotaka, M. Sasaibe, Y. Myashita, and K. Goto, Present status of bacterial α-amylases. Assay methods and some enzymic properties, *Denpun Kagaku*, 27 (1980) 151–157.
537. M. Aurangzeb, M. A. Qadeer, and J. Iqbal, Production of raw starch hydrolyzing amylolytic enzymes by *Streptococcus bovis* PCSIR-7B, *Pak. J. Sci. Ind. Res.*, 35 (1992) 520–523.
538. S. Okada and S. Kitahata, Hydrolysis of α-1,6-glucosidic bond by *Streptococcus bovis* α-amylase (in Japanese), *Denpun Kagaku*, 30 (1983) 199.
539. Y. S. Kwak, T. Akiba, and T. Kudo, Purification and characterization of α-amylase from hyperthermophilic archeon *Thermococcus profundis* which hydrolyses both 1,4 and 1,6 glucosidic linkages, *J. Ferm. Bioeng.*, 86 (1998) 363–367.
540. S. T. Joergensen, Cloning and expression of *Pyrococcus* α-amylase gene and use of enzyme in starch degradation, EU Patent WO 9419454 (1994); *Chem. Abstr.*, 121 (1994) 223663.
541. K. H. Park, S. J. Sung, H. S. Lee, C. S. Park, and Y. R. Kim, Hyperthermophilic enzyme from *Pyrococcus furiosus* having α-amylase and cyclodextrin-hydrolyzing activities and use for production of maltose syrup from starch at high temperature, South Korean Patent, 2006065811 (2006); *Chem. Abstr.*, 146 (2006) 22542.
542. Y. Takahashi, N. Nakamura, and M. Yamamoto, Thermostable α-amylase manufacture with *Thermomonospora* for use in maltose manufacture, Jpn. Kokai Tokkyo Koho, 02113886 (1990); *Chem. Abstr.*, 114 (1991) 41043.
543. R. S. Mishra and R. Maheshwari, Amylases of the thermophilic fungus *Thermomyces lanuginosus:* Their purification, properties, action on starch and response to heat, *J. Biosci.*, 21 (1996) 653–672.
544. W. Dierchen, Bakterieller Stärkeabbau in Abhängigkeit von natürliche Stickstoffquellen und proteolytischen Enzymen, *Starch/Stärke*, 5 (1953) 133–138.
545. L. Buranakarl, K. Ito, K. Izaki, and H. Takahashi, Purification and characterization of a raw starch-digestive amylase from non-sulfur purple photosynthetic bacterium, *Enzyme Microb. Technol.*, 10 (1988) 173–179.
546. Hirokoshi and T. Hayashi, Manufacture of amylose H with *Bacillus* for manufacture of maltoologosaccharides, Jpn. Kokai Tokkyo Koho 62,208,278 (1987); *Chem. Abstr.*, 109 (1988) 21663c.
547. B. K. Diderichsen, H. Outtrup, M. Schulein, and B. E. Norman, Chimeric α-amylases, EU Patent EP 252,666 (1988); *Chem. Abstr.*, 108 (1988) 200890a.
548. A. Viksoe-Nielsen, C. Andersen, and J. Liu, Production of a starch hydrolysate using chimeric amylases, EU Patent, EP055961 (2007); *Chem. Abstr.*, 148 (2007), 77270.
549. S. B. Ghosh and A. K. Chandra, Action pattern and substrate specificity of a thermostable α-amylase from *Bacillus apiarius* CMBL 152, *Ann. Microbiol. Enzimol.*, 39(Pt. 2), (1989) 195–202.
550. S. Fukuyama, T. Matsui, C. L. Soong, E. Allain, A. Vikso-Nielsen, H. Udagawa, Y. Liu, D. J. Ye, W. Wu, L. N. Andersen, and S. Landvik, Chimeric α-amylases comprising catalytic and carbohydrate-binding modules and the use for starch processing, EU Patent WO 2006069290 (2006); *Chem. Abstr.*, 145 (2006) 98570.
551. R. G. Von Tigerstrom and S. Stelmaschuk, Purification and partial characterization of an amylase from *Lysobacter brunescens*, *J. Gen. Microbiol.*, 133 (1987) 3437–3443.
552. G. Antranikian, P. Zablowski, and G. Gottschalk, Conditions for the overproduction and *thermohydrosulfuricum* DSM 567, *Appl. Microbiol. Biotechnol.*, 27 (1987) 75–81.
553. H. Melasniemi and M. Korhola, α-Amylase-pullulanase from *Clostridium thermhydroosulfuricum*, EU Patent EP 258050 (1988); *Chem. Abstr.*, 109 (1988) 168991.
554. B. A. Annous and H. P. Blaschek, Isolation and characterization of α-amylase derived from starch-grown *Clostridium acetobutylicum* ATCC 824, *J. Ind. Microbiol.*, 13 (1994) 10–16.
555. W. W. Johnston and A. M. Wyne, The amylase of *Clostridium acetobutylicum*, *J. Bact.*, 30 (1935) 491–501.

556. D. J. D. Hockenhull and D. Herbert, The amylase and maltase of *Clostridium acetobutylicum*, *Biochem. J.*, 39 (1945) 102–106.
557. J. F. Shaw, F. P. Lin, S. C. Chen, and H. C. Chen, Purification and properties of an extracellular α-amylase from *Thermus* sp, *Bot. Bull. Acad. Sinica*, 36 (1995) 195–200.
558. B. Punpeng, Y. Nakata, M. Goto, Y. Teramoto, and S. Hayashida, A novel raw starch digesting yeast α-amylase from *Lipomyces starkeyi* HN-606, *J. Ferm. Bioeng.*, 73 (1992) 108–111.
559. I. Sa-Correia and N. Van Uden, Production of biomass and amylases by the yeast *Lipomyces kononenkoae* in starch-limited continuous culture, *Eur. J. Appl. Microbiol. Biotechnol.*, 13 (1981) 24–28.
560. J. H. Lee, S. O. Lee, G. O. Lee, E. S. Seo, S. S. Chang, S. K. Yoo, D. W. Kim, D. F. Day, and D. Kim, Transglycosylation reaction and raw starch hydrolysis by novel carbohydrolases from *Lipomyces starkeyi*, *Biotechnol. Bioprocess Eng.*, 8 (2003) 106–111.
561. J. J. Wilson and W. M. Ingledew, Isolation and characterization of *Schwanniomyces alluvius* amylolytic enzymes, *Appl. Environ. Microbiol.*, 44 (1982) 301–307.
562. B. Simoes-Mendes, Purification and characterization of the extracellular amylases of the yeast *Schwanniomyces alluvius*, *J. Microbiol.*, 30 (1984) 1163–1170.
563. G. B. Calleja, S. R. Levy-Rick, A. Nasim, and C. V. Lusena, Extracellular amylases of starch-fermenting yeast: pH effect on export and residence time in the periplasm, *CRC Crit. Rev. Biotechnol.*, 5 (1987) 177–184.
564. A. M. Sills, P. S. J. Zygora, and G. G. Stewart, Characterization of *Schwanniomyces castelli* mutants with increased productivity of amylases, *Appl. Microbiol. Biotechnol.*, 20 (1984) 124–128.
565. M. Richter, P. Klingenberg, D. Körner, H. Ruttloff, and F. Schierbaum, Thermitase with α-amylolytic activity, East German Patent, 225,711 (1985); *Chem. Abstr.*, 104 (1985) 49867t.
566. S. K. Soni, I. K. Sandhu, K. S. Bath, U. C. Banerjee, and P. R. Patnaik, Extracellular amylase production by *Saccharomycopsis capsularis* and its evaluation for starch saccharification, *Folia Microbiol. (Prague)*, 41 (1996) 243–248.
567. K. Hamada, H. Watanabe, R. Nakajima, Y. Tanaka, M. Hirata, and T. Tahara, *Saccharomyces* for manufacturing of maltotriose-high composition, Jpn. Kokai Tokkyo Koho 2007215495 (2007); *Chem. Abstr.*, 147 (2007) 276335.
568. K. Myrbäck and B. Örtenblad, Wirkung der Amylasen auf einige Stärkeabbauprodukte, *Skand. Arch. Physiol.*, 80 (1938) 334–340.
569. A. Schulz, Untersuchungen über den Abbau von Getreide Stärken durch Pilzamylase, *Angew. Chem.*, 62 (1950) 294.
570. Y. Pomeranz, Liquefying action of pancreatic, cereal, fungal, and bacterial α-amylase, *J. Food Sci.*, 28 (1963) 149–155.
571. C. W. Brabender and B. Pagenstedt, Influence of cereal and fungal amylases on the gelatinization characteristics of wheat starch, *Bakers Dig.*, 27 (1953) 17–22.
572. J.V. Bhat and N. Mahomed, Amylolytic activity of certain genera of molds, *J. Univ. Bombay*, 18B, Sci. 27 (1950) 1–2.
573. H. Okazaki, Amylases of various species of molds.I. Classification of molds into five types after the manner of potato hydrolysis, *J. Agr. Chem. Soc. Jpn.*, 24 (1950) 88–96.
574. N. Crowley, Degradation of starch by strains of Group A streptococci having related antigens, *J. Gen. Microbiol.*, 4 (1950) 156–170.
575. T. T. Hansen, Industrial applications possibilities for an acid-stable alpha amylase from *Aspergillus niger*, *Progr. Biotechnol.*, 1 (1985) 211–216.
576. H. L. Maslow and W. C. Davson, The effect of the hydrogen-ion concentration upon the starch-liquefying activity of the amylase of *Aspergillus oryzae*, *J. Biol. Chem.*, 68 (1926) 83–93.

577. S. P. Pak, S. Y. Jan, and S. U. Sin, Effect of *Aspergillus oryzae* 38 enzymes on structural changes of maize starch granules, *Choson Minjujuui Inmin Konghwaguk KwahagwonTongbo*, (3) (1985) 40–43; *Chem. Abstr.*, 103 (1985) 137512h.
578. J. Starka, Dynamics of starch hydrolysis by *Aspergillus niger* and *Aspergillus oryzae* amylases (in Czekh), *Preslia*, 27 (1955) 154–169.
579. D. A. Fairbarn, F. G. Pries, and J. R. Stark, Extracellular amylase synthesis by *Streptomyces limosus*, *Enzyme Microb. Technol.*, 8 (1986) 89–92.
580. K. Wako, S. Hashimoto, S. Kubomura, A. Yokota, K. Aikawa, and J. Kamaeda, Purification and some properties of a maltotriose-producing amylase, *J. Jpn. Soc. Starch Sci.*, 26 (1979) 175–181.
581. M. T. Hang, S. Furuyoshi, T. Yagi, and S. Yamamoto, Purification and characterization of raw starch-digesting α-amylase from *Streptomyces thermocyaneoviolaceus* IFO 14271, *Oyo Toshitsu Kagaku*, 43 (1996) 487–497.
582. H. E. M. McMahon, C. T. Kelly, and W. M. Fogarty, High maltose-producing amylolytic system of *Streptomyces* sp, *Biotechnol. Lett.*, 21 (1999) 23–26.
583. R. Cruz, E. L. de Souza, E. H. E. Hoffmann, M. Z. Bellini, V. D. Cruz, and C. R. Vieira, Relationship between carbon source, production and patterns action of α-amylase from *Rhizopus* sp, *Rev. Microbiol.*, 28 (1997) 101–105.
584. S. C. Peixoto, J. A. Jorge, H. F. Terenzi, and M. T. M. Polizeli, *Rhizopus microsporus* var. *rhizodopoformis* a thermotolerant fungus with potential for production of thermostable amylases, *Int. Microbiol.*, 6 (2003) 269–273.
585. J. J. Marshall, Starch-degrading enzymes derived from *Cladosporium resinae*, US Patent 4,234,686 (1980); *Chem. Abstr.*, 94 (1980) 119482.
586. B. V. McCleary and M. A. Anderson, Hydrolysis of α-D-glucans and α-D-gluco- oligosaccharides by *Cladosporum resinae* glucoamylases, *Carbohydr. Res.*, 86 (1980) 77–96.
587. G. Moulin and P. Galzy, Etude de l'α-amylase de la proi *Pichia burtonii* BOIDIN, *Z. Allgem. Mikrbiol.*, 18 (1978) 269–274.
588. Z. Jurić, Hydrolysis of various carbohydrates by amylolytic systems of some *Aspergillus* species (in Serbo-Croatian), *Mikrobiologija*, 3 (1966) 91–99.
589. S. I. Pronin and E. I. Blinnikova, The thermostability and regeneration of amylase from *Aspergillus niger* (in Russian), *Arch. Sci. Biol. (USSR)*, 41(3), (1936) 11–14.
590. V. M. Hanrahan and M. L. Caldwell, A study of the action of taka amylase, *J. Am. Chem. Soc.*, 75 (1953) 2191–2197.
591. K. Ahlborg and K. Myrbäck, Grenzdextrine und Stärke. II. Takadistase und Maisstärke, *Biochem. Z.*, 297 (1938) 172–178.
592. K. Myrbäck, Trisccharide unter den Abbauprodukten der Stärke, *Biochem. Z.*, 297 (1938) 179–183.
593. S. A. Barinova, Effect of temperature and the concentration of enzyme and starch on the saccharification of starch by amylase from *Aspergillus oryzae* (in Russian), *Mikrobiologija*, 13 (1944) 82–87.
594. R. V. Feniksova and G. K. Ermoshina, Products of starch hydrolysis by α-amylase of *Aspergillus oryzae* 3-9-15, *Prikl. Biokhim. Mikrobiol.*, 7(1), (1971) 24–29.
595. G. Tomita and S. S. Kim, Substrate effect on heat inactivation of taka-amylase A, *Nature*, 205 (1965) 46–48.
596. G. Tomita and S. S. Kim, Effects of substrate on heat inactivation of taka-amylase A, *Z. Naturforsch.*, B22 (1967) 294–300.
597. M. S. Levin, The saccharification of starch paste by the amylase of *Aspergillus oryzae* (in Russian), *Tr. Ukr. Nauchn. Issled. Inst. Pishch. Prom.* (1), (1954) 18–29.
598. K. Nishikawa, Takadiastase, *Biochem. Z.*, 188 (1927) 386–404.
599. H. Nakamura, Amylase protecting substances. I. Introduction and preliminary tests (in Japanese), *J. Soc. Chem. Ind. Jpn.*, 33(Suppl.), (1930) 521–522.

600. I. D. Collins, Quantitative hydrolysis of starch by buffered taka-diastase, *Science*, 66 (1927) 430–431.
601. T. Kitano, Taka-amylase. I. The amount of glucose formed from starch by the action of taka diastase (in Japanese), *J. Soc. Chem. Ind. Jpn.*, 38(Suppl.), (1935) 376–381.
602. C. E. Mangels and J. J. Martin, Jr., Effect of diffeent buffers and type of substrate on the diastases of wheat starch, *Cereal Chem.*, 12 (1935) 256–268.
603. R. Yamamoto, S. Yoshimura, and R. Tanaka, Kaoliang as the source of starch-making, milling, and ame manufacturing. IV. Manufacture of ame from cleaned kaoliang (in Japanese), *J. Agr. Chem. Soc. Jpn.*, 12 (1936) 503–511.
604. M. Ninomiya, S. Kataoka, and R. Yamamoto, The velocities of liquefaction and saccharification of stach (in Japanese), *J. Agr. Chem. Soc. Jpn.*, 12 (1936) 531–540.
605. R. Yamamoto and S. Matsumoto, The maltose-dextrin ratio in the saccharified products of starch by malt-diastase (in Japanese), *J. Agr. Chem. Soc. Jpn.*, 12 (1936) 541–547.
606. R. Yamamoto, S. Matsumoto, and S. Arami, Effects of heat-treatment under pressure on the composition and saccharification of kaoliang (in Japanese), *J. Agr. Chem. Soc. Jpn.*, 12 (1936) 548–554.
607. D. Das, J. Roy, and G. P. Sen, Potentiality of *Aspergilus awamori* (NRRL 3112) for alpha-amylase and glucoamylase production and their effect on saccharification of potato starch, *Indian Chem. Eng. A*, 37 (1995) 125–131.
608. D. N. Klimovskii and V. I. Rodzevich, Amylolytic enzymes and *Aspergillus* (in Russian), *Mikrobiologiya*, 19 (1950) 60–64.
609. T. Matsubara, Y. Ben Anmar, T. Anindyawati, S. Yamamoto, K. Ito, M. Iizuka, and N. Minamiura, Degradation of raw starch granules by a-amylase purified from culture of *Aspergillus awamori* KT-11, *J. Biochem. Mol. Biol.*, 37 (2004) 422–428.
610. N. Ramaseh, K. R. Sreekantiah, and V. S. Murthy, Purification and characterization of a thermophilic α-amylase of *Aspergillus niger* van Tieghem, *Starch/Stärke*, 34 (1982) 274–279.
611. Y. Minoda and K. Yamada, Acid-stable α-amylase of black *Aspergilli*, *Agric. Biol. Chem.*, 27 (1963) 806–811.
612. Y. Minoda, M. Arai, Y. Torigoe, and K. Yamada, Acid-stable α-amylase of black *Aspergilli*. III. Separation of acid-stable α-amylase and acid-unstable α-amylase from the same mold amylase preparation, *Agric. Biol. Chem.*, 32 (1968) 110–113.
613. Y. Minoda, M. Arai, and K. Yamada, Acid-stable α-amylase of black *Aspergilli*. V. Amino acid composition and amino-terminal amino acid, *Agric. Biol. Chem.*, 33 (1969) 572–578.
614. M. Arai, T. Koyano, H. Ozawa, Y. Minoda, and K. Yamada, Acid-stable α-amylase from black *Aspergilli*. IV. Some physicochemical properties, *Agric. Biol. Chem.*, 32 (1968) 507–513.
615. H. P. Heldt-Hansen and K. Aunstrup, Acid-stable α-amylase composition and its use, EU Patent EP 138,428 (1985); *Chem. Abstr.*, 103 (1985) 5044x.
616. P. Ducroo, J. J. M. Labout, and B. Noordam, Enzyme product and its use in the saccharification of starch, EU Patent EP 140,410 (1985); *Chem. Abstr.*, 103 (1985) 84145g.
617. V. Planchot, P. Colonna, D. J. Gallant, and P. Bouchet, Extensive degradation of native starch granules by alpha-amylase from *Aspergillus fumigatus*, *J. Cereal Sci.*, 21 (1995) 163–171.
618. J. Abe, F. W. Bergmann, K. Obata, and S. Hizukuri, Raw-starch degrading enzyme of *Aspergillus* K-27 (in Japanese), *Denpun Kagaku*, 32 (1985) 128–135.
619. A. M. Omemu, I. Akpan, M. O. Bankole, and O. D. Teniola, Hydrolysis of raw tyuber starches by amylase of *Aspergillus niger* AM07 isolated from the soil, *Afr. J. Biotechnol.*, 4 (2005) 19–25.
620. E. M. Doyle, C. T. Kelly, and W. M. Fogarty, The high maltose-producing α-amylase of *Penicillium expansum*, *Appl. Microbiol. Biotechnol.*, 30 (1989) 492–496.
621. T. Shambe and O. Ejembi, Production of amylase and cellulose: Degradation of starch and carboxymethylcellulose by extracellular enzymes from four fungal species, *Enzyme Microb. Technol.*, 9 (1987) 308–312.

622. V. V. Michelena and F. J. Castillo, Production of amylase by *Aspergillus foetidus* on rice flour medium and characterization of the enzyme, *J. Appl. Bacteriol.*, 56 (1984) 395–407.
623. W. J. Middlehoven and C. M. Muylwijk, High-affinity amylolytic enzymes produced by the mold *Trichoderma harzianum*, *Appl. Microbiol. Biotechnol.*, 23 (1985) 400–403.
624. O. Ya. Rashba and I. D. Varbanets, Splitting of some polysaccharides by different species of *Fusarium* (in Ukrainian), *Mikrobiol. Zhurn. (Kiev)*, 32 (1970) 392–394.
625. S. J. Chary and S. M. Reddy, Starch degrading enzymes of two species of *Fusarium*, *Folia Microbiol. (Prague)*, 30 (1985) 452–477.
626. S. Takao, H. Sasaki, K. Kurosawa, M. Tanida, and Y. Kamagata, Production of a raw starch saccharyfying enzyme by *Corticum rolfsii*, *Agric. Biol. Chem.*, 50 (1986) 1979–1987.
627. J. Zemek, L. Kuniak, and L. Marvanova, α-amylase, Czechoslovak Patent 230,889 (1986); *Chem. Abstr.*, 107 (1987) 5574t.
628. J. H. Do and S. D. Kim, Properties of amylase produced from higher funi *Ganoderma lucidum*, *Sanop Misaengmul Hakhoechi*, 13 (1985) 173–178. *Chem. Abstr.*, 104 (1985) 2589p.
629. R. De Mot and H. Verachtert, Purification and characterization of extracellular α-amylase and glucoamylase from the yeast *Candida antaertica* CBS 6678, *Eur. J. Biochem.*, 164 (1987) 643–654.
630. S. L. Turchi and T. Becker, Improved purification of α-amylase isolated from *hizomucor pusillus* by affinity chromatography, *Curr. Microbiol.*, 15 (1987) 203–205.
631. A. K. Gosh and S. Sengupta, Multisubstrate specific amylase from mushroom *Termitomyces clypeatus*, *J. Biosci.*, 11 (1987) 275–285.
632. J. Augustin, J. Zemek, O. Fassatiova, and L. Kuniak, Production of α-amylase by microscopic fungi, *Folia Microbiol. (Prague)*, 26 (1981) 142–146.
633. A. M. Sills, M. E. Sauder, and G. G. Stewart, Amylase activity in certain yeasts and a fungal species, *Dev. Ind. Microbiol.*, 24 (1983) 295–303.
634. E. Dubreucq, H. Boze, D. Nicol, G. Moulin, and P. Galzy, Kinetics of the α-amylase of *Schwanniomyces castelii*, *Biotechnol. Bioeng.*, 33 (1989) 369–373.
635. A. B. Pasari, R. A. Korus, and R. C. Heimsch, Kinetics of the amylase system of *Schwanniomyces castelii*, *Enzyme Microb. Technol.*, 10 (1988) 156–160.
636. G. A. Somkuti and D. H. Steinberg, Thermoacidophilic extracellular α-amylase of *Mucor pusillus*, *Dev. Ind. Microbiol.*, 21 (1980) 327–337.
637. O. L. Franco, D. J. Rigden, F. R. Melo, and M. F. Grossi-De-Sá, Plant α-amylase inhibitors and their interaction with insect α-amylases, *Eur. J. Biochem.*, 269 (2002) 397–412.
638. L. P. Kouame, E. D. Ahipo, S. L. Niamke, F. A. Kouame, and A. Kamenan, Synergism of cockroach (*Periplaneta americana*) α-amylase and α-glucosidase hydrolysis of starch, *Afr. J. Biotechnol.*, 3 (2004) 529–533.
639. M. Ben Elarbi, H. Khemiri, T. Jridi, and J. Ben Hamida, Purification and characterization of alpha-amylase from safflower (*Corthanus tinctorius*) germinating seeds, *C. R. Biol.*, 332 (2009) 426–432.
640. S. E. M. De-Almeida, K. Mizuta, and R. Giglio, *Pycnoporus sanguineus*: A novel source of alpha-amylase, *Mycol. Res.*, 101 (1997) 188–190.
641. B. Schwermann, K. Pfau, B. Liliensiek, M. Schleyer, T. Fischer, and E. P. Baker, Purification, properties and structural aspects of a thermoacidophilic α-amylase for *Alicyclobasillus acidocaldarius* cyclodextrins, *Eur. J. Biochem.*, 226 (1994) 981–991.
642. G. Feller, E. Narinx, J. L. Arpigny, Z. Zekhini, J. Swings, and C. Gerday, Temperature dependence of growth, enzyme secretion and activity of psychrophilic Antarctic bacteria, *Appl. Microbiol. Biotechnol.*, 41 (1994) 477–479.
643. G. Feller, T. Lohnienne, C. Deroanne, C. Libioulle, J. Vanbrrumrn, and C. Gerday, Purification, characterization and nucleotide sequence of the thermolabile alpha-amylase from Antarctic psychrotroph *Alteromonas haloplanctis* A23, *J. Biol. Chem.*, 267 (1992) 5217–5221.

644. H. Sugita, A. Kuruma, and Y. Deguchi, Purification and some properties of an α-amylase from anaerobic bacterium isolated from coastal sediment, *Biosci. Biotechnol. Biochem.*, 61 (1997) 1757–1759.
645. H. Melasniemi, Purification and some properties of the extracellular α-amylase–pullulanase produced by *Clostridium thermosulfuricum*, *Biochem. J.*, 250 (1988) 813–818.
646. E. W. Boyer, M. B. Ingle, and G. D. Mercer, Isolation and characterization of unusual bacterial amylase, *Starch/Stärke*, 31 (1979) 166–170.
647. E. S. Demirkan, B. Mikami, M. Adachi, T. Higasa, and S. Utsumi, α-amylase from *B.* amyloliquefaciens: Purification, characterization, raw starcg degradation and expression in *E. coli*, *Process Biochem.*, 40 (2005) 2629–2636.
648. J. A. Milner, J. D. Martin, and A. Smith, Two-stage inocula for the production of alpha-amylase by *Bacillus amyloliquefaciens*, *Enzyme Microb. Technol.*, 21 (1997) 382–386.
649. D. Gangadharan, S. Sivaramnakrishna, K. M. Nampoorthiri, and A. Pandey, Solid culturing of *Bacillus* amyloliquefaciens for alpha amylase production, *Food Technol. Biotechnol.*, 44 (2006) 269–274.
650. P. Hiller, D. A. J. Wase, A. N. Emery, and G. L. Solomon, Instability of α-amylase production and morphological variation in continuous culture of *Bacillusamyloliquefaciens* is associated with plasmid loss, *Process Biochem.*, 32 (1997) 51–59.
651. C. S. Lin, S. C. Hong, C. Y. Cheng, and C. L. Chen, Design of a continuous rotating annular bioreactor for immobilized-cell fermentations, *J. Chem. Technol. Biotechnol.*, 69 (1997) 433–437.
652. A. Tonkova, V. Ivanova, E. Dobreva, M. Stefanova, and D. Spassova, Thermostable α-amylase production by immobilized *Bacillus licheniformis* cells in agar gel and calcium alginate beads, *Appl. Microbiol. Biotechnol.*, 41 (1994) 517–522.
653. M. Stefanova, A. Tonkova, E. Dobreva, and D. Spassova, Agar gel immobilization of *Bacillus brevis* cells for production of thermostable α-amylase, *Folia Microbiol.*, 43 (1998) 42–44.
654. O. Ariga, H. Tyofuku, I. Minegishi, T. Hattori, Y. Ano, and M. Nagura, Efficient production of recombinant enzymes using PVA encapsulated bacteria, *Ferment. Bioeng.*, 84 (1997) 553–557.
655. S. Ebisu, M. Mori, H. Takagi, K. Kadowaki, H. Yamagata, H. Tsukagoshi, and S. Udaka, Production of a fungal protein. Taka amylase A by protein producing *Bacillus brevis* HDPS1, *J. Ind. Microbiol. Biotechnol.*, 11 (1993) 83–88.
656. L. G. Loginova, S. Ya. Karpukhina, and E. P. Guzhova, α-amylases of thermophilic bacteria (in Russian), *Izv. Akad. Nauk SSSR Biol.*, 4 (1967) 595–599.
657. A. Harder, B. Noordam, and A. M. Brekelmans, Kinetis of isomaltose formation by amyloglucosidase and purification of disaccharide by fermentation of undesired by-products, *Ann. N. Y. Acad. Sci.*, 413 (1983) 340–351.
658. K. R. Babu and T. Satynarayana, α-amylase production by thermophilic *Bacillus coagulans* in solid state fermentation, *Process Biochem.*, 30 (1995) 305–309.
659. C. T. Kelly, D. J. Bolton, and W. M. Fogarty, Bi-phasic production of α-amylase of *Bacillus flavothermus* in batch fermentation, *Biotechnol. Lett.*, 19 (1997) 675–677.
660. D. J. Bolton, C. T. Kelly, and W. M. Fogarty, Purification and characterization of the α-amylase from of *Bacillus flavothermus*, *Enzyme Microb. Technol.*, 20 (1997) 340–343.
661. A. Bandopadhyay, S. K. Seri, and S. C. Pal, Extracellular amylose synthesis by *Bacillus globisporus* BH-1b, *Acta Biotechnol.*, 14 (1994) 97–194.
662. S. A. El-Aassar, S. H. Omar, M. K. Gouda, A. M. Ismail, and A. F. Abdelfattah, Purification of α-amylase from *Bacillus lentus* cultures, *Appl. Microbiol. Biotechnol.*, 38 (1992) 312–314.
663. F. J. Morgan and F. G. Priest, Characterization of a thermostable α-amylase from *Bacillus licheniformis* NCIB 6346, *J. Appl. Microbiol.*, 50 (1981) 107–114.

664. F. Qin, J. Man, B. Xu, M. Hu, M. Gu, Q. Liu, and C. Wei, Structural properties of hydrolyzed high-amylose rice starch by α-amylose from *Bacillus licheniformis*, *J. Agric. Food Chem.*, 59 (2011) 12667–12673.
665. E. Dobreva, A. Tonkova, V. Ivanova, M. Stefanova, and L. Kabaivanova, Immobilization of *Bacillus licheniformis* cells, properties of thermostable α-amylase, *J. Ind. Microbiol. Biotechnol.*, 20 (1998) 166–170.
666. S. Decordt, K. Vanhoof, J. Hu, G. Maesmans, M. Hendrickx, and P. Tobback, Thermostability of soluble and immobilized α-amylase from *Bacillus licheniformis*, *Biotechnol. Bioeng.*, 40 (1992) 396–402.
667. A. R. Varlan, W. Sansen, A. VanLoey, and M. Hendrickx, Covalent enzyme immobilization on paramagnetic polyacrolein beads, *Biosens. Bioelectron.*, 11 (1996) 443–448.
668. M. V. Ramesh and B. K. Lonsane, Purification of thermostable alpha-amylase produced by *Bacillus licheniformis* M27 under solid state fermentation, *Process Biochem.*, 24 (1989) 176–178.
669. I. C. Kim, J. H. Cha, J. R. Kim, S. Y. Jang, B. C. Seo, T. K. Cheong, D. S. Lee, and K. H. Park, Catalytic properties of the cloned amylase from *Bacillus licheniformis*, *J. Biol. Chem.*, 267 (1992) 22108–22114.
670. M. V. Ramesh and B. K. Lonsane, Characteristics and novel features of thermostable α-amylase produced by *Bacillus licheniformis* M27 under solid state fermentation, *Starch/Stärke*, 42 (1990) 233–238.
671. P. J. Brumm, R. E. Hebeda, and W. M. Teague, Purification ad characterization of the commercialized, cloned *Bacillus megaterium* α-amylase. I. Purification and hydrolytic properties, *Starch/Stärke*, 43 (1991) 319–323.
672. G. Mamo, B. A. Gashe, and A. Gassesse, A highly thermostable amylase from a newly isolated thermophilic *Bacillus* sp. WN 11, *J. Appl. Microbiol.*, 86 (1999) 557–560.
673. L. Lin, C. Chyau, and W. H. Hsu, Production and properties of a raw starch degrading amylase from thermophilic and alkaliphilic *Bacillus* sp. TS-23, *Biotechnol. Appl. Biochem.*, 28 (1998) 61–68.
674. L. M. Hamilton, C. T. Kelly, and W. M. Fogarty, Production and properties of the raw starch digesting α-amylase of *Bacillus* sp. IMD-435, *Process Biochem.*, 35 (1999) 27–31.
675. X. D. Liu and Y. Xu, A novel raw starch digesting α-amylase from a newly isolated *Bacillus* sp. YX-1: Purification and characterization, *Bioresour. Technol.*, 99 (2008) 4315–4320.
676. R. Vieira de Carvalho, T. L. Ribeiro Correa, J. C. Matos da Silva, L. Ribeiro Coutinho de Oliverira Mansur, and M. L. Leal Martins, Properties of an amylase from thermophilic Bacillus sp., *Braz. J. Microbiol.*, 39 (2008) No. 1.
677. M. Ben Ali, M. Mezghani, and S. A. Bejar, A thermostable α-amylase producing maltohexaose from new isolated *Bacillus* sp. US100: Study of activity and molecular cloning of the corresponding gene, *Enzyme Microb. Technol.*, 24 (1999) 584–589.
678. G. B. Manning and L. L. Campbell, Thermostable α-amylase of *Bacillus stearothermophilus*. I. Crystallization and some general properties, *J. Biol. Chem.*, 236 (1961) 2952–2957.
679. L. L. Campbell and G. B. Manning, Thermostable α-amylase of *Bacillus stearothermophilus*. II. Physical properties and molecular weight, *J. Biol. Chem.*, 236 (1961) 2962–2965.
680. L. L. Campbell and P. D. Cleveland, Thermostable α-amylase of *Bacillus stearothermophilus*. III. Amino acid composition, *J. Biol. Chem.*, 236 (1961) 2966–2969.
681. K. Ogasahara, A. Imanishi, and T. Isemura, Studies on thermophilic alpha-amylase from *Bacillus stearothermophilus*. Thermal stability of thermophilic alpha-amylase, *J. Biochem.*, 67 (1970) 77–80.
682. K. Yutani, Molecular weight of thermostable α-amylase from *B. stearothermophilus*, *J. Biochem.*, 74 (1973) 581–584.

683. Y. Yutani, I. Sasaki, and K. Ogasahara, Comparison of thermostable α-amylases from *B. stearothermophilus* grown at different temperatures, *J. Biochem.*, 74 (1973) 573–577.
684. E. M. Eegelseer, J. Schocher, U. B. Sleyter, and M. Sara, Evidence that N-terminal S-layer protein fragment triggers the release of a cell associated high-molecular weight amylase in *Bacillus stearothermophilus* ATCC1298, *J. Bacteriol.*, 178 (1996) 5602–5609.
685. R. D. Wind, R. M. Buitelaar, G. Eggink, H. J. Huizing, and L. Dijkhuizen, Characterization of a new *Bacillus stearothermophilus* isolate: A high thermostable α-amylase producing strain, *Appl. Microbiol. Biotechnol.*, 41 (1994) 155–162.
686. J. L. Marco, L. A. Bataus, F. F. Valencia, C. J. Ulho, S. Astolfi-Filho, and C. R. Felix, Purification and characterization of truncated *Bacillus subtilis* α-amylase produced by *Escherichia coli*, *Appl. Microbiol. Biotechnol.*, 44 (1996) 746–752.
687. T. J. G. Garcia Salva and I. O. Moraes, Effect of carbon source on α-amylase production from *Bacillus subtilis* BA-04, *Rev. Microbiol.*, 26 (1995) 46–51.
688. G. C. Uguru, D. Robb, J. A. Akinyanju, and A. Sani, Purification, characterization and mutagenic enhancement of a thermoactive α-amylase from *Bacillus subtilis*, *J. Ind. Microbiol. Biotechnol.*, 19 (1997) 273–279.
689. I. Spencer-Martins and N. Van Uden, Extracellular amylolytic system of the yeast *Lipomyces kononenkoae*, *Eur. J. Appl. Microbiol. Biotechnol.*, 6 (1979) 241–250.
690. S. Hayashida, Y. Teramoto, and T. Inoue, Production and characteristics of raw potato starch digesting α-amylase from *Bacillus subtilis* 61, *Appl. Environ. Microbiol.*, 54 (1988) 1516–1522.
691. S. K. Lee, Y. B. Kim, and G. E. Ji, Purification of amylase secreted from *Bifidobacterium adolescentis*, *J. Appl. Microbiol.*, 83 (1997) 267–272.
692. K. Ratanakhanokchai, J. Kaneko, Y. Kamio, and K. Izaki, Purification and properties of maltotetraose- and maltotriose-producing amylose from *Chloroflexus aurantiacus*, *Appl. Environ. Microbiol.*, 58 (1992) 2490–2494.
693. V. Paquet, C. Croux, G. Goma, and P. Soucaille, Purification and characterization of the alpha amylase from *Clostridium acetobutylicum* ATCC824, *Appl. Environ. Microbiol.*, 57 (1991) 212–218.
694. A. Chojecki and H. P. Blaschek, Effect of carbohydrate source on α-amylase and glucoamylase formation by *Clostridium acetobutylicum* SA-1, *J. Ind. Microbiol. Biotechnol.*, 1 (1986) 63–67.
695. B. Ensley, J. J. McHugh, and L. L. Barton, Effect of carbon sources on formation of α-amylase and glucoamylase by *Clostridium acetobutylicum*, *J. Gen. Appl. Microbiol.*, 21 (1975) 51–59.
696. T. Tanaka, E. Ishimoto, Y. Shimomura, M. Taniguchi, and S. Oi, Purification and some properties of raw starch binding amylase of *Clostridium butylicum* isolated from mesophilic methane sludge, *Agric. Biol. Chem.*, 51 (1987) 399–405.
697. N. J. Shih and R. G. Labbe, Purification and characterization of an extracellular alpha amylase from *Clostridium perfringens* type A, *Appl. Environ. Microbiol.*, 61 (1995) 1776–1779.
698. M. V. Swamy and G. Seenayya, Thermostable pullulanase and α-amylase activity from *Clostridium thermosulfurogenes* SV9. Optimization conditions for enzymaticproduction, *Process Biochem.*, 31 (1996) 157–162.
699. M. V. Swamy and G. Seenayya, *Clostridium thermosulfurogene* SV9. A thermophilic amylases producer, *Indian J. Microbiol.*, 36 (1996) 181–184.
700. C. L. Jeang, Y. H. Lee, and L. W. Chang, Purification and characterization of a raw starch digesting amylase from a soil bacterium *Cytophaga* Cytophaga sp, *Biochem. Mol. Biol. Int.*, 35 (1995) 549–557.
701. E. P. Delahaye, M. J. Foglietti, C. Andrieux, I. Chardonloriaux, O. Szyli, and P. Raibaud, Identification and properties of an alpha-amylase from a strain of *Eubacterium* sp. isolated from rat intestinal tract, *Comp. Biochem. Physiol. A Comp. Physiol.*, 98 (1991) 351–354.

702. S. Patel, R. Bagai, and D. Madamwar, Stabilization of alkalophilic α-amylase by calcium alginate immobilization, *Biocat. Biotrans.*, 14 (1996) 147–155.
703. M. J. Coronado, C. Vargas, J. Hofemeister, A. Ventosa, and J. J. Nieto, Production and biochemical characterization of α-amylase from the moderate halophile *Halomonas meridiana*, *FEMS Microbiol. Lett.*, 183 (2000) 67–71.
704. C. L. Fergus, The production of amylase by some thermophilic fungi, *Mycologia*, 61 (1969) 1171–1177.
705. A. Burgess-Cassler and S. Imam, Partial purification and comparative characterization of α-amylase secreted by *Lactobacillus amylovorus*, *Curr. Microbiol.*, 61 (1991) 207–213.
706. M. O. Ilori, O. O. Amud, and O. Omidji, Purification and properties of a glucose-forming amylase of *Lactobacillus brevis*, *World J. Microbiol. Biotechnol.*, 11 (1995) 595–596.
707. M. O. Ilori, O. O. Amud, and O. Omidji, Purification and properties of an α-amylase produced by a cassava-fermenting strain of *Micrococcus luteus*, *Folia Microbiol.*, 42 (1997) 445–449.
708. S. Sen and S. L. Chakraborty, Amylose frm *Lactobacillus cellobiosus* D-39 isolated from vegetable waste: Characteristics of immobilized enzyme in whole cell, *Enzyme Microb. Technol.*, 9 (1987) 112–116.
709. F. Giraud, L. Gosselin, B. Marin, J. L. Parada, and M. Raimbault, Purification and characterization of an extracellular amylase activity from *Lactobacillus plantarum* strain A6, *J. Appl. Bacteriol.*, 75 (1993) 276–282.
710. A. I. Adeleye, Purification and properties of α-amylase from *Micrococcus various*, *Basic Microbiol.*, 30 (1990) 717–722.
711. M. Hasegawa and K. Sakurai, Process for producing neurainidase, US Patent 5258290 (1993).
712. L. G. Loginova, I. A. Tsaplina, E. P. Guzhova, E. V. Boltyanskaya, I. D. Kovolevskaya, and L. M. Seregina, Thermophilic actinomycetes of the species *Micromonospora vulgaris* active producers of amylolytic and proteolytic enzymes (in Russian), *Mikrobiologiya*, 39 (1970) 784–789.
713. M. E. Farezvidal, A. Fernandezvivas, F. Gonzalez, and J. M. Arias, Properties and significance of an α-amylase produced by *Myxococcus coralloides*, *J. Appl. Bacteriol.*, 78 (1995) 14–19.
714. A. Stevens, M. Gunasekaran, and P. Sangan, Purification and properties of amylase from *Nocardia asteroids*, *J. Gen. Appl. Microbiol.*, 40 (1994) 319–329.
715. Y. Kimura, M. Ogata, M. Yoshida, and T. Nanakui, Stability of immobilized maltotetraose forming amylase from *Pseudomonas stutzeri*, *Biotechnol. Bioeng.*, 33 (1989) 845–855.
716. S. N. Freer, Purification and characterization of the extracellular alpha amylase from *Streptococcus bovis* JB1, *Appl. Environ. Microbiol.*, 59 (1993) 1398–1402.
717. H. E. M. McMahon, C. T. Kelly, and W. M. Fogarty, Thermostability of three α-amylases of Streptomyces sp. IMD 2679, *J. Ind. Microbiol. Biotechnol.*, 22 (1999) 96–99.
718. D. Primarin and Y. Ohta, Some enzyme properties of raw starch digesting amylases from *Streptomyces* sp. No.4, *Starch/Stärke*, 52 (2000) 28–32.
719. R. R. Ray, D. Jana, and G. Nanda, Immobilization of beta-amylase drim *Bacillus megaterium* B_6 into gelatin film bycross-linking, *J. Appl. Bacteriol.*, 79 (1995) 157–162.
720. Z. T. Abramov, I. A. Tsaplina, V. M. Stepanov, and K. G. Loginova, Purification and properties of α-amylase from *Thermoctinomyces vulgaris* strain (in Russian), *Biokhimiya*, 51 (1986) 100–106.
721. O. Hesse, G. Hansen, W. E. Hohne, and D. Körner, Eine thermostabile alpha-Amylase aus *Thermoactinomyces vulgaris*. Reinigung und Charakteriziening, *Biomed. Biochim. Acta*, 50 (1991) 225–232.
722. A. M. Allam, A. M. Hussein, and A. M. Ragab, Amylase of the thermophilic actinomycete *Thermomonospora vulgaris*, *Z. Allgem. Mikrobiol.*, 15 (1975) 393–399.
723. B. S. Collins, C. T. Kelly, W. M. Fogarty, and E. M. Doyle, The high maltose-producing α-amylase of the thermophilic qctinomycete *Thermomonospora curvata*, *Appl. Microbiol. Biotechnol.*, 39 (1993) 31–35.

724. F. Schnutzenberger and R. Carnell, Amylose production by *Thermomonospora curvata*, *Appl. Environ. Microbiol.*, 34 (1977) 234–236.
725. J. L. Glymph and F. J. Stutzenberger, Production, purification and characterization of alpha amylase from *Thermomonospora curvata*, *Appl. Environ. Microbiol.*, 34 (1977) 391–397.
726. J. E. Bush and F. J. Stutzenberger, Amylolytic activity of *Thermomonospora fusca*, *World J. Microbiol. Biotechnol.*, 13 (1997) 637–642.
727. Y. C. Chung, T. Kobayashi, H. Kanai, T. Akiba, and T. Kudo, Purification and properties of extracellular amylose from the hyperthermophilic archeon *Thermococcus profundus* DT 5432, *Appl. Environ. Microbiol.*, 61 (1995) 1502–1506.
728. Y. Muramatsu, K. Takahashi, and N. Nakamura, Characterization of a maltogenic amylase of *Thermononospora viridis* and application in branched cyclodextrin production, *Starch/Stärke*, 45 (1993) 99–104.
729. W. Liebl, I. Stemplinger, and P. Ruille, Properties and gene structure of *Thermatoga maritima* α-amylase Amy 1, a putative lipoprotein of a hyperthermophilic bacterium, *J. Bacteriol.*, 179 (1997) 941–948.
730. M. C. V. Egas, M. S. da Costa, D. A. Covan, and E. M. V. Pires, Extracellular α-amylase from *Thermus filiformis* Ork A2: Purification and biochemical characterization, *Exothermophiles*, 2 (1998) 23–32.
731. V. R. Linardi, C. M. M. Andrade, M. M. Figueira, M. C. Andrade, and A. V. V. Souza, Characterization of the amylolytic system of *Candida* strain, *Folia Microbiol.*, 38 (1993) 281–284.
732. H. Onishi, Candida tsukubaensis sp, *Antonie van Leeuwenhoek J. Microbiol.*, 38 (1972) 365–367.
733. R. De Mot, E. Van Oudendijck, and H. Verachtert, Purification and characterization of an extracellular glucoamylase from the yeast *Candida tsukubaensis* CBS6389, *Antonie van Leeuwenhoek J. Microbiol.*, 51 (1985) 275–287.
734. R. De Mo and H. Verachtert, Purification and characterization of extracellular amylolytic enzymes from yeast *Filobasidum capsuligenum*, *Appl. Environ. Microbiol.*, 50 (1985) 1474–1482.
735. A. S. Narayanan and E. R. B. Shanmugasundaram, Studies on amylase from *Fusarium vasinefectum*, *Arch. Biochem. Biophys.*, 118 (1967) 317–322.
736. A. B. Tsiomenko, D. S. Musaev, V. V. Lupashin, and I. S. Kulaev, Secreted α-amylase of the basidomycetous yeast *Filobasidium capsuligenum:* Isolation, purification and properties (in Russian), *Biokhimiya*, 57 (1992) 287–303.
737. J. A. Prieto, B. R. Bort, J. Martinez, F. Randez-Gil, C. Buesa, and P. Sanz, Purification and characterization of a new α-amylase of intermediate thermal stability from the yeast *Lipomyces kononenkoae*, *Biochem. Cell Biol.*, 73 (1995) 41–49.
738. G. Moulin and P. Galzy, Study of an amylase and its regulation in *Lipomyces starkeyi*, *Agric. Biol. Chem.*, 43 (1977) 1165–1171.
739. C. T. Kelly, M. E. Moriarty, and W. M. Fogarty, Thermostablr extracellular α-amylase and α-glucosidase of *Lipomyces starkeyi*, *Appl. Microbiol. Biotechnol.*, 22 (1985) 352–358.
740. V. N. Thanh, Lipomyces orientalis sp. nov., a yeast species isolated from soil in Vietnam, *Int. J. Syst. Evol. Microbiol.*, 56 (2006) 2009–2013.
741. L. M. P. De Moraes, S. Astolfi-Filho, and C. J. Ulhao, Purification and some properties of an α-amylase glucoamylase fusion protein from *Saccharomyce cerevisiae*, *World J. Microbiol. Biotechnol.*, 15 (1999) 561–564.
742. H. Ruttloff, A. Taufel, and F. Zickler, Glucoamylase aus *Endomycopsis bispora*. I. Zur Produkten des Enzymes in Schüttelkultur, *Z. Allgem. Mikrobiol.*, 19 (1979) 195–201.
743. H. Ebertova, Amylolytic enzymes of *Endomycopsis capsularis*. II. A study of properties of isolated α-amylase, amyloglucosidase and maltose transglucosidase, *Folia Microbiol.*, 11 (1966) 422–438.

744. J. Fukumoto, Y. Tsujisaka, and M. Araki, Studiem of amylose of Endomycopsis, *Kagaku to Kogyo*, 34 (1981) 423–427.
745. A. I. Stepanov, V. P. Afanaseva, G. V. Zaitseva, A. P. Mednokova, and I. B. Lupandina, Regulation of biosythesis of amylolytic enzymes from *Endomyces fibuligera* strain 20-9 (in Russian), *Prikl. Biokhim. Mikrobiol.*, 11 (1975) 682–685.
746. H. Gasparik, E. Hostinova, and J. Zelinka, Production of extracellular amylase by *Endomycopsis fibuligera* on comlex starch substrate, *Biologia (Bratislava)*, 40 (1985) 1174–1176.
747. C. Laluce, M. C. Bertolini, J. R. Ernandes, A. V. Martini, and A. Martini, New amylolytic yeast strain for starch and extrin fermentation, *Appl. Environ. Microbiol.*, 54 (1988) 2447–2451.
748. A. M. Yurkov, M. Kemler, and D. Bergerov, Assesment of yeast diversity in soils under different management regimes, *Fungal Ecol.*, 5 (2012) 24–35.
749. A. Strasser, F. B. Martens, J. Dohmen, and C. P. Hollenberg, Amylolytic enzymes producing microorganisms constructed by recombinant DNA technology and their use for fermentation process, US Patent 5,100,794 (1992).
750. R. De Mot and H. Verachtert, Secretion of α-amylase amd multiple froems of glucoamylase by the yeast *Trichosporon pullulans*, *Can. J. Microbiol.*, 32 (1986) 47–51.
751. I. I. Perevozchenko and A. S. Tsyperovich, Comparative investigation of the properties of α-amylases of malt fungi of the genus *Aspergillus*, *Appl. Biochem. Microbiol.*, 8 (1972) 12–18.
752. R. S. Bhella and I. Tosaar, Purification and some properties of extracellular α-amylase from *Aspergillus awamori*, *Can. J. Microbiol.*, 31 (1985) 149–153.
753. B. N. Okolo, F. S. Ire, L. L. Ezeogu, C. U. Anyanwu, and J. F. C. Odibo, Purification and some properties of a novel raw starch-digesting amylase from *Aspergilluscarbonarius*, *J. Sci. Food Agric.*, 81 (2001) 329–336.
754. P. O. Olutiola, α-Amylolytic activity of *Aspergillus chevalieri* from mouldy maize seeds, *Indian Phytopathol.*, 35 (1982) 428–433.
755. D. Siedenberg, S. R. Gerlach, A. Czwalinna, and K. Schugerl, Production of xylanase of *Aspergillus awamori* on complex medium in stirred tank and airlift tower loop reactors, *J. Biotechnol.*, 56 (1997) 205–216.
756. S. L. Khoo, A. A. Amirul, M. Kamaruzaman, N. Nazalan, and M. N. Azizan, Purification and characterization of alpha amylase from *Aspergillus flavus*, *Folia Microbiol.*, 39 (1994) 392–398.
757. Y. D. Hang and E. E. Woodams, Baked-bean waste: A potential substrate for producing fungal amylase, *Appl. Environ. Microbiol.*, 33 (1977) 1293–1294.
758. C. Planchot and P. Colonna, Purification and characterization of extracellular alpha-amylose from *Aspergillus fumigatus*, *Carbohydr. Res.*, 272 (1995) 97–109.
759. C. M. Dominigues and R. M. Peralta, Production of amylase by soil fungi and partial biochemical characterization of amylase of a selected strain (*Aspergillus fumigatus* Fresenius), *Can. J. Microbiol.*, 39 (1993) 681–685.
760. C. E. Goto, Production of amylase by *Aspergillus fumnigatus* utilizing α-methyl-D-glucoside a synthetic analogue of maltose as substitute, *FEMS Microbiol. Lett.*, 167 (1998) 139–142.
761. D. Alazard and J. F. Baldensperger, Amylolytic enzymes from *Aspergillus henneberge (A. niger* group): Purification and characterization of amylose from solid and liquid cultures, *Carbohydr. Res.*, 107 (1982) 231–241.
762. M. Ohnishi, K. Iwata, T. Tomita, U. Nishikawa, and K. Hiromi, Kinetic properties of *Rhizopus* glucoamylase and *Bacillus* α-amylase which are immobilized on on Cellulofine, *Starch/Stärke*, 42 (1990) 486–489.
763. B. N. Okolo, L. L. Ezeogu, and C. L. Mba, Production of raw starch digesting amylase from *Aspergillus niger* grown on native starch sources, *J. Sci. Food Agric.*, 69 (1995) 109–115.
764. M. Arai, Y. Minoda, and K. Yamada, Acid-stable α-amylase of black *Aspergilli*. VI. Carbohydrate and metal content, *Agric. Biol. Chem.*, 33 (1969) 922–928.

765. K. N. Varalakshmi, B. S. Kumundinu, B. N. Nandini, J. Solomon, R. Suhas, B. Makesh, and A. P. Kavitha, Production and charaktrisation of α-amylase from *Aspergillus niger* JGI24 isolated in Bangalore, *Pol. J. Microbiol.*, 58 (2009) 29–36.
766. K. Hoshino, M. Katagiri, M. Taniguchi, T. Sasakura, and M. Fuijii, Hydrolysis of starchy materials by repeated use of an amylase immobilized on a novel thermoresponsive polymer, *J. Ferment. Bioeng.*, 77 (1994) 407–412.
767. O. Bhumibhamon, Production of amyloglucosidase by submerged culture, *Thai J. Agric. Sci.*, 16 (1983) 173–184.
768. N. Ramachandran, K. R. Skreekantiah, and V. S. Murthy, Studies on the thermophilic amylolytic enzymes of a strain *Aspergillus niger*, *Starch/Stärke*, 30 (1978) 272–275.
769. N. Ramachandran, K. R. Skreekantiah, and V. S. Murthy, Influence of media composition on the production of alpha-amylase and amyloglucosidase by a strain of *Aspergillus niger*, *Starch/Stärke*, 31 (1979) 134–138.
770. M. Klich, Identification of Common *Aspergillus* Species, (2002) Centraalbureau voor Schimmelcultures, Utrecht.
771. M. Yabuki, N. Ono, K. Hoshino, and S. Fujui, Rapid induction of alpha amylase by nongrowing mycelia of *Aspergillus oryzae*, *Appl. Environ. Microbiol.*, 34 (1977) 1–6.
772. I. Jodal, L. Kandra, J. Harangi, P. Nanasi, and J. Szejtli, Hydrolysis of cyclodextrin by *Aspergillus oryzae* α-amylase, *Starch/Stärke*, 36 (1984) 140–143.
773. C. H. Nguyen, R. Tsurumizu, T. Sato, and M. Takeuchi, Taka-amylase A in conidia of *Aspergillus oryzae* RIB40, *Biosci. Biotechnol. Biochem.*, 63 (2005) 2035–2041.
774. V. M. Harahan and M. L. Caldwell, Additional studies of the properties of taka amylase, *J. Am. Chem. Soc.*, 75 (1953) 4030–4034.
775. Y. Minoda, T. Koyano, M. Arai, and K. Yamada, Acid-stable α-amylase of black *Aspergilli*. II. Some general properties, *Agric. Biol. Chem.*, 32 (1968) 104–109.
776. N. Suetsugu, S. Koyama, K. I. Takeo, and T. Kuge, Kinetic studies on the hydrolysis of α-, β- and γ-cyclodextrin by taka amylase A, *J. Biochem.*, 76 (1974) 57–63.
777. M. A. Murado, M. P. Gonzalez, A. Torrado, and L. M. Pastrana, Amylose production by solid state culture of *Aspergillus oryzae* on polyurethane foams. Some mechanistic approaches from empirical model, *Process Biochem.*, 32 (1997) 35–42.
778. S. Sivuramakrishnan, D. Gaugadharan, K. M. Nampoothiri, C. R. Soccol, and A. Pandey, Alpha amylase production by *Aspergillus oryzae* employing solid state fermentation, *J. Sci. Ind. Res.*, 66 (2007) 621–626.
779. C. T. Chan, M. S. Tang, and C. F. Lin, Purification and properties of alpha-amylase from *Aspergillus oryzae* ATCC 76080, *Biochem. Mol. Biol. Int.*, 36 (1995) 185–193.
780. O. Juhasz, V. Lukačova, V. Vollek, and B. Skarka, Purification and characteristics of the amylose system of *Aspergillus oryzae*, *Biologia* (1991) 627–634.
781. H. Toda, K. Kondo, and K. Narita, The complete amino acid sequence iof taka-amylase A, *Proc. Jpn. Acad., B*, 58 (1982) 208–212.
782. T. Harada, Preparation of *Aspergillus oryzae* enzymes, *Ind. Eng. Chem.*, 23 (1931) 1424–1427.
783. T. Saganuma, N. Tahara, K. Kitahara, T. Nagahama, and K. Inuzuka, N-terminal sequence of amino acids and some properties of an acid-stable α-amylase from citric acid koji (*Aspergillus usami* var.), *Biosci. Biotech. Biochem.*, 60 (1996) 177–179.
784. H. Lefuji, M. Chino, M. Kato, and Y. Kimura, Raw starch digesting and thermostable α-amylase from the yeast *Cryptococcus* sp. S-2: Purification, characterization, cloning and sequencing, *Biochem. J.*, 318 (1996) 989–996.
785. R. Sadhukhan, S. K. Roy, and S. L. Chakraborty, Immobilization of α-amylase from *Mycellophthora thermophila* D-14 (ATCC48104), *Enzyme Microb. Technol.*, 15 (1993) 801–804.

786. W. Hu, M. E. Persia, and L. Kung, Jr., Short communication: In vitro ruminal fermentability of a modified corn cultivar expressing a thermotolerant alpha amylase, *J. Dairy Sci.*, 99 (2010) 4846–4849.
787. C. T. Zenin and Y. K. Park, Purification and characterization of acid α-amylase from *Pacecilomyces* sp, *J. Ferment. Technol.*, 61 (1983) 199–201.
788. R. Haska and Y. Ohta, Purification and properties of the raw starch digesting amylase from *Penicillium brunneum* No. 24, *Starch/Stärke*, 46 (1994) 480–485.
789. E. M. D. Siqueira, K. Mizuta, and J. R. Giglio, *Pycnoporus sanguineus*: A novel source of α-amylase, *Mycol. Res.*, 101 (1997) 188–190.
790. E. L. de Souza, E. H. E. Hoffmann, V. M. Castilho, V. A. de Lima, and M. Z. Bellini, Production and characterization of amylase from *Rhizopus* sp, *Arq. Biol. Technol.*, 39 (1996) 831–839.
791. T. Shimazaki, S. Hara, and M. Sato, Production, purification and some properties of extracellular amylase of *Schizophyllum commune*, *J. Ferment. Technol.*, 62 (1984) 165–170.
792. F. J. C. Odibo, N. Okafor, and B. U. Okafor, Purification and immobilization of *Scytolidium* sp. α-Amylase and its general properties, *J. Gen. Appl. Microbiol.* (1992) 1–11.
793. J. A. Schellart, F. M. W. Visser, T. Zandstva, and W. J. Middlehover, Starch degradation by the mold *Trichoderma viride*. I. The mechanism of degradation, *J. Microbiol. Serol.*, 42 (1976) 229–238.
794. B. Jensen, J. Olson, and K. Allerman, Purification of extracellular amylolytic enzymes from the thermophilic fungus *Thermomyces lanuginosus*, *Can. J. Microbiol.*, 34 (1988) 218–223.
795. B. Jensen and J. Olsen, Physicochemical properties of purified alpha-amylase from the thermophilic fungus *Thermomycs lanuginosus*, *Enzyme Microb. Technol.*, 14 (1992) 112–116.
796. E. A. Barnett and C. L. Fergus, The relation of extracellular amylase, mycelium and time in some thermophilic and mesophilic *Humicola* species, *Mycopathol. Mycol. Appl.*, 44 (1971) 131–141.
797. F. Canganella, C. M. Andrade, and G. Antranikian, The characterization of amylolytic and pullolytic enzymes *from Thermophilic archea* and from a new *Fervidobacterium* species, *Appl. Microbiol. Biotechnol.*, 42 (1994) 239–245.
798. K. A. Laderman, B. R. Davis, H. C. Krutsch, M. S. Lewis, Y. V. Griko, P. L. Privalov, and C. B. Anfinsen, The purification and characterization of extremely thermostable alpha-amylase from the hyperthermophilic archaebacterium *Pyrococcus furiosus*, *J. Biol. Chem.*, 268 (1993) 24394–24401.
799. R. Koch, A. Spreinat, K. Lemke, and G. Antranikian, Purification and properties of a hyperthermoactive α-amylase from archaeobacterium *Pyrococcus woesei*, *Arch. Microbiol.*, 155 (1991) 572–578.
800. C. Haseltine, M. Rolfsmeier, and P. Blum, The glucose effect and regulation of alpha amylase synthesis in the hyperthermophilic archaeon, *Sulfolobus solfataricus*, *J. Bacteriol.*, 178 (1996) 945–950.
801. C. Sjoholm and G. Antranikian, Thermococcus amylase and pullulanase, EU Patent WO 9523852 (1995).
802. T. Kobayashi, Y. S. Kwak, T. Akiba, T. Kudo, and K. Horikoshi, *Thermococcus profundus* sp. nov. A new hyperthermophilic archeon isolated from a deep-sea hydrothermal vent, *Syst. Appl. Microbiol.*, 17 (1994) 232–236.
803. S. Varavinit, N. Chaokasem, and S. Shobsngob, Immobilization of thermostable alpha-amylase, *Sci. Asia*, 28 (2002) 247–251.
804. G. Dey, V. Nagpal, and R. Banerjee, Immobilization of α-amylase from *Bacillus circulans* GRS 313 on coconut fiber, *Appl. Biochem. Biotechnol.*, 102–103 (2002) 303–314.
805. S. D. Shewale and A. B. Pandit, Hydrolysis of soluble starch using *Bacillus licheniformis* α-amylase immobilized on superporous CELBEADS, *Carbohydr. Res.*, 342 (2007) 997–1008.
806. J. P. Chen, Y. M. Sun, and D. H. Chu, Immobilization of α-amylase to a compatible temperature-sensitive membrane for starch hydrolysis, *Biotechnol. Prog.*, 14 (1998) 473–478.

807. M. I. G. Siso, M. Graber, J. S. Condoret, and D. Combes, Effect of diffusional resistance on the action pattern of immobilized alpha-amylase, *J. Chem. Technol. Biotechnol.*, 48 (1990) 185–200.
808. B. T. Hofreiter, K. L. Smiley, J. A. Boundy, C. L. Swanson, and R. J. Fecht, Novel modification of cornstarch by immobilized α-amylase, *Cereal Chem.*, 55 (1978) 995–1006.
809. Z. Konsuola and M. Liakopoulou-Kyriakides, Starch hydrolysis by the action of an entrapped in alginate capsules α-amylase from *acillus subtilis*, *Process Biochem.*, 41 (2006) 343–349.
810. T. S. Lai and M. J. Lan, Some properties of covalently bound immobilized α-amylase and the effect of α-amylolysis on starch saccharification, *J. Chin. Biochem. Soc.*, 5(1), (1976) 7–12.
811. A. L. Khurana and C. T. Ho, Imnnobilization of amylases on silica support to study breakdown products of potato starch by HPLC, *J. Liq. Chromatogr.*, 12 (1989) 1669–1677.
812. K. Hoshino, M. Taniguchi, Y. Netsu, and M. Fujii, Repeated hydrolysis of raw starch using amylase immobilized on a reversibly soluble-insoluble carrier, *J. Chem. Eng. Jpn.*, 22 (1989) 54–59.
813. L. Cong, R. Kaul, U. Dissing, and B. Mattiasson, A model study on Eudargit and polyethyleneimine as soluble carriers of α-amylase for repeated hydrolysis of starch, *J. Biotechnol.*, 42 (1995) 75–84.
814. H. Rozie, W. Somers, K. Van't Riet, F. M. Romouts, and J. Visser, Crosslinked potato starch as an affinity adsorbent for bacterial α-amylase, *Carbohydr. Polym.*, 15 (1991) 349–365.
815. A. Konieczny-Molenda, A. Walaszek, P. Kochanowski, R. Bortel, and P. Tomasik, Immobilization of alpha-amylase on polymeric supports, *Chem. Eng. J.*, 146 (2009) 515–519.
816. T. Kimura, M. Ogata, H. Kobayashi, M. Yoshida, K. Oishi, and T. Nakakuki, Continuous production of maltotetraose using a dual immobilized enzyme system abd maltoteraose-forming amylase and pullulanase, *Biotechnol. Bioeng.*, 36 (1990) 790–796.
817. P. Q. Flor and Y. Hayashida, Continuous production of high glucose syrup by immobilized amylase, *J. Biotechnol. Bioeng.*, 25 (1983) 1973–1980.
818. J. P. Lenders and R. R. Crichton, Thermal stabilization of amylolytic enzymes by covalent coupling to soluble polysaccharide, *Biotechnol. Bioeng.*, 26 (1984) 1343–1351.
819. M. Shimada, F. Kokubun, and Y. Nakamura, Immobilized α-amylase with high activity, *Sen'i Gakkaishi*, 40 (1984) T406–T410.
820. Z. Zhao, Continuous production of glucose with immobilized amylase on spongy microporous polystyrene, *Zhongguo Nangzao*, 1(5), (1982) 26–30. *Chem. Abstr.*, 99 (1983) 120669r.
821. D. H. Strumeyer, A. Constantinides, and J. Freudenberger, Preparation and characterization of α-amylase immobilized on collagen membranes, *J. Food Sci.*, 39 (1974) 498–502.
822. G. Ongen, G. Yilmaz, R. O. J. Jongboom, and H. Fell, Encapsulation of α-amylase in a starch matrix, *Carbohydr. Polym.*, 50 (2002) 1–5.
823. Y. Quan, C. Zhao, and Y. Liu, Immobilization of α-amylase by microencapsulation, *Jiangsu Shiyou Huagong Xueyuan Xuebao*, 9(2), (1997) 33–36. 52; *Chem. Abstr.*, 128 (1998) 125530.
824. J. F. Kennedy and C. A. White, Stability and kinetic properties of a magnetic immobilized α-amylase K, *Starch/Stärke*, 31 (1979) 375–381.
825. Inoue-Japax Research Inc., Enzyme reaction under magnetic field, Jpn. Kokai Tokkyo Koho 80,148,090 (1980); *Chem. Abstr.*, 94 (1980) 116880g.
826. Tajwan Sugar Corp., Production of refined maltose from tapioca starch, *Tajwan Sugar*, 27(5), (1980) 165–166.
827. Nikken Chemicals Co., Hydrolyzed starches, Jpn. Kokai Tokkyo Koho 81 52,000 (1981); *Chem. Abstr.*, 95 (1981) 99656q.
828. E. Lee, The action of sweet potato β-amylase on glycogen and amylopectin. Formation of a novel limit dextrin, *Arch. Biochem. Biophys.*, 146 (1971) 488–492.
829. K. Myrbäck and G. Neumuller, Amylases and the hydrolysis of starch and glycogen, in J. Sumner and K. Myrbäck (Eds.), *The Enzymes*, Vol. 1, Pt. 1.
830. G. A. van Klinkenberg, The separation and action of two malt amylases; the relation of starch to glycogen, *Proc. Acad. Sci. Amsterdam*, 34 (1931) 893–905.

831. K. H. Meyer, P. Bernfeld, P. Guertler, and G. Noelting, Recherches sur l'amidon. XXXIX. L'amidon de mais, *Helv. Chim. Acta*, 31 (1948) 108–110.
832. H. Henriksnas and T. Lovgren, Chain length distribution of starch hydrolyzate after α- or β-amylase action, *Biotechnol. Bioeng.*, 20 (1978) 1303–1307.
833. K. Myrbäck and E. Sihlbom, Wirkung der Malzamylasen auf niedrigmolekulare Hydrolyseprodukten der Stärke, *Arkiv Kemi*, 1(1), (1949) 1–16.
834. C. G. Bohnenkamp and P. J. Reilly, Use of immobilized glucoamylase-β-amylase and fungal amylase mixtures to produce high maltose syrup, *Biotechnol. Bioeng.*, 22 (1980) 1753–1758.
835. M. Abdullah, E. Lee, and W. Whelan, Requirement of rabbit-muscle glycogen phosphorylase for primer, *Biochem. J.*, 97 (1965) 10P.
836. J. Fukumoto, T. Yamamoto, and K. Ichikawa, Bacterial saccharogenic amylase, *Symp. Enzyme Chem. Jpn.*, 7 (1952) 104–105.
837. S. Miyoshi, M. Higashihara, and S. Okada, Action of bacterial β-amylase on raw starch (in Japanese), *Denpun Kagaku*, 33 (1986) 238–243.
838. M. Higashihara, S. Miyoshi, and S. Okada, Action of β-amylase on raw starch (in Japanese), *Denpun Kagaku*, 34 (1987) 106–112.
839. O. F. Stamberg and C. H. Bailey, Action of wheat amylases on soluble starch, *J. Biol. Chem.*, 126 (1938) 479–488.
840. B. Filipowitz, Influence of proteins and protein derivatives on the enzymic hydrolysis of starch, *Biochem. J.*, 25 (1931) 1874–1884.
841. C. S. Hanes, Action of the 2 amylases of barley, *Can. J. Res.*, 13B (1935) 185–208.
842. V. D. Martin, N. M. Naylor, and R. M. Hixon, Action of β-amylase from soybeans on various starches, *Cereal Chem.*, 16 (1939) 565–573.
843. M. Samec, Über die Wirkung von α-Malzamylasen am einige Stärkesubstanzen. 5. Über enzymatische Amylolysis in der von M. Samec und E. Waldschmidt-Leite sogenannten Untersuchungreihe, *Z. Physiol. Chem.*, 248 (1937) 117–128.
844. A. Tychowski, Untersuchungen über die Stärke Grenzabbau von Stärken verschiedener Herkunft bei Einwirkung verschiedener Getreideamylasen undUntersuchung der Hydrolyseprodukte, *Biochem. Z.*, 291 (1937) 138–158.
845. A. Tychowski, Hydrolyse der Stärke mit β-Amylase, *Biochem. Z.*, 291 (1937) 247–253.
846. K. H. Meyer, M. Wertheim, and P. Bernfeld, Recherches sur l'amidon. XI. Sur la dextrine residuelle de l'amidon de mais (erythrogranulose), *Helv. Chim. Acta*, 24 (1941) 212–216.
847. C. S. Hanes, Enzymic degradation of starch, *Dept. Sci. Ind. Res., Rept. Food Invest. Board* (1936) 140.
848. S. Peat, S. J. Pirt, and W. J. Whelan, Enzymic synthesis and degradation of starch. XV. β-Amylase and the constitution of amylose, *J. Chem. Soc.* (1952) 705–713.
849. J. M. G. Cowie, I. T. Fleming, C. T. Greenwood, and D. J. Manners, Physicochemical studies on starches. IX. The mechanism of the β-amylolysis of amylose and the nature of the β-limit dextrins, *J. Chem. Soc.* (1958) 697–702.
850. K. Myrbäck and W. Thorsell, Hydrolyse longkettiger Moleküle; enzymatisch Spaltung der Amylose, *Svensk Kim. Tids.*, 54 (1942) 50–60.
851. J. Fukumoto and W. Thorsell, Studies on bacterial amylase. VI. Liquefaction and saccharification (in Japanese), *J. Agr. Chem. Soc. Jpn.*, 20 (1944) 23–26.
852. D. N. Klimovskii and V. I. Rodzevich, Hydrolysis of starch by amylase from different sources (in Russian), *Biokhimiya*, 14 (1949) 26–34.
853. S. Ueda, Phosphoric acid in starch in relation to amylase action (in Japanese), *Kagaku to Seibutsu*, 4 (1966) 701–704.
854. T. N. Evreinova, T. A. Shubert, and M. N. Nestyuk, Coacervates and enzymes. Protein–carbohydrate coacervates and β-amylase (in Russian), *Dokl. Akad. Nauk SSSR*, 105 (1955) 137–140.

855. E. Macovschi, S. S. Vasu, and M. Cirsteanu, Action of urease and β-amylase on urea and on starch in the presence of certain glucoprotein coacervates (in Roumanian), *Studii Cerc. Acad. Rep. Pop. Romie, Inst. Biochim. Studii Cerc. Biochim.*, 1 (1958) 297–311.
856. W. J. Olson, B. A. Burkhart, and A. D. Dickson, Inactivation of α- and β-amylase in aqueous malt extracts, *Cereal Chem.*, 20 (1943) 126–138.
857. G. Jiang and S. Zhang, Manufacture and quality control of sweet potato maltose, *Shipin Kexue*, 152 (1992) 8–11. *Chem. Abstr.*, 118 (1993) 100696.
858. T. Baba and K. Kainuma, Partial hydrolysis of sweet potato starch with β-amylase, *Agric. Biol. Chem.*, 51 (1987) 1365–1371.
859. M. Z. H. Bhuiyan and J. M. V. Blanshard, Studies on the biochemical morphology of maize starch, *Bangladesh J. Sci. Ind. Res.*, 22 (1987) 81–88.
860. Higashihara, T. Maruo, and Y. Hatanaka, Studies on retrogradation of raw corn starch treated with bacterial β-amylase, *Kagaku to Kogyo*, 66 (1992) 101–105.
861. M. Batsalova, Susceptibility of wheat starch to the action of β-amylase (in Bulgarian), *Nauch. Tr. Vissh. Selskostop. Inst. Vasil Kolarov, Plovdiv*, 25(1), (1980) 61–68.
862. B. Örtenblad and K. Myrbäck, Amylasewirkung auf oxidierte Stärke, *Biochem. Z.*, 307 (1941) 129–131.
863. A. Rowe and C. Weill, Inhibition of β-amylase by ascorbic acid, *J. Am. Chem. Soc.*, 81 (1959) 921–924.
864. J. Vandor and I. F. Laszlo, β-Diastase. I. Correlation of diastase stability with temperature and pH, *Mag. Kem. Folyoirat*, 56 (1950) 373–377.
865. A. W. Rowe and C. E. Weill, The inhibition of β-amylase by ascorbic acid. II, *Biochim. Biophys. Acta*, 65 (1962) 245–250.
866. E. Purgatto, F. M. Lajolo, J. R. Oliveira de Nascimento, and B. R. Cordenunei, Inhibition of β-amylase activity, starch degradation and sucrose formation by indole-3-acetic acid during banana ripening, *Planta*, 212 (2001) 823–828.
867. M. Nakamura, Failure of cyanide to inhibit β-amylase, *J. Agr. Chem. Soc. Jpn.*, 26 (1952) 497–502.
868. U. Misra and D. French, Inhibition and action pattern of β-amylase in presence of maltose, *Biochem. J.*, 77 (1960) 1P.
869. J. Baker, Properties of amylases from midgets of larvae *Sitophilus zeamais* and *Sitophilus granaries*, *Insect Biochem.*, 13 (1983) 421–428.
870. C. O. Bjoerling, α- und β-Amylasen, *Svensk. Farm. Tids.*, 39 (1935) 453–457, 470–474, 485–492, 501–504.
871. K. V. Giri, Further studies on the hydrolysis of starch by sweet-potato amylase, *J. Indian Chem. Soc.*, 15 (1938) 249–262.
872. K. Myrbäck, Über Grenzdextrine und Stärke, *Biochem. Z.*, 297 (1938) 160–171.
873. Y. Tokuoka, Koji amylase. IX. Existence of β-amylase, *J. Agr. Chem. Soc. Jpn.*, 13 (1937) 586–594.
874. S. Yamakawa, M. Yoshimoto, T. Kumagaya, S. Tada, K. Tokimura, H. Tonoue, K. Ikeda, and M. Ichiki, The preparation of β-amylase extract from sugarcane and its use in maltose syrup production, Jpn. Kokai Tokkyo Koho 20050461110 (2005); *Chem. Abstr.*, 142 (2005) 235494.
875. E. Sarikaya, T. Higasa, M. Adachi, and R. Mikami, Comparison of degradation abilities of α- and β-amylases on raw starch granules, *Process Biochem.*, 35 (2000) 711–715.
876. M. V. Hedge, F. Roy, and P. N. Joshi, A new method for the preparation of β-amylase from sweet potato, *Prep. Biochem.*, 9 (1979) 71–84.
877. P. Colman and B. Matthews, Symmetry, molecular weight and crystallographic data for sweet potato β-amylase, *J. Mol. Biol.*, 60 (1971) 163–164.
878. M. Samec, Enzymatische Amylolysis. IV. Wirkung der β-Amylase auf einige Stärkesubstanzen, *Z. Physiol. Chem.*, 236 (1935) 103–118.

879. O. E. Stamberg, Wheat starch. IV. Fractionation and amylase hydrolysis, *Cereal. Chem.*, 17 (1940) 372–378.
880. J. M. Newton, F. F. Farley, and N. M. Naylor, The use of soybean β-amylase to follow the modification of starch, *Cereal Chem.*, 17 (1940) 342–355.
881. P. C. Lu and S. C. Ma, Action of wheat beta amylase on lotus-rhizome starch, *J. Chin. Chem. Soc.*, 10 (1943) 194–199. *Chem. Abstr.*, 38 (1944) 3299.
882. Y. F. Ma, J. K. Ellington, D. E. Evans, S. J. Logue, and P. Langridge, Removal of the four C-terminal glycine-rich repeats enhances the thermostability and substrate binding affinity of barley β-amylase, *Biochemistry*, 39 (2000) 13350–13355.
883. H. Volker, R. Buckow, and D. Knorr, Catalytic activity of β-amylase from barley in different pressure/temperature domains, *Biotechnol. Prog.*, 21 (2005) 1632–1638.
884. P. Bernfeld and P. Guertler, Recherches sur l'amidon. XXXVIII. Methode perfectionnee de degradation β-amylatique de l'amylose et de l'amylopectin, *Helv. Chim. Acta*, 31 (1948) 106–108.
885. J. Voss, Über den enzymatischen Abbau von Stärken, *Starch/Staerke*, 4 (1952) 192–200.
886. M. Kato, K. Hiromi, and Y. Morita, Purification and kinetic studies of wheat bran β-amylase. Evaluation of subsite affinities, *J. Biochem.*, 75 (1974) 563–576.
887. R. Shinke and N. Mugibayashi, Types of amylases in rice grains, *Agr. Biol. Chem.*, 37 (1973) 2437–2438.
888. D. C. Doehlert, S. H. Duke, and L. Anderson, β-Amylases from alfalfa (*Medicago sativa*. L) roots, *Plant Physiol.*, 69 (1982) 1096–1102.
889. J. E. Kruger, Relation between free and latent beta-amylases in wheat, *Cereal Chem.*, 47 (1970) 79–85.
890. M. Hoshino, Y. Hirose, K. Sano, and K. Mitsugi, Adsorption of microbial β-amylase on starch, *Agric. Biol. Chem.*, 39 (1975) 2415–2416.
891. J. Fukumoto, T. Yamamoto, and K. Ichikawa, Starch hydrolysis by bacterial amylases of two different types, *Symp. Enzyme Chem. Jpn.*, 8 (1953) 40–46.
892. P. S. Rao and K. V. Giri, Further studies on mechanism of β-amylase inhibition by vitamin C, *Proc. Indian Acad. Sci.*, 28 (1948) 71–82.
893. R. Geoffroy, Starch and diastase in flour, *Ind. Agr. Aliment.*, 69 (1952) 411–420. *Chem. Abstr.*, 47 (1953) 34866.
894. H. Outtrup and B. E. Norman, Properties and application of a thermostable maltogenic amylase produced by a strain of *Bacillus* modified by recombinant-DNA techniques, *Starch/Stärke*, 36 (1984) 405–411.
895. A. V. Gasparyan, V. A. Abelyan, and E. K. Afrikyan, Some characteristics of β-amylase produced by *Bacillus polymyxa* (n Russian), *Biokhimiya*, 57 (1992) 856–861.
896. C. B. Sohn, S. M. Lee, M. H. Kim, J. H. Ko, K. S. Kim, J. E. Chang, Y. K. Ahn, and C. H. Kim, Purification and characterization of β-amylase from *Bacillus polymyxa* No. 26-1, *J. Food Sci.*, 61 (1996) 230–234.
897. R. Ahmad, J. Iqbal, M. A. Baig, R. H. Jafri, M. A. Qadee, and J. Khan, Raw starch digestion by *Bacillus polymyxa* PCSIR 90 beta-amylase, *Pak. J. Sci.*, 47 (1995) 85–91.
898. S. Ueda and J. J. Marshall, Raw starch digestion by *Bacillus polymyxa* β-amylase, *Starch/Stärke*, 32 (1980) 122–125.
899. R. R. Rani, S. C. Jana, and G. Nanda, Saccharification of indigenous starches by β-amylase of *Bacillus megaterium*, *World J. Microbiol. Biotechnol.*, 10 (1994) 691–693.
900. R. Shinke, T. Nanmori, K. Aoki, H. Nishira, K. Nishikawa, K. Yamane, and K. Nishida, Biomass utilization by microbial agents. III. Raw starch digestion by microbial and plant β-amylases, *Kobe Daigaku Nogakubu Kenkyu Hokoku*, 16 (1984) 309–316.
901. Y. Takasaki and Y. Takahara, Adsorption of bacterial β-amylase, Jpn. Kokai Tokkyo Koho, 76 70,873 (1976); *Chem. Abstr.*, 85 (1976) 141332b.

902. Y. I. Henis, T. Yaron, R. Lamed, J. Rishpon, E. Sahar, and E. Katchalski-Katzir, Mobility of enzymes on insoluble substrates: The β-amylase–starch gel system, *Biopolymers*, 27 (1988) 123–138.
903. H. Okazaki, Starch saccharyfying enzyme from *Aspegillus oryzae* (in Japanese), Japanese Patent 49 ('56).
904. H. Okazaki, Amylases of various species of molds. VII. Purification and properties of saccharogenic amylase of *Aspergillus oryzae*, *Nippon Nogei Kagaku Kaishi*, 28 (1954) 48–51.
905. H. Okazaki, Amylases of various species of molds. VIII. Purification and properties of saccharogenic amylase of *Aspergillus oryzae*, *Nippon Nogei Kagaku Kaishi*, 28 (1954) 51–56.
906. H. Okazaki, Amylases of various species of molds. IX. Preparation and properties of saccharogenic amylase of *Aspergillu oryzae*. 3, *Nippon Nogei Kagaku Kaishi*, 29 (1955) 181–185.
907. H. Okazaki, Amylases of various species of molds. X. Amylase and maltase actions of saccharogenic amylase of *Aspergillus oryzae*. 1, *Nippon Nogei Kagaku Kaishi*, 29 (1955) 273–277.
908. H. Okazaki, Amylases of various species of molds. XI. Amylase and maltase actions of saccharogenic amylase, *Nippon Nogei Kagaku Kaishi*, 29 (1955) 277–283.
909. I. Yamazaki, S. Ueda, S. Hayashida, and S. Kayashima, Black koji amylases. V. Effect of phosphorus in potato starch on amylase actions, *Hakko Kagaku Zasshi*, 47 (1969) 753–758.
910. G. J. Shen, B. C. Saha, Y. E. Lee, L. Bhatnagar, and J. G. Zeikus, Purification and characterization of a novel thermostable β-amylase from *Clostridium thermosulphurogenes*, *Biochem. J.*, 254 (1988) 835–840.
911. J. G. Zeikus and H. H. Hyun, Thermostable β-amylase, EU Patent WO 86 01,832 (1986); *Chem. Abstr.*, 104 (1986) 223649k.
912. B. C. Saha, G. J. Shen, and J. G. Zeikus, Behavior of a novel thermostable β-amylase on raw starch, *Enzyme Mikrob. Technol.*, 9 (1987) 598–601.
913. M. Hagishihara and S. Miyoshi, Raw starch digestion with β-amylase. III. Raw starch digestion with β-amylase, effect of debranching enzymes (in Japanese), *Kagaku to Kogyo*, 63 (1989) 363–367.
914. R. R. Ray and R. Chakraverty, Extracellular β-amylase from *Syncephalastrum racemosum*, *Mycol. Res.*, 102 (1998) 1563–1567.
915. K. Martensson, Preparation of an immobilized two-enzyme system, β-amylase-pullulanase, on an acryslic copolymer for the conversion of starch to maltose. I. Preparation and stability of immobilized β-amylase, *Biotechnol. Bioeng.*, 16 (1974) 567–577.
916. P. Vretblad and R. Axen, Preparation and properties of an immobilized barley β-amylase, *Biotechnol. Bioeng.*, 15 (1973) 783–794.
917. K. Martensson, Preparation of an immobilized two-enzyme system, β-amylase-pullulanase, on an acryslic copolymer for the conversion of starch to maltose. II. Coupling of the enzymes and use in a packed bed column, *Biotechnol. Bioeng.*, 16 (1974) 579–591.
918. T. Noda, S. Furuta, and S. Suda, Sweet potato β-amylase immobilized on chitosan beads and its application in the semi-continuous production of maltose, *Carbohydr. Polym.*, 44 (2001) 189–195.
919. F. Shiraishi, K. Kawakami, T. Kojima, A. Yuasa, and K. Kusunoki, Maltose production from soluble starch by β-amylase and debranching enzyme immobilized on ceramic monlith (in Japanese), *Kagaku Kogaku Ronbunshu*, 14 (1988) 288–294.
920. T. Harada, Enzymic conversion of starch into maltose syrup using β-amylase and *Pseudomonas isoamylse*, *Methods Carbohydr. Chem.*, 10 (1994) 245–248.
921. F. Shiraishi, Maltose production from soluble starch by β-amylase and debranching enzyme immobilized on ceramic monolith (in Japanese), *Kagaku Kogaku*, 54 (1990) 398–400.
922. V. Ramesh and C. Singh, Studies on kinetics and activation of doluble and immobilized sweet potato β-amylase, *J. Mol. Catal.*, 10 (1981) 341–355.
923. A. O. Kolawole, J. O. Ajele, and R. Sirdeshmukh, Purification and characterization of alkaline stable β-amylase in malted African millet (*Eleusine coracana*) seeds, *Process Biochem.*, 46 (2011) 2178–2186.

924. Y. Yamasaki, β-Amylase in germinating millet seeds, *Phytochemistry*, 64 (2003) 935–939.
925. M. L. Fan, Partial purification and properties of potato beta-amylase, *Taiwania*, 20 (1975) 132–138.
926. C. P. Sandiford, R. D. Tee, and A. J. Taylor, The role of cereal and fungal amylases in cereal flour hypersensitivity, *Clin. Exp. Allergy*, 24 (1994) 549–557.
927. A. Balls, M. Walden, and R. Thompson, A crystalline β-amylase from sweet potato, *J. Biol. Chem.*, 173 (1948) 9–19.
928. Y. Morita, S. Aibara, H. Yamashita, F. Yagi, T. Suganuma, and K. Hiromi, Crystallization and preliminary x-ray investigation of soybean β-amylase, *J. Biochem.*, 77 (1975) 343–351.
929. G. Chapman, J. Pallas, and J. Mendicino, The hydrolysis of maltodextrins by a β-amylase isolated from leaves of *Vicia fabia*, *Biochim. Biophys. Acta*, 276 (1972) 491–507.
930. A. Kohno, T. Nanmori, and R. Shinke, Purification of β-amylase from alfalfa (*Medicago sativa* L.) seeds, *J. Biochem.*, 105 (1989) 232–233.
931. K. Subbramaiah and R. Sharma, Characterization of β-amylase from *Synopsis alba* cotyledons, *Phytochemistry*, 29 (1990) 1417–1419.
932. H. S. Kwan, K. H. So, K. Y. Chan, and S. C. Cheng, Purification and properties of β-amylase from *Bacillus circulans* S31, *World J. Microbiol. Biotechnol.*, 10 (1994) 597–598.
933. E. Berka, E. Berka, M. Rey, and P. Ramaiya, Bacillus licheniformis chromosome, US Patent Appl. 20100064393 (2010).
934. D. E. Hensley, K. L. Smaley, J. A. Boundy, and A. L. Lagoda, Beta-amylase production by *Bacillus polymyxa* on a corn steep-starch-salts medium, *Appl. Environ. Microbiol.*, 39 (1980) 678–680.
935. M. V. Swamy, M. Sai Ram, and G. Senayya, β-Amylase from *Clostridium thermocellum* SS8 – Thermophilic anaerobic, cellulolutic bacterium, *Lett. Appl. Microbiol.*, 18 (1994) 301–304.
936. H. H. Hyun and J. G. Zeikus, General biochemical characterization of thermostable extracellular β-amylase from *Clostridium thermohydrosulfirogenes*, *Appl. Environ. Microbiol.*, 49 (1985) 1162–1167.
937. P. R. M. Reddy, M. V. Swamy, and G. Seenyya, Purification and characterization of thermostable β-amylase and pullulanase from high yielding *Clostridium thermosulfurogenes* SV2, *World J. Microbiol. Biotechnol.*, 14 (1998) 89–94.
938. A. Nipkow, G. Shen, and J. Zeikus, Continuous production of thermostable β-amylase with *Clostridium thermosulfurogenes*. Effect of culture conditions and metabolite levels on enzyme synthesis and activity, *Appl. Environ. Microbiol.*, 55 (1989) 689–694.
939. W. M. Fogarty and C. T. Kelly, Recent advances in microbial amylases, In: W. M. Fogarty and C. T. Kelly, (Eds.) *Microbial Enzymes and Biotechnology*, 2nd edn. Elsevier Science Publ, London, 1990, pp. 71–132.
940. F. J. C. Odibo, N. Okafor, M. U. Tom, and C. A. Oyeka, Purification and characterization of a β-amylase of *Hendresonula toruloidea*, *Starch/Stärke*, 44 (1992) 192–195.
941. R. R. Ray and R. Chakravorty, Extracellular β-amylase from *Syncephalastrum racemosum*, *Mycol. Res.*, 102 (1998) 1563–1567.
942. R. N. Tharanathan, S. V. Paramahans, and J. A. K. Tareen, In vitro digestibility of native starch granules of samsi and sanwa, *Starch/Stärke*, 35 (1983) 235–236.
943. S. Hizukuri, Raw starch digesting activity and kinetic properties of glucoamylases (in Japanese), *Denpun Kagaku*, 34 (1987) 98–105.
944. J. J. Kelly and D. H. Alpers, Properties of human intestinal glucoamylase, *Biochim. Biophys. Acta*, 315 (1973) 113–122.
945. H. Masuda, M. Murata, T. Takahashi, and S. Sugawara, Purification and properties of glucoamylase from sugar beet cells in suspension culture, *Plant Physiol.*, 88 (1988) 172–177.
946. I. D. Fleming, Amyloglucosidase: α-1,4-gluucans glucohydrolase, In: J. A. Radley, (Ed.), *Starch and Its Derivatives*, 4th edn. Chapman Hall Ltd., London, 1968, p. 498.

947. D. J. Wijbeng, G. Beldman, A. Veen, and D. J. Binnema, Production of native starch degrading enzymes by a *Bacillus lentus* strain, *Appl. Microbiol. Biotechnol.*, 35 (1991) 180–184.
948. H. H. Hyun and J. G. Zeikus, General biochemical characterization of thermostable pullulanase and glucoamylase from *Clostridium thermohydrosulfuricum*, *Appl. Environ. Microbiol.*, 49 (1985) 1168–1173.
949. U. Specka, F. Meyer, and G. Antranikian, Purification and properties of a thermoactive glucoamylase from *Clostridium thermosaccharolyticum*, *Appl. Environ. Microbiol.*, 57 (1991) 2317–2323.
950. M. O. Ilori, O. O. Amund, and O. Omidjii, Effect of carbon and nitrogen sources on glucoamylase production in *Lactobacillus brevis*, *Folia Microbiol.*, 41 (1996) 339–340.
951. P. M. Taylor, E. J. Napier, and L. D. Fleming, Some properties of a glucoamylase produced by the thermophilic fungus *Humicola huminosa*, *Carbohydr. Res.*, 61 (1978) 301–308.
952. H. Bender, A bacterial glucoamylase degrading cyclodextrins, *Eur. J. Biochem.*, 115 (1985) 287–291.
953. H. Iizuka and S. Mineki, Studies on the genus *Monascus*, *J. Gen. Appl. Microbiol.*, 23 (1977) 217–230.
954. B. K. Dalmia and Z. L. Nikolov, Characterization of glucoamylase adsorption to raw starch, *Enzyme Microb. Technol.*, 13 (1991) 982–990.
955. S. Towprayoon, Y. Fujio, and S. Ueda, Effect of adsorption of an inhibitory factor on raw starch hydrolysis by glucoamylase, *World J. Microbiol. Biotechnol.*, 6 (1990) 400–403.
956. S. Towprayoon, B. C. Saha, Y. Fujio, and S. Ueda, Some characteristic of a raw starch digestion inhibitory factor from *Aspergillus niger*, *Appl. Microbiol. Biotechnol.*, 29(948), (1988) 289–291.
957. W. M. Fogarty and C. P. Benson, Purification and characterization of a thermophilic amyloglucosidase from *Aspergillus niger*, *Eur. J. Appl. Microbiol. Biotechnol.*, 18 (1983) 271–278.
958. J. S. Smith and D. R. Lineback, Hydrolysis of native wheat and corn starch granules by glucoamylases from *Aspergillus niger* and *Rhizopus nivea*, *Starch/Stärke*, 28 (1976) 243–249.
959. N. Arai and S. Amano, Some rpoperties of commercial preparations of glucoamylases. (in Japanese), *Denpun Kagaku*, 27 (1980) 146–150.
960. K. Tsekova, A. Vicheva, and A. Tzekova, Enhanced thermostability of glucoamylase from *Aspergillus niger*, *Dokl. Bulg. Akad. Nauk*, 50(7–8), (1997) 53–56.
961. G. Konieczny-Janda, Enzymic modification of starch. Glucoamylase as an example as starch treatment, *GBF Monogr. Ser.*, 11 (1988) 143–157.
962. B. C. Saha and S. Ueda, Inhibition of raw starch digestion by one glucoamylase from black *Aspergillus* at high enzyme concentration. III. Raw starch digestion inhibitory factor in onbe glucoamylase preparation of black *Aspergillus*, *Agric. Biol. Chem.*, 47 (1983) 2773–2779.
963. B. C. Saha and S. Ueda, Production and characteristics of inhibitory factor of raw starch digestion from *Aspergillus niger*, *Appl. Microbiol. Biotechnol.*, 27 (1984) 341–346.
964. B. C. Saha and S. Ueda, Raw starch degradation by glucoamylase II of black *Aspergillus* (in Japanese), *Denpun Kagaku*, 31 (1984) 8–13.
965. Y. Xu, Q. Su, Y. He, and Y. Li, Studies on the high-yielding glucoamylase mutants *Aspergillus niger* UV_{11}-33and *A. niger* UV_{11}-22. II. Fermentative conditions in the laboratory and enzyme characteristics of mutants, *Gongye Weishengwu*, 17(1), (1987) 20–25. *Chem. Abstr.*, 106 (1987) 194718r.
966. J. A. Smith and J. R. Frankiewicz, Glucoamylase by fermentation, Brazilian Patent 77 04,255 (1978); *Chem. Abstr.*, 89 (1978) 127775z.
967. C. B. Sohn and Y. J. Park, Studies on the raw starch saccharyfying enzyme from *Aspergillus niger* and its mutants, *Nongop Kisul Yongu Pogo (Chungnam Taehakkyo)*, 10(1), (1983) 166–185. *Chem. Abstr.*, 100 (1983) 47545e.
968. H. Guan, Z. Yan, and S. Zhang, Purification and properties of raw starch-digesting glucoamylase from an *Aspergillus niger* mutant, *Shengwu Huaxue Yu Shengwu Wuli Xuebao*, 25 (1993) 53–460. *Chem. Abstr.*, 122 (1995) 26404.

969. S. Ueda, R. Ohba, and S. Kano, Fractionation of the glucoamylase system from black koji mold and the effects of adding isoamylse and α-amylase on amylolysis by glucoamylase fractions, *Starch/Stärke*, 26 (1974) 374–378.
970. M. Dreimane and R. Zeltsin, Use of amylolytic enzymes of the yeasts *Endomycopsis fibuligera* R313 for saccharyfication of starch-containing raw material (in Russian), In: D. Ya. Kreslinya, (Ed.), *Fermentatsiya*, Zinantne, Riga, 1974, pp. 18–21.
971. A. R. Kusnadi, C. Ford, and Z. L. Nikolaev, Functional starch binding domain of *Aspergillus* glucoamylase I in *Eschericia coli*, *Gene*, 127 (1999) 193–197.
972. S. K. Soni, M. V. Rao, and D. Das, Studies on glucoamylase produced from *Aspergillus awamori* (NRRL-3112) and their effect on saccharification of potato starch, *Indian J. Exp. Biol.*, 33 (1995) 957–961.
973. S. Hayashida and M. Goto, The mechanism for digestion of crystalline carbohydrates by amylase and cellulose based on the theory of the affinity site (in Japanese), *Gakugei Zasshi, Kyushu Daigaku Nogakubu*, 47 (1993) 73–83.
974. M. Goto, K. Tanigawa, W. Kanlayakrit, and S. Hayashida, The mechanism of binding of glucoamylase I from *Aspergillus awamori* var. kawachi to cyclodextrin and raw starch, *Biosci. Biotechnol. Biochem.*, 58 (1994) 49–54.
975. S. Hayashida, Selective submerged productions of three types of glucoamylases from by a black koji mold, *Agric. Biol. Chem.*, 39 (1975) 2093–2099.
976. O. V. Grabovska, O. M. Maidanets, and N. I. Shtangeeva, Intensification of enzymic hydrolysis of starch by nechanical homogenization (in Ukrainian), *Tsukor Ukr.* (1–2), (2005) 47–48, 52.
977. P. Q. Flor and S. Hayashida, Production and characteristics of raw starch-digesting glucoamylase O from a protease-negative, glycosidase-negative *Aspergillus awamori* var. kawchi mutant, *Appl. Environ. Microbiol.*, 45 (1983) 905–912.
978. K. B. Svenson and M. R. Sierks, Enzymatic saccharification of starch, using a genetically engineered glucoamylase, EU Patent WO 9200381 (1992); *Chem. Abstr.*, 116 (1992) 126987
979. Y. Hata, H. Ishida, Y. Kojima, E. Ichikawa, A. Kawata, K. Suginami, and S. Imayasu, Comparison of two glucoamylases produced by *Aspergillus oryzae* in solid-state culture (koji) and in submerged culture, *J. Ferment. Bioeng.*, 84 (1997) 532–537.
980. T. S. Yu, T. H. Kim, and C. Y. Joo, Purification and characteristics of glucoamylase in *Aspergillus oryzae* NR 3-6 isolated from traditional Korean Nuruk, *J. Microbiol. (Seoul)*, 37(2), (1999) 80–85.
981. C. H. Song, H. U. Bong, and K. Y. Jin, Effect of temperature on the reaction rate in the enzymatic hydrolysis of starch by glucoamylase, *Choson Minjujuul Inmin Konghwaguk Kwahagwon Tongbo* (2), (2005) 37–39. *Chem. Abstr.*, 143 (2005) 265683.
982. C. H. Song, H. U. Bong, and K. Y. Jin, Effect of the concentrations of substrate and enzyme on the reaction rate in hydrolysis by glucoamylase, *Choson Minjujuul Inmin Konghwaguk Kwahagwon Tongbo* (1), (2005) 46–48. *Chem. Abstr.*, 143 (2005) 417972.
983. L. Ray and S. K. Majumdar, Enzymic hydrolysis of starch to glucose, and immobilization of amyloglucosidase in calcium alginate gel, *Res. Ind.*, 39 (1994) 109–112.
984. P. Manjunath and M. R. R. Rao, A comparative studies on glucoamylase from three fungal sources, *J. Biosci.*, 1 (1979) 409–425.
985. T. Takahashi, N. Inokuchi, and M. Irie, Purification and characterization of a glucoamylase from *Aspergillus saitoi*, *J. Biochem. (Tokyo)*, 89 (1981) 125–134.
986. T. Takahashi, Y. Ikegami, M. Irie, and E. Nakao, Different behavior towards raw starch of two glucoamylases from *Aspergillus saitoi*, *Chem. Pharm. Bull.*, 38 (1990) 2780–2783.
987. A. Ghose, B. S. Chatterjee, and A. Das, Characterization of glucoamylase from *Aspergillus terreus* 4, *FEMS Microbiol. Lett.*, 66 (1990) 345–349.
988. T. Noda, Y. Takahata, T. Nagata, and M. Monma, Digestibility of sweet potato raw starches by glucoamylase, *Starch/Stärke*, 44 (1992) 32–35.

989. B. G. Na and C. H. Yang, Purification and characterization of glucoamylase from *Aspergillus shirousamii*, *Hanguk Saenghwa Hakhoechi*, 18(3), (1985) 222–231. *Chem. Abstr.*, 104 (1985) 2639b.
990. T. Takahashi, K. Kato, Y. Ikegami, and M. Irie, Different behavior towards raw starch of three forms of glucoamylase from *Rhizopus* sp, *J. Biochem. (Tokyo)*, 98 (1985) 663–671.
991. W. N. Hou and M. J. Chung, Hydrolysis of various substrated by two forms of the purified glucoamylase from *Rhizopus oryzae*, *Hanguk Sikpum Kwahakhoechi*, 16 (1984) 398–402. *Chem. Abstr.*, 102 (1985) 91961c.
992. A. Tanaka and S. Takeda, Random attack as the hydrolytic reaction mode of oligosaccharides by *Rhizopus* glucoamylase, *Biosci. Biotechnol. Biochem.*, 59 (1995) 1372–1373.
993. S. Nahar, F. Hossein, B. Feroza, and M. A. Halim, Production of glucoamylase by *Rhizopus* sp. in liquid culture, *Pak. J. Bot.*, 40 (2008) 1693–1698.
994. B. C. Saha and S. Ueda, Raw starch adsorption elution and digestion behavior of glucoamylase of *Rhizopus niveus*, *J. Ferment. Technol.*, 61 (1983) 67–72.
995. H. Kondo, H. Nakatani, R. Matsumo, and K. Hiromi, Product distribution in amylase catalyzed hydrolysis of amylose. Comparison of experimental results and theoretical predictions, *J. Biochem.*, 87 (1980) 1053–1070.
996. H. Ishigami, Raw starch-digesting amylase from *Chalaraparadoxa*, *Denpun Kagaku*, 34 (1987) 66–74.
997. K. Kainuma, H. Ishigami, and S. Kobayashi, Isolation of a novel raw starch-digestive amylase from a strain of black moild *Chalara paradoxa*, *Denpun Kagaku*, 32 (1985) 136–141.
998. National Institute of Food Research, Saccharification of starch by *Chalara* enzyme, Jpn. Kokai Tokkyo Koho 59,140,896 (1984); *Chem. Abstr.*, 101 (1984) 209096j.
999. M. Monma and K. Kainuma, Studies on the novel raw starch-digesting amylase obtained from *Chalara paradoxa*. XIII. Preparation of *Chalara paradoxa* glucoamylase modified by fluorescein-isothiocyanate and binding to starch, *Agr. Biol. Chem.*, 52 (1988) 2087–2089.
1000. M. Monma, Studies on properties of *Chalara paradoxa* raw starch digesting amylase (in Japanese), *Denpun Kagaku*, 38 (1991) 45–50.
1001. H. Ishigami, H. Hashimoto, and K. Kainuma, Novel raw starch-degrading amylase obtained from *Chalara paradoxa*. II. Determination of optimum culture conditions for the *Chalara paradoxa* production, *Denpun Kagaku*, 32 (1985) 189–196.
1002. Y. Marlida, N. Saari, Z. Hassan, S. Radu, and J. Bakar, Purification and characterization of sago starch-degrading glucoamylase from *Acremonium* sp. endophytic fungus, *Food Chem.*, 71 (2000) 221–227.
1003. H. Boze, G. Moulin, and P. Galzy, Influence of culture conditions on the cell yield and amylase biosynthesis in continuous culture by *Schwanniomyces castellii*, *Arch. Microbiol.*, 148 (1987) 162–166.
1004. H. Boze, J. B. Guyot, G. Moulin, and P. Galzy, Kinetics of the amyloglucosidase of *Schwanniomycs castellii*, *Yeast*, 5(Spec. Issue), (1989) S117–S121.
1005. F. Sato, M. Okuyama, H. Nakao, H. Mori, S. Chiba, and S. Chiba, Glucoamylase originating from *Schwanniomyces occidentalis* is a typical α-glucosidase, *Biosci. Biotechnol. Biochem.*, 69 (2005) 1905–1913.
1006. I. M. Gracheva, A. I. Sadova, V. P. Gaidenko, and T. V. Brovarets, Optimum conditions for starch enzymic hydrolysis by *Endomycopsis* species 20-9 enzymes (in Russian), *Prikl. Biokhim. Mikrobiol.*, 5 (1969) 282–288.
1007. L. A. Nakhapetyan and I. I. Manyailova, Hydrolysis of concentrated starch solutionsnusing immobilized glucoamylase (in Russian), *Ferment. Spirt. Prom.* (2), (1976) 41–43.
1008. H. Iizuka and S. Mineki, Studies on the genus *Monascus*. II. Substrate specificity of two glucoamylases obtained from *Monascus kaoliang* F-1, *J. Gen. Appl. Microbiol.*, 24(3), (1978) 185–192.
1009. T. E. Abraham, V. P. Sreedharan, and S. V. Ramakrishna, Development of an alternative route for the hydrolysis of cassava flour, *Starch/Stärke*, 41 (1989) 472–476.

1010. S. Z. Zhang, Comparative studies on two forms of glucoamylase from *Monascus rubiginosus* Sato, *Proc. Symp. Acids, Proteins*, 1979 (1980) 402–412. *Chem. Abstr.*, 95 (1981) 57046k.
1011. Y. S. Wang, J. W. Sun, and S. Z. Zhang, Substrate specificity of glucoamylase from *Monascus* sp, *Wei Sheng Wu Hsueh Pao*, 20 (1980) 398–406. *Chem. Abstr.*, 94 (1980) 43228p.
1012. A. Kaji, M. Sato, M. Kobayashi, and T. Maruo, Acid-stable glucoamylase produced in a medium containing sucrose by *Corticum rolfsi* (in Japanese), *Nippon Nogei Kagaku Kaishi*, 50 (1976) 509–517.
1013. M. Sawada, K. Kurosawa, H. Sasaki, and S. Takao, Starch saccharification with *Corticium* glucoamylase, Jpn. Kokai Tokkyo Koho 62,126,989 (1987); *Chem. Abstr.*, 107 (1987) 152902m.
1014. H. J. Kim and S. M. Byun, Purification and characterization of acid-stable glucoamylase, *Hanguk Saenghwa Hakhoe Chi*, 10(3), (1977) 207–218. *Chem. Abstr.*, 90 (1978) 50370z.
1015. Y. Takeda, H. E. H. Mauui, M. Tanida, S. Takao, and S. Chiba, Purification and substrate specificity of gklucoamylase of *Paecilomyces varioti* AHU 9417, *Agric. Biol. Chem.*, 49 (1985) 1633–1641.
1016. B. R. Nielsen, R. I. Nielsen, and J. Lehmbeck, Thermostable glucoamylase of *Talaromyces emersonii* and its use in starch saccharification. US Patent 6255084 (2001); *Chem. Abstr.*, 135 (2001) 89139.
1017. B. R. Nielsen, R. I. Nielsen and J. Lehmbeck, Thermostable glucoamylase of *Talaromyces emersonii* and its use in starch saccharification, EU Patent WO 9928448 (1999); *Chem. Abstr.*, 131 (1999) 375649.
1018. D. M. Katkocin, N. S. Word, and S. S. Yang, Thermostable glucoamylase, EU Patent EP 135,138 (1985); *Chem. Abstr.*, 102 (1985) 202606n.
1019. V. Horvathova, K. Slajsova, and E. Sturdik, Evaluation of the glucoamylase Glm from *Saccharomycopsis fibuligera* IFO 011 in hydrolysing corn starch, *Biologia (Bratislava)*, 59 (2004) 361–365.
1020. Y. Yamasaki and Y. Suzuki, Purification and properties of α-glucosidase and glucoamylase from *Lentinus edodes* (Berk.) Sing, *Agric. Biol. Chem.*, 42 (1978) 971–980.
1021. D. R. Krauze, C. J. Wood, and D. J. Maclean, Glucoamylase (exo-1,4-α-D-glucan glucanohydrolase EC 3.2.1.3) is the major starch-degrading enzyme secreted by the phytopathogenic fungus *Colletotrichum gloeosporioides*, *J. Gen. Microbiol.*, 137 (1991) 2463–2468.
1022. V. V. Joutsjoki, E. E. M. Parkkinen, and T. K. Torkkeli, A novel glucoamylase preparation for grain mash saccharification, *Biotechnol. Lett.*, 15 (1993) 277–282.
1023. T. L. Baldwin, B. S. Bower, G. K. Chotani, N. Dunn-Coleman, O. Lantero, Jr., S. E. Lantz, M. J. Pepsin, J. K. Shetty, B. A. Strohm, and H. Wang, Expression of granular starch-hydrolyzing glucoamylase from filamentous fungi *Trichoderma* for producing glucose syrup from granular starch substrates, EU Patent WO 2005052148 (2005); *Chem. Abstr.*, 143 (2005) 42784.
1024. J. J. M. Labout, Conversion of liquified starch into glucose using a novel glucoamylase system, *Starch/Stärke*, 37 (1985) 157–161.
1025. D. Caratti de Lima and Y. K. Park, Production of glucose sirup from cassava starch using fungal enzymes (in Portugese), *Ecletica Quim.*, 2 (1977) 61–67.
1026. F. Shierbaum and M. Richter, Glukoamylasezusatz zum Stärkehydrolysaten, *Ernaehrungsforsch.*, 13 (1968) 399–410.
1027. Z. Boskov and J. Jakovljević, Study of enzyme preparation used for production of starch hydrolysate (in Serbo-Croatian), *Technika (Belgrade)*, 24 (1969) 1078–1080.
1028. S. V. Durmishidze, G. I. Kvesitadze, and G. N. Kokonashvili, Glucoamylase of *Aspergillus awamori* (in Russian), *Dokl. Akad. Nauk SSR*, 217 (1974) 470–471.
1029. A. Kimura and J. F. Robyt, Reaction of enzymes with starch granules enhanced reaction of glucoamylase with gelatinized starch granules, *Carbohydr. Res.*, 288 (1996) 233–240.
1030. L. Ludvig and T. Ferenczi, Increasing the yield of crystalline dextrose (in Hungarian), *Szeszipar*, 21 (4), (1973) 139–141.

1031. T. A. Ladur, N. G. Gulyuk, Z. M. Borodina, D. M. Pikhalo, E. G. Bondar, V. G. Shorniko, and G. D. Malyi, Industrial tests of production technology for crystalline glucose with the acidic-enzymic hydrolysis of starch (in Russian), *Sakh. Prom.* (11), (1977) 71–74.
1032. A. Sroczyński and K. Budryn, Effect of initial enzymic hydrolysis of starch on its saccharification with glucoamylase (in Polish), *Roczn. Technol. Chem. Zywn.*, 24 (1974) 279–285.
1033. A. M. Madgavkar, Y. T. Shah, and J. T. Cobb, Hydrolysis of starch in a membrane reactor, *Biochem. Bioeng.*, 19 (1977) 1719–1726.
1034. Y. Hattori, Saccharification enzyme of *Endomyces* (in Japanese), *Dempunto Gijutsu Kenkyu Kaiho*, 23 (1961) 28–33.
1035. T. Cayle and F. Viebrock, The application of Mathew's formula in enzymic starch conversion, *Cereal Chem.*, 43 (1966) 237–244.
1036. H. Ohnishi, H. Sakai, and T. Ohta, Purification and some properties of glucoamylase from *Clostridium* sp. G005, *Agric. Biol. Chem.*, 55 (1991) 1901–1902.
1037. H. H. Hyun and J. G. Zeikus, Regulation and genetic enhancement of glucoamylase and pullulanase production in *Clostridium thermohydrosulfuricum*, *J. Bacteriol.*, 164 (1985) 1146–1152.
1038. A. Oren, A thermophilic amyloglucosidase from *Halobacterium sodomense*, a halophilic bacterium from the Dead Sea, *Curr. Microbiol.*, 8 (1983) 225–230.
1039. D. M. Bui, I. Kunze, S. Forster, T. Wartmann, and C. Horstmann, Cloning and expression of an *Arxula adeninivorans* glucoamylase gene in *Saccharomyces cerevisiae*, *Appl. Microbiol. Biotechnol.*, 44 (1996) 610–619.
1040. H. Furuta, T. Arai, H. Hama, N. Shiomi, A. Kondo, and H. Fukuda, Bioconversion of maleic acid to fumaric acid by *Pseudomonas alcaligenes* strain XD-1, *J. Ferment. Bioeng.*, 84 (1997) 169–171.
1041. I. Yamashita and S. Fukui, Molecular cloning of a glucoamylase producing gene in the yeast *Saccharomyces*, *Agric. Biol. Chem.*, 47 (1983) 2689–2692.
1042. B. Hulseweh, U. M. Dahlems, J. Dohmen, A. W. M. Strasser, and C. P. Hollenberg, Characterization of the active sites of *Schwanniomyces occidentalis* glucoamylase by in vitro mutagenesis, *Eur. J. Biochem.*, 244 (1997) 128–133.
1043. *DSM Catalogue of strains*, (1989) 167.
1044. H. K. Wang, E. W. Swain, and C. W. Haseltine, Glucoamylase of *Amylomyces rouxii*, *J. Food Sci.*, 49 (1984) 1210–1211.
1045. M. L. Chiquetto, M. C. R. Facciotti, B. V. Kilkian, and W. Schmidell, Influence of carbon and nitrogen source on glucoamylase production by *Aspergillus* in batch process, *Biotechnol. Lett.*, 14 (1992) 465–468.
1046. A. B. Ariff and C. Webb, The influence of different fermenter configurations and modes of operation on glucoamylase production by *Aspergillus awamori*, *Asia Pac. J. Mol. Biol.*, 4 (1996) 183–195.
1047. J. G. Silva, H. J. Nascimento, V. F. Soares, and E. P. S. Bon, Glucoamylase isoenzymes tailoring through medium composition, *Appl. Biochem. Biotechnol.*, 63–65 (1997) 87–96.
1048. M. C. R. Queiroz, M. C. R. Facciotti, and W. Schmidell, Rheological changes of *Aspergillus awamori* broth during amyloglucosidase production, *Biotechnol. Lett.*, 19 (1997) 167–170.
1049. Y. Yamasaki, Y. Suzuki, and J. Ozawa, Purification and properties of two forms of glucoamylase from *Penicillium oxalicum*, *Agric. Biol. Chem.*, 41 (1977) 755–762.
1050. V. K. Morya and D. Yadov, Isolation and screening of different isolates of *Aspergillus* for amylase production, *Internet J. Microbiol.*, 7(2009) http://dx.doi.org/10 5580/21f8.
1051. E. Yoshino and S. Hayashida, Enzymic modification of glucoamylase of *Aspergillus awamori* var. kawachi, *J. Ferment. Technol.*, 56 (1978) 289–294.
1052. A. R. Shah and M. Madamwar, Xylanase production by a newly isolatd *Aspergillus foetidus* strain and its characterization, *Process Biochem.*, 40 (2005) 1763–1777.
1053. P. Selvakumar, L. Ashakumary, and A. Pandey, Biosynthesis of glucoamylase from *Aspergillus niger* by solid state fermentation using tea waste as the basis of a solid substrate, *Bioresour. Technol.*, 65 (1998) 83–85.

1054. P. Selvakumar, L. Ashakumary, A. Helen, and A. Pandey, Purification and characterization of glucoamylase produced by *Aspergillus miger* in solid state fermentation, *Lett. Appl. Microbiol.*, 23 (1996) 403–406.
1055. A. Pandey, P. Selvakumar, and L. Ashakumary, Performance of a column bioreactor for glucoamylase synthesis by *Aspergillus niger* in SSF, *Process Biochem.*, 31 (1996) 43–46.
1056. A. Pandey, L. Ashakumary, and P. Selvakumar, Copra – A novel substrate for solid state fermentation, *Bioresour. Technol.*, 51 (1995) 217–220.
1057. A. Pandey, L. Ashakumary, P. Selvakumar, and K. S. Vijaylakshmi, Influence of water activity on growth and activity of *Aspergillus niger* for glucoamylase production in solid state fermentation, *World J. Microbiol. Biotechnol.*, 10 (1994) 485–486.
1058. A. Pandey, P. Selvakumar, and L. Ashakumary, Glucoamylase production by *Aspergillus niger* on rice bran is improved by adding nitrogen source, *World J. Microbiol. Biotechnol.*, 10 (1994) 348–349.
1059. A. Pandey and S. Radhakrishnan, The production of glucoamylase by *Aspergillus niger* NCINI, *Process Biochem.*, 28 (1993) 305–309.
1060. A. Pandey and S. Radhakrishnan, Packed bed column bioreactor for producing enzymes, *Enzyme Microb. Technol.*, 14 (1992) 486–488.
1061. D. Singh, J. S. Dahiya, and P. Nigam, Simultaneous raw starch hydrolysis and ethanol fermentation by glucoamylase from *Rhizotonia solani* and *Saccharomyces cerevisiae*, *J. Basic Microbiol.*, 35 (1995) 117–121.
1062. B. Stoffer, T. P. Frandsen, P. K. Busk, P. Shneider, I. Svedsen, and B. Svensson, Production, purification and characterization of the catalytic domain of glucoamylase from *Aspergillus niger*, *Biochem. J.*, 292 (1993) 197–201.
1063. A. Pandey, Production of starch saccharifying enzyme (glucoamylase) in solid cultures, *Starch/Stärke*, 44 (1992) 75–77.
1064. A. A. Amirul, S. L. Khoo, M. N. Nazalan, M. S. Razip, and M. N. Azizan, Production and properties of two forms of glucoamylase from *Aspergillus niger*, *Folia Microbiol.*, 41 (1996) 165–174.
1065. K. Ono, K. Shintani, and S. Shikata, Comparative studies of various molecular species in *Aspergillus niger* glucoamylase, *Agric. Biol. Chem.*, 52 (1988) 1699–1706.
1066. K. Venkataraman, P. Manjunath, and M. R. R. Raghavandra, Glucoamylases of *Aspergillus niger* NRRL330, *Indian J. Biochem. Biophys.*, 12 (1975) 107–110.
1067. M. N. Miah and S. Ueda, Multiplicity of glucoamylase of *Aspergillus oryzae*. II. Enzymatic and physicochemical properties of three forms of glucoamylase, *Starch/Stärke*, 29 (1977) 235–239.
1068. A. Gromada, J. Fiedurek, and J. Szczodrak, Isoglucose production from raw starchy materials based on two-stage enzymatic system, *Pol. J. Microbiol.*, 57 (2008) 141–148.
1069. T. Kobayashi, T. Nagamura, and I. Endo, An effective production method of glucoamylase in a novel solid-state culture system, *Ann. N. Y. Acad. Sci.*, 613 (1990) 846–855.
1070. Y. Imai, M. Suzuki, M. Masamoto, K. Nagayasu, and M. Kishimoto, Glucoamylase production of *Aspergillus oryzae* in fed-batch culture using a statistical regression model, *J. Fermen. Bioeng.*, 78 (1994) 310–314.
1071. C. Kneck, Production of glucoamylase using *Aspergillus phoenicus* immobilized in calcium alginate beads, *Appl. Microbiol. Biotechnol.*, 35 (1991) 466–468.
1072. D. R. Lineback and W. E. Burmann, Purification of a glucoamylase from *Aspergillus phoenicis*, *Carbohydr. Res.*, 14 (1970) 341–343.
1073. M. Naczk, R. M. Myhara, and F. Sahidi, Effects of processing on the oligosaccharides of oilseed and legume protein meals, *Food Chem.*, 45 (1992) 193–197.
1074. A. B. Ghosh, B. Chaterjee, and A. Das, Production of glucoamylase by 2-deoxy-D-glucose resistant mutant of *Aspergillus terreus*, *Indian J. Technol.*, 29 (1991) 529–533.
1075. S. Ali, Z. Hossain, S. Mahmood, and R. Alam, Induction of glucoamylase productivity by nonstarchy carbohydrates in *Aspergillus terreus*, *World J. Microbiol. Biotechnol.*, 6 (1990) 19–21.

1076. A. Ghosh, B. Chatterjee, and A. Das, Purification and characterization of *Aspergillus terreus* NA 170 mutant, *J. Appl. Bacteriol.*, 71 (1991) 162–164.
1077. G. Mamo and A. Geessese, Production of raw starch digesting amyloglucosidase by *Aspergillus* sp, GP-21 in solid state fermentation, *J. Ind. Microbiol. Biotechnol.*, 22 (1999) 622–626.
1078. M. Krzechowska and H. Urbanek, Isolation and some properties of glucoamylase from *Cephalosporum charticola* Lindau, *Appl. Environ. Micobiol.*, 30 (1975) 163–166.
1079. S. Poulivong, L. Cai, H. Chen, E. H. C. McKenzie, K. Abdelsalam, E. Choukeatirote, and K. D. Hyde, *Colletotrichum gloeosporioides* is not a common pathogen on tropical fruits, *Fungal Div.*, 44 (2010) 33–43.
1080. J. King, The glucoamylase of *Coniophora cerebella*, *Biochem. J.*, 105 (1967) 577–583.
1081. Y. Nagasaka, K. Kurosawa, A. Yokota, and F. Tomita, Purification and properties of the raw starch digesting glucoamylase from *Cortocium rolfsii*, *Appl. Microbiol. Biotechnol.*, 50 (1998) 323–330.
1082. A. Kaji, M. Sato, M. Kobayashi, and T. Murao, Acid-stable glucoamylase produced in medium containing sucrose by *Corticium rolfsii*, *J. Agric. Chem. Soc. Jpn.*, 50 (1976) 509–511.
1083. T. Fujui and Z. Nikuni, Preparation and properties of glucoamylase from *Endomyces* sp. IFO 0111, *Agric. Biol. Chem.*, 33 (1969) 884–885.
1084. I. M. Gracheva, T. A. Lushchik, Y. A. Tyrsin, and E. E. Pinchukova, Effect of the nature of the precipitating agent and its concentration on the precipitation of enzymes from the culture of liquid of *Endomycopsis* species 20-9 (in Russian), *Biokhimiya*, 42 (1977) 1803–1805.
1085. I. Ferrero, C. Rossi, N. Marmiroli, C. Donnini, and P. P. Puglisi, Effect of chloramphenicol, antimycin A and hydroxamate on the morphogenetic development of the dimorphic ascomycete *Endomycopsis capsularis*, *Antonie Van Leeuwenhoek*, 47 (1981) 311–323.
1086. C. Gonzalez, O. Delgado, M. Baigori, C. Abate, and L. I. C. De Figuero Allieri, Ethanol production from native cassava starch by mixed culture of *Endomycopsis fibuligera* and *Zymomonas mobilis*, *Acta Biotechnol.*, 18 (1998) 149–155.
1087. O. V. S. Reddy and S. C. Basappa, Direct fermentation of cassava starch to ethanol by mixed cultures of *Endomycopsis fibuligera* and *Zymomonas mobilis*. Symergism and limitations, *Biotechnol. Lett.*, 18 (1996) 1315–1318.
1088. N. Ohno, T. Juln, S. Song, S. Uchiyama, H. Shinoyama, A. Ando, and T. Fujii, Purification and properties of amylases extracellularly produced by an imperfect fungus *Fusidium* sp. BX-1 in a glycerol medium, *Biosci. Biochem. Biotechnol.*, 56 (1992) 465–469.
1089. Y. Yamasaki and H. Kanno, Purification and properties of glucoamylase from *Mucor javanicus*, *Agric. Biol. Chem.*, 55 (1991) 2429–2430.
1090. S. J. Rasmussen-Wilson, J. S. Palas, V. J. Wolf, C. S. Taft, and C. P. Selitrennikoff, Expression of a plant protein by *Neurospora crassa*, *Appl. Environ. Microbiol.*, 63 (1997) 3488–3493.
1091. Y. Nagasaka, N. Muraki, A. Kimura, K. Kitamoto, A. Yokota, and F. Tomita, Study on glucoamylase from *Corticium rolfsii*, *J. Appl. Glycosci.*, 46 (1999) 169–178.
1092. M. Michelin, E. Ruller, R. J. Ward, L. A. Moraes, J. A. Jorge, H. F. Ferenz, and M. L. Polizeli, Purification and biochemical characterization of a thermostable extracellular glucoamylase produced by thermotolerant fungus *Paecolomyces variotii*, *J. Ind. Microbiol. Biotechnol.*, 35 (2008) 17–25.
1093. R. Nandi and S. K. Mukherjee, Effect of various organic compounds on synthesis of glucoamylase by an isolated strain of *Penicillium italicum*, *Indian J. Exp. Biol.*, 27 (1989) 1101–1102.
1094. H. Sun, X. Ge, and W. Zhang, Production of novel raw starch digesting glucoamylase by Penicillium sp. X-1 under solid state fermentation and its use in direct hydrolysis of raw starch, *World J. Microbiol. Biotechnol.*, 23 (2007) 603–613.
1095. A. Yukhi, T. Watanabe, and K. Matsuda, Purification and properties of saccharogenic amylase from *Piricularia oryzae*, *Starch/Stärke*, 29 (1977) 265–269.
1096. C. R. Soccol, B. Marin, M. Reaimbault, and J. M. Lebault, Breeding and growth of Rhizopus in raw cassava by solid state fermentation, *Appl. Microbiol. Biotechnol.*, 41 (1994) 330–336.

1097. Y. Fujio and H. Morita, Improved glucoamylase production by *Rhizopus* sp. A-11 using metal-ion supplemented liquid medium, *J. Ferment. Bioeng.*, 82 (1996) 554–567.
1098. J. Pazur and S. Okada, Properties of the glucoamylase from *Rhizopus delemar*, *Carbohydr. Res.*, 4 (1967) 371–379.
1099. C. R. Soccol, I. Iloki, B. Marin, and M. Raimbault, Comparative production of alpha-amylase, glucoamylase and protein enrichment of raw and cooked cassava by *Rhizopus* strains in submerged and solid state fermentations, *J. Food Sci. Technol.*, 31 (1994) 320–323.
1100. M. D. Chiarello, C. R. Soccol, S. C. Stertz, L. Furlanetto, J. D. Fontana, and N. Krieger, Biodegradation of cassava crude starch granules during solid state fermentation by *Rhizopus* glucoamylase, *Arq. Biol. Technol.*, 40 (1997) 771–785.
1101. K. Watanabe and F. Fukimbara, The composition of saccharogenic amylase from *Rhizopus javanicus* and the isolation of glycopeptides, *Agric. Biol. Chem.*, 37 (1973) 2755–2761.
1102. S. Moriyama, R. Matsuno, and T. Kamikubo, Influence of dielectric constants and ligand binding on thermostability of glucoamylase, *Agric. Biol. Chem.*, 41 (1977) 1985–1993.
1103. K. Ikasari and D. A. Mitchell, Protease production by *Rhizopus oligosporus* in solid state fermentation, *World J. Microbiol. Biotechnol.*, 10 (1994) 320–324.
1104. H. Morita, M. Matsunaga, K. Mizuno, and Y. Fujio, A comparison of raw starch digesting glucoamylase production in liquid and solid cultures of *Rhizopus* strains, *J. Gen. Appl. Microbiol.*, 44 (1998) 211–216.
1105. The UK National Culture Collection, In: D. Smith, M. J. Ryan, and J. G. Day, (Eds.). *UKNCC Biological Resource: Properties, Maintenance and Management,* Pineapple Planet Design Studio, Ltd., Old Town, Swindon, UK, 2001.
1106. S. Kumar and T. Satyanarayana, Purification and kinetics of a roaw starch hydrolyzing thermostable and neutral glucoamylase of the thermophilic mold *Thermomucor indicae-seudatricae*, *Biotechnol. Prog.*, 19 (2003) 936–944.
1107. V. B. Rao, N. V. S. Sastry, and P. V. S. Rao, Purification and characterization of thermostable glucoamylase from thermophilic fungus *Thermomyces lanuginosus*, *Biochem. J.*, 193 (1981) 379–382.
1108. A. Subrahmanyam, A. S. K. Mangallum, and K. S. Gopalakrishnan, Amyloglucosidase production by *Torula thermophila* (thermophilic fingus), *Indian J. Exp. Biol.*, 15 (1977) 495–500.
1109. R. Watts, J. Dahiya, K. Chaudhary, and P. Tauro, Isolation and characterization of a new antifungal metabolite of *Trichoderma resei*, *Plant Soil*, 107 (1988) 81–84.
1110. J. A. Schellart, E. J. F. van Arem, M. A. J. S. Van Boekel, and W. J. Middlehoven, Starch degradation by the mould *Trichoderma viride*. II. Regulation of enzyme synthesis, *Antonie van Leeuwenhoek J. Microbiol.*, 42 (1976) 239–244.
1111. J. P. Zhou, J. Q. Wei, W. Z. Gong, Z. Y. Wang, and H. Yang, Study of immobilized glucoamylase and industrial application, *Gansu Gongye Daxue Xuebao*, 29(4), (2003) 74–76. *Chem. Abstr.*, 140 (2004) 373998.
1112. K. Kusunoki, K. Kawakami, K. Kato, and F. Shiraishi, Hydrolysis of soluble starch on immobilized glucoamylase using ceramic honeycomb carriers, *Kenkyu Hokoku Asahi Garasu Kogyo Gijutsu Shoreikai*, 39 (1981) 293–302. *Chem. Abstr.*, 97 (1982) 184294f.
1113. Y. Y. Lee, D. D. Lee, and G. T. Tsao, Continuous production of glucose from starch hydrolysates by immobilized glucoamylase, *Proc. 3rd Int. Biodegradation Symp.*, 1975, pp. 1021–1032.
1114. S. V. Yelchits, L. A. Ivanova, L. A. Golubeva, and L. A. Melnichenko, Effectiveness of the hydrolysis of starch by immobilized glucoamylase (in Russian), *Pishch. Prom.* (4), (1981) 47–48.
1115. R. P. Rohrbach, M. J. Maliarik, and T. P. Malloy, High sugar syrups, US Patent 4,511,654 (1985); *Chem. Abstr.*, 103 (1985) 21489b.
1116. CPC International Inc., Method using glucoamylase immobilized on porous alumina, *Res. Discl.*, 193 (1980) 195–197.

1117. M. Abdulla and F. C. Armbruster, Method using glucoamylase immobilized on porous aluminum, US Patent 4,226,937 (1980); *Chem. Abstr.*, 94 (1980) 3009m.
1118. H. H. Weetall, W. P. Vann, W. H. Pitcher, Jr., D. D. Lee, Y. Y. Lee, and G. T. Tsao, Scale-up studies on immobilized purified glucoamylase covalently coupled to porous ceramic support, *Inf. Chim.*, 155 (1976) 153–158, 161–165.
1119. F. Toldra, N. B. Jansen, and G. T. Tsao, Hydrolysis of maltose and cornstarch by glucoamylase immobilized in porous glass fibers and beads, *Process Biochem.*, 27(3), (1992) 177–181.
1120. S. H. Cho and Z. U. Kim, Characteristics and applications of immobilized glucoamylase, *Hanguk Nonghwa Hakhoechi*, 28(4), (1985) 233–238. *Chem. Abstr.*, 105 (1986) 168007t.
1121. F. B. Song, H. Y. Chen, J. B. Yu, and S. Z. Li, Studies on performance of glucoamylase immobilized on molecular sieve, *Fudan Xuebao, Ziran Kexueban*, 39 (2000) 363–367, 373; *Chem. Abstr.*, 134 (2000) 177432.
1122. G. Sanjay and S. Sugunan, Glucoamylase immobilized on montmorillonite: Characterization and starch hydrolysis activity in a fixed bed reactor, *Catal. Commun.*, 6 (2005) 525–530.
1123. G. A. Kovalenko, L. V. Perminova, G. V. Plaksin, T. V. Chuenko, O. V. Komova, and N. A. Rudina, Immobilized glucoamylase: A biocatalyst of dextrin hydrolysis, *Appl. Biochem. Microbiol.*, 42 (2006) 145–149.
1124. A. C. Chakrabarti and K. B. Storey, Co-immobilization of amyloglucosidase and pullulanase for enhanced starch hydrolysis, *Appl. Microbiol. Biotechnol.*, 33 (1990) 48–50.
1125. A. Wójcik, J. L. Łobarzewski, A. Błaszczyńska, and J. Fiedurek, Silica gels activated by BCl_3 and aliphatic diamines as supports for glucoamylase immobilization, *Biotech. Bioeng.*, 30 (1987) 983–986.
1126. J. Łobarzewski, A. Paszczyński, T. Wolski, and J. Fiedurek, Keratin and polyamide coated inorganic matrices as supports for glucoamylase immobilization, *Biochem. Biophys. Res. Commun.*, 121 (1984) 220–228.
1127. M. Barta, J. Kučera, V. Vojcišek, and K. Čulik, Continuous production of glucose and products of its enzymic conversion, Czechoslovak Patent 216,646 (1984); *Chem. Abstr.*, 103 (1985) 86538m.
1128. G. M. Zanin and F. F. de Mores, Stability of immobilized amyloglucosidase in the process of cassava starch saccharification, *Appl. Biochem. Biotechnol.*, 51(52), (1995) 253–262.
1129. H. Maeda, L. F. Chen, and G. T. Tsao, Glucoamylase immobilized on diethylaminoetylcellulose beads in column reaction, *J. Ferment. Technol.*, 57 (1979) 238–243.
1130. M. Tomar and K. A. Prabhu, Immobilization of glucoamylase on DEAE-cellulose activated with chloride compounds, *Enzyme Microb. Technol.*, 7 (1985) 557–560.
1131. V. Arasaratnam, T. Murugapoopthy, and K. Balasubramanian, Effect of pH on preparation and performance of physically immobilized amyloglucosidase on DEAE-cellulose, *Starch/Stärke*, 46 (1994) 146–149.
1132. B. Szajani, G. Klamar, and L. Ludvig, Preparation, characterization and laboratory application of an immobilized glucoamylase, *Enzyme Microb. Technol.*, 7 (1985) 488–492.
1133. G. Klamar, B. Szajani, and L. Laszlo, Preparation of immobilized glucoamylase and its use in the starch industry (in Hungarian), *Szeszipar*, 32(2), (1984) 41–43.
1134. J. Popa and C. Beldie, Comparison between the hydrolysis of different substrates in the presence of immobilized amyloglucosidase on acrylic support, *Progr. Catal.*, 3 (1994) 97–102.
1135. N. Ohtani, T. Ishidao, Y. Iwai, and Y. Arai, Hydrolysis of starch or pullulan by glucoamylase or pullulanase immobilized on poly(N-isopropyloacrylamide) gel, *Colloid Polym. Sci.*, 277 (1999) 197–202.
1136. T. Handa, S. Goto, T. Akino, T. Mizukura, and Y. Ayukawa, The hydrolysis of soluble starch in granular immobilized enzyme columns (in Japanese), *Dempun Kagaku*, 24(1), (1977) 1–8.
1137. W. Błaszków and T. Miśkiewicz, Use of starch as a substrate for immobilized enzymes: α-amylase, glucoamylase and glucose isomerase (in Polish), *Przem. Ferm. Owoc.-Warz.* (1), (1980) 6–8.

1138. H. Sahashi, T. Hibino, T. Okada, H. Ishizuka, and M. Ito, Conversion of insoluble substrates to a pharmaceutical or food product with immobilized enzymes, Jpn. Kokai Tokkyo Koho 63129987 (1988); *Chem. Abstr.*, 110 (1989) 22316.
1139. D. Gembicka, I. Nowacka, and J. Janicki, Application of glucoamylase bound with DEAE-cellulose for hydrolysis of starch by continuous method, *Acta Aliment. Pol.*, 1(1), (1975) 33–38.
1140. N. B. Milosavić, R. M. Prodanović, S. M. Jovanović, V. M. Maksimović, and Z. M. Vujcić, Characterization and use of amyloglucosidase immobilized on the macroporous copolymer of ethylene glycol dimethylacrylate and glycidyl methacrylate under simulated industrial conditions (in Serbo-Croatian), *Hem. Ind.*, 58 (2004) 493–498.
1141. Y. K. Park and D. C. Lima, Continuous conversion of starch to glucose by amyloglucosidase-resin complex, *J. Food Sci.*, 38 (1973) 358–359.
1142. M. Bellal, J. Boudrant, and C. Cheftel, Tubular enzyme, reactor using glucoamylase adsorbed onto an anionic resin. Application to maltodextrin hydrolysis, *Ann. Technol. Agric.*, 27 (1978) 469–488.
1143. P. N. Nahete and V. Shankar, Immobilization of amyloglucosidase on polystyrene anion exchange resin. III. Product analysis, *Food Biotechnol.*, 6 (1992) 127–134.
1144. I. D. Ruadze, N. A. Zherebtsov, Yu. I. Slepokurova, V. F. Selemenev, I. V. Shkutina, and O. F. Stoyanova, Preparation and characterization of immobilized *Aspergillus awamori* 466 glucoamylase, *Appl. Biochem. Microbiol.*, 37 (2001) 178–183.
1145. V. F. Selemenev, O. F. Stoyanova, I. V. Shkutina, I. D. Ruadze, N. A. Zeretsov, Yu. I. Slepokurova, and V. B. Smirnov, Immobilization of glucoamylase on the sorbent MKhDE 100 or the ion exchanger ANKB-2 or ANKB-35 for preparation of glucose from starch in food industry, Russian Patent 2181770 (2002) 348421; *Chem. Abstr.*, 137 (2002) 348421.
1146. J. Krauze and B. Wawrzyniak, Properties of gel-entrapped glucoamylase, *Starch/Stärke*, 40 (1988) 314–319.
1147. D. Y. Schafhauser and K. B. Story, Co-immobilization of amyloglucosidase and pullulanase onto granular chicken bone for enhanced starch degradation, *Biotechnol. Appl. Biochem.*, 17 (1993) 103–113.
1148. N. Miwa and K. Otomo, Continuous hydrolysis of high molecular weight compounds by immobilized enzymes, Jpn. Kokai Tokkyo Koho, 75,116,684 (1975); *Chem. Abstr.*, 84 (1975) 29181e.
1149. R. Torres, B. C. Pessela, C. Mateo, C. Ortez, M. Fuentes, J. M. Guisan, and R. Fernandez-Lafuente, Reversible immobilization of glucoamylase by ionic adsorption on sephabeads coated with polyethyleneimine, *Biotechnol. Prog.*, 20 (2004) 1297–1300.
1150. K. J. Raju, C. Ayyanna, D. V. S. Padmasree, S. S. Rao, and R. J. Rao, Bioconversion of tapioca starch to glucose using amyloglucosidase immobilized in hen egg white, *Indian Chem. Eng.*, 33(2), (1991) 58–61.
1151. D. Uttapap, Y. Koba, and A. Ishizaki, Recycle use of immobilized glucoamylase by tangential flow filtration unit, *Biotechnol. Bioeng.*, 33 (1989) 542–549.
1152. I. Roy and M. N. Gupta, Hydrolysis of starch by a mixture of glucoamylase and pullulanase entrapped individually in calciu alginate beads, *Enzyme Microb. Technol.*, 31 (2004) 26–32.
1153. T. Kaneko, T. Ohno, and N. Ohisa, Purification and characterization of a thermostable raw starch degrading amylase from a *Streptomyces* sp. isolated from milling factory, *Biosci. Biotechnol. Biochem.*, 69 (2005) 1073–1081.
1154. Y. Marlida, N. Saari, S. Radu, and F. A. Bakar, Production of α-amylase degrading raw starch by *Giberella pulicaris*, *Biotechnol. Lett.*, 22 (2000) 95–97.
1155. W. A. Davies, Distribution of maltase in plants. I. The functions of maltase in starch degradation and the influence on the amyloclastic activity of plant materials, *Biochem. J.*, 10 (1916) 31–48.
1156. S. Chiba, N. Hibi, and T. Shimomura, Substrate specificity of a neutral α-glucosidase, *Agric. Biol. Chem.*, 40 (1976) 1813–1817.

1157. H. Nakai, T. Ito, S. Tanizawa, K. Matsubara, T. Yamamoto, M. Okuyama, H. Mori, S. Chiba, Y. Sano, and A. Kimura, Plant α-glucosidase: Molecular analysis of rice α-glucosidase and degradation mechanism of starch granules in germination stage, *J. Appl. Glycosci.*, 53 (2006) 137–142.
1158. M. A. Taylor, H. A. Ross, D. McRae, D. Stewart, I. Roberts, G. Duncan, F. Wright, S. Millan, and H. V. Davies, A potato α-glucosidase gene encodes a glucoprotein processing α-glucosidase II-like activity. Demonstration of enzyme activity and effects of down regulation in transgenic plants, *Plant J.*, 24 (2000) 305–316.
1159. Z. Wierzchowski, Maltase aus Stärke, *Biochem. Z.*, 50 (1913) 209–219.
1160. S. Chiba and T. Shimomura, Substrate specificity of flint corn α-glucosidase, *Agric. Biol. Chem.*, 39 (1975) 1041–1047.
1161. N. Takahashi, T. Shimomura, and S. Chiba, Studies on α-glucosidase of rice. I. Isolation and some properties of α-glucosidase I and α-glucosidase II, *Agric. Biol. Chem.*, 35 (1971) 2015–2024.
1162. N. Takahashi and T. Shimomura, α-Glucosidase in rice. II. Action of rice α-glucosidase on maltose and starch, *Agric. Biol. Chem.*, 37 (1973) 67–74.
1163. S. Murata, H. Matsui, S. Chiba, and T. Shimomura, Substrate specificity of α-glucosidase II in rice seeds, *Agric. Biol. Chem.*, 43 (1979) 2131–2135.
1164. Z. Sun and C. A. Henson, Degradation of native starch granules by barley α-glucosidase, *Plant Physiol.*, 94 (1990) 320–327.
1165. H. Matsui, S. Chiba, and T. Shimomura, Substrate specificity of an α-glucosidase in sugar beet seeds, *Agric. Biol. Chem.*, 42 (1978) 1855–1860.
1166. M. J. Sissons and A. W. MacGregor, Hydrolysis of barley starch granules by α-glucosidases from malt, *J. Cereal. Sci.*, 19 (1994) 161–169.
1167. Y. Suzuki and K. Uchida, Three forms of α-glucosidase from Welsh onion (*Allium fistolosum* L.), *Agric. Biol. Chem.*, 48 (1984) 1343–1345.
1168. T. S. M. Pirttijarvi, G. Wahlstrom, F. A. Rainey, P. E. J. Saris, and M. S. Salkinoja-Salonen, Inhibition of bacilli in industrial starches by nisin, *J. Ind. Microbiol. Biotechnol.*, 26 (2001) 107–114.
1169. Agency of Industrial Science and Technology, Amylase G3 and α-glucosidase, Jpn. Kokai Tokkyo Koho 59 17,983 (1984); *Chem. Abstr.*, 101 (1984) 37218g.
1170. M. Kawai, S. Mori, S. Hirose, and H. Tsuji, Heat-stable debranching enzyme, Bacillus and process for producing same and manufacture of glucose- or maltose-containing liquids from starch, EU Patent EP 558036 (1993); *Chem. Abstr.*, 119 (1993) 198567.
1171. A. Zdziebło and J. Synowiecki, New source of the thermostable α-glucosidase suitable for single step starch processing, *Food Chem.*, 79 (2002) 485–491.
1172. A. Martino, C. Schraldi, S. Fusco, I. Di Lemia, T. Costabile, P. Pellicano, M. Marotta, M. Generoso, J. Van der Oost, C. W. Sensen, R. L. Charlebois, M. Moracci, M. Rossi, and M. De Rosa, Properties of recombinant α-glucosidase from *Sulfolobus solfataricus* in relation to starch processing, *J. Mol. Catal. B*, 11 (2001) 787–794.
1173. F. Duchiron, E. Legin, C. Ladrat, H. Gentelet, and G. Barbier, New thermostable enzymes for crop fractionation, *Ind. Crops Prod.*, 6 (1997) 265–270.
1174. K. Gray, J. B. Garrett, N. M. Aboushadi, R. Knowles, E. O'Donoghue, and E. Waters, Thermostable α-glucosidase genes from environmental bacteria and their use for hydrolysis of maltooligosaccharides and liquefied starch in food processing and dental care products, EU Patent WO 2004085615 (2004); *Chem. Abstr.*, 141 (2004) 326804.
1175. Y. Yamasaki, T. Miyaki, and Y. Suzuki, α-glucosidase from *Mucor javanicus*. II. Properties of crystalline enzyme, *Bull. Ohara Inst. Agric. Biol. Okayama Univ.*, 17(3), (1978) 123–141. *Chem. Abstr.*, 89 (1978) 125052n.
1176. S. P. Amarakone, H. Ishigami, and K. Kainuma, Conversion of oligosaccharides formed during starch hydrolysis by a dual enzyme system, *Denpun Kagaku*, 31 (1984) 1–7.

1177. H. Tanaka, H. Fukuda, and M. Kuroda, Enzymic manufacture of oligosaccharides, Jpn. Kokai Tokkyo Koho 08023990 (1996); *Chem. Abstr.*, 124 (1996) 287218.
1178. I. Kobayashi, M. Tokuda, H. Hashimoto, T. Konda, H. Nakano, and S. Kitahata, Purification and characterization of a new type of α-glucosidase from *Paecilomyces lilacinus* that has transglycosylation activity to produce α-1,3- and α1,2-linked oligosaccharides, *Biosci. Biotechnol. Biochem.*, 67 (2003) 29–35.
1179. G. M. Zanin and F. F. de Moraes, Thermal stability and energy deactivation of free and immobilized amyloglucosidase in the saccharification of liquefied cassava starch, *Appl. Biochem. Biotechnol.*, 70–72 (1998) 383–394.
1180. K. Kanaya, S. Chiba, and T. Shimomura, Changes in the hydrolytic activities of buckwheat α-glucosidase by acetylation with N-acetylimidazole, *Agric. Biol. Chem.*, 42 (1978) 1887–1893.
1181. G. de Burlet, C. Vannier, J. Giudicelli, and P. Sudaka, Neutral alpha-glucosidase from human kidneys: Molecular and immunological properties, relationship with intestinal glucoamylase, *Biochemie*, 61 (1980) 1177–1183.
1182. N. Takahashi and T. Shimomura, Biochemical studies on α-glucosidase from buckwheat (*Fagopyrum essculentum* Moench). II. Purification of buckwheat α-glucosidase and some properties on maltose and soluble starch, *Agric. Biol. Chem.*, 32 (1968) 929–939.
1183. P. Blum, Hyperthermophilic alpha-amylase and its use, US Patent Appl. 09/376,343 (2003).
1184. M. Alonso-Benito, A. Polo, B. Gonzalez, M. Fernando-Lobato, and J. Sanz-Aparicio, Structural and kinetic analysis of *Schwanniomyces occidentalis* invertase reveals a new oligomerization pattern and the role of its supplementary domain in substrate binding, *J. Biol. Chem.*, 285 (2010) 13930–13941.
1185. H. R. Constantino, S. H. Brown, and R. M. Kelly, Purification and characterization of an alpha-glucosidase from hyperthermophilic aerobacterium *Pyrococcus furiosus* exhibiting a temperature optimum of 105 to 115°C, *J. Bacteriol.*, 172 (1990) 3654–3660.
1186. B. Linke, A. Ruediger, G. Wittenberg, P. L. Jorgensen, and G. Antranikian, Production of heat-stable pullulanase and α-glucosidase from extremely thermophilic archaeon *Pyrococcus woesei*, *Proc. DECHEMA Biotechnol. Conf.*, 1992, pp. 161–163.
1187. I. Di Lernia, A. Morana, A. Ottombrino, S. Fusco, M. Rossi, and M. de Rosa, Enzymes from *Sulfolobus shibatae* for the production of trehalose and glucose from starch, *Extremophiles*, 2 (1998) 409–416.
1188. M. Rolfsmeier and P. Blum, Purification and characterization of a maltase from the extremely thermophilic crenarchaete *Sulfolobus solfataricus*, *J. Bacteriol.*, 177 (1995) 582–585.
1189. H. A. Ernst, L. Lo Leggio, M. Willemoes, G. Leonard, P. Blum, and S. Larsen, Structure of the *Sulfolobus solfataricus* α-glucosidase. Implications for domain conservation and substrate recognition in GH31, *J. Mol. Biol.*, 358 (2006) 1106–1124.
1190. E. Legin-Copinet, Mise en evidence et etude des propprietes d'une α-glucosidase thermostable du nouvel archaeon hyperthermophile *Thermococcus hydrothermalis* AL662 isole d'un ecosysteme hydrothermal sous-marin, *Ph.D. Theses Univ. Reims, France*, 1997.
1191. K. Piller, R. M. Daniel, and H. H. Petach, Properties and stabilization of an extracellular α-glucosidase from the extremely thermophilic archaebacteria *Thermococcus* strain AN1: Enzyme activity at 130°C, *Biochem. Biophys. Acta*, 1292 (1996) 197–205.
1192. K. Stenholm, S. Home, M. Lauro, M. Perttula, and T. Suortti, Hydrolysis of barley starch by malt limit dextrinase, *Proc. 26th. Eur. Brewery Conv.*, 1997, pp. 283–290. *Chem. Abstr.*, 129 (1998) 548859.
1193. D. Byrom and S. H. Collins, Debranching enzyme from *Streptomyces* and its use in starch saccharification, EU Patent EP 242,075 (1987); *Chem. Abstr.*, 108 (1988) 166130r.
1194. T. Kobayashi and K. Yamanouchi, The limit dextrinase activity of molds, *J. Agric. Chem. Soc. Jpn.*, 27 (1953) 180–186.

1195. B. Maruo and T. Kobayashi, Enzymic scission of the branch links to amylopectin, *Nature*, 167 (1951) 606–607.
1196. P. N. Hobson, W. J. Whelan, and S. Peat, Enzymic synthesis and degradation of starch. XIV. R-enzyme, *J. Chem. Soc.* (1951) 1451–1459.
1197. S. Peat, W. J. Whelan, and G. J. Thomas, Enzymic synthesis and degradation of starch. XXII. Evidence of multiple branching in waxy maize starch – Correction, *J. Chem. Soc.* (1956) 3025–3030.
1198. Y. Endo, N. Tanaka, and M. Kakimi, Malktohexaose-high sugar manufacture with pullulanase Amano, Jpn. Kokai Tokkyo Koho 06335394 (1994); *Chem. Abstr.*, 122 (1995) 131178.
1199. I. Maeda, Z. Nikuni, H. Taniguchi, and M. Nakamura, Purification of the debranching enzyme (R-enzyme) from malted barley and the role of the enzyme in the digestion of starch granules during the germination of barley seeds, *Carbohydr. Res.*, 61 (1978) 309–320.
1200. I. Maeda, N. Jimi, H. Taniguchi, and M. Nakamura, Purification of R-enzyme from malted barley and its role in in vitro digestion of barley starch granules (in Japanese), *Denpun Kagaku*, 26(2), (1979) 117–127.
1201. W. F. Line, V. K. Chaudhary, E. Chicoye, and R. J. Mizerak, Production of dextrose and maltose syrups using an enzyme derived from rice, US Patent 4734364 (1988); *Chem. Abstr.*, 109 (1988) 209694.
1202. S. Ge, S. Yang, and S. Zhang, Synergistic effects of pullulanase and β-amylase and the linkage between sugars and peptide chain of pullulanase, *Shengwu Huaxue Yu Shengwu Wuli Jinzhan*, 58 (1984) 29–32. *Chem. Abstr.*, 101 (1984) 225795s.
1203. S. Ge, S. J. Yang, and S. Z. Zhang, Studies on pullulanase from *Acetobacter aerogenes*. I. Purification and some properties, *Wei Sheng Wu Hsueh Pao*, 20(4), (1980) 415–420. *Chem. Abstr.*, 94 (1980) 43152j.
1204. R. Koch and G. Antranikian, Action of amylolytic and pullulytic enzymes from various thermophiles on linear and branched glucose polymers, *Starch/Stärke*, 42 (1990) 397–403.
1205. B. C. Saha, G. J. Shen, K. C. Srivastava, L. W. LeCureux, and J. G. Zeikus, New thermostable α-amylases-like pullulanase from thermophilic Bacillus sp. 3183, *Enzyme Microb. Technol.*, 11 (1989) 760–764.
1206. B. E. Norman, A novel debranching enzyme for application in the glucose syrup industry, *Starch/Stärke*, 34 (1982) 340–346.
1207. Y. Takasaki, Pullulanase-like enzyme processing α-amylase method for its production and its use in starch saccharification, EU Patent EP 188,049 (1986); *Chem. Abstr.*, 105 (1986) 151574c.
1208. A. Lappalainen, M. L. Niku-Paavola, T. Suorti, and K. Poutanen, Purification and characterization of *Bacillus acidopullulyticus* pullulanase fro enzymic starch modification, *Starch/Stärke*, 43 (1991) 477–482.
1209. Y. Takasaki, T. Tsuruta, S. Hayashi, and K. Imada, Reaction conditions for saccharification of liquified starch by a maltotriose-producing amylase from *Microbacterium imperiale* (in Japanese), *Hakko Kogaku Kaishi*, 70 (1992) 255–258.
1210. A. R. Plant, H. W. Morgan, and R. M. Daniel, A highly stable pullulanase from *Thermus aquaticus* YT-1, *Enzyme Microb. Technol.*, 8 (1986) 668–672.
1211. A. R. Plant, R. M. Clemens, R. M. Daniel, and H. W. Morgan, Purification and preliminary characterization of an extracellular pullulanase from *Thermoanaerobium* Tok6-B1, *Appl. Microbiol. Biotechnol.*, 26 (1987) 427–433.
1212. S. H. Brown and R. M. Kelly, Characterization of amylolytic enzymes having both α-1,4 and α-1,6 hydrolytic activity from the thermophilic archea *Pyrococcus furiosa* and *Thermococcus litoralis*, *Appl. Environ. Microbiol.*, 59 (1993) 2614–2621.
1213. H. Gantelet and F. Duchiron, A new pullulanase from hyperthermophilic archeon for starch hydrolysis, *Biotechnol. Lett.*, 21 (1999) 71–75.

1214. J. G. Zeikus and H. H. Hyun, Thermostable starch converting enzymes, EU Patent WO 86 01,831 (1986); *Chem. Abstr.*, 105 (1986) 5194n.
1215. T. Kimura and K. Horikoshi, Manufacture of maltotetraose with *Micrococcus* species from amylose, amylopectin or starch, Jpn. Kokai Tokkyo Koho 01218598 (1989); *Chem. Abstr.*, 112 (1990) 19975.
1216. Y. Takasaki, Thermostable pullulanase its production with *Microbacterium* and its use in saccharification of starch, EU Patent EP 557637 (1993); *Chem. Abstr.*, 119 (1993) 198568.
1217. J. J. Marshall, Starch degrading enzymes derived from *Cladosporium resinae*, US Patent 4,211,842 (1980); *Chem. Abstr.*, 93 (1980) 103601c.
1218. L. Furegon, A. D. B. Peruffo, and A. Curioni, Immobilization of rice limit dextrinase on γ-alumina beads and its possible use in starch processing, *Process Biochem.*, 32 (1997) 113–120.
1219. E. Nebesny, Changes of carbohydrates and molecular structure of dextrins during enzymatic hydrolysis of starch with maltogenase participation, *Starch/Stärke*, 42 (1990) 432–436.
1220. H. S. Lee, K. R. Shockley, G. J. Shut, S. B. Comes, C. I. Montero, M. R. Johnson, C. J. Chou, S. L. Bridger, N. Wigner, S. D. Brehm, F. E. Jenny, Jr., D. A. Comfort, R. M. Kelly, and M. W. W. Adams, Transcriptional and biochemical analysis of starch metabolism in the hyperthermophilic aercheon *Pyrococcus furiosus*, *J. Bacteriol.*, 188 (2006) 2115–2125.
1221. A. Ruediger, P. L. Jorgensen, and G. Antranikian, Isolation and characterization of a heat-stable pullulanase from the hyperthermophilic archeon *Pyrococcus woesei* after cloning and expression of its gene in *Escherichia coli*, *Appl. Environ. Microbiol.*, 61 (1995) 567–575.
1222. H. Gantelet and F. Duchiron, Purification and properties of thermoactive and thermostable pullulanase from *Thermococcus hydrothermalis*, a new hyperthermophilic archeon isolated from a deepsea hydrothermal vent, *Appl. Microbiol. Biotechnol.*, 49 (1998) 770–777.
1223. C. Dong, C. Vielle, and J. G. Zeikus, Cloning, sequencing and expression of the gene encoding amylopullulanase from *Pyrococcus furiosus* and biochemical characterization of the recombinant enzyme, *Appl. Environ. Microbiol.*, 63 (1997) 3577–3584.
1224. S. Okada, T. Imanaka, T. Kuriki, K. Yoshikawa, and H. Hidaka, Manufacture of panose-containing starch with Bacillus enzyme, Jpn. Kokai Tokkyo Koho 01171493 (1989); *Chem. Abstr.*, 112 (1990) 117362.
1225. T. Kuriki, M. Yanase, H. Takata, T. Imanaka, and S. Okada, Highly branched oligosaccharide produced by the transglycosylation reaction of neopullulanase, *J. Ferment. Bioeng.*, 76 (1993) 184–190.
1226. M. J. Yebra, J. Arroyo, P. Sanz, and J. A. Prieto, Characterization of novel neopullulanase from *Bacillus polymyxa*, *Appl. Biochem. Biotechnol.*, 68 (1997) 113–120.
1227. B. Maruo, T. Kobayashi, Y. Tsukano, and N. Yamada, The mechanism of the action of amylosynthase, *J. Agric. Chem. Soc. Jpn.*, 25 (1951) 34–37.
1228. M. J. Kaczmarek and H. Rosenmund, The action of human pancreatic and salivary isoamylases on starch and glycogen, *Clin. Chim. Acta*, 79 (1977) 69–73.
1229. R. Nakagawa and S. Kobayashi, Analysis of degradation products from amylose, amylopectin and starch by isoamylase, *Hokkaidoritsu Kogyo Shikenjo Hokoku*, 289 (1990) 79–88. *Chem. Abstr.*, 117 (1992) 65400.
1230. A. Kimura and J. F. Robyt, Reaction of enzymes with starch granules: Reaction of isoamylase swith native and gelatinized granules, *Carbohydr. Res.*, 287 (1996) 255–261.
1231. J. T. Lai, S. C. Wu, and H. S. Liu, Investigation on the immobilization of *Pseudomonas* isoamylase onto polysaccharide matrixes, *Bioprocess Eng.*, 18(3), (1998) 155–161.
1232. W. P. Chou, P. M. Wang, and W. S. Chu, Preservation of isoamylase adsorbed onto raw corn starch, *Biotechnol. Tech.*, 13 (1999) 259–261.
1233. S. Hisaka, J. Abe, Y. Takahashi, and N. Nakamura, Isoamylase its manufacture with *Flavobacterium odoratum* and manufacture of sugars with isoamylase, Jpn. Kokai Tokkyo Koho 05227959 (1993); *Chem. Abstr.*, 119 (1993) 269192.

1234. S. Hizukuri, T. Kozuma, H. Yoshida, J. Abe, Y. Takahashi, M. Yamamoto, and N. Nakamura, Properties of *Flavobacterium odoratum* KU isoamylase, *Starch/Stärke*, 48 (1996) 295–300.
1235. N. Tsutsumi, H. Bisgard-Frantzen, and A. Svendsen, Starch conversion process using thermostable isoamylases from *Rhodothermus* or *Sulfolobus*, EU Patent WO 9901545 (1999); *Chem. Abstr.*, 130 (1999) 106939.
1236. K. Ohata, M. Shima, and T. Harada, Actions of isoamylase and glucoamylase on starches from different origins after heating at various temperatures and before heating in aqueous suspensions (in Japanese), *Kobe Joshi Daigaku Kiyo*, 21 (1988) 251–257. *Chem. Abstr.*, 109 (1989) 148099d.
1237. M. Kubota, T. Nakada and S. Saikai, Novel maltotetraose-producing amylase and its fermentative manufacture and use, Jpn. Kokai Tokkyo Koho 63240784 (1988); *Chem. Abstr.*, 111 (1989) 513774.
1238. Y. Takasaki, Pullulanase-amylase complex enzyme from *Bacillus subtilis*, *Agric. Biol. Chem.*, 51 (1987) 9–16.
1239. K. Takahashi, A. Totsuka, T. Nakakuki, and N. Nakamura, Production and application of maltogenic amylase by a strain of *Thermomonospora viridis* TF-35, *Starch/Stärke*, 44 (1992) 96–101.
1240. S. Peat, G. J. Thomas, W. J. Whelan, and N. Nakamura, Z-enzyme, *J. Chem. Soc.* (1952) 722–733.
1241. S. Peat, S. J. Pirt, and W. J. Whelan, Enzymic synthesis and degradation of starch. XVI. The purification and properties of β-amylase, *J. Chem. Soc.* (1952) 714–722.
1242. J. M. G. Cowie, I. D. Fleming, C. T. Greenwood, and D. J. Manners, α-1,4-Glucosans. VII. Enzymic degradation and molecular structure of amylase, *J. Chem. Soc.* (1957) 4430–4437.
1243. I. A. Popadich, A. I. Mirovich, N. I. Mironova, and O. E. Stolbikova, Effect of heat treatment on the activity of amylolytic enzymes in flour (in Russian), *Khlebopek. Kondter. Prom.* (5), (1975) 10–11.
1244. V. I. Rodzevich, N. S. Mazur, and O. I. Kapustina, Malt enzymes supplementing the action of amylases during the hydrolysis of starch (in Russian), *Ferment. Spirt. Prom.* (3), (1974) 22–24.
1245. T. Oguma, T. Kurokawa, K. Tobe, S. Kitao, and M. Kobayashi, Purification and some propertie of glucotransferase from Arthrobacter globiformis T-3044, *Oyo Toshitsu Kagaku*, 43 (1996) 73–78. *Chem. Abstr.*, 124 (1996)290496.
1246. J. P. Petzel and P. A. Hartman, A note on starch hydrolysis and β-glucuronidase activity among flavobacteria, *J. Appl. Bacteriol.*, 61 (1986) 421–426.
1247. S. Bielecki, Enzymatic conversions of carbohydrates, In: P. Tomasik, (Ed.), *Chemical and Functional Properties of Food Saccharides*, CRC Press, Boca Raton, 2004, pp. 131–157.
1248. M. Mizuno, T. Tonozuka, K. Ichikawa, S. Kamitori, A. Nishikawa, and Y. Sakano, Three dimensional structure of glucodextrinase a glycoside hydrolase family 15 enzyme, *Biologia (Bratislava)*, 60(Suppl. 16), (2005) 171–176.
1249. X. Sun, X. Wang, and X. Zhang, Investigation on kinetics properties of α-a1,4-glucan lyase from red seaweed *G. lemanaeformis*, *Qingdao Haiyang Daxue Xuebao*, 30 (2000) 230–236. *Chem. Abstr.*, 133 (2000) 146795.
1250. M. Seiji, Digestion of starch by α-limit dextrinase (in Japanese), *J. Biochem. Jpn.*, 40 (1953) 519–525.
1251. S. A. Barker, E. J. Bourne, I. A. Wilkinson, and S. Peat, Enzymic synthesis and degradation of starch. VI. Properties of purified P- and Q-enzymes, *J. Chem. Soc.* (1950) 84–92.
1252. S. A. Barker, E. J. Bourne, I. A. Wilkinson, and S. Peat, Enzymic synthesis and degradation of starch. VII. The mechanism of Q-enzyme action, *J. Chem. Soc.* (1950) 93–99.
1253. S. Peat, W. J. Whelan, and M. J. Bailey, Enzymic synthesis and degradation of starch. XVIII. Minimum chain length for Q-enzyme action, *J. Chem. Soc.* (1953) 1422–1427.
1254. M. Grześkowiak, R. Kaczmarek, and M. Remiszewski, Effect of different factors on the degree of saccharification and the sugar composition in starch hydrolysates produced using a fungal amylolytic preparation (in Polish), *Zesz. Probl. Post. Nauk Roln.*, 159 (1974) 265–273.

1255. I. Maeda, S. Kiribuchi, and M. Nakamura, Digestion of barley starch granules by the combined action of α- and β-amylases purified from barley and barley malt, *Agric. Biol. Chem.*, 42 (1978) 259–267.
1256. L. Słomińska and G. Starogardzka, Application of a multi-enzyme complex in the utilization of potato pulp, *Starch/Stärke*, 39 (1987) 121–125.
1257. J. Fischer, G. Wahl, W. Kempa, G. Schwachula, A. Schellenberger, W. Hettwer, H. W. Mansfeld, K. Haeupke, and W. Hohaus, Manufacture of starch hydrolyzates with immobilized enzymes, East German Patent 264457 (1989); *Chem. Abstr.*, 112 (1990) 19966.
1258. K. Myrbäck, Zur Kenntnis der Stärkespaltung durch pflanzliche Amylase. I, *Biochem. Z.*, 285 (1936) 290–293.
1259. Z. M. Borodina and T. A. Ladur, Liquefaction of starch during enzymic production of glucose (in Russian), *Sakh. Prom.* (2), (1974) 53–58.
1260. S. Ueda, B. C. Saha, and Y. Koba, Direct hydrolysis of raw starch, *J. Microbiol. Sci.*, 1(1), (1984) 21–24.
1261. J. O. Carroll, T. R. Swanson, and P. C. Trackman, Starch liquefaction with alpha amylase mixtures, US Patent 4933279 (1990); *Chem. Abstr.*, 114 (1991) 60471.
1262. P. Rona and J. Hefter, Gleichzeitig Wirkung von Speichelpankreas und Malzamylase auf Stärke, *Biochem. Z.*, 217 (1930) 113–124.
1263. J. Fukumoto, Studies on the production of bacterial amylose. I. (in Japanese), *J. Agr. Chem. Soc. Jpn.*, 19 (1943) 309–314.
1264. H. Okazaki, Amylases from fungi (in Japanese), *J. Agr. Chem. Soc. Jpn.*, 31 (1950) 309–314.
1265. N. G. Kursheva and A. F. Fedorov, Hydrolysis of starch during the separate and combined action of α- and β-amylases from barley (in Russian), *Ferment. Spirt. Prom.*, 34(3), (1968) 18–20.
1266. S. Ueda, Amylolytic system of black-koji molds. II. Raw starch digestibility of the saccharogenic amylase fraction and its interaction with dextrinogenic amylase fraction, *Bull. Agr. Chem. Soc. Jpn.*, 21 (1957) 284–290.
1267. I. A. Preece and M. Shadaksharaswamy, Reducing-group production from starch by the action of α- and β-amylases of barley malt, *Biochem. J.*, 44 (1949) 270–274.
1268. Y. Otani and S. Takahashi, Saccharification of the mixtures of starch, soluble starch, dextrin and maltose by amylase. 3, *J. Ferment. Technol.*, 33 (1955) 467–470.
1269. D. E. Briggs, Salts and the degradation of starch by mixtures of α- and β-amylase, *Enzymologia*, 26 (1963) 355–364.
1270. H. Behmenburg, Production of maltose and glucose through enzymic breakdown of carbohydrates in flour, German Patent 802,334 (1950); *Chem Abstr.*, 45 (1941) 9769.
1271. W. Witt and J. J. Sauter, In vitro degradation of starch grains by phosphorylases and amylases from polar wood, *J. Plant Physiol.*, 146 (1995) 35–40.
1272. A. Spreinat and G. Antranikian, Analysis of the amylolytic enzyme system of *Clostridium thermosulfurogenes* EM1: Purification and synergistic action of pullulanases and maltohexaose forming α-amylase, *Starch/Stärke*, 44 (1992) 305–312.
1273. V. B. Patil and N. B. Patil, Biomass conversion: Synergistic use of immobilized α-amylase and amyloglucosidase for rapid and maximum conversion of starch into glucose, *Indian J. Chem. Technol.*, 7(2), (2000) 47–50.
1274. M. Miranda, I. G. Siso, P. Gonzalez, M. A. Murad, and J. Miron, Amylolysis in systems with α-amylase and glucoamylase. A comparative studies of six producers of evaluation, *Biotechnol. Tech.*, 1 (1987) 195–200.
1275. A. Sroczyński and T. Pierzgalski, Hydrolysis of polysaccharides and oligosaccharides during saccharification of starch by the enzyme-enzyme method (in Polish), *Zesz. Nauk. Politech. Lodz. Chem. Spoz.*, 24 (1974) 59–81.
1276. A. Karakatsanis and M. Liakopoulou-Kyriakides, Comparative study of hydrolysis of various starches by alpha-amylase and glucoamylase in PEG-dextran and PEG-substrate aqueous two phase systems, *Starch/Stärke*, 50 (1998) 349–353.

1277. A. Karakatsanis, M. Liakopoulou-Kyriakides, and M. Stamatoutis, Hydrolysis of various starches by synergistic action of α-amylase and glucoamylase in aqueouos two-phase impeller-agitated systems, *Starch/Stärke*, 49 (1997) 194–199.
1278. V. Arasaratnam and K. Balasubramanian, Synergistic action of α-amylase and glucoamylase on raw corn, *Starch/Stärke*, 45 (1993) 231–233.
1279. M. Fujii and Y. Kawamura, Synergistic action of α-amylase and glucoamylae on hydrolysis of starch, *Biotechnol. Bioeng.*, 27 (1985) 260–265.
1280. H. Nagamoto and H. Inoue, Hydrolysis of soluble starch by complex amylase system. (in Japanese), *Kagaku Kogaku Ronbunshu*, 10 (1984) 698–706. *Chem. Abstr.*, 102 (1984) 130351a.
1281. M. Fujii and Y. Kawamura, Kinetics of hydrolysis of starch by mixed endo- and exo- enzyme systems (in Japanese), *Kagaku Kogaku*, 47 (1983) 786–788.
1282. E. Nebesny, Carbohydrate compositions and molecular structure of dextrins in enzymatic high conversion starch syrups, *Starch/Stärke*, 41 (1989) 431–435.
1283. A. Poli, E. Eposito, L. Lama, P. Orlando, G. Nicolaus, F. de Appolonia, A. Gambacorta, and B. Nicolaus, *Anoxybacillus amylolyticum* sp. Nov. a thermophilic amylase producing bacterium isolated from Mount Rittman (Antarctida), *Syst. Appl. Microbiol.*, 29 (2006) 300–307.
1284. C. Artom and L. Cioglia, The digestion of starch by pancreatic and enteric amylases (in Italian), *Boll. Soc. Ital. Biol. Sper.*, 7 (1932) 1385–1389.
1285. L. Coglia, The combined digestion of starch by the enteric juice and by amylase of different origin (in Italian), *Boll. Soc. Ital. Biol. Sper.*, 8 (1933) 489–492.
1286. W. W. Pigman, Extent of hydrolysis of starches by amylases in the presence and absence of yeasts, *J. Res. Natl. Bur. Standards*, 33 (1944) 105–120.
1287. A. K. Balls and S. Schwimmer, Digestion of raw starch, *J. Biol. Chem.*, 156 (1944) 203–210.
1288. Y. Narita and A. Okamoto, Manufacture of glucose-transferred products from starch hydrolyzates with α-amylasee and α-glucosidase, Jpn. Kokai Tokkyo Koho 04148693 (1992); *Chem. Abstr.*, 117 (1992) 190294.
1289. A. R. J. R. Verplaetse, K. F. P. Koenraad, F. R. G. M. Van Lancker, and A. M. P. Destexhe, Glucose syrup with specific properties and syrup preparation by using amylase mixture, EU Patent WO 9812342 (1998); *Chem. Abstr.*, 128 (1998) 256679.
1290. Y. Miyashita, S. Hirooka, T. Hirohashi, and T. Kimura, Manufacture of high-conversion hydrolyzed starch syrup containing isomaltose, Jpn. Kokai Tokkyo Koho, 63287496 (1988); *Chem. Abstr.*, 110 (1989) 153065.
1291. S. Schwimmer, The role of maltase in the enzymolysis of raw starch, *J. Biol. Chem.*, 161 (1945) 219–234.
1292. H. H. Sato and Y. K. Park, Production of maltose from starch by simultaneous action of β-amylase and Flavobacterium isoamylase, *Starch/Stärke*, 32 (1980) 352–355.
1293. E. Nebesny, Combined enzymic starch hydrolysis, *Starch/Stärke*, 41 (1989) 266–270.
1294. K. Ciesielski, J. Bryjak, and I. Zbyciński, Modeling of starch hydrolysis in two-enzyme system using artificial neural networks (in Polish), *Inz. Chem. Proc.*, 25 (2004) 801–806.
1295. L. Słomińska and G. Starogardzka, Method for enzymatic saccharification of liquid starch milk, Polish Patent 152541 (1991); *Chem. Abstr.*, 115 (1991) 254311.
1296. M. Chiba, S. Mori, H. Tsuji, and R. Oya, Yield-increased glucose manufacture with glucoamylase and mutarotase, Jpn. Kokai Tokkyo Koho 03139289 (1991); *Chem. Abstr.*, 116 (1992) 19754.
1297. N. Zherebtsov, Yu. Serbulov, N. Derkanosova, A. Yakovle, and S. Fomina, Optimization of the enzymic hydrolysis of wheat flour (in Russian), *Khleboprodukty* (9), (1991) 36–38.
1298. H. Okazaki, Joint actions of mold dextrinogenic saccharogenic and maltose- oligosaccharide-transglucosidic enzyme upon starch (in Japanese), *Symp. Enzyme Chem. Jpn.*, 11 (1956) 215–222.

1299. H. Okazaki, Action of a dextrinogenic amylase a glucose-producing amylase and transglucosidase upon starch in the presence and absence of yeast, *Arch. Biochem. Biophys.*, 63 (1956) 322–333.
1300. T. Yamanobe and Y. Takasaki, Amylase from Bacillus effective for production of maltose. IV. Treatment of starch with α-amylase after digestion with β-amylase and pullulanase from *Bacillus cereus* var. mycoides, *Biseibutsu Kogyo Gijutsu Kenkyusho Kenkyu Hokoku*, 50 (1978) 51–62. *Chem. Abstr.*, 89 (1978) 58335s.
1301. T. Yamanobe and Y. Takasaki, Amylases from Bacillus effective for production of maltose. III. Effect of conditions in liquefaction of starch on production of maltose by β-amylae and pullulanase of *Bacillus cereus* var. mycoides, *Biseibutsu Kogyo Gijutsu Kenkyusho Kenkyu Hokoku*, 50 (1978) 41–49. *Chem. Abstr.*, 89 (1978) 58336t.
1302. T. Yamanobe and Y. Takasaki, Studies on amylases from Bacillus effective for production of maltose. III. Effect of conditions in liquefaction of starch on production of maltose by β-amylae and pullulanase of *Bacillus cereus* var. mycoides. (in Japanese), *Hakko Kagaku Kaishi*, 56 (1978) 272–279.
1303. T. Yamanobe and Y. Takasaki, Studies on amylases from Bacillus effective for production of maltose. IV. Further treatment of starch hydrolysates produced by β-amylase and pullulanase of Bacillus sp. by α-amylase of *Bacillus subtilis* and *Aspergillus oryzae* (in Japanese), *Hakko Kagaku Kaishi*, 56 (1978) 280–286.
1304. T. Yamanobe and Y. Takasaki, Amylase from Bacillus effective for production of maltose. VIII. Production of maltose from starch of various origin by β-amylase and and pullulanase of *B. cereus* var, mycoides, *Biseibutsu Kogyo Gijutsu Kenkyusho Kenkyu Hokoku*, 52 (1979) 17–23. *Chem. Abstr.*, 89 (1978) 144997r.
1305. T. Yamanobe and Y. Takasaki, Production of maltose from starch of various origins by β-amylase and pullulanase of *Bacillus cereus* var.mycoides (in Japanese), *Nippon Nogei Kagaku Kaishi*, 53(3), (1979) 77–80.
1306. D. Mikulikova, M. Benkova, and J. Kraič, The potential of common cereals to form retrograded resistant starch, *Czech. J. Genet. Plant Breed.*, 43 (2006) 95–102.
1307. X. Cao and Y. Liu, Application of debranching enzyme in the process of saccharification (in Japanese), *Nianjiu*, 30(4), (2003) 78–79. *Chem. Abstr.*, 143 (2004) 228393.
1308. K. N. Thompson, N. E. Lloyd, and J. A. Johnson, Dextrose using mixed immobilized systems, US Patent 4,102,745 (1978); *Chem. Abstr.*, 90 (1978) 20842n.
1309. R. L. Starnes, C. L. Hoffman, V. M. Flint, P. C. Trackman, D. J. Duhart, and D. M. Katkocin, Reactions of glucoamylases in the biomass conversion of sucrose, *ACS Symp. Ser.*, 460 (1991) 384–393.
1310. H. K. Nielsen, Novel bacteriolytic enzymes and cyclodextrin glycosyl transferases for the food industry, *Food Technol.*, 45 (1991) 102–104.
1311. H. Bender, Branched saccharides formed by the action of his-modified cyclodextrin glycosyltransferase from *Klebsiella pneumonia* M 5 aI on starch, *Carbohydr. Res.*, 222 (1991) 239–244.
1312. U. Schoenebaum, M. Kunz, and K. D. Vorlop, Enzymic hydrophilation of functionalized carbohydrates of technical interest by immobilized cyclodextrin glucosyltransferase (CGT), *Proc. 8th Bioechnol. DECHEMA Meet.*, 4B (1990) 759–762.
1313. T. N. Kaper, B. Talik, T. J. Ettema, H. Bos, M. J. E. C. van der Maarel, and L. Dijkhuizen, Maylomaltase of *Pyrobaculum aerophilum* IM2 produces thermoreversible starch gels, *Appl. Environ. Microbiol.*, 71 (2005) 5098–5106.
1314. Y. Takasaki, Enhancement of glucose yield in saccharification with glucoamylase and glucosyltransferase, Jpn. Kokai Tokkyo Koho, 01202294 (1989); *Chem. Abstr.*, 112 (1990) 75353.
1315. M. Coogan and R. R. B. Russell, Starch hydrolysis by a glucosyltransferase-like enzyme produced by *Streptococcus sanguis*, *Zentrbl. Bakteriol. Suppl.*, 22 (1992) 247–248.

1316. E. Kim, S. Ryu, H. Bae, N. Huong, and S. Lee, Biochemical characterisation of a glycogen branching enzyme from *Streptococcus mutans*: Enzymatic modification of starch, *Food Chem.*, 110 (2008) 979–984.
1317. I. Przylas, K. Tomoo, Y. Terada, T. Takaha, K. Fujii, W. Saenger, and N. Straeter, Crystal structure of amylomaltase from *Thermus aquaticus*, a glycosyltransferase catalyzing the production of large cyclic glucans, *J. Mol. Biol.*, 296 (2000) 873–886.
1318. B. N. Gawande, A. Goel, A. Y. Patkar, and S. N. Nene, Puridfication and properties of novel raw starch degrading cyclomaltodextrin glucanotransferase from *Bacillus fermus*, *Appl. Microbiol. Biotechnol.*, 51 (1999) 504–509.
1319. B. N. Gawande and A. Y. Patkar, Purification and properties of a novel raw starch degrading cyclodextrin glycosyltransferase for *Klebsiella pneumonia* AS-22, *Enzyme Microb. Technol.*, 28 (2001) 733–743.
1320. K. Y. Lee, Y. R. Kim, K. H. Park, and H. G. Lee, Effects of α-glucotransferase treatment on the thermoreversibility and freeze-thaw stability of rice starch, *Carbohydr. Polym.*, 63 (2006) 347–354.
1321. W. H. Peterson, S. W. Scott, and W. S. Thompson, Aus Stärke und Cellulose durch Bakterien gebildeter reduzierender Zucker, *Biochem. Z.*, 219 (1930) 1–6.
1322. O. v. Friedrichs, Stärke und ihre hydrolytische Abbauprodukten, *Ark. Kemi. Min. Geol.*, 5 (1913) No.2.
1323. A. A. Imshenetskii, Biochemical activity of thermophilic bacteria (in Russian), *Izv. Akad. Nauk SSR*, 30 (1941) 671–674.
1324. C. M. Zhang, X. W. Huang, W. Z. Pan, J. Zhang, K. B. Wie, H. P. Klenk, S. K. Tang, W. J. Li, and K. Q. Zhang, *Anoxybacillus tengchongenses* sp. nov. and *Anoxybacillus eryuanensis* sp. nov. facultatively anaerobic, alkalitolerant bacteria from hot springs, *Int. J. Syst. Evol. Microbiol.*, 61 (2011) 118–122.
1325. B. Ollivier, R. A. Mah, T. J. Ferguson, D. R. Boone, J. L. Garcia, and R. Robinson, Emendation of the gene *Thermobacteroides*: *Thermobacteroides proteolyticus* sp. Nov., a proteolytic acetogen from a methanogenic enrichment, *Int. J. Syst, Bacter.*, 35 (1985) 425–428.
1326. M. J. E. C. van der Maarel, A. Veen, and D. J. Wijbenga, *Paenibacillus granivorans* sp. Nov. a new Paenibacillus species which degrades native potato starch granules, *Syst. Appl. Microbiol.*, 23 (2000) 344–348.
1327. A. M. Balayan, V. A. Abelyan, and L. S. Markosyan, Enzymic hydrolysis of starch with thermophilic bacillus (in Russian), *Biol. Zh. Arm.*, 41 (1988) 108–113.
1328. N. Yoshigi, T. Chikano, and M. Kamimura, Studies on production of maltopentaose by *B. cereus* NY-14. I. Characterization of maltopemtaose producing bacterium and its cultural conditions, *Agric. Biol. Chem.*, 49 (1985) 2379–2384.
1329. G. E. Meyers, Paraformaldehyde-resistant starch-fermenting bacteria in starch-base drilling mud, *Appl. Microbiol.*, 10 (1962) 418–421.
1330. I. A. Makrinov, A new microorganism bringing about the fermentation of starch and pectous materials (in Russian), *Arch. Sci. Biol. (Skt. Petersburg)*, 18 (1915) No. 5.
1331. T. Tanaka, Y. Shimomura, M. Taniguchi, and S. Oi, Raw starch-utilizing anaerobe from mesophilic methane sludge and its production of amylase, *Agric. Biol. Chem.*, 51 (1987) 591–592.
1332. G. J. Walker and P. M. Hope, Degradation of starch granules by some amylolytic bacteria from rumen of sheep, *Biochem. J.*, 90 (1964) 398–408.
1333. M. A. Cotta, Intersaction of ruminal bacteria in the production and utilization of maltooligosaccharides from starch, *Appl. Environ. Microbiol.*, 58 (1992) 48–54.
1334. A. R. Abou Akkada, P. N. Hobson, and B. H. Howard, Carbohydrate fermentation by rumen oligotrich protozoa of the genus *Entodinium*, *Biochem. J.*, 73 (1959) 44P–45P.
1335. B. K. Mukerji and D. P. Johari, Transformation of starch by Rhizobium species species, *Indian J. Agr. Sci.*, 38 (1968) 964–970.

1336. K. K. Suk, G. T. Veeder, and D. D. Richey, Polysaccharide and bacterial fermentation for its preparation, US Patent 4,286,059 (1981); *Chem. Abstr.*, 95 (1981) 167148c.
1337. S. Yoshida and H. Morishita, Biodegradation of polysaccharides with strains of *Pseudomonas* sp, *Seikatsu Esei*, 34 (1990) 119–123. *Chem. Abstr.*, 113 (1990) 168718.
1338. P. Cros, Fermentative manufacture of polysaccharides from starch in the presence of amylolytic enzymes, EU Patent EP 319372 (1989); *Chem. Abstr.*, 112 (1990) 97002.
1339. T. Tadokoro, N. Takasugi, and T. Sasaki, The chemical change of starch by *Bacterium cassavanum* (in Japanese), *J. Chem. Soc. Jpn.*, 62 (1941) 1255–1256.
1340. T. Tadokoro and T. Sasaki, Chemical change of starch by *Bacterium cassavanum* (in Japanese), *J. Chem. Soc. Jpn.*, 63 (1942) 751–752.
1341. S. Nowak, A. Heinath, and J. Wesenberg, Enzymic saccharification of cereal starch using yeast, East German Patent 296100 (1991); *Chem. Abstr.*, 116 (1992) 150469.
1342. K. H. Meyer and P. Bernfeld, Reccherches sur l'amidon.XVI. Sur la degradation des hydrates de carbone du groupe de l'amidon per le suc de Lebedew dialyse, *Helv. Chim. Acta*, 24 (1941) 1400–1403.
1343. D. K. Apar and B. Oezbek, Corn, rice and wheat starch hydrolysis by using various alpha-amylase enzymes at temperature 40°C, *J. Eng. Nat. Sci.* (2), (2004) 55–67.
1344. Y. C. Pan and W. C. Lee, Production of high-purity isomalto-oligosaccharides syrup by the enzymatic conversion of transglucosidase and fermentation of yeast cells, *Biotechnol. Bioeng.*, 89 (2005) 794–804.
1345. M. R. Dhawale and W. M. Ingledew, Starch hydrolysis by depressed mutants of *Schwanniomyces castellii*, *Biotechnol. Lett.*, 5 (1983) 825–830.
1346. H. Sahara, A. Kodaka, H. Hisada, Y. Hata, A. Kawato, Y. Abe, and A. Ueda, Glycosylase cell surface expressing yeast for use in polysaccharide degradation, Jpn. Kokai Tokkyo Koho 2005245335 (2005); *Chem. Abstr.*, 143 (2005) 280526.
1347. K. J. Ma, L. L. Lin, H. R. Chien, and W. H. Hsu, Efficient utilization of starch by a recombinant strain of *Saccharomyces cerevisiae* producing glucoamylase and isoamylase, *Biotechnol. Appl. Biochem.*, 31 (2000) 55–59.
1348. C. R. Soccol, S. C. Stertz, M. Raimbault, and L. I. Pinheiro, Biotransformation of solid waste from cassava starch production by *Rhizopus* in solid state fermentation. II. Scale-up in different bioreactors, *Braz. Arq. Biol. Technol.*, 38 (1995) 1319–1326.
1349. Y. Otani, S. Takahash, and M. Inage, The degradation products of potato starch by the amylase from submerged culture of fungi (in Japanese), *Hakko Kagaku Zasshi*, 36 (1958) 235–241.
1350. Y. Otani, S. Takahashi, and E. Yamamoto, The degradation products of potato starch by the β-amylase from submerged culture of fungi (in Japanese), *Hakko Kagaku Zasshi*, 36 (1959) 241–245.
1351. S. K. Garg and H. W. Doelle, Optimization of cassava starch conversion to glucose by *Rhizopus oligosporus*, *J. Appl. Microbiol. Technol.*, 5 (1989) 297–305.
1352. S. K. Garg and H. W. Doelle, Optimization of physiological factors for direct saccharification of cassava starch to glucose by *Rhizopus oligosporus* 145F, *Biotechnol. Bioeng.*, 33 (1989) 948–954.
1353. H. Tokuda, T. Soga, and K. Nakanishi, Hydrolysis of raw starch by *Rhizopus achamydosporus* immobilized with the gas phase cultivation method, *Nippon Jozo Kyokashi*, 92 (1997) 911–914. *Chem. Abstr.*, 128 (997) 812373.
1354. C. R. Soccol, S. C. Stertz, M. Raimbault, and L. I. Pinheiro, Biotransformation of solid waste from cassava starch production by Rhizopus in solid state fermentation. II Optimization of the culture conditions and growth kinetics, *Braz. Arq. Biol. Tecnol.*, 38 (1995) 1311–1318.
1355. D. E. Johnson, G. E. N. Nelson, and A. Ciegler, Starch hydrolysis by conidia of *Aspergillus wentii*, *Appl. Microbiol.*, 16 (1968) 1678–1683.
1356. S. Nishimura, Abbau der Stärke durch der Amylase der *Aspergillus oryzae* und der Malz mit Berücksichtig der Grenzabbaues, *Wochschr. Brau.*, 44 (1927) 533–535.

1357. O. Tanabe, T. Nakamura, H. Tsuru, T. Uchida, and Y. Yamaguchi, Liquefaction and saccharification of starch by molds, *J. Agr. Chem. Jpn.*, 23 (1950) 229–230.
1358. O. Tanabe and H. Tsuru, Liquefaction and saccharification of starch by molds (in Japanese), *J. Agr. Chem. Soc. Jpn.*, 23 (1950) 230.
1359. R. V. Feniksova and A. A. Shilova, Glucoamylase as the basic enzyme in an amylolytic complex of molds ensuring the complete saccharification of starch (in Russian), *Ferment. Spirt. Prom.*, 36(8), (1970) 16–18.
1360. F. Boas, Untersuchungen über Stärkewirkung und Bildung löslicher Stärke bei Schimmelpilze, *Zentr. Bakt. Parasitenk., II Abt.*, 56 (1921) 7–11.
1361. I. Nakajima and J. Kato, The digestion of raw starch by *Aspergillus oryzae* (in Japanese), *J. Ferment. Technol.*, 28 (1950) 220–221.
1362. E. Pinto da Soilva Bon and N. Monteiro, Jr., Fermentative hydrolysis of starch from palms into glucose syrup, Brazilian Patent 2002004544 (2004); *Chem. Abstr.*, 143 (2005) 328042.
1363. K. Christa, M. Soral-Śmietana, and G. Lewandowicz, Buckwheat starch: Structure, functionality and enzyme in vitro susceptibility upon the roasting process, *Int. J. Food Sci. Nutr.*, 60(Suppl. 4), (2009) 140–154.
1364. J. A. Smith and J. R. Frankiewicz, Glucoamylase, culture medium for its production and its use for the hydrolysis of starch, German Patent 2,554,850 (1976); *Chem. Abstr.*, 85 (1976) 92205h.
1365. S. Ueda, Amylolytic system of black koji molds. VI. Action of black koji amylase system on raw starches, especially glutinous and non-glutinous starches (in Japanese), *Nippon Nogei Kagaku Kaishi*, 32 (1958) 648–650.
1366. B. Drews, H. Specht, and E. Rothenbach, Die amylolytische Aktivität von *Aspergillus niger* mit submerser Zuchtung in verschiedener Schleupen, *Biochem. Z.*, 322 (1952) 380–387.
1367. S. Ueda, Amylolytic system of the black koji molds. III. Relation between the susceptibility to digestion and efficiency of adsorption of black koji amylase to raw starch (in Japanese), *Nippon Nogei Kagaku Kaishi*, 31 (1957) 898–902.
1368. S. Ueda, Amylolytic system of the black koji molds. IV. Comparative study of various black koji molds on raw starch digestion (in Japanese), *Nippon Nogei Kagaku Kaishi*, 31 (1957) 902–904.
1369. M. Burger and K. Beran, Mechanism of the action of maltase of *Aspergillus niger*. I. Effect of temperature on the activation of starch hydrolysis by fungal enzyme preparation, *Chem. Listy*, 50 (1956) 133–139.
1370. C. Li, H. Lu, and H. Li, Compound ferment bacteria and producing thereof, Chinese Patent 135057 (2002); *Chem. Abstr.*, 139 (2003) 130422.
1371. R. A. Koutinas, K. Belafi-Bako, A. Kabiri-Badr, A. Toth, L. Gubicza, and C. Webb, Enzymatic hydrolysis of polysaccharides: Hydrolysis of starch by an enzyme complex from fermentation by *Aspergillus awamori*, *Food Bioprod. Proc.*, 79(C1), (2001) 41–45.
1372. A. E. Creedy, P. Jowett, and T. K. Walker, Formation of D-cellobiose from starch and other substrates by *Acetobacter*, *Chem. Ind.* (1954) 1297–1298.
1373. S. A. Filippov, S. P. Avakyants, and S. P. Bartenev, Conditions for the immobilization of cells of microscopic fungi and their use for enzymic hydrolysis of starch (in Russian), *Ferment. Spirt Prom.* (2), (1928) 30–31.
1374. N. Cochet, A. Marcipar, and J. M. Lebeault, Starch degradation by immobilized filamentous fungi, *Cell Immobiliises Colloq.* (1979) 211–227. *Chem. Abstr.*, 94 (1980) 154972w.
1375. S. L. Slaughter, P. R. Ellis, and P. J. Butterworth, An investigation of the action of porcine pancreatic α-amylase on native and gerlatinized starches, *Biochim. Biophys. Acta Gen. Subj.*, 1525 (2001) 29–36.
1376. Y. Higuchi, A. Ohashi, H. Imachi, and H. Harada, Hydrolytic activity of alpha-amylase in anaerobic digested sludge, *Water Sci. Technol.*, 52(1–2), (2005) 259–266.
1377. G. J. Walker and P. M. Hope, Action of some α-amylases on starch granules, *Biochem. J.*, 86 (1963) 52–62.

1378. E. J. Bourne, D. A. Sitch, and S. Peat, Enzymic synthesis and degradation of starch. III. The role of carbohydrate activators, *J. Chem. Soc.* (1949) 1448–1457.
1379. E. A. Sym, Einfluß des kolloidalen Zustandes der Stärke und Amylaselösung auf die Geschwindigkeit der Amylolyse, *Biochem. Z.*, 251 (1932) 116–132.
1380. P. Petit, and Richard, Influence des mode de dissolution de l'amylase sur la saccharification de l'amidon, *Compt. Rend.*, 181 (1925) 575–577.
1381. T. Omori, The influence of chloride ion on the salivary amylase (in Japanese), *J. Biochem. Jpn.*, 14 (1931) 339–342.
1382. L. Ambard and S. Trautmann, Rolle der H^+ und Cl^- Ionen bei der Aktievirung der Amylase, *Compt. Rend. Soc. Biol. Paris*, 112 (1933) 1532–1533. *Chem. Zentr.*, 1934, I, 3220.
1383. S. Ueda, Amylase adsorption on raw starch and its relation to raw starch digestion (in Japanese), *Dempun Kagaku*, 25 (1978) 124–131.
1384. K. A. Chandrokar and N. P. Badenhuizen, How meaningful are determinations of glucosyltransferas activation in starch-enzyme complexes, *Starch/Stärke*, 18 (1966) 91–95.
1385. Y. G. Ann, T. Anindyawati, K. Ito, M. Iizuka, and N. Minamuira, Stabilization of amylolytic enzymes by modification with periodate-oxidized starch, *Hanguk Sikpum Yongyang Hakhoechi*, 11 (1998) 561–564. *Chem. Abstr.*, 132 (1999) 136568.
1386. K. Nagai, Investigations into the nature of starch. I. Ultramicroscopic studies on the fermentive processs of starch (in Japanese), *Acta Schol. Med. Univ. Imp. Kyoto*, 7 [IV] (1925) 569–575.
1387. K. Myrbäck, Constitution of cozymase, *Skand. Arch. Physiol.*, 77 (1937) 58–59. *Chem. Abstr.*, 28 (1937) 4750.
1388. A. Hock, Verdauung von verschiedene Stärkeartes, *Biedermanns Zentr. B. Tiereraehr.*, 10 (1938) 3–20.
1389. F. Nowotny, The influence of malt and barley amylase on natural not pasted starch (in Polish), *Pol. Agr. Forest Ann.*, 45 (1938) 1–36.
1390. N. V. Bhide and D. L. Sahasrabuddhe, Comparative disintegration of flours and starches with hydrochloric acid, diastase and pancreatin, *Poona Agr. Coll. Mag.*, 35 (1943) 17–27.
1391. E. Masłyk, W. Leszczyński, and A. Gryszkin, Modification-induced changes in potato starch susceptibility to amylolytic enzyme attack, *Pol. J. Food Nutr. Sci.*, 12(1), (2003) 54–56.
1392. J. Fukumoto, T. Yamamoto, and K. Ichikawa, Starch hydrolysis by bacterial amylases (in Japanese), *Symp. Enzyme Chem. Jpn.*, 8 (1953) 40–45.
1393. I. A. Popadich, N. N. Tregubov, A. I. Mirovich, N. I. Mironova, and E. D. Bondareva, Action of enzymic preparations on different types of starch (in Russian), *Izv. Vyssh. Uchebn. Zaved. Pishch. Tekhnol.* (2), (1975) 34–37.
1394. A. Sroczyński and T. Pierzgalski, Transformation of carbohydrates during enzymic solubulization of starch (in Polish), *Zesz. Nauk. Politech. Lodz. Chem. Spoz.*, 23 (1975) 171–198.
1395. G. J. Woo and J. D. McCord, Bioconversion of unmodified native starches by *Pseudomonas stutzeri* maltohydrolase: Effect of starch type, *Appl. Microbiol. Biotechnol.*, 38 (1993) 586–591.
1396. S. Sringam, Comparison of enzymic hydrolyses of four kinds of starch, *Kasetsart J. Nat. Sci.*, 28 (1994) 264–272. *Chem. Abstr.*, 123 (1995) 167916.
1397. M. V. Ramesh and B. K. Lonsane, End product profiles of starch hydrolysis by bacterial alpha-amylases at different temperature and pH values, *Biotechnol. Lett.*, 11 (1989) 649–652.
1398. E. Herzfeld and R. Klinger, Berichtigung und Erganzung zu unser Arbeit "Zur Chemie der Polysaccharide", *Biochem. Z.*, 112 (1920) 55–60.
1399. J. O'Sullivan, Influence of the temperature, concentration, duration of the mash and slackness of malt on the starch products of the extract of malt, *J. Soc. Chem. Ind.*, 39 (1920) 22T–27T.
1400. P. Rosendal, B. H. Nielsen, and N. K. Lange, Stability of bacterial α-amylase in the starch liquefaction process, *Starch/Stärke*, 31 (1979) 368–372.
1401. W. Windisch, Untersuchungen auf Diastase, *Wochschr. Brau.*, 42 (1925) 136–138.

1402. W. Windisch, Mechanismus de Wirkung von Amylase, *Wochschr. Brau.*, 42 (1925) 148–150, 153–156, 159, 166–167, 171–173.
1403. F. Chodat and M. Philia, Phänomenon voon Ambard.(Adsorption von Amylase durch Stärke), *Compt. Rend. Seances Soc. Phys. Hist. Nat. Geneve*, 41 (1924) 118–122. *Chem. Zentr.*, 1924, II, 2850.
1404. Ya. P. Barmenkov, The electrokinetic potential of starch and its role in enzymic hydrolysis (in Russian), *Biokhimyia*, 3 (1938) 740–750.
1405. J. L. Baker and H. F. E. Hulton, Action of diastase on starch granules, I, *J. Chem. Soc.*, 105 (1913) 1529–1536.
1406. M. Lisbonne and E. Vulquin, Inactivation of malt diastase by electrical dialysis and activation by electrolytes, *Compt. Rend. Soc. Biol.*, 72 (1914) 936–938. *Chem. Abstr.*, 10 (1914) 639.
1407. W. Banks and C. T. Greenwood, Starch degrading enzymes. X. Theories of the stepwise degradation of linear polysaccharides: Influence of the molecular weight distribution of the substrate on the apparent mechanism, *Starch/Stärke*, 20 (1968) 316–319.
1408. J. C. Curvelo-Santana, A. F. Libranz, and E. Tambourgi, Approach model for simulation of the starch hydrolysis by ?-amylase and alcohol production from manioc root starch, *Chem. Prod. Process Model.*, 4 (2009) 1394–2659.
1409. Y. Kozaki and H. Inoue, Adsorption of α-amylase onto raw starch (in Japanese), *Hakko Kogaku Kaishi*, 62 (1984) 111–117.
1410. A. Hoschke, E. Laszlo, and J. Hollo, A study of the role of histidine side-chains at the active centre of amylolytic enzymes, *Carbohydr. Res.*, 81 (1980) 145–156.
1411. A. Hoschke, E. Laszlo, and J. Hollo, A study of the role of tyrosine groups at the active centre of amylolytic enzymes, *Carbohydr. Res.*, 81 (1980) 157–166.
1412. B. Svensson, H. Jespersen, M. R. Sierks, and E. A. MacGregor, Sequence homology between putative raw-starch binding domains from different starch degrading enzymes, *Biochem. J.*, 264 (1989) 309–311.
1413. S. Hayashida, Y. Teramoto, T. Inoue, and S. Mitsuiki, Occurrence of an affinity site apart from the active site on the raw starch-digesting but non-raw-starch adsorbable *Bacillus subtilis* 65 α-amylase, *Appl. Environ. Microbiol.*, 56 (1990) 2584–2586.
1414. R. Rodriguez-Sanoja, B. Ruiz, J. P. Guyot, and S. Sanchez, Starch binding domain affects catalysis in two *Lactobacillus* α-amylases, *Appl. Environ. Microbiol.*, 71 (2005) 297–302.
1415. N. Juge, J. Nohr, M. F. Le Gal-Coeffet, B. Kramhoft, C. S. M. Furniss, V. Planchot, D. B. Archer, G. Williamson, and B. Svensson, The activity of barley α-amylase on starch granules is enhanced by fusion of a starch binding domain from *Aspergillus niger* glucoamylase, *Biochim. Biophys. Acta*, 1764 (2006) 275–284.
1416. Anonymus, Study on maltotetraose forming amylase structure (in Japanese), *Hyogo-kennitsu Kogyo Gijutsu Senta Kenkyu Hokokusho*, 9 (1999) 55. *Chem. Abstr.*, 133 (2000) 307081.
1417. R. Rodriguez Sanoja, J. Morlon-Guyot, J. Jore, J. Pintado, N. Juge, and J. P. Guyot, Comparative characterization of complete and truncated forms of *Lactobacillus amylovorus* α-amylase and role of the C-terminal direct repeats in raw-starch binding, *Appl. Environ. Microbiol.*, 66 (2000) 3350–3356.
1418. S. Hayashida, S. Kunisaki, M. Nakao, and P. Q. Flor, Evidence for raw starch-affinity site on *Aspergillus awamori* glucoamylase I, *Agric. Biol. Chem.*, 46 (1982) 83–89.
1419. J. Abe, The structure and action of fungal glucoamylases (in Japanese), *Denpun Kagaku*, 36 (1989) 51–57.
1420. Y. Hata, K. Kitamoto, K. Gomi, C. Kumagai, and G. Tamura, Functional elements of the promoter region of the *Aspergillus oryzae* galA gene encoding glucoamylase, *Curr. Genetics*, 22 (1992) 85–91.

1421. T. Takahashi, N. Muroi, M. Irie, and Y. Ikegami, Different binding behavior to chitin of multiple forms of glucoamylase from *Aspergillus saitoi* and *Rhizopus* sp, *Chem. Pharm. Bull.*, 39 (1991) 2387–2390.
1422. N. J. Belshaw and G. Williamson, Specificity of the binding domain of glucoamylase 1, *Eur. J. Biochem.*, 211 (1993) 717–724.
1423. L. Chen, Deletion analysis of the starch binding domain from Aspergillus glucoamylase, *Diss. Abstr., Int. B*, 53 (1993) 5072.
1424. R. Fagerstroem, Evidence for a polysaccharide-binding domain in *Hormoconis resinae* glucoamylase P: Effects of its proteolytic removal on substrate specificity and inhibition by β-cyclodextrin, *Microbiology (Reading, UK)*, 140 (1994) 2399–2407.
1425. M. Goto, T. Semimaru, K. Furukawa, and S. Hayashda, Analysis of the raw starch-binding domain by mutation of glucoamylase from *Aspergillus awamori* var. kawachi expressed in *Saccharomyces cerevisae*, *Appl. Environ. Microbiol.*, 60 (1994) 3926–3930.
1426. K. Sorimachi, M. F. Le Gal-Coeffet, G. Williamson, D. B. Archer, and M. P. Williamson, Solution structure of the granular starch binding domain of *Aspergillus niger* glucoamylase, *Structure (London)*, 5 (1997) 647–681.
1427. K. Kimura, S. Kataoka, A. Nakamura, T. Takano, S. Kobayashi, and K. Yamane, Functions of the carboxy terminal region of cyclodextrin glucanotransferase of alkalophilic *Bacillus* sp. #1011: Relation to catalyzing activity and pH stability, *Biochem. Biophys. Res. Commun.*, 161 (1989) 1273–1279.
1428. J. R. Villette, F. S. Krzewinski, P. J. Looten, P. J. Sicard, and S. J. L. Bouquelet, Cyclomaltodextrin glucanotransferase from *Bacillus circulans* E192. IV. Evidence for a raw starch-binding site and its interaction with a β-cyclodextrin copolymer, *Biotechnol. Appl. Biochem.*, 16 (1992) 57–63.
1429. D. Penninga, B. A. van der Veen, R. M. A. Knegtel, S. A. F. T. van Hilum, H. J. Rozenboom, K. H. Kalk, B. W. Dijkstra, and L. Dijkhuizen, The raw starch binding domain of cyclodextrin glycosyltransferase from *Bacillus circulans* strain 251, *J. Biol. Chem.*, 271 (1996) 32777–32784.
1430. K. Ohdan, T. Kuriki, H. Takata, and S. Okada, Cloning of the cyclodextrin glucanotransferase gene from alaklophilic Bacillus sp. A2-5a and analysis of the raw starch-binding domain, *Appl. Microbiol. Biotechnol.*, 53 (2000) 430–434.
1431. A. R. Plant, R. M. Clemens, H. W. Morgan, and R. M. Daniel, Active site and substrate specificity of *Thermoanaerobium* Tok6-B1 pullulanase, *Biochem. J.*, 246 (1987) 537–541.
1432. G. Francis, L. Brennan, S. Stretton, and T. Ferenci, Genetic mapping of starch and lambda-receptor sites in maltoporin. Identification of substitutions causing direct and indirect effects on binding sites by cysteine mutagenesis, *Mol. Microbiol.*, 5 (1991) 2293–2301.
1433. N. A. Zhrebtsov, I. D. Ruadze, and A. N. Yakovlev, Mechanism of acid catalyzed and enzymic hydrolysis of starch (in Russian), *Prikl. Biokhim. Mikrobiol.*, 31 (1995) 599–603.
1434. M. Halpern and Y. Leibowitz, Stereochemical mechanism of starch hydrolysis by β-amylase, *Bull. Res. Counc. Isr.*, A6 (1957) 131–132.
1435. M. Halpern and Y. Leibowitz, The stereochemical mechanism of starch hydrolysis by α-amylase, *Bull. Res. Counc. Isr.*, A8 (1959) 41.
1436. M. Halpern and J. Leibowitz, Mechanism of the glucosidic bond cleavage in the enzymic hydrolysis of starch, *Biochim. Biophys. Acta*, 36 (1959) 29–34.
1437. M. Samec, Change in molecular structure in the enzymic degradation of starch (in Serbo-Croatian), *Kem. Zbornik* (1951) 7–9.
1438. C. S. Hanes, Studies on plant amylases. I. Effect of starch concentration upon the velocity of hydrolysis by the amylase of germinated barley, *Biochem. J.*, 26 (1932) 1406–1421.
1439. S. Fajbergerówna, Structure and enzymic reactions. X. Influence of viscosity on the state of aggregation of the dispersed phase (in Polish), *Acta Biol. Exp. (Warsaw)*, 6 (1931) 143–172.

1440. J. Berkson and L. B. Flexner, The rate of reaction between enzyme and substrate, *J. Gen. Physiol.*, 11 (1928) 433–457.
1441. I. Stein, The saccharification process and its effect on amylase activity (in Slovak), *Chem. Zvesti*, 4 (1950) 225–268.
1442. V. A. von Klinkerberg, Über die Spezifiät der Amylasen. II. Die enzymatische Analyse von Stärke und Glykogen, *Z. Physiol. Chem.*, 212 (1932) 173–195.
1443. L. M. Marchal and J. Tramper, Hydrolytic gain during hydrolysis reaction: Implications and coreaction procedures, *Biotechnol. Tech.*, 13 (1999) 325–328.
1444. O. Holmbergh, Adsorption von α-Amylase, *Biochem. Z.*, 266 (1933) 203–215.
1445. J. Chełkowski and T. Zegar, Hydrolysis of starch with bacterial α-amylase (in Polish), *Przem. Spoz.*, 32 (1978) 298–300.
1446. G. Rafaeli, Studies on behavior of amylase in the presence of crude starch (in Italian), *Arch. Ital. Biol.*, 80 (1928) 161–166.
1447. W. O. James and M. Cattle, The physiological importance of the mineral elements in plants. VI. The influence of potassium chloride on the rate of diastatic hydrolysis of starch, *Biochem. J.*, 27 (1933) 1805–1809.
1448. M. S. Shulman and L. M. Vainer, Amylase adsorption by starch (in Russian), *Prikl. Biokhim. Mikrobiol.*, 3 (1967) 717–722.
1449. L. M. Vainer and M. S. Shulman, Preparation of amylase adsorbed on modified starch (in Russian), *Prikl. Biokhim. Mikrobiol.*, 5 (1969) 364–366.
1450. R. M. Sandstedt and S. Ueda, α-Amylase adsorption on raw starch and its relation to raw starch digestion, *Denpun Kogyo Gakkaishi*, 17 (1969) 215–228.
1451. V. M. Leloup, P. Colonna, and G. Marchis-Mouren, Mechanism of the adsorption of pancreatic alpha-amylse onto starch crystallites, *Carbohydr. Res.*, 232 (1992) 367–374.
1452. V. M. Leloup, P. Colona, and S. G. Ring, α-Amylase adsorption on starch crystallites, *Biotechnol. Bioeng.*, 38 (1991) 127–134.
1453. V. Leloup, P. Colonna, and S. Ring, α-Amylase adsorption on crystalline starch, *Food Hydrocoll.*, 5 (1991) 225–227.
1454. B. Chatterjee, A. Ghosh, and A. Das, Starch digestion and adsorption by beta-amylase of *Emericella nidulans* (*Aspergillus nidulans*), *J. Appl. Bacteriol.*, 72 (1992) 208–213.
1455. S. Medda, B. C. Saha, and S. Ueda, Raw starch adsorption and elution behavior of glucoamylase I of black *Aspergillus*, *J. Ferment. Technol.*, 60 (1982) 261–264.
1456. R. F. Tester and M. D. Sommerville, The effect of non-starch polysaccharide on the extent of gelatinization, swelling and α-amylase hydrolysis of maize and wheat starches, *Food Hydrocoll.*, 17 (2003) 41–64.
1457. R. Koukiekolo, V. Desseaux, Y. Moreau, G. Marchis-Mouren, and M. Santimone, Mechanism of porcine pancreatic α-amylase. Inhibition of amylose and maltopentaose hydrolysis by α-, β- and γ-cyclodextrins, *Eur. J. Biochem.*, 268 (2001) 841–848.
1458. P. M. Wojciechowski, A. Kozioł, and A. Noworyta, Iteration model of starch hydrolysis by amylolytic enzymes, *Biotechnol. Bioeng.*, 75 (2001) 530–539.
1459. P. M. Wojciechowski, Modelling of enzymic depolymerization using Monte Carlo method (in Polish), *Wiad. Chem.*, 55 (2001) 473–490.
1460. J. Bryjak and L. J. Królikowski, Modeling of enzymic liquefaction of starch (in Polish), *Inz. Chem. Process.*, 25 (2004) 745–750.
1461. T. R. Swanson, J. O. Carroll, R. A. Britto, and D. J. Duhart, Development and field confirmation of a mathematical model for amyloglucosidase/pullulanase saccharification, *Starch/Stärke*, 38 (1986) 382–387.
1462. D. K. Apar and B. Ozbek, α-Amylase inactivation during corn starch hydrolysis process, *Process Biochem.*, 39 (2004) 1877–1892.

1463. V. Komolprasert and R. Y. Ofoli, Starch hydrolysis kinetics of *Bacillus licheniformis* α-amylase, *J. Chem. Technol. Biotechnol.*, 51 (1991) 209–223.
1464. E. K. Pye and L. B. Wingard, *Enzyme Engineering*, Vol. 2, (1973) Plenum Press, New York.
1465. B. M. Schoonees, Starch hydrolysis using α-amylase. A laboratory evaluation using response surface methodology, *Proc. 78th Annu. Congr. S. Afr. Sugar Technol. Assoc.*, 2003, pp. 427–439. *Chem. Abstr.*, 142 (2004) 431837.
1466. L. Pincussen and T. Oya, Fermente und Licht. Versuche über den Einfluß der temperatur bei der Lichtwirkung, *Biochem. Z.*, 207 (1929) 410–415.
1467. E. Semmens, Hydrolysis of starch grains by light polarized by small particles, *Nature*, 117 (1926) 821–822.
1468. E. C. C. Baly and E. S. Semmens, Selective photochemical action of polarized light. I. The hydrolysis of starch, *Proc. Roy Soc. (London)*, 97B (1924) 250–253.
1469. A. E. Navez and B. B. Rubenstein, Starch hydrolysis as affected by light, *J. Biol. Chem.*, 95 (1932) 645–660.
1470. M. Fiedorowicz, A. Konieczna-Molenda, G. Khachatryan, H. Staroszczyk, and P. Tomasik, Polarized light – A powerful tool in physical and enzymatic processing of polysaccharides, In: E. S. Lazos, (Ed.), *Proc. 5th Int. Congr. Food Technology, Thessaloniki*, Vol. 2, Hellenic Association of Food Technologists, Thessaloniki, 2007, pp. 181–190.
1471. M. Fiedorowicz and G. Khachatryan, Effect of illumination with the visible polarized and nonpolarized light on α-amylolysis of starches of different botanical origin, *J. Agric. Food Chem.*, 51 (2003) 7815–7819.
1472. A. Konieczna-Molenda, M. Fiedorowicz, and P. Tomasik, The polarized light-induced enzymatic formation and degradation of biopolymers, *Macromol. Symp.*, 272 (2008) 117–124.
1473. M. Fiedorowicz, A. Konieczna-Molenda, and G. Khachatryan, Light stimulated enzymatic processes of selected polysaccharides, In: P. Tomasik, V. P. Yuryev, and E. Bertoft, (Eds.) *Starch. Progress in Basic and Applied Science,* Polish Society of Food Technologists, Malopolska Branch, Cracow, 2007, pp. 191–199.
1474. M. Fiedorowicz, A. Molenda-Konieczny, G. Khachatryan, and P. Tomasik, Light stimulated enzymatic reactions of polysaccharides, Pol. Patent, PL 211549 (2012).
1475. E. S. Semmens, Hydrolysis of starch grains by polarized infrared radiation, *Nature*, 163 (1949) 371.
1476. K. Fuke, T. Yukimasa, H. Yaku, and H. Oka, Speeding up biopolymer mediated reactions with microwave radiation, Jpn. Kokai, Tokkyo Koho, 2003265175 (2003); *Chem. Abstr.*, 139 (2003) 257745.
1477. H. Zhang, M. Zhang, X. Shen, and X. Lu, Optimization of parameters for preparation of corn retrograded starch by microwave – Enzyme method, *Shipin Kexue*, 28 (2007) 237–240. *Chem. Abstr.*, 149 (2008) 199102.
1478. E. Hirakawa, Method for control of enzymatic activities with magnetic field and ist application to manufacture and processing of foods manufacture of pharmaceuticals enzyme detergents and treatment of waste, Jpn. Kokai Tokkyo Koho 2005218437 (2005); *Chem. Abstr.*, 143 (2005) 1892124.
1479. Y. Li, S. Guo, L. Li, and M. Cai, Enhancement of starch enzymolysis and ultrafiltration in membrane reactor by using an electric field (I), *Zhengzhou Liangshi Xueyuan Xuebao*, 20(4), (1999) 28–32. *Chem. Abstr.*, 133 (2000) 57937.
1480. R. Qian, Effect of gamma radiation on α-amylase and amyloglucosidase, *Wuxi Qinggongye Xueyuan Xuebao*, 11 (1992) 192–196. *Chem. Abstr.*, 120 (1994) 321755.
1481. E. L. King, Unusual kinetic consequences of certain enzyme catalysis mechanism, *J. Phys. Chem.*, 60 (1956) 1378–1381.
1482. V. I. Kovalenko, L. A. Danilova, A. Kolpakchi, and K. A. Kalunyants, Kinetics of enzymic hydrolysis in the malting process (in Russian), *Ferment. Spirt. Prom.* (5), (1984) 31–34.

1483. S. Adachi, K. Hashimoto, and K. Mukai, A rate expression of liquefaction of starch by α-amylase, *Nihama Kogyo Koto Senmon Gakko Kiyo Rikogaku-hen*, 24 (1988) 77–83. *Chem. Abstr.*, 109 (1989) 75600y.
1484. M. Ulmann and M. Richter, Rate constant for the hydrolysis of potato starch by diastase, *Monatsber. Deut. Akad. Wiss. Berlin*, 1 (1959) 123–133. *Chem. Abstr.*, 54 (1960) 3554.
1485. H. Sobue and K. Matsumoto, Kinetics of enzyme action. I. Action of amylase on starch (in Japanese), *J. Chem. Soc. Jpn. Ind. Chem. Sect.*, 52 (1949) 331–332.
1486. I. A. Popadich, N. N. Tregubov, A. I. Mirovich, N. I. Mironova, and E. D. Bondareva, Hydrolysis of starches of different origins by some amylolytic enzyme preparations (in Russian), *Ferment. Spirt. Prom.* (7), (1975) 33–36.
1487. A. Marc, J. M. Engasser, M. Moll, and F. Flayeux, A kinetic model of starch hydrolysis by α- and β-amylases during mashing, *Biotechnol. Bioeng.*, 25 (1983) 481–486.
1488. J. T. Park and J. E. Rollings, Effects of substrate branching characteristics on kinetics of enzymatic depolymerization of mixed linear and branched polysaccharides: II. Amylose/amylopectin α-amylolysis, *Biotechnol. Bioeng.*, 44 (1994) 792–800.
1489. J. T. Park and J. E. Rollings, Effects of substrate branching characteristics on kinetics of enzymatic depolymerization of mixed linear and branched polysaccharides: II. Amylose/glycogen α-amylolysis, *Biotechnol. Bioeng.*, 46 (1995) 36–42.
1490. S. I. Pronin, Kinetics of starch hydrolysis by pancreatic amylase. III, (in Russian), *Biokhimyia*, 5 (1940) 648–657.
1491. J. W. Van Dyk, J. T. King, and M. L. Caldwell, The study of the kinetics of the hydrolysis of waxy maize starch by crystalline pancreatic amylase from swine, *J. Am. Chem. Soc.*, 78 (1956) 3343–3345.
1492. J. W. Van Dyk and M. L. Caldwell, The kinetics of the hydrolysis of waxy maize starch by crystalline pancreatin amylase from swine, *J. Am. Chem. Soc.*, 78 (1956) 3345–3350.
1493. G. Orestano, Comparison of the enzymic liquefaction and saccharification of starch. II. Pancreatic amylase (in Italian), *Bull. Soc. Chim. Biol.*, 14 (1932) 1531–1551.
1494. S. I. Pronin, Dependence of the kinetics of the enzymic hydrolysis of starch on the enzyme concentration (in Russian), *Byul. Vsesoyuz. Inst. Eksp. Med.*, 3 (1937) 419–422.
1495. C. Artom and G. Orestano, Comparison of the enzymic liquefaction and saccharification of starch. I. Soy bean amylase (in Italian), *Bull. Soc. Chim. Biol.*, 13 (1931) 516–541.
1496. S. L. Slaughter, P. J. Butterworth, and P. R. Ellis, Mechanism of the action of porcine pancreatic α-amylase on native and heat treated starches from various botanical sources, *Roy. Chem. Soc. Spec. Publ.*, 27 (2001) 110–115.
1497. G. Zhang, Y. Shi, Y. Wei, and S. Ouyang, Kinetics of buckwheat starch enzymatic hydrolysis with fungal α-amylase, *Nongye Gongcheng Xuebao*, 23(5), (2007) 42–46. *Chem. Abstr.*, 149 (2008) 265870.
1498. E. A. Chistyakova and A. R. Sapronov, Kinetics of the enzymic hydrolysis of barley starch (in Russian), *Ferment. Spirt. Prom.* (5), (1972) 33–35.
1499. T. Kitano, The constants of the reaction velocities of the amylolytic and maltatic actions (in Japanese), *J. Chem. Soc. Ind. Jpn.*, 38(Suppl), (1935) 381–385.
1500. W. E. Henderson and W. M. Teague, A kinetic model of *Bacillus stearothermophilus* α-amylase under process conditions, *Starch/Stärke*, 40 (1988) 412–418.
1501. E. Raabe and D. Knorr, Kinetics of starch hydrolysis with *Bacillus amyloliquefaciens* α-amylase under high hydrostatic pressure, *Starch/Stärke*, 48 (1996) 409–414.
1502. T. Heitmann, E. Wenzig, and A. Mersmann, Characterization of three different potato starches and kinetics of their enzymic hydrolysis by an α-amylase, *Enzyme Microb. Technol.*, 20 (1997) 259–267.
1503. S. Liu and H. Tan, Kinetic model of enzyme-catalyzed saccharification of corn starch, *Chongqinq Daxue Xuebao, Ziran Kexueban*, 20(1), (1997) 109–111. *Chem. Abstr.*, 127 (1997) 249621.

1504. S. Zhong, L. Zhang, and Q. Hu, Kinetic and mathematical simulation of degrading potato starch with α-amylase, *Zhongguo Liangyou Xuebao*, 12(2), (1997) 34–38. *Chem. Abstr.*, 127 (1997) 462067.
1505. P. Reinikainen, T. Suortti, J. Olkku, V. Malkki, and P. Linko, Extrusion cooking in enzymatic liquefaction of wheat starch, *Starch/Stärke*, 38 (1986) 20–26.
1506. Continental Engineering Ingeniuersbureau voor de Process Industrie N.V. Hydrolysis of starch, Dutch Patent 6,504,840 (1966); *Chem. Abstr.*, 66 (1966) 56988h.
1507. D. Paolucci-Jeanjean, M. P. Belleville, N. Zakhia, and G. M. Rios, Kinetics of cassava starch hydrolysis with Termamyl enzyme, *Biotechnol. Bioeng.*, 68 (2000) 71–77.
1508. K. H. Meyer and P. Bernfeld, Die Konstitution des Amylopektins, *Compt. Rend. Soc. Phys. Hist. Nat. Geneve*, 57 (1940) 89–91.
1509. K. H. Meyer and J. Press, Recherches sur l'amidon, IX. Degradation per la β-amylase toi d'action des masses, *Helv. Chim. Acta*, 24 (1941) 50–58.
1510. W. D. Claus, Kinetics of β-amylase action in 20% starch pastes at elevated temperature, *Cereal Chem.*, 24 (1947) 59–69.
1511. D. M. Belenkii, Kinetics of the hydrolysis of different substrates by rabbit liver α-amylase (in Russian), *Dokl. Akad. Nauk SSSR*, 205 (1972) 230–232.
1512. F. Shiraishi, K. Kawamaki, and K. Kusunoki, Kinetics of condensation of glucose into maltose and isomaltose in hydrolysis of starch by glucoamylase, *Biotechnol. Bioeng.*, 27 (1985) 498–502.
1513. L. P. Pashchenko, N. A. Zherebtsov, and A. V. Zubchenko, Kinetics of starch hydrolysis by glucoamylases of *Aspergillus awamori* and *Rhizopus delemar* B (in Russian), *Izv. Vyssh. Uchebn. Zaved., Pishch. Tekhnol.* (2), (1975) 79–82.
1514. V. L. Yarovenko, B. M. Nakhmanovich, T. V. Fedorova, and E. M. Belov, Kinetics of the enzymic hydrolysis of barley starch (in Russian), *Ferment. Spirt. Prom.* (2), (1974) 8–11.
1515. N. G. Cherevko, Study of the kinetics of enzymic hydrolysis of starch (in Ukrainian), *Visn. Lviv. Politekh, Inst.*, 111 (1977) 28–30.
1516. H. Tatsumi and H. Katano, Kinetic analysis of enzymic hydrolysis of raw starch by glucoamylase using amperometric glucose sensor, *Chem. Lett.*, 33 (2004) 692–693.
1517. H. Tatsumi and H. Katano, Kinetics of the surface hydrolysis of raw starch by glucoamylase, *J. Agric. Food Chem.*, 53 (2005) 8123–8127.
1518. M. Miranda, M. A. Murado, A. Sanroman, and J. M. Lema, Mass transfer control of enzymic hydrolysis of polysaccharides by glucoamylase, *Enzyme Microb. Technol.*, 13 (1991) 142–147.
1519. A. Sanroman, M. A. Murado, and J. M. Lema, The influence of substrate structure on the kinetics of the hydrolysis of starch by glucoamylase, *Appl. Biochem. Biotechnol.*, 59 (1996) 329–336.
1520. S. Zhong, S. Liu, and L. Zhang, Kinetics of saccharification of potato starch with glucoamylase, *Zhongguo Liangyou Xuehui*, 13(1), (1998) 21–25, 44; *Chem. Abstr.*, 129 (1998) 148235.
1521. R. de Lima Camargo Giordano and W. Schmidell, Kinetic study of starch enzymic hydrolysis: Glucose inhibition (in Portuguese), *Braz. Rev. Microbiol.*, 20 (1989) 376–381.
1522. K. Kusunoki, K. Kawakami, F. Shiraishi, K. Kato, and M. Kai, A kinetic experiment for hydrolysis of soluble starch by glucoamylase, *Biotechnol. Bioeng.*, 24 (1982) 347–354.
1523. S. Adachi, K. Hashimoto, and M. Ueyama, An empirical rate expression for saccharfication of starch by glucoamylase (in Japanese), *Nihama Kogyo Koto Senmon Gakko Kiyo Rikogaku-hen*, 22(1), (1986) 97–102. *Chem. Abstr.*, 105 (1987) 77541w.
1524. S. Adachi, K. Hashimoto, and M. Ueyama, A kinetic model for saccharification of liquified starch by glucoamylase (in Japanese), *Nihama Kogyo Koto Senmon Gakko Kiyo Rikogaku-hen*, 22(2), (1986) 39–44. *Chem. Abstr.*, 105 (1987) 99395d.
1525. K. Okumura and K. Suga, A kinetic study of the hydrolysis of amylose by glucoamylase (in Japanese), *Kagaku Kogaku Ronbunshu*, 11 (1985) 147–154.
1526. M. Matsumura, J. Hirata, S. Ishii, and J. Kobayashi, Kinetics of saccharification of raw starch by glucoamylase, *J. Chem. Technol. Biotechnol.*, 42 (1988) 51–67.

1527. G. M. A. Van Beynum, J. A. Roels, and R. Van Tilburg, Equilibriun relationships in the degradation of starch by an amyloglucosidase, *Biotechnol. Bioeng.*, 22 (1980) 643–649.
1528. J. Kučera, Continuous hydrolysis of soluble starch by immobilized amyloglucosidase (EC 3.2.1.3), *Collect. Czech. Chem. Commun.*, 41 (1976) 2978–2986.
1529. K. N. Waliszewski, M. Garcia Alvarado, and J. De la Cruz Medina, Kinetics of enzymic hydrolysis of cassava flour starch – Optimization and modeling, *Int. J. Food Sci. Technol.*, 27 (1992) 465–472.
1530. A. H. Fahmy, K. M. ElSahy, R. M. Attia, and S. H. Ahmed, Effect of native and derivatized potato starches on the kinetics of amyloglucosidase, *Chem. Mikrobiol. Technol. Lebensm.*, 9 (1985) 28–30.
1531. O. Gaouar, C. Aymard, N. Zakhia, and G. M. Rios, Kinetic studies on the hydrolysis of soluble and cassava starches by maltogenese, *Starch/Stärke*, 49 (1997) 231–237.
1532. H. Okazaki and S. Saito, Rates of the secondary phase of starch hydrolysis by the joint actions of crystalline taka-α-amylase and saaccharogenic amylase system prepared from taka diastase (in Japanese), *J. Agr. Chem. Soc. Jpn.*, 26 (1952) 447–453.
1533. F. Shiraishi, K. Kawakami, A. Yuasa, T. Kojima, and K. Kusunoki, Kinetic expression for maltose production from soluble starch by simultaneous use of β-amylase and debranching enzymes, *Biotechnol. Bioeng.*, 30 (1987) 374–380.
1534. K. M. Bendetskii and V. L. Yarovenko, Kinetics of the enzymic hydrolysis of starch (in Russian), *Ferment. Spirt. Prom.* (6), (1975) 39–42.
1535. A. R. Sapronov, E. A. Chistyakova, and A. B. Starobinskaya, Kinetics of the enzymic saccharifiction of potato and barley starches (in Russian), *Ferment. Spirt. Prom.* (5), (1975) 20–22.
1536. M. Fukui, S. Murakami, Y. Yamada, T. One, and T. Nakamura, A kinetic equation for hydrolysis of polysacchrides by mixed exo- and endoenzyme systems, *Biotechnol. Bioeng.*, 23 (1981) 1393–1398.
1537. J. Zhou, L. Huang, and J. Zhou, Kinetics on coaction of α-amylase and glucoamylase in starch hydrolysis, *Zhongguo Tiaoweipin* (7), (1997) 10–12, 32; *Chem. Abstr.*, 128 (1998) 192848.
1538. C. Akerberg, G. Zacchi, N. Torto, and L. Gorton, A kinetic model for enzymatic wheat starch saccharification, *J. Chem. Technol. Biotechnol.*, 75 (2000) 306–314.
1539. H. Wang, J. Y. Han, S. S. Zhong, and H. Y. Chang, Hydrolysis kinetics of potato starch with two enzymes in one step, *Gaoxiao Huaxue Gongcheng Xuebao*, 15 (2001) 446–452. *Chem. Abstr.*, 136 (2001) 120118.
1540. W. Yao and H. Yao, Kinetics of mixed enzyme hydrolyzing raw starch, *Gongye Weishengwu*, 35(4), (2005) 15–18, 24; *Chem. Abstr.*, 146 (2006) 495778.
1541. L. Pastrana, Influence of retrogradation of potato starch on the parameters of Michaelis equation for amylase (in Portuguese), *Affinidad*, 59 (2002) 65–69.
1542. D. M. Lu and L. S. Yang, Synergism kinetics of α-amylase and glucoamylase during hydrolysis of microwave modified tapioca starch, *Jingxi Huagong*, 21 (2004) 768–771. *Chem. Abstr.*, 143 (2005) 228221.
1543. T. R. Johnson, J. O. Carroll, R. A. Britto, and D. J. Dubart, Development and field confirmation of a mathematical model for amyloglucosidase/pullulanase sacchrification, *Starch/Stärke*, 38 (1986) 382–387.
1544. T. A. Kovaleva, Kinetic and thermodynamic aspects of catalysis of polysaccharides by native and immobilized amylases (in Russian), *Biofizika*, 45 (2000) 439–444.
1545. J. E. Sawicki, Kinetic model for the simultaneous hydrolysis if starch by α-amylase, β-amylase and maltase, *US Nat. Tech. Inform. Serv., PB Rep.*, No. 220626/6 (1973); *Chem. Abstr.*, 79 (1973) 112698w.
1546. E. L. King and C. Altman, A schematic method of deriving the rate laws for enzyme catalyzed reactions, *J. Phys. Chem.*, 60 (1956) 1375–1378.

1547. P. Gonzalez-Tello, F. Camacho, E. Jurado, and E. M. Gaudix, A simple method for obtaining kinetic equations to describe the enzymic hydrolysis of biopolymers, *J. Chem. Technol. Biotechnol.*, 67 (1996) 286–290.
1548. H. Nakatani, Monte Carlo simulation of multiple attack mechanism of alpha-amylase, *Biopolymers*, 39 (1996) 665–669.
1549. P. Wojciechowski, E. Zienkiewicz, A. Kozioł, and A. Noworyta, Model analysis of the kinetics of starch hydrolysis by endo- and exo-amylolytic enzymes (in Polish), *Zesz. Nauk. Polit. Lodz, Inz. Chem. Process.* (26), (1999) 167–172.
1550. K. Martensson, Preparation of immobilized two-enzyme system, β-amylase–pullulanase, to an acrylic polymer for the conversion of starch to maltose. III. Process kinetics on continuous reactors, *Biotechnol. Bioeng.*, 16 (1974) 1567–1587.
1551. R. M. Costa and F. X. Malcata, Multisubstrate Michaelis-Menten kinetics. Explicit dependence of substrate concentration on time for batch reactors, *Bioprocess Biosyst. Eng.*, 10 (1994) 155–159.
1552. T. V. Subramanian, Process hydrolysis of starch with enzyme catalysis, *Biotechnol. Bioeng.*, 22 (1980) 661–665.
1553. W. T. M. Sanders, M. Geerink, G. Zeeman, and G. Lettinga, Anaerobic hydrolysis kinetics of particulate substrates, *Water Sci. Technol.*, 41 (2000) 17–24.
1554. K. Honda, M. Yamashita, and J. Kuboi, Enzymic hydrolysis in de-gassed water for improving efficiency, Jpn. Kokai Tokkyo Koho 07274959 (1995); *Chem. Abstr.*, 124 (1995) 28119.
1555. M. Nomura, Y. Ishii, and T. Yomo, Activation of α-amylase by exposure in high oxygen atmosphere, Jpn. Kokai Tokkyo Koho 200732554 (2007); *Chem. Abstr.*, 148 (2007) 48248.
1556. S. H. Choi, C. J. Kim, and S. K. Lee, Attrition effects of beads on enzymic saccharification of raw starch, *Hanguk Nonghwa Hakkoechi*, 32 (1989) 374–377. *Chem. Abstr.*, 113 (1990) 113730.
1557. L. X. Yang and Y. S. Xu, Study of destruction of amylase in liquefaction of starch, *Jiangsu Shiyou Huagong Xueyuan Xuebao*, 14(3), (2002) 22–24. *Chem. Abstr.*, 138 (2002) 339908.
1558. J. H. Walkup and J. G. Leech, Continuous starch enzymic conversion process, US Patent 3,149,049 (1964).
1559. A. Kamperman, N. C. Laane, and D. J. M. Schmedding, Recovery of enzymes used for polysaccharide modification, EU Patent EP 574066 (1993); *Chem. Abstr.*, 120 (1994) 105125.
1560. R. H. Clark, F. L. Fowles, and P. T. Black, Activation of amylase, *Trans. Roy. Soc. Can., III*, [3] 25 (1931) 99–105.
1561. L. A. Underkoefler, L. J. Denault, and E. F. Hore, Enzymes in the starch industry, *Starch/Stärke*, 17 (1965) 179–184.
1562. G. A. Jeffreys, Enzymic hydrolysis of starch, US Patent 2,370,665 (1945).
1563. M. Samec and P. Dolar, Enzymatische Amylolysis. VIII. Aufganphase der Stärkehydrolyse, *Z. Physiol. Chem.*, 259 (1939) 204–212.
1564. B. Örtenblad and K. Myrbäck, Über Grenzdextrine und Stärke, Über den Mechanismus der Spaltung durch Dextrinogenamylase, *Biochem. Z.*, 315 (1943) 233–239.
1565. E. J. Bourne, N. Haworth, A. Macey, and S. Peat, Amylolytic degradation of starch. A revision of the hypothesis of sensitization, *J. Chem. Soc.* (1948) 924–930.
1566. K. Babenko, Rate of depolymerization of starch in the system soluble starch–amylase–sodium chloride (in Polish), *Med. Dosw. Spol.*, 22 (1937) 106–116.
1567. T. Watanabe, S. Kawamura, and K. Matsuda, Sugar composition of hydrol from the enzymic hydrolysate of starch (in Japanese), *Nippon Nogei Kagaku Kaishi*, 40 (1966) 306–310.
1568. M. Pompejus, S. Freyer, M. Lohscheidt, O. Zelder, and M. Boy, Fermentative production ofc fine hemicals from starch hydrolyzates, EU Patent WO 2005116228; *Chem. Abstr.*, 144 (2005) 36384.
1569. J. J. Holthuis, P. H. Kerkhoven, and J. Mulder, Möglichkeiten und Vorteile der Vakuumreihe in der Kartoffelstärkeindustrie, *Starch/Stärke*, 33 (1981) 14–17.

1570. J. Hollo and E. Laszlo, Der Verlauf der Stärkeabbaues bei der Mälzung und während des Maischens in der Bierbrauerai, *Starch/Stärke*, 24 (1972) 226–229.
1571. J. Gunkel, M. Voetz, and F. Rath, Effect of the malting barley variety (*Hordeum vulgare* L.) on thermostability, *J. Inst. Brew.*, 108 (2002) 355–361.
1572. M. A. Pozen, Some observations on the mashing process, *Wallerstein Labs. Comm.*, 3(10), (1940) 199–206.
1573. G. He and L. Cao, The relationship between protein content and malting quality of wheat, *Zhongguo Liangyou Xuebao Bianjibu*, 17(6), (2002) 32–34. *Chem. Abstr.*, 140 (2003) 110350.
1574. J. Bartfay, Alcohol manufacture from starch-containing and similar material (in Hungarian), *Szeszipar*, 22 (1974) 101–106.
1575. P. Limonov and A. Loseva, Loss of starch during the malting process (in Russian), *Spirt Vodoch. Prom.*, 14(8), (1937) 8–10.
1576. R. P. Ellis, The use of high amylose barley for the production of whiskey malt, *J. Inst. Brew. (London)*, 82(5), (1976) 280–281.
1577. W. L. Mose, Diastase from the technical aspect, *Chem. Eng. Mining Rev.*, 15 (1923) 328–330.
1578. T. Fukimbara and K. Muramatsu, The action of mold amylse on alcoholic fermentation. I. The saccharogenic enzyme of taka-diastase and alcoholic fermentation with the saccharyfying agents of different ratios of enzymes, *J. Agr. Chem. Soc. Jpn.*, 27 (1953) 398–402.
1579. T. Fukimbara and I. Toshihiko, The action of mold amylse on alcoholic fermentation. II. Correlation between residual dextrin value of mold amylase and alcohol yield, *J. Agr. Chem. Soc. Jpn.*, 27 (1955) 62–66.
1580. F. G. Krivopalov and L. E. Sinelnikova, Influence of salt treatment of starch on its fermentative activity (in Russian), *Izv. Vyssh. Uchebn. Zaved., Pishch. Tekhnol.* (6), (1961) 33–36.
1581. M. Kugimiya and T. Fujimura-Ito, Rupture and damage of cell particles in dry mashed potatoes on the market (in Japanese), *Nippon Shokuhin Kagaku Kogaku Kaihi*, 43 (1996) 946–950.
1582. W. Windisch, P. Kolbach, and E. Schild, Über den Stärkeabbau beim Maischen, *Wochschr. Brau.*, 49 (1932) 289–295, 298–303.
1583. P. Kolbach and G. W. Haase, Stärkeabbau in gekochter Maische, *Wochschr. Brau.*, 56 (1939) 105–109, 113–118, 124–128, 134–136, 140–143.
1584. E. Bertoft and H. Henriknas, Starch hydrolysis in malting and mashing, *J. Inst. Brew.*, 89 (1983) 279–282.
1585. E. Briess, Starch conversion in mashes or doughs, US Patent 2,275,836 (1941).
1586. J. P. Casey, Starch paste. US Patent 2,472,790 (1949).
1587. E. Allan, K. S. Wenge, and H. Bisgard-Frantzen, Processes for producing a fermentation product such as ethanol, from milled starch without gelatinization using glucoamylase from *Athelia rolfsii* and acod α-amylase, EU Patent WO 2005069840 (2005); *Chem. Abstr.*, 143 (2005) 192412.
1588. M. R. Zakirova, Enzymic hydrolysis of proteins during bioconversion of starch-based raw materials (in Russian), *Uzbek. Kim. Zhurn.* (1), (2005) 42–45.
1589. R. H. Hopkins, Action and properties of β-amylase, *Rev. Ferm. Ind. Aliment.*, 10 (1955) 199–202.
1590. E. K. Y. Shih and O. J. Banasik, Amylolytic digestion of barley and malt starch during experimental mashing, *Brew. Dig.*, 35(12), (1960) 48–60.
1591. E. H. Muslin, C. B. Karpelenia, and C. A. Henson, The impact of thermostable α-glucosidase on the production of fermentable sugars during mashing, *J. Am. Soc. Brew. Chem.*, 61(3), (2003) 142–145.
1592. O. B. Joergensen, Barley malt alpha-aglucosidase. V. Degradation of starch and dextrins, *Acta Chem. Scand.*, 18 (1964) 1975–1978.
1593. H. J. Pieper and T. Senn, Pretreatment of starch-containing raw materials for alcohol fermentation, German Patent 3638529 (1988); *Chem. Abstr.*, 110 (1988) 6390.

1594. Y. Otani, S. Takahashi, and N. Kaneko, Improvement of alcohol mash. XIV. The composition of the culture media appropriate for the production of fungal amylase. 3. (in Japanese), *Hakko Kagaku Zasshi*, 36 (1958) 59–68.
1595. D. L. Goode, H. M. Ulme, and E. K. Arendt, Model studies to understand the effects of amylase additions and pH adjustment on the rheological behavior of simulated brewery mashes, *J. Inst. Brew.*, 111 (2005) 153–164.
1596. S. Trojaborg, O. B. Joergensen, C. Visigalli, and R. G. Martinez, Thermoascus aurantiacus thermostable β-glucanase and its use in starch pretreatment and liquefaction, EU Patent WO 2008023060 (2008); *Chem. Abstr.*, 148 (2008) 306597.
1597. A. V. Stepanov and A. Kuzin, Preparation of soluble starch by enzymic action (in Russian), *Khim. Farm. Prom.*, 111 (1932) 321–325.
1598. S. Zhao, S. Xiong, Z. Zhang, J. Shi, and L. Guo, Method for manufacturing water-soluble starch, Chinese Patent 20070124 (2007); *Chem. Abstr.*, 146 (2007) 253931.
1599. P. A. Singer and H. Perlstein, Starch conversion, US Patent 1.548,637 (1925).
1600. J. F. Walsh, S. M. Kinzinger, and A. H. Goodman, Nonbitter starch-conversion products, US Patent 2,192,213 (1940).
1601. The American Diamalt Co., Starch conversion products, British Patent 382,517 (1931).
1602. J. Wachman and G. J. Sippel, Starch conversion product, US Patent 1,916,872 (1932).
1603. G. V. Galkina, Highly hydrolyzed products from double hydrolysis of starch (in Russian), *Tr. Tsentr. Nauchn.-Issled. Inst. Krakhmal. Patochn. Prom.* (6), (1963) 25–39.
1604. W. G. Kingma, Erfahrungen bei der kontinuierlichen Verflüssigung von Stärke mit Säure und α-Amylase, *Starch/Stärke*, 17 (1965) 284–287.
1605. S. Bhargava, H. Bisgard-Frantzen, H. Frisner, A. Vikso-Nielsen, and M. Johal, Enzymic starch liquefaction process for improved ethanol production, EU Patent WO 2005092015 (2005); *Chem. Abstr.*, 143 (2005) 345492.
1606. A. L. Nugey, Enzymic gelatinization of raw starch in the manufacture of malt beverages, US Patent 2,951,762 (1960).
1607. S. Bhargava and S. Kohl, Enzymic liquefaction of starch-containing material, EU Patent WO 2006052787 (2006); Chem. Abstr., 144 (2006) 449492.
1608. Miles Laboratories Inc., Hydrolysis of starch, Dutch Patent 6,514,659 (1966).
1609. J. Han, Saccharification-free method for producing liquid starch syrup, Chinese Patent 1793384 (2006); *Chem. Abstr.*, 145 (2006) 169296.
1610. T. Maezawa, Y. Hayakawa, and M. Okubo, Liquefaction of starch. Criticism of experimental liquefaction method for starch (in Japanese), *Dempunto Gijutsu Kenkyu Kaiho*, 30 (1964) 74–81.
1611. T. Komaki, Enzymic liquefaction and saccharification of starch. VI. Preparation and properties of insoluble starch particles remaining in saccharified liquid starch after treatment with bacterial α-amylase and glucoamylase, *Agr. Biol. Chem. (Tokyo)*, 32 (1968) 123–129.
1612. L. Ludvig and M. Varga, Semi-industrial starch degradation with Hungarian enzyme preparations (in Hungarian), *Szeszipar*, 16 (1968) 22–26.
1613. S. Slott and G. B. Madsen, Liquefying starch, US Patent 3,912,590 (1975); *Chem. Abstr.*, 83 (1975) 207826h.
1614. M. E. Carr, L. T. Black, and M. O. Bagby, Continuous enzymic liquefaction of starch for saccharification, *Biotechnol. Bioeng.*, 24 (1982) 2441–2449.
1615. T. Komaki and N. Taji, Enzymic liquefaction and saccharification of starch. VIII. Liquefying conditions of corn starch by bacterial alpha-amylase, *Agr. Biol. Chem. (Tokyo)*, 32 (1968) 860–872.
1616. H. W. Leach, R. E. Hebeda, and D. J. Holik, Enzymic hydrolysis of granular starch, US Patent 3,922,196 (1975); *Chem. Abstr.*, 84 (1975) 87947v.

1617. G. S. Cattell and I. S. Daoud, Method and apparatus for the continuous liquefaction of starch, German Patent 2,930,614 (1980); *Chem. Abstr.*, 92 (1980) 148888r.
1618. P. H. E. Postier, Method and apparatus for continuous manufacture of starch paste, French Patent 2,322,925 (1977); *Chem. Abstr.*, 87 (1977) 203374u.
1619. P. C. Roger Carcassonne-Leduc, Continuous diastatic (partial) conversion of starch in grain flours, US Patent 2,841,497 (1958).
1620. H. W. Leach, R. E. Hebeda, D. J. Holik, E. A. Huske, D. K. Przybylski, R. Walon, and E. F. Berghmans, Conversion of granular starch into a soluble hydrolysate, German Patent, 2,417,639 (1974); *Chem. Abstr.*, 83 (1975) 62355e.
1621. C. W. Hinman and W. J. Nelson, Low-dextrose-equivalent starch hydrolyzates, S. Afr. Rep. Patent 68 07,248 (1968); *Chem. Abstr.*, 72 (1969) 22778y.
1622. CPC International Inc., Water-soluble low-dextrose-equivalent starch hydrolyzates, French Patent 2,116,770 (1972); *Chem. Abstr.*, 78 (1973) 99446v.
1623. F. Ueno, Y. Hanno, M. Ogura, Y. Takanayagi, and Y. Mochizuki, Enzymic hydrolysis of starch, Jpn. Kokai Tokkyo Koho, 74 19,049 (1974); *Chem. Abstr.*, 82 (1974) 123364x.
1624. F. Hattori, M. Nakamura, N. Taji, M. Nojiri, T. Nakai, and K. Kusai, Hydrolysis of starch, Jpn. Kokai Tokkyo Koho 76 44,652 (1976); *Chem. Abstr.*, 85 (1976) 61428x.
1625. Union Starch and Refining Co., Inc., Starch liquefaction, British Patent 1,200,817 (1970); *Chem. Abstr.*, 74 (1970) 4884s.
1626. Miles Laboratory Inc., Starch liquefaction process, British Patent 1,266,647 (1972); Chem. Abstr., 77 (1972) 21939t.
1627. M. L. Salo, I. Korhonen, and U. R. Lehtonen, Improving the fermentability of starch by hydrothermal processing. I. Studies of pure starches and cereal meals (in Finnish), *Maataloustieteelinen Aikak*, 42 (1970) 154–164.
1628. C. Veit, C. Felby, and C. C. Fuglsang, Secondry starch liquefaction in fermentation ethanol production, EU Patent WO 2002038787 (2002); *Chem. Abstr.*, 136 (2002) 385041.
1629. Y. Ji, H. Wang, Q. Dong, and X. Wang, Discussion on double enzyme process in alcohol production, *Henan Huagong*, (7) (2001) 20, 43; *Chem. Abstr.*, 137 (2002) 19437.
1630. B. H. Nielsen and P. Rosendal, Application of low-temperature liquefaction in production of ethanol from starch, *Proc. 4th Int. Symp. Alcohol Fuel Technol., Sao Paulo, 1980*, 1 (1981) 51–55. *Chem. Abstr.*, 95 (1981) 172552j.
1631. A. Sippel, Die Zeitgesetz der diastatischen Verflüssigung von Stärkekleister, *Biochem. Z.*, 314 (1943) 227–231.
1632. C. Mitchinson and L. P. Solheim, Method for liquefying starch, EU Patent WO 9638578 (1996); *Chem. Abstr.*, 126 (1997) 73841.
1633. R. Haga, M. Ishida, and M. Katsurayama, Liquefying starch. EU Patent EP 180,952 (1986); *Chem. Abstr.*, 105 (1986) 41287d.
1634. B. C. Park and D. J. Lim, Formation of PEG/dextran aqueous two-phase system for starch hydrolysis using α-amylase, *Sanop Misaengmul Hakkoechi*, 20 (1992) 190–195; *Chem. Abstr.*, 119 (1993) 26817.
1635. M. Richter, F. Schierbaum, and S. Augustat, Starch hydrolysis products, German Patent 2,365,850 (1976); *Chem. Abstr.*, 85 (1975) 179380d.
1636. R. G. P. Walon, Hydrolysis of starch dispersion with high solids content, Belgian Patent 841,714 (1976); *Chem. Abstr.*, 87 (1977) 83201m.
1637. R. G. P. Walon, Starch hydrolysis at high dry substances, US Patent 4,235,965 (1980); *Chem. Abstr.*, 94 (1981) 86073b.
1638. A. L. Morehouse and P. A. Krome Sander, Low d.e. starch hydrolyzates, Canadian Patent, 1,208,586 (1986); *Chem. Abstr.*, 105 (1985) 151821f.
1639. N. C. Holt, K. V. Rachlitz, and C. Bos, Starch hydrolyzates, German Patent 2,334,290 (1974); *Chem. Abstr.*, 81 (1974) 12087u.

1640. A. E. Staley, Diluted starch hydrolyzates, Jpn. Kokai Tokkyo Koho 75,100,251 (1975); *Chem. Abstr.*, 84 (1975) 137556q.
1641. DDS-Kroyer A/S, Starch hydrolysate preparation by enzymic hydrolysis, Belgian Patent 827,182 (1975); *Chem. Abstr.*, 84 (1975) 134128r.
1642. D. Medvecki, J. Zapletal, and J. Burianek, Dried starch hydrolyzates, Czechoslovak Patent 192,663 (1981); *Chem. Abstr.*, 96 (1982) 141442q.
1643. L. E. Coker and K. Venkatasubramanian, Low D.E. maltodextrins, US Patent 4,447,532 (1984); *Chem. Abstr.*, 101 (1984) 53679p.
1644. P. R. Witt, Liquefaction of starch, US Patent 4,540,663 (1985); *Chem. Abstr.*, 103 (1985) 176974e.
1645. T. Yoshida, Y. Ishige, M. Matsudaira, and T. Takahashi, Branched dextrins and straight-chain oligosaccharides, Jpn. Kokai Tokkyo Koho 61,205,494 (1986); *Chem. Abstr.*, 106 (1986) 118179a.
1646. T. Arami and F. Arami, Enzymic manufacture of dextrin from starch in high yields, Jpn. Kokai Tokkyo Koho 10215893 (1998); *Chem. Abstr.*, 129 (1998) 174730.
1647. W. Zhao, Y. Bi, and Z. Li, Preparation for α-amylase-degraded starch by deep-processing corn, Chinese Patent 1903882 (2007); *Chem. Abstr.*, 146 (2007) 231297.
1648. X. Hao, S. Wang, and L. Cui, Enzymic hydrolysis of corn starch by α-amylase, *Shipin Kexue*, 27 (2008) 141–143. *Chem. Abstr.*, 147 (2007) 142094.
1649. G. Xu, Z. Liu, and Z. Chen, Study on preparation of maltotriose from wheat starch, *Zhengzhou Gongcheng Xueyuan Xuebao*, 23(3), (2002) 1–4. *Chem. Abstr.*, 139 (2003) 306944.
1650. D. K. Apar, M. Turhan, and B. Ozbek, Enzymatic hydrolysis of starch by using a sonifier, *Chem. Eng. Commun.*, 193 (2006) 1117–1126.
1651. Q. Gao, L. Huang, N. Huang, and T. Huang, Liquefaction of cereal flour with thermostable α-amylase, *Zhongguo Liangyou Xuebao*, 16(6), (2001) 1–4. *Chem. Abstr.*, 137 (2002) 369026.
1652. M. J. Ruiz, C. I. Samchez, R. G. Torres, and D. R. Molin, Enzymatic hydrolysis of cassava starch for production of bioethanol with a Colombian wild yeast strain, *J. Braz. Chem. Soc.*, 22 (2011) 2337–2342.
1653. C. V. Anoop, A. K. Suresh, and V. A. Juvekar, An enzymatic hydrolysis process for the production of liquefacts (maltodextrins) from tapioca starch, Indian Patent 190472 (2003); *Chem. Abstr.*, 145 (2006) 187207.
1654. P. O. Serrano and C. M. L. Franco, Annealing and enzymatic hydrolysis of cassava starch (in Portugese), *Braz. J. Food Technol.*, 8 (2005) 220–232.
1655. A. Sołoducha and K. W. Szewczyk, Rheological properties of starch suspensions subjected to enzymatic liquefaction (in Polish), *Inz. Aparat. Chem.*, 44(4S), (2005) 97–98.
1656. F. Schierbaum, M. Richter, S. Augustat, and S. Radosta, Production, properties and use of gel-forming products of starch hydrolysis (in Russian), *Sakh. Prom.* (2), (1978) 66–68.
1657. F. Li, L. Wang, S. Tang, A. Li, Z. Zhang, and Y. Ou, Enzymatic preparation of rice starch maltodextrin with a low dextrose equivalent, *Shipin Yu Fajiao Gongye*, 32(5), (2006) 71–73. *Chem. Abstr.*, 148 (2007) 378276.
1658. S. Rogols and R. L. High, Enzymatic hydrolysis of wheat starch. I. A rapid method for colorimetrically estimating amylase activity, *Starch/Stärke*, 15 (1963) 1–4.
1659. G. V. Galkina, Enzymic hydrolysis of starch-containing material for the preparation of glucose products (in Russian), *Tr. Tsentr. Nauchn.-Issled. Inst. Krakhmal. Patochn. Prom.* (6), (1963) 48–62.
1660. H. R. Kathrein, Treatment and use of enzymes for the hydrolysis of starch, US Patent 3,108,928 (1963).
1661. H. Barfoed, Enzymatische Methoden zur Stärkeverflüssigung, *Starch/Stärke*, 19 (1967) 291–295.
1662. A. L. Morehouse, R. C. Malzahn, and J. T. Day, Starch hydrolyzates, German Patent 1,955,392 (1971); *Chem. Abstr.*, 75 (1971) 119377s.

1663. A. L. Morehouse and P. A. Krone, Low dextrose equivalent starch hydrolyzates and their manufacture, US Patent 4,684,410 (1987); *Chem. Abstr.*, 107 (1987) 153175p.
1664. Y. Y. Linko, P. Saarinen, and M. Linko, Starch conversion by soluble and immobilized α-amylase, *Biotechnol. Bioeng.*, 17 (1975) 153–165.
1665. Hayashibara Biochemical Laboratories Inc., Starch hydrolyzates by an immobilized enzyme, Jpn. Kokai Tokkyo Koho 80,124,498 (1980); *Chem. Abstr.*, 94 (1981) 101346p.
1666. T. Maezawa, S. Hayakawa, M. Okubo, and F. Shimbori, Liquefying properties of starch used for enzyme-hydrolyzed glucose production. VI. Effects of fermentation and acid treatment on the liquefaction of starch (in Japanese), *Nippon Nogei Kagaku Kaishi*, 41 (1967) 365–369.
1667. T. Komaki, Studies on enzymic liquefaction and saccharification of starch. VII. On the content of insoluble starch particles in some types of starch and increase of these materials by treatment under several conditions, *Agr. Biol. Chem. (Tokyo)*, 32 (1968) 314–319.
1668. R. W. Kerr, Solid total hydrolyzate of starch, US Patent 3,311,542 (1967); *Chem. Abstr.*, 67 (1967) 3992u.
1669. V. A. Yakovenko, N. V. Romenskii, and L. V. Masenko, Acidic and enzymic hydrolysis of corn starch (in Russian), *Izv. Vyssh. Uchebn. Zaved., Pishch. Tekhnol.* (6), (1967) 20–22.
1670. W. Nierle, Versuche zur enzymatischen Stärkeverflüssigung, *Starch/Stärke*, 19 (1967) 389–393.
1671. H. Xie, Manufacture of corn starch sugars by enzyme hydrolysis, Chinese Patent 101117647 (2008); *Chem. Abstr.*, 148 (2008) 287104.
1672. D. Wong, A. K. Uppalanchi, X. Gan, and B. Zhong, Enhanced enzymatic hydrolysis of starch by addition of manganese ions, US Patent 2008102497 (2008); *Chem. Abstr.*, 148 (2008) 470174.
1673. H. Kamasaka, T. Kuriki, and S. Okada, Modification of starch with enzymes modified starch and foods containing it, Jpn. Kokai Tokkyo Koho 2001103991 (2001); *Chem. Abstr.*, 134 (2001) 310108.
1674. L. Słomińska, M. Mączyński, L. Jarosławski, and W. Grajek, Method for enzymic starch hydrolysis, Polish Patent 191661 (2006); *Chem. Abstr.*, 146 (2006) 80443.
1675. L. E. Cameron and R. H. M. Stouffs, Starch hydrolysis, EU Patent; 136087; *Chem. Abstr.*, 102 (1984) 5186995f.
1676. J. Wang, A. Zeng, Z. Liu, X. Yuan, and S. Wu, Hydrolysis of raw starch granules by glucoamylase and product inhibition during the hydrolysis, *Trans. Tianjin Univ.*, 11 (2005) 199–203. *Chem. Abstr.*, 145 (2005) 29756.
1677. A. Hersiczky, P. Bende, and J. Ruff, Enzymatic degradation of starch, Hungarian Patent 15,486 (1978); *Chem. Abstr.*, 90 (1979) 70618q.
1678. J. W. Beishuizen, Enzymatic production of specialized sugar syrups from New Zealand-grown maize and wheat, *Chem. N.Z.*, 46(6), (1982) 125–128.
1679. M. H. David and H. Gunther, Enzymic conversions, EU Patent EP 164,933 (1986); *Chem. Abstr.*, 104 (1986) 128248v.
1680. L. M. S. Gome and I. O. Moraes, Effect of the feed method on the enzymatic hydrolysis of sorghum starch (in Portuguese), *Rev. Braz. Eng.*, 4 (1986) 17–24.
1681. G. Q. He, C. Q. Liu, Q. H. Chen, H. Ruan, and Y. I. Wu, Hydrolysis conditions for porous starch manufacturing from early rice by α-amylase and glucoamylase, *Zhongguo Shuidao Kexue*, 20 (2006) 227–230. *Chem. Abstr.*, 146 (2006) 357289.
1682. F. S. Kaper, J. Aten, M. A. Reinders, P. Dijkstra, and A. J. Suvee, A method of making and applying beta-limit dextrin containing starch hydrolysates, EU Patent EP 242,913 (1987); *Chem. Abstr.*, 108 (1987) 73845n.
1683. N. Nakamura and K. Horikoshi, Manufacture of sugars from starch with heat-resistant pullulanase, Jpn. Kokai Tokkyo Koho, 63 42,696 (1988); *Chem. Abstr.*, 109 (1988) 53236g.
1684. N. Tanaka and M. Kakimi, Manufacture of starch sugars with immobilized enzymes, Jpn. Kokai Tokkyo Koho 07000192 (1995); *Chem. Abstr.,* 122 (1995) 158791.

1685. T. Okada, T. Hibino, H. Ishizuka, M. Ito, H. Sahashiand, and T. Saiga, Manufacture of malto-oligosaccharides from starches and sucrose with cyclodextrin glucanotransferase using ultrafiltration membranes, Jpn. Kokai Tokkyo Koho 01179698 (1989); *Chem. Abstr.*, 112 (1990) 53775.
1686. L. Henderson and C. Constable, Enzymic liquefaction of starch, EU Patent WO 2005086640 (2005); *Chem. Abstr.*, 143 (2005) 284847.
1687. B. A. Paulson, D. S. Power, S. W. Sandra, J. K. Shetty, and D. E.Ward, Starch liquefaction using a mixture of *Buttiauxella* phytase and α-amylase, EU Patent WO 2008097620 (2008); *Chem. Abstr.*, 149 (2008) 265501.
1688. T. Miyamoto, T. Nakajima, M. Asano, and M. Nagai, Enzymic manufacture of branched-chain oligosaccharide-high sugar, Jpn. Kokai Tokkyo Koho 63291588 (1988); *Chem. Abstr.*, 111 (1989) 113794.
1689. K. A. Ram and K. Venkatasubramanian, Enhancement of starch conversion efficiency with free and immobilized pullulanase and α-1,4-glucosidase, *Biotechnol. Bioeng.*, 24 (1982) 355–369.
1690. K. Hayashida, K. Kunimoto, F. Shiraishi, K. Kawakami, and Y. Arai, Enzymic hydrolysis of soluble starch in a polyethylene glycol-dextran aqueous two-phase system, *J. Ferment. Bioeng.*, 69 (1990) 240–243.
1691. M. K. Morsi, S. M. Mohsin, and A. A. Hussein, Effect of added sugars during liquefaction and saccharification on the percent hydrolysis and its chemical composition on enzyme hydrolysis, *Egypt J. Food Sci.*, 6(1–2), (1978) 1–12.
1692. N. Dey, R. Soni, and S. K. Soni, A novel thermostable α-amylase from thermophilic Bacillus sp. SN-1 and its application in the liquefaction of sorghum starch for ethanol fermentation, *Asian J. Microbiol. Biotechnol. Environ. Sci.*, 4 (2002) 159–164.
1693. W. C. Muller and F. D. Miller, Hydrolysis of starch and fermentable hydrolyzates therefrom, US Patent 4,356,266 (1982).
1694. W. C. Muller and F. D. Miller, Hydrolysis of starch and fermentable hydrolysates obtained therefrom, US Patent 4,266,027 (1981); *Chem. Abstr.*, 95 (1981) 26921n.
1695. S. Ueda and Y. Fujio, Raw starch digestion and ethanol fermentation of starch materials without cooking for saving energy, *Proc. 3rd Pacific Chem. Eng. Congr., 1983*, 3 (1983) 421–424. *Chem. Abstr.*, 101 (1984) 189683t.
1696. R. F. Larson and A. W. Turner, Nonretrograded thinned hydrolyzates, US Patent 3,783,100 (1974); *Chem. Abstr.*, 80 (1974) 119316z.
1697. D. P. Langlois, Starch conversion products, US Patent 3,804,716 (1974); *Chem. Abstr.*, 81 (1974) 65510h.
1698. Agency of Industrial Science and Technology, Enzymic maltohexaose production, Jpn. Kokai Tokkyo Koho 58 63,392 (1983); *Chem. Abstr.*, 99 (1983) 103705g.
1699. Agency of Industrial Science and Technology, Oligosaccharide with amylase G4.5, Jpn. Kokai Tokkyo Koho 58 170,492 (1983); *Chem. Abstr.*, 100 (1983) 33267d.
1700. Agency of Industrial Science and Technology, Oligosaccharide by reaction of amylase G4.5, Jpn. Kokai Tokkyo Koho 58 170,491 (1983); *Chem. Abstr.*, 99 (1983) 33270z.
1701. Y. Takasaki, Manufacture of maltotetraose, Jpn. Kokai Tokkyo Koho 62 25,993 (1986); *Chem. Abstr.*, 107 (1986) 57475r.
1702. M. Karube, K. Hayashi, and T. Morita, Enzymic manufacture of α-maltopentaose in water-hydrophobic solvent system, Jpn. Kokai Tokkyo Koho, 03080093 (1991); *Chem. Abstr.*, 116 (1992) 5332.
1703. M. Karube, K. Hayashi, and T. Morita, Enzymic manufacture of oligosaccharides in water-hydrophobic solvent system, Jpn. Kokai Tokkyo Koho, 03080092 (1992); *Chem. Abstr.*, 116 (1992) 5331.
1704. R. L. Antrim, Branched starches and branched starch hydrolyzates, US Patent 2002065410 (2002); *Chem. Abstr.*, 136 (2002) 403419.

1705. F. Deleyn, C. Alken, F. Derez, D. Mauro, D. Provost, M. O. Stalin, and B. Vanhemelrijck, Process for the production of maltodextrins, EU Patent WO 2006047176 (2006); *Chem. Abstr.*, 144 (2006) 431656.
1706. J. Milner, Glucose polymers, US Patent 3,928,135 (1975); *Chem. Abstr.*, 84 (1975) 155681f.
1707. K. W. Kim, H. S. Ji, T. D. Kim, and M. H. Do, Preparation method of branched dextrins and oligosaccharides, S. Korean Patent 139262 (1998); *Chem. Abstr.*, 140 (2004) 110197.
1708. O. V. Grabovska, O. M. Maidanets, N. I. Shatangeeva, and V. O. Miroshnik, Optimization of technological parameters for enzymic liquefaction of starch (in Ukrainian), *Tsukor Ukr.* (6), (2004) 31–33.
1709. K. Ahlborg and K. Myrbäck, Grenzdextrine und Stärke. II. Takadiastase und Maizestärke, *Biochem. Z.*, 297 (1938) 172–178.
1710. J. Xie, Shortening of saccharification time in enzymic glucose production, *Shipin Yu Fajiao Gongye* (5), (1938) 9–14. *Chem. Abstr.*, 101 (1984) 70902y.
1711. L. Słomińska, Enzymic modification of low conversion starch products, *Starch/Stärke*, 41 (1989) 180–183.
1712. J. L. Bidrawn, F. G. Ewing, B. H. Landis, and H. R. Wheeler, Continuous enzyme conversion of cornstarch, *Tappi*, 48(9), (1965) 101A–104A.
1713. V. B. Prasnjak and E. N. Moskalova, Method for production of starch syrup, EU Patent WO 2006130124 (2006).
1714. Z. Xu, H. Zhang, and Y. Han, Enzymatic preparation of high maltose syrups from starch, *Zhongguo Shipin Xuebao*, 5(1), (2006) 37–42. *Chem. Abstr.*, 145 (2006) 488004.
1715. L. Wallerstein and O. P. Gray, Dextrose, US Patent 2,583,451 (1952).
1716. T. Tominaga, M. Niimi, and H. Sugihara, Starch saccharification with enzyme complex, Jpn. Kokai Tokkyo Koho 69 01,360 (1969); *Chem. Abstr.*, 71 (1969) 2152r.
1717. Y. Otani and S. Takahashi, Saccharification of the mixtures of starch, soluble starch, dextrin and maltose by amylase. 2, *J. Ferm. Technol. Jpn.*, 33 (1955) 465–467.
1718. K. K. K. Kroeyer, Enzymic hydrolysis of starch, French Patent 1,450,301 (1966); *Chem. Abstr.*, 66 (1966) 67009r.
1719. T. A. Ladur, N. G. Gulyuk, and R. M. Karpenko, Process for manufacture of glucose-containing syrups, USSR Patent 1.337,412 (1987); *Chem. Abstr.*, 108 (1987) 7819f.
1720. G. V. Galkina, Effect of various factors on enzymic hydrolysis to obtain highly sweetened syrups (in Russian), *Tr. Vses. Nauchn.-Issled. Ind. Krakhmaloprod.* (7), (1964) 87–101.
1721. A. B. Ariff, M. A. Sobri, S. Shioya, and M. I. A. Karim, Continuous production of a generic fermentation medium from sago starch, *Biotechnol. Sust. Util. Biol. Res. Tropic*, 12 (1998) 278–287.
1722. Daikin Kogyo Co., Ltd., Direct saccharification of starch by *Aspergillus fumigatus* K27, Jpn. Kokai Tokkyo Koho 60 83,595 (1985); *Chem. Abstr.*, 103 (1985) 159127p.
1723. H. Goss and H. W. Maurer, Continuous process for manufacturing starch degradation products, British Patent 1,028,212 (1966).
1724. H. Goss and H. W. Maurer, Continuous manufacture of starch hydrolysates, German Patent 1,190,890 (1965); *Chem. Abstr.*, 62 (1965) 16495.
1725. K. Reczey, E. Laszlo, and J. Hollo, Comparison of enzymic saccharification of starch and cellulose from technological and economic aspects, *Starch/Stärke*, 38 (1986) 306–310.
1726. S. Morita, H. Tamaya, S. Hakamata, T. Suzuki, and K. Sato, Manufacture of starch syrups, Jpn. Kokai Tokkyo Koho 60,199,400 (1985); *Chem. Abstr.*, 104 (1985) 131844s.
1727. P. Ducroo, Improvements relating to the production of glucose syrups and purified starches from wheat and other cereal starches containing pentosans, EU Patent EP 228,732 (1987); *Chem. Abstr.*, 108 (1987) 4704d.

1728. L. Huang, Z. Xu, M. Zhao, Q. Wen, Y. Zhou, J. Luo, and S. Xu, Method for preparing nitrogen-containing pure syrup by enzymic hydrolysis of wheat B-starch, Chinese Patent 101049133 (2007); *Chem. Abstr.*, 147 (2007) 468257.
1729. G. Collazos Hernandez, Process for preparation of maltodextrin,glucose, maltose and dextrose from cassava starch, EU Patent WO 2004007739 (2004); *Chem. Abstr.*, 140 (2004) 113143.
1730. C. V. Anoop, A. K. Suresh, and V. A. Juvekar, An enzymatic hydrolysis process for the production of glucose syrups from tapioca starch, Indian Patent 189622 (2003); *Chem. Abstr.*, 143 (2005) 385967.
1731. G. D. Thevenot, Saccharyfying starchy materials, US Patent 1,547,845 (1925).
1732. A. R. Ling and D. R. Nanji, Liquefying and saccharyfying starch, US Patent 1,548,721 (1925).
1733. A. R. Ling and D. R. Nanji, Starch, Canadian Patent 261,214 (1926).
1734. R. Lecoq and S. Wary, Malt flour and malting of cooked and raw starch, *J. Pharm. Chim.*, 30 (1924) 231–236.
1735. J. Bartfay and M. Varga, Preparation of crystalline glucose by enzymatic hydrolysis of starch. II. (in Hungarian), *Szeszipar*, 15(3), (1967) 78–85.
1736. K. K. K. Kroyer, Manufacture of products containing glucose and maltose from starch and polysaccharides, Dutch Patent 6.612,585 (1967); *Chem. Abstr.*, 67 (1967) 55331x.
1737. T. Tamura, S. Shibata, N. Kawase, F. Saeki, and S. Suzuki, Optimum conditions of enzymic conversion of starch to glucose and effects of continuous liquefaction by acid in the process (in Japanese), *Dempunto Gijutsu Kenkyu Kaiho*, 24 (1961) 45–56.
1738. M. Kujawski and F. Nowotny, Influence of starch liquefaction method on the source of its enzymic saccharification (in Polish), *Roczn. Technol. Chem. Zywn.*, 19 (1970) 103–121.
1739. T. Arai, Acid liquefaction and enzymic saccharification of cornstarch (in Japanese), *Dempunto Gijutsu Kenkyu Kaiho*, 30 (1964) 65–73.
1740. M. Hirao and M. Mitsuhashi, Starch syrup manufacture, French Patent, 2,066,516 (1971); *Chem. Abstr.*, 77 (1972) 7643k.
1741. J. Bartfay and M. Varga, Preparation of crystalline glucose by enzymic hydrolysis of starch. I. (in Hungarian), *Szeszipar*, 14(3), (1966) 83–88.
1742. T. L. Hurst, Starch conversion syrups, US Patent 3,490,922 (1970); *Chem. Abstr.*, 72 (1970) 123251b.
1743. Les Usines de Melle, Saccharyfying starch, French Patent 849,463 (1939).
1744. W. G. Kingma, Starch conversion processes, *Process Biochem.*, 1 (1966) 49–52.
1745. M. Huchette, F. Devos, and E. Requin, Continuous hydrolysis of starch, French Patent 1,391,011 (1965).
1746. M. Huchette and F. Devos, Continuous hydrolysis of starch, French Patent 1,391,082 (1965).
1747. Z. Boskov and J. Jakovljević, Feasibility of continuous enzymic saccharification starch (in Serbo-Croatian), *Hem. Ind.*, 26(5), (1972) 182–185.
1748. R. R. Barton and C. E. Land, Jr., Sugar syrups, Belgian Patent 644,301 (1964).
1749. F. C. Armbruster, E. R. Kooi, and C. F. Harjes, Low dextrose equivalent starch hydrolysates, S. African Rep. Patent 67 07,132 (1968); *Chem. Abstr.*, 70 (1969) 48867k.
1750. T. Gou, Y. Yang, J. Jiang, and G. Li, Production of sugar from crude corn starch by saccharification and fermentation, Chinese Patent 1205360 (1999); *Chem. Abstr.*, 132 (2000) 307366.
1751. J. O. Carrol, T. R. Swanson, and P. C. Trackman, Alpha-amylase mixtures for starch liquefaction, EU Patent EP 252,730 (1988); *Chem. Abstr.*, 109 (1988) 53235f.
1752. F. G. Ewing, J. L. Bidrawn, B. H. Landis, H. R. Wheeler, and M. V. Dennis, Continuous starch conversion, US Patent 3,371,018 (1968); *Chem. Abstr.*, 68 (1968) 115861c.
1753. Y. Takasaki, Starch saccharification by a pullulanase-like compound enzyme from *Bacillus*, Jpn. Kokai Tokkyo Koho 61 162,197 (1986); *Chem. Abstr.*, 105 (1986) 224629m.

1754. J. Krebs, Enzymic starch-degradation products, German Patent 1,113,430 (1960).
1755. Y. Otani and S. Takahashi, Improvement of alcohol mash. IV. Saccharification of the mixtures of starch soluble starch, dextrin and maltose by amylase. I. (in Japanese), *J. Ferm. Technol.*, 33 (1955) 413–418.
1756. A. Shimizu and H. Inoue, Starch hydrolysis of immobilized glucoamylase on collagen membrane, (in Japanese), *Kagaku Kogaku Ronbunshu*, 7 (1981) 635–642.
1757. T. T. Hansen and S. Rugh, High conversion syrups using immobilized α-amylase, EU Patent EP 157,638 (1985); *Chem. Abstr.*, 104 (1985) 4667m.
1758. P. Wang, V. Singh, H. Xue, D. B. Johnston, K. D. Rausch, and M. E. Tumbleson, Comparison of raw starch hydrolyzing enzyme with conventional liquefaction and saccharification enzymes in dry-grind corn processing, *Cereal Chem.*, 84 (2007) 10–14.
1759. H. Maeda, S. Miyado, and H. Suzuki, Water-insoluble enzymes. 2. Continuous saccharification of liquefied starch (in Japanese), *Hakko Kyokaishi*, 28 (1970) 391–397.
1760. S. Maruo, N. Tachikake, and Y. Ezure, Saccharides preparation from starch or its derivatives with exo-type enzyme, EU Patent WO 9310256 (1993); *Chem. Abstr.*, 119 (1993) 70571.
1761. L. P. Hayes, thinning and saccharification of starch pastes with glucoamylase, US Patent 3,806,415 (1974); *Chem. Abstr.*, 81 (1974) 65511j.
1762. R. H. M. Stouffs and H. P. Venker, High-dextrose starch hydrolyzates using immobilized glucoamylase, EU Patent EP 110,574 (1984); *Chem. Abstr.*, 101 (1984) 71055m.
1763. Nippon Shiryo Kogyo Co., Ltd., Saccharification of starch, Dutch Patent 6,515,054 (1966); *Chem. Abstr.*, 67 (1967) 3990s.
1764. J. T. Garbutt and A. M. Hanson, Treatment and use of enzymes for hydrolysis of starch, US Patent 3,268,417 (1966).
1765. T. Tominaga, T. Nimi, and H. Sugihara, Saccharification of starch, Jpn. Kokai Tokkyo Koho 68 23,560 (1968); *Chem. Abstr.*, 70 (1968) 89032c.
1766. Z. Boskov and J. Jakovljević, Feasibility of continuous enzymic saccharification of starch (inSerboCroatian), *Technika (Belgrade)*, 27 (1972) 960–963.
1767. B. L. Dwiggins, C. E. Pickens, and C. W. Niekamp, Raw starch saccharification, EU Patent EP 176,297 (1986); *Chem. Abstr.*, 104 (1986) 184843x.
1768. K. K. K. Kroeyer, Enzymic hydrolysis of starch, French Patent 1,516,475 (1968); *Chem. Abstr.*, 70 (1968) 89033d.
1769. P. Van Twisk and G. Tegge, Enzymatischer Abbau von Stärke in trocken enkeimten Mais, *Starch/Stärke*, 20 (1968) 319–324.
1770. M. Seidman and C. L. Royal, Enzymatically hydrolyzing starch with thermostable alpha-amylase, US Patent 3,551,293 (1970); *Chem. Abstr.*, 74 (1971) 100818u.
1771. C. A. Vecher and M. V. Levitskaya, Change in the carbohydrate complex of potato pulp during its enzymic hydrolysis (in Russian), *Vestsi Akad. Navuk Belorus SR, Ser. Biyal. Navuk* (3), (1973) 108–111.
1772. D. Schwengers, C. Bos, and E. Andersen, Refined starch hydrolysate from cereals, German Patent 2,803,030 (1978); *Chem. Abstr.*, 89 (1978) 131463p.
1773. Ajinomoto Co., Direct saccharification of milo, Inc., Jpn. Kokai Tokkyo Koho 81,137,895 (1981); *Chem. Abstr.*, 96 (1981) 67268u.
1774. System Agricultural Center Co., Ltd., Concentrated sugar solution from starchy materials, Jpn. Kokai Tokkyo Koho 58 141,794 (1983); *Chem. Abstr.*, 100 (1983) 33251u.
1775. M. Kurimoto, Starch hydrolyzates of high purity, German Patent 2,046,471 (1972); *Chem. Abstr.*, 78 (1973) 86262g.
1776. M. Kurimoto, Saccharification of starch, French Patent 2,062,411 (1971); *Chem. Abstr.*, 77 (1971) 7642j.
1777. M. C. Cadmus, L. G. Jayko, D. E. Hansley, H. Gasdorf, and K. L. Smiley, Enzymic production of glucose syrup from grains and its use in fermentation, *Cereal Chem.*, 43 (1966) 658–669.

1778. A. W. Turner, W. C. Mussulman, and T. L. Hurst, Process of the production of starch syrup, Belgian Patent 633,645 (1963).
1779. Miles Laboratories, Inc., Noncrystallizing high-dextrose-equivalent syrups, British Patent 1,071,538 (1967); *Chem. Abstr.*, 67 (1967) 55330w.
1780. A. Sroczyński, T. Pierzgalski, and K. Nowakowska, The production of high-saccharified syrups by two-stage enzymic hydrolysis of potato starch. II. The effect of starch concentration on its saccharification with minimum amounts of glucoamylase preparations and at an unchanged pH during the whole process of hydrolysis, *Acta Aliment. Pol.*, 4(1), (1978) 63–70.
1781. P. W. Thompson, Continuous starch hydrolysis, German Patent 3.012,143 (1980); *Chem. Abstr.*, 94 (1980) 3016m.
1782. R. Borriss, D. Schmidt, E. Gorges, H. Ruttloff, A. Taeufel, P. Lietz, P. Steffen, W. Heinig, and U. Gerges, Enzyme mixture for saccharification of grain mash, East German Patent, 152,142 (1981); *Chem. Abstr.*, 96 (1982) 216057d.
1783. H. Anger, M. Richter, B. Kettlitz, R. Schirner, G. Haeusler, and T. Roick, Preparation of cereal starch hydrolyzates containing at least 95% glucose, East German Patent 298,431 (1992); *Chem. Abstr.*, 117 (1992) 29134.
1784. Y. Nakajima, T. Adachi, J. Ito, and H. Mizuno, Saccharification of starch in high maltose yields, Jpn. Kokai Tokkyo Koho 74 55,857 (1974); *Chem. Abstr.*, 81 (1974) 137833r.
1785. K. Kitahara, M. Kurushima, and T. Abe, The enzymic saccharification in the fermentation industries (in Japanese), *Bull. Res. Inst. Food Sci., Kyoto Univ.*, 3 (1950) 72–79.
1786. S. Barghava, M. Johal, and H. Frisner, Enzymic starch liquefaction and saccharification processes, EU Patent WO 2006028897 (2006); *Chem. Abstr.*, 144 (2006) 272819.
1787. R. G. P. Walon, Maltose syrup with high maltose content from starch, Belgian Patent 810,205 (1974); *Chem. Abstr.*, 82 (1974) 60400h.
1788. B. C. Saha and J. G. Zeikus, Improved method for preparing high-maltose conversion syrup, *Biotechnol. Bioeng.*, 34 (1989) 299–303.
1789. V. Arasaratnam, K. Thayanathan, and K. Balasubramanian, Sugar syrup (DE 50-70) from corn flour, *Starch/Stärke*, 50 (1998) 95–98.
1790. S. Sakai and N. Tsuyama, Saccharified starch product with maltose as the chief constituent, German Patent 2,532,078 (1976); *Chem. Abstr.*, 84 (1976) 178211s.
1791. F. J. Zucker, H. Houben, and G. Ostwald, Alcohol from starch containing materials, German Patent 2,944,482 (1981); *Chem. Abstr.*, 95 (1981) 26920m.
1792. M. Kusakabe, K. Takahashi, and Z. Yoshino, Polysaccharides for alcoholic beverages. Jpn. Kokai Tokkyo Koho, 63,109,792 (1988); Chem. Abstr., 109 (1989) 127347t.
1793. B. A. Ustinnikov, A. N. Lazareva, and V. L. Yarovenko, Hydrolysis of starch during the separate and combined action of α-amylase and glucoamylase applicable to conditions of alcohol production (in Russian), *Ferm. Spirt. Prom.*, 37(2), (1971) 13–17.
1794. T. Oya and N. Yokoi, Maltose syrup production, Jpn. Kokai Tokkyo Koho 79 92,637 (1977); *Chem. Abstr.*, 91 (1979) 173702n.
1795. Agency of Science and Technology Hokkaido Sugar Co., Ltd., Saccharification of starch, Jpn. Kokai Tokkyo Koho 82 65,199 (1980); *Chem. Abstr.*, 97 (1982) 11155g.
1796. K. B. Martensson, Enzymic production of starch conversion product having a high maltose content, US Patent 3,996,107 (1976); *Chem. Abstr.*, 86 (1976) 53943s.
1797. R. G. P. Walon, Maltose-containing starch hydrolyzate and recovery of crystallized maltose, German Patent 2,855,392 (1979); *Chem. Abstr.*, 91 (1979) 93338h.
1798. B. E. Norman, Saccharification of starch hydrolyzate, British Patent 2,074,167 (1981); *Chem. Abstr.*, 96 (1982) 1212001g.
1799. Y. Takasaki, High maltose preparation, Jpn. Kokai Tokkyo Koho 60 186,296 (1985); *Chem. Abstr.*, 104 (1984) 128245s.

1800. J. J. Caboche, Process for producing a syrup rich in maltose. EU Patent 1016728 (2000); *Chem. Abstr.*, 133 (2000) 73249.
1801. Corn Products Co., Enzymic conversion of starch to high maltose syrup, British Patent 1,144,950 (1969); *Chem. Abstr.*, 70 (1969) 89034e.
1802. Kyorin Pharmaceutical Co., Ltd., Highly purified maltose, Jpn. Kokai Tokkyo Koho 81 64,794 (1981); *Chem. Abstr.*, 95 (1981) 167165f.
1803. S. Shiratori, Separation of maltose from saccharified starch preparations, Jpn. Kokai Tokkyo Koho 81, 127,097 (1981); *Chem. Abstr.*, 96 (1981) 21589w.
1804. M. Kurimoto, Manufacture of maltose, Jpn. Kokai Tokkyo Koho 74 102,854 (1974); *Chem. Abstr.*, 83 (1975) 43683f.
1805. S. Kobayashi, Production of maltose from Ipomoea (sweet potato) starch, *Baiomasu Henkan Keikaku Kenkyu Hokoku* (19), (1989) 78–85. *Chem. Abstr.*, 115 (1991) 7189.
1806. K. K. K. Kroeyer, Conversion of starch and other polysaccharides to dextrose and maltose, Belgian Patent 650,378 (1965).
1807. Hayashibara Co., Ltd., High-purity maltose, French Patent 1,569,499 (1969); *Chem. Abstr.*, 72 (1969) 45262q.
1808. M. Mitsuhashi and K. Masuda, Production of very pure maltose from starch, French Patent 2,012,831 (1970); *Chem. Abstr.*, 74 (1970) 14345f.
1809. M. Hirao and J. Ogasawara, Maltose from starches, S. African Rep. Patent 69 08,247 (1970); *Chem. Abstr.*, 75 (1971) 38205g.
1810. M. H. Do, H. S. Ji, S. C. An, and K. W. Kim, Process for preparing high purity of maltose by enzymic saccharification, S. Korean Patent 2000021208 (2000); *Chem. Abstr.*, 136 (2002) 85785.
1811. W. Murayama, S. Koizumi, and N. Kobayashi, Maltose, Jpn. Kokai Tokkyo Koho 76 98,346 (1976); *Chem. Abstr.*, 85 (1976) 194482g.
1812. M. Mitsuhashi, M. Hirao, and K, Sugimoto, Maltitol, French Patent 2,000,580 (1969); *Chem. Abstr.*, 72 (1970) 80625g.
1813. H. Konishi, T. Miyamoto, T. Nakajima, and J. Nashimoto, Maltose having a high purity, Jpn. Kokai Tokkyo Koho 78 118,532 (1978); *Chem. Abstr.*, 90 (1978) 40518q.
1814. Meiji Confectionery Co. Ltd., Maltose. Dutch Patent 74 06,921 (1974); *Chem. Abstr.*, 83 (1975) 145754x.
1815. T. Yabuki, T. Hanatani, and T. Yamada, Preparing maltose, Jpn. Kokai Tokkyo Koho 76 101,141 (1976); *Chem. Abstr.*, 86 (1976) 53936s.
1816. Meiji Seika Kaisha Ltd., Maltose, Jpn. Kokai Tokkyo Koho 80 77,896 (1980); *Chem. Abstr.*, 93 (1980) 170018z.
1817. Z. Yoshino, Maltose powder production, Belgian Patent 900,901 (1984); *Chem. Abstr.*, 102 (1984) 222445n.
1818. R. P. Rohrbach, M. J. Maliarik, T. P. Malloy, G. J. Thompson, K. T. Lin, and D. W. Penner, Glucose or maltose from starch, S. African Rep. Patent 84 07,452 (1985); *Chem. Abstr.*, 104 (1985) 90870b.
1819. R. P. Rohrbach, M. J. Maliarik, T. P. Malloy, G. J. Thompson, K. T. Lin, and D. W. Penner, Manufacture of glucose and maltose by continuous starch hydrolysis, Roumanian Patent 92,103 (1987); *Chem. Abstr.*, 108 (1987) 148896r.
1820. M. Niimi, Y. Hario, K. Kataura, Y. Ishii, and K. Kato, Enzymic manufacture of highly pure maltose and its hydrolyzate, Jpn. Kokai Tokkyo Joho 02092296 (1990); *Chem. Abstr.*, 113 (1990) 96168.
1821. J. L. Baker, Maltose. British Patent 564,895 (1944).
1822. H. C. Gore, Maltose from starch, British Patent 226,812 (1923).
1823. E. Kusubara, Maltose from starch, Jpn. Patent 3027 ('57) (1957).
1824. R. Walon, A. Reeve, and J. Dinsdale, Maltose from starch, Belgian Patent, 862,614 (1978); *Chem. Abstr.*, 89 (1978) 165294z.

1825. T. Yoshida and K. Yoritomi, Maltose from starch, German Patent 2,416,987 (1975); *Chem. Abstr.*, 82 (1975) 112222y.
1826. W. F. Line, V. K. Chaudhary, E. Chicoye, and R. J. Mizerak, Dextrose and maltose syrups produced using an enzyme derived from rice, EU Patent EP 127,291 (1984); *Chem. Abstr.*, 102 (1985) 111867g.
1827. S. Pedersen and H. Vang Hendricksen, Method for production of maltose and/or enzymatically modified starch, EU Patent WO 2001016349 (2001); *Chem. Abstr.*, 134 (2001) 236335.
1828. Y. Takasaki, Enzymic production of maltose. Jpn. Kokai Tokkyo Koho 62 126,991 (1987); *Chem. Abstr.*, 107 (1987) 152933x.
1829. Y. Takasaki, Enzymic manufacture of maltose, Jpn. Kokai Tokkyo Koho 62 126,994 (1987); *Chem. Abstr.*, 107 (1987) 152932w.
1830. Y. Takasaki, Enzymic production of maltose, Jpn. Kokai Tokkyo Koho 62 126,993 (1987); *Chem. Abstr.*, 107 (1987) 152864a.
1831. Y. Takasaki, Enzymic production of maltose, Jpn. Kokai Tokkyo Koho 62 126,992 (1987); *Chem. Abstr.*, 107 (1987) 152865b.
1832. M. Li, J. W. Kim, and T. L. Peeples, Amylase partitioning and extractive bioconversion of starch using thermoseparating aqueous two-phase systems, *J. Biotechnol.*, 93 (2002) 15–26.
1833. VEB Maisan Werke, D-Glucose and hydrolysis products from starch by acid or enzymic catalysis, Belgian Patent 902,257 (1985); *Chem. Abstr.*, 104 (1985) 67821c.
1834. G. Thompson, K. F. Lin, and D. W. Penner, Glucose or maltose from starch, US Patent 4,594,322 (1986); *Chem. Abstr.*, 105 (1986) 113633p.
1835. D. C. Young, Glucose manufacture by polysaccharide hydrolysis, US Patent 4,664,717 (1987); *Chem. Abstr.*, 107 (1987) 219419e.
1836. Y. Wang, L. Wang, Y. Wang, B. Cui, and C. Chen, Liquefying process in producing glucose process, Chinese Patent 1389572 (2003); *Chem. Abstr.*, 141 (2003) 55977.
1837. E. Stern, Starch degradation products, German Patent 523,349 (1924).
1838. E. E. Fisher, Enzyme hydrolysis of carbohydrates, US Patent 3,720,583 (1973); *Chem. Abstr.*, 79 (1973) 20609q.
1839. D. F. Rentshler, D. P. Langlois, R. F. Larson, L. H. Alverson, and R. W. Ligget, Crystalline glucose preparation by enzymic hydrolysis of starch, US Patent 3,039,935 (1962).
1840. J. K. Kumar, R. Ravi, and T. Karunanthi, From starch to glucose. Hydrolysis of tuberous root by *Bacillus cereus*, *Chem. Eng. World*, 39(11), (2004) 50–52.
1841. B. Selmi, D. Marion, J. M. P. Cornet, J. P. Douzals, and P. Gervais, Amyloglucosidase hydrolysis of high-pressure and thermally gelatinized corn and wheat starches, *J. Agric. Food Chem.*, 48 (2000) 2629–2633.
1842. R. Lopez-Ulibarri and G. M. Hall, Continuous hydrolysis of cassava flour starch in an enzymic membrane reactor, *BHR Group Conf. Ser. Publ.*, 3 (1993) 95–105. *Chem. Abstr.*, 120 (1994) 190036.
1843. D. Schwengers, Glucose from starch-rich plant parts, German Patent 2,255,666 (1972); *Chem. Abstr.*, 82 (1974) 1984d.
1844. H. S. Lee, W. G. Lee, S. W. Park, H. Lee, and H. N. Chang, Starch hydrolysis using enzyme in supercritical carbon dioxide, *S. Korea Biotechnol. Tech.*, 7 (1993) 267–270.
1845. H. S. Lee, Y. W. Ryu, and C. Kim, Hydrolysis of starch by α-amylase and glucoamylase in supercritical carbon dioxide, *S. Korea J. Microbiol. Biotechnol.*, 4 (1994) 230–232.
1846. M. Larsson, V. Arasaratnam, and B. Mattiasson, Integration of bioconversion and downstream processing starch hydrolysis in an aqueous two-phase system, *Biotechnol. Bioeng.*, 33 (1989) 758–766.
1847. Y. Hattori and S. Iida, Enzymic production of glucose from starch, Jpn. Kokai Tokkyo Koho 63 13,048 (1959).
1848. P. M. Sinclair, Enzymes convert starch to dextrose, *Chem. Eng.*, 72(18), (1965) 90–92.

1849. L. E. Jackson and M. Seidman, Enzymic hydrolysis of granular starch directly to glucose, EU Patent EP 171,218 (1986); *Chem. Abstr.*, 104 (1985) 147190m.
1850. T. Furukawa, S. Shinohara, A. Kato, and T. Noguchi, Enzymic hydrolysis of starch, Jpn. Kokai Tokkyo Koho 69 01,358 (1964); *Chem. Abstr.*, 70 (1968) 107708j.
1851. S. Gargova and M. Beshkov, Application of the enzyme preparation glucoamylase obtained from *Aspergillus niger* 16/132 for maize starch hydrolysis (in Bulgarian), *Acta Microbiol. Bulg.*, 10 (1982) 40–46.
1852. C. Cojocaru and M. Gazea, Preparation of glucose by enzymic hydrolysis of starch. Isolation of amyloglucosidase-producing mold strain (in Roumanian), *Ind. Aliment. (Bucharest)*, 22 (1971) 254–255.
1853. D. Pomeroy, N. O. Sousa, I. Marques da Silva, and H. L. Martelli, Production of a fermented preparation with α-amylase and amyloglucosidase activity (in Portuguese), *Rev. Quim. Ind. (Rio de Janeiro)*, 56 (1987) 5–9.
1854. N. Salomon, Glucose production from starch, Israeli Patent 21,408 (1968); *Chem. Abstr.*, 69 (1968) 107758e.
1855. Academia Sinica, Institute of Microbiology, Taipei, Study on preparation of glucose for injection by acid enzyme method, *Yiyao Gongye*, (2) (1983) 10–16; *Chem. Abstr.*, 99 (1983) 86538k.
1856. Dorr-Oliver, Inc., Enzymic convesion of starch to glucose, Dutch Patent 6,401,495 (1964).
1857. A. L. Wilson, Enzymic hydrolysis of starch, Belgian Patent 657,627 (1965).
1858. T. Omaki, Dextrose, Jpn. Kokai Tokkyo Koho 70 37,816 (1970); *Chem. Abstr.*, 75 (1971) 78410b.
1859. H. Meyer, G. Wahl, H. D. Teuerkauf, F. Schierbaum, M. Richter, and H. Ruttloff, Sccharification of starch polysccharides and their decomposition products or starch-containing raw materials, East German Patent 119,341 (1976); *Chem. Abstr.*, 86 (1976) 57101p.
1860. M. Agrawal, S. Pradeep, K. Chandraraj, and S. N. Gummadi, Hydrolysis of starch by amylase from Bacillus sp. KCA102: A statistical approach, *Process Biochem.*, 40 (2005) 2499–2507.
1861. M. Ito and Y. Fukushima, Manufacture of glucose by saccharification process having system for controlling enzyme feeding and retention time, Jpn. Kokai Tokkyo Koho 62 272,987 (1987); *Chem. Abstr.*, 109 (1988) 21679n.
1862. Denki Kagaku Kogyo K. K., Enzymic glucose production from starch, Jpn. Kokai Tokkyo Koho 82 132,893 (1982); *Chem. Abstr.*, 97 (1982) 196851y.
1863. K. L. Smiley, Continuous conversion of starch to glucose with immobilized glucoamylase, *Biotechnol. Bioeng.*, 13 (1971) 309–317.
1864. D. D. Lee, P. J. Reilly, and E. V. Collins, Jr., Pilot plant production of glucose from starch with soluble α-amylase and immobilized glucoamylase, *Enzyme Eng.*, 3 (1975) 525–530.
1865. V. Krasnobajew, PAG-B-Immobilized glucoamylase: Glucose production from soluble starch, *Prepr. 1st Eur. Congr. Biotechnol., Frankfurt/Main* (1978) 111–113. *Chem. Abstr.*, 91 (1978) 89507v.
1866. H. Sahashi, M. Ito, T. Okada, H. Ishizuka, and T. Saiga, Continuous manufacture of glucose and immobilized glucoamylase, Jpn. Kokai Tokkyo Koho 01144990 (1989); *Chem. Abstr.*, 111 (1989) 213360.
1867. K. L. Smiley, Continuous conversion of modified starch to glucose by immobilized glucoamylase, *Proc. Int. Symp. Convers. Manuf. Foodst. Microorg.*, 1971 (1972) 79–86. *Chem. Abstr.*, 82 (1975) 108197b.
1868. R. E. Hebeda, D. J. Holik, and H. W. Leach, Dextrose from starch, Belgian Patent 861,799 (1978); *Chem. Abstr.*, 89 (1978) 61340v.
1869. M. Abdullah and F. C. Armbruster, Method using glucoamylase immobilized on porous alumina, US Patent 4,226,937 (1980); *Chem. Abstr.*, 94 (1980) 3009m.
1870. D. D. Lee, Continuous production of glucose from dextrin by glucoamylase immobilized on porous silica, *Rep. Kans. State Univ., Inst. Syst. Des. Optim.*, 66 (1975) 51–62. *Chem. Abstr.*, 85 (1975) 175501w.

1871. K. Imata and K. Asano, Preparation of dextrose solution, Jpn. Kokai Tokkyo Koho 71 10,018 (1971); *Chem. Abstr.*, 76 (1971) 32880x.
1872. L. S. Khvorova, N. R. Andreev, N. D. Lukin, T. V. Lapidus, and N. N. Nyunina, Method of glucose production, Russian Patent 2314351 (2008); *Chem. Abstr.*, 148 (2008) 102223.
1873. S. Sakai and H. Akai, Glucose, French Patent 2,504,150 (1982); *Chem. Abstr.*, 98 (1982) 70359w.
1874. Y. Takasaki, Manufacture of glucose by enzymic saccharification of starch, Jpn. Kokai Tokkyo Koho 06062882 (1994); *Chem. Abstr.*, 121 (1994) 33297.
1875. T. L. Hurst, Enzyme system: Mixture of amylo-1,6-glucosidase and glucoamylase suitable for converting starch into sugar, German Patent 1,943,096 (1970); *Chem. Abstr.*, 73 (1970) 119231y.
1876. Agency of Industrial Science and Technology, Promotion of glucose production, Jpn. Kokai Tokkyo Koho 81 23,894 (1979); *Chem. Abstr.*, 95 (1981) 59942y.
1877. Agency of Industrial Science and Technology, Promotion of glucose production, Jpn. Kokai Tokkyo Koho 81 23,895 (1979); *Chem. Abstr.*, 95 (1981) 59943z.
1878. B. Noordam, A method for increasing monosaccharide levels in the saccharification of starch and enzymes useful thereof, EU Patent WO 9613602 (1996); *Chem. Abstr.*, 125 (1996) 56376.
1879. F. Takahashi, Y. Sakai, S. Kobayashi, and S. Wakabayashi, Continuous saccharification of starch with glucoamylase, Jpn. Kokai Tokkyo Koho 01191693 (1989); *Chem. Abstr.*, 112 (1990) 117400.
1880. P. Bruckmayer, Process and apparatus for making glucose from a starch solution, EU Patent EP 1671704 (2006); *Chem. Abstr.*, 145 (2006) 64780.
1881. Y. Takasaki, Manufacture of maltotetraose, Jpn. Kokai Tokkyo Koho 62 25,992 (1987); *Chem. Abstr.*, 107 (1987) 76430p.
1882. T. Morita, H. J. Lim, and I. Karube, Enzymic hydrolysis of polysaccharides in water-immiscible organic solvent biphasic system, *J. Biotechnol.*, 38 (1995) 253–261.
1883. K. Wako, S. Hashimoto, J. Kanaeda, S. Kubomura, C. Takahashi, H. Tonda, S. Makino, S. Shiotsu, and K. Mimura, Starch sugar containing maltotriose as a major component, Jpn. Kokai Tokkyo Koho 76 09,739 (1976); *Chem. Abstr.*, 85 (1976) 80017d.
1884. S. Hashimoto, K. Wako, J. Kanaeda, S. Kubomura, C. Takahashi, H. Toda, S. Makino, S. Shiotsu, and K. Mimura, Starch sugar containing high amounts of maltotriose, Jpn. Kokai Tokkyo Koho 76 110,049 (1976); *Chem. Abstr.*, 86 (1976) 45038k.
1885. M. Hirata, Y. Tanaka, T. Tanba, and H. Naito, Manufacture of maltotriose from liquified starch, Jpn. Kokai Tokkyo Koho 2005058110 (2005); *Chem. Abstr.*, 142 (2005) 238784.
1886. Agency of Industrial Science and Technology, Maltohexaose production by *Bacillus* amylase, Jpn. Kokai Tokkyo Koho 57 146,591 (1982); *Chem. Abstr.*, 98 (1982) 33076f.
1887. F. E. Knudsen and J. Karkalas, Enzymatisch hergestellte Stärkehydrolysate, *Starch/Stärke*, 21 (1969) 284–291.
1888. G. Tegge and G. Richter, Zur enzymatischen Hydrolyse verschidener Stärkearten, *Starch/Stärke*, 38 (1986) 329–335.
1889. W. Windisch, Über der verschiedenen Rohfruchtmaterialen bei der diastatischen Lösung und Verzuckerung, *Wochschr. Brau.*, 39 (1922) 13–14.
1890. Y. Suzuki, K. Onishi, T. Takaya, and H. Fuwa, Comparative susceptibility to pancreatin of starch granules from different plant sources (in Japanese), *Denpun Kagaku*, 26 (1979) 182–190.
1891. M. Loes, Vergleichende Untersuchungen über den enzymatischen Abbau der verschiedenen Stärkearten, *Ann. Pediatr.*, 181 (1953) 1–16.
1892. K. H. Meyer, E. Preiswerk, and R. Jeanloz, Recherches sur l'amidon. XV. La cinetique de la degradation de l'amidon non degrade de pommes de terre et de mass par la β-amylase, *Helv. Chim. Acta*, 24 (1941) 1395–1400.
1893. J. C. Valetudie, P. Colonna, B. Bouchet, and D. J. Gallant, Hydrolysis of tropical tuber starches by bacterial and pancreatic α-amylases, *Starch/Stärke*, 45 (1993) 270–276.

1894. V. Pomeranz and J. A. Shellenberger, Effect of alpha-amylases on pregelatinized starches as measured by the amylograph, *J. Sci. Food Agr.*, 14 (1963) 145–148.
1895. H. W. Leach and T. J. Schoch, Structure of the starch granule. II. Action of various amylases on granular starches, *Cereal Chem.*, 38 (1961) 34–46.
1896. K. Yoshida, H. Hattori, and H. Murukami, Fluctuation of α-amylase activation of an enzyme preparation acting on various lots of Merck's soluble starch and its adjustant (in Japanese), *Nippon Jozo Kyokai Zasshi*, 76 (1981) 856–858. *Chem. Abstr.*, 96 (1981) 64637c.
1897. V. F. Milovskaya, Resistance of starch from different sorts of wheat to the hydrolyzing action of amylase (in Russian), *Biokhimiya*, 5 (1940) 589–595.
1898. D. J. Atkins and J. F. Kennedy, A comparison of the susceptibility of two common grades of wheat starch to enzymic hydrolysis and the resultant oligosaccharide product spectra, *Starch/Stärke*, 37 (1985) 421–427.
1899. K. Kanenaga, A. Harada, and T. Harada, Actions of various amylases on starch granules from different plant origins heat at 60°C in their aqueous suspensions, *Chem. Express*, 5 (1990) 465–468.
1900. T. B. Darkanbaev and T. N. Kostyukova, Enzymic hydrolyzability of starch from various varieties of wheat (in Russian), *Izv. Akad. Nauk Kazakh. SSR Biol.* (10), (1955) 94–99.
1901. H. Kuehl, Diastatische Abbau der Weizen- und Roggenstärke im Mehl, *Muehlenlab.*, 3 (1933) 65–70.
1902. E. Nebesny, Changes of carbohydrate compositions during enzymatic hydrolysis of starch of various origin, *Starch/Stärke*, 45 (1993) 426–429.
1903. H. C. Sherman and J. C. Baker, Experiments upon starch as substrate for enzyme action, *J. Am. Chem. Soc.*, 38 (1916) 1885–1904.
1904. K. Myrbäck and B. Persson, Products of potato starch degradation by malt α-amylase, *Arkiv Kemi*, 5 (1953) 365–378.
1905. R. L. Lohmar, Jr., F. B. Weakley, and G. E. Lauterback, Controlled degradation of waxy cornstarch by malt α-amylase, *Cereal Chem.*, 33 (1956) 198–206.
1906. E. Nebesny, Changes of carbohydrates and molecular structure of dextrins during enzymic liquefaction of starch, *Starch/Stärke*, 44 (1992) 398–401.
1907. J. Blom, T. Schmith, and B. Schwartz, Abbau von Kartoffelstärke durch alpha-Amylasen aus Malz und aus *Bacillus subtilis*. Differenzierung der Abbauprodukten, *Acta Chim. Scand.*, 6 (1952) 591–598.
1908. E. Nebesny, Carbohydrate compositions and molecular structure of dextrins in enzymic high-maltose syrups, *Starch/Stärke*, 42 (1990) 437–444.
1909. E. Nebesny, T. Pierzgalski, and S. Brzeziński, Changes in carbohydrate composition during enzymic hydrolysis of starch with mycolase participation, *Starch/Stärke*, 48 (1996) 263–266.
1910. A. S. Vecher and M. V. Levitskaya, Acidic and enzymic saccharification of the starch in potato pulp (in Russian), *Vestsi Akad. Navuk BSSR, Ser Biyal. Navuk* (5), (1974) 117–120.
1911. L. M. Marchal, J. Jonkers, G. T. Franke, C. D. De Gooijer, and J. Tramper, The effect of process conditions on the α-amylolytic hydrolysis of amylopectin potato starch. An experimental design approach, *Biotechnol. Bioeng.*, 62 (1999) 348–357.
1912. Y. Han, H. Zhang, and Z. Xu, Production of high maltose syrup using potato starch influencing factors of starch liquefaction, *Shipin Gongye Keji*, 25(2), (2004) 57–59. *Chem. Abstr.*, 143 (2005) 438820.
1913. Z. Xu, Y. J. Han, and H. W. Zhang, Influencing factors of liquefaction of potato starch, *Huaxue Yu Nianhe* (1), (2004) 20–22. *Chem. Abstr.*, 141 (2004) 53222.
1914. Y. Wang, C. Zhang, and J. Yin, Saccharification of red potato starch with enzymolysis, *Huaxue Shije*, 29 (1988) 417–419. *Chem. Abstr.*, 110 (1989) 73853.

1915. R. Sakamoto, T. Kimura, N. Kuwatani, Y. Kawashima, and N. Arai, Enzymic manufacture of phosphorylated oligosaccharides and preparation of their calcium salt, Jpn. Kokai Tokkyo Koho 10084985 (1998); *Chem. Abstr.*, 128 (1998) 307590.
1916. N. Absar, I. S. M. Zaidul, S. Takigawa, N. Hashimoto, C. Matsuura-Endo, H. Yamauchi, and T. Noda, Enzymatic hydrolysis of potato starch containing different amounts of phosphorus, *Food Chem.*, 112 (2009) 57–62.
1917. T. Posternak and H. Pollaczek, De la protection contre hydrolyse enzymatique exercee par les groupes phosphoryles. Etude de la degradation enzymatique d'un peptide et d'un polyose phosphoryles, *Helv. Chim. Acta*, 24 (1941) 921–930.
1918. L. J. McBurney and M. D. Smith, Effect of esterified phosphorus in potato starch on the action pattern of salivary α-amylase, *Cereal Chem.*, 42 (1965) 161–167.
1919. S. J. Thannhauser, S. Z. Sorkin, and B. N. F. Boncoddo, The amylolytic and phosphatase activity of liver tissue in von Gierke's disease, *J. Clin. Invest.*, 19 (1940) 681–683.
1920. M. Samec, Studien über Pflanzkolloide. IV. Die Versuchungen des Phosphorgehaltes bei den Zustandänderung und dem diastatische Abbau der Stärke, *Koll. Chem. Beihefte*, 6 (1914) 23–54.
1921. K. Myrbäck and B. Kihlberg, Grenzdextrine und Stärke. XIX. Phosphorsäure in Stärkedextrin, *Biochem. Z.*, 315 (1943) 250–258.
1922. V. V. Kapustina and M. S. Dudkin, Enzymic hydrolysis of millet starch (in Russian), *Ferment. Spirt. Prom.* (3), (1972) 34–36.
1923. M. S. Pisareva, V. I. Rodzevich, and V. L. Yarovenko, Improvement in the starch hydrolysis process by acid phosphatase from *Aspergillus awamori* strain 22 (in Russian), *Ferment. Spirt. Prom.* (2), (1978) 38–41.
1924. T. Fukui, Y. Uta, and Z. Nikuni, Sweet potato starch difficultly liquefable by amylase. The viscogram and temperature curve properties (in Japanese), *Denpun Kogyo Gakkaishi*, 11 (1964) 25–28.
1925. Y. Komai and T. Tominaga, Varieties of tapioca and sago starch granules with regard to their behavior with respect to α-amylase (in Japanese), *Dempunto Gijutsu Kenkyu Kaiho*, 30 (1964) 20–30.
1926. M. Shin and S. Ahn, Action of crude amylolytic enzymes extracted from sweet potatoes and purified amylases on sweet potato starches, *Hanguk Shikpum Kwahakhoechi*, 18 (1986) 431–436. *Chem. Abstr.*, 107 (1986) 57659d.
1927. V. Hagenimana, R. E. Simard, and L. P. Vezina, Method for the hydrolysis of starchy materials by sweet potato endogenous amylases, US Patent 5525154 (1996); *Chem. Abstr.*, 125 (1996) 410964.
1928. P. L. C. Rupp and S. J. Schwartz, Characterization of the action of *Bacillus subtilis* alpha amylase on sweet potato starch amylose and amylopectin, *J. Food Biochem.*, 12 (1988) 191–204.
1929. Y. Hu, T. Zhang, W. Kang, D. Wang, and X. Yuan, New process for treatment of preserved sweet potato by enzyme, Chinese Patent 1451288 (2003); *Chem. Abstr.*, 142 (2005) 46270.
1930. G. R. P. Moore, L. Rodrigues de Canto, R. E. Amante, and V. Soldi, Cassava and corn starch in maltodextrin production, *Quim. Nova*, 28 (2005) 596–600.
1931. R. Soto Ibanez, Production of glucose by enzymic hydrolysis of cassava starch, *Rev. Boliv. Quim.*, 8 (1989) 17–21. *Chem. Abstr.*, 113 (1990) 4600.
1932. N. de Queiroz Araujo, I. E. de Castro, R. Falcao dos Reis, T. Bravo da Costa Ferreira, H. Ferreira de Castro, J. de Oliveira Barbosa, P. Suzzi, and F. de Arauno Costa, Enzymic hydrolysis of manioc (in Portuguese), *Inf. INT Rio de Janeiro*, 8(9), (1975) 42–52. *Chem. Abstr.*, 88 (1975) 103315.
1933. S. Yamamoto, T. Kaneko, and Y. Mino, Degradation of starch by α- and β-amylases from tap root of alfalfa, *Grassland Sci.*, 43 (1997) 14–17.
1934. N. M. Kim, J. S. Lee, and B. H. Lee, Enzymatic hydrolysis of Korean ginseng starch and characteristics of produced maltooligosaccharides, *J. Ginseng Res.*, 24(1), (2000) 41–45.

1935. G. V. Chowdary, S. Hari Krishna, and G. Hanumatha Rao, Optimization of enzymic hydrolysis of mango kernel starch by response surface methodology, *Bioprocess Eng.*, 23 (2000) 681–685.
1936. P. V. Hung and N. Morita, Physicochemical properties and enzymatic digestibility of starch from edible canna (*Canna* edulis) grown in Vietnam, *Carbohydr. Polym.*, 61 (2005) 314–321.
1937. H. Anger, M. Richter, B. Kettlitz, H. Kalb, H. Neurnberger, H. Kraetz, and P. Schulz, One-step enzymic manufacture of maltose- and maltotriose-rich hydrolyzate from starch, East German Patent 285112 (1990); *Chem. Abstr.*, 114 (1991) 183885.
1938. S. Rogols and R. G. Hyldon, Enzymic hydrolysis of wheat starch. II. The effect of pH on the degradation products of wheat and corn starch substrates and their role in titanium dioxide retention, *Starch/Stärke*, 15 (1963) 364–367.
1939. I. W. Dadswell and J. F. Gardner, The relation of α-amylase and susceptible starch to diastatic activity, *Cereal Chem.*, 24 (1947) 79–99.
1940. M. Gibiński, M. Pałasiński, and P. Tomasik, Properties of defatted oat starch, *Starch/Stärke*, 45 (1993) 354–357.
1941. G. Tegge, Verarbeitungstechnische Eigenschaften von Inland Mais. 3. Direktverzuckerung, *Starch/Stärke*, 24 (1972) 375–379.
1942. Z. Bebić and M. Radosavljević, Studies of enzyme hydrolysis of maize starch (in Serbo-Croatian), *Arh. Poljopr. Nauke*, 46 (1985) 267–278.
1943. O. V. Fursova, L. A. Anikeeva, V. A. Kuzovlev, A. A. Khakimshanov, and Z. S. Khaidarova, Enzymic hydrolysis of cereal grain starch granules (in Russian), *Prikl. Biokhim. Mikrobiol.*, 26 (1990) 371–378.
1944. K. Balasubramanian and S. Mahendran, Corn malt amylases for saccharification of starch in corn, *J. Microb. Biotechnol.*, 5 (1990) 42–46.
1945. J. Yang and S. Yiang, Maltodextrin preparation by enzymic liquefaction of starch with α-amylase, Chinese Patent 1038308 (1989); *Chem. Abstr.*, 113 (1990) 76653.
1946. A. Rotsch, Besonderheiten bei der diastatischen Verzukerung von Maismehl, *Starch/Stärke*, 2 (1950) 78–81.
1947. C. Gerard, P. Colonna, A. Buleon, and V. Planchot, Amylolysis of maize mutant starches, *J. Sci. Food Agric.*, 81 (2001) 1281–1287.
1948. G. Jiang and Q. Liu, Characterization of residues from partially hydrolyzed potato and high amylose corn starches by pancreatic α-amylase, *Starch/Stärke*, 54 (2002) 527–533.
1949. M. Yamamoto, S. Ushiro, N. Nakamura, and S. Hizukuri, Digestion of raw corn starch by the raw starch digesting amylase of *Aspergillus* sp. K-27 (in Japanese), *Denpun Kagaku*, 35 (1988) 235–243.
1950. B. C. Saha and R. J. Bothast, Starch conversion by amylases from *Aureobasidium pullulans*, *J. Ind. Microbiol.*, 12 (1993) 413–416.
1951. A. Golc-Wondra and L. Vitez, Carbohydrates in food industry – Enzyme hydrolysis of starch (in Slovenian), *Nova Proizvod.*, 36 (1985) 81–87.
1952. S. H. Imam, A. Burgess-Cassler, G. L. Cote, S. H. Gordon, and F. L. Baker, A study of cornstarch granule digestion by an unusually high molecular weight α-amylase, *Curr. Microbiol.*, 22 (1991) 365–370.
1953. T. Okemoto, A. Iwamoto, K. Hara, H. Hashimoto, and S. Kobayashi, Manufacture of maltopentaose syrups with α-amylase, Jpn. Kokai Tokkyo Koho 04045794 (1992); *Chem. Abstr.*, 117 (1992) 46743.
1954. M. Matsudaira and T. Yoshida, Manufacture of nonreducing branched dextrin and nonreducing linear oligosaccharides with α-amylase from starch, Jpn. Kokai Tokkyo Koho 01056705 (1989); *Chem. Abstr.*, 111 (1989) 132619.
1955. J. W. Lim, S. H. Mun, and M. Shin, Action of α-amylase and acid on resistant starches prepared from normal maize starch, *Food Sci. Biotechnol.*, 14 (2005) 32–38.
1956. V. Arasarantnam and K. Balasubramanian, The hydrolysis of starch in wet and dry milled corn, *Int. Sugar J.*, 96 (1994) 368–370.

1957. L. V. Kaprelyants, L. V. Tarakhtii, and I. V. Styngach, Enzymic hydrolysis of wheat starch with various amylases (in Russian), *Biotekhnologi* (6), (1991) 50–52.
1958. M. Richter, H. Anger, B. Kettlitz, F. Schierbaum, P. Klingenberg, D. Korner, H. Kraetz, and P. Schulz, Preparation of maltose-maltotriose with α-amylase from *thermoactinomyces,* East German Patent 287732 (1991); *Chem. Abstr.*, 115 (1991) 7009.
1959. M. Remiszewski, Y. Y. Linko, M. Leisola, and P. Linko, Enzymic conversion of wheat starch and grain, *Proc. 2nd Int. Congr. Food Process Eng., 1979*, 2 (1980) 237–245. *Chem. Abstr.*, 94 (1980) 119676f.
1960. E. Legin, A. Copinet, and F. Duchiron, A single step high temperature hydrolysis of wheat starch, *Starch/Stärke*, 50 (1998) 84–89.
1961. S. D. Textor, G. A. Hill, D. G. MacDonald, and E. St Denis, Cold enzyme hydrolysis of wheat starch granules, *Can. J. Chem. Eng.*, 76 (1998) 87–93.
1962. E. Nebesny, J. Rosicka, and T. Pierzgalski, Enzymic hydrolysis of wheat starch into glucose, *Starch/Stärke*, 50 (1998) 337–341.
1963. E. Nebesny, J. Rosicka, and D. Sucharzewska, The effect of process conditions of enzymic hydrolysis on the properties of wheat starch hydrolyzates, *Zywn. Technol. Jakosc*, 5(4, Suppl), (1998) 181–189.
1964. D. P. Atkins and J. F. Kennedy, The influence of pullulanase and α-amylase upon the oligosaccharide product spectra of wheat starch hydrolyzates, *Starch/Stärke*, 37 (1985) 126–131.
1965. J. Holm and I. Bjoerck, Effects of thermal processing of wheat on starch. II. Enzymic availability, *J. Cereal Sci.*, 8 (1988) 261–268.
1966. A. Leon, E. Duran, and C. Benedito De Barber, Firming of starch gels and amylopectin retrogradation as related to dextrin production by α-amylase, *Z. Lebensm. Unters. Forsch. A.*, 205 (1997) 131–134.
1967. J. E. Kruger and B. A. Marchylo, A comparison of the catalysis of starch components by isoenzymes from two major groups of germinated wheat α-amylases, *Cereal Chem.*, 62 (1985) 11–18.
1968. S. Hizukuri, Y. Takeda, and T. Matsubayashi, Effect of phosphorus in starch granules on raw starch digestion by bacterial α-amylase (in Japanese), *Denpun Kagaku*, 26 (1979) 112–116.
1969. E. Berliner and R. Ruter, Diastase und Weizenmehl, *Z. Ges. Muehlenw.*, 5 (1928) 134–140, 156–162.
1970. J. G. Malloch, Studies on the resistance of wheat do diastatic action, *Can. J. Res.*, 1 (1929) 111–147.
1971. P. Gonzales Tello, F. Camacho Rubio, A. Robles Medina, and L. J. Morales Villena, Hidrolisi de harina de trigo con α-amilasa *de A. oryzae*, *Rev. Agroquim. Tecnol. Aliment.*, 30 (1990) 51–58.
1972. Z. Dai, Y. Yin, and Z. Wang, Comparison of starch accumulation and enzyme activity in grains of wheat culivars differing in kernel type, *Plant Growth Regul.*, 17 (2008) 153–162.
1973. A. Faraj, T. Vasanthan, and R. Hoover, The influence of α-amylase-hydrolyzed barley starch fractions on the viscosity of low and high purity barley β-glucan concentrates, *Food Chem.*, 96 (2006) 56–65.
1974. K. Poutanen, M. Lauro, T. Suortti, and K. Autio, Partial hydrolysis of gelatinized barley and waxy barley starches by alpha-amylase, *Food Hydrocoll.*, 10 (1996) 269–275.
1975. M. Lauro, A. Lappalainen, T. Suortti, K. Autio, and K. Poutanen, Modification of barley starch by α-amylase and pullulanase, *Carbohydr. Polym.*, 21 (1993) 151–152.
1976. M. Lauro, S. G. Ring, V. J. Bull, and K. Poutanen, Gelation of waxy barley starch hydrolyzates, *J. Cereal Sci.*, 26 (1997) 347–354.
1977. E. Bertoft, R. Manelius, P. Myllarinen, and A. H. Shulman, Characterization of dextrins solubilized by α-amylae from barley starch granules, *Starch/Stärke*, 52 (2000) 160–163.
1978. Y. Y. Linko, A. Lindroos, and P. Linko, Soluble and immobilized enzyme technology in bioconversion of barley starch, *Enzyme Mirob. Technol.*, 1 (1979) 273–278.

1979. D. H. Ahn and H. N. Chang, Liquefaction and saccharification of starch using α-amylase and immobilized glucoamylase, *Sanop Misaengmul Hakkoechi*, 19 (1991) 497–503. *Chem. Abstr.*, 119 (1993) 26818.
1980. K. V. Kobelev, R. A. Kolcheva, K. A. Kalunyants, and I. A. Popadich, Kinetics of the enzymic hydrolysis of an insoluble barley starch (in Russian), *Ferment. Spirt. Prom.* (8), (1981) 28–31.
1981. J. H. Li, T. Vasanthan, R. Hoover, and B. G. Rossnagel, Starch from hull-less barley. V. In vitro susceptibility of waxy, normal and high-amylose starches towards hydrolysis by alpha-amylases and amyloglucosidase, *Food Chem.*, 84 (2004) 621–632.
1982. S. You and M. S. Izydorczyk, Comparison of the physicochemical properties of barley starches after the partial hydrolysis of and acid alcohol hydrolysis, *Carbohydr. Polym.*, 69 (2007) 489–502.
1983. F. J. Warth and D. B. Darabsett, The fractional liquefaction of starch, *Mem. Dept. Agr. India, Chem. Ser.*, 3 (1914) 135–147.
1984. K. P. Basu and S. Mukherjee, Biochemical investigations of different varieties of Bengal rice. III. Enzymic digestability of rice starch and protein: Action of salivary and pancreatic amylase as well as pepsin and trypsin, *Indian J. Med. Res.*, 23 (1936) 777–787.
1985. N. N. Ivanov, M. M. Kurgatnikov, and V. A. Kirsanova, Differences in the structure of starch as demonstrated by the use of diastase (in Russian), *Enzymologia*, 4 (1937) 163–168.
1986. D. K. Apar and B. Oezbek, α-Amylase inactivation during rice starch hydrolysis, *Process Biochem.*, 40 (2005) 1367–1379.
1987. K. H. Meyer and M. Fuld, Recherches sur l'amidon. XVII. L'amidon du riz collant, *Helv. Chim. Acta*, 24 (1941) 1404–1407.
1988. S. Wang, P. Zhang, and P. Wang, Production of extreme high maltose syrup and high protein rice flour by whole enzymic method, *Wuxi Qinggong Daxue Xuebao*, 17(2), (1998) 28–33. *Chem. Abstr.*, 130 (1998) 124090.
1989. J. H. Auh, H. Y. Chae, Y. R. Kim, K. H. Shim, S. H. Yoo, and K. H. Park, Modification of rice starch by selective degradation of amylose using alkalophilic *Bacillus* cyclomaltodextrinase, *J. Agric. Food Chem.*, 54 (2006) 2314–2319.
1990. V. K. Griffin and J. R. Brooks, Production and size distribution of rice maltodextrins hydrolyzed from milled rice flour using heat-stable alpha-amylase, *J. Food Sci.*, 54 (1989) 190–193.
1991. B. N. Okolo, L. I. Ezeogu, and K. E. Ugwuanyi, Amylolysis of sorghum starch as influenced by cultivar, germination time and gelatinization temperature, *J. Inst. Brew.*, 103 (1997) 371–376.
1992. H. Guzman-Maldonado, O. Paredes-Lopez, and J. Dominiquez, Optimization of an enzymatic procedure for the hydrolytic depolymerization of amaranthus starch by response surface methodology, *Lebensm. Wiss. Technol.*, 26 (1993) 28–33.
1993. G. Zhang and B. R. Hamaker, Low α-amylase starch digestibility of cooked sorghum flours and the effect of protein, *Cereal Chem.*, 75 (1998) 710–713.
1994. A. P. Barba de la Rosa, O. Paredes Lopez, A. Carabez Trejo, and C. Ordorica Falomir, Enzymic hydrolysis of amaranth flour differential scanning calorimetry and scanning electron microscopy studies, *Starch/Stärke*, 41 (1989) 424–428.
1995. A. M. Zając, Direct hydrolysis of milled triticale grain. I. Liquefaction, *Pol. J. Food Nutr. Sci.*, 2(4), (1993) 33–39.
1996. L. Słomińska, E. Łozińska, and A. Grześkowiak-Przywecka, Direct hydrolysis of starch sources, In: V. P. Yuryev, P. Tomasik, and H. Ruck, (Eds.) *Starch; From Starch Containing Sources to Isolation of Starches and Their Application,* Nova Science Publ. Inc., New York, 2004, pp. 57–64.
1997. K. Lewen, D. Fletcher, R. S. Dickmann, G. Yang, J. B. Holder, L. A. Wilson, M. R. Andrews, and K. R. Vadlamani, β-Glucan containing oat based cereal base prepared by enzymic hydrolysis, EU Patent WO 2000030457 (2000); *Chem. Abstr.*, 132 (2000) 347013.
1998. W. J. Wang, A. D. Powell, and C. G. Oates, Pattern of enzymic hydrolysis in raw sago starch: Effect of processing history, *Carboohydr. Polym.*, 26 (1995) 91–97.

1999. A. Arbakariya, B. Asbi, and N. R. Norjehan, Rheological behavior of sago starch during liquefaction and saccharification, *Ann. N. Y. Acad. Sci.*, 613 (1990) 610–613.
2000. N. Haska and Y. Ohta, Glucose production from treated sago starch granules by raw starch digesting amylase from *Penicillium brunneum*, *Starch/Stärke*, 43 (1991) 102–107.
2001. S. Govindasamy, C. G. Oates, and H. A. Wong, Characterization of changes of sago starch components during hydrolysis by thermostable alpha-amylase, *Carbohydr. Polym.*, 18 (1992) 89–100.
2002. S. Gorinstein, C. G. Oates, S. M. Chang, and C. Y. Lii, Enzymatic hydrolysis of sago starch, *Food Chem.*, 49 (1994) 411–417.
2003. B. Cheirslip, A. H-Kittikun, and S. Shioya, Kinetic study on enzymatic hydrolysis of sago starch, *Biotechnol. Sust. Util. Biol. Res. Tropics*, 17 (2004) 43–47.
2004. C. W. Wong, S. K. S. Muhammed, M. H. Dzulkifly, N. Saari, and H. M. Ghazali, Enzymatic production of linear long-chain dextrin from sago (Metroxylon sagu) starch, *Food Chem.*, 100 (2007) 774–780.
2005. K. Bujang, D. Salwani, A. Adeni, and P. Jolhiri, Effects of starch concentration and pH on enzymic hydrolysis of sago starch, *Biotechnol. Sust. Util. Biol. Res. Tropics*, 14 (2000) 32–35.
2006. Z. I. Xie and X. H. Wu, Study on optimum liquefaction conditions of chestnut starch, *Shipin Kexue*, 24(10), (2003) 62–66. *Chem. Abstr.*, 143 (2005) 345688.
2007. C. Lopez, A. Torrado, P. Fucinos, N. P. Guerra, and L. Pastrana, Enzymic inhibition and thermal inactivation in the hydrolysis of chestnut puree with an amylase mixture, *Enzyme Microb. Technol.*, 39 (2006) 252–258.
2008. C. Lopez, A. Torrado, P. Fucinos, N. P. Guerra, and L. Pastrana, Enzymatic hydrolysis of chestnut puree: Process optimization using mixtures of α-amylase and glucoamylase, *J. Agric. Food Chem.*, 52 (2004) 2907–2914.
2009. C. Lopez, A. Torrado, N. P. Guerra, and L. Pastrana, Optimization of solid-state enzymatic hydrolysis of chestnut using mixtures of αp-amylase and glucoamylase, *J. Agric. Food Chem.*, 53 (2005) 989–995.
2010. D. Mu, M. Kang, Y. Li, L. Zhou, and W. Zhao, Study on starch hydrolysis by enzymes during processing chestnut beverage, *Shipin Gongye Keji*, 23(11), (2002) 72–75. *Chem. Abstr.*, 141 (2004) 224368.
2011. T. E. Panasyuk and M. S. Dudkin, Enzymic hydrolysis of pea and bean starch (in Russian), *Izv. Vyssh. Uchebn. Zaved., Pishch. Tekhnol.* (4), (1968) 29–31.
2012. E. Bertoft, Z. Qin, and R. Manelius, α-Amylase action on granular pea starch, *Proc. 7th Int. Symp. Plant Polym. Carbohydr. Berlin 1992*, 2 (1995) 199–202.
2013. S. Peat, W. J. Whelen, J. R. Turvey, and K. Morgan, The structure of isolichenin, *J. Chem. Soc.* (1961) 623–629.
2014. P. Karrer, Isolichenin und Stärkeabbau, *Z. Physiol. Chem.*, 148 (1925) 62–64.
2015. S. Nishimura, Starch. II. Amylose and amylopectin (in Japanese), *J. Agr. Chem. Soc. Jpn.*, 9 (1933) 767–770.
2016. H. Ninomyia, Effect of salivary amylase on different kinds of starch and the different kinds of starch and the different effects of salivary and pancreatic amylase on amylopectin and amylose (in Japanese), *J. Biochem. Jpn.*, 31 (1940) 429–435.
2017. Y. Otani and S. Takahashi, Improvement of alcoholic mash. XI. Liquefaction and saccharification of starch by the bacterial and fungal amylases separately and in combination (in Japanese), *Hakko Kagaku Zasshi*, 36 (1958) 453–457.
2018. R. M. Attia and A. H. Fahmy, Effect of bacterial α-amylase on the physical properties of rice starch, *Agric. Res. Rev.*, 52 (1974) 115–122.
2019. K. Kainuma, S. Matsunaga, M. Itagawa, and S. Kobayashi, New enzyme system — β-amylase–pullulanase to determine the degree of gelatinization and retrogradation of starch or starch products (in Japanese), *Denpun Kagaku*, 28 (1981) 235–240. *Chem. Abstr.*, 96 (1982) 67325k.

2020. P. J. Brumm, Non-random cleavage of starch and the low dextrose equivalent starch conversion products produced thereby, Canadian Patent 2109368 (1994); *Chem. Abstr.*, 121 (1994) 203550.
2021. K. Myrbäck and L. G. Sillen, Branched molecules. II. Application of our models to starch and glycogen with specific regard to enzymic saccharification, *Svensk. Kem. Tids.*, 55 (1943) 311–323. *Chem. Abstr.*, 39 (1944) 2247.
2022. K. Myrbäck, The constitution and enzymic degradation of starch and glycogen, *Svensk. Kemi Tids. (The Svedberg Mem. Vol.)*, 56 (1944) 508–522.
2023. R. Sutra, Sur l'action des amylases dans ses rapports avec la constitution de l'amidon, *Bull. Soc. Chim. Biol.*, 30 (1948) 439–449.
2024. J. A. Rendleman, Jr., Hydrolytic action of α-amylase on high-amylose starch of low molecular mass, *Biotechnol. Appl. Biochem.*, 31 (2000) 171–178.
2025. D. L. Fox and R. Craig, Enzymic reactions in heavy water. II. Deuterium and the hydrolysis of starch, *Proc. Soc. Exptl. Biol. Med.*, 33 (1935) 206–209.
2026. Y. Komai and K. Mifune, Enzymic hydrolysis of starch in relation to the degree of oxidation (in Japanese), *Denpun Kogyo Gakkaishi*, 11 (1964) 32–44.
2027. H. E. Horn and B. A. Kimball, Maltodextrins of improved stability prepared by enzymic hydrolysis of oxidized starch, US Patent 3,974,034 (1976); *Chem. Abstr.*, 85 (1976) 162289h.
2028. A. S. H. El-Saied, R. M. Attia, and A. H. Fahmy, Studies on kinetics of amyloglucosidase as affected by native and derivatized corn starch, *Food Chem.*, 15 (1984) 45–50.
2029. W. Ziese, Specifität der Amylasen, *Z. Physiol. Chem.*, 235 (1935) 235–245.
2030. R. L. Conway and L. F. Hood, Pancreatic alpha-amylase hydrolysis products of modified and unmodified tapioca starches, *Starch/Stärke*, 28 (1976) 341–343.
2031. B. M. N. M. Azemi and M. W. Kensington, Distribution of partial digestion products of hydroxypropyl derivatives of maize, waxy maize and high amylose maize starch, *Starch/Stärke*, 46 (1994) 440–443.
2032. M. A. Quilez and R. Clotet, Effect of the acylation of corn starch on the kinetics of enzymic and chemical hydrolysis, *Rev. Agroquim. Tecnol. Aliment.*, 31 (1991) 551–557. *Chem. Abstr.*, 116 (1991) 213150.
2033. A. Copinet, V. Coma, J. P. Onteniente, and Y. Couturier, Enzymatic degradation of native and acetylated starch-based extruded blends, *Packag. Technol. Sci.*, 11 (1998) 69–81.
2034. A. Copinet, V. Coma, E. Legin, Y. Couturier, and J. C. Prudhomme, Hydrolysis of acetylated starches with thermostable alpha-amylase, *Starch/Stärke*, 49 (1997) 492–498.
2035. H. Derradji-Serghat, G. Bureau, Y. Couturier, and J. C. Prudhomme, Aerobic biodegradatiom of acetylated starch, *Starch/Stärke*, 51 (1999) 362–368.
2036. H. Serghat-Derradji, A. Copinet, G. Bureau, and Y. Couturier, Aerobic degradation of extruded polymer blends with native starch as major component, *Starch/Stärke*, 51 (1999) 369–375.
2037. A. Rajan and T. E. Abraham, Enzymatic modification of cassava starch by bacterial lipase, *Bioprocess Biosyst. Eng.*, 29(1), (2006) 65–71.
2038. G. Hamdi and G. Ponchel, Enzymatic degradation of epichlorohydrin crosslinked starch micorspheres by α-amylase, *Pharm. Res.*, 16 (1999) 867–875.
2039. R. Manelius, K. Nurmi, and E. Bertoft, Characterization of dextrins obtained by enzymatic treatment of cationic potato starch, *Starch/Stärke*, 57 (2005) 291–300.
2040. R. Manelius, K. Nurmi, and E. Bertoft, Enzymic and acidic hydrolysis of cationized waxy maize starch granules, *Cereal Chem.*, 77 (2000) 345–353.
2041. B. Zhang, Y. Liang, L. Yang, S. Yu, X. Zeng, and X. Fu, Preparation of degraded starch by enzymolysis, Chinese Patent 1390855 (2003); *Chem. Abstr.*, 141 (2004) 262314.
2042. H. Selek, G. Ponchel, S. Kas, and A. A. Hincal, Preparation and enzymic degradation of the crosslinked degradable starch microspheres by pancreatin (in Slovenian), *Farm. Vestn (Ljubljana)*, 50 (1999) 318–319.

2043. T. Takaya, Y. Sugimoto, E. Imo, Y. Tominaga, N. Nakatani, and H. Fuwa, Degradation of starch granules by α-amylase of fungi, *Starch/Stärke*, 30 (1978) 289–293.
2044. A. Kimura and J. F. Robyt, Reaction of enzymes with starch granules: Kinetics and products of the reaction with glucoamylase, *Carbohydr. Res.*, 277 (1995) 87–107.
2045. G. Orestano, Liquefaction and saccharification of starch by malt amylase (in Italian), *Boll. Soc. Ital. Biol. Sper.*, 7 (1932) 259–262.
2046. I. A. Popadich, I. S. Shub, N. I. Mironova, and E. D. Bondareva, Effect of wheat starch grain size on the tendency toward degradation by amylolytic preparations (in Russian), *Izv. Vyssh. Ucheb. Zaved., Pishch. Tekhnol.* (3), (1972) 24–26.
2047. F. Delpeuch and J. C. Favier, Characteristics of starches from from tropical food plants. Alpha amylase hydrolysis swelling and solubility patterns, *Ann. Technol. Agric.*, 29 (1980) 53–67.
2048. T. Maezawa, S. Hayakawa, M. Okubo, and F. Shinbori, Liquefying properties of starch for enzyme hydrolyzed glucose production. VII. Effects of the treatment of chlorinating reagent and differences between large starch granules and small ones (in Japanese), *Nippon Nogei Kagaku Kaishi*, 41 (1967) 422–427.
2049. A. W. MacGregor and J. E. Morgan, Hydrolysis of barley starch granules by alpha-amylase from barley malt, *Cereal Foods World*, 31 (1986) 688–693.
2050. R. G. Swanson, The susceptibility of gelatinized starches to the malt amylase, *Brew. Dig.*, 13(10), (1938) 185T–188T.
2051. M. S. Buttrose, Difference in x-ray diffraction type of two woody starches and the relation of this to acid and enzyme breakdown, *Starch/Stärke*, 18 (1966) 122–126.
2052. H. Tatsumi, H. Katano, and T. Ikeda, Kinetic analysis of glucoamylase-catalyzed hydrolysis of starch granules from various botanical sources, *Biosci. Biotechnol. Biochem.*, 71 (2007) 946–950.
2053. S. H. Imam, S. H. Gordon, A. Mohamed, R. Harry-O'kuru, B. S. Chiou, G. M. Glenn, and W. J. Orts, Enzyme catalysis of insoluble cornstarch granules: Impact on surface morphology, properties and biodegradability, *Polym. Degr. Stab.*, 91 (2006) 2894–2900.
2054. B. J. D. Meeuse and B. N. Smith, Amylolytic breakdown of some raw algal starches, *Planta*, 57 (1962) 624–635.
2055. A. W. MacGregor and D. L. Balance, Hydrolysis of large and small starch granules from normal and waxy barley cultivars by α-amylase from barley malt, *Cereal Chem.*, 57 (1980) 397–402.
2056. B. W. Kong, J. I. Kim, M. J. Kim, and J. C. Kim, Porcine pancreatic α-amylase hydrolysis of native starch granules as a function of granule surface area, *Biotechnol. Progr.*, 19 (2003) 1162–1166.
2057. H. Le Corvaisier, The action of the mineralization (ash) of starch on the liquefaction und hydrolysis of starch paste by malt diastase, *Bull. Soc. Sci. Bretagne*, 15(Suppl.), (1938) 1–74. *Chem. Abstr.*, 34 (1940) 44709.
2058. T. Noda, Y. Takahata, and T. Nagata, Factors relating to digestibility of raw starch by amylase, *Denpun Kagaku*, 40 (1993) 271–276.
2059. A. Stevnebo, S. Sahlstroem, and B. Svihus, Starch structure and degree of starch hydrolysis of small and large starch granules from barley varieties with varying amylose content, *Animal Feed Sci. Technol.*, 130 (2006) 23–38.
2060. C. M. L. Franco and C. F. Ciacco, Factors which affect the enzymatic degradation of natural starch granules. Effect of the size of granules, *Starch/Stärke*, 44 (1992) 422–426.
2061. H. Fuwa, Y. Sugimoto, M. Tanaka, and D. V. Glover, Susceptibility of various starch granules to amylases as seen by scanning electron microscope, *Starch/Stärke*, 30 (1978) 186–191.
2062. H. Fuwa, Y. Sugimoto, T. Takaya, and Z. Nikuni, Scanning electron microscopy of starch granules with or without amylase attack, *Carbohydr. Res.*, 70 (1979) 233–238.
2063. U. R. Bhat, S. V. Paramahans, and R. N. Tharanathan, Scanning electron microscopy of enzyme-digested starch granules, *Starch/Stärke*, 35 (1983) 261–265.

2064. Y. Fukai, E. Takaki, and S. Kobayashi, Changes in starch granules after enzymic treatment. II. Changes in three kinds of starch granules after enzymic treatment, *Nippon Nogei Kagaku Kaishi*, 68 (1994) 793–800.
2065. P. Tomasik, C. H. Schilling, and S. Shepardson, Microscopic imaging of starch granule envelope, In: V. P. Yuryev, P. Tomasik, and H. Ruck, (Eds.). *From Starch Containing Sources to Isolation of Starches and Their Applications,* Nova Science Publ, New York, 2004.
2066. E. Bertof and H. Henriksnas, Initial stages in α-amylolysis of barley starch, *J. Inst. Brew.*, 88 (1982) 261–267.
2067. M. Lauro, T. Suortti, K. Autio, P. Linko, and K. Poutanen, Accessibillity of barley starch granules to α-amylase during different phases of gelatinization, *J. Cereal Sci.*, 17 (1993) 125–136.
2068. M. Lauro, P. M. Forsell, M. T. Suortti, S. H. D. Hulleman, and K. S. Poutanen, α-amylolysis of large barley starch granules, *Cereal. Sci.*, 76 (1999) 925–930.
2069. M. Lauro, α-Amylolysis of starch, *VTT Publ.*, 433 (2001) 1–92. *Chem. Abstr.*, 136 (2001) 166285.
2070. Y. Ikawa and H. Fuwa, Changes in some properties of starch granules of maize having amylose extender gene by amylase attack, *Starch/Stärke*, 32 (1980) 145–149.
2071. H. K. Sreenath, Studies on starch granules digestion by α-amylase, *Starch/Stärke*, 44 (1992) 61–63.
2072. Y. Liang, B. S. Zhang, L. S. Yang, and D. W. Gao, Study on α-amylase degraded activity of corn starch with non-crystallized granule state under different conditions, *Shangqiu Shifan Xueyuan Xuebao*, 21(5), (2005) 96–101. *Chem. Abstr.*, 145 (2005) 251047.
2073. M. I. Smirnova, Biochemical study of maize grains (in Russian), *Zhurn. Prikl. Botan., Genetik*, 25 (1931) 329–343.
2074. M. Sujka and J. Jamroz, Starch granule porosity and its change by means of amylolysis, *Int. Agrophys.*, 21 (2007) 107–113.
2075. M. Sujka and J. Jamroz, α-Amylolysis of native potato and corn starches – SEM, AFM, nitrogen and iodine sorption investigations, *LWT- Food Sci. Technol.*, 42 (2009) 1219–1224.
2076. T. Fukui, M. Fujii, and J. Nikuni, Digestion of raw starch granules by amylase, particularly rice starch granules, *Nippon Nogei Kagaku Kaishi*, 38 (1964) 262–266.
2077. Y. Fukai, E. Takaki, and S. Kobayashi, Characteristic change of various starch granules by enzymic treatment. I. Characteristic change of rice starch granules by enzymic treatment (in Japanese), *Denpun Kagaku*, 40 (1993) 263–269.
2078. E. Bertoft, R. Manelius, and Z. Qin, Studies of the structure of pea starches. I. Initial stages in α-amylolysis of granular smooth pea starch, *Starch/Stärke*, 45 (1993) 215–220.
2079. P. Wuersch, S. Del Vedovo, and B. Koellreutter, Cell structure and starch nature as key determinants of the digestion rate of starch in legumes, *Am. J. Clin. Nutr.*, 43 (1986) 25–29.
2080. Y. Liang, B. Zhang, L. Yang, and D. Gao, Structure characteristics and α-amylase degradability of tapioca starch in noncrystallized granule state, *Shipin Gongye Keji*, 25(9), (2004) 49–51. *Chem. Abstr.*, 144 (2005) 275920.
2081. Y. Liang, B. S. Zhang, L. S. Yang, and D. W. Gao, Study on degradation characteristics of potato starch in noncrystalline granular state by microorganisms, *Jingxi Huagong*, 20 (2003) 361–363. 380; *Chem. Abstr.*, 140 (2003) 237427.
2082. Y. Liang, B. S. Zhang, L. S. Yang, and D. W. Gao, Study of α-amylase degradation activity of potato starch in noncrystalline granular state, under different conditions, *Jingxi Huagong*, 21 (2004) 763–767. *Chem. Abstr.*, 143 (2005) 288293.
2083. T. Sasaki, J. Hayashi, N. Ishida, and K. Kainuma, Swelling, solubilization and its morphological changes of sweet potato starch granules by urea and pullulanase (in Japanese), *Nippon Shokuhin Kogyo Gakkaishi*, 27 (1980) 489–497. *Chem. Abstr.*, 94 (1980) 82460c.
2084. R. Manelius, Z. Qin, A. K. Avall, H. Andtfolk, and E. Bertoft, The mode of action of granular wheat starch by bacterial α-amylase, *Starch/Stärke*, 49 (1997) 142–147.

2085. R. M. Sandstedt, Photomicrographic studies on wheat starch. III. Ezymic digestion and granule structure, *Cereal Chem.*, 32(Suppl.), (1955) 17–47.
2086. J. A. Popadich, F. A. Lysyuk, S. E. Traubenberg, and I. S. Shub, Micrographic and chromatographic studies of wheat starch grains during enzymic hydrolysis (in Russian), *Sakh. Prom.* (1), (1975) 60–66.
2087. O. V. Fursov, L. A. Anikieeva, B. K. Idgeev, and T. B. Darkanbaev, Susceptibility of wheat grain starch to α-amylase (in Russian), *Izv. Akad. Nauk Kaz. SSR Biol.* (6), (1982) 16–19.
2088. N. H. Thomson, M. J. Miles, S. G. Ring, P. R. Shewry, and A. S. Tatham, Real time imaging of enzymic degradation of starch granules by atomic force microscopy, *J. Vacuum Sci. Technol.*, B12 (1994) 1565–1568.
2089. F. R. Rosenthal and T. Nakamura, Structure of *Leguminosae* starches. I. Solubility in dimethyl sulfoxide and by enzymatic action, *Starch/Stärke*, 24 (1972) 152–158.
2090. Y. Sugimoto, Scanning electron microscopic observation of starch granules attacked by enzymes (in Japanese), *Denpun Kagaku*, 27 (1980) 28–40.
2091. K. Nagai, Untersuchungen über die Natur der Stärke, *Acta Schol. Med. Univ. Imp. Kyoto*, 7 [IV] (1925) 569–575; *Chem. Zentr.*, 1926, II, 1014.
2092. P. T. Boekestein, Adsorption of amylases by starch grain, *Acta Brevia Neerland., Physiol. Pharmacol. Microbiol.*, 2 (1932) 132–134.
2093. R. A. Gortner and C. Hamalainen, Protein films and the susceptibility of raw starch to diastatic attack, *Cereal Chem.*, 17 (1940) 378–383.
2094. Y. Koba, B. C. Saha, and S. Ueda, Adsorption on and digestion of raw starch by malt and bacterial alpha-amylase, *Denpun Kagaku*, 33 (1986) 199–201.
2095. R. N. Tharathan and U. R. Bhat, Scanning electron microscopy of chemically and enzymically treated black gram (*Phaseolus* mungo) and ragi (*Eleusine* coracana) starch granule, *Starch/Stärke*, 40 (1989) 378–382.
2096. V. Planchot, P. Colonna, and A. Buleon, Enzymatic hydrolysis of α-glucan crystallites, *Carbohydr. Res.*, 298 (1997) 319–326.
2097. Y. K. Kim and J. F. Robyt, Enzyme modification of starch granules: In situ reaction of glucoamylase to give complete retention of D-glucose inside granules, *Carbohydr. Res.*, 318 (1999) 129–134.
2098. H. Fredriksson, I. Bjorck, R. Andersson, H. Liljeberg, J. Silverio, A. C. Eliasson, and P. Åman, Studies of α-amylase degradation of retrograded starch gels from waxy maize and high-amylopectin potato, *Carbohydr. Polym.*, 43 (2000) 81–87.
2099. F. E. Volz and P. E. Ramstad, Effects of various physical treatments upon the amyloclastic susceptibility of starch, *Food Res.*, 17 (1952) 81–92.
2100. P. Leman, A. Bijttebier, H. Groesaert, G. E. Vandeputte, and J. A. Delcour, Influence of amylases on the rheological and molecular properties of partially damaged wheat starch, *J. Sci. Food Agric.*, 86 (2006) 1662–1669.
2101. I. A. Farhat, J. Protzmann, A. Becker, B. Valies-Pamies, R. Naele, and S. E. Hill, Effect of extended conversion and retorgradation on the digestibility of potato starch, *Starch/Stärke*, 53 (2001) 431–436.
2102. Nagase & Co., Adsorption of amylase on starch, French Patent 1,368,631 (1964).
2103. M. S. Shulman and L. M. Vainer, Sorption of amylase on starch (in Russian), *Tr. Vses. Nauchn.-Issled. Inst. Prod. Brosheniya*, 19 (1970) 174–181.
2104. J. H. Lin and Y. H. Chang, Effects of type and concentration of polyols on the molecular structure of corn starch kneaded with pullulanase in a farinograph, *Food Hydrocoll.*, 20 (2006) 340–347.
2105. S. De Cordt, M. Hendrickx, G. Maesmans, and P. Tobback, The influence of polyalcohols and carbohydrates on the thermostability of α-amylase, *Biotechnol. Bioeng.*, 43 (1994) 107–114.
2106. Y. H. Lee and K. H. Jo, Saccharification of uncooked starches in an attrition-coupled reaction system, *Sanop Misaengmul Hakkoechi*, 14(1), (1986) 29–36. *Chem. Abstr.*, 104 (1986) 205506g.

2107. M. Kumakura and I. Kaetsu, Pretreatment of starch raw materials and their enzymic hydrolysis by immobilized glucoamylase, *Enzyme Microb. Technol.*, 5 (1983) 199–203.
2108. N. Cao, Q. Xu, J. Ni, and L. F. Chen, Enzymic hydrolysis of corn starch after extraction of corn oil with ethanol, *Appl. Biochem. Biotechnol.*, 57(58), (1996) 39–48.
2109. Y. Ishii, M. Nomura, and T. Yomo, A simple activation method for α-amylase, Jpn. Kokai Tokkyo Koho 2007014258 (2007); *Chem. Abstr.*, 146 (2007) 137592.
2110. F. M. Porodko, E. L. Rosenfeld, and I. Ya. Soloveichik, The effect of the culinary treatment of potatoes upon the accessibility of the starch to amylase (in Russian), *Vopr. Pitaniya*, 8 (1939) 109–122.
2111. M. T. Tollier and A. Guilbot, Über die Wirkung der α- und β-Amylasen auf bestrahlte Stärke, *Starch/Stärke*, 18 (1966) 305–310.
2112. Y. Tanaka, K. Kawashima, T. Hayashi, and K. Umeda, On the increase of α-amylase affinity of starch granule by γ-irradiation (in Japanese), *Shokuhin Sogo Kenkyushio Kenkyu Hokoku* (34), (1980) 147–152. *Chem. Abstr.*, 94 (1981) 82441x.
2113. T. Kume and N. Tamura, Change in digestibility of raw starch by gamma-irradiation, *Starch/Stärke*, 39 (1987) 71–74.
2114. C. Zheng, Y. Wang, and A. Du, Effect of irradiation on enzymic hydrolysis of sweet potato starch,, *Huaxue Shije*, 30(3), (1989) 128–129. *Chem. Abstr.*, 111 (1989) 76510.
2115. R. Hayashi and A. Hayashida, Increased amylase digestibility of pressure treated starch, *Agr. Biol. Chem.*, 53 (1989) 2543–2544.
2116. L. Wang, Y. Chen, D. Shen, and Y. Meng, Studies on saccharification and filtration characteristics of extruded corn as adjunct to brew beer under different processing conditions, *Zhongguo Liangyou Xuebao*, 14 (1999) 44–48. *Chem. Abstr.*, 132 (1999) 150880.
2117. W. Krzyżaniak, W. Białas, A. Olesienkiewicz, T. Jankowski, and W. Grajek, Characteristics of oligosaccharides produced by enzymatic hydrolysis of potato starch using mixture of pullulanase and alpha-amylase, *e-J. Pol. Agric. Univ.*, 6(2), (2003).
2118. Y. Ayano and N. Ishida, Digestibility and amylogram of starch monophosphate (in Japanese), *Eiyogaku Zasshi*, 22 (1964) 117–120.
2119. M. Wootton and M. A. Chaudhry, In vitro digestion of hydroxypropyl derivatives of wheat starch. I. Digestibility and action pattern using porcine pancreatic α-amylase, *Starch/Stärke*, 33 (1981) 135–137.
2120. W. R. Fetzer and R. M. Hamilton, Conditioning of starch for enzymic conversion, US Patent 2,720,465 (1955).
2121. R. W. Best, R. A. Doughty, L. F. Van der Burgh, and A. E. Staley, Bleached starch composition of improved enzyme convertibility, US Patent 3,709,788 (1973); *Chem. Abstr.*, 78 (1973) 113037w.
2122. E. Pozerski, Verdaulichkeit von Rohstärke, *Bull. Soc. Sci. Hyg. Aliment.*, 21 (1933) 1–29. *Chem. Zentr.*, 1933, II, 1388.
2123. C.B. Duryea, Making maltose, US Patent 1,110,756 (1913).
2124. A. Sroczyński, M. Boruch, and T. Pierzgalski, Obtaining thermally modified starch for enzymic hydrolysis (in Polish), *Roczn. Technol. Zywn.*, 16 (1969) 7–19.
2125. A. Sreenivasan, Die Aufgangphase der α-diastatische Stärkehydrolyse. 8. Über enzymatische Amylolyse, *Biochem. Z.*, 301 (1939) 210–218.
2126. S. Hagiwara, K. Esaki, K. Nishiyama, S. Kitamura, and T. Kuge, Effect of microwave irradiation on potato starch granules (in Japanese), *Denpun Kagaku*, 33(1), (1986) 1–9.
2127. M. F. Dull, R. G. Swanson, and J. D. Solomon, The susceptibility of the starches of germinating barley and malt to diastase, *Brew. Dig.*, 19(7), (1944) 37–39. 90T–92T.
2128. M. R. Kweon and M. S. Shin, Comparison of enzyme resistant starches formed during heat-moisture treatment and retrogradation of high-amylsoe corn starches, *Hanguk Nonghwa Hakkoechi*, 40 (1997) 508–513. *Chem. Abstr.*, 128 (1998) 139870.

2129. H. Anger and G. Stoof, Effect of hydrothermal treatment of starches on their enzymic degradation in vitro and in vivo, *Proc. 7th Int. Symp. Plant Polymeric Carobohydr. Res., Berlin 1992* (1995) 119–122. *Chem. Abstr.*, 126 (1997) 32203.
2130. P. Cairns, V. Leloup, M. J. Miles, S. G. Ring, and V. J. Morris, Resistant starch. An x-ray diffraction study into the effect of enzymic hydrolysis on amylose gels in vitro, *J. Cereal Sci.*, 12 (1990) 203–206.
2131. C. M. L. Franco, C. F. Ciacco, and D. Q. Tavares, Effect of the heat-moisture treatment on the enzymic susceptibility of corn starch granules, *Starch/Stärke*, 47 (1995) 223–228.
2132. I. Maruta, Y. Kurahashi, R. Takano, K. Hayashi, K. Kudo, and S. Hara, Enzymic digestibility of reduced pressurized, heat-moisture treated starch, *Food Chem.*, 61 (1998) 163–165.
2133. G. Stoof, H. Anger, D. Schmiedl, and W. Bergthaller, Hydrothermische Behandlung von Stärke in Gegenwart von α-Amylase. III. Väranderung der rheologischen Eigenschaften von Weizenstärke durch hydrothermisch-enzymatische Behandlung, *Starch/Stärke*, 49 (1997) 225–231.
2134. G. Stoof, D. Schmiedl, H. Anger, and W. Bergthaller, Hydrothermische Behandlung von Stärke in Gegenwart von α-Amylase. IV. Änderung der Eigenschaften von Kartoffelstärke durch hydrothermisch-enzymatische Behandlung, *Starch/Stärke*, 50 (1998) 108–114.
2135. H. Anger, M. Richter, B. Kettlitz, and S. Radosta, Hydrothermal treatment of starch in presence of α-amylase. I. Hydrolysis of starch with alpha-amylase in heterogenous phase, *Starch/Stärke*, 46 (1994) 182–186.
2136. M. Kurakake, Y. Tachibana, K. Masaki, and T. Komaki, Adsorption of α-amylase on heat-moisture treated starch, *J. Cereal Sci.*, 23 (1996) 163–168.
2137. S. I. Pronin and G. B. Balashova, Effect of time of saccharification on observed activity of rye amylase after its incomplete thermal inactivation (in Russian), *Dokl. Akad. Nauk SSSR*, 113 (1957) 866–868.
2138. I. A. Popadich, A. I. Mirovich, N. S. Martynenko, and M. V. Babkina, Effect of some additives on the thermostability of amylolytic enzymic preparations (in Russian), *Prikl. Biokhim. Mikrobiol.*, 10 (1974) 729–733.
2139. L. M. Marchal, A. M. J. Van de Laar, E. Goetheer, E. B. Schimmelpennink, J. Bergsma, H. H. Beeftink, and J. Tramper, Effect of temperature on the saccharide composition obtained after α-amylolysis of starch, *Biotechnol. Bioeng.*, 63 (1999) 344–355.
2140. M. Sujka, K. O. Udeh, and J. Jamroz, α-Amylolysis of native corn, potato, wheat and rice granule, *Ital. J. Food Sci.*, [4], 18 (2006) 433–439.
2141. S. I. Pronin, The temperature coefficients of enzymic hydrolysis of natural starches (in Russian), *Biokhimyia*, 5(1), (1940) 65–73.
2142. S. Trautmann and L. Ambard, Wirkung der Temperatur auf die Verdauung der Stärke durch Amylase, *Ann. Physiol. Physicochim. Biol.*, 9 (1933) 707–712. *Chem. Zentr.,* 1934, I, 3353.
2143. T. Kitano, The optimum hydrogen ion concentration in the action of taka diastase (in Japanese), *J. Soc. Chem. Ind. Jpn.*, 38(Suppl), (1935) 385–388.
2144. T. Kitano, The change in the amylolyic and maltatic activities of taka diastase solution on standing (in Japanese), *J. Soc. Chem. Ind. Jpn.*, 38(Suppl.), (1935) 447–449.
2145. J. L. Baker and H. F. E. Hulton, The hydrolysis of potato starch paste by malt amylase at different temperature, *J. Inst. Brew.*, 43 (1937) 301–307.
2146. A. D. Kiliç Apar and B. Ozbek, α-Amylase inactivation by temperature during starch hydrolysis, *Process Biochem.*, 39 (2004) 1137–1144.
2147. H. Okazaki, Discussion on the mechanism of starch decomposition (1), *J. Agr. Chem. Soc. Jpn.*, 31 (1950) 121–126.
2148. L. Lin and J. Kang, Effect of starch and calcium on thermostability of microbial α-amylase, *Shipin Kexue (Beijing)*, 104 (1988) 5–9. *Chem. Abstr.*, 110 (1989) 131108.

2149. K. Myrbäck and N. O. Johansson, Über die Wirkungsweise der Amylasen und ihre Affinität zu verschiedenen Substraten, *Arkiv Kemi*, A20 (1945) No. 6.
2150. G. S. Eadie, Effect of substrate concentration on the hydrolysis of starch by the amylase of germinated barley, *Biochem. J.*, 20 (1926) 1016–1023.
2151. I. Ganchev and K. Tsekova, Kinetics of hydrolysis of starch by glucoamylase of *Aspergillus niger* strain B 77. III. Enzyme hydrolysis of starch by the preparation of glucamyl (in Bulgarian), *Acta Microbiol. Bulg.*, 18 (1986) 49–55.
2152. R. F. Tester and M. D. Sommerville, Swelling and enzymatic hydrolysis of starch in low water system, *J. Cereal Sci.*, 33 (2001) 193–203.
2153. R. Drapron and A. Guilbot, Contribution a l'etude des reactions enzymatiques dans les milieux biologiques peu hydrates, *Ann. Technol. Agr.*, 11 (1962) 275–371.
2154. R. Drapron and A. Guilbot, Contribution a l'etude des reactions enzymatiques dans les milieux biologiques peu hydrates, *Ann. Technol. Agr.*, 11 (1962) 175–196.
2155. F. Kiermeier and E. Coduro, Über den distatischen Stärkeabbau in lufttrockenen Substanzen, *Biochem. Z.*, 325 (1954) 280–287.
2156. D. Devdariani and L. Tskhvediani, Activity of amylase in the presence of a substrate and crystalloid in air-dried material (in Russian), *Tr. Gruz. Inst. Subtrop. Khoz.*, 142 (1969) 497–500.
2157. D. Devdariani, Activity of amylase in air-dried state (in Russian), *Tr. Gruz. Inst. Subtrop. Khoz.*, 142 (1969) 583–587.
2158. D. Devdariani and M. Nodiya, Activity of amylase in the presence of a substrate in air-dried material (in Russian), *Tr. Gruz. Inst. Subtrop. Khoz.*, 142 (1969) 639–642.
2159. C. Yook and J. F. Robyt, Reaction of alpha-amylases with starch granules in aqueous suspension giving product in solution and in a minimum amount water giving products inside the granule, *Carbohydr. Res.*, 337 (2002) 1113–1117.
2160. M. E. Van der Veen, S. Veelaert, A. J. Van der Goot, and R. M. Boom, Starch hydrolysis under low water conditions. A conceptual process design, *J. Food Eng.*, 75 (2006) 178–186.
2161. R. G. Griskey and T. Richter, Effect of pressure of the alpha amylase-ctalyzed hydrolysis of starch, *Biotechnol. Bioeng.*, 6 (1964) 469–471.
2162. G. P. Talwar, E. Barbu, J. Basset, and M. Macheboeuf, Studies on the enzymatic hydrolysis of amidon under increased pressure, *Bull. Soc. Chim. Biol.*, 33 (1951) 1793–1804.
2163. T. Matsumoto, Studies on enzyme reaction under high pressure. I, *Repts. Ind. Res. Center Shiga Pref.* 7 (1972) 59–65; *Chem. Abstr.*, 120 (1994) 161740.
2164. T. Matsumoto, Maltopentaose manufacture with amylase under high pressure, Jpn. Kokai Tokkyo Koho 08009988 (1996); *Chem. Abstr.*, 124 (1996) 200302.
2165. M. R. A. Gomez, R. Clark, and D. A. Ledward, Effects of high pressure on amylases and starch in wheat and barley flours, *Food Chem.*, 63 (1998) 363–372.
2166. Y. T. Liang, J. L. Morrill, F. R. Anstaett, A. D. Dayton, and H. D. Pfost, Effect of pressure moisture and cooking time on susceptibility of corn or sorghum grain starch to enzymic attack, *J. Dairy Sci.*, 53 (1970) 336–341.
2167. G. Weemanes, S. De Cordt, K. Goossens, L. Ludikhuyze, and M. Hemdrickx, High pressure, thermal and combined pressure–temperature stabilities of α-amylases from *Bacillus* species, *Biotechnol. Bioeng.*, 50 (1996) 49–56.
2168. M. Wallerstein, Thermal stabilization of α-amylase, US Patent 905,029 (1909).
2169. J. Hsiu, E. H. Fischer, and E. A. Stein, Alpha-amylases as calcium metalloenzymes. II. Calcium and the catalytic activity, *Biochemistry*, 3 (1964) 61–65.
2170. F. Uyar, Z. Baysal, and M. Dogru, Purification and some characterization of an extracellular alpha amylase from thermotolerant *Bacillus subtilis*, *Ann. Microbiol.*, 53 (2003) 315–322.
2171. B. L. Vallee, E. A. Stein, W. N. Sumerwell, and E. H. Fischer, Metal content of α-amylases of various origin, *J. Biol. Chem.*, 234 (1959) 2901–2905.

2172. R. Manelius and E. Bertoft, The effect of Ca^{2+} ions on the α-amylolysis of granular starches from oats and waxy-maize, *J. Cereal Sci.*, 24 (1996) 139–150.
2173. J. V. Giri and J. G. Shirkhande, Studies in salt activation. I. Influence of neutral salts on the enzyme hydrolysis of starch, *J. Indian Chem. Soc.*, 12 (1935) 273–286.
2174. N. Aghani, G. Feller, C. Gerday, and R. Haser, Structural basis of alpha amylase activation by chloride, *Protein Sci.*, 11 (2002) 1435–1441.
2175. G. Baumgarten, Structure and enzyme reactions. IX. Action of salts in the systems: Starch-amylase-proteins, *Biochem. J.*, 26 (1932) 539–542.
2176. K. Harada, Hydrolysis of starch by amylase (in Japanese), *J. Biochem. Jpn.*, 4 (1924) 123–137.
2177. Y. Takasaki, Stabilization of amylase with aluminum salts, Jpn. Kokai Tokkyo Koho 01104173 (1989); *Chem. Abstr.*, 112 (1990) 213092.
2178. W. Neugebauer, Starch degrading enzymes, US Patent 2,081,670 (1937).
2179. W. M. Clifford, The effect of halogen salts on salivary and pancreatic amylase, *Biochem. J.*, 30 (1936) 2049–2053.
2180. J. Burger, Examination of the amylase of potato (in Hungarian), *Magyar Chem. Folyoirat*, 34 (1928) 120–128, 135–140, 150–154.
2181. G. Radaeli, Verhalten der Amylase der Stärke gegenüber, *Pathologica*, 20 (1928) 269–279. *Chem. Zentr.*, 1929, II, 53.
2182. F. J. McClure, A review of fluorine and its physiological effects, *Physiol. Rev.*, 13 (1933) 277–300.
2183. J. R. Broeze, Einwirkung der Ptyalins auf Stärke. II. Elektrolyteeinflüsse, *Biochem. Z.*, 231 (1931) 365–384.
2184. A. Zlatarov, Enzyme chemistry of heavy metals. V. The influence of cadmium salts on the salivary amylase (in Bulgarian), *Ann. Univ. Sofia II, Fac. Phys-Math*, [2] 31 (1935) 1–4.
2185. R. S. Potter, The effect of certain substances on the action of malt amylase on starch, *J. Chem. Soc. Ind.*, 59 (1940) 45–47.
2186. T. Sabalitschka and R. Crzellitzer, Sublimatmikrokonzentrationen und Stärkeabbau durch Söichelamylase, *Mikrochemie*, 25 (1938) 225–227.
2187. J. Fischer, The effect of heavy metals on *Aspergillus niger*. I. Starch decomposition, *Planta*, 32 (1942) 395–413.
2188. W. Windisch and Derz, Reaktivität und Stärkeverzuckerung. Phosphaten, *Wochschr. Brau.*, 30 (1914) 533–537.
2189. V. I. Rodzevich, M. S. Pisareva, N. S. Mazur, and N. A. Kalashnikova, Saccharification of starchy raw material with enzymic preparations, USSR Patent 460,292 (1975); *Chem. Astr.*, 83 (1975) 26297r.
2190. H. C. Sherman and J. A. Walker, Influence of certain electrolytes upon the course of the hydrolysis of starch by malt amylase, *J. Am. Chem. Soc.*, 39 (1917) 1476–1493.
2191. M. Kleinman, I. H. Evans, and E. A. Bevan, Sodium phosphate enhancement of starch hydrolysis by a diastatic strain of *Saccharomyces cerevisiae*, *Biotechnol. Letts.*, 10 (1988) 825–828.
2192. S. E. Charm and B. L. Wong, Enzyme inactivation with shearing, *Biotechnol. Bioeng.*, 12 (1970) 1103–1109.
2193. A. W. MacGregor, R. J. Weselake, R. D. Hill, and J. E. Morgan, Effect of an α-Amylase inhibitor from barley kernels on the formation of products during the hydrolysis of amylose and starch granules by α-amylase II from malted barley, *J. Cereal Sci.*, 4 (1986) 125–132.
2194. U. Olsson, Vergiftungserscheinigungen an Malzamylase und Beitrage zur Kenntnis der Stärkeveflüssigung, *Z. Physiol. Chem.*, 126 (1923) 29–99.
2195. E. Heusch, Influence of phenols on the amylolytic action of aqueous extract of malt (in Italian), *Arch. Farm. Sper.*, 16 (1913) 308–316.
2196. T. Kitano, The course of the amylolytic and maltatic action with addition of alcohol and various salts (in Japanese), *J. Soc. Chem. Ind. Jpn.*, 38(Suppl.), (1935) 447.

2197. T. G. Hanna and J. Lelievre, Effect of lipid on the enzymic degradation of wheat starch, *Cereal Chem.*, 52 (1975) 697–701.
2198. E. Nebesny, J. Rosicka, and M. Tkaczyk, Influence of conditions of maize starch enzymic hydrolysis on physicochemical properties of glucose syrups, *Starch/Stärke*, 56 (2004) 132–137.
2199. S. S. Despande and M. Cheryan, Natural oxidant extraction from fenugreek (*Trigonella foenumgraceum*) for ground beef patties, *J. Food Sci.*, 49 (1984) 516–520.
2200. T. C. Crowe, S. A. Seligman, and L. Copeland, Inhibition of enzymic digestion of amylose by free fatty acids in vitro contributes to resistant starch formation, *J. Nutr.*, 130 (2000) 2006–2008.
2201. T. Maezawa, S. Hayakawa, M. Okuba, and F. Shimbori, Liquefying properties of starch for enzyme-hydrolyzed glucose production. VIII. Effects of fatty acids on liquefaction of starch (in Japanese), *Nippon Nogei Kagaku Kaishi*, 42 (1968) 211–215.
2202. S. Kende, Die Wirkung der Seifen auf den Fermentativen, *Biochem. Z.*, 82 (1917) 9–30.
2203. T. Isawa, Action of sodium oleate on amylolytic fermentation (in Japanese), *Kyoto Igaku Zasshi*, 18 (1921) 49–82.
2204. H. C. Gore and C. N. Frey, Augmentic enzymic starch conversion, US Patent 2,351,954 (1944).
2205. V. Borissovskii and N. Vvedenskii, Passievirung der Fermentwirkung. Einfluß organischen Fettsäure auf der Speichels, *Biochem. Z.*, 219 (1930) 72–78.
2206. K. Myrbäck and B. Person, Enzyme-substrate compounds and activity-pH-curves of the amylase. 3. pH-curves of barley beta-amylase, *Arkiv Kemi*, 5 (1952) 177–185.
2207. Y. Hara and M. Honda, The inhibition of α-amylase by tea polyphenols, *Agr. Biol. Chem.*, 54 (1990) 1939–1945.
2208. Z. Ghalanbor, N. Ghaemi, S. A. Marashi, M. Amanlou, M. Habibi-Rezaei, K. Khajeh, and B. Ranjbar, Binding of Tris to *Bacillus licheniformis* α-amylase can affect its starch hydrolysis activity, *Protein Peptide Lett.*, 15 (2008) 212–214.
2209. K. M. Benedetski and V. L. Yarovenko, Action of α-amylase from *Bacillus subtilis* on starch (in Russian), *Biokhimyia*, 38 (1973) 568–572.
2210. O. Holmbergh, Adsorption von α-Amylase aus Malz an Stärke, *Biochem. Z.*, 258 (1933) 134–140.
2211. H. Euler and K. Josephson, Zur Kenntnis der Aciditätbedingungen der enzymatischen Rohzuckerspaltung, *Z. Physiol Chem.*, 155 (1926) 1–30.
2212. M. Al Kazaz, V. Desseaux, G. Marchis-Mouren, E. Prodanov, and M. Santimone, The mechanism of porcine pancreatic alpha-amylase. Inhibition of maltopentaose hydrolysis by acarbose, maltose and maltotriose, *Eur. J. Biochem.*, 252 (1998) 100–107.
2213. H. Yamagishi, The course of saccharification by *Aspergillus oryzae* (in Japanese), *Sci. Repts. Tohoku Imp. Univ.*, [4] 3 (1928) 179–204.
2214. C. Reid, The inhibition of starch hydrolysis in the presence of glucose, *J. Physiol.*, 75 (1932) 10P–11P.
2215. C. A. Reddy and M. M. Abouzied, Glucose feedback inhibition of amylase activity in *Aspergillus* sp. and release of this inhibition when cocultured with *Saccharomyces cerevisiae*, *Enzyme Microb. Technol.*, 8 (1986) 659–664.
2216. K. Uehara and S. Mannen, Interaction of sweet potato β-amylase with its reaction product, maltose, *J. Biochem.*, 85 (1979) 105–113.
2217. S. Freiberger, Structure and enzymic reactions. IX. The systems amylase–starch–gelatin and urease–urea–gelatin, *Biochem. J.*, 25 (1931) 705–712.
2218. E. Ohlsson, Einfluß der Citrate auf der enzymatische Stärkehydrolyse, *Arch. Int. Pharmacodynamie*, 37 (1930) 98–107. *Chem. Zentr.*, 1930, II, 2909.
2219. E. Ohlsson, Einwirkung von Chinaalkaloiden auf amylolytische Eznyme, *Arch. Int. Pharmacodynamie*, 37 (1930) 108–114. *Chem. Zentr.*, 1930, II, 2909.
2220. W. Błaszków, Effect of an extract from buckwheat endosperm on the destructurizing activity of barley malt α-amylase (in Polish), *Pr. Nauk. Wyzsz. Szk. Ekon. Wroclaw*, 52 (1974) 33–41.

2221. H. Weenen, R. J. Hamer, and R. A. De Wijk, Semi-solid food products and methods for their production based on inhibiting amylase-induced starch breakdown, EU Patent WO 2005070227 (2005); *Chem. Abstr.*, 143 (2005) 171825.
2222. V. E. Kudryashova, A. K. Gladilin, A. V. Vakurov, F. Heitz, A. V. Levashov, and V. V. Mozhayev, Enzyme polyelectrolyte complexes in water–ethanol mixtures. Negatively charged groups artificially introduced into α-chymotrypsin provided additional activation and stabilization effects, *Biotechnol. Bioeng.*, 55 (1997) 267–277.
2223. R. Mukerjea, G. Slocum, R. Mukerjea, and J. F. Robyt, Significant differences in the activities of α-amylases in the absence and presence of polyethylene glycol assayed on eight starches solubilized by two methods, *Carbohydr. Res.*, 341 (2006) 2049–2054.
2224. J. F. Robyt, Stabilization and activation of nine starch degrading enzymes and significant differences in the activities of α-amylases assayed on eight different starches, *ACS Symp. Ser.*, 972 (2007) 61–72.
2225. C. Hatanaka, T. Haraguchi, and T. Mitsutake, Production of high purity glucose by semi-batch membrane reactor, *Maku*, 14 (1989) 344–351. *Chem. Abstr.*, 112(1990) 137413.
2226. F. Takao, Y. Sugimoto, and H. Fuwa, Effect of bile salts on degradation of various starch granules by pancreatic α-amylase (in Japanese), *Dempun Kagaku*, 25 (1978) 12–18.
2227. R. R. Ray, S. C. Jana, and G. Nanda, Adsorption of β-amylase from *Bacillus carotarum* B6 onto different native starches, *Letts. Appl. Microbiol.*, 19 (1994) 454–457.
2228. G. Zhang and B. R. Hamaker, SDS sulfite increases enzymic hydrolysis of native sorghum starches, *Starch/Stärke*, 51 (1999) 21–25.
2229. F. Caujolie and J. Molinier, Einfluß der aliphatischer Amine und ihre Chlorhydrate auf der amylolytische Wirkung von Speichel und Pankreas, *Bull. Sci. Pharmacol.*, 37 (1930) 290–297. *Chem. Zentr.*, 1930, I, 3317.
2230. F. Caujoli and J. Molinier, Einfluß der aliphatische Amine und ihrer Chlorhydrate auf der Verzuckerung der Stärke durch die Speichel, *Bull. Sci. Pharmacol.*, 37 (1930) 351–355. *Chem. Zentr.*, 1930, I, 1303.
2231. F. Caujolie and J. Molinier, Einfluß der aliphatische Amine und ihre Chlorhydrate auf der Verzuckerung der Stärke durch Malzextrakt, *Bull. Sci. Pharmacol.*, 37 (1930) 355–357. *Chem. Zentr.*, 1930, I, 1303.
2232. H. C. Sherman and N. M. Naylor, Influence of some organic compounds upon the hydrolysis of starch by salivary and pancreatic amylases, *J. Am. Chem. Soc.*, 44 (1922) 2957–2966.
2233. Y. Pomeranz and F. P. Mamaril, Effect of *N*-ethylmaleimide on the starch liquefying enzyme from soy flour, *Nature*, 203 (1964) 863–864.
2234. R. L. Antrim, G. Mitchinson, and L. P. Solheim, Amylase – Aqueous starch mixture and its liquefaction at low pH, EU Patent WO 9628567 (1996); *Chem. Abstr.*, 125 (1996) 303700.
2235. H. Nakamura, Protecting action of proteases (in Japanese), *J. Soc. Chem. Ind. Jpn.*, 33(Suppl.), (1930) 523–524.
2236. H. Nakamura, Protective action of proteins and their digestion products (in Japanese), *J. Soc. Chem. Ind. Jpn.*, 33 Suppl (1930) 524–526.
2237. H. Pringsheim, H. Borchardt, and H. Hupfer, Über Glutathion als Aktivator der fermentativer Stärkeverzuckerung, *Biochem. Z.*, 238 (1931) 476–477.
2238. C. Wunderly, Die Steuerung eines enzymatischen Abbaues durch einen anderen, *Helv. Chim. Acta*, 23 (1940) 414–428.
2239. H. Ninomyia, Salivary amylase. II. Hydrolysis of salivary amylase by protease and its carbohydrate and iron content (in Japanese), *J. Biochem. Jpn.*, 31 (1940) 421–428.
2240. N. Haska and Y. Ohta, Effect of cellulose addition on hydrolysis of sago starch granules by raw starch digesting amylase *Penicilium brunneum* No. 24, *Starch/Stärke*, 45 (1993) 237–241.
2241. H. Chrempinska, Structure and enzyme reactions. X. Action of salts on the systems amylase–astarch–proteins, *Biochem. J.*, 25 (1931) 1555–1564.

2242. H. C. Sherman and M. L. Caldwell, Influence of arginine, histidine and cystine upon the hydrolysis of starch by purified pancreatic amylase, *J. Am. Chem. Soc.*, 43 (1921) 2469–2476.
2243. H. C. Sherman and F. Walker, Influence of certain amino acids upon the enzyme hydrolysis of starch, *J. Am. Chem. Soc.*, 43 (1921) 2461–2469.
2244. H. C. Sherman and M. L. Caldwell, Action of amino acid in protecting amylase from inactivation by mercury, *J. Am. Chem. Soc.*, 44 (1922) 2923–2926.
2245. H. C. Sherman and M. L. Caldwell, Influence of lysine upon the hydrolysis of starch by purified pancreatic amylase, *J. Am. Chem. Soc.*, 44 (1922) 2926–2930.
2246. D. Narayanamurti and C. V. R. Ayyar, Influence of amino acids on the hydrolysis of starch by cumbu amylase, *J. Indian Chem. Soc.*, 8 (1931) 645–650.
2247. S. Nishimura, The liquefaction of starch by the enzyme in yeast autolyzate (in Japanese), *Bull. Agr. Chem. Soc. Jpn.*, 4 (1928) 126.
2248. S. Nishimura, Stärke verflüssigendes Enzym in Trockenhefeautolysaten, *Biochem. Z.*, 223 (1930) 161–170.
2249. K. Myrbäck and S. Myrbäck, Über die Grenzverzuckerung der Stärke und das sogenannte Komplement der Amylose, *Svensk. Kem. Tids.*, 45 (1933) 230–236.
2250. A. I. Oparin, T. N. Evreinova, T. A. Schubert, and M. N. Nestyuk, Coacervates and enzymes. Protein–carbohydrate coacervates and α-amylase (in Russian), *Dokl. Akad. Nauk SSSR*, 104 (1955) 581–583.
2251. M. W. Strickberger, *The Evolution*, (2000) Jones and Bartlett Publ, London, Chapter 4.
2252. H. A. Campbell, F. Hollis, Jr., and R. V. MacAllister, Improved methods for evaluating starch specific uses, *Food Technol.*, 4 (1950) 492–496.
2253. Y. Zhang and X. Wang, Enzymatic production of glucose, *Shipin Gongye Keji*, 27(11), (2006) 122–123, 126; *Chem. Abstr.*, 149 (2008) 30736.
2254. H. Xie, Method for production of sugar powder from corn starch, Chinese Patent 101126109 (2008); *Chem. Abstr.*, 148 (2008) 357648.
2255. G. P. Closset, Y. T. Shah, and J. T. Cobb, Analysis of membrane reactor performance for hydrolysis of starch by glucoamylase, *Biotechnol. Bioeng.*, 15 (1973) 441–445.
2256. E. Tachauer, J. T. Cobb, and Y. T. Shah, Hydrolysis of starch by a mixture of enzymes in a membrane reactor, *Biotechnol. Bioeng.*, 16 (1974) 545–550.
2257. J. J. Marshall and W. J. Whelan, New approach to the use of enzymes in starch technology, *Chem. Ind. (London)* (25), (1971) 701–702.
2258. S. P. O'Neill, P. Dunnill, and M. D. Lilly, Comparative study of immobilized amyloglucosidase in packed bed reactor and a continuous feed stirred tank reactor, *Biotechnol. Bioeng.*, 13 (1971) 337–352.
2259. S. Chen, Y. Liu, and P. Yu, Study on column reactor of chitosan-immobilized amylase, *Fujian Shifan Daxue Xuebao, Ziran Kexueban*, 12(4), (1996) 75–79. *Chem. Abstr.*, 127 (1997) 46739.
2260. A. Sanroman, R. Chamy, M. J. Nunez, and J. M. Lema, Enzymic hydrolysis of starch in a fixed-bed pulsed-flow reactor, *Appl. Biochem. Biotechnol.*, 28–29 (1991) 527–538.
2261. D. M. G. Freire and G. L. Sant'anna, Jr., Hydrolysis of starch with immobilized glucoamylase. A comparison between two types of expanded-bed reactor, *Appl. Biochem. Biotechnol.*, 26 (1990) 23–34.
2262. M. Liakopoulou-Kyriakides, A. Karakatsanis, and M. Stamatoudis, Enzymic hydrolysis of starch in agitated PEG-dextran aqueous two-phase system, *Starch/Stärke*, 48 (1996) 291–294.
2263. G. A. Kovalenko, S. V. Sukhinin, A. V. Simakov, L. V. Perminova, O. V. Khomova, and O.Yu Borovtsova, Rotating inertial bioreactor for heterogenous biocatalytic processes. I. Fermentation hydrolysis of starch (in Russian), *Biotekhnologiya* (1), (2004) 83–90.
2264. G. A. Kovalenko, S. V. Sukhinin, L. V. Perminova, A. V. Simakov, O. V. Khomova, V. V. Khomov, O. Yu. Borovtsova, and T. V. Chuenko, A process and reactor for enzymic hydrolysis of carbohydrates, Russian Patent 2245925 (2005); *Chem. Abstr.*, 142 (2005) 179153.

2265. M. M. Meahger and D. D. Grafelman, Method for liquefaction of cereal grain starch substrate and apparatus thereof, US Patent 5981237 (1999); *Chem. Abstr*, 131 (1999) 298729.
2266. K. H. Jo and Y. H. Lee, Enhancing mechanism of the saccharification of uncooked starch in an agitated bead reaction system, *Sanop Misaengmul Hakhoechi*, 14 (1986) 407–413. *Chem. Abstr.*, 106 (1986) 136987s.
2267. Y. H. Lee and J. S. Park, Direct conversion of raw starch to maltose in an agitated bead enzyme reactor using fungal α-amylase, *Sanop Misaengmul Hakhoechi*, 19 (1991) 290–295. *Chem. Abstr.*, 119 (1993) 47440.
2268. A. N. Emery and J. P. Cardoso, Parameter evaluation and performance studies in a fluidized-bed immobilized enzyme reactor, *Biotechnol. Bioeng.*, 20 (1978) 1903–1929.
2269. J. M. S. Cabral, J. M. Novais, and J. P. Cardoso, Modeling of immobilized glucoamylase reactions, *Ann. N. Y. Acad. Sci.*, 413 (1983) 535–541.
2270. J. M. S. Cabral, J. M. Novais, J. P. Cardoso, and J. F. Kennedy, Design of immobilized glucoamylase reactors using a simple kinetic model for the hydrolysis of starch, *J. Chem. Technol. Biotechnol.*, 36 (1986) 247–254.
2271. G. M. Zanin, L. M. Kambara, L. P. V. Calsavara, and F. F. De Moraes, Performance of fixed bed reactors and immobilized enzyme, *Appl. Biochem. Biotechnol.*, 45–46 (1994) 627–640.
2272. G. M. Zanin, L. M. Kambara, L. P. V. Calsavara, and F. F. De Moraes, Modeling fixed and fluidized reactors for cassava starch saccharification with immobilized enzyme, *Appl. Biochem. Biotechnol.*, 63–65 (1997) 527–540.
2273. Yu. A. Ramazanov, V. I. Kislykh, and I. P. Kosyuk, Method and bioreactor for preparing molasses from starch by simplified accelerated enzymic hydrolysis, Russian Patent 2283349 (2006); *Chem. Abstr.*, 145 (2006) 294882.
2274. J. M. Novais, J. M. S. Cabral, and J. P. Cardoso, Kinetic model and comparison of continuous reactor for starch hydrolysis by immobilized amyloglucosidase (in Portuguese), *Actas 9th Simp. Iberoam. Catal.*, 1, (1984) 397–406. *Chem. Abstr.*, 102 (1987) 4446x.
2275. X. D. Wang, Studies on maize starch fermentation with immobilized yeast in a fluidized-bed bioreactor, *Xian Shifan Daxue Xuebao Ziran Kexueban*, 29(1), (2004) 98–101. *Chem. Abstr.*, 141 (2004) 189710.
2276. C. Schiraldi, A. Martino, T. Costabile, M. Generoso, M. Marotta, and M. De Rosa, Glucose production from maltodextrins employing a thermophilic immobilized cell biocatalyst in a packed-bed reactor, *Enzyme Microb. Technol.*, 34 (2004) 415–424.
2277. C. Nolasco-Hipolito, G. Kobayashi, K. Sonomoto, and A. Ishizaki, Synchronized fresh cell bioreactor system for continuous L-(+)-lactic acid production using *Lactococcus lactis* IO-1 in hydrolyzed sago starch, *Biotechnol. Sust. Util. Biol. Resources Tropics*, 16 (2003) 51–56.
2278. D. Darnoko, M. Cheryan, and W. E. Artz, Saccharification of cassava starch in an ultrafiltration reactor, *Enzyme Microb. Technol.*, 11 (1989) 154–159.
2279. M. Ito, T. Hibino, T. Okada, H. Ishizuka, H. Sahashi, and T. Saiga, Membrane reactors for enzymic reactions, Jpn. Kokai Tokkyo Koho 01141587 (1989); *Chem. Abstr.*, 112 (1990) 137490.
2280. M. Nakajima, K. Iwasaki, H. Nabetani, and A. Watanabe, Continuous hydrolysis of soluble starch by free β-amylase and pullulanase using ultrafiltration membrane reactor, *Agric. Biol. Chem.*, 54 (1990) 2793–2799.
2281. K. A. Sims and M. Cheryan, Hydrolysis of liquefied corn starch in a membrane reactor, *Biotechnol. Bioeng.*, 39 (1992) 960–967.
2282. Y. Sahashi, H. Ishizuka, and K. Hibino, Characteristics of glucoamylase immobilized on membrane and performance of membrane reactor (in Japanese), *Kagaku Kogaku Ronbunshu*, 20 (1994) 47–53.
2283. D. Uttapap, Y. Koba, and A. Ishizaki, Continuous hydrolysis of starch in membrane and connected immobilized enzyme reactor, *J. Fac. Agr. Kyushu Univ.*, 33 (1989) 167–175. *Chem. Abstr.*, 111 (1989) 35766.

2284. N. Nemestothy, L. Ribeiro, A. Thoth, K. Belafi-Bako, L. Gubicza, and C. Webb, Kinetics of enzymatic hydrolysis of polysaccharides, *Mededelinge Fac. Landbouw. Toegepaste Biol. Wettensch. Univ. Gent*, 66(3a), (2001) 191–194.
2285. G. Catapano and E. Klein, Performance of the dialytic reactor with product inhibited enzyme reactions: A model study, *Bioseparation*, 4 (1994) 201–211.
2286. H. Li, Z. Li, C. Huang, J. Wei, G. Tang, and H. Wu, Study on continuous process of producing maltooligosaccharide with system of membrane reaction, *Zhongguo Liangyou Xuebao*, 13(5), (1998) 49–52. *Chem. Abstr.*, 130 (1998) 97108.
2287. L. Słomińska, W. Grajek, A. Grześkowiak, and M. Gocałek, Enzymic starch saccharification in an ultrafiltration membrane reactor, *Starch/Stärke*, 50 (1998) 390–396.
2288. D. Paolucci-Jeanjean, M. P. Belleville, and G. M. Rios, A comprehensive study of the loss of enzyme activity in a continuous membrane reactor – Application to starch hydrolysis, *J. Chem. Technol. Biotechnol.*, 76 (2001) 273–278.
2289. U. Pankiewicz, J. Wierciński, A. Trejgel, L. Słomińska, A. Grześkowiak, L. Jarosławski, and R. Zielonka, Enzymatic hydrolysis of maltodextrin into maltose syrup in a continuous membrane reactor, *Pol. J. Food Nutr. Sci.*, 11(1), (2002) 19–22.
2290. D. Paolucci-Jeanjean, M. P. Belleville, M. P. Zakhia, and G. M. Rios, Continuous recycled membrane reactor for starch conversion, *Rec. Progr. Genie Procedes*, 13 (1999) 291–298.
2291. A. Grześkowiak-Przywecka and L. Słomińska, Saccharification of potato starch in an ultrafiltration reactor, *J. Food Eng.*, 79 (2007) 539–545.
2292. G. J. Woo and J. D. McCord, Maltotetraose production using *Pseudomonas stutzeri* exo- α-amylase in a membrane recycle bioreactor, *J. Food Sci.*, 56 (1991) 1019–1023.
2293. G. Bayramoglu, M. Yilmaz, and M. Y. Arica, Immobilization of a thermostable α-amylase onto reactive membranes: Kinetics characterization and application to continuous starch hydrolysis, *Food Chem.*, 84 (2004) 591–599.
2294. J. Y. Houng, J. Y. Chiou, and K. C. Chen, Production of high maltose syrup using an ultrafiltration reactor, *Bioprocess Eng.*, 8 (1992) 85–90.
2295. H. Fukuda and Y. Endo, Manufacture of maltoologosaccharides from α-amylase-treated starch with reverse osmosis membranes, Jpn. Kokai Tokkyo Koho, 03133387 (1991); *Chem. Abstr.*, 115 (1991) 254346.
2296. M. Hakoda, T. Chiba, and K. Nakamura, Characteristics of electro-ultrafiltration bioreactor (in Japanese), *Kagaku Kogaku Ronbunshu*, 17 (1991) 470–476.
2297. M. Ohnishi, C. Ishimoto, and J. Seto, The properties of the multistage bioreactor constructed by the LB technique, *Bull. Chem. Soc. Jpn.*, 64 (1991) 3581–3584.
2298. M. Onda, Y. Lvov, K. Ariga, and T. Kunitake, Methods for preparing an ultrathin film reactor with immobilized protein and using it for chemical reactions, German Patent 19642882 (1997); *Chem. Abstr.*, 126 (1997) 327748.
2299. M. Engasser, J. Caumon, and A. Marc, Hollow fiber enzyme reactors for maltose and starch hydrolysis, *Chem. Eng. Sci.*, 35 (1980) 99–105.
2300. A. Marc, C. Burel, and J. M. Engasser, Hollow fiber enzymic reactors for the hydrolysis of starch and sucrose, *Proc. Int. Symp. Util. Enzymes Technol. Aliment.* (1982) 35–39. *Chem. Abstr.*, 98 (1984) 67921t.
2301. Y. H. Ju, W. J. Chen, and C. K. Lee, Starch slurry hydrolysis using α-amylase immobilized on a hollow-fiber reactor, *Enzyme Microb. Technol.*, 17 (1995) 685–688.
2302. A. Grześkowiak-Przywecka and L. Słomińska, Continuous potato starch hydrolysis process in a membrane reactor with tubular and hollow-fiber membranes, *Desalination*, 184 (2005) 105–112.
2303. M. Ito, T. Okada, T. Hibino, H. Sahashi, H. Ishizuka, and T. Saiga, Hollow fiber-type bioreactor for continuous enzymic reactors, Jpn. Kokai Tokkyo Koho, 02109986 (1990); *Chem. Abstr.*, 113 (1990) 150900.

2304. J. Wei, Hollow fiber enzymic membrane reactor and catalyst kinetics. I. Hollow fiber enzymic membrane reactor of multienzyme system, *Shuichuli Jishu*, 21 (1995) 342–346. *Chem. Abstr.*, 124 (1996) 230025.
2305. A. Taki, O. Ueshima, and M. Okada, Studies on production of maltose by membrane separation technique, *Shokuhin Sangyo Senta Gijutsu Kenkyu Hokoku*, 3 (1979) 13–23. *Chem. Abstr.*, 93 (1980) 148430f.
2306. L. Hakkarainen, P. Linko, and J. Olkku, State vector model for Contherm scraped surface heat exchanger used as an enzyme reactor in wheat starch conversions, *J. Food Eng.*, 4 (1985) 135–153.
2307. H. Akdogan, High moisture food extrusion, *Int. J. Food Sci. Technol.*, 34 (1999) 195–207.
2308. V. Komolprasert and R. Y. Ofoli, A dispersion model for prediction of the extent of starch liquefaction by *Bacillus licheniformis* α-amylase during reactive extrusion, *Biotechnol. Bioeng.*, 37 (1991) 681–690.
2309. H. Chouvel, J. Boudrant, and J. C. Cheftel, Continuous enzymic hydrolysis of starch cereal flours in an extruder-boiler, *Proc. Int. Symp. Util. Enzymes Technol. Aliment.* (1982) 171–176. *Chem. Abstr.*, 98 (1983) 105786n.
2310. J. Burianek, J. Zapletal, and E. Jurisova, Modified starch and flour by combined extrusion and enzymic degradation, Czechoslovak Patent 225,322 (1984); *Chem. Abstr.*, 103 (1985) 213772k.
2311. D. Shen, Extrusion-processig method, apparatus and saccharification method of starch syrup materials added with enzyme, Chinese Patent 101230406 (2008); *Chem. Abstr.*, 149 (2008) 269844.
2312. M. E. Camire and A. L. Camire, Enzymic starch hydrolysis of extruded potato peels, *Starch/Stärke*, 46 (1994) 308–311.
2313. Y. Y. Linko, H. Makela, and P. Linko, A novel process for high-maltose syrup production from barley starch, *Ann. N. Y. Acad. Sci.*, 413 (1983) 352–354.
2314. R. Y. Ofoli, V. Komolprasert, B. C. Saha, and K. A. Berglund, Production of maltose by reactive extrusion of carbohydrates, *Lebensm. Wiss. Technol.*, 23 (1990) 262–266.
2315. L. Roussel, A. Vieille, I. Billet, and J. C. Cheftel, Sequential heat gelatinization and enzymic hydrolysis of corn starch in an extrusion reactor, *Lebensm. Wiss. Technol.*, 24 (1991) 449–458.
2316. W. Krzyżaniak, T. Jankowski, and W. Grajek, Optimization of the enzymatic hydrolysis of potato starch combined with extrusion (in Polish), *Zywnosc (Food)*, 12(1), (2005) 48–62.
2317. D. Curić, D. Karlovćc, B. Tripalo, and D. Ježek, Enzymic conversion of corn starch in twin-screw extruder, *Chem. Biochem. Eng. Quart.*, 12(2), (1998) 63–71.
2318. I. Hayakawa, K. Sakamoto, H. Hagita, and Y. Furio, Production of hydrolysable starchy material in highly concentrated substrate by a two-stage extrusion cooking method, *Nippon Shokuhin Kogyo Gakkaishi*, 38 (1991) 945–953. *Chem. Abstr.*, 116 (1992) 172654.
2319. P. E. Barker, N. J. Ajongwen, M. T. Shieh, and G. Ganetsos, Simulated counter-current chromatographic bioreactor-separator, *Stud. Surface Sci. Catal.*, 80 (1993) 35–44.
2320. S. F. D'Souza and B. S. Kubal, A cloth-strip bioreactor with immobilized glucoamylase, *J. Biochem. Biophys. Methods*, 51 (2002) 151–159.
2321. V. L. Varella, B. R. V. Concone, J. T. Senise, and P. P. Doin, Continuous enzymic cooking and liquefaction of starch using the technique of direct resistive heating, *Biotechnol. Bioeng.*, 26 (1984) 654–657.
2322. D. A. Sirotti and A. H. Emery, Diffusion and reaction in an immobilized-enzyme starch saccharification catalyst, *Appl. Biochem. Biotechnol.*, 9 (1984) 27–39.
2323. J. H. Walkup and J. G. Leech, Mechanism of enzyme conversion of starch, *Symp. Ser. Chem. Eng. Progr.*, 64(86), (1968) 1–5.
2324. L. Yang and F. Luo, Development of techniques for continuous starch liquefying and ejecting enzyme method, *Shipin Gongye Keji* (4), (1995) 3–10. *Chem. Abstr.*, 124 (1996) 143925.
2325. S. Neuman, Degraded starch products, British Patent 579,702 (1946).

2326. F. Starzyk, C. Y. Lii, and P. Tomasik, Light absorption, transmission and scattering in potato starch granule, *Pol. J. Food Nutr. Sci.*, 10(4), (2001) 27–34.
2327. G. H. Zheng and R. S. Bhatty, Enzyme-assisted wet separation of starch from other seed components of hull-less barley, *Cereal Chem.*, 75 (1998) 247–250.
2328. W. Van der Ham and A. Lehmussari, A process for producing starch from cereals, EU Patent EP 267,637 (1988); *Chem. Abstr.*, 109 (1988) 56929w.
2329. D. B. Johnston and V. Singh, Enzymatic milling of corn: Optimization of soaking, grinding and enzyme incubation steps, *Cereal. Chem.*, 81 (2004) 626–632.
2330. K. Kroeyer, Ist es notwendig Stärke hergestellen um Stärkesirup, Dextrose und Total Sugar zu erhalten? Stärkehaltigen Rohstoffen mittels Enzymen und die Weitereverarbeitung der Extraktionprodukte, *Starch/Stärke*, 18 (1966) 311–316.
2331. L. Słomińska, D. Wiśniewska, and J. Niedbach, The influence of additional enzyme treatment of corn and wheat starches on filtration properties of hydrolysates, *e-J. Pol. Agric. Univ.*, 7(2), (2005).
2332. J. A. Moheno-Perez, H. D. Almeida-Dominguez, and S. O. Serna-Saldivar, Effect of fiber degrading enzymes on wet milling and starch properties of different types of sorghum and maize, *Starch/Stärke*, 51 (1999) 16–20.
2333. J. D. Steinke and L. A. Johnson, Steeping maize in the presence of multiple enzymes. I. Static batchwise steeping, *Cereal Chem.*, 68 (1991) 7–12.
2334. T. Minakawa, Manufacture of starch using an enzyme, Japanese Patent 5026 ('57) (1957).
2335. J. Cao, Technology for producing rice starch by enzyme, Chinese Patent 1552885 (2004); *Chem. Abstr.*, 143 (2005) 129549.
2336. F. Schierbaum, S. Radosta, M. Richter, B. Kettlitz, and C. Gernat, Studies on rye starch properties and modifications. I. Composition and properties of rye starch granules, *Starch/Stärke*, 43 (1991) 331–339.
2337. T. Verwimp, G. E. Vandeputte, K. Marrant, and J. A. Delcour, Isolation and characterization of rye starch, *J. Cereal Sci.*, 39 (2004) 85–90.
2338. A. M. Dos Mohd, M. N. Islam, and B. M. Noor, Enzymic extraction of native starch from sago (*Metroxylon sagu*) waste residue, *Starch/Stärke*, 53 (2001) 639–643.
2339. T. Hidaka, H. Hyodo, S. Hamakawa, and K. Otsubo, Application of commercial cellulose for sweet potato starch manufacture (in Japanese), *Denpun Kogyo Gakkaishi*, 13(3), (1966) 89–93.
2340. N. Toyama, N. Fujii, and K. Ogawa, Production of sweet potato starch using a cell-separating enzyme and cellulose. II. Enzymic isolation of starch from raw starch waste made by a continuous juicer and from sweet potato powder prepared by the dry milling process (in Japanese), *Hakko Kogaku Zasshi*, 44 (1966) 731–740.
2341. Y. Hamazaki and H. Kawahara, Production of sweet potato starch by microbial enzymes. II. Properties of the starch obtained by the enzyme *Clostridium acetobuylicum* S-1. (in Japanese), *Denpun Kogyo Gakkaishi*, 13(4), (1966) 105–108.
2342. S. Sato and S. Oka, Preparation of starch from wheat flour by pepsin treatment (in Japanese), *Denpun Kogyo Gakkaishi*, 14(1), (1967) 13–19.
2343. A. Caraban, V. Jascanu, A. Fodor, T. Dehelean, and O. Stanasel, The influence of some food additives on the hydrolysis reaction of starch with amylase from wheat flour, *Acta Univ. Cibiniensis Ser. F.*, 8(2), (2005) 69–74.
2344. Q. Huang, X. Fu, F. Luo, X. He, and L. Li, Method for pretreating starch, Chinese Patent 1911966 (2007); *Chem. Abstr.*, 146 (2007) 298045.
2345. A. Liu, Q. Chen, B. Qi, and G. Lin, Studies on α-amylase and glucoamylase application in refrigeration of food, *Shipin Kexue (Beijing)*, 25(4), (2004) 115–118. *Chem, Abstr.*, 143 (2005) 285003.
2346. X. Fang, S. Huang, and S. Liu, Preparation of microporous starch with α-amylase, *Shipin Gongye Keji*, 26(1), (2005) 60–62. *Chem. Abstr.*, 145 (2005) 105492.

2347. Q. Zhou, T. Zhang, X. Liu, J. Kan, and Z. Chen, Preparation of microporous cassava starch granules,, *Shipin Gongye Keji*, 26(10), (2005) 127–129. *Chem. Abstr.*, 146 (2006) 523697.
2348. W. Yao, G. Liang, and H. Yao, Researches on porous starch. II. Optimization of porous starch production process, *Zhongguo Liangyou Xuebao*, 20(4), (2005) 21–24, 28; *Chem. Abstr.*, 148 (2007) 217119.
2349. Y. Fu, Y. Xu, H. Liu, X. Liu, and J. Kan, Study on comparing enzyme preparation process of microporous starch granules, *Shipin Gongye Keji* (3), (2006) 24–27. *Chem. Abstr.*, 147 (2007) 165146.
2350. T. Shibuya and H. Chaen, Amylose granules and their preparation, British Patent 2247242 (1992); *Chem. Abstr.*, 116 (1992) 216699.
2351. T. Ishii, N. Hasegawa, M. Katsuro, K. Suzuki, and M. Koishi, Powdery pharmaceutical preparation and method of manufacturing it, EU Patent WO 9517909 (1995); *Chem. Abstr.*, 123 (1995) 179507.
2352. H. N. Dunning and E. H. Borochoff, Increasing the fat absorptivity of starch by enzymic hydrolysis, US Patent 3,596,666 (1971); *Chem. Abstr.*, 75 (1971) 128683r.
2353. L. Huang, R. Yu, X. Yang, Q. Wen, N. Huang, H. Wang, and L. Xiong, Preparation of porous starch through high-temperature enzymolysis method, Chinese Patent 1661028 (2005); *Chem. Abstr.*, 144 (2006) 329913.
2354. T. Yoshida and M. Matsudaira, Preparation of adsorbents from branched dextrin, Jpn. Kokai Tokkyo Koho, 63130139 (1988); *Chem. Abstr.*, 109 (1988) 213109.
2355. K. Ishii and S. Oka, Properties of porous polymer films prepared by amylolytic elimination of starch dispersed in methyl methacrylate-2-ethylhexyl acrylate copolymer films, *Kobunshi Ronbunshu*, 48 (1991) 163–169. *Chem. Abstr.*, 114 (1991) 186805.
2356. A. Ishii and S. Oka, Jpn. Kokai Tokkyo Koho 03223350 (1991); Manufacture of porous films from starch-polymer blends, *Chem. Abstr.*, 117 (1992) 9396.
2357. R. Deinhammer and C. Andersen, Methods for preventing, removing, reducing or disrupting biofilm by using bacterial α-amylase, EU Patent WO 2006031554 (2006); *Chem. Abstr.*, 144(2006) 272819.
2358. X. Chen, S. Zou, and R. Zeng, Study on improvement of quality of starch film using pullulanase, *Shipin Gongye Keji*, 23(10), (2002) 20–22. *Chem. Abstr.*, 139 (2003) 163812.
2359. F. Baum and C. Wardsack, Production of starch-free pectin preparations, *Ernährungsforsch.*, 5 (1960) 539–550. *Chem. Abstr.*, 55 (1961) 55844.
2360. W. R. Mueller-Stoll and H. Schroeder, The selective degradation of starch in crude pectin solutions by means of enzymes from the mycellia of molds, *Zentr. Bakteriol. Parasitenk. Abt. II*, 115(3), (1962) 297–313.
2361. W. Rzędowski and D. Ostaszewicz, Amylolytic preparations for removing starch from pectin solutions (in Polish), *Pr. Inst. Lab. Bad. Przem. Spoz.*, 13(4), (1963) 23–29.
2362. Y. Yamada, K. Boki, and M. Takahashi, Characterizaation of starches from roots of *Panax* ginseng C.A. Meyer and *Panax notoginseng* (Burk.) F.H. Chen, *J. Appl. Glycosci.*, 52 (2005) 351–357.
2363. Y. L. Niu, P. Zhang, C. P. Song, S. P. Li, W. F. Wang, and C. R. Gong, Preliminary experiments on enzymic degradation of starch in Henan flue-cured tobacco, BF2, C3F and X2L, *Yancao Keji* (3), (2005) 26–28, 32; *Chem. Abstr.*, 144 (2005) 84327.
2364. L. F. Chen, Hydrolysis protocol for *Pueraria*, *Dalian Quinggongye Xueyuan Xuebao*, 18 (1999) 234–237. *Chem. Abstr.*, 132 (1999) 221714.
2365. Y. Dugenet, Process for fractioning cereal bran and fractions thus obtained, French Patent 2874930 (2006); *Chem. Abstr.*, 144 (2006) 273245.
2366. M. Watanabe, J. Watanabe, K. Sonoyama, and S. Tanabe, Novel method for producing hypoallergenic wheat flour by enzymic fragmentation of the constituent allergens and its application to food processing, *Biosci. Biotechnol. Biochem.*, 64 (2000) 2663–2667.
2367. K. Hamar, Utilization of starch derivatives obtained by hydrolytic degradation in the paper industry (in Hungarian), *Papirip. Magy. Grafika*, 10 (1966) 219–225.

2368. L. E. Cave and F. R. Adams, Continuous enzyme converting of starch for paper surface treatment, *Tappi*, 51(Pt. 1; 11), (1968) 109A–112A.
2369. W. Nachtergaele and J. Van der Meeren, New starch derivatives for the coating industry, *Starch/Stärke*, 39 (1987) 135–141.
2370. R. A. Gale, Enzymic conversion of corn starch, *Paper Mill*, 63 (1940) No.43, 13.
2371. R. W. Kerr, Modified starch for use in paper coating and the like, US Patent 2,380,848 (1945).
2372. H. H. Schopmeyer and H. A. Kaufmann, Enzyme-thinned starch size, US Patent 2,364,590 (1944).
2373. A. Nacu and I. Keneres-Ursu, Production of enzyme-degraded starch (in Roumanian), *Celul. Hirlie*, 18 (1969) 201–206.
2374. M. L. Cushing and C. W. Turner, Enzyme-converted starches as coating adhesives. Variables affecting performance, *Tappi*, 41 (1958) 345–349.
2375. W. L. Craig, Starch-containing coating compositions, US Patent 606,278 (1950).
2376. Schweizerische Ferment A-G. Enzymic degradation of starch to dextrin for paper coating, French Patent 1,506,047 (1967); *Chem. Abstr.*, 70 (1968) 5341s.
2377. H. Konerth, N. Jalba, and G. Iliescu, Enzymic hydrolysis of starch and its use for surface treatment of paper (in Roumanian), *Celul. Hirtie*, 19 (1970) 316–323.
2378. A. F. Ottinger and P.R. Graham, Modifiying starches with copolymers and enzymes for use in sizing paper, US Patent 3,436,309 (1969); *Chem. Abstr.*, 70 (1969) 116415p.
2379. H. Yamada and Y. Baba, Enzyme-degraded cationized starch paper strengthening agents with good disintegrability and manufacture of disintegrable paper or fiber sheets using the agents, Jpn. Kokai Tokkyo Koho 11012979 (1999); *Chem. Abstr.*, 130 (1999) 126479.
2380. H. Yamada and Y. Baba, Manufacture of recyclable paper containing enzyme-degraded cationic starch and carboxymethylcellulose with high tensile strength in the dry state and good disintegrability, Jpn. Kokai Tokkyo Koho 11107188 (1999); *Chem. Abstr.*, 130 (1999) 313395.
2381. Y. Kojima and S. L. Yoon, Improved enzymatic hydrolysis of waste paper by ozone pretreatment, *J. Mater. Cycles Waste Manag.*, 10 (2008) 134–139.
2382. D. Schönbohm, A. Blüher, and G. Banik, Enzymes in solvent conditioned poultices for the removal of starch-based adhesives from iron gall ink corroded, *Restaurator*, 25 (2004) 267–281.
2383. N. F. Schink and R. E. McConiga, Equipment for the production of starch hydrolysates, French Patent 1,364,190 (1964).
2384. W. Jülicher, Enzymic desizing of fabrics, *Melliland Textilber.*, 33 (1952) 511–516. *Chem. Abstr.*, 46 (1952) 89015.
2385. L. Dantal, Enzymic reactions and their application for textiles, *Ind Textile*, 69 (1952) 450–456, 504–507.
2386. Mead Corp., High-solids starch adhesives, British Patent 1,026,444 (1966).
2387. E. L. Speakman, Dried starch-enzyme blend used to prepare adhesives, US Patent 3,544,345 (1970); *Chem. Abstr.*, 74 (1970) 65920x.
2388. A. M. Aboul-Enein, R. M. Abdel-Azize, A. M. Youssef, and M. A. Dawod, Improving efficiency of bacterial alpha-amylase during desizing of textiles, *Egypt J. Food Sci.*, 15 (1987) 91–104.
2389. N. Nakamura, K. Uno, M. Matsuzawa, and S. Nagatomo, Starch adhesives for paper, Jpn. Kokai, Tokkyo Koho, 79 43,246 (1979); *Chem. Abstr.*, 91 (1978) 93335e.
2390. J. J. Salama, Aqueous colloidal composition from starch, Swiss Patent 557,866 (1975); *Chem. Abstr.*, 82 (1974) 158112r.
2391. E. P. Gillan, Enzyme conversions of cornstarch, *Paper Trade J.*, 115(14), (1942) 109–111.
2392. M. A. Hanson and R. D. Gilbert, New look at desizing with enzymes, *Text. Chem. Color.*, 6 (1974) 262–265.
2393. W. Hahn, A. Seitz, M. Riegels, R. Koch, and M. Pirkotsch, Enzymic mixtures and process for the desizing of starch-containing textiles, EU Patent EP 789075 (1997); *Chem. Abstr.*, 127 (1997) 553434.

2394. F. A. Bozich, Jr., Method for producing high solid dextrin adhesives, US Patent 4921795 (1990); *Chem. Abstr.*, 113 (1990) 42710.
2395. A. Svendsen, T. V. Borchert, and H. Bisgard-Frantzen, Recombinant alpha-amylase mutants and their use in textile desizing, starch liquefaction and washing, EU Patent WO 9741213 (1997); *Chem. Abstr.*, 128 (1997) 20052.
2396. T. de Souza Rocha, A. P. De Almeida Carneiro, and C. M. Landi Franco, Effect of enzymatic hydrolysis on some physicochemical properties of root and tuber granular starches, *Cienc. Tecnol. Aliment. Campinas*, 30 (2010) 544–551.
2397. N. A. Ibrahim, M. El-Hossamy, M. S. Morsy, and B. M. Eid, Optimization and modification of enzymatic desizing of starch-size, *Polym-Plast. Technol. Eng.*, 43 (2004) 519–538.
2398. V. V. Safonov, I. A. Sidorenko, V. V. Baev, and Yu. V. Novichkova, Application conditions of amylosubtilin G 3X α-amylase for desizing of cotton fabrics (in Russian), *Biotekhnologiya* (2), (1991) 61–64.
2399. V. I. Lebedeva, A. V. Cheshkova, B. N. Melnikov, V. P. Gavrilova, and N. I. Sverdlova, Enzyme-containing compositions for removing starch size from cellulose-containing fabrics, USSR Patent 1776707 (1992); *Chem. Abstr.*, 121 (1994) 233014.
2400. K. Opwis, D. Knittel, A. Kele, and E. Schollmeyer, Enzymatic recycling of starch-containing desizing liquors, *Starch/Stärke*, 51 (1999) 348–353.
2401. W. Wei, L. Yang, and X. Liang, Study on properties of α-amylase in starch adhesive, *Zhanjie*, 19(5), (1998) 9–10, 13; *Chem. Abstr.*, 130 (1998) 239093.
2402. H. Yuki, Purification of dampening water containing starch by enzyme in offset printing, Jpn. Kokai Tokkyo Jkoho 07088468 (1995); *Chem. Abstr.*, 123 (1995) 44452.
2403. K. K. Yaoi Kagaku Kogyo, Starch glue releasing agents, Jpn. Kokai Tokkyo Koho 59 24,770 (1984); *Chem. Abstr.*, 101 (1984) 25321x.
2404. I. Bertoti, B. Kallai, and T. A. Toth, Amylase-promoted starch hydrolysis for animal food and glue manufacture, Hungarian Patent 38,955 (1986); *Chem. Abstr.*, 107 (1987) 25057q.
2405. 2. M. Berghgracht, Preparation of hydrolytic starch derivatives, Belgian Patent 667,071 (1965).
2406. J. L. Eden, Y. C. Shi, R. J. Nesiewicz and J. Wieczorek, Jr., High-solids storage-stable maltodextrin-based adhesives, US Patent 5932639 (1999); *Chem. Abstr.*, 131 (1999) 131453.
2407. A. Yoshizawa and T. Kitazawa, Paste for gummed tape from hydrolyzed starch, US Patent 3,770,672 (1973); *Chem. Abstr.*, 80 (1973) 61368z.
2408. H. Wegner, Reinigungsmöglichkeiten für eiweisshaltige Stärken auf enzymatischem Wege, *Starch/Stärke*, 3 (1951) 254–255.
2409. P. L. Weegels, J. P. Marseille, and A. M. B. Voorpostel, Enzymes as a processing aid in the separation of wheat into starch and gluten, *Proc. 4th Int Workshop gluten proteins, 1990*, (1990) 199–203; *Chem. Abstr.*, 116 (1992) 57882.
2410. P. L. Weegels, J. P. Marseille, and R. J. Hamer, Enzymes as a processing aid in the separation of wheat flour into starch and gluten, *Starch/Stärke*, 44 (1992) 44–48.
2411. S. A. Frederix, C. M. Courtin, and J. A. Delcour, Impact of endoxylanases with different substrate selectivity on gluten-starch separation, *Proc. 3rd Eur. Symp. Enzyme Grain Processing, Leuven* 2002, pp. 247–253. *Chem. Abstr.*, 142 (2004) 410020.
2412. F. David, Starch hydrolysis with increased dry weight, Czechoslovak Patent 200,995 (1979); *Chem. Abstr.*, 98 (1983) 162745.
2413. Z. Wang, Quick-acting dirt-removing cleaning agent, Chinese Patent 1046554 (1990); *Chem. Abstr.*, 115 (1991) 52313.
2414. M. Linderer, G. Wildbrett, and G. Enderle, Investigation of mechanical dishwashing. III. Starch removal from various formulations, *SOFW J.*, 121 (1995) 160, 162–164, 166–167.
2415. B. Kottwitz, K. H. Maurer, R. Breves, I. Schmidt, A. Weber, A. Hellebrandt, and L. Polanyi, Bacillus cyclodextrin glucanotransferase/α-amylase and gene and application of enzyme, especially in cleaning agents, EU Patent WO 2002044350 (2002); *Chem. Abstr.*, 137 (2002) 17128.

2416. C. Mitchinson, C. Requadt, T. Ropp, L. P. Solheim, C. Ringer, and A. Day, Bacillus α-amylase mutant recombinant production, improved low pH starch liquefaction, thermal stability and activity and use as laundry detergent or dishwashing detergent, EU Patent WO 9639528 (1996); *Chem. Abstr.*, 126 (1997) 105762.
2417. B. Kottwitz, H. D. Speckmann, and K. H. Maurer, Laundry detergents containing amylase and dye transfer inhibitor and their use, EU Patent WO 9963035 (1999); *Chem. Abstr.*, 132 (1999) 24155.
2418. B. Kottwitz, H. D. Speckmann, K. H. Maurer, and C. Nitsch, Detergents containing amylase and alkali percarbonate and their use, EU Patent WO 9963036 (1999); *Chem. Abstr.*, 132 (1999) 24156.
2419. B. Kottwitz, H. D. Speckmann, K. H. Maurer, and C. Nitsch, Detergents containing amylase and percarboxylic acid and their use, EU Patent WO 9963037 (1999); *Chem. Abstr.*, 132 (1999) 24157.
2420. B. Kottwitz, H. D. Speckmann, K. H. Maurer, and C. Nitsch, Detergents containing amylase and acetonitirle quarternary ammonium derivative and their use, EU Patent WO 9963038 (1999); *Chem. Abstr.*, 132 (1999) 24158.
2421. B. Kottwitz, H. D. Speckmann, K. H. Maurer, and C. Nitsch, Detergents containing amylase and their use, EU Patent WO 9963039 (1999); *Chem. Abstr.*, 132 (1999) 24159.
2422. B. Kottwitz, H. D. Speckmann, K. H. Maurer, and C. Nitsch, Detergents containing amylase and protease and their use, EU Patent WO 9963040 (1999); *Chem. Abstr.*, 132 (1999) 24160.
2423. C. C. Barnett, C. Mitchinson, and S. D. Power, An improved cleaning composition containing *Bacillus licheniformis* α-amylase mutants with improved thermal stability and oxidation resistance, EU Patent WO 9605295 (1996); *Chem. Abstr.*, 125 (1996) 4407.
2424. H. D. Speckmann, B. Kottwitz, C. Nitsch, K. H. Maurer, H. Blum, and L. Zuchner, Detergents containing amylase and bleach-activating transition metal compounds and their use, EU Patent WO 9963041 (1999); *Chem. Abstr.*, 132 (1999) 24161.
2425. J. Slade, Enzymic method for removing contaminants from ion exchange and fractionation resin, EU Patent WO 9804344 (1998); *Chem. Abstr.*, 128 (1998) 129823.
2426. M. Tokida, Y. Asayama, Y. Masumoto, and K. Endo, Enzyme treatment of starch for coating color – Effects of pullulanase, *Kami Parupu Kenkyu Happyokai Koen Yoshiu 63rd*, (1996) 144–147; *Chem. Abstr.*, 125 (1996) 250808.
2427. W. C. Chung and J. J. Kasica, Enzymically debranched starches as tablet excipients, EU Patent EP 499648 (1992); *Chem. Abstr.*, 117 (1992) 239833.
2428. W. C. Chung and J. J. Kasica, Enzymically debranched starches as tablet excipients, Canadian Patent 2032385 (1992); *Chem. Abstr.*, 118 (1992) 27490.
2429. W. C. Chung and J. J. Kasica, Enzymically debranched starches as tablet excipients, US Patent 5468286 (1995); *Chem. Abstr.*, 124 (1995) 66616.
2430. P. A. Truter, E. D. Dilova, T. L. Van der Merwe, and L. Thilo, Polysaccharides and enzymes for making quick-release compositions, EU Patent WO 9901108 (1999); *Chem. Abstr.*, 130 (1999) 1000695.
2431. E. K. Blue, C. W. Chiu, Z. Hussain, H. Shah, P. Trubiano, and D. Boyd, Manufacture of glucoamylase-converted starch derivatives as emulsifying and encapsulating agents, EU Patent EP 913406 (1999); *Chem. Abstr.*, 130 (1999) 326514.
2432. E. K. Blue, C. W. Chiu, Z. Hussain, R. Jeffcoat, H. Shah, P. Trubiano, and D. Boyd, Use of enzymically converted starch derivatives as an encapsulating agent, EU Patent EP 922449 (1999); *Chem. Abstr.*, 131 (1999) 49473.
2433. L. Tuovinen, S. Peitonen, M. Likola, M. Hotakainen, M. Lahtela-Kakkonen, A. Poso, and K. Jarvinen, Drug release from starch-acetate microparticles and films with and without incorporated α-amylase, *Biomaterials*, 25 (2004) 4355–4362.
2434. X. X. Li, L. Chen, and L. Lin, Digestibility of acetated tapioca starches for controlled drug release carriers, *Gongneng Gaofenzi Xuebao*, 16 (2003) 561–564. *Chem. Abstr.*, 141 (2004) 42729.

2435. S. Kio, N. Tanaka, N. Yago, K. Suzuki, Y. Endo, K. Nagatsuka, and H. Fukuda, Preparation of starch-based microcapsules, Jpn. Kokai Tokkyo Koho, 01159047 (1989); *Chem. Abstr.*, 111 (1989) 234712.
2436. Z. Luo, F. Luo, and X. He, Method for preparing starch-based immobilized enzyme carrier, Chinese Patent 1966708 (2007); *Chem. Abstr.*, 147 (2007) 51711.
2437. H. G. Bazin, F. W. Barresi, and J. Wang, Enzymatically modified hydrophobic starch, EU Patent WO 2002024938 (2002); *Chem. Abstr.*, 136 (2002) 278225.
2438. M. Kawano, K. Mura, and W. Tanimura, Changes in bound fatty acids during enzymic conversion of corn starch, *Tokio Nogo Daigaku Nogaku Shuho*, 33 (1988) 1–8. *Chem. Abstr.*, 110 (1989) 93545.
2439. P. C. Trubiano and A. G. Makarious, Modified starch–enzyme delivery system with increased bioavailability, EU Patent EP 1484055 (2004); *Chem. Abstr.*, 142 (2004) 28173.
2440. K. Poutanen, Enzymes: An important tool in the improvement of the quality of cereal foods, *Trends Food Sci. Technol.*, 8 (1997) 300–306.
2441. K. Takami, A. Ito, S. Hamada, I. Kimura, and T. Kooryama, Cooking Indica rice with α-amylase and starch, Jpn. Kokai Tokkyo Koho 0732615 (1995); *Chem. Abstr.*, 124 (1996) 174267.
2442. I. S. Dogan, Dynamic rheological properties of dough as affected by amylase from various sources, *Nahrung*, 46 (2002) 399–403.
2443. K. M. Hardiner, K. Maninder, and G. S. Bains, Effect of cereal, fungal and bacterial α-amylase on the rheological and breadmaking properties, *Nahrung*, 27 (1983) 609–618.
2444. K. M. Hardiner, K. Maninder, and G. S. Bains, Interaction of amyloglucosidase and α-amylase supplements in breadmaking, *Nahrung*, 28 (1984) 837–849.
2445. N. Chamberlain, T. H. Collins, and E. E. McDermott, α-Amylase and bread properties, *J. Food Technol.*, 16 (1981) 127–152.
2446. P. Van Hung, M. Yamamori, and N. Morita, Formation of enzyme-resistant starch in bread as affected by high-amylose wheat flour substitutions, *Cereal Chem.*, 82 (2005) 690–694.
2447. J. F. Conn, A. J. Johnson, and B. S. Miller, An investigation of commercial fungal and bacterial α-amylase preparations in baking, *Cereal Chem.*, 27 (1950) 191–205.
2448. J. Krebs, B. Sproessler, H. Uhlig, and R. Schmitt, Enzymes for bakery products, German Patent 2,227,368 (1974); *Chem. Abstr.*, 82 (1974). 71839.
2449. T. Jimenez and M. A. Martinez-Anaya, Amylases and hemicellulases in breadmaking. Degradation by-products and potential relationship with functionality, *Food Sci. Technol. Int.*, 7 (2001) 5–14.
2450. S. Sahlstroem and E. Braathen, Effects of enzyme preparations for baking, mixing time and resting time on bread quality and bread staling, *Food Chem.*, 58 (1996) 75–80.
2451. T. T. Valjakka, J. G. Ponte, Jr., and K. Kulp, Studies on a raw-starch digesting enzyme. II. Replacement of sucrose in white pan bread, *Cereal Chem.*, 71 (1994) 145–149.
2452. O. Silberstein, Heat-stable bacterial α-amylase in baking. Application to white bread, *Bakers Dig.*, 38 (1964) 66–72.
2453. M. L. Martin and R. C. Hoseney, A mechanism of bread firming. II. Role of starch hydrolyzing enzymes, *Cereal. Chem.*, 68 (1991) 496–503.
2454. A. E. Leon, E. Duran, and C. Beneditto de Barber, Utilization of enzyme mixtures to retard bread crumb firming, *J. Agric. Food Chem.*, 50 (2002) 1416–1419.
2455. M. A. Martinez-Anaya and T. Jimenez, Functionality of enzymes that hydrolyse starch and non-starch polysaccharide in breadmaking, *Lebensm. Unters. Forsch.*, A 205 (1997) 209–214.
2456. K. Yoshida, Bread with slight aroma and sweetness and its manufacture, Jpn. Kokai Tokkyo Koho 2002345392 (2002); *Chem. Abstr.*, 137 (2002) 384186.
2457. D. Curić, J. Dugum, and I. Bauman, The influence of fungal α-amylase supplementation on amylolytic activity and baking quality of flour, *Int. J. Food Sci. Technol.*, 37 (2002) 673–680.
2458. V. Horvathova and K. Valachova, Use of yeast glucoamylase Glm in bakery technology (in Slovak), *Bull. Potravin. Vysk.*, 44 (2005) 261–271.

2459. H. Ogasawara and S. Egawa, Manufacture of bread containing rice flour to which α-amylase is added, Jpn. Kokai Tokkyo Koho, 2004135608 (2004); *Chem. Abstr.*, 140 (2004) 374293.
2460. S. Sasaki, H. Hagiwara, and M. Moriyama, Bakery quality-improving sols and gels, their manufacture and foods containing them, Jpn. Kokai Tokkyo Koho 2008000133 (2008); *Chem. Abstr.*, 148 (2008) 120724.
2461. M. Jahnke, Microwave baking additive containing gelling gum and enzyme components, US Patent 2003206994 (2003); *Chem. Abstr.*, 139 (2003) 337264.
2462. R. Geoffroy, Amylolysis in bread doughs, *Compt. Rend. Acad. Agr. France*, 41 (1955) 270–271. *Chem. Abstr.*, 46 (1955) 51993.
2463. H. Xiang, L. Chen, X. Li, Y. Pang, L. Li, and L. Yu, Study on the crystal structure and enzymatic resistance of kneaded starch, *Shipin Gongye Keji*, 27(2), (2006) 84–86. *Chem. Abstr.*, 147 (2007) 405453.
2464. C. C. Walden, Action of flour amylase during oven baking, *Baker's Dig.*, 33 (1959) 24–31.
2465. L. Eynard, N. Guerrieri, and P. Cerletti, Modifications of starch dyring baking: Studied through reactivity with amyloglucosidase, *Cereal. Chem.*, 72 (1995) 594–597.
2466. M. Siljestrom, I. Bjork, A.-C. Eliasson, C. Lonner, M. Nyman, and N.-G. Asp, Effect on polysaccharides during baking and storage of bread – In vitro and in vivo studies, *Cereal Chem.*, 65 (1988) 1–8.
2467. L. H. Johansen and S. Lorenzen, Enzymic modification in manufacture of masa dough, US Patent 2003165593 (2003); *Chem. Abstr.*, 139 (2003) 229744.
2468. A. Svendsen, C. Andersen, T. Spendler, A. Viksoe-Nielsen, and H. Östdal, Hybrid enzymes containing endoamylase catalytic domain fused to carbohydrate binding domain for use in starch saccharification and baking, EU Patent WO2006066596 (2006); *Chem. Abstr.*, 145 (2006) 98562.
2469. P. Leman, H. Goesaert, G. E. Vandeputte, B. Lagrain, and J. A. Delcour, Maltogenic amylase has a non-typical impact on the molecular and rheological properties of starch, *Carbohydr. Polym.*, 62 (2005) 205–213.
2470. F. Kiermeier and E. Coduro, Der Einfluß des Ëssergehaltes auf Enzymreaktionen in wasserarmen Lebensmitteln. III. Über Enzyme Reaktionen in Mehl, Teg und Brot, *Z. Lebensm. Untersuch. Forsch.*, 102 (1955) 7–12.
2471. M. Rimbault, General and microbial aspects of solid substrate fermentation, *E. J. Biotechnol.*, 1 (1998) No. 3.
2472. J. Tapodo, F. Hirschberg, E. Laszlo, and A. Hoschke, Stable sweet cakes, Hungarian Patent 11,663 (1976); *Chem. Abstr.*, 85 (1976) 191088x.
2473. P. Nicolas and C. E. Hansen, Flour based food product comprising fermentable alpha-amylase, EU Patent EP 1415539 (2004); *Chem. Abstr.,* 140 (2004) 356319.
2474. T. Sugimoto, S. Takeshi, and T. Matsusue, Emulsifier-free oil and fat composition containing sugar-degrading enzymes for bread making, Jpn. Kokai Tokkyo Koho 09233993 (1997); *Chem. Abstr.*, 127 (1997) 602577.
2475. K. H. Meyer and W. F. Gonon, Rercherches sur l'amidon. 50. La degradation l'amylose per les α-amylases, *Helv. Chim. Acta*, 34 (1951) 294–307.
2476. K. H. Meyer and W. F. Gonon, Rercherches sur l'amidon. 50. La degradation l'amylopectin per les α-amylases, *Helv. Chim. Acta*, 34 (1951) 308–316.
2477. S. Hug-Iten, F. Escher, and B. Conde-Petit, Structural properties of starch in bread and bread model systems: Influence of an antistaling α-amylase, *Cereal Chem.*, 78 (2001) 421–428.
2478. M. Haros, C. M. Rosell, and C. Benedito, Effect of different carbohydrases on fresh bread texture and bread staling, *Eur. Food Res. Technol.*, 215 (2002) 425–430.
2479. R. D. Dragsdorf and E. Varriano-Marsdorf, Bread staling: X-ray diffraction studies on bread supplemented with α-amylase from different sources, *Cereal Chem.*, 57 (1980) 310–314.

2480. H. R. Palacios, P. B. Schwarz, and B. L. D'Appolonia, Effect of α-amylases from different sources on the firming of concentrated wheat starch gels: Relationship to bread staling, *J. Agric. Food Chem.*, 52 (2004) 5987–5994.
2481. H. R. Palacios, P. B. Schwarz, and B. L. D'Appolonia, Effect of α-amylases from different sources on the retrogradation and recrystallization of concentrated wheat starch gels: Relationship to bread staling, *J. Agric. Food Chem.*, 52 (2004) 5978–5986.
2482. Y. Champenois, G. Della Valle, V. Planchot, A. Buleon, and P. Colonna, Influence of α-amylase on bread staling and on retrogradation of wheat starch models, *Sci. Aliment.*, 19 (1999) 471–486.
2483. P. Leman, H. Goesaert, and J. A. Delcour, Residual amylopectin structures of amylase-treated wheat starch slurries reflect amylase mode of action, *Food Hydrocoll.*, 23 (2009) 153–164.
2484. T. A. Siswoyo, N. Tanaka, and N. Morita, Effect of lipase combined with α-amylase on retrogradation of bread, *Food Sci. Technol. Res.*, 5 (1999) 356–361.
2485. K. M. Kragh, Amylase in baking, *Proc. 3rd Eur. Symp. Enzyme in Grain Processing, Louvain Belgium, 2002*, (2003) 221–223; *Chem. Abstr.*, 142 (2004) 372760.
2486. F. E. Volz and P. E. Ramstad, Effect of emulsifiers on the enzyme susceptibility of starch during staling of bread, *Cereal Chem.*, 28 (1951) 118–122.
2487. S. S. Jackel, A. S. Schultz, F. D. Schoonover, and W. E. Schaeder, The decrease in susceptibility of bread crumb starch to β-amylase during staling, *Cereal Chem.*, 29 (1952) 190–199.
2488. R. R. Bergkvist and K. O. Claesson, Process for the enzymic degradation of whole flour carbohydrates to produce a foodstuff, the foodstuff and its use, EU Patent EP 231,729 (1987); *Chem. Abstr.*, 107 (1987) 153198y.
2489. I. P. Batsalov and L. Ya, Auerman, Carbon dioxide-amylase complex in wheat (in Bulgarian), *Nauch. Tr. Vyssh. Inst. Khranit. Vkus. Prom. Plovidiv*, 8(Pt. 2), (1962) 169–181.
2490. J. Elms, P. Beckett, P. Griffin, P. Evans, C. Sams, M. Roff, and A. D. Curran, Job categories and their effect on exposure to fungal alpha-amylase and inhalable dust in the U.K. baking industry, *AIHA J.*, 64 (2003) 467–471.
2491. R. I. Nicholson and M. Horsley, The removal of starch from cane juice, *Int. Sugar J.*, 60 (1958) 260–267.
2492. T. Z. Chin and T. Y. Lin, The removal of starch from cane juice by an enzymic process, *Tajwan Sugar*, 7 (1960) 21–24.
2493. R. Bretschneider and J. Čopikova, Enzymic hydrolysis of some saccharides (in Czekh), *Listy Cukrov.*, 85 (1969) 188–191.
2494. Y. K. Park, I. S. H. Martens, and H. H. Sato, Enzymic removal of starch from sugarcane juice during sugarcane processing, *Process Biochem.*, 20(2), (1985) 57–59.
2495. J. Bruijn and R. P. Jennings, Enzymic hydrolysis of starch in cane juice, *Proc. Annu. Congr. S. Afr. Sugar Technol. Assoc.*, 42 (1968) 45–52.
2496. M. I. Egorova, L. I. Belayeva, S. I. Kazakova, and A. I. Chugunov, Removal of starch from raw cane juice (in Russian), *Pishch. Prom.*, (10), (2004) 58–60.
2497. P. Hidi, I. G. R. Burgess, and R. H. Holdgate, Process for reducing the starch content of raw rugar solutions, French Patent 1,520,947 (1968); *Chem. Abstr.*, 71 (1969) 23134b.
2498. P. L. Turecek, F. Pittner, and F. Brinkner, Degradation of polysaccharides by immobilized depolymerizing enzymes, *Int. J. Food Sci. Technol.*, 25 (1990) 1–8.
2499. B. Jędrychowska and R. Sawicka-Żukowska, Hydrolysis of starch in apple juices and the effectiveness of selected enzyme preparations (in Polish), *Przem. Ferm. Owoc-Warzyw.*, 34(1), (1990) 15–17.
2500. V. F. Tarytsa, V. F. Karazhiya, and T. A. Zubak, Use of enzymes in apple juice processing (in Russian), *Pishch. Prom.*, 1(9), (1990) 17–18.
2501. M. E. Carrin, L. N. Ceci, and J. E. Lozano, Characterization of starch in apple juice and its degradation by amylases, *Food Chem.*, 87 (2004) 173–178.

2502. G. L. Baker, Enzymic hydrolysis of starch in pectic extractions from apple pomace, *Bull 192 (Annu. Rept. 1934)* (1935) 27.
2503. M. Furuta, Y. Takada, H. Suenaga, and S. Ohta, Development of starch in apple juice and its degradation by amylases, *Kenkyu Hokoku Fukuoka-ken Kogyo Gijutsu Senta*, 2 (1991) 95–100. *Chem. Abstr.*, 119 (1993) 224730.
2504. G. L. Baker, Fruit juices. IX. The role of pectin 5. The enzymic hydrolysis of starch in the presence of pectin in pectic extracts and in apple pomace, *Univ. Del. Agr. Exp. Sta., Bull.*, 204(1936) 89pp.
2505. A. Totsuka, T. Nanakuki, T. Unno, M. Arima, K. Tanabe, K. Yamada, N. Katsumata, and A. Watanabe, Maltooligosaccharide compositions for use as sweetening agents and their preparation from astarch with α-amylase and debranching enzymes, Jpn. Kokai Tokkyo Koho 05038265 (1993); *Chem. Abstr.*, 119 (1993) 27066.
2506. A. Totsuka, T. Nakakuki, T. Unno, M. Fuse, M. Suga, and K. Kato, Reduced starch sugar compositions containing maltohexitol and maltoheptitol for use in foods and beverages, Jpn. Kokai Tokkyo Koho 05038272 (1993); *Chem. Abstr.*, 119 (1993) 48070.
2507. S. Sakai, Sweetened starch products, German Patent 2,532,005 (1976); *Chem. Abstr.*, 85 (1976) 162285d.
2508. M. Kubota, T. Nakada, and S. Sakai, Enzymic preparation of sweetening agent maltotetraose from starch, Jpn. Kokai Tokkyo Koho 08205865 (1996); *Chem. Abstr.*, 125 (1996) 220221.
2509. T. Adachi, T. Nakamura, and H. Hidaka, Enzymic hydrolysis of starch for candy production, Jpn. Kokai Tokkyo Koho 79 11,954 (1979); *Chem. Abstr.,* 91 (1979) 173694m.
2510. Meiji Seika Kaisha Ltd., Starch hydrolysis used as a sweetener, French Patent 2,396,079 (1979); *Chem. Abstr.*, 91 (1979) 122432s.
2511. C. Takahashi, J. Kanaeda, S. Kubomura, S. Hashimoto, and H. Toda, Production of starch sugars with high maltose content by a new enzyme, Jpn. Kokai Tokkyo Koho 75 125,046 (1975); *Chem. Abstr.*, 84 (1975) 104098p
2512. S. Hirooka, N. Yamazaki, and T. Hatori, Enzymic preparation of saccharified health product from starch, Jpn. Kokai Tokkyo Koho 2008000100 (2008); *Chem. Abstr.,* 148 (2008) 77846.
2513. C. Wei, Method for preparing fresh sweet and waxy corn by converting amylopectin with α-amylase, Chinese Patent 101040678 (2007); *Chem. Abstr.*, 147 (2007) 447522.
2514. D. P. Langlois and J. K. Dale, Starch conversion syrup, US Patent 2,202,609 (1940).
2515. A. Kovago, J. Petroczy, I. Petroczy, G. Torok, and J. Feher, Water-soluble sugars and sugar mixtures by enzymic hydrolysis of natural products containing $\geq$ 20 wt% starch, Hungarian Patent 17,516 (1979); *Chem. Abstr.*, 92 (1980) 144985x.
2516. J. M. Quarles and R. J. Alexander, Hydroxypropyl starch hydrolysate products as low-calorie sweetener, EU Patent WO 9203936 (1992); *Chem. Abstr.*, 117 (1992) 25097.
2517. S. Okada, S. Kitahata, and H. Ishibashi, Enzymic manufacture of oligosaccharides as mild sweeteners, Jpn. Kokai Tokkyo Koho 63202396 (1988); *Chem. Abstr.*, 110 (1989) 133959.
2518. H. Katsuragi, K. Shimodi, E. Kimura, and H. Hamada, Synthesis of capsaicin glycosides and 8-nordihydrocapsaicin glycosides as potential weight loss formulations, *Biochem. Imnsights*, 3 (2010) 35–39.
2519. D. P. Langlois, Application of enzymes to corn-sirup production, *Food Technol.*, 7 (1953) 303–307.
2520. M. Hirao and J. Ogasawara, Enzyme-converted starch syrups, S. African Rep. Patent 69 08,248 (1970); *Chem. Abstr.*,74 (1971) 100813p.
2521. K. Sugimoto, M. Hirao, M. Mitsuhashi, and J. Ogasawara, Starch sirup of low viscosity, German Patent 1,916,726 (1970); *Chem. Abstr.*, 74 (1970) 4885t.
2522. R. E. Heady, High-dextrose-equivalent starch hydrolyzate sirup, US Patent 3,535,123 (1970); *Chem. Abstr.*, 74 (1970) 14349k.
2523. A. Sroczyński, T. Pierzgalski, and K. Nowakowska, Manufacture of sirups by enzymic hydrolysis of starch (in Polish), *Zesz. Nauk. Politech. Lodz, Ser. Chem. Spoz.*, 28 (1976) 55–72.

2524. A. Sroczyński, T. Pierzgalski, and K. Nowakowska, Production of high saccharified syrups by two-stage enzymic hydrolysis of potato starch. I. The effect of glucoamylse concentration on reducing-sugar content increase at constant and variable pH values in hydrolysis of starch, *Acta Aliment. Pol.*, 3 (1977) 107–114.

2525. A. Sroczyński, T. Pierzgalski, and K. Nowakowska, Manufacture of maltose sirups by enzymic hydrolysis of starch (in Polish), *Zesz. Nauk. Politech. Lodz, Ser. Chem. Spoz.*, 28 (1976) 73–77.

2526. M. Hirao and M. Mitsuhashi, Recovery of amylose sirup by saccharification of starch with an acid and (or) enzyme, German Patent 2,013,863 (1970); *Chem. Abstr.*, 74 (1970) 14348j.

2527. J. G. Zeikus and B. C. Saha, Preparing high conversion syrups and other sweetener by thermostable β-amylase and other thermostable enzymes, US Patent 4,647,538 (1989); *Chem. Abstr.*, 111 (1989) 132913.

2528. P. J. Reilly, Production of glucose and fructose with immobilized glucoamylase and glucose isomerase, *Rep. US Natl. Sci. Fund., Res. Appl. Natl. Needs*, NSF/RA-760032 (1975) PB-265 548, 80–84; *Chem. Abstr.*, 88 (1977) 35835v.

2529. T. Kimura and T. Nakakuki, Maltotetraose, a new saccharide of tertiary property, *Starch/Stärke*, 42 (1990) 151–157.

2530. T. Spendler and J. B. Nielsen, Enzymatic treatment of starchy food products for shortening the tempering step, EU Patent WO 2003024242 (2003); *Chem. Abstr.*, 138 (2003) 237261.

2531. H. Jian, Q. Gao, and S. Liang, Preparation of resistant starch with enzymes process, *Shipin Yu Fajiao Gongye*, 28(5), (2002) 6–9. *Chem. Abstr.*, 137 (2002) 365546.

2532. Y. C. Shi, X. Cui, and A. M. Birkett, Resistant starch prepared by isoamylase debranching of low amylose starch, EU Patent EP 1362869 (2003); *Chem. Abstr.*, 139 (2003) 364027.

2533. M. Henley and C. W. Chiu, Amylase-resistant starch product from debranched high-amylose starch, US Patent 5409542 (1995); *Chem. Abstr.*, 123 (1995) 173309.

2534. C. W. Chiu, M. Henley, and P. Altieri, Preparation of an amylase-resistant starch by enzymic and thermal processing, EU Patent EP 564893 (1993); *Chem. Abstr.*, 119 (1993) 248177.

2535. F. Qiu, W. Zhao, and C. Tao, Resistant starch-technology for the production of special dietary fiber, *Shipin Gongye Keji*, 26(1), (2005) 102–103. *Chem. Abstr.*, 144 (2005) 149366.

2536. N. Tanaka, N. Kago, H. Fukuda, and Y. Doi, Non-digestible starch and its manufacture, Jpn. Kokai Tokkyo Koho 11255802 (1999); *Chem. Abstr.*, 131 (1999) 215797.

2537. K. Ookuma, T. Hanno, K. Inada, I. Matsuda, and Y. Katsuta, Dietary fiber-high dextrins preparation from potato starch, Jpn. Kokai Tokkyo Koho 05176719 (1993); *Chem. Abstr.*, 119 (1993) 158864.

2538. K. Ookuma, T. Hanno, K. Inada, I. Matsuda, and Y. Katsuta, Preparation of low-calorie filling agents from potato starch, Jpn. Kokai Tokkyo Koho 05176716 (1993); *Chem. Abstr.*, 119 (1993) 158862.

2539. G. Dongowski and G. Stoof, Investigations of potato pulp as a dietary fiber source. Compositions of potato pulp after influence of pectinases and cellulases and enzymic degradation of starch, *Starch/Stärke*, 45 (1993) 234–238.

2540. D. Liu and Y. Zuo, Extraction of dietary fiber from potato dregs by enzymatic hydrolysis, *Shipin Gongye Keji*, 26(5), (2005) 90–92. *Chem. Abstr.*, 145 (2006) 123298.

2541. Q. Li, Method for manufacturing dietary fiber powder from wheat bran by multi-enzyme hydrolysis, Chinese Patent 1718107 (2006); *Chem. Abstr.*, 145 (2006) 45259.

2542. K. Yamada, M. Matsuda, and M. Sugiyama, Manufacture of indigestible starch-rich wheat flour for diet, Jpn. Kokai Tokkyo Koho 2004187598 (2004); *Chem. Abstr.*, 141 (2004) 88261.

2543. H. S. Guraya, C. James, and E. T. Champagne, Effect of enzyme concentration and storage temperature on the formation of slowly digestible starch from cooked debranched rice starch, *Starch/Stärke*, 53 (2001) 131–139.

2544. Q. Gao, Z. Luo, and L. Yang, Method for manufacturing resistant starch with amylase combined with heat-moisture treatment, Chinese Patent 1973688 (2007); *Chem. Abstr.*, 147 (2007) 94688.
2545. D. Huang, X. Kou, C. Li, and S. Wang, Method for manufacturing resistant maltodextrin, Chinese Patent 1808017 (2007); *Chem. Abstr.*, 146 (2007) 276399.
2546. P. Snow and K. O'Dea, Factors affecting the rate of hydrolysis of starch in food, *Am. J. Clin. Nutr.*, 34 (1981) 2721–2727.
2547. J. Kodet and S. Šterba, KMS – Enzymatically modified starches (in Czech), *Prum. Potravin*, 29 (1978) 438–439.
2548. Y. Fukai, E. Tagaki, T. Matsuzawa, and S. Shimada, Enzyme-modified starch for preparation of food, Jpn. Kokai Tokkyo Koho 06022800 (1994); *Chem. Abstr.*, 120 (1994) 297213.
2549. C. W. Chiu, D. P. Huang, J. J. Kasica, and Z. F. Xu, Use of amylase-treated low-viscosty starch in foods, US Patent 5599569 (1997); *Chem. Abstr.*, 126 (1997) 185283.
2550. Z. F. Xu, J. L. Senkeleski, J. P. Zallie, and P. J. Hendrikx, Use of enzymatically treated starches as viscosifiers and their use in food products, EU Patent EP 898901 (1999); *Chem. Abstr.*, 130 (1999) 196066.
2551. I. Kanbara and H. Miyamoto, β-Amylase in preparation of starchy foods, Jpn. Kokai Tokkyo Koho 199220228 (1992); *Chem. Abstr.*, 116 (1992) 254448.
2552. L. Chedid, D. P. Huang, and P. Baytan, Flavored popping corn with low or no fat, US Patent 5754287 (1998); *Chem. Abstr.*, 129 (1998) 15526.
2553. K. Baba and N. Kurushima, Manufacture of food additives containing α-amylase inhibitor, Jpn. Kokai Tokkyo Koho 08182475 (1996); *Chem. Abstr.*, 125 (1996) 220214.
2554. B. Meschonat, H. Herrman, R. Spannägl, V. Sander, G. Konieczny-Janda, and M. Sommer, Enzyme granules for mixing into feed, EU Patent WO 9742837 (1997); *Chem. Abstr.*, 128 (1997) 3408.
2555. H. Herrman and R. Spannägl, Enzyme granulate for use in food technology, EU Patent WO 9742839 (1997); *Chem. Abstr.*, 128 (1997) 34045.
2556. I. Hayasaka and E. Yoshida, Enzymic preparation of canned starchy beverages, Jpn. Kokai Tokkyo Koho 01257432 (1989); *Chem. Abstr.*, 113 (1990) 57825.
2557. Y. Fukai, E. Tagaki, T. Matsuzawa, and S. Shimada, Enzyme-treated starch granules for food, Jpn. Kokai Tokkyo Koho 05123116 (1993); *Chem. Abstr.*, 119 (1993) 94127.
2558. S. King, R. Luis, H. Kistler, and A. Laurenz, Acidic amyloceus fermented composition for reducing the caloric content of foods containing fats and oils, EU Patent EP 663153 (1995); *Chem. Abstr.*, 123 (1995) 142379.
2559. N.Yu. Sharova, Enzyme-containing food additive for hydrolysis of plant carbohydrates and proteins, Russian Patent 2294368 (2007); *Chem. Abstr.*, 146 (2007) 273046.
2560. M. L. Dondero, M. W. Montgomery, L. A. McGill, and D. K. Law, Preparation of a potato hydrolyzate using α-amylase, *J. Food Sci.*, 43 (1978) 1698–1701.
2561. S. Xu, Z. Wang, and R. Yang, Manufacture and application of maltodextrin with low dextrose equivalent value by spraying liquefaction and enzymolysis, Chinese Patent 1528910 (2004); *Chem. Abstr.*, 143 (2005) 284827.
2562. K. Nojiri, T. Takahashi, S. Ideie, and S. Igarashi, Manufacture of oligosaccharides by glucosidase as bifidobacterium growth promoter in health foods, Jpn. Kokai Tokkyo Koho 62,205,793 (1987); *Chem. Abstr.*, 108 (1987) 93093u.
2563. S. Endo, M. Tanaka, and Y. Goto, Enzymic additives to rice cooking, Jpn. Kokai Tokkyo Koho 08140600 (1996); *Chem. Abstr.*, 125 (1996) 166262.
2564. M. Watanabe, Taste improvement of rice with polysaccharide-hydrolyzing enzymes and protein-hydrolyzing enzymes, Jpn. Kokai Tokkyo Koho 04258263 (1992); *Chem. Abstr.*, 118 (1993) 6035.
2565. H. Takahashi, Protein-, phosphorus-, and potassium-low processed rice and their manufacture, Jpn. Kokai Tokkyo Koho 11332482 (1999); *Chem. Abstr.*, 132 (1999) 2979.

2566. S. Okada, K. Yoshikawa, H. Hidaka, A. Onoe, M. Katayama, and K. Inoe, Manufacture of aging-prevented starch with amylase, Jpn. Kokai Tokkyo Koho 01153099 (1989); *Chem. Abstr.*, 111 (1989) 231067.
2567. T. Nishimoto, K. Hino, T. Okura, H. Chaen, S. Fukuda, and T. Miyaki, Enzymic preparation of antiaging branched starch, EU Patent WO 2006112222 (2006); *Chem. Abstr.*, 145 (2006) 395656.
2568. H. Y. Chae, C. R. Choi, Y. R. Kim, H. G. Lee, T. W. Moon, J. H. Park, K. H. Park, and M. S. Shin, Thermostable α-glucanotransferase which reconstructs structure of starch in food, enhances storage of food and improves properties and taste of food, S. Korean Patent 2005041069 (2005); *Chem. Abstr.*, 145 (2006) 330585.
2569. N. Yamada, T. Ogawa, A. Maruyama, T. Okamoto, and H. Wakabayashi, Process for production of starch-containng food and enzyme preparation for modification of starch-containing food, EU Patent, WO 20070627 (2008); *Chem. Abstr.*, 148 (2008) 77750.
2570. H. Weenen, R. A. De Wijk, and R. J. Hamer, Improved semi-solid food products and methods for their production based on inhibiting amylase induced starch breakdown, EU Patent EP 1557094 (2005). *Chem. Abstr.*, 143 (205) 152339.
2571. Q. Gao and Y. Yang, The improvement of starch film with isoamylases, *Shipin Kexue*, 165 (1993) 8–12. *Chem. Abstr.*, 120 (1994) 75970.
2572. T. Yunusov and S. R. Gregory, Enzymic processing of grains and oil seeds for production of highly soluble food products, EU Patent WO 2004006691 (2004); *Chem. Abstr.*, 140 (2004) 110388.
2573. J. Fang, H. Chen, Z. Li, and H. Liu, Studies on hydrolysis of wheat B-starch by enzyme and preparation of caramel pigment from hydrolysis, *Zhongguo Tiaoweipin* (11), (2006) 46–49. *Chem. Abstr.*, 147 (2007) 520948.
2574. S. Guttapadu, S. A. Vinayak, V. P. Pradyumnarai, and B. A. Gopalarathanam, Enzymically modified whole meal cereal flour with enhanced taste and flavor for use in chapattis, Indian Patent 2002MU00946 (2004); *Chem. Abstr.*, 147 (2007) 501385.
2575. D. Zhan, Method for producing wheat food with enzyme, Chinese Patent 101040683 (2007); *Chem. Abstr.*, 147 (2007) 447527.
2576. I. Hayasaka and M. Kobayashi, Preparation of bean jam having good preservability, Jpn. Kokai Tokkyo Koho, 01235557 (1989); *Chem. Abstr.*, 112 (1990) 157046.
2577. Hohnen Oil Co., Ltd., Starch hydrolyzate for coating of frozen fried foods, Jpn. Kokai Tokkyo Koho Jpn. Kokai Tokkyo Koho 80 85,376 (1980); *Chem. Abstr.*, 93 (1980) 130971y.
2578. D. W. Harris and J. A. Little, Enzymic fragmentation of amylose for use as fat substitutes, EU Patent EP 529892 (1993); *Chem. Abstr.*, 118 (1993) 168048.
2579. M. H. Do, K. I. Lee, and H. S. Ji, Manufacture of starch hydrolyzates for use as bulking agents, S. Korean Patent 152349 (1998); *Chem. Abstr.*, 142 (2004) 73700.
2580. X. Zhou, L. Zhang, and J. Ou, Optimization of technology for enzyme processing of instant edible kudzu powder, *Shipin Kexue*, 26(8), (2005) 238–241. *Chem. Abstr.*, 145 (2006) 487907.
2581. K. Shimazu, M. Magara, Y. Tateno, K. Tanaka, and K. Kato, Enzymic preparation of high-purity maltitol, Jpn. Kokai Tokkyo Koho, 07123994 (1995); *Chem. Abstr.*, 123 (1995) 81761.
2582. M. Yang, Y. Yu, and F. Gao, Fat substitute research. I. Hydrolysis of potato starch by different amylases to produce low DE value maltodextrin, *Shipin Kexue*, 26 (2005) 149–155. *Chem. Abstr.*, 145 (2006) 247856.
2583. N. Dunn-Coleman, J. K. Shetty, G. Duan, A. Sung, and Y. Qian, High-purity rice protein concentrate obtained by enzymic hydrolysis and solubilization of starch component with α-amylase and gluco-amylase, EU Patent WO 2005082155 (2005); *Chem. Abstr.*, 143 (2005) 285338.
2584. S. Pedersen and H. Vang Hendricksen, Maltogenic amylase-modified starch derivatives, EU Patent WO 2001016348 (2001); *Chem. Abstr.*, 134 (2001) 192313.
2585. V. V. Aleksandrov, N. V. Afanasyev, and V. S. Gryuner, Effect of the method of starch hydrolysis on the properties of sirup (in Russian), *Tr. Vses. Nauch-Issled. Inst. Kondit. Prom.* (9), (1953) 49–72.

2586. G. Skrede, O. Herstad, S. Sahlstrom, A. Holck, E. Slinde, and A. Skrede, Effects of lactic acid fermentation on wheat and barley carbohydrate compositionand production performance in the chicken, *Animal Feed Sci. Technol.*, 105 (2003) 135–148.
2587. A. T. Adesogan, M. B. Salawu, A. B. Ross, D. R. Davies, and A. E. Brooks, Effect of *Lactobacillus buchneri, Lactobacillus fermentum, Leuconostoc mesenteroides* inoculants or a chemical additive on the fermentation aerobic stability and nutritive value of crimped wheat grains, *J. Dairy Sci.*, 86 (2003) 1789–1796.
2588. Q. Gao and Y. Yang, Study on improving starch film with isoamylase, *Huanan Ligong Daxue Xuebao, Ziran Kexueban*, 21(2), (1993) 62–68.
2589. H. Kalinowska, M. Turkiewicz, and S. Bielecki, Enzymes of new generation in food manufacture (in Polsh), *Przem. Spoz.* (10), (2000) 3–5.
2590. H. Kalinowska, M. Turkiewicz, and S. Bielecki, Enzymes of new generation in food manufacture. II. Production of health-promoting food components (in Polish), *Przem. Spoz.* (12), (2000) 8.
2591. R. D. Preignitz, Food products containing α-amylase and process, EU Patent EP 105,051 (1984); *Chem. Abstr.*, 101 (19834) 109311g.
2592. M. Isaksen, K. Kragh, and T. Gravesen, Animal feed additive containing amylase for resistant starch degradation, EU Patent WO 2003049550 (2003); *Chem. Abstr.*, 139 (2003) 35786.
2593. A. C. Ivy, C. R. Schmidt, and J. M. Beazell, Effectiveness of malt amylase on the gastric digestion of starch, *J. Nutr.*, 12 (1936) 59–83.
2594. S. Steenfeldt, M. Hammershoj, A. Mullertz, and J. J. Fris, Enzyme supplementation of wheat-based diets for broilers. 2. Effect on apparent metabolizable energy content and nutrient digestibility, *Animal Feed Sci. Technol.*, 75 (1998) 45–64.
2595. Y. B. Wu, V. Ravindran, and W. H. Hendriks, Influence of exogenous enzyme supplementation on energy utilization and nutrient digestibility of cereals for broilers, *J. Sci. Food Agric.*, 84 (2004) 1817–1822.
2596. B. Haberer, E. Schulz, and G. Flachowsky, Effects of β-glucanase and xylanase supplementation in pigs fed a diet rich in nonstarch polysaccharides. Disappearance and disappearance rate of nutrients including the nonstarch polysaccharides in stomach and small intestine, *J. Animal Physiol Animal Nutr.*, 78 (1998) 95–103.
2597. J. Gdala, H. N. Johansen, K. E. Bach Knudsen, I. H. Knap, P. Wagner, and O. B. Jorgensen, The digestibility of carbohydrates, protein and fat in the small and large intestinen of piglets fed non-supplemented and enzyme supplemented diets, *Animal Feed Sci. Technol.*, 65 (1997) 15–33.
2598. W. Hackl, G. Bolduan, and M. Beck, Investigation about efficiency of several enzymes or enzyme mixtures and their combination in the mixed feed in growing pigs, *Proc. 5th Symp. Vitamine, Zusatzschtoffe in der Ernährung Mensch und Tier, Jena 1995* (1995) 376–381. *Chem. Abstr.*, 127 (1997) 49619.
2599. H. A. Herrmann, G. Konieczny-Janda, B. Meschonat, V. Sander, M. Sommer, and R. Spannagel, Enzyme pre-granules for granular fodder. Hungarian Patent 9903667 (2000); *Chem. Abstr.*, 146 (2007) 440695.
2600. C. A. Miller, Hyperthermophilic bacterial glycosidic bond hydrolyzing enzyme sequences and use thereof in animal feeds and related grain-based enzyme-releasing matrix and method of preparation, EU Patent WO 2003072717 (2003); *Chem. Abstr.*, 139 (2003) 226483.
2601. N. O. Ankrah, G. L. Campbell, R. T. Tyler, B. G. Rossnagel, and S. R. T. Sokhansanj, Hydrothermal and β-glucanase effects on the nutritional and physical properties of starch in normal and waxy hull-less barley, *Animal Feed Sci. Technol.* 81 (1999) 205–219.
2602. X. Meng, B. A. Slominski, C. M. Nyachoti, L. D. Campbell, and W. Guenter, Degradation of cell wall polysaccharides by combinations of carbohydrates enzymes and their effect on nutrient utilization and broiler chicken performance, *Poultry Sci.*, 84 (2005) 37–47.

2603. M. Ishida, R. Haga, and M. Katsurayama, Hydrolysis with amylase-starch complex or the like, Jpn. Kokai Tokkyo Koho 61,219,393 (1986); *Chem. Abstr.*, 107 (1986) 152890f.
2604. L. M. Prescott, J. P. Harley, and D. A. Klein, *Microbiology*, 6th edn. (2005) McGraw Hill, New York.
2605. J. T. Kim and W. S. Lee, High functional fermented starch products oreoared by order-fermentation of microorganisms and preparation thereof, S. Korean Patent 2002024197 (2002); *Chem. Abstr.*, 142 (2004) 197020.
2606. L. Stryer, *Biochemistry*, (1981) W. H. Freeman and Company, San Francisco.
2607. J. Błażewicz, Estimation f the usability of triticale malts in brewing industry, *Pol. J. Food Nutr. Sci.*, 2(1), (1993) 39–45.
2608. J. Błażewicz, B. Foszczyńska, and J. Kiersnowski, Possibility of replacing in brewery barley malts with triticale malts (in Polish), *Zesz. Nauk. AR Wroclaw, Ser. Technol. Zywn.*, 8 (1995) 39–46.
2609. Z. Włodarczyk and J. Jarniewicz, Attempts to saccharify potato starch mashes with a crude mold preparation obtained by submerged cultivation (in Polish), *Przem. Ferm. Rolny*, 20(10), (1976) 9–13.
2610. L. Thomas, T. Senn, and H. J. Pieper, Bioethanol aus Triticale, *GIT Fachzschr. Lab.*, 35 (1991) 1087–1088, 1092.
2611. D. K. Roy, Role of limit dextrinase on alcohol yield from starch, *Ann. Biochem. Exptl. Med. (India)*, 12 (1952) 115–118.
2612. A. W. MacGregor, S. L. Bazin, and S. W. Schroeder, Effect of starch hydrolysis products on the determination of limit dextrinase and limit dextrinase inhibitors in barley and malt, *J. Cereal Sci.*, 35 (2002) 17–28.
2613. G. Thevenot, Saccharification of raw cereal mashes (in brewing), *West. Brew.*, 48 (1917) 124–127.
2614. H. J. Pieper and T. Senn, New pressureless process for enzymic conversion of rye starch by dispersion of the whole grain below gelatinization temperature, *Proc. 3rd Eur. Congr. Biotechnol.*, 2 (1984) 233–238. *Chem. Abstr.*, 105 (1984) 22982k.
2615. R. Muller, A mathematical model of the formation of fermentable sugars from starch hydrolysis during high-temperature mashing, *Enzyme Microb. Technol.*, 27 (2000) 337–344.
2616. C. Brandam, X. M. Meyer, J. Proth, P. Strehaiano, and H. Pingaud, A new reaction scheme for the starch hydrolysis and temperature policy influence during mashing, *Food Sci. Biotechnol.*, 11 (2002) 40–47.
2617. S. Bhargava and M. Johal, Enzymic liquefaction process, EU Patent WO 2006017294 (2006); *Chem. Abstr.*, 144 (2006) 211233.
2618. G. E. Gill, B. J. Savage, and S. Bhargava, Enzymic starch liquefaction and saccharification for improved starch fermentation, EU Patent WO 2007035730 (2007); *Chem. Abstr.*, 146 (2007) 357402.
2619. B. M. Nakhmanovich, V. L. Yarovenko, A. P. Levchik, O. P. Maslova, and L. A. Levchik, Saccharification of starch-containing raw material fore the production of alcohol, USSR Patent, 299,539 (1971); *Chem. Abstr.*, 75 (1971) 87094d.
2620. B. M. Nakhmanovich, V. L. Yarovenko, A. P. Levchik, L. A. Orlova, A.Ya Pankratov, and G. G. Azarova, Saccharification of starch media by cultures of mold fungi under fermentation conditions (in Russian), *Izv. Vyssh. Uchebn. Zaved., Pishch. Prom.* (5), (1974) 74–77.
2621. R. H. Hopkins and D. Kulka, Enzymes involved in promoting dextrin fermentation, *Proc. Eur. Brew. Conv. Copenhagen* (1957) 182–193. *Chem. Abstr.*, 52 (1959) 19013b.
2622. B. A. Ustinnikov, S. V. Pykhova, and A. N. Lazareva, Fermentative hydrolysis of starch in the spirit industry (in Russian), *Tr. Vses, Nauchn.-Issled. Inst. Ferm. Spirt. Prom.* (16), (1965) 139–149.
2623. J. Zhao, H. Luo, and S. Wu, Application of saccharyfying enzyme in the production of fen-flavor Xiaoqu liquor, *Shipin Keji* (3), (2006) 94–95. *Chem. Abstr.*, 147 (2007) 165120.
2624. T. Ishido and M. Yamada, Alcoholic food and drink and its manufacture, Jpn. Kokai Tokkyo Koho, 2007282561 (2007); *Chem. Abstr.*, 147 (2008) 468368.

2625. E. Otto and J. Escovar-Kousen, Ethanol production by simultaneous saccharification and fermentation (SSF), EU Patent WO 2004046333 (2004); *Chem. Abstr.*, 141 (2004) 5891.
2626. T. Sugano and I. Ito, Jpn. Saccharides for the manufacture of alcohol, Japanese Patent 71 24,057 (1971); *Chem. Abstr.*, 77 (1972) 73703f.
2627. C. Sjoeholm and G. Antranikian, Fervidobacterium amylase and pullulanase, fermentation and sweetener and ethanol production from starch, EU Patent WO 9523850 (1995); *Chem. Abstr.*, 123 (1995) 283752.
2628. C. Sjoeholm and G. Antranikian, Staphylothermus amylase, EU Patent WO 9523851 (1995); *Chem. Abstr.*, 123 (1995) 283753.
2629. N. G. Kursheva and A. F. Fedorov, Saccharification of starch and acceleration of fermentation (in Russian), *Ferm. Spirt. Prom.*, 36(7), (1970) 14–17.
2630. Kyowa Hakko Kogy Co., Ltd., Saccharification of starches, Jpn. Kokai Tokkyo Koho 59,179,093 (1984); *Chem. Abstr.*, 102 (1984) 44368e.
2631. N. W. Lützen, Enzyme technology in the production of alcohol – recent process development, *Proc. 6th Int. Ferm. Symp. Adv. Biotechnol. 1980*, 2 (1981) 5–21. *Chem. Abstr.*, 96 (1981) 120829f.
2632. S. Fang and M. Zhu, Study on alcoholic fermentation of raw starch without cooking, *Shipin Yu Fajiao Gongye* (2), (1988) 13–19. *Chem. Abstr.*, 109 (1989) 127311b.
2633. O. J. Lantero, M. Li, and J. K. Shetty, Process for conversion of granular starch to ethanol, US Patent 2007281344 (2007); *Chem. Abstr.*, 148 (2007) 9526.
2634. S. Pedersen and H. S. Olsen, Production of ethanol from granular starch, EU Patent, WO 200606582 (2006); *Chem. Abstr.*, 145 (2006) 61528.
2635. H. W. Doelle, Conversion of starch hydrolysates to ethanol using *Zymomonas mobilis*, EU Patent WO 87 02,706 (1987); *Chem. Abstr.*, 107 (1987) 152929a.
2636. J. H. Lee, R. J. Pagan, and P. L. Rogers, Continuous simultaneous saccharification and fermentation of starch using *Zymomonas mobilis*, *Biotechnol. Bioeng.*, 25 (1983) 659–663.
2637. Soc. Francaise des Disatilleries de l'Indo-Chine, Conversion of starch and starchy materials into sugar and alcohols, French Patent 459,634 (1912); *Chem. Abstr.*, 8 (1914) 22056.
2638. Soc. Francaise des Disatilleries de l'Indo-Chine, Saccharyfying starch for the manufacture of alcohol and wine, French Patent 459,815 (1912); Chem. Abstr., 8 (1914) 22057.
2639. V. L. Yarovenko, B. M. Nakhmanovich, V. V. Yarovenko, and L. A. Orlova, Intesification of ethanol fermentation of starch media (in Russian), *Prikl. Biokhim. Mikrobiol.*, 12 (1976) 509–514.
2640. V. L. Yarovenko, B. A. Ustinnikov, I. S. Salmanova, S. V. Pykhova, and A. N. Lazareva, Continuous production of alcohol from a starch raw material, USSR Patent 276,888 (1973); *Chem. Abstr.*, 81 (1974) 2413h.
2641. V. A. Marinchenko, Treatment of starch raw material before alcohol fermentation, USSR Patent, 403,721 (1973); *Chem. Abstr.*, 81 (1975) 62120v.
2642. F. W. Bai, W. A. Anderson, and M. Moo-Young, Ethanol fermentation technologies from sugar and starch feedstock, *Biotechnol. Adv.*, 26 (2008) 89–105.
2643. G. B. Calleja, C. V. Lusena, I. A. Veliky, and F. Moranelli, Two-step process for direct conversion of starch to ethanol with yeast, Canadian Patent 1238592 (1988); *Chem. Abstr.,* 109 (1988) 228777.
2644. M. Banerjee, S. Debnath, and S. L. Majumdar, Production of alcohol from starch by direct fermentation, *Biotechnol. Bioeng.*, 32 (1988) 831–834.
2645. M. Taniguchi, K. Hoshino, Y. Netsu, and M. Fujii, Repeated simultaneous sccharification and fermentation of raw starch by a combination of a reversibly soluble-insoluble amylase and yeast cells, *J. Chem. Eng. Jpn.*, 22 (1989) 313–314.
2646. J. Hariantono, A. Yokota, S. Takao, and F. Tomita, Ethanol production from raw starch by simultaneous fermentation using *Schizosaccharomyces pombe* and raw starch saccharyfying enzyme from *Corticium rolfsii*, *J. Ferm. Bioeng.*, 71 (1991) 367–369.

2647. Y. W. Ryu, J. H. Seong, and S. H. Ko, Direct alcohol fermentation of starch by *Schwanniomyces castelii*, *Proc. Asia-Pac. Biochem. Eng. Conf.*, 2001 (1992) 743–746. *Chem. Abstr.*, 119 (1993) 115424.
2648. A. Strasser, F. B. Martens, J. Dohmen, and C. P. Hollenberg, Saccharification of starch for brewing using *Saccharomyces* expressing amylase genes from *Schwanniomyces,* US Patent 5100794 (1992) ; *Chem. Abstr.*, 117 (1992) 232774.
2649. H. Fukuda, A. Kondo, A. Tanaka, M. Ueda, and E. Satoh, Yeast expressing rice glucoamylase and yeast α-amylase for ethanol fermentation, EU Patent WO 2003016525 (2000); *Chem. Abstr.*, 138 (2003) 203804.
2650. T. S. Khaw, Y. Katakura, K. Ninomiya, C. Moukamnerd, A. Kondo, M. Ueda, and S. Shioya, Enhacement of ethanol production by promoting surface contact between starch granules and arming yeast in direct ethanol fermentation, *J. Biosci. Bioeng.*, 103 (2007) 95–97.
2651. H. Tanaka, H. Kurosawa, and H. Murakami, Ethanol production from starch by a coimmobilized culture system of *Aspergillus awamori* and *Zymomonas mobilis*, *Biotechnol. Bioeng.*, 28 (1986) 1761–1768.
2652. H. Kurosawa, N. Nomura, and H. Tanaka, Ethanol production from starch by a coimmobilized mixed culture system of *Aspergillus awamori* and *Saccharomyces cerevisiae*, *Biotechnol. Bioeng.*, 33 (1989) 716–723.
2653. N. Fujii, T. Oki, A. Sakurai, S. Suye, and M. Sakakibara, Ethanol production from starch by immobilized *Aspergillus awamori* and *Saccharomyces pastorianus* using cellulose carriers, *J. Ind. Microbiol. Biotechnol.*, 27 (2001) 52–57.
2654. R. L. C. Giordano, P. C. Hirano, L. R. B. Gonçalves, and N. W. Schmidell, Study of biocatalyst to produce ethanol from starch, *Appl. Biochem. Biotechnol.*, 84–86 (2000) 643–654.
2655. P. Chen, C. Jiang, C. Li, and X. Chu, Ethanol fermentation from starch by coimmobilized yeast and *Aspergillus niger* cells, *Gongye Weishengwu*, 20(8), (1991) 8–13. *Chem. Abstr.*, 114 (1991) 227446.
2656. J. Nowak and H. Roszak, Co-immobilization of *Aspergillus niger* and *Zymomonas mobilis* for ethanol production from starch, *Pol. J. Food Nutr. Sci.*, 6(3), (1997) 65–70.
2657. S. C. Choi, S. W. Lee, S. K. Park, C. K. Sung, B. S. Shon, and N. K. Sung, Ethanol production from raw starch by co-immobilized mixed *Rhizopus japonicus* and *Zymomonas mobilis*, *Hanguk Shikpum Yongyang Kwahak Hoechi*, 25 (1996) 708–714. *Chem. Abstr.*, 126 (1997) 58920.
2658. M. Y. Sun, N. P. Nghiem, B. H. Davison, O. F. Webb, and P. R. Bienkowski, Production of ethanol from starch by co-immobilized *Zymomonas mobilis* – Glucoamylase in a fluidized bed reactor, *Appl. Biochem. Biotechnol.*, 70-72 (1998) 429–439.
2659. C. H. Kim and S. K. Rhee, Process development for simultaneous starch saccharification and ethanol fermentation by *Zymomonas mobilis*, *Process Biochem.*, 28(5), (1993) 331–339.
2660. N. Chithra and A. Baradarajan, Studies on coimmobilization of amyloglucosidase and *Saccharomyces cerevisiae* for direct conversion of starch to ethanol, *Process Biochem.*, 24(6), (1989) 208–211.
2661. N. Chithra and A. Baradarajan, Direct conversion of starch hydrolyzate to ethanol using a coimmobilizate of amyloglucosidase and *Saccharomyces cerevisiae* in batch stirred tank reactor, *Bioprocess Eng.*, 7 (1992) 265–267.
2662. W. He, P. Chen, and B. Chen, Fermented beverage of starch crops and its preparation method, Chinese Patent CN 101254009 (2008); *Chem. Abstr.*, 149 (2008) 401171.
2663. L. V. A. Reddy, O. V. S. Reddy, and S. C. Basappa, Potentiality of amylolytic yeasts for direct fermentation of starchy substrates to ethanol, *Indian J. Microbiol.*, 45 (2005) 1–15.
2664. E. T. Oner, Optimization of ethanol production from starch by an amylolytic nuclear petite *Saccharomyces cerevisae* strain, *Yeast*, 23 (2006) 849–856.
2665. J. Sun and C. Sun, Manufacture of white spirits with microbial enzyme preparation, Chinese Patent 1062924 (1992); *Chem. Abstr.,* 118 (1993) 37922.

2666. C. H. Horn, J. C. Du Preez, and S. G. Kilian, Fermentation of grain sorghum starch by cocultivation of *Schwanniomyces occidentalis* and *Saccharomyces cerevsiae*, *Biores. Technol.*, 42 (1992) 27–31.
2667. K. D. Kirby and C. J. Mardon, Solid phase fermentation for intermediate scale ethanol production, *Proc. 4th Int. Symp. Alcohol Fuels Technol., 1980*, 1 (1981) 13–19. *Chem. Abstr.*, 95 (1981) 172548n.
2668. G. Saucedo-Castaneda, B. K. Lonsane, J. M. Navarro, S. Roussos, and M. Raimbaut, Potential of using a single fermentor for biomass build-up starch hydrolysis and ethanol production. Solid state fermentation system involving *Schwanniomyces castellii*, *Appl. Biochem. Biotechnol.*, 36 (1992) 47–61.
2669. H. S. Olsen, Production of ethanol from starch, EU Patent WO 2008135547 (2008); *Chem. Abstr.*, 149 (2008) 531527.
2670. K. S. Wenger, E. Allain, S. M. Lewis, J. M. Finck, and D. L. Roth, Improved distillation by addition of amylase and protease to mashes, EU Patent WO 2005099476 (2005); *Chem. Abstr.*, 143 (2005) 420966.
2671. N. Dunn-Coleman, O. J. Lantero, S. E. Lantz, M. J. Pepsin, and J. K. Shetty, Heterologous expression of an *Aspergillus kawachi* acid-stable α-amylase and application in granular starch hydrolysis, US Patent 2005266543 (2005); *Chem. Abstr.*, 144 (2005) 1309.
2672. A. Molhant, Saccharyfying and fermenting amyloceus material, US Patent 1,134,281 (1914).
2673. K. Myrbäck, B. Örtenblad, and K. Ahlborg, Vergärung von Grenzdextrinen Stärke und Disacchariden, *Enzymologia*, 3 (1937) 210–219.
2674. H. Leopold and B. Buka, Amyloverfahrung, *Zentr. Bakt. Parasitenk. II Abt.*, 97 (1938) 353–386.
2675. A. I. Zhyryanov, Determination of starch fermentation by *Vibrio comma* (in Russian), *Biol. Kultiv. Mikroorg.*, (1969) 170–173.
2676. T. Imamura, M. Kawamoto, and Y. Takaoka, Characteristic of main mash injected by a killer yeast in sake brewing and the nature of its killer fraction (in Japanese), *Hakko Kagaku Zasshi*, 52 (1974) 293–299.
2677. S. Ueda, Ethanol fermentation of starch materials without cooking (in Japanese), *Denpun Kagaku*, 29 (1982) 123–130.
2678. S. W. Lee, M. Yajima, and M. Tanaka, Use of food additives to prevent contamination during fermentation using a co-immobilized mixed culture system, *J. Ferm. Bioeng.*, 75 (1993) 389–391.
2679. S. W. Lee, Y. U. Cho, H. C. Kim, S. K. Park, and N. K. Sung, Effect of Neupectin-L on ethanol production from raw starch using a co-immobilized *Aspergillus awamori* and *Zymomonas mobilis*, *Hanguk Nonghwa Hakhoechi*, 40 (1997) 89–94. *Chem. Abstr.*, 127 (1999) 80206.
2680. A. Purr and M. Reichel, Saccharification of starch, German Pat. 2,227,976 (1974); *Chem. Abstr.*, 81 (1975) 51505y.
2681. D. Wuester-Botz, A. Aivasidis, and C. Wandrey, Continuous ethanol production by *Zymomonas mobilis* in a fluidized bed reactor. II. Process development for the fermentation of hydrolyzed B-starch without sterilization, *Appl. Microbiol. Biotechnol.*, 39 (1993) 685–690.
2682. W. I. Cho, Process for preparing high quality wine by adding amylase which decomposes starch material s to be easily fermented to improve availability of raw materials of wine and fermentation efficiency, S. Korean Patent 2005017954 (2005); *Chem. Abstr.*, 145 (2006) 270080.
2683. B. S. Enevoldsen, Degradation of starch by amylase in beer brewing, *Dempun Kagaku*, 25 (1978) 89–99.
2684. J. H. Thorup, J. G. M. Heathcote, and M. B. M. Fenton, Liquefaction of starch-based biomass, EU Patent WO 2008135775 (2008); *Chem. Abstr.*, 149 (2008) 531547.
2685. M. Larson and B. Mattiason, Continuous conversion of starch to ethanol using a combination of an aqueous two-phase system and an ultrafiltration unit, *Ann. N. Y. Acad. Sci.*, 434 (1984) 144–147.

2686. A. Ramirez, M.J. Boudarel, D. Ghozlan, N. Conil, and H. Blachere, Process for continuous ethanol manufacture by simultaneous hydrolysis and fermentation of starch-containing substances, French Patent 2609046 (1988); *Chem. Abstr.*, 110 (1988) 73886.
2687. K. Hoshino, M. Taniguchi, H. Marumoto, and M. Fujii, Repeated batch conversion of raw starch to ethanol using amylase immobilized on a reversible soluble-autoprecipitating carrier and flocculating yeast cells, *Agr. Biol. Chem.*, 53 (1989) 1961–1967.
2688. K. Hoshino, M. Taniguchi, H. Marumoto, and M. Fujii, Continuous alcohol production from raw starch using a reversibly soluble-autoprecipitating amylase and flocculation yeast cell, *J. Ferm. Bioeng.*, 69 (1990) 228–233.
2689. S. Fukushima and K. Yamade, Continuous alcohol fermentation of starch materials with a novel immobilized cell/enzyme bioreactor, *Ann. N. Y. Acad. Sci.*, 434 (1984) 148–151.
2690. K. Yamade and S. Fukushima, Continuous alcohol production from starchy materials with novel immobilized cell/enzyme bioreactor, *J. Ferm. Bioeng.*, 67 (1989) 97–101.
2691. B. Guan, Method and apparatus for continuous enzymic starch saccharification and culture of alcohol-fermenting yeast, Chinese Patent CN 1069066 (1993); *Chem. Abstr.*, 119 (1983) 70539.
2692. P. Feng, Q. Xu, Y. Zhao, and Y. Huang, Continuous fermentation to produce ethanol using *Schizosaccharomycs pombe* yeast flocs, *Shengwu Gongcheng Xuebao*, 5 (1989) 70–76. *Chem. Abstr.*, 111 (1990) 22146.
2693. T. Hisayasu, S. Asano, M. Yamadaki and K. Kida, Preparation of flocculable *Saccharomyces* hybrid by protoplast fusion and manufacture of ethanol with it, Jpn. Kokai Tokkyo Koho 63 44,880 (1988); *Chem. Abstr.*, 109 (1988) 53237h.
2694. V. L. Yarovenko, B. A. Ustinnikov, S. V. Pykhova, A. N. Lazareva, L. S. Salmanova, and A. P. Levchik, Continuous alcohol production from starch raw material, USSR Patent 316,719 (1971); *Chem. Abstr.*, 76 (1971) 32885c.
2695. V. L. Yarovenko, B. M. Nakhmanovich, L. A. Levchik, and A. P. Levchik, Alcohol from a starch raw material, USSR Patent 384,863 (1973); *Chem. Abstr.*, 80 (1973) 2327n.
2696. L. I. Navrodskaya, B. A. Ustinnikov, E. A. Maksimova, and V. M. Dobriyan, Complex saccharification of starchy raw material by bacterial α-amylase, glucoamylase and cytase (in Russian), *Ferment. Spirt. Prom.* (3), (1977) 26–29.
2697. T. S. Li, K. Li, and Y. Q. Liu, Preliminary study on alcoholic continuous fermentation by co-immobilization of saccharifying enzymes and yeasts, *Zhengzhou Gongye Daxue Xuebao*, 22(3), (2000) 8. *Chem. Abstr.*, 136 (2001) 354232.
2698. Y. Shimizu, Preparation and application of immobilized yeast for ethanol production, Jpn. Kokai Tokkyo Koho 59,220,186 (1984); *Chem. Abstr.*, 102 (1984) 183752p.
2699. S. Debnath, M. Bannerjee, and S. K. Majumdar, Production of alcohol from starch by immobilized cells of *Saccharomyces diastaticus* in batch and continuous process, *Process Biochem.*, 25(2), (1990) 43–46.
2700. P. Gunsekaran, T. Karmakaran, and M. Kasthuriban, Fermentation pattern of *Zymomonas mobilis* strains in different substrates – Comparative studies, *J. Biosci.*, 10 (1986) 181–186.
2701. D. Gombin, M. Toth, G. Klamar, and B. Szajani, Ethanol production from thinned starch using solid-phase biocatalysts (in Hungarian), *Magyar Kemiai Folyoirat*, 100 (1994) 509–512.
2702. J. M. Segraves, W. G. Allen, J. Harris, and K. L. Pratt, Valley "Ultra-Thin" – A novel alpha amylase for true pH 4.5 liquefaction, *Int. Sugar J.*, 107 (2005) 616–618, 620-621.
2703. T. Senn, Examination in starch degradation using technical enzyme preparations in bioethanol production, *Proc. DECHEMA Biotechnol. Conf., 1992*, 1993(5A), (1993) 155–160.
2704. Y. Koba, B. Feroza, F. Yusaku, and S. Ueda, Preparation of Koji from corn hulls for alcoholic fermentation without cooking, *J. Ferment. Technol.*, 64 (1986) 175–178.
2705. W. L. Owen, Production of industrial alcohol from grain by amylo process, *Ind. Eng. Chem.*, 25 (1933) 87–89.

2706. R. V. Feniksova, Application of molds during the manufacturing of alcohol (in Russian), *Nauchn. Cht. Sbornik* (1952) 99–113. *Ref. Zhurn.,* (1954) No. 32705.
2707. J. S. Horn, The amylo saccharification process, *Chem. Eng.,* 24 (1918) 204.
2708. F. Taufel, N. A. Ruttloff, V. L. Yarovenko, B. A. Ustinnikov, P. Lietz, and P. Steffen, Improvement of crude alcohol production using glucoamylase prepared from *Endomycopsis bispora* (in Bulgarian), *Khranit. Prom.* (3), (1979) 32–34.
2709. F. Taufel, N. A. Ruttloff, V. L. Yarovenko, P. Lietz, B. A. Ustinnikov, and P. Steffen, Improved production of raw alcohol using glucoamylase from *Endomycopsis bispora* (in Polish), *Przem. Spoz.,* 33 (1979) 134–136.
2710. S. W. Lee, T. Ebata, Y. C. Liu, and H. Tanaka, Co-immobilization of thee strains of microorganisms and its application in ethanol production from raw starch under in sterile conditions, *J. Ferm. Bioeng.,* 75 (1993) 36–42.
2711. Y. Mori and T. Inaba, Ethanol production from starch in a pervaporation membrane bioreactor using *Clostridium thermohydrosulfuricum, Biotechnol. Bioeng.,* 36 (1990) 849–853.
2712. R. Rasap and A. Glemza, Purification of β-amylase from *Bacillus polymyxa* no. 3 on corn starch (in Russian), *Prikl. Biokhim. Mikrobiol.,* 17 (1981) 702–707.
2713. D. Xiao, H. Zhao, S. Zhao, K. Xu, Y. Chen, L. Du, and B. Guo, Method for increasing alcohol content by adding *Saccharomyces cerevisiae* powder in high-concentration mash fermentation of amyloid material, Chinese Patent 1661024 (2005); *Chem. Abstr.,* 144 (2009) 349024.
2714. D. Xiao, Y. Chen, L. Du, and N. Shen, Method for increasing ethanol resistance of *Saccharomyces cerevisiae,* Chinese Patent 1944657 (2007); *Chem. Abstr.,* 146 (2009) 461054.
2715. A. Kondo, H. Shigechi, M. Abe, K. Uyama, T. Matsumoto, S. Takahashi, M. Ueda, A. Tanaka, M. Kishimoto, and H. Fukuda, High-level ethanol production from starch by a flocculent *Saccharomyces cerevisiae* strain displaying cell-surface glucoamylase, *Appl. Microbiol. Biotechnol.,* 58 (2002) 291–296.
2716. T. Sakai, Direct ethanol production from starch, Jpn. Kokai Tokkyo Koho, JP 59 45,874 (1984); *Chem. Abstr.,* 101 (1984) 22038z.
2717. W. C. Muller and F. D. Miller, Hydrolysis of starch and continuous fermentation of the sugars obtained from it to provide ethanol, US Patent 4,243,750 (1981); *Chem. Abstr.,* 94 (1981) 207163s.
2718. N. Hammond and J. Prevost, Novel stillage treatment process for ethanol fermentation, US Patent 2006194296 (2006); *Chem. Abstr.,* 145 (2007) 247666.
2719. C. S. Cooper, J. W. Spouge, G. G. Stewart, and J. H. Bryce, The effect of distillery backset on hydrolytic enzymes in mashing and fermentation, In: J. H. Bryce and G. G. Stewart, (Eds.) *Distilled Spirits,* Nottingham University Press, Nottingham, 2004, pp. 79–88.
2720. S. Ueda and Y. Koba, Novel alcoholic fermentation of raw starch without cooking by using black *Aspergillus* amylase, *Proc. 6th Int. Symp. Adv. Biotechnol., 1980* (2), (1981) 169–174. *Chem. Abstr.,* 96 (1981) 120830z.
2721. V. L. Yarovenko, S. V. Pykhova, B. A. Ustinnikov, A. N. Lazareva, and D. M. Makeev, Enzymic hydrolysis of starch in continuous alcoholic fermentation (in Russian), *Ferm. Spirit Prom.,* 31(1), (1965) 5–10.
2722. G. B. Borglum, Starch hydrolysis for ethanol production, *Prepr. Pap. Am. Chem. Soc., Div. Fuel Chem.,* 25 (1980) 264–269.
2723. V. L. Rodzevich, N. S. Mazur, and B. A. Ustinnikov, Reserve of alcohol production efficiency (in Russian), *Ferm. Spirt. Prom.,* 3 (1982) 22–24.
2724. G. Konieczny-Janda, Progress in enzymatic saccharification of wheat starch, *Starch/Stärke,* 43 (1991) 308–315.
2725. S. Kunisaki and N. Matsumoto, Application of a raw starch-digestive enzyme preparation for alcoholic fermentation with noncooking (in Japanese), *Denpun Kagaku,* 32 (1985) 152–161.

2726. Japan Tax Administration Agency, Manufacture of alcohol from potatoes, Jpn. Kokai Tokkyo Koho, 82,12,990 (1982); *Chem. Abstr.*, 96 (1982) 216068h.
2727. Novo Industri A/S, Manufacture of ethanol by fermentation, Jpn. Kokai Tokkyo Koho 81,121,487 (1981); *Chem. Abstr.*, 96 (1982) 216069j.
2728. H. Mueller and F. Mueller, Alcohol from grain, French Patent 2,442,887 (1980); *Chem. Abstr.*, 94 (1980) 63796a.
2729. H. Bergmann, E. Arendt, G. Welz, V. Schierz, and M. Prause, Ethanol by fermentation, East German Patent, 148,063 (1981); *Chem. Abstr.*, 95 (1981) 167164e.
2730. T. I. Prosvetova, E. A. Dvadtsatova, V. L. Yarovenko, N. Ya. Vasileva, B. M. Nakhmanovich, B. A. Ustinnikov, V. F. Shamrin, S. I. Karaichev, S. V. Pykhova et al., Saccharification of starch raw material in alcohol production, USSR Patent 840,099 (1981); *Chem. Abstr.,* 95 (1981) 167163d.
2731. W. C. Mueller and F. D. Miller, Fermentable sugar from the hydrolysis of starch derived from dry milled cereal grains, US Patent 4,448,881 (1984); *Chem. Abstr.*, 101 (1984) 37216e.
2732. S. Bhargava, H. Frisner, H. Bisgard-Ftantzen, and J. W. Tams, Production of ethanol from enzymatically hydrolyzed starch, EU Patent WO 2005113785 (2005); *Chem. Abstr.*, 144 (2005) 5525.
2733. M. P. T. Smith and G. A. Crabb, Ethanol fermentation using enzyme hydrolyzed starch, EU Patent WO 2005118827 (2005); *Chem. Abstr.,*144 (2005) 21927.
2734. E. D. Faradzheva, N. I. Gunkina, S. I. Savin, and V. A. Malkin, Method for production of alcohol from starchy raw material, Russian Patent 2113486 (1998); *Chem. Abstr.*, 132 (2000) 346716.
2735. B. A. Ustinnikov. P. S. Efremov, V. A. Polyakov, S. I. Gromov, V. N. Polikashin, Yu. V. Polikashin, and V. N. Sidorin, Method of production of ethyl alcohol from starch-containing raw, Russian Patent 2156806 (2000); *Chem. Abstr.*, 136 (2002) 4798.
2736. M. Aurangzeb, M. A. Quadeer, and J. Iqbal, Ethanol fermentation of raw starch, *Pak. J. Sci Ind. Res.*, 35(4), (1992) 162–164.
2737. A. K. Biswas and D. Das, An improved process for the preparation of ethanol from starchy material, Indian Patent 188562 (2002); *Chem. Abstr.*, 148 (2008) 333854.
2738. M. Qiao, K. Zhang, Z. Yu, and W. Zhou, Method for manufacturing fermentation ethanol from starch with increased yield, Chinese Patent 1814763 (2006); *Chem. Abstr.*, 145 (2006) 229437.
2739. M. H. Malfait, G. Moulin, and P. Galzy, Ethanol inhibition of growth, fermentation and starch hydrolysis in *Schwanniomyce castelli*, *J. Ferm. Technol.*, 64 (1986) 279–284.
2740. R. Buethner, R. Bod, and D. Birnbaum, Alcoholic fermentation of starch by *Arxula adeninivorans*, *Zentrbl. Mikrobiol.*, 147 (1992) 225–230.
2741. J. G. Zeikus and H. H. Hyun, Co-culture production of alcohol and enzymes, EU Patent WO 80 01, 833 (1984); *Chem. Abstr.*, 105 (1984) 5204r.
2742. Y. H. Kim and J. H. Seu, Culture conditions for glucoamylase production and ethanol productivity of heterologous transformant of *Sccharomyces cerevisae* by glucoamylase gene of *Saccharomyces diastaticus*, *Sanop Misaengmul Hakhoechi*, 16 (1988) 494–498. *Chem. Abstr.*, 110 (1988) 191180.
2743. G. Verma, P. Nigam, D. Singh, and K. Chaudhary, Bioconversion of starch to ethanol in a single-step process by coculture of amylolytic yeasts and *Saccharomyces cerevisae* 21, *Biores. Technol.*, 72 (2000) 261–266.
2744. G. Verma and K. Chaudhary, Direct fermentation of starch to ethanol by mixed culture of *Saccharomyces cerevisae* and amylolytic yeasts, *Asian J. Microbiol. Biotechnol. Environ. Sci.*, 5 (2003) 593–597.
2745. G. Verma and K. Chaudhary, Direct fermentation of starch to ethanol by mixed culture of *Saccharomyces cerevisae* and amylolytic yeasts, *Asian J. Microbiol. Biotechnol. Environ. Sci.*, 5 (2003) 523–527.
2746. L. O. Ingram, D. S. Beall, G. F. Burchhardt, W. V. Guimaraes, K. Ohta, B. E. Wood, and K. T. Shanmugam, Ethanol manufacture from oligosaccharides with microbial hosts expressing

foreign genes for alcohol dehydrogenase, pyruvate decarboxylase and polysaccharides, US Patent 5424202 (1995); *Chem. Abstr.*, 123 (1995) 141888.
2747. H. H. Hyun and H. H. Hyun, Manufacture of thermostable amylolytic enzyme and/or ethanol with catabolite repression-resistant *Clostridium* mutants, US Patent 4737459 (1988); *Chem. Abstr.*, 109 (1988) 188827.
2748. V. L. Dos Santos, W. V. Guimaraes, E. G. De Barros, and E. F. Araujo, Fermentation of maltose and starch by *Klebsiella oxytoca* P2, *Biotechnol. Lett.*, 20 (1998) 1179–1182.
2749. V. L. Dos Santos, E. F. Araujo, E. G. De Barros, and W. V. Guimaraes, Fermentation of starch by *Klebsiella oxytoca* P2 containing plasmids with α-amylase and pullulanase genes, *Biotechnol. Bioeng.*, 65 (1998) 673–676.
2750. M. G. G. Ribeiro do Santos, H. Aboutboul, J. B. Faria, A. C. G. Schenberg, and W. Schmidell, Genetic improvement of *Saccharomyces* for ethanol production from starch, *Yeast*, 5(Spec. Issue), (1989) 11–15.
2751. Y. Nakamura, T. Sawada, and K. Yamaguchi, Breeding and cultivation of glucoamylase-producing yeast with inactivation of MAT locus, *J. Chem. Eng. Jpn.*, 32 (1999) 424–430.
2752. B. Yamada, S. Mitsuiki, K. Furukawa, and M. Saka, Alcohol fermentation with recombinant *Saccharomyces cerevisiae* harboring glucoamylase gene from starch (in Japanese), *Kyushu Sangyo Daigaku Kogakubu Kenkyu Hokoku*, 38 (2001) 173–176.
2753. R. Fitzsimon, L. M. Pepe de Moraes, F. Sineriz, and M. E. Lucca, Evaluation of starch fermentation products by amylolytic recombinant *Saccharomyces cerevisae* strains, *Braz. Arch. Biol. Technol.*, 49 (Spec. Issue), (2006) 153–158.
2754. A. M. Knox, J. C. du Preez, and S. G. Kilian, Starch fermentation characteristics of *Saccharomyces cerevisiae* strains transformed with amylase genes from *Lipomyces konenkoae* and *Saccharomycopsis fibuligera*, *Enzyme Microbiol. Technol.*, 34 (2004) 453–460.
2755. T. Borchert, S. Danielsen, and E. Allain, Methods for engineering fusion proteins of glucoamylase or α-amylase starch binding and catalytic domains of *Aspergillus niger, Talaromyces emersonii* and *Althera rolfsii* for production of ethanol in *Saccharomyces cerevisaie*, EU Patent WO 2005045018 (2005); *Chem. Abstr.*, 142 (2005) 480912.
2756. E. T. Oner, S. G. Oliver, and B. Kirdar, Production of ethanol from starch by respiration-deficient recombinant *Saccharomyces cerevisiae*, *Appl. Environ. Microbiol.*, 71 (2005) 6443–6445.
2757. M. M. Altintas, B. Kirdar, Z. I. Oensan, and K. O. Uelgen, Cybernetic modeling of growth and ethanol production in recombinant *Saccharomyces cerevisiae* strain secreting bifunctional protein, *Process Biochem.*, 37 (2002) 1439–1445.
2758. I. S. Pretorius, R. R. C. Otero, and P. Van Rensburg, Recombinant *Saccharomyces cerevisiae* strains expressing α-amylase and glucoamylase genes from *Lipomyces konenkoae* and *Saccharomycopsis fibuligera* and their use for starch degradation to ethanol, S. African Patent 20044006714 (2005); *Chem. Abstr.*, 147 (2007) 321415.
2759. R. E. Hebeda and C. R. Styrlund, Starch hydrolysis products as brewing adjuncts, *Cereal Foods World*, 31 (1986) 685–687.
2760. K. M. Shahani and B. A. Friend, Fuel alcohol production from whey and grain mixtures, *Prepr. Pap. Am. Chem. Soc., Div. Fuel Chem.*, 25 (1980) 281–284.
2761. C. Compagno, D. Porro, C. Smeraldi, and B. M. Ranzi, Fermentation of whey and starch by transformed *Saccharomyces cerevisae* cells, *Appl. Microbiol. Biotechnol.*, 43 (1995) 822–825.
2762. Y. Chen, Z. Tang, W. Yu, and X. Chen, Fermentation of ethanol from sugarcane juice and starch material, Chinese Patent 101070549 (2007); *Chem. Abstr.*, 148 (2008) 9519.
2763. B. L. Maiorella, C. R. Wilke, and H. W. Blanch, Alcohol production and recovery, *Adv. Biochem. Eng.*, 20 (1981) 43–49.
2764. L. R. Lindeman and C. Rocchiccioli, Ethanol in Brasil. Brief summary of the state of industry, *Biotechnol. Bioeng.*, 21 (1979) 1107–1119.

2765. C. R. Wilke, R. D. Yang, A. F. Scamanna, and R. P. Freitas, Raw materials evaluation and process development studies for conversion of biomass to sugars and ethanol, *Biotechnol. Bioeng.*, 23 (1981) 163–183.
2766. M. P. Cereda and O. F. Vilpoux, Cassava fermentation in Latin America: Fermented starch an overview, *Microbiol. Biotechnol. Horticult.*, 1 (2006) 397–419.
2767. M. Matsumoto, O. Fujkushi, M. Miyanaga, K. Kakihara, E. Nakajima, and M. Yoshizumi, Industrialization of a noncooking system for alcoholic fermentation from grain, *Agric. Biol. Chem.*, 46 (1982) 1549–1558.
2768. B. A. Ustinnikov, A. N. Lazareva, and V. L. Yarovenko, Improvement of the fermentation process using an *Aspergillus batatae* 61 culture with increased glucoamylase activity (in Russian), *Ferm. Spirt. Prom.*, 7 (1972) 16–18.
2769. K. Kotarska, G. Kłosowski, and B. Czupryński, Characterization of technological features of dry yeast (strain I-7-43) preparation, product of electrofusion between *Saccharomyces cerevisiae* and *Saccharomyces diastaticus*, in industrial application, *Enzyme Microb. Technol.*, 49 (2011) 38–43.
2770. B. Y. Jeon, D. H. Kim, B. K. Na, D. H. Ahn, and D. H. Park, Production of ethanol directly from potato starch by mixed culture of *Saccharomyces cerevisiae* and *Aspergillus niger* using electrochemical reactor, *J. Microbial. Biotechnol.*, 18 (2008) 545–551.
2771. M. L. Lazić, S. Rasković, M. Stanković, and V. B. Vekjović, Enzymic hydrolysis of potato starch and ethanol production (in Serbo-Croatian), *Hem. Ind.*, 58 (2004) 322–326.
2772. X. Xiong, H. Zhou, X. Tan, Q. Li, and Y. Hu, Method for producing ethanol with potato, Chinese Patent 101085991 (2007); *Chem. Abstr.*, 148 (2007) 99225.
2773. H. Li and Y. Zhao, Method for producing ethanol from microwave-heated potato, Chinese Patent 101168745 (2008); *Chem. Abstr.*, 148 (2008) 536246.
2774. S. N. Mandavilli, Performance characteristics of an immobilized enzyme reactor producing ethanol from starch, *J. Chem. Eng. Jpn.*, 33 (2000) 886–890.
2775. B. Y. Jeon, S. J. Kim, D. H. Kim, B. K. Na, D. H. Park, H. T. Tran, R. Zhang, and D. H. Ahn, Development of a serial bioreactor system for direct ethanol production from starch using *Aspergillus niger* and *Saccharomyces cerevisiae*, *Biotechnol. Bioprocess Eng.*, 12 (2007) 566–573.
2776. W. Słowiński, J. A. Bernat, M. Remiszewski, and L. Słomińska, Saccharification of potato starch waste with different types of enzymes, *Proc. 1st Symp. Bioconvers. Cellul Subst. Energy, Chem. Microb. Protein 1977* (1978) 435–447. *Chem. Abstr.*, 90 (1978) 20819k.
2777. L. Chen, F. He, X. Li, and L. Li, Study on biodegradability of micronized potato starch, *Shipin Gongye Keji*, 22(3), (2001) 16–18. *Chem. Abstr.*, 136 (2001) 163859.
2778. J. Nowak, M. Łasik, T. Miśkiewicz, and Z. Czarnecki, Biodegradation of high temperature wastewater from potato starch industry, *Proc. 1st Int. Conf. Waste Management Environment, Cadiz, Spain, 2002* (2002) 655–663. *Chem. Abstr.*, 138 (2003) 389963.
2779. K. Kubota, Shochu production from uncooked raw materials, Jpn. Kokai Tokkyo Koho 60,232,083 (1985); *Chem. Abstr.*, 104 (1986) 107925y.
2780. S. Asai, A. Makino, and A. Makino, Preparation of shochu materials from sweet potato, Jpn. Kokai Tokkyo Koho, 01257470 (1989); *Chem. Abstr.*, 112(1990) 156724.
2781. K. Kubota, Y. Matsumura, and T. Yamamoto, Enzymic saccharification for alcohol fermentation of various uncooked cereal grains (in Japanese), *Denpun Kagaku*, 32 (1985) 169–176.
2782. T. E. Abraham, C. I. Krishnaswamy, and S. V. Ramakrishna, Effect of hydrolysis conditions of cassava on the oligosaccharide profile and alcohol fermentation, *Starch/Stärke*, 39 (1987) 237–240.
2783. R. Amutha and P. Gunasekaran, Production of ethanol from liquefied cassava starch using co-immobilized cells of *Zymomonas mobilis* and *Saccharomyces diastaticus*, *J. Biosci. Bioeng.*, 92 (2001) 560–564.

2784. Y. C. Ho and H. M. Ghazali, Alcohol production from cassava starch by co-immobilized *Zymomonas mobilis* and immobilized glucoamylase, *Pertanika*, 9 (1986) 235–240.
2785. K. D. Nam, M. H. Choi, W. S. Kim, H. S. Kim, and B. H. Ryu, Simulraneous saccharification and alcohol fermentation of unheated starch by free immobilized and coimmobilized systems of glucoamylase and *Saccharomyces cerevisiae*, *J. Ferment. Technol.*, 66 (1988) 427–432.
2786. O. V. S. Reddy and S. C. Basappa, Selection and characterization of *Endomycopsis fibuligera* strains for one-step fermentation of starch to ethanol, *Starch/Stärke*, 45 (1993) 187–194.
2787. K. Zhang and H. Feng, Fermentation potentials of *Zymomonas mobilis* and its application in ethanol production from low-cost raw sweet potato, *Afr. J. Biotechnol.*, 9 (2010) 6122–6125.
2788. V. Del Bianco, N. Araujo, A. Miceli, P. C. Souza e Silva, and J. A. Burle, Manioc alcohol by continuous fermentation (in Portuguese), *Inf. Inst. Nac. Tecnol. Rio de Janeiro*, 9(10), (1976) 20–26.
2789. F. P. De Franca, D. S. Mano, and S. G. F. Leite, Production of ethanol from manioc flour by strains of *Zymomonas* species (in Portuguese), *Rev. Latinoam. Microbiol.*, 28 (1986) 313–316.
2790. W. Xue, X. Dang, and X. Li, Study on technology of cassava ethanol fermentation, *Niangiju*, 32(4), (2005) 39–40. *Chem. Abstr.*, 46 (2006) 227761.
2791. L. Bringhenti, C. Cabello, and L. H. Urbano, Alcoholic fermentation of hydrolyzed starchy substrate enriched with sugar cane molasses (in Potuguese), *Ciencia Agrotec.*, 31 (2007) 429–432.
2792. M. Lewandowska, W. Bednarski, M. Adamczak, J. Tomasik, and S. Michalak, Ethanolic fermentation of starch mash with lactose by free, immobilized, coimmobilized yeasts and galactosidase, *Meded. Fac. Landbouw. Toegep. Biol. Wetensch. (Gent Univ.)*, 65(3a), (2000) 335–338.
2793. Y. Fujio, P. Suyanadona, P. Attasampunna, and S. Ueda, Alcoholic fermentation of raw cassava starch by *Rhizopus koji* without cooking, *Biotechnol. Bioeng.*, 26 (1984) 314–319.
2794. Y. Fujio, M. Ogata, and S. Ueda, Ethanol fermentation of raw cassava starch with *Rhizopus koji* in a gas circulation type fermentor, *Biotechnol. Bioeng.*, 27 (1985) 1270–1273.
2795. A. Abe, Y. Oda, K. Asano, and T. Sene, *Rhizopus delamer* is a proper name for *Rhizopus oryzae* fumaric-malic acid producer, *Mycologia*, 99 (2007) 714–722.
2796. P. Atthasampunna, P. Somchai, A. Eur-Aree, and S. Artjariyasripong, Production of fuel ethanol from cassava, *MIRCEN J. Appl. Microbiol. Biotechnol.*, 3 (1987) 135–142.
2797. B. H. Ryu and K. D. Nam, Large-scale alcohol fermentation with cassava slices at low temperature, *Sanop Misaengmul Hakhoechi*, 15 (1987) 75–79. *Chem. Abstr.* 107 (1988) 132561w.
2798. C. G. Teixera, Ethyl alcohol from cassava. Use of fungi in the hydrolysis of starch (in Portuguese), *Bragantia*, 10 (1950) 277–286.
2799. S. Ueda, C. T. Zenin, C. D. Monteiro, and Y. K. Park, Production of ethanol from raw cassava starch by a nonconventional fermentation method, *Biotechnol. Bioeng.*, 23 (1981) 291–299.
2800. S. Ueda and Y. Koba, Alcoholic fermentation of raw starch without cooking by using black-koji amylase, *J. Ferment. Technol.*, 58 (1980) 237–242.
2801. S. K. Rhee, G. M. Lee, Y. T. Han, Z. A. M. Yusof, M. H. Han, and K. J. Lee, Ethanol production from cassava and sago starch using *Zymomonas mobilis*, *Biotechnol. Lett.*, 6 (1984) 615–620.
2802. S. Srikanta, S. A. Jaleel, and K. R. Sreekantiah, Production of ethanol from tapioca (*Manihot esculenta*, Crantz), *Starch/Stärke*, 39 (1987) 132–135.
2803. Y. Wu and Z. Sun, Determination of physiological-biochemical characteristics of a thermoduric yeast and its ethanol fermentation tests, *Gongye Weishengwu*, 21(2), (1991) 10–14, 5; *Chem. Abstr.*, 115 (1991) 134137.
2804. M. Bae and J. M. Lee, Ethanol fermentation of raw cassava starch (II), *Sanop Misaengmul Hakhoechi*, 12 (1984) 261–264. *Chem. Abstr.*, 103 (1985) 159081u.
2805. C. Balagopalan, Ethanol manufacture by fermentation of hydrolyzed cassava starch, Indian Patent 160, 842 (1987); *Chem. Abstr.*, 109 (1988) 53223a.

2806. N. K. Aggarwal, P. Nigam, D. Singh, and B. S. Yadav, Process optimization for the production of sugar for the bioethanol industry from tapioca, a non-conventional source of starch, *World J. Microbiol. Biotechnol.*, 17 (2001) 783–787.
2807. P. Gao and F. Zhou, Study on increasing alcoholic yield of sweet potato by using cellulase, *Shipin Yu Faxiao Gongye* (4), (1982) 26–31. *Chem. Abstr.*, 98 (1982) 3501y.
2808. J. W. Chua, N. Fukui, Y. Wakabayashi, T. Yoshida, and H. Taguchi, Enzymic hydrolysis of sweet potato for energy-saving production of ethanol, *J. Ferment. Technol.*, 62 (1984) 123–130.
2809. P. Chen and C. Li, Breeding of raw starch degradation enzyme producing mutant and ethanol fermentation, *Weishengwuxue Tongbao*, 18 (1991) 268–270. *Chem. Abstr.*, 117 (1992) 46644.
2810. Ueda Chemical Industry Co., Ltd., Hankyu Kyoeibussan Co., Ltd., Mitsui Engineering and Shipbuilding Co., Ltd., Saccharification of uncooked raw starch materials, Jpn. Kokai Tokkyo Koho 82 18, 991 (1982); *Chem. Abstr.*, 97 (1982) 22101p.
2811. B. C. Saha and S. Ueda, Alcoholic fermentation of raw sweet potato by nonconventional method using *Endomycopsis fibuligera* glucoamylase preparation, *Biotechnol. Bioeng.*, 25 (1983) 1181–1186.
2812. Y. Yamamoto, Y. Matsumura, K. Kakutani, and K. Uenakai, maceration of uncooked sweet potato for alcoholic fermentation (in Japanese), *Denpun Kagaku*, 29 (1982) 117–122.
2813. Wuxi Institute of Light Industry, An experimental report on application of new technology – Uncooked alcoholic fermentation from dried sweet potato in alcoholic production, *Shipin Yu Fajiao Gongye* (5), (1985) 1–8. *Chem. Abstr.*, 104 (1985) 49793r.
2814. B. Yu, F. Zhang, P. Wang, and Y. Zheng, Scale-up research on alcoholic fermentation from dried sweet potato by immobilized yeast, *Shipin Yu Fajiao Gongye* (2), (1994) 1–7. *Chem. Abstr.*, 121 (1994) 228899.
2815. G. Shi, L. Zhang, and K. Zhang, Efficient economic and clean ethanol production, *Proc. 3rd Int. Conf. Food Sci. Technol., Davis CA, 1997* (1999) 68–75. *Chem. Abstr.*, 131 (1999) 285483.
2816. J. S. Epchtein, C. A. Gonsalves, T. Igarashi, and J. Finguerut, Continuous fermentation of sweet potato for alcohol production, Brasilian Patent 80 07,164 (1982); *Chem. Abstr.*, 97 (1982) 196858f.
2817. H. Yan, Method for producing starch sugar from diascorea by enzyme liquefaction and acid saccharification and extracting fuel ethanol by residue-free rapid fermentation, Chinese Patent 1966699 (2007); *Chem. Abstr.*, 147 (2008) 51706.
2818. N. K. Maldakar and S. N. Maldakar, Ethanol fermentation of glucose derived from *Diascorea sativa* tubers, Indian Patent 185247 (2000); *Chem. Abstr.*, 141 (2004) 348887.
2819. J. C. Scheerens, M. J. Kopplin, I. R. Abbas, J. M. Nelson, A. C. Gathman, and J. W. Berry, Feasibility of enzymic hydrolysis and alcoholic fermentation of starch contained in buffalo gourd (*Cucurbita foetidissima*) roots, *Biotechnol. Bioeng.*, 29 (1987) 436–444.
2820. R. W. Kerr and N. F. Schink, Fermentability of cornstarch products – relation to starch structure, *Ind. Eng. Chem.*, 33 (1941) 1418–1421.
2821. I. Y. Han and M. P. Steinberg, Simultaneous hydrolysis and fermentation of raw dent and high lysine corn and their starches, *Biotechnol. Bioeng.*, 33 (1989) 906–911.
2822. J. Fang and S. Fang, Alcoholic fermentation of raw starch without yeast, *Shipin Yu Fajiao Gongye* (5), (1992) 7–12. *Chem. Abstr.*, 118 (1993) 100479.
2823. I. Y. Han and M. P. Steinberg, Solid-state yeast fermentation of raw corn with simultaneous koji hydrolysis, *Proc. 8th Biotechnol. Bioeng. Symp., 1986*, 17 (1986) 449–462. *Chem. Abstr.*, 106 (1987) 212508d.
2824. K. Yoshizawa, H. Hamaki, and N. Okumura, Studies on grain alcohol. I. Digestion of corn starch by amylase, *Nippon Jozo Kyokai Zasshi*, 75 (1980) 765–767. *Chem. Abstr.*, 94 (1981) 63720w.
2825. Kyowa Hakko Kogyo Ltd., Manufacture of ethanol by fermentation, Jpn. Kokai Tokkyo Koho 81 15,691 (1981); *Chem. Abstr.*, 94 (1981) 190333x.

2826. Institute of Microbiology, Peking Alcohol Plant and Tientsin Alcohol Plant, Application of glucoamylase produced by *Aspergillus niger* mutant A.S.3.4309 in alcohol industry, *Shih Pin Yu Fa Hsiao Kung Eh* (2), (1981) 63–67. *Chem. Abstr.*, 95 (1981) 22924t.
2827. I. Y. Han and M. P. Steinberg, Amylolysis of raw corn by *Aspergillus niger* for simultaneous ethanol fermentation, *Biotechnol. Bioeng.*, 30 (1987) 225–232.
2828. D. Vadehra, A process for the conversion of starch based agricultural products into alcohol, Indian Patent 158,345 (1986); *Chem. Abstr.*, 107 (1987) 132668m.
2829. Z. Chi and Z. Liu, High concentration ethanol fermentation from starch in corn with *Saccharomyces* sp. H0, *Shengwu Gongcheng Xuebao*, 10 (1994) 130–134. *Chem. Abstr.*, 121(1994) 228928.
2830. G. Kłosowski, B. Czupryński, and M. Wolska, Characteristics of alcoholic fermentation with the application of *Saccharomyces cerevisiae* yeast as-4 strain and I-7-23 fusant with amylolytic properties, *J. Food Eng.*, 76 (2006) 500–505.
2831. D. Inlow, J. McRae, and A. Ben-Bassat, Fermentation of corn starch to ethanol with genetically engineered yeast, *Biotechnol. Bioeng.*, 32 (1988) 227–234.
2832. D. H. Lee, J. S. Choi, J. U. Ha, S. C. Lee, and Y. I. Hwang, Ethanol fermentation of corn starch by a recombinant *Saccharomyces cerevisiae* having glucoamylse and α-amylase activities, *J. Food Sci. Nutr.*, 6 (2001) 206–210.
2833. G. Mencinicopshi, C. Mosu, E. Balauta, and M. Ferdes, Amyloglucosidase and yeast co-immobilization for ethanol production from corn starch, *Sti. Technol. Aliment. (Bucharest)*, 1(1), (1993) 4–12.
2834. J. E. McGhee, M. E. Carr, and G. St. Julian, Continuous bioconversion of starch to ethanol by calcium alginate immobilized enzymes and yeasts, *Cereal Chem.*, 61 (1984) 446–449.
2835. Z. Chi, J. Liu, and P. Xu, High-concentration ethanol production from cooked corn starch by using middle-temperature cooking process, *Shengwu Gongcheng Xuebao*, 11 (1995) 228–232. *Chem. Abstr.*, 124 (1996) 173564.
2836. H. S. Olsen, S. Pedersen, R. Beckerich, C. Veit, and C. Felby, Improved fermentation process, EU Patent WO 2002074895 (2002); *Chem. Abstr.*, 137 (2002) 246620.
2837. H. Yan, Method for extracting fuel ethanol from corn with high yield and rapid speed by enzyme liquefaction and acid saccharification main fermentations and extraction, Chinese Patent 1966698 (2007); *Chem. Abstr.*, 147 (2007) 51705.
2838. K. Wei, J. F. Li, X. X. Qin, and G. T. Ma, Study on alcohol fermentation of uncooked corn starch, *Guangxi Nongye Shengwu Kexue*, 26 (2007) 253–255. *Chem. Abstr.*, 149 (2008) 400907.
2839. M. S. Krishnan, N. P. Nghiem, and B. H. Davison, Ethanol production from corn starch in a fluidized bed bioreactor, *Appl. Biochem. Biotechnol.*, 77–79 (1999) 359–372.
2840. Z. J. Bebić, J. B. Jakovljević, and J. Baraš, Corn starch hydrolysates as fermentation substrates for ethanol production (in Serbo-Croatian), *Hem. Ind.*, 54(1), (2000) 5–9.
2841. L. Mojović, S. Nikolić, M. Rakin, and M. Vukasinovich, Production of bioethanol from corn meal hydrolyzates, *Fuel*, 85 (2006) 1750–1755.
2842. B. S. Dien, N. N. Nichols, P. J. O'Brian, L. B. Iten, and R. J. Bothast, Enhancement of ethanol yield from the corn dry grind process by conversion, *ACS Symp. Ser.*, 887 (2004) 63–77.
2843. B. S. Dien, N. Nagle, K. B. Hicks, V. Singh, R. A. Moreau, M. P. Tucker, N. N. Nichols, D. B. Johnston, M. A. Cotta, Q. Nguyen, and R. J. Bothast, Fermentation of quick fiber produced from a modified corn-milling process into ethanol and recovery of corn fiber oil, *Appl. Biochem. Biotechnol.*, 113–115 (2004) 937–949.
2844. N. Voča, T. Krička, V. Janusič, and A. Matin, Bioethanol production from corn kernel grown with different cropping intensities, *Cereal Res. Commun.*, 35(2, Pt. 2), (2007) 1309–1312.
2845. J. Lu, H. Jin, L. Hou, J. Piao, and Q. Zhang, Ferment enhancing agent for preparing ethanol fuel, Chinese Patent 1353190 (2002); *Chem. Abstr.*, 138 (2003) 168890.

2846. Y. Takahashi, New ethanolic beverages from corn grits, Jpn. Kokai Tokkyo Koho 62 55068 (1987); *Chem. Abstr.*, 107 (1987) 5773s
2847. Y. K. Park, H. H. Sato, E. M. San Martin, and C. F. Ciacco, Production of ethanol from extruded degermed corn grits by a nonconventional fermentation method, *J. Ferment. Technol.*, 65 (1987) 469–473.
2848. L. Wang, X. Ge, and W. Zhang, Improvement of ethanol yield from raw corn flour by *Rhizopus* sp, *World J. Microbiol. Biotechnol.*, 23 (2007) 461–465.
2849. K. Sato, S. Miyazaki, N. Matsumoto, K. Yoshizawa, and K. Nakamura, Pilot-scale solid-state ethanol by gas circulation using moderately thermophilic yeast, *J. Ferment. Technol.*, 66 (1988) 173–180.
2850. S. Baba and T. Iida, Brewing of a new type of sake from corn starch, *Nagano-ken Shokuhin Kogyo Shikenjo Kenkyu Hokoku*, 15 (1987) 1–5. *Chem. Abstr.*, 109 (1987) 5209d.
2851. S. R. Parekh and M. Wayman, Performance of a novel continuous dynamic immobilized cell bioreactor in ethanolic fermentation, *Enzyme Microbiol. Technol.*, 9 (1987) 406–410.
2852. K. Mikuni, M. Monma, and K. Kainuma, Alcohol fermentation of corn starch digested by *Chalara paradoxa* amylase without cooking process, *Biotechnol. Bioeng.*, 29 (1987) 729–732.
2853. J. Wang and S. Wang, Method for producing propanol, butanol, and ethanol from starch emulsion after extracting gluten from wheat starch, Chinese Patent 101230360 (2008); *Chem. Abstr.*, 149 (2009) 286968.
2854. P. Monceaux and E. Segard, Production of ethanol and various byproducts from cereals, French Patent 2,586,032 (1987); *Chem. Abstr.*, 107 (1987) 152928z.
2855. E. Grampp, B. Sprössle, and H. Uhlig, Starch degradation for the manufacture of ethanol beverages, German Patent 2,153,151 (1973); *Chem. Abstr.*, 79 (1973) 16933y.
2856. S. Hisaku, K. Nakanishi, and K. Obata, Alcoholic fermentation from starch without steaming, Jpn. Kokai Tokkyo Koho 61 12,292 (1986); *Chem. Abstr.*, 105 (1984) 41294d.
2857. T. Montesinos and J. M. Navarro, Production of alcohol from raw wheat flour by amyloglucosidase and *Saccharomyces cerevisiae*, *Enzyme Microb. Technol.*, 27 (2000) 362–370.
2858. K. Peglow, M. Tischler, K. W. Echtermann, W. Hirte, and G. Schulz, Use of specific enzyme complex for pressure-less starch decomposition process in grain distillation, East German Patent 236,109 (1986); *Chem. Abstr.*, 106 (1986) 3862a.
2859. S. Sharma, M. Pandey, and B. Saharan, Fermentation of starch to ethanol by an amylolytic yeast *Saccharomyces diastaticus* SM-10, *Indian J. Exp. Biol.*, 40 (2002) 325–328.
2860. M. Wyman, S. Chen, R. S. Parekh, and S. R. Parekh, Comparative performance of *Zymomonas mobilis* and *Saccharomyces cerevisiae* in alcoholic fermentation of saccharified wheat starch B in a continuous dynamic immobilized biocatalyst bioreactor, *Starch/Stärke*, 40 (1988) 270–275.
2861. K. Suresh, N. Kiransree, and L. V. Rao, Production of ethanol by raw starch hydrolysis and fermentation of damaged grains of wheat and sorghum, *Bioprocess Eng.*, 21 (1999) 165–168.
2862. P. Lerminet, Ethanol production from starch, French Patent 2,586,701 (1987); *Chem. Abstr.*, 107 (1987) 152926x.
2863. T. E. Mudrak, L. V. Kislaya, V. A. Marinchenko, and E. V. Zelinskaya, Two-stage procedure for the fermentation of wort prepared from starch-containing materials by thermotolerant yeast strains (in Russian), *Ferment. Spirt. Prom.*, 6 (1987) 24–25.
2864. E. F. Torres and J. Baratti, Ethanol production from wheat flour by *Zymomonas mobilis*, *J. Ferment. Technol.*, 66 (1988) 167–172.
2865. T. Montesinos and J. M. Navarro, Role of the maltose in the simultaneous-saccharification-fermentation process from raw wheat starch and *Saccharomyces cerevisiae*, *Bioprocess Eng.*, 23 (2000) 319–322.

2866. S. K. Soni, A. Kaur, and J. K. Gupta, A solid state fermentation based bacterial α-amylase and fungal glucoamylase system and its suitability for the hydrolysis of wheat starch, *Process Biochem.*, 39 (2003) 185–192.
2867. J. Wesenberg, M. Heide, K. Laube, S. Nowak, J. Mücke, and P. Lietz, Ethanol fermentation with mixed yeast population, East German Patent 253,044 (1988); *Chem. Abstr.*, 109 (1988) 53240d.
2868. K. Stecka, J. Milewski, A. Miecznikowski, B. Łączyński, M. Cieślak, and M. Błażejczak, Method of obtaining ethyl alcohol, Polish Patent 159970 (1993); *Chem. Abstr.*, 122 (1995) 79236.
2869. E. Rzepka, K. M. Stecka, E. Badocha, and M. Zbieć, The use of *Zymomonas mobilis* KKP/604 in normal pressure technology of ethanol production (in Polish), *Pr. Inst. Lab. Bad. Przem. Spoz.*, 56 (2001) 124–137.
2870. J. Kučerova, Usability of triticale into bioethanol producing through the use of technical enzymes (in Czekh), *Acta Univ. Agric. Silvicult. Mend. Brunensis*, 54(4), (2006) 33–38.
2871. K. Kirste, W. Freyberg, G. Trimde, E. Kersten, G. Nebelung, K. W. Echtermann, U. Zimare, K. Göck, J. Wessenberg et al., Saccharification of triticale mash, East German Patent 235,667 (1986); *Chem. Abstr.,* 106 (1986) 3861z.
2872. Japan National Tax Administration Agency, Manufacture of alcohol, Jpn. Kokai Tokkyo Koho 82 12,991 (1982); *Chem. Abstr.*, 96 (1982) 216055b.
2873. Okura Shuzo Co., Ltd., Saccharified rice starch for sake production, Jpn. Kokai Tokkyo Koho 58 43 43,780 (1983); *Chem. Abstr.*, 98 (1983) 214149c.
2874. D. Y. Corredor, S. Bea, and D. Wang, Pretreatment and enzymatic hydrolysis of sorghum bran, *Cereal. Chem.*, 84 (2007) 61–66.
2875. R. Hosting, *A Dictionary of Japanese Food*, 3rd edn. (2003) Tuttle Publ, Tokyo.
2876. F. Wang, C. Pan, Y. Xiong, J. Dong, S. Zhao, X. Zheng, X. Wei, and J. Hu, Study on production of beer with bifid growth factors by glucoside conversion with enzymes, *Niangjiu*, 29(6), (2004) 87–89. *Chem. Abstr.*, 141 (2004) 224026.
2877. K.K. Naigai Shokuhin Koki, Production of saccharified liquid for sake brewing, Jpn. Kokai Tokkyo Koho 58 00,882 (1983); *Chem. Abstr.*, 98 (1983) 141876a.
2878. T. Suda, Microorganism material for manufacture of fertilizer from rice bran, Jpn. Kokai Tokkyo Koho 2001261473 (2001); *Chem. Abstr.,* 135 (2001) 241528.
2879. N. Dey, R. Soni, and S. K. Soni, A novel thermostable alpha amylase from thermophilic *Bacillus* sp. SN-1, *Asian J. Microbiol. Biotechnol. Environ. Sci.*, 4 (2002) 159–164.
2880. X. Zhan, D. Wang, S. R. Bean, X. Mo, X. S. Sun, and D. Boyle, Ethanol production from supercritical-fluid-extrusion cooked sorghum, *Ind. Crops Prod.*, 23 (2006) 304–310.
2881. P. Thammarutwasik, Y. Koba, and S. Ueda, Alcoholic fermentation of sorghum without cooking, *Biotechnol. Bioeng.*, 28 (1986) 1122–1125.
2882. G. Negrut, L. Table, G. Giurgiulescu, S. D. Matu, I. Farcasescu, I. Grancea, A. Dragulescu, M. Mracec, and A. Goie, Manufacture of ethanol from *Sorghum saccharatum*, Roumanian Patent 90,724 (1987); *Chem. Abstr.*, 107 (1987) 174433m.
2883. D. Ma, C. Sun, and P. Gao, Ethanol production with simultaneous saccharification and fermentation (SSF) of sweet sorghum, *Gongye Weishengwu*, 17(4), (1987) 5–9. *Chem. Abstr.*, 107 (1987) 174322z.
2884. A. Garcia, III and P. Grilione, Ethanol production characteristics for a respiratory deficient mutant yeast strain, *Trans. ASAE*, 25 (1982) 1396–1399.
2885. P. S. O, D. J. Cha, and H. W. Suh, Alcohol fermentation of naked barley without cooking, *Sanop Misaengmul Hakhoechi*, 14 (1986) 415–420. *Chem. Abstr.*, 106 (1986) 136951a.
2886. K. Morimoto, K. Yoshioka, and N. Hashimoto, Enzyme saccharification technology for brewing barley. IV. A saccharification enzyme preparation designed for solving the problems caused by enzymatic saccharification of barley, *Monatsschr. Brauwiss.*, 41(2), (1989) 85–89.

2887. G. Kłosowski, B. Czupryński, K. Kotlarska, and M. Wolska, Study of using amaranthus grain in agricultural distilleries using pressureless starch release technology (in Czech), *Kvasny Prum.*, 48 (2002) 302–308.
2888. X. Wu, D. Wang, S. R. Bean, and J. P. Wilson, Ethanol production from pearl millet using *Saccharomyces cerevisiae*, *Cereal Chem.*, 83 (2006) 127–131, 316.
2889. V. V. R. Bandaru, S. R. Somalanka, D. R. Mendu, N. R. Madicherla, and A. Chityala, Optimization of fermentation conditions for the production of ethanol from sago starch by co-immobilized amyloglucosidase and cells of *Zymomonas mobilis* using response surface methodology, *Enzyme Microbiol. Technol.*, 38 (2006) 209–214.
2890. G. M. Lee, C. H. Kim, K. J. Lee, Z. A. M. Yusof, M. H. Han, and S. K. Rhee, Simultaneous saccharification and ethanol fermentation of sago starch using immobilized *Zymomonas mobilis*, *J. Ferment. Technol.*, 64 (1986) 293–297.
2891. S. K. Rhee, G. M. Lee, C. H. Kim, Z. Abidin, and M. H. Han, Simultaneous sago starch hydrolysis and ethanol production by *Zymomonas mobilis* and glucoamylase, *Proc. 8th Biotechnol. Bioeng. Symp.*, 117 (1986) 481–493. *Chem. Abstr.*, 107 (1987) 5679r.
2892. C. H. Kim, G. M. Lee, Z. Abidin, M. H. Han, and S. K. Rhee, Immobilization of *Zymomonas mobilis* and amyloglucosidase for ethanol production from sago starch, *Enzyme Microbiol. Technol.*, 10 (1988) 426–430.
2893. C. H. Kim, Z. Abidin, C. C. Ngee, and S. K. Rhee, Pilot-scale fermentation by *Zymomonas mobilis* from simultaneously saccharified sago starch, *Biores. Technol.*, 40(1), (1992) 1–6.
2894. C. G. Lee, C. H. Kim, and S. K. Rhee, A kinetic model and simulation of starch saccharification and simultaneous ethanol fermentation by amyloglucosidase and *Zymomonas mobilis*, *Bioprocess Eng.*, 7 (1992) 335–341.
2895. A. Ishizaki and S. Tripetchkul, Continuous ethanol production by *Zymomonas mobilis* from sago starch hydrolyzate and natural rubber waste, *Acta Hortic.*, 389 (1995) 131–145.
2896. N. Haska and Y. Ohta, Alcohol fermentation from sago starch granules using raw sago starch-digesting amylases from *Penicillium brunneum* No. 24 and *Saccharomyces cerevisiae* No. 33, *Starch/Stärke*, 45 (1993) 241–244.
2897. D. C. Ang, S. Abd-Aziz, H. M. Yusof, M. I. A. Karim, A. B. Arif, K. Uchiyama, and S. Shioya, Direct fermentation of sago starch to ethanol using recombinant yeast, *Biotechnol. Sust. Util. Biol. Resour. Tropics*, 16 (2003) 19–24.
2898. K. Bujang, D. S. A. Adeni, and A. Ishizaki, Effects of pH on production of ethanol from hydrolyzed sago starch utilizing *Zymomonas mobilis*, *Biotechnol. Sust. Util. Biol. Res. Tropics*, 14 (2000) 20–26.
2899. N. N. Nichols, B. S. Dien, Y. V. Wu, and M. A. Cotta, Ethanol fermentation of starch from field peas, *Cereal Chem.*, 82 (2005) 554–558.
2900. J. Yang, W. W. Zheng, J. H. Li, and D. H. Zhou, Preparation of instant whole mung bean beverage using enzyme hydrolysis method, *Nanchang Daxue Xuebao Gongkeban*, 28 (2006) 115–118. *Chem. Abstr.*, 146 (2007) 481036.
2901. Y. Jiang, L. Jiang, and D. M. Kong, Application of saccharifying enzyme – α-amylase method in preparation of Baiju (white spirit) alcoholic beverage, *Jilin Gongxueyuan Xuebao, Ziran Kexueban*, 21(2), (2000) 47–49. *Chem. Abstr.*, 134 (2001) 325476.
2902. M. A. Murado, L. Pastrana, J. A. Velazquez, J. Miron, and M. Pilar Gonzalez, Alcoholic chestnut fermentation in mixed culture. Compatibility criteria between *Aspergillus oryzae* and *Saccharomyces cerevisiae*, *Biores. Technol.*, 99 (2008) 7255–7263.
2903. S. Maeda and T. Matsumoto, Bioconversion of organic waste to fermentable sugars, *Kenkyu Hokoku – Kanagawa-ken Sangyo Gijutsu Sogo Kenkyushu*, 5 (1999) 23–27. *Chem. Abstr.*, 132 (2000) 346683.

2904. L. Davis, P. Rogers, J. Pearce, and P. Peiris, Evaluation of Zymomonas-based ethanol production from a hydrolyzed waste starch stream, *Biomass Bioenergy*, 30 (2006) 809–814.
2905. H. Fukuda, K. Iwase, A. Ito, and Y. Okamoto, Composite microorganism group and its use as a means to degrade organic waste material, Jpn. KokaiTokkyo Koho JP 2002058471 (2002); *Chem. Abstr.*, 136 (2002) 196948.
2906. I. I. Erokhin, N. N. Zmeeva, and E. V. Kuznetsov, Enhanced complex enzyme preparation for hydrolysis of vegetable waste, among them food waste comprising α-amylase, glucoamylase, xylanase and cellulose, Russian Patent 2238319 (2004); *Chem. Abstr.*, 141 (2004) 327650.
2907. J. Gao, B. S. Hooker, and R. S. Skeen, Genetically engineered microorganisms useful in expression of bioproducts from starch waste, EU Patent WO 2000050610 (2000); *Chem. Abstr.*, 133 (2000) 203816.
2908. S. Suzuki and H. Honda, Alcohol fermentation from starch waste materials by solid-state fermentation, Jpn. Kokai Tokkyo Koho 2005065695 (2005); *Chem. Abstr.*, 142 (2005) 260136.
2909. N. Raman and C. Pothiraj, One-step process for starch fermentation employing the starch digestion yeast *Saccharomyces diastaticus*, *J. Indian Chem. Soc.*, 85 (2008) 452–454.
2910. Licher Privatbrauerai Ihring-Melchior G.m.b.H, Process for fermantative production of ethanol from renewable raw material, Germany, German Patent 102006047782 (2008); *Chem. Abstr.*, 148 (2008) 401462.
2911. R. J. M. P. De Baynast de Septifontaines, F. E. M. E. Brouard, G. A. J. M. Yvon, J. L. A. G. Baret, and C. Leclerc, Saccharification of starchy vegetable for alcohol fermentation, French Patent 2637294 (1990); *Chem. Abstr.*, 113 (1990) 113889.
2912. S. M. Lewis, Method for producing ethanol using raw starch, US Patent 7,842,484 (2010).
2913. J. J. Rusek, M-L. R. Rusek, and J. D. Ziulkowsky, High-octane automotive fuel and aviation gasoline manufactured by fermentation of renewable carbohydrate feedstock, US Patent 2008244961 (2008); *Chem. Abstr.*, 149 (2008) 429091.
2914. R. J. Wallace, S. J. A. Wallace, N. McKain, V. L. Nsereko, and G. F. Hartnell, Influence of supplementary fibrolytic enzymes on the fermentation of corn and grass silages by mixed ruminal microorganisms in vitro, *J. Animal Sci.*, 79 (2001) 1905–1916.
2915. Y. Zhou and Y. Zhou, Method for producing ethanol from wild plant starch such as corn starch, Chinese Patent 101085992 (2007); *Chem. Abstr.*, 148 (2007) 99226.
2916. N. Suzuki, K. Matsunaga, H. Moriyama, H. Onoda, H. Ohata, K. Matsumoto, and S. Matsunaga, Ethanol manufacture from marine plant and algae, Jpn. Kokai Tokkyo Koho, 2003310288 (2003); *Chem. Abstr.*, 139 (2003) 322394.
2917. N. Hirayama and R. Ueda, Ethanol production method with microalgae and device, Jpn. Kokai Tokkyo Koho 2000228993 (2000); *Chem. Abstr.*, 133 (2000) 149816.
2918. A. Azzini and T. J. Barreto di Menezes, Alcohol from bamboo by enzymic hydrolysis of starch, cellulose and pentosan, Brazilian Patent 80 01,361 (1981); *Chem. Abstr.*, 96 (1981) 67266s.
2919. A. Azzini and T. J. Barreto di Menezes, Alcohol from bamboo by enzymic hydrolysis of starch and acid hydrolysis of pentosane and cellulose, Brazilian Patent 80 01,362 (1981); *Chem. Abstr.*, 96 (1981) 67267t.
2920. G. Amin, R. De Mot, K. Van Dijk, and H. Verachtert, Direct alcoholic fermentation of starchy biomass using amylolytic yeast strains in batch and immobilized cell system, *Appl. Microbiol. Biotechnol.*, 22 (1985) 237–245.
2921. C. Felby, J. Larsen, H. Joergensen, and J. Vibe-Pedersen, Enzymatic hydrolysis of biomass having a high dry matter content, EU Patent WO 2006056838 (2006); *Chem. Abstr.*, 144 (2006) 487360.
2922. L. Lindet, The byproducts of alcoholic fermentation, *Bull. Assoc. Chim. Sucr. Dist.*, 35 (1917) 232–236. *Chem. Abstr.*, 12 (1918) 10600.

2923. S. P. Gulyaev, Alcoholic fermentation of starchy and sugary materials (in Russian), USSR Patent 112,341 (1958).
2924. A. Lee, Evaluation of saccharifying methods for alcoholic fermentation of starchy substrates, *Iowa State Coll. J. Sci.*, 30 (1956) 403–404.
2925. G. Kłosowski and B. Czupryński, Kinetics of acetals and esters formation during alcoholic fermentation of various starchy raw materials with application of yeasts *Saccharomyces cerevisiae*, *J. Food Eng.*, 72 (2006) 242–246.
2926. G. Kłosowski and B. Czupryński, Acetal and ester formation during alcoholic fermentation of various starchy raw materials with application of yeast *Saccharomyces cerevisiae* As-4 and D-1 (in Polish), *Pr. Inst. Lab. Bad. Przem. Spoz.*, 59 (2004) 126–137.
2927. E. Bauer, H.-E. Jacob, and H. Berg, Electrostimulation of CO_2 production in yeast cells, *Stud. Biophys.*, 119 (1986) 137–140.
2928. E. Bauer, H.-H. Grosse, H. Berg, and H.-E. Jacob, Method for stimulation of fermentation process, EU Patent: WP AGl B263 8133 (1984); *Chem. Abstr.*, 108 (1985) 203334.
2929. K. Nakanishi, H. Tokuda, T. Soga, T. Yoshinaga, and M. Takeda, Effect of electric current on growth and alcohol production by yeast cells, *J. Ferm. Bioeng.*, 85 (1998) 250–253.
2930. N. I. Danilyak, Model of hydrolysis of starch-cellulose substrates using carbohydrase systems (in Russian), *Ferment. Spirt. Prom.* (6), (1987) 32–34.
2931. L. A. Rovinskii and V. L. Yarovenko, Simulation of the kinetics of enzyme hydrolysis of starch in standard apparatus used in alcohol production (in Russian), *Ferm. Spirt. Prom.* (2), (1977) 14–17.
2932. L. A. Rovinskii and V. V. Ivanov, Calculation of the hydrolysis of starch in alcohol fermentation (in Russian), *Ferm. Spirt. Prom.* (7), (1979) 30–33.
2933. V. L. Yarovenko and B. M. Nakhmanovich, Kinetics of continuous alcohol fermentation during the reprocessing of raw starch (in Russian), *Izv. Vyssh. Ucheb. Zaved. Pishch. Prom.* (1), (1970) 82–85.
2934. F. Parisi, A. Converti, M. Del Borghi, P. Perego, M. Zilli, and G. Ferraiolo, Continuous alcohol production from starch hydrolyzate, *Biotechnol. Bioeng. Symp.*, 17 (1986) 379–389.
2935. G. Zhong and X. Li, Study on ethanol fermentation and its kinetics from starch hydrolyzate by immobilized yeast cells, *Weishengwuxue Tongbao*, 21 (1994) 19–22. *Chem. Abstr.*, 121 (1994) 253802.
2936. A. Converti, C. Bargagliotti, C. Cavanna, C. Nicolella, and M. Del Borghi, Evaluation of kinetic parameters and thermodynamic quantities of starch hydrolyzate alcohol fermentation by *Saccharomyces cerevisiae*, *Bioprocess Eng.*, 15(2), (1996) 63–69.
2937. L. R. B. Goncalves, R. C. Giordano, and R. L. C. Giordano, A bidisperse model to study the production of ethanol from starch using enzyme and yeast coimmobilized in pectin gel, *Proc. 9th Eur. Bioenergy Conf. Copenhagen, 1996*, 3 (1996) 1566–1571; *Chem. Abstr.*, 127 (1997) 296138.
2938. A. D. Kroumov, A. N. Modenes, and M. C. de Araujo Tait, Development of new unstructured model for simultaneous saccharification and fermentation of starch to ethanol by recombinant strain, *Biochem. Eng. J.*, 28 (2006) 243–255.
2939. E. Rzepka, E. Badocha, and K. M. Stecka, Selected kinetic aspects of the alcoholic fermentation of mashes prepared under atmospheric pressure (in Polish), *Pr. Inst. Lab. Bad. Przem. Spoz.*, 54 (1999) 32–49.
2940. V. L. Yarovenko and B. M. Nakhmanovich, Kinetics of product synthesis in continuous alcoholic fermentation, *Pure Appl. Chem.*, 36 (1973) 397–405.
2941. C. Weizmann, *Trial and Error*, (1963) Hamilton, London.
2942. H. B. Speakman, Gas production during the acetone and butyl alcohol fermentation of starch, *J. Biol. Chem.*, 43 (1920) 401–411.

2943. H. B. Speakman, Biochemistry of the acetone and butyl alcohol fermentation of starch by *Bacillus granulobacter pectinovorum, J. Biol. Chem.*, 41 (1920) 319–343.
2944. C. L. Gabriel, Butanol fermentation process, *Ind. Eng. Chem.*, 20 (1928) 1063–1067.
2945. D. A. Legg and M. T. Walton, Ethyl alcohol fermentation, US Patent 2,132,358 (1938).
2946. J. Mueller, Butyl acetonic fermentation, US Patent 2,132,039 (1938).
2947. M. I. Zalesskaya and F. M. Kinzburgskaya, Starch hydrolysis in acetone-butanol fermentation. I. Amylolytic enzymes of acetone-butanol bacteria (in Russian), *Tr. Vses. Nauch. Issled. Inst. Spirt. Prom.* (3), (1954) 140–148.
2948. M. I. Zalesskaya and F. M. Kinzburgskaya, Starch hydrolysis in acetone-butanol fermentation. II. Changes in the carbohydrate composition of the medium at different stages of acetone-butanol fermentation in Russian, *Tr. Vses. Nauch. Issled. Inst. Spirt. Prom.* (3), (1954) 148–155.
2949. A. Afschar and K. Schäller, Continuous fermentation of acetone and butanol from starch-containing wastes, German Patent 4006051 (1991); *Chem. Abstr.*, 116 (1992) 127018.
2950. H. R. Badr, Starch utilization and solvent synthesis by *Clostridium acetobutylicum* for batch and continuous acetone-butanol fermentation, *Diss. Abstr. Int.*, B 52 (1991) 1151.
2951. T. C. Ezeji, N. Quareshi, and H. P. Blaschek, Continuous butanol fermentation and feed starch retrogradation: Butanol fermentation sustainability using *Clostridium beijerinckii* BA101, *J. Biotechnol.*, 115 (2005) 179–187.
2952. S. Onuma, M. Naganum, and H. Oiwa. Isopropyl alcohol manufacture from corn, Jpn. Kokai Tokkyo Koho 61 67,493 (1984); *Chem. Abstr.*, 105 (1986) 5206t.
2953. T. Hanai, S. Atsumi, and J. C. Liao, Engineered synthetic pathway for isopropanol production in *Escherichia coli, Appl. Environ. Microbiol.*, 73 (2007) 7814–7818.
2954. T. Jojima, M. Inui, and H. Yukawa, Production of isopropanol by metabolitically engineered *Escherichia coli, Appl. Microbiol. Biotechnol.*, 77 (2008) 1219–1224.
2955. G. A. Adams, Production and properties of 2,3-butanediol. VII. Fermentation of wheat by *Aerobacillus polymyxa* under aerobic and anaerobic conditions, *Can. J. Res.*, 24F (1946) 1–11.
2956. G. A. Adams and J. D. Leslie, Production and properties of 2,3-butanediol. VIII. pH control in *Aerobacillus olymyxa* fermentations and its effects on products and their recovery, *Can. J. Res.*, 24F (1946) 12–28.
2957. S. B. Fratkin and G. A. Adams, Production and properties of 2,3-butanediol. IX. The effect of various nutrient materials on the fermentation of starch by *Aerobacillus polymyxa, Can. J. Res.*, 24F (1946) 29–38.
2958. P. Perego, A. Converti, A. Del Borghi, and P. Canepa, 2,3-Butanediol production by *Enterobacter aerogenes* selection of the optimal conditions and application to food industry residues, *Bioprocess Eng.*, 23 (2000) 613–620.
2959. D. Liu, K. Cheng, H. Liu, R. Lin, and J. Hao, Method for manufacturing 1,3-propanediol and 2,3-butanediol from crude starch containing materials by microbial fermentation, Chinese Patent 1710086 (2005); *Chem. Abstr.*, 145 (2006) 375463.
2960. Y. Zheng, Z. Hu, Z. Liu, and Y. Shen, Application of *Bacillus licheniformis* B-05571 in manufacturing 1,3-dihydroxyacetone, Chinese Patent 101037659 (2007); *Chem. Abstr.*, 47 (2007) 425643.
2961. L. J. Wickerham, L. B. Lockwood, O. G. Pettijohn, and G. E. Ward, Starch hydrolysis and fermentation by the yeast *Endomycopsis fibuligera, J. Bact.*, 48 (1944) 413–427.
2962. Y. L. Huang, Z. Wu, L. M. Zhang, C. Cheung, and S. T. Yang, Production of carboxylic acids from hydrolyzed corn meal by immobilized cell fermentation in a fibrous-bed bioreactor, *Bioresource Technol.* 82 (2002) 51–59.
2963. Y. C. Tsai, M. C. Huang, S. F. Lin, and Y. C. Su, Method for the production of itaconic acid using *Aspergillus terreus* solid state fermentation, US Patent 6171831 (2001); *Chem. Abstr.*, 134 (2001) 85176.

2964. A. Steinbüchel, *Biopolymers*, (2002) J. Wiley-VCH, Weinheim.
2965. A. Suryani, N. Atifah, E. Hambali, K. Syamsu, and R. Meliawati, Production of poly(3-hydroxybutyrate) by fed-batch culture of *Ralstonia eutropha* on sago starch hydrolysate as the carbon source, *Biotechnol. Sust. Util. Biol. Tesources Tropics*, 17 (2004) 84–90.
2966. D. Rusendi and J. D. Sheppard, Hydrolysis of potato processing waste for the production of poly-β-hydroxybutyrate, *Bioresource Technol.*, 54 (1995) 191–196.
2967. S. A. Waksman and D. Kirsh, Butyric acid and butyl alcohol fermentation of hemicellulose- and starch-rich materials, *Ind. Eng. Chem.*, 25 (1933) 1036–1041.
2968. C. Coolhaas, Zur Kenntnis der Dissimilation fettsäurer Salze und Kohlehydrate durch thermophile Bakterien, *Zentr. Bakt. Parasitenk, II Abt.*, 75 (1928) 344–360.
2969. C. Weizmann and H. M. Spiers, Fermenting starchy materials, British Patent 164,366 (1918).
2970. X. Zhang and H. Wang, Method for producing microbial polyunsaturated fatty acid from wastewater of starch processing, Chinese Patent 1686858 (2005); *Chem. Abstr.*, 145 (2006) 123161.
2971. N. Campbell and J. Reece, *Biology*, 7th edn. (2005) Benjamin.
2972. R. J. Siezen,, J. Kok,, T. Abee,, and G. Schaafsma, (Eds.) Lactic Acid Bacteria, Genetics, Metabolism and Applications. *Proc. 74th Symp. Lactic Acid Bacteria, Egmond aan Zee, The Netherlands*, Kluwer Academic Publ, Dordrecht, 2002.
2973. H. Steinbök, Zur Frage des Stärkeabbauvermogens einige Milchsäurebakterien. *Milchwiss. Ber.*, 12 (1962) 73–106.
2974. E. Doumer, Starches as indispensable complement in the lactic ferment treatment, *J. Pharm. Chim.*, 20 (1919) 188–190.
2975. J. S. Lee, G. Y. Heo, J. W. Lee, Y. J. Oh, J. A. Park, Y. H. Park, Y. R. Pyun, and J. S. Ah, Analysis of kimchi microflora using denaturing gradient gel electrophoresis, *Int. J. Food Microbiol.*, 102 (2005) 143–150.
2976. M. Kim and J. Chun, Bacterial community structure in kimchi, a Korean fermented vegetable food as revealed by 16S tRNA gene analysis, *Int. J. Food Microbiol.*, 103 (2005) 91–96.
2977. O. Meyerhof, Über die enzymatische Milchsäurebildung in Muskelextract. I, *Biochem. Z.*, 178 (1926) 395–418.
2978. O. Meyerhof, Über die enzymatische Milchsäurebildung in Muskelextract. II. Die Spaltung der Hexosediphosphorsäure, *Biochem. Z.*, 178 (1926) 462–490.
2979. T. Viljava and H. Koivikko, Method for preparing pure lactic acid, EU Patent WO 9704120 (1997); *Chem. Abstr.*, 126 (1997) 170493.
2980. Y. J. Wee, H. O. Kim, J. S. Yun, and H. W. Ryun, Pilot-scale lactic acid production via batch culturing of *Lactobacillus* sp. RKY2 using corn steep liquor as a nitrogen source, *Food Technol. Biotechnol.*, 44 (2006) 293–298.
2981. P. Yin, N. Nishina, Y. Kosakai, K. Yahiro, Y. Park, and M. Okabe, Enhanced production of L(+)-lactic acid from corn starch in a culture of *Rhizopus oryzae* using an air-lift bioreactor, *J. Ferment. Bioeng.*, 84 (1997) 249–253.
2982. S. Miura, T. Arimura, M. Hoshino, M. Kojima, L. Dwiarti, and M. Okabe, Optimization and scale-up of L-lactic acid fermentation by mutant strain *Rhizopus* sp, MK-96-1196 in airlift bioreactors, *J. Biosci. Bioeng.*, 96 (2003) 65–69.
2983. B. Jin, L. P. Huang, and P. Lant, *Rhizopus arrhizus* – A producer for simultaneous saccharification and fermentation of starch waste materials to L(+) lactic acid, *Biotechnol. Lett.*, 25 (2003) 1983–1987.
2984. X. J. Li, L. J. Pan, S. T. S. P. Jiang, L. H. Pan, and Z. Zheng, Production of L-lactic acid from sweet potato starch with *Rhizopus oryzae*, *Hefei Gongye Daxue Xuebao Ziran Kexueban*, 27 (2004) 127–130. *Chem. Abstr.*, 142 (2005) 409784.
2985. D. C. Sheu, K. J. Duan, C. Y. Wu, and M. C. Yu, L-lactic acid production by *Rhizopus oryzae*, US Patent 2006234361 (2006); *Chem. Abstr.*, 145 (2006) 395653.
2986. K. Fukushima, K. Sogo, S. Miura, and Y. Kimura, Production of D-lactic acid by bacterial fermentation of rice starch, *Macromol. Biosci.*, 4 (2004) 1021–1027.

2987. C. Vishnu, G. Seenayy, and G. Reddy, Direct conversion of starch to L(+)-lactic acid by amylase-producing *Lactobacillus amylophilus* GV6, *Bioprocess Eng.*, 23 (2000) 155–158.
2988. M. Altaf, B. J. Naveena, M. Venkateshwar, E. V. Kumar, and G. Reddy, Single step fermentation of starch to (+) lactic acid by *Lactobacillus amylophilus* G6V in SSF using inexpensive nitrogen sources to replace peptone and yeast extract – Optimization by RSM, *Process Biochem.*, 41 (2006) 465–472.
2989. G. Aguilar, J. Morlon-Guyot, B. Trejo-Aguilar, and J. P. Guyot, Purification and characterization of an extracellular α-amylase produced by *Lactobacilllus manihotivoran* LMG 18910^T, an amylolytic lactic acid bacterium, *Enzyme Microb. Technol.*, 27 (2000) 406–413.
2990. J. P. Guyot, M. Calderon, and J. Morton-Guyot, Effect of pH control on lactic acid fermentation of starch by *Lactobacillus manihotivorans* LMG 18010T, *J. Appl. Microbiol.*, 88 (2000) 176–182.
2991. S. Jurcoane, M. Paraschiv, A. D. D. Ionescu, M. Vlad, V. Dodan, F. Soltuzu, L. Nicola, A. Avram, S. Bogdanescu, G. Raitaru and E. Iancu, Lactic acid fermentation by *Lactobacillus delbrueckii* in submerged culture, Roumanian Patent 115177 (1999); *Chem. Abstr.*, 134 (2001) 161997.
2992. W. Soetaert, K. Buchholz, and E. J. Vandamme, Production of D-mannitol and D-lactic acid from starch hydrolysates by fermentation with *Leuconostoc mesenteroides*, *Compt. Rend. Acad. Agric. France*, 80 (1994) 119–126.
2993. K. Kitahara and H. Ishida, Production of lactic acid by direct fermentation from starch (in Japanese), *Bull. Res. Inst. Food Sci., Kyoto Univ.*, 2 (1949) 27–31.
2994. M. Yokoyama, Lactic acid production, Jpn. Kokai Tokkyo Koho 76 88,691 (1976); *Chem. Abstr.*, 85 (1976) 175613j.
2995. L. Ray, G. Mukherjee, and S. K. Majumdar, Production of lactic acid from potato fermentation, *Indian J. Exp. Biol.*, 29 (1991) 681–682.
2996. T. R. Schamala and K. R. Sreekantiah, Fermentation of starch hydrolysates by *Lactobacillus plantarum*, *J. Ind. Microbiol.*, 3 (1988) 175–178.
2997. R. Anuradha, A. K. Suresh, and K. V. Venkatesh, Simultaneous saccharification and fermentation of starch to lactic acid, *Process Biochem.*, 35 (1999) 367–375.
2998. K. I. Ago, M. Azuma, and K. Takahashi, Liquefaction for the lactic acid fermentation from rice, *J. Chem. Eng. Jpn.*, 40 (2007) 529–533.
2999. W. C. Lee, Production of D-lactic acid by bacterial fermentation of rice, *Fibers Polym.*, 8 (2007) 571–578.
3000. M. Sumiyoshi, K. Natsui, and Y. Nomura, Production of lactic acid from starch by simultaneous saccharification and fermentation, *Biotech. Sust. Util. Biol. Resources Tropics*, 17 (2004) 7–12.
3001. C. Figueroa, A. M. Davila, and J. Pourquie, Original properties of ropy strains of *Lactobacillus plantarum* isolated from the sour cassava starch fermentation, *J. Appl. Microbiol.*, 82 (1997) 68–72.
3002. K. J. Zielińska, K. M. Stecka, A. H. Miecznikowski, and A. M. Suterska, Degradation of raw potato starch by an amylolytic strain of *Lactobacillus plantarum* C, *Progr. Biotechnol.*, 17 (2000) 187–192.
3003. S. H. Panda and R. C. Ray, Direct conversion of raw starch to lactic acid by *Lactobacillus plantarum* MTCC 1407 in semi-solid fermentation using sweet potato (*Ipomea batatas* L) flour, *J. Sci. Ind. Res.*, 67 (2008) 531–537.
3004. J. Pintado, M. Raimbault, and J. P. Guyot, Influence of polysaccharides on oxygen dependent lactate utilization by an amylolytic *Lactobacillus plantarum* strain, *Int. J. Food Microbiol.*, 98 (2005) 81–88.
3005. P. H. Billard, D. Montet, and J. Pourquie, Lactic fermentation of cassava starch enriched in yeast proteins, *Microb. Aliment. Nutr.*, 15 (1997) 315–321.
3006. S. H. Panda, M. Parmanick, and R. C. Ray, Lactic acid fermentation of sweet potato (*Ipomoea batatas* L.) into pickles, *J. Food Proc. Preserv.*, 31 (2007) 83–101.
3007. J. L. Parada, E. Zapata, S. V. de Fabrizio, and A. Martinez, Microbiological and technological aspects of cassava-starch fermentation, *World J. Microbiol. Biotechnol.*, 12 (1996) 53–56.

3008. L. J. Forney and C. A. Reddy, Fermentative conversion of potato processing wastes into crude protein feed supplement by lactobacilli, *Dev. Ind. Microbiol.*, 18 (1976) 135–143.
3009. K. J. Zielińska, K. M. Stecka, A. H. Miecznikowski, and A. M. Suterska, Degradation of raw potato starch by the amylases of lactic acid bacteria (in Polish), *Pr. Inst. Lab. Bad. Przem. Spoz.*, 55 (2000) 22–29.
3010. J. Yan, R. Bajpai, E. Iannotti, M. Popovic, and R. Mueller, Lactic acid fermentation from enzyme-thinned starch with immobilized *Lactobacillus amylovorus*, *Chem. Biochem. Eng. Quart.*, 15(2), (2001) 59–63.
3011. M. Morita and Y. Yokota, Pretreatment of fermentation feed for lactic acid production – Liquefaction of potato starch in lactic acid solution (in Japanese), *Kagaku Kogaku Ronbunshu*, 22 (1996) 938–940. *Chem. Abstr.*, 125 (1996) 112854.
3012. Zhejiang Institue of Technology, Lactic acid fermentation by immobilized cells. I., *Zhejiang Gongxueyuan Xuebao* (3), (1991) 9–15. *Chem. Abstr.*, 116 (1992) 126882.
3013. L. K. Nakamura and C. D. Crowell, *Lactobacillus amylophilus* a new starch-hydrolyzing species from swine waste-corn fermentation, *Dev. Ind. Microbiol.*, 20 (1979) 531–540.
3014. D. M. Flores, K. Shibata, J. L. Tagubase, G. Kobayashi, and K. Sonomoto, Screening, isolation and identification of potent lactic acid bacteria for direct lactic acid fermentation of sago starch, *Biotech. Sust. Util. Biol. Resources Tropics*, 17 (2004) 13–19.
3015. S. Sirisansaneeyakul, P. Mekvichitsaeng, K. Kittikusolthum, S. Pattaragulwanit, M. Laddee, S. Bhuwapathanapun, and A. Ishizaki, Lactic acid production from starch hydrolysates using *Lactococcus lactis* IO-1, *Thai J. Agric. Sci.*, 33 (2000) 53–64.
3016. C. Nolasco-Hipolito, E. Crabbe, G. Kobayashi, K. Sonomoto, and A. Ishizaki, pH-dependent continuous lactic acid fermentation by *Lactococcus lactis* IO-1 using hydrolyzed sago starch, *J. Fac. Agric. Kyushu Univ.*, 44 (2000) 367–375.
3017. A. Ishizaki, C. Nolasco-Hipolito, G. Kobayashi, and K. Sonomoto, Continuous L-lactic acid production using *Lactococcus lactis* IO-1 from hydrolyzed sago starch, *Biotech. Sust. Util. Biol. Resources Tropics*, 14 (2000) 49–59.
3018. T. Sato, H. Nawa, S. Sasaki, and K. Fukuda, Saccharification of sago palm by wood-rotting fungi and fermenting into lactic acid, *JIRCAS Work. Rep.*, 39 (2005) 144–147. *Chem. Abstr.*, 143 (2005) 476478.
3019. K. Hofvendahl, C. Akerberg, G. Zacchi, and B. Hahn-Hagerdal, Simultaneous enzymatic wheat starch saccharification and fermentation to lactic acid by *Lactococcus lactis*, *Appl. Microbiol. Biotechnol.*, 52 (1999) 163–169.
3020. M. Calderon Santoyo, G. Loiseau, R. Rodriguez Sanoja, and J. P. Guyot, Study on starch fermentation at low pH by *Lactobacillus fermentum* Ogi E1 reveals uncoupling between growth and α-amylase production at pH 4.0, *Int. J. Food Microbiol.*, 80 (2002) 77–87.
3021. G. Diaz Ruiz, J. P. Guyot, F. Ruiz Teran, J. Morlon-Guyot, and C. Wacher, Microbial and physiological characterization of weakly amylolytic but fast-growing lactic acid bacteria: A functional role in supporting microbial diversity in pozol, a Mexican fermented maize beverage, *Appl. Environ. Microbiol.*, 69 (2003) 4367–4374.
3022. K. Richter, U. Becker, R. Berger, I. Rühlemann, W. Hohaus, E. Mendow, and B. Kupfer, Use of potato hydrolyzates as substrate for fermentative manufacture of lactic acid, East German Patent 266590 (1989); *Chem. Abstr.*, 112 (1990) 97036.
3023. K. Hoshino, M. Taniguchi, H. Marumoto, K. Shimizu, and M. Fujii, Continuous lactic acid production from raw starch in a fermentation system using a reversibly soluble-autoprecipitating amylase and immobilized cells of *Lactobacillus casei*, *Agric. Biol. Chem.*, 55 (1991) 479–485.
3024. Y. Ohkouchi and Y. Inoue, Direct production of (+) lactic acid from starch and food waste using *Lactobacillus manihotivorans* LMG18011, *Bioresource. Technol.*, 97 (2006) 1554–1562.
3025. L. Kung, Jr., and N. K. Ranjit, The effect of *Lactobacillus buchneri* and other additives on the fermentation and aerobic stability of barley silage, *J. Dairy Sci.*, 84 (2001) 1149–1155.

3026. N. K. Ranjit, C. C. Taylor, and L. Kung, Jr., Effect of *Lactobacillus buchneri* 40788 on fermentation, aerobic stability and nutritive value of maize silage, *Grass Forage Sci.*, 57 (2002) 73–81.
3027. A. T. Adesogan and M. B. Salawu, Effect of applying formic acid, heterolactic bacteria or homolactic and heterolactic bacteria on the fermentation of bi-crops of peas and wheat, *J. Sci. Food Agric.*, 84 (2004) 983–992.
3028. S. Yang, Experimental research on the production of citric acid from maize starch, *Huagong Yejin*, 12 (1991) 150–155. *Chem. Abstr.*, 116 (1992) 5221.
3029. T. K. Nguyen, L. Martinkova, L. Seichert, and F. Machek, Citric acid production by *Aspergillus niger* using media containing low concentration of glucose or corn starch, *Folia Microbiol. (Prague)*, 37 (1992) 433–441.
3030. W. Lesniak, J. Pietkiewicz, and W. Podgorski, Citric acid fermentation from starch and dextrose syrups by a trace metal resistant mutant of *Aspergillus niger*, *Biotechnol. Lett.*, 24 (2002) 1065–1067.
3031. I. U. Haq, S. Ali, and J. Iqbal, Direct production of citric acid from raw starch by *Aspergillus niger*, *Process Biochem.*, 38 (2003) 921–924.
3032. E. Gąsiorek and W. Leśniak, The effect of natural raw material addition on the yield of citric acid biosynthesis by solid substrate fermentation (in Polish), *Zywnosc (Food)*, 10(1), (2003) 48–58.
3033. S. Rugsaseel, K. Kirimura, and S. Usami, Selection of mutants of *Aspergillus niger* showing enhanced productivity of citric acid from starch in shaking culture, *J. Ferment. Bioeng.*, 75 (1993) 226–228.
3034. S. Moruya and K. S. Janari, Production of citric acid from starch-hydrolysate by *Aspergillus niger*, *Microbiol. Res.*, 155 (2000) 37–44.
3035. N. Yu. Sharova and T. A. Nikiforova, Method for production of citric acid alpha-amylase and glucoamylase, Russian Patent 2266950 (2004); *Chem. Abstr.*, 144 (2005) 68684
3036. M. Calderon, G. Loiseau, and J. P. Guyot, Fermentation by *Lactobacillus fermentum* Ogi E1 of different combinations of carbohydrates occurring naturally in cereals; consequences on growth energetics and α-amylase production, *Int. J. Food Microbiol.*, 80 (2003) 161–169.
3037. T. A. Nikiforova, E. P. Nazarets, L. N. Mushnikova, I. N. Voronova, A. V. Galkin, and T. A. Pozdnyakova, Method of citric acd production (in Russian), Russian Patent 2132384 (1999).
3038. N. Yu. Sharova, L. N. Mushnikova, T. A. Pozdnyakova, and T. A. Nikiforova, Method of citric acid production, Russian Patent 2186850 (2002); *Chem. Abstr.*, 138 (2003) 135947.
3039. F. C. Prado, L. P. S. Vanderberghe, A. L. Wojciechowski, L. A. Rodrigues-Leon, and C. R. Soccol, Citric acid production by solid-state fermentation on a semi-pilot scale using different percentages of treated cassava bagasse, *Braz. J. Chem. Eng.*, 22 (2005) 547–555.
3040. Y. Yamada and P. Yukhpan, Genera and species in acetic acid bacteria, *Int. J. Food Microbiol.*, 125 (2008) 15–24.
3041. J. Cleerwerck and P. de Vos, Polyphasic taxonomy of acetic acid bacteria. An overview of currently applied methodology, *Int. J. Food Microbiol.*, 125 (2008) 2–14.
3042. P. Raspor and D. Goranović, Biotechnological applications of acetic acid bacteria, *Crit. Rev. Biotechnol.*, 28 (2008) 101–124.
3043. D. Zuo, A. Li, G. Li, and X. Xu, Production of vinegar, *Zhongguo Tiaoweipin* (12), (1999) 4–8. *Chem. Abstr.*, 133 (2000) 42436.
3044. N. Jiang, Q. Wang, A. Shen, P. He, and D. Lu, Method for producing pyruvic acid with *Trichosporum cutaneum* by fermentation, Chinese Patent 1468964 (2004); *Chem. Abstr.,* 142 (2004) 36996.
3045. S. Anastassiadis and H. J. Rehm, Continuous gluconic acid production by the yeast-like *Aureobasidium pullulans* in a cascading operation of two bioractors, *Appl. Microbiol. Biotechnol.* 73 (2006) 541–548.
3046. S. Anastassiadis and I. G. Morgunov, Gluconic acid production, *Recent Pat. Biotechnol.* 1 (2007) 167–180.

3047. Y. C. Su, W. H. Liu, and L. Y. Jang, Studies on microbial production of sodium gluconate and glucono-δ-lactone from starch, *Proc. Natl. Sci. Counc. (Taiwan)*, Pt. 2, 10 (1977) 143–160; *Chem. Abstr.*, 88 (1978) 61102y.
3048. V. O. Singh and R. P. Singh, Utilization of agro-food by-products for gluconic acid production by *Aspergillus niger* ORS-4 under surface culture cultivation, *J. Sci. Ind. Res.* 61 (2002) 356–360.
3049. M. Mączyński, P. Gzyl, M. Remiszewski, Z. Antecki, G. Kaczmarowicz, S. Walisch, W. Wnuk, and A. Sokołowski, Method of obtaining sodium gluconate through a microbial processes, Polish Patent 154,789 (1993); *Chem. Abstr.*, 122 (1995) 104053.
3050. Z. Wang, L. Wu, M. Zhou, and Y. Ding, Duoenzyme saccharification in glutamic acid fermentation, *Zhongguo Tiaoweipin* (9), (1991) 12–14. *Chem. Abstr.*, 116 (1992) 233817.
3051. A. B. Ariff, R. Mohamed, P. C. X. Sim, M. M. Saileh, and M. I. A. Mohamed, Production of kojic acid by *Aspergillus flavus link* using starch as a carbon source, *Biotechnol. Sust. Util. Biol. Resources Tropics*, 11 (1997) 296–302.
3052. M. Rozfarizan, A. B. Ariff, H. Shimizu, M. A. Hassan, M. I. A. Karim, and S. Shioya, Direct fermentation of sago starch to kojic acid by *Aspergillus niger*, *Biotechnol. Sust. Util. Biol. Resources Tropics*, 14 (2000) 242–257.
3053. R. I. Santamaria, G. Del Rio, G. Saab, M. E. Rodriguez, X. S. Oberon, and A. Lopez- Munguia, Alcoholysis reactions from starch with α-amylases, *FEBS Letts.*, 452 (1999) 346–350.
3054. N. Nakanishi and Y. Hidaka, Manufacture of ethyl α-glucoside with α-glucosidase of *Aspergillus* species, Jpn. Kokai Tokkyo Koho 2002017396 (2002); *Chem. Abstr.*, 136 (2002) 84803.
3055. J. Larsson, D. Svensson, and P. Aldercreutz, α-Amylase catalyzed synthesis of alkyl glucosides, *J. Mol. Catal.*, B 37(1–6), (2005) 84–87.
3056. W. Ma, Q. Ban, and J. Wu, Amylase-catalyzed synthesis of alkyl glucosides, Chinese Patent 101168761 (2008); *Chem. Abstr.*, 148 (2008) 536242.
3057. J. Y. Park, S. O. Lee, and T. H. Lee, Synthesis of 1-*O*-benzyl-α-glucoside and 1-*O*-α – Maltoside by trnasglycosylation of α-amylase from soluble starch in aqueous solution, *Biotechnol. Lett.*, 21 (1999) 81–86.
3058. A. Taki, O. Uejima, K. Ogawa, and Y. Usui, Manufacture of genipin malto-oligosaccharides from malto-oligosaccharides and geniposide with amylase, Jpn. Kokai Tokkyo Koho, 03058791 (1991); *Chem. Abstr.*, 115 (1991) 181548.
3059. Y. C. Shin and M. S. Byun, A novel maltotetraose-forming alkaline α-amylase from an alkalophoilic *Bacillus* strain GM 8901. In: K. H. Park, J. Robyt, and Y. D. Choi, (Eds.) *Enzymes for Carbohydrate Engineering*, Elsevier Sci. B.V, Amsterdam, 2000, pp. 61–81.
3060. G. J. Woo and J. D. McCord, Bioconversion of starches into maltotetraose using *Pseudomonas stutzeri* maltotetrahydrolase in a membrane recycle bioreactor: Effect of multiple enzyme systems and mass balance study, *Enzyme Microb. Technol.*, 16 (1994) 1016–1020.
3061. J. Schmidt and J. Michael, Starch metabolism in *Pseudomonas stutzeri*, I. Studies on maltotertraose-forming amylase, *Biochim. Biophys. Acta*, 566 (1979) 88–99.
3062. J. Y. Park, K. S. Kim, Y. H. Kim, S. K. Kim, J. I. Lee, C. W. Seo, and H. H. Kim, Enzymic preparation of benzyl- and phenyl-α-D-glucopyranoside derivatives, S. Korean Patent 167489 (1999); *Chem. Abstr.*, 142 (2004) 73501.
3063. S. Mori and M. Goto, Glycosides manufacture with γ-cyclodextrin glucanotransferase, Jpn. Kokai Tokkyo Koho 09009987 (1997); *Chem. Abstr.*, 126 (1997) 185093.
3064. C. P. Chiu, Y. H. Lee, H. Y. Chiang, and H. T. Tsai, Studies on the yield of coupling sugar from starch using alkalophilic Bacillus CGTase, *Jiemian Kexue Huizhi*, 14(1), (1991) 13–19. *Chem. Abstr.*, 116 (1992) 19685.
3065. K. Washino and M. Iwata, Preparation of quercetin 3-*O*-glycoside and method for modification of water-sparingly soluble flavonoid using glycosides, Jpn. Kokai Tokkyo Koho 07010898 (1995); *Chem. Abstr.*, 122 (1995) 265924.

3066. S. Maruo, H. Yamashita, K. Miyazaki, H. Yamamoto, Y. Kyotani, H. Ogawa, M. Kojima, and Y. Ezure, A novel and efficient method for enzymic synthesis of high purity maltose using moranoline(1-deoxynojirimycin), *Biosci. Biotechnol. Biochem.* 56 (1992) 1406–1409.
3067. M. Sugiyama, Y. Ezure, M. Kojima, and K. Katsunori, Enzymic manufacture of moranoline derivatives, Jpn. Kokai Tokkyo Koho 62267292 (1987); *Chem. Abstr.*, 109 (1988) 209662.
3068. K. Koizumi, Y. Okada, Y. Kubota, and T. Utamura, Pharmaceutical compositions containing α, β or γ-cyclodextrins to improve water-solubility of drugs, Jpn. Kokai Tokkyo Koho, 63027440 (1988); *Chem. Abstr.*, 110 (1989) 121385.
3069. S. Kobayashi, W. Suzuki, N. Watanabe, and R. Ooya, Glucosylcyclodextrin enzymic preparation, Jpn. Kokai Tokkyo Koho 08056691 (1996); *Chem. Abstr.*, 124 (1996) 287225.
3070. T. Nakada and M. Kubota, Enzymic manufacture of pharmaceutical α-glycosyl derivatives of catecholamines, EU Patent EP 564099 (1993); *Chem. Abstr.*, 121 (1994) 7422.
3071. T. G. Kim, D. C. Park, H. D. Shin, and Y. H. Lee, Preparation method of glucosyl xylitol using intermolecular transglycosylation reaction of cyclodextrin glucanotransferase, S. Korean Patent 2000040009 (2000); *Chem. Abstr.*, 137 (2002) 93947.
3072. P. Degn, K. L. Larsen, J. O. Duus, B. O. Pedersen, and W. Zimmermann, Two-step enzymatic synthesis of maltooligosaccharide esters, *Carbohydr. Res.* 329 (2000) 57–63.
3073. T. Watanabe, M. Kuwabara, T. Koshijima, M. Ueda, and M. Nakajima, Preparation of oligosaccharides by enzymic transglycosidation of cellololigosaccharides with starch etc, Jpn. Kokai Tokkyo Koho, 06271596 (1994); *Chem. Abstr.*, 122 (1995) 56402.
3074. M. Karube and T. Morita, Manufacture of high polymerization degree oligosaccharides from polysaccharides, Jpn. Kpkai Tokkyo Koho 06121693 (1994); *Chem. Abstr.*, 121 (1994) 155930.
3075. Y. Sakano, R. Sakamoto, N. Uotsu, K. Hosoya, and T. Kimura, Enzymie manufacture of heterooligosaccharides from starch or starch hydrolyzate, Jpn. Kokai Tokkyo Koho, 2005278457; *Chem. Abstr.*, 143 (2005) 385290.
3076. K. Funane, K. Mizuno, K. Terasawa, Y. Kitamura, T. Baba, and M. Kobayashi, Reaction of rice branching enzyme RBEI on various kinds of starches analyzed by high-performance anion exchange chromatography, *J. Appl. Glycosci.*, 46 (1999) 453–457.
3077. T. Sawai and E. J. Hehre, A novel amylase (*Candida* transglucosyl-amylase) that catalyzed glucosyl transfer from starch to dextrin, *J. Biol. Chem.*, 237 (1962) 2047–2052.
3078. M. J. E. C. van der Maarel, G. J. W. Euverink, D. J. Binnema, H. T. P. Bos, and J. Bergsma, Amylomaltase from the hyperthermophilic bacterium *Thermus thermophilum*: Enzyme characteristics and application in the starch industry, *Mededelingen –Fac. Landdbouwk. ToegepasteBiol. Wetensch. Univ. Gent*, 65(3a), (2000) 231–234.
3079. M. J. E. C. van der Maarel, I. Capron, G. J. W. Euverink, H. T. Bos, T. Kaper, D. J. Binnema, and P. A. M. Steenken, A novel thermoreversible gelling product made by enzymic modification of starch, *Starch/Stärke*, 57 (2005) 465–472.
3080. T. Miyake and H. Tsuchiya, Enzymic preparation of glycosylglycirrhizin as calorie-low sweetener, Jpn. Kokai Tokkyo Koho 02219595 (1980); *Chem. Abstr.*, 114 (1991) 60704.
3081. I. Yamamoto, N. Muto, and T. Miyake, Alpha-glycosyl-1-ascorbic acid and its preparation and use in foods, pharmaceuticals and cosmetics, EU Patent EP 398484 (1990); *Chem. Abstr.*, 114 (1991) 183859.
3082. Y. H. Lee, H. D. Shin, and T. K. Kim, Enzymic transglycosylation method of L-ascorbic acid in heterogenous system by using insoluble starch as glycosyl donor which allows of enhancement of yield and reaction rate of L-ascorbic acid, inhibition of production of by-products and easy removal of insoluble remaining glycosyl donor, S. Korean Patent 2005061007 (2005); *Chem. Abstr.*, 145 (2006) 375460.

3083. T. Shibuya, H. Chaen, and S. Sakai, Lactoneotrehalose and its preparation and use, EU Patent EP 480640 (1992); *Chem. Abstr.*, 117 (1992) 27048.
3084. R. A. Gross, Preparation of acylated sugars via enzymic esterification of monosaccharides and polysaccharides with (R) -3-hydroxybutyric acid, EU Patent WO 2006012490 (2006); *Chem. Abstr.*, 144 (2006) 171193.
3085. K. Mukai, K. Tsusaki, M. Kubota, S. Fukusa, and T. Miyake, Process for producing 2-O-α- D-glucopyranosyl-L-ascorbic acid, EU Patent WO 2004013344; *Chem. Abstr.*, 140 (2004) 180245.
3086. Y. Zhang, Maltodextrin modified by crosslinking-enzymolysis or crosslinking-esterification-enzymolysis its preparation method and application, Chinese Patent 1594364 (2005); *Chem. Abstr.*, 144 (2006) 305116.
3087. Z. Lubiewski, J. Le Thanh, L. Stendera, and G. Lewandowicz, Enzymatic hydrolysis of starch sodium octenyl succinate in a recirculation membrane reactor (in Polish), *Zywnosc (Food)*, 14(5), (2007) 9–22.
3088. A. Copinet, C. Bilard, J. P. Onteniente, and Y. Couturier, Enzymic degradation and deacetylation of native and acetylated starch-based extruded blends, *Polym, Degr. Stab.* 71 (2001) 203–212.
3089. S. Chakraborty, B. Sahoo, I. Teraoka, L. M. Miller, and R. A. Gross, Enzyme-catalyzed regioselective modification of starch nanoparticles, *ACS Symp. Ser.*, 900 (2005) 246–265.
3090. S. Chakraborty, B. Sahoo, I. Teraoka, L. M. Miller, and R. A. Gross, Enzyme-catalyzed regioselective modification of starch nanoparticles, *Macromolecules*, 38 (2005) 61–68.
3091. A. Rajan, V. S. Prasad, and T. E. Abraham, Enzymatic esterification of starch using recovered coconut oil, *Int. J. Biol. Macromol.*, 39 (2006) 265–272.
3092. N. V. Thoai, J. Roche, and M. J. Silhol-Bernere, Dephosphorylation and the action of amylases on starch phosphates and glycogen phosphates, *Compt. Rend. Acad. Sci.*, 223 (1946) 931–933.
3093. F. von Falkenhausen, Darstellung und enzymatische Spaltung von Amylophosphorsäuren, *Biochem. Z.*, 253 (1932) 152–160.
3094. Y. Yang and X. Chen, Study on edible wrapping film made by modified starch compound by enzymic esterification, *Shipin Gongye Keji* (1), (1997) 28–30. *Chem. Abstr.*, 126 (1997) 329863.
3095. K. H. Meyer and P. Bernfeld, Recherches sur l'amidon. XXII. L'action de la phosphorylase de pommes de terre, *Helv. Chim. Acta*, 25 (1942) 404–405.
3096. K. H. Meyer and P. Bernfeld, Recherches sur l'amidon.XXI. Sur les enzymes amylolytiques de la levuse, *Helv. Chim. Acta*, 25 (1942) 399–403.
3097. P. N. Hobson, W. J. Whelan, and S. Peat, The enzyme synthesis and degradation of starch. XII. The mechanism of synthesis of amylopectin, *J. Chem. Soc.* (1951) 596–598.
3098. S. Shibuya, S. Suzuki, and M. Nishida, Enzymic phosphorolysis of starch and inulin (in Japanese), *J. Chem. Soc. Jpn., Pure Chem. Sect.*, 73 (1952) 168–170.
3099. A. Kumar and G. G. Sanwal, Kinetics of starch phosphotylase from young banana leaves, *Phytochemistry*, 27 (1988) 983–988.
3100. S. J. Kelly and L. G. Butler, Enzymic approaches to production of sucrose from starch, *Biotechnol. Bioeng.*, 22 (1980) 1501–1507.
3101. S. Daurat-Lerroque, L. Hammar, and W. J. Whelan, Enzyme reactors for the continuous synthesis of sucrose from starch, *J. Appl. Biochem.*, 4 (1982) 133–152.
3102. A. Kamogawa and T. Fukui, Inhibition of α-glucan phosphorylase by bisulfite competition at the phosphate binding site, *Biochim. Biophys. Acta*, 302 (1973) 158–166.
3103. D. K. Myers and N. K. Matheson, Affinity purification and kinetics of pea cotyledon starch phosphorylases, *Phytochemistry*, 30 (1991) 1079–1087.
3104. Y. Sato and M. Hirao, α-Glucose 1-phosphate, German Patent 2,031,269 (1971); *Chem. Abstr.*, 74 (1970) 100816s.
3105. T. Imamura, S. Kayane, T. Kurosaki, M. Morishita, and M. Tanigaki, Method for purifying sugar phosphates or their salts, US Patent Appl. 07/012,496 (1987).

3106. C. S. Hanes, Breakdown and synthesis of starch by an enzyme system from pea seeds, *Proc. Roy. Soc. (London)*, B128 (1940) 421–450.
3107. S. Pavgi-Upadhye and A. Kumar, Immobilization of starch phosphorylase from Bengal gram seeds: Production of glucose 1-phosphate, *Genet. Eng. Biotechnol.*, 16 (1996) 145–151.
3108. G. P. R. de Chatelperon, C. M. T. A. Bonissol, G. M. S. Paris, and G. F. J. Blachere, preparation of α-glucose-6-phosphate, French Patent 1,379,668 (1964).
3109. Y. Nishimura and K. Tsukamoto, Formation of glucose 6-phosphate from starch (in Japanese), *Eiyo to Shokuryo*, 19(1), (1966) 30–39.
3110. N. V. Thoai, J. Roche, and E. Danzas, Nonhydrolyzing phosphorylation of starch and glycogen with alkaline phosphates, *Compt Rend. Acad. Sci.*, 222 (1946) 259–261.
3111. G. Sarzana and F. Cacioppo, The phosphoirylation of starch by enzyme in various animal tissues (in Italian), *Biochim Terap. Sper.*, 25 (1938) 359–365.
3112. P. Ostern, J. A. Guthke, and B. Umschweif, Enzymatische Phosphorilierung der Stäerke, *Enzymologia*, 3 (1937) 5–9.
3113. M. Samec, Dephosphorylation of potato starch by brain phosphatase (in Italian), *Atti Accad. Ital. Rend. Cl. Sci. Fis. Mat. Natur.*, 3 (1941) 128–131.
3114. T. Iwata, T. Funaguma, and A. Hara, Purification and some properties of two phosphorylases from *Typha latifolia* pollen, *Agric. Biol. Chem.*, 52 (1988) 407–412.
3115. A. Alissandrolos, N. Baudendistel, S. L. Flitsch, B. Hanes, and P. J. Halling, Lipase-catalyzed acylation of starch and determination of the degree of substitution by methanolysis and GC, *BMC Biotechnol.*, 10 (2010) 82.
3116. S. D. Watson and B. I. Pietschke, The effect of sulfide on α-glucosidases: Implications for starch degradation in anaerobic bioreactors, S. Afr *Chemosphere*, 65 (2006) 159–164.
3117. M. Kitagawa and Y. Taguchi, Apparatus and method for anaerobic digestion of starch particle-containing wastewaters, Jpn. Kokai Tokkyo Koho, 2002224686 (2002); *Chem. Abstr.*, 137 (2002) 144798.
3118. S. H. Bhosale, M. B. Rao, and V. V. Deshpande, Molecular and industrial aspects of glucose isomerase, *Microbiol. Rev.*, 60 (1996) 280–300.
3119. S. H. Bok, M. Seideman, and P. W. Wopet, Selective isolation of acidophilic *Streptomyces* strain for glucose isomerase production, *Appl. Environ. Microbiol.*, 47 (1984) 1213–1215.
3120. R. O. Horwath and R. M. Irbe, Enzyme for the preparation of fructose, EU Patent EP 97,973 (1984); *Chem. Abstr.*, 100 (1983) 101628z.
3121. K. Dekker, H. Yamagata, K. Sakaguchi, and S. Udaka, Xylose (glucose) isomerase gene from the thermophile *Thermus thermophilus*: Cloning, sequencing and comparison with other thermostable xylose isomerases, *J. Bacteriol.*, 173 (1991) 3078–3083.
3122. C. Lee, L. Bhatnagar, B. C. Saha, Y. E. Lee, M. Takagi, T. Imanaka, M. Bagdasarian, and J. G. Zeikus, Cloning and expression of the *Clostridium thermosulfurogenes* glucose isomerase gene in *Escherichia coli* and *Bacillus subtilis*, *Appl. Environ. Microbiol.*, 56 (1990) 2638–2643.
3123. L. F. Rojas, J. C. Mazo, C. Sanchez, R. Rios, and C. Figueroa, Effects of Ca^{2+} on the catalytic activity of immobilized glucose isomerase during production of high fructose syrup from yucca starch, *Colombia Aliment. Equip. Tecnol.*, 184 (2003) 49–54.
3124. G. Weidenbach, D. Bonse, and G. Richter, Advances in the enzymic processing of the starch industry, *Proc. BioTech*, 83 (1983) 1017–1027. *Chem. Abstr.*, 100 (1984) 123115g.
3125. R. G. P. Walon, Levulose from granuar starch, US Patent 4,009,074 (1977); *Chem Abstr.*, 86 (1977) 137969t.
3126. R. V. Macallister, N. E. Lloyd, R. G. Dworschack, and W. Nelson, Jr., Fructose containing syrup, British Patent 1,267,119 (1972); *Chem. Abstr.*, 77 (1972) 7647q.
3127. R. E. Hebeda, H. W. Leach, and R. G. P. Walon, Preparation of levulose from granular starch, British Patent 1,450,671 (1976); *Chem. Abstr.*, 86 (1976) 70137z.

3128. CPC International Inc., Conversion of starch into levulose, Dutch Patent 75 03,239 (1975); *Chem. Abstr.*, 86 (1976) 53916k.
3129. R. E. Hebeda, H. W. Leach, and R. G. Walon, Levulose from granular starch, Belgian Patent 820,243 (1975); *Chem. Abstr.*, 83 (1974) 166151p.
3130. A. Emery, V. W. Rodwell, H. C. Lim, P. C. Wankat, F. E. Regmer, and D. R. Schneider, Enzymic production of fructose, *Natl. Sci. Found., Res. Appl. Natl. Needs*, [*Rep.*] NSF/RA (US)-760032 (1975) 73–79; *Chem. Abstr.*, 88 (1975) 20518v.
3131. Y. Ghali, R. M. Attia, M. Roushdi, and M. Alaa, El-Din, Studies on the combined action of amylases and glucose isomerase on starch and its hydrolysates. II. Immobilization of glucoamylase, *Starch/Stärke*, 32 (1980) 303–308.
3132. A. Lindroos, Y. Y. Linko, and P. Linko, Barley starch conversion by immobilized glucoamylase and glucose isomerase, *Proc. 2nd Conf. Food Process Eng., 1979*, 2 (1980) 92–102; *Chem. Abstr.*, 94 (1980) 138021e.
3133. W. Błaszków and T. Miśkiewicz, Studies on the glucose fructose syrup production from pure glucose and potato starch on columns with immobilized enzymes, glucose isomerase, α-amylase and glucoamylase (in Polish), *Pr. Nauk. Akad. Ekon. Im. Oskara Langego Wroclaw*, 167 (1980) 5–13, 15.
3134. H. Mueller, Glucose-fructose mixtures from grains, Swiss Patent 623,357 (1981); *Chem. Abstr.*, 95 (1981) 95797p.
3135. H. Müller, Glucose and fructose from starch-containing plant products, Swiss Patent 622,028 (1981); *Chem. Abstr.*, 95 (1981) 26915p.
3136. L. Lin, J. Chen, K. Yang, J. Zhao, B. Xie, and X. Chen, Study on enzymic conversion of *Cana edulis* starch to fructose, *Shipin Kexue*, 39 (1984) 9–15. *Chem. Abstr.*, 103 (1985) 21372b.
3137. L. Zhu, J. Xie, and R. Derjani, Study on the production of maltodextrins and fructose syrup from banana starch, *Henan Gongye Daxue Xuebao, Ziran Kexueban*, 26(6), (2005) 53–56. *Chem. Abstr.*, 147 (2006) 71516.
3138. G. Tegge and R. Richter, Hirse und Bruchreis als Basismaterial der Glucosegewinnung, *Starch/Stärke*, 34 (1982) 386–390.
3139. T. Kaneko, S. Takahash, and K. Saito, Characterization of acid-stable glucose isomerase from *Streptomyces* sp. and development of single-step process for high-fructose corn sweetener (HFCS) production, *Biosci. Biotechnol. Biochem.*, 64 (2000) 940–947.
3140. Y. Wang, L. Wang, W. Li, and, A. Si, Process for manufacturing crystalline fructose from corn starch, Chinese Patent 1876845 (2006); *Chem. Abstr.*, 146 (2006) 192611.
3141. N. H. Aschengreen, B. Helwig Nielsen, P. Rosendal, and J. Ostergaard, Liquefaction, saccharification and isomerization of starches from sources other than maize, *Starch/Stärke*, 31 (1979) 64–66.
3142. R. I. Antrim, W. Colill, and B. J. Schnyder, Glucose isomerase production of high fructose syrups, *Appl. Biochem. Bioeng.*, 2 (1979) 97–155.
3143. R. E. Hebeda, H. W. Leach, and R. G. Philip, Production of fructose from starch, USSR Patent 688,138 (1979); *Chem. Abstr.*, 92 (1980) 20577w.
3144. Y. Takasaki, Conversion of starch into a D-fructose-D-glucose mixture, Japan Kokai Tokkyo Koho 73 22,643 (1973); *Chem. Abstr.*, 79 (1973) 51839u.
3145. K. Parker, M. Salas, and V. C. Nwosu, High fructose corn syrup. Production, uses and public health concerns, *Biotechnol. Mol. Biol. Rev.*, 5 (2010) 71–78.
3146. H. Müller, A mixture of glucose and fructose from raw cereals, Swiss Patent 622,552 (1981); *Chem. Abstr.*, 95 (1981) 95793j.
3147. E. Junko Tomotani and M. Vitolo, Production of high-fructose syrup using immobilized invertase in a membrane reactor, *J. Food Eng.*, 80 (2006) 662–667.
3148. N. M. Khalid and P. Markakis, Production of high-fructose syrup from cassava starch, *Proc. Symp. 2nd Int. Flavor Conf.*, 1 (1981) 319–326; *Chem. Abstr.*, 95 (1981) 202253g..

3149. H. P. G. Knapik and W. H. Mueller, Staged immobilized amyloglucosidase and immobilized glucose isomerase in producing fructose from thinned starch, US Patent 4,582,803 (1986); *Chem. Abstr.*, 104 (1986) 223909v.
3150. K. Yamanaka, The utilization of sugar isomerases. I. Basal conditions for glucose isomerase reaction. (in Japanese), *Nippon Nogei Kagaku Kaishi*, 37 (1963) 231–236.
3151. M. Boruch and E. Nebesny, Die Wirkung von Glucoseisomerase auf Oligosaccharide in Stärkehydrolysaten, *Starch/Stärke*, 31 (1979) 345–347.
3152. G. L. Shukla and K. A. Prabhu, Production of high fructose syrup as sweetening agent from corn and bagasse, *Proc. 49th Annu. Conv. Sugar Technol. Assoc. India*, pp. G41–G48. *Chem. Abstr.*, 107 (1987) 153036u.
3153. Y. Takeda, T. Kiyofuji, and S. Hizukuri, Improved co-immobilizate of glucoamylase and glucose isomerase. Continuous production of fructose syrup directly from liquefied starch, *Denpun Kagaku*, 38 (1991) 375–378.
3154. X. Cha, H. Zhou, W. Li, and J. Shen, Effects of microenvironment on immobilized enzymes – Effects of the property of carriers ionic groups, *Shengwu Huaxue Zazhi*, 4 (1988) 558–565. *Chem. Abstr.*, 110 (1989) 110816.
3155. S. Yabuki, T. Hanatani, and S. Yamada, Fructose separation from invert sugar or glucose isomeriation mixtures, Jpn. Kokai Tokkyo Koho 76 86,143 (1976); *Chem. Abstr.*, 85 (1976) 162293e.
3156. H. J. Shin, Y. J. Cho, and B. J. Jun, Method for producing fructose-6-phosphate from starch using thermostable enzyme derived from *Thermus* sp, with high efficiency, S. Korean Patent 2006059622 (2006); *Chem. Abstr.*, 146 (2006) 96305.
3157. T. Kurimura and Y. Endo, Formation of fructose in the enzymatic hydrolysis of starch (in Japanese), *Dempunto Gijutsu Kenkyu Kaiho*, 27 (1963) 13–17.
3158. A. Moradian and S. A. Benner, A biomimetic biotechnological process for converting starch to fructose: Thermodynamic and evolutionary considerations in applied enzymology, *J. Am. Chem. Soc.*, 114 (1992) 6980–6987.
3159. R. Walon, Maltulose-containing syrups, US Patent 4,217,413 (1979); *Chem. Abstr.*, 93 (1980) 184621c.
3160. R. G. Walon, Syrups containing maltulose, Belgian Patent 873,615 (1979); *Chem. Abstr.*, 91 (1979) 89910w.
3161. M. Boruch and E. Nebesny, The periodic isomerization of glucose in starch hydrolysates using a soluble and insoluble preparations of glucose isomerase, *Acta Aliment. Pol.*, 30(4), (1980) 215–226.
3162. M. Boruch and E. Nebesny, Continuous isomerization of glucose in starch hydrolysates with immobilized glucose isomerase, *Acta Aliment. Pol.*, 35(4), (1985) 414–425.
3163. C. Y. Lin, C. C. Chang, and C. H. Hung, Fermentative hydrogen production from starch using natural mixed cultures, *Int. J. Hydrogen Energy*, 33 (2008) 2445–2453.
3164. R. Oztekin, I. K. Kapdan, F. Kargi, and H. Argun, Optimization of media composition for hydrogen gas production from hydrolyzed wheat starch by dark fermentation, *Int. J. Hydrogen Energ*, 33 (2008) 4083–4090.
3165. O. Hirayama, Y. Kumada, and Y. Naruse, Hydrogen production from starch by a bacterium isolated soil, *Kinki Daigaku Nogakubu Kiyo*, 32 (1999) 53–60. *Chem. Abstr.*, 131 (1999) 202121.
3166. J. H. Shin, J. H. Yoon, E. K. Ahn, M. S. Kim, J. S. Sim, and T. H. Park, Fermentative hydrogen production by newly isolated *Enterobacter asburiae* SNU-1, *Int. J. Hydr. Energ.*, 32 (2007) 192–199.
3167. B. Fabiano and P. Perego, Thermodynamic study and optimization of hydrogen production by *Enetrobacter aerogenes*, *Int. J. Hydrogen Energ.*, 27 (2002) 149–156.
3168. S. Liu, W. Zhang, and H. Ma, Hydrogen fermentation using immobilized microorganisms, Chinese Patent 1789414 (2006); *Chem. Abstr.*, 145 (2006) 165618.

3169. Y. C. Lo, S. D. Chen, C. Y. Chen, T. I. Huang, C. Y. Lin, and J. S. Chang, Combining enzymatic hydrolysis and dark-photo fermentation process for hydrogen production from starch feedstock. A feasibility study, Int. J. Hydrogen Energ., 33 (2008) 5224–5233.
3170. H. Yokoi, A. Saitsu, H. Uchida, J. Hirose, S. Hayashi, and Y. Takasaki, Microbial hydrogen production from sweet potato starch residue, J. Biosci. Bioeng., 91 (2001) 58–63.
3171. E. Lopez, N. Arrien, J. Antonanzas, E. Egizabal, M. Belsue, and J. M. Valero, Produccion de hidrogeno a partir de residuos mediante fermetacion, Rev. Ing. Quim. (Madrid), 37 (2005) 175–181.
3172. T. Kanai, H. Imanaka, A. Nakajima, K. Uwamori, Y. Omori, T. Fukui, H. Atomi, and T. Imanaka, Continuous hydrogen production by thermophilic archeon Thermococcus kodakaraeusis, J. Biotechnol., 116 (2005) 271–282.
3173. S. A. Van Ooteghem, S. K. Beer, and P. C. Yue, Hydrogen production by the thermophiolic bacterium Thermatoga neapolitana, Appl. Biochem. Biotechnol., 98–100 (2002) 177–189.
3174. A. Tabuchi, T. Mandai, T. Shibuya, S. Fukada, T. Sugimot, and M. Kurimoto, Formation of trehalose from starch by novel enzymes, Oyo Toshitsu Kagaku, 42 (1995) 401–406. Chem. Abstr., 124 (1996) 143710.
3175. T. Shibuya, T. Sugimoto, and T. Miyake, Energy-supplementing saccharide source and its uses, Canadian Patent 2119070 (1994); Chem. Abstr., 122 (1995) 131146.
3176. K. Maruta, M. Kubota, and T. Sugimoto, Sulfolobus acidocaldarus thermostable enzyme forms trehalose-containing non-reducing saccharide from reducing amyloceus saccharide and recombinant enzyme use in saccharification and sweetener or syrup manufacture, Canadian Patent 2154307 (1996); Chem. Abstr., 124 (1996) 252731.
3177. K. Mukai, A. Tabuchi, T. Nakada, T. Shibuya, H. Chaen, S. Fukuda, M. Kurimoto, and Y. Tsujisaka, Production of trehalose from starch by the thermostable enzymes from Sulfolobus acidocaldarius, Starch/Stärke, 49 (1997) 26–30.
3178. M. Yoshida, N. Nakamura, and K. Horikoshi, Production of trehalose from starch by maltose phosphorylase and trehalose phosphorylase from a strain of Plesiomonas, Starch/Stärke, 49 (1997) 21–26.
3179. Y. Inoue, G. Nomura, and S. Miyoshi, Trehalose syrup enzymic manufacture from liquefied starch, Jpn. Kokai Tokkyo Koho 2001197898 (2001); Chem. Abstr., 135 2001) 136485.
3180. L. Lama, B. Nicolaus, A. Trincone, P. Morzillo, M. De Rosa, and A. Gambacorta, Starch conversion with immobilized thermophilic archeobacterium Sulfolobus solfataricus, Biotechnol. Letts, 12 (1990) 431–432.
3181. X. Wang and Z. Chi, Manufacture of trehalose by Saccharomycopsis fibuligera fermentation, Chinese Patent 1740333 (2006); Chem. Abstr., 144 (2006) 310625.
3182. S. Kobayashi and N. Shibuya, Enzymic manufacture of neotrehalose and centose, Jpn. Kokai Tokkyo Koho 63216492 (1988); Chem. Abstr., 110 (1989) 152832.
3183. M. A. Hassan, Y. Shirai, A. Kubota, M. I. A. Karim, K. Nakanishi, and K. Hashimoto, Effect of oligosaccharides on glucose consumption by Rhodobacter spheroides in polyhydroxyalkanoate production from enzymatically treated crude sago starch, Biotechnol. Sust. Util. Biol. Resources Tropics, 13 (1999) 184–193.
3184. Q. Wu, H. Huang, J. Chen, and G. Chen, Application of Bacillus cereus as strain in synthesis of polyhydroxy butyrate, Chinese Patent 1238382 (1999); Chem. Abstr., 133 (2000) 149275.
3185. Y. Kawabata, K. Toeda, T. Takahashi, N. Shibamoto, and M. Kobayashi, Preparation of highly branched starch by glycogen branching enzyme Neurospora crassa N2-44 and its characterization, J. Appl. Glycosci., 49 (2002) 273–279.
3186. K. Kruus, M. L. Niku-Paavola, and L. Viikari, Laccase – A useful enzyme for modification of biopolymers, Proc. 1st & 2nd Int. Conf. Biopolymer Technol., Coimbra (1999) & Naples (2000), pp. 255–261. Chem. Abstr., 139 (2003) 18851.

3187. S. Mathew and P. Adlercreutz, Mediator facilitated laccase catalyzed oxidation of granular potato starch and the physico-chemical characterization of the oxidized products, *Biores. Technol.*, 100 (2009) 3576–3584.
3188. S. Rogols and R. L. High, fermentable starch compositions, US Patent 3,652,295 (1972); *Chem. Abstr.*, 77 (1973) 47018s.
3189. W. Z. Hassid and R. M. McCready, Molecular constitution of enzymatically synthesized starch, *J. Am. Chem. Soc.*, 63 (1941) 2171–2173.
3190. T. Minagawa, Amylosynthease. XXIV. The velocity 3 (in Japanese), *J. Agr. Chem. Soc. Jpn.*, 10 (1934) 550–553.
3191. S. Nishimura, Enzymatisch Synthese der höhren Dextrine, *Biochem. Z.*, 225 (1930) 264–266.
3192. B. Maruo and T. Kobayashi, The enzymic formation and degradation of starch. I. Examination of amylosynthease (in Japanese), *J. Agr. Chem. Soc. Jpn.*, 23 (1949) 115–120.
3193. B. Maruo and T. Kobayashi, The enzymic formation and degradation of starch. II. Examination of amylosynthease (in Japanese), *J. Agr. Chem. Soc. Jpn.*, 23 (1949) 120–123.
3194. B. Maruo, T. Kobayashi, Y. Tsukano, and N. Yamada, The enzymic formation and degradation of starch. VIII. The enzyme transformation of amylopectin into amylose (in Japanese), *J. Agr. Chem. Soc. Jpn.*, 24 (1951) 347–349.
3195. M. Nakamura, The enzymic formation and degradation of starch. XIV. Potato phosphorylase (in Japanese), *J. Agr. Chem. Soc. Jpn.*, 26 (1952) 260–267.
3196. M. Nakamura, The enzymic formation and degradation of starch. XV Potato phosphorylase 2 (in Japanese), *J. Agr. Chem. Soc. Jpn.*, 26 (1952) 267–272.
3197. B. Maruo, The enzymic formation and degradation of starch. III. Dry preparation of potato phosphorylase, *J. Agr. Chem. Soc. Jpn.*, 23 (1950) 271–274.
3198. K. F. Gottlieb, P. M. Bruinenberg, J. B. Schotting, and D. J. Binnema, Process for preparing chain-extended starch, EU Patent EP 590736 (1994); *Chem. Abstr.*, 121 (1994) 55999.
3199. M. Nakamura, The enzymic formation and degradation of starch. XVII. Lima bean phosphorylase (in Japanese), *J. Agr. Chem. Soc. Jpn.*, 27 (1953) 70–75.
3200. M. Rongine de Fekete, L. F. Leloir, and C. E. Cardin, Mechanism of starch biosynthesis, *Nature*, 187 (1960) 918–919.
3201. E. F. Recondo and L. F. Leloir, Adenosine diphosphate glucose and starch synthesis, *Biochem. Biophys. Res. Communs.*, 6 (1961) 85–89.
3202. A. I. Oparin, T. N. Evreiinova, T. I. Larinova, and I. M. Davydova, Synthesis and degradation of starch in coacervate droplets (in Russian), *Dokl. Akad. Nauk SSSR*, 143 (1962) 980–983.
3203. K. Hayashibara, Starch gel containing oligosaccharides having fructose terminals, Jpn. Kokai Tokkyo Koho 80 39,304 (1970); *Chem. Abstr.*, 94 (1980) 64088h.
3204. AVEBE B.A., Enzymic production of amylose from starch, Dutch Patent 86 00,936 (1987); *Chem. Abstr.*, 108 (1987) 73848r.
3205. S. Peat, W. J. Whelan, and W. R. Rees, Enzymic synthesis and degradation of starch. XX. Disproportionating enzyme (D-enzyme) of the potato, *J. Chem. Soc.* (1956) 44–53.
3206. S. Peat, W. J. Whelan, and G. W. F. Kroll, Enzymic synthesis and degradation of starch. XXI. Dextrins synthesized by D-enzyme, *J. Chem. Soc.* (1956) 53–55.
3207. Y. K. Cho and K. H. Park, Production of red bean starch granule with cellulose from *Fusarium monioliforme*, *Hanguk Nonghwa Hakhoechi*, 29(1), (1986) 44–50. *Chem. Abstr.*, 106 (1986) 17195x.
3208. K. Yamamoto, K. Yoshikawa, and S. Okada, Effective dextrin production from starch by dextrin dextrinase with debranching enzyme, *J. Ferment. Bioeng.*, 76 (1993) 411–413.
3209. C. V. A. Cynter, M. Hang, J. De Jersey, B. Patel, P. A. Inkemann, and S. Hamilton, Isolation and characterization of a thermostable dextranase, *Enzyme Microb. Technol.*, 20 (1997) 242–247.

3210. J. Pongjanta, A. Utaipatanacheep, O. Navikul, and K. Piyachamkwon, Enzymes-resistant starch (RS III) from pullulanase–Debranched high amylose rice starch, *Kasetsart J. (Nat. Sci.)*, 42 (2008) 198–205.
3211. W. Sun, N. Jiang, Y. Ren, C. Xu, and T. Chunxi, Preparation of pullulan by fermentation, Chinese Patent 1216781 (1999); *Chem. Abstr.*, 132 (2000) 307365.
3212. A. Villiers, Sur la transformation de la fecule en dextrine par la ferment butyrique, *Compt. Rend. Acad. Sci.*, 435 (1891) 536–538.
3213. F. Schardinger, Thermophilic bacteria from various foods and milk and the products formed when these bacteria are cultivated in media containing carbohydrate, *Z. Untersuch. Nahrung Genussm.*, 6 (1903) 865–880. *Chem. Abstr.*, 2 (1906) 116347.
3214. F. Schardinger, Formation of crystalline nonreducing polysaccharides from starch by microbial action, *Zentr. Bakteriol. Parasitenk. Abt. II.*, 22 (1908) 98–103. *Chem. Abstr.*, 3 (1909) 4804.
3215. F. Schardinger, The formation of crystalline polysaccharides (dextrins) from starch pest. by mioc. ovrddot organisms, *Zentr. Bakteriol. Parasitenk.* Abt. II., 29 (1911) 188–197; *Chem. Abstr.*, 5 (1911) 14969.
3216. K. Freudenberg, E. Plankenhorn, and H. Knauber, Über Schardinger Dextrine aus Stärke, *Ann.*, 558 (1947) 1–10.
3217. K. Freudenberg, E. Plankenhorn, and H. Knauber, Schardinger dextrins from starch, *Chem. Ind.* (1947) 731–735.
3218. Kiangsu Institute of Food Fermentation, A new method for the production of cyclodextrin by Bacillus, *Shih Pin Yu Fa Hsiao Kung Yeh* (6), (1980) 1–9. *Chem. Abstr.*, 95 (1981) 130942f.
3219. P. Pongsawasdi and M. Yagisawa, Screening and identification of cyclomaltodextrin glucanotransferase-producing bacterium, *J. Ferment. Technol.*, 65 (1987) 463–467.
3220. Y. Kato, J. Nomura, K. Mikuni, K. Hara, H. Hashimoto, T. Nakajima, and S. Kobayashi, Chemical characterization of dextrins obtained from potato starch by treatment with *Bacillus macerans* enzyme, *J. Ferment. Bioeng.*, 68 (1989) 14–18.
3221. H. Hashimoto, K. Hara, K. Kainuma, and S. Kobayashi, α-Rich cyclodextrin manufacture, Jpn. Kokkai Tokkyo Koho JP 60,188,088 (1984); *Chem. Abstr.*, 104 (1985) 49874t.
3222. C. P. Yang and G. D. Lei, Production of cyclodextrin. II. Preparation of soluble starch for cyclodextrin (CD) production and the miniaturization of β-CD crystal, *Tatung Hsueh Pao*, 14 (1984) 85–94. *Chem. Abstr.* 102 (1985) 205684x.
3223. E. B. Tilden and S. Hudson, Conversion of starch to crystalline dextrins by the action of a new type of amylase separated from cultures of *Aerobacillus macerans*, *J. Am. Chem. Soc.*, 61 (1939) 2900–2902.
3224. W. S. McCleanhan, E. B. Tilden, and C. S. Hudson, Study on the products obtained from starch by the action of amylase of *Bacillus macerans*, *J. Am. Chem. Soc.*, 64 (1942) 2139–2144.
3225. R. W. Kerr, Significance of the degradation of starch by macerans amylase, *J. Am. Chem. Soc.*, 65 (1943) 188–193.
3226. S. Akiya and T. Watanabe, A crystalline decomposition product of starch by a Bacillus. III. A new strain of *Bacillus macerans* (in Japanese), *J. Pharm. Soc. Jpn.*, 70 (1950) 572–576.
3227. S. Akiya and T. Watanabe, A crystalline decomposition product of starch by a Bacillus. IV. A new hexasaccharide (in Japanese), *J. Pharm. Soc. Jpn.*, 70 (1950) 576–578.
3228. K. Yamamoto, Z. Z. Zhang, and S. Kobayashi, Cycloamylose (cyclodextrin) glucanotransferase degrades intact granules of potato raw starch, *J. Agric. Food Chem.*, 48 (2000) 962–966.
3229. R. W. Kerr, Action of macerans enzyme on a component of corn starch, *J. Am. Chem. Soc.*, 64 (1942) 3044.
3230. G. Seres and J. Szejtli, Enzymic formation of cyclodextrins from starch, *Hung. Annu. Meet. Biochem.*, 18 (1978) 169–171.

3231. M. Gottvaldova, H. Hrabova, V. Sillinger, and J. Kučera, Biospecific sorption of cyclodextrin glucosyltransferase on physically modified starch, *J. Chromatogr. Biomed. Appl.*, 427 (1988) 331–334.
3232. C. Moriwaki, F. M. Pelissari, R. A. C. Gonçalves, J. E. Gonçalves, and G. Matioli, Immobilization of *Bacillus fermus* strain 37 in inorganic matrix for cyclodextrin production, *J. Mol. Catal., B*, 49 (2007) 1–7.
3233. M. Samec and F. Cernigoj, Über die Hydrolyse von Abbauprodukten der Stärke durch *Bacillus macerans* und sein Enzym, *Berichte*, 75B (1942) 1758–1760.
3234. K. Myrbäck and L. G. Gjörling, Starch cleavage by *Bacillus macerans*, *Arkiv Kemi, A*, 20 (1945) 1–13.
3235. K. Myrbäck and E. Willstädt, On the action of *Bacillus macerans* amylase, *Acta Chem. Scand.*, 3 (1949) 91–93.
3236. J. S. Choe, J. H. Chon. J. C. Han, S. H. Lim, and K. T. Jung, Direct fermentation process for cyclodextrin production by using microorganism, S. Korean Patent 9614108 (1996). *Chem. Abstr.*, 133 (2000) 119040.
3237. L. Li and Y. Sun, Production of cyclodextrins directly from raw starch, *Shipin Yu Fajiao Gongye*, 26 (4), (2000) 21–24. *Chem. Abstr.*, 134 (2001) 161951.
3238. Y. Suzuki, H. Iwasaki, and F. Kamimoto, Chloropircin inclusion complexes, Jpn. Kokai Tokkyo Koho, 75,89,306 (1975); *Chem. Abstr.*, 84 (1976) 16737.
3239. S. Yano, T. Miyauchi, H. Hitaka, and M. Sawata, Cyclodextrin (in Japanese), Jpn. Kokai Tokkyo Koho JP 7,109,224 (1971).
3240. C. P. Yang and C. H. Hung, Cyclodextrin production. I. Reaction conditions for cyclodextrin formation from soluble starch by cyclodextrin glucosyltransferase, *Tatung Hsueh Pao*, 14 (1984) 95–102. *Chem. Abstr.*, 102 (1985) 205685y.
3241. M. A. Hassan, Y. Shirai, A. Kubota, M. I. A. Karim, K. Nakanishi, and K. Hashimoto, Effect of oligosaccharides on glucose consumption by *Rhodobacter sphaeroides* in polyhydroxyalkanoate production from enzymatically treated crude sago starch, *J. Ferment. Bioeng.*, 86 (1998) 57–61.
3242. E. K. C. Yu, H. Aoki, and M. Misawa, Specific α-cyclodextrin production by a novel thermostable cyclodextrin glycosyltransferase, *Appl. Microbiol. Biotechnol.*, 28 (1988) 377–379.
3243. N. Nakamura and K. Horikoshi, Purification and properties of cyclodextrin glycosyltransferase of an alkalophilic Bacillus sp, *Agric. Biol. Chem.*, 40 (1976) 935–941.
3244. K. Horikoshi and N. Nakamura, Cyclodextrin. US Patent 4,135,977 (1979); *Chem. Abstr.*, 90 (1980) 136264e.
3245. AVEBE G.A. Cyclodextrin, Dutch Patent 81 04,410 (1983); *Chem. Abstr.*, 99 (1983) 20925c.
3246. N. Szerman, I. Schroh, A. L. Rossi, A. M. Rosso, N. Krymkiewicz, and S. A. Ferrarotti, Cyclodextrin production by cyclodextrin glucosyltransferase from *Bacillus circulans* DF 9R, *Bioresource Technol.*, 98 (2007) 2886–2891.
3247. G. R. Cucolo, H. F. Alves-Prado, E. Gomes, and R. da Silva, Optimization of CGTase production by Bacillus sp. subgroup alkalophilus E16 in submerged fermentation of cassava starch (in Portuguese), *Braz. J. Food Technol.*, 9 (2006) 201–208.
3248. E. J. Vandamme, C. Declecq, and I. Debonne, Dynamics of the *Bacillus circulans* var. alkalophilus cyclodextrin glucosyltransferase fermentation, *Proc. 3rd Eur. Congr. Biotechnol.*, 1 (1984) 327–332. *Chem. Abstr.*, 105 (1986) 5097h.
3249. P. Mattson, T. Korpela, S. Paavilainen, and M. Makela, Enhanced conversion of starch to cyclodextrins in ethanolic solutions by *Bacillus circulans* var alkalophilus cyclomaltodextrin glucanotransferase, *Appl. Biochem. Biotechnol.*, 30 (1991) 17–28.
3250. R. Wadetwar, K. Upadhye, S. Bakhle, S. Deshpande, and V. Nagulwar, Production of β-cyclodextrin: Effect of pH, time and additives, *Indian J. Pharm. Sci.*, 68 (2006) 520–523.
3251. N. Burhan and V. Beschkov, Kinetic study of cyclodextrin production by crude cyclodextrin gluconotransferase, *Bulg. Chem. Communs.*, 38 (2006) 121–125.

3252. L. Dijkhuizen, B. W. Dijkstra, C. Andersen, and C. Van der Osten, Cyclomaltodextrin glucanotransferase variants with altered specificity for cyclodextrin synthesis, EU Patent WO 9633267 (1996); *Chem. Abstr.*, 126 (1996) 3787.

3253. P. T. Mattson, M. J. Makela, and T. K. Korpela, Procedure for producing maltodextrins and maltose syrups containing cyclodextrins with immobilized cyclomaltodextrin glucanotransferase, EU Patent WO 9109963 (1991); *Chem. Abstr.*, 115 (1991) 134196.

3254. K. Horikoshi, Purification and characterization of cyclomaltodextrin glucanotransferase with an acid pH optimum from *Bacillus coagulan*, EU Patent EP 327099 (1989); *Chem. Abstr.*, 111 (1989) 170081.

3255. T. Kaneko, M. Yoshida, M. Yamamoto, N. Nakamura, and K. Horikoshi, Production of cyclodextrins by simultaneous action of two CGTases from three strains of Bacillus, *Starch/Stärke*, 42 (1990) 277–281.

3256. K. Akimaru, T. Yagi, and S. Yamamoto, Cyclomaltodextrin glucanotransferase-producing moderate thermophile *Bacillus coagulans*, *J. Ferment. Bioeng.*, 71 (1991) 63–65.

3257. I. H. Higuti, P. A. Da Silva, and A. J. Do Nascimento, Studies on alkalophilic CGTase producing bacteria and effect of starch on cyclodextrin glucosyltrnasferase activity, *Braz. Arch. Biol. Technol.*, 47 (2004) 135–138.

3258. K. Tomita, T. Tanaka, Y. Fujita, and K. Nakanishi, Some factors affecting the formation of γ-cyclodextrin using cyclodextrin glycosyl transferase from *Bacillus* sp.AL-6, *J. Ferment. Bioeng.*, 70 (1990) 190–192.

3259. G. Matioli, G. M. Zanin, and F. F. De Moraes, Enhancement of selectivity for producing γ-cyclodextrin, *Appl. Biochem. Biotechnol.*, 84–86 (2000) 955–962.

3260. A. Geol and S. Nene, A novel cyclomaltodextrin glucanotransferase from *Bacillus firmus* that degrades raw starch, *Biotechnol. Lett.*, 17 (1995) 411–416.

3261. D. G. Yim, H. H. Sato, Y. H. Park, and Y. K. Park, Production of cyclodextrin from starch by cyclodextrin glucosyltransferases from *Bacillus firmus* and characterization of purified enzyme, *J. Ind. Microbiol. Biotechnol.*, 18 (1997) 402–405.

3262. Hayashibara Biochemical Laboratories Inc., Enzymic conversion of starch to β-cyclodextrin, French Patent 2,154,396 (1973); *Chem. Abstr.*, 79 (1973) 113990x.

3263. S. Kitahata and S. Okada, Cyclodextrin glucosyltransferase. II. Action of cyclodextrin glucosyltransferase from *Bacillus megaterium* strain No. 5 on starch, *Agric. Biol. Chem.*, 38 (1974) 2413–2417.

3264. S. Kitahata and S. Okada, Cyclodextrin glucosyltransferase. III. Transfer action of cyclodextrin glucosyltransferase on starch, *Agric. Biol. Chem.*, 39 (1975) 2185–2191.

3265. S. Kitahata, M. Kubota, and S. Okada, Industrial production of cyclodextrin (in Japanese), *Kagaku to Kogyo (Osaka)*, 60 (1986) 335–338.

3266. S. Mori, T. Mase, and R. Ooya, Preparation of cyclodextrin glucanotransferase with *Bacillus megaterium* and its use for manufacturing γ-cyclodextrin, Jpn. Kokai Tokkyo Koho 06113843 (1994); *Chem. Abstr.*, 121 (1995) 77245.

3267. I. Pishtiyski and B. Zhekova, Effect of different substrates and their preliminary treatment on cyclodextrin production, *World J. Microbiol. Biotechnol.*, 22 (2006) 109–114.

3268. Y. D. Lee and H. S. Kim, Enzymic production of cyclodextrins from unliquified corn starch in an attrition bioreactor, *Biotechnol. Bioeng.*, 37 (1991) 795–801.

3269. J. T. Kim, Y. D. Lee, and H. S. Kim, Production of cyclodextrins from unliquified cornstarch using cyclodextrin glucotransferase in a membrane reactors, *Ann. N. Y. Acad. Sci.*, 672 (1992) 552–557.

3270. K. Horikoshi and T. Kato, Microbial production of γ-cyclodextrin synthease and its use in manufacture of γ-cyclodextrin, Jpn. Kokai Tokkyo Koho 62 25,976 (1987); *Chem. Abstr.*, 107 (1987) 57466p.

3271. K. Koizumi, T. Utamura, M. Sato, and Y. Yagi, Isolation and characterization of branched cyclodextrins, *Carbohydr. Res.*, 153 (1986) 55–67.
3272. M. Sato, H. Nagano, Y. Yagi, and T. Ishikura, Increasing γ-cyclodextrin yield, Jpn. Kokai Tokkyo Koho 60,227,693 (1985); *Chem. Abstr.*, 104 (1985) 128250q.
3273. H. C. Chung, S. H. Yoon, M. J. Lee, M. J. Kim, K. S. Kweon, I. W. Lee, J. W. Kim, B. H. Oh, H. S. Lee, V. A. Spiridonova, and K. H. Park, Characterization of a thermostable cyclodextrin glucanotransferase isolated from *Bacillus stearothermophilus* ET1, *J. Agric. Food Chem.*, 46 (1998) 952–959.
3274. M. Aikawa, Y. Sawaguchi, Y. Murata, and K. Sasaki, Selective manufacture of β-cyclodextrin with cyclodextrin glucanotransferase, Jpn. Kapki Tokkyo Koho 03083592 (1991); *Chem. Abstr.*, 116 (1992) 5333.
3275. V. A. Abelyan, S. I. Dikhtyarev, and E. G. Afrikyan, Some new properties of thermophilic cyclodextrin-transferase of Bacillus sp. (in Russian), *Biokhimya*, 56 (1991) 1583–1590.
3276. K. Horikoshi, N. Nakamura, and M. Matsuzawa, Maltosugars containing cyclodextrin, Jpn. Kokai Tokkyo Koho, 78,52,693 (1978); *Chem. Abstr.*, 89 (1979) 147188p.
3277. K. Horikoshi, Production and industrial applications of β-cyclodextrin, *Process Biochim.*, 14 (1979) 26–28, 30.
3278. K. H. Chen and C. L. Jeang, Effect of the substrate and enzyme concentration on the production of cyclodextrin using response surface methodology, *Taiwan Nongye Huaxue Yu Shipin Kexue*, 38 (2000) 367–371. *Chem. Abstr.*, 134 (2001) 328130.
3279. C. L. Jeang, Y. L. Chang, and H. Y. Sung, Production of cyclodextrins (5). Purification and characterization of γ-cyclodextrin forming enzyme from *Bacillus subtilis* no. 313, *Zhongguo Nongye Huaxue Huizi*, 29 (1991) 502–512. *Chem. Abstr.*, 117 (1992) 85691.
3280. M. Shimashita, H. Sada, S. Takahashi, and N. Miyazaki, Enzymic manufacture of cyclodextrin from crosslinked starch, Jpn. Kokai Tokkyo Koho 02227087 (1990); *Chem. Abstr.*, 114 (1991) 99995.
3281. S. Mori, T. Mase, and R. Ooya, Preparation of cyclodextrin glucanotransferase with *Brevibacterium* and its use for manufacturing γ-cyclodextrin, Jpn. Kokai Tokkyo Koho 06113842 (1994); *Chem. Abstr.*, 121 (1995) 77244.
3282. S. Mori, T. Mase, and T. Ohya, *Brevibacterium* cyclodextrin glucanotransferase process for producing the enzyme and process for producing gamma-cyclodextrin, EU Patent 614971 (1994); *Chem. Abstr.*, 121 (1994) 249505.
3283. R. L. Starnes, Conversion of liquified starch to cyclodextrins using *Clostridium* cyclodextrin, EU Patent WO 9109962 (1991); *Chem. Abstr.*, 115 (1991) 134232.
3284. J. H. Lee, K. H. Choi, J. Y. Choi, Y. S. Lee, I. B. Kwon, and J. H. Yu, Enzymic production of α-cyclodextrin with cyclomaltodextrin glucanotransferase of *Klebsiella oxytoca* 19-1, *Enzyme Microb. Technol.*, 14 (1992) 1017–1020.
3285. J. Y. Choi, J. H. Lee, K. H. Cho, and I. B. Kwon, *Klebsiella oxytoca* No. 19-1 capable of producing α-cyclodextrin, US Patent US 5492829 (1996); *Chem. Abstr.*, 124 (1996) 230175.
3286. H. Bender, Kinetische Untersuchungen der durch die Cyclodextin glucotransferase krystalyesierten (1-4)-α-D-glucopyranosyl transferreaktione insbesondere der Cyclisierungsreaktion mit Amylose, Amylopektin und gesamt Stärke als Substrate, *Carbohydr. Res.*, 78 (1980) 147–162.
3287. B. Gawande and A. Patkar, Alpha-cyclodextrin production using cyclodextrin glucosyltransferase from *Klebsiella pneumoniae* AS-22, *Starch/Stärke*, 53 (2001) 75–83.
3288. H. F. Bender, Cyclooctaamylose, German Patent DE 3,317,064 (1984); *Chem. Abstr.*, 102 (1985) 26775f.
3289. E. Flaschel, J. P. Landert, D. Spiesser, and A. Renken, The production of α-cyclodextrin by enzymatic degradation of starch, *Ann. N. Y. Acad. Sci.*, 434 (1984) 70–77.
3290. H. Bender, Studies of the mechanism of the cyclisation reaction catalysed by the wild type and a truncated α-cyclodextringlucotransferase from *Klebsiella pmeumonia* strain M5 aI and the

β-cyclodextrin glucosyltransferase from *Bacillus circulans* strain 8, *Carbohydr. Res.*, 206 (1990) 257–267.
3291. Y. Yagi, K. Kourio, and T. Inui, Cyclodestrins by reacting starch with an amylase from *Micrococcus* microorganisms, EU Patent. 17,242 (1980); *Chem, Abstr.*, 94 (1980) 45606j.
3292. K. L. Larsen, L. Duedahl-Olsen, L. H. Pedersen, and W. Zimmermann, Production of cyclomaltonanoase (δ-cyclodextrin) by various cyclodextrin glycosyltrnasferases, *Proc. 9th Symp. on Cyclodextrins, Santiago de Compostela, May/June 1998*, pp. 89–92. *Chem. Abstr.*, 132 (1999) 206996.
3293. R. E. Vollu, F. F. da Mota, E. A. Gomes, and L. Seldin, Cyclodextrin production and genetic characterization of cyclodextrin glucanotransferase of *Paenibacillus graminis*, *Biotechnol. Lett.*, 30 (2008) 929–935.
3294. R. L. Starnes, P. C. Trackman, and D. M. Katkocin, Manufacture and use of thermostable cyclodextrin glycosyl transferase, EU Patent, WO 8903421 (1989); *Chem. Abstr.*, 112 (1990) 156726.
3295. T. J. Kim, B. C. Kim, and H. S. Lee, Production of cyclodextrin using raw corn starch without a pretreatment, *Enzyme Microb. Technol.*, 20 (1997) 506–509.
3296. L. Słomińska, A. Szostek, and A. Grześkowiak, Studies on enzymatic continuous production of cyclodextrins in an ultrafiltration membrane bioreactor, *Carbohydr. Polym.*, 50 (2002) 423–428.
3297. H. Takada, T. Takabane, T. Kuriki, and S. Okada, Cyclic glucan synthesis with thermostable branching enzyme for use in food production, Jpn. Kokai Tokkyo Koho 2000316581 (2000); *Chem. Abstr.*, 133 (2000) 331188.
3298. Y. Yoshimura, S. Okada, and S. Kitahata, Manufacture of branched cyclodextrin, Jpn. Kokai Tokkyo Koho 62 03,795 (1987); *Chem. Abstr.*, 107 (1987) 57464m.
3299. Y. Yoshimura, S. Okada, and S. Kitahata, Manufacture of branched cyclodextrin, Jpn. Kokai Tokkyo Koho JP 62 06, 696 (1987); *Chem. Abstr.*, 107 (1987) 57465n.
3300. T. Kaneko, N. Nakamura, and K. Horikoshi, Kinetic characterization of chimeric cyclomaltodextrin glucanotransferase from genes of two alkalophilic Bacillus, *Starch/Stärke*, 42 (1990) 354–358.
3301. G. Schmid, δ-Cyclodextrin glycosyltransferase of Bacillus for γ-cyclodextrin manufacture, German Patent DE 4009822 (1991); *Chem. Abstr.*, 116 (1992) 101791.
3302. K. Yamane and K. Kimura, Cyclomaltodextrin glucanotransferase (CGTase) muteins for enhanced preparation of γ-cyclodextrin, Jpn. Kokai Tokkyo Koho 05041985 (1993); *Chem. Abstr.*, 119 (1994) 23687.
3303. J. B. Hwang and S. H. Kim, Cyclodextrin production from potato starch with *Bacillus stearothermophilus* cyclomaltodextrin glucanotransferase, *Sanop Misaengmul Hakhoechi*, 20 (1992) 344–347. *Chem. Abstr.*, 119 (1993) 26785.
3304. T. Uozumi, H. Masaki, A. Nakamura, and K. Shin, Cyclodextrin glucanotransferase (CGTase) of Bacillus and its mutants for enhanced manufacture of γ-cyclodextrin (CD), Jpn. Kokai Tokkyo Koho 05219948 (1993); *Chem. Abstr.*, 120 (1994) 72411.
3305. G. E. Schulz and G. Parsiegla, Engineering of cyclodextrin glycosyltransferase deletion mutants for improved production of gamma-cyclodextrin, US Patent US 6472192 (2002); *Chem. Abstr.*, 137 (1994) 334622.
3306. Q. Qi, Z. Wang, and P. Wang, Method for producing β-cyclodextrin with yeast, Chinese Patent CN 1803854 (2006); *Chem. Abstr.*, 145 (2006) 190854.
3307. Y. H. Lee and D. C. Park, Enzymatic synthesis of cyclodextrin in an heterogenous enzyme reaction system containing insoluble extruded starch, *Sanop Misaengmul Hakhoechi*, 19 (1991) 514–520. *Chem. Abstr.*, 119 (1993) 47514.
3308. N. Nakamura and K. Horikoshi, Production of Schardinger β-dextrin by soluble and immobilized cyclodextrin glycosyltransferase of an alkalophilic Bacillus sp, *Biotechnol. Bioeng.*, 19 (1977) 87–99.
3309. K. Yoritomi and T. T. Yoshida, Cyclodextrin, Jpn. Kokai Tokkyo Koho 76,136,889 (1976); *Chem. Abstr.*, 86 (1977) 137968s.

3310. Y. Sawaguchi, M. Aikawa, K. Kikuchi, and K. Sasaki, Effective production of γ-cyclodextrin and/or α-glucosylglycylrrhizin with cyclodextrin glucanotransferase, Jpn. Kokai Tokkyo Koho 02255095 (1990); *Chem. Abstr.*, 114 (1991) 141647.
3311. C. S. Su and C. P. Yang, A novel method for continuous production of cyclodextrin using an immobilized enzyme system, *J. Chem. Technol. Biotechnol.*, 48 (1990) 313–323.
3312. B. Xie et al., Beta-cyclodextrin production, Chinese Patent 1054072 (1991); *Chem. Abstr.*, 116 (1992) 88184.
3313. Y. H. Lee and D. C. Park, Direct synthesis of cyclodextrin in a heterogenous enzyme reaction system containing insoluble extruded starch, *Proc. Asia Pacific Biochem. Conf.*, pp. 127–129. *Chem. Abstr.*, 119 (1993) 137457.
3314. W. Sieh and A. Hedges, Immobilized enzyme for removal of residual cyclodextrin, US Patent US 5565226 (1996); *Chem. Abstr.*, 125 (1996) 274271.
3315. G. Seres, M. Jarai, S. Piukovich, G. Szigetvari, M. Gabanyi, and J. Szejtli, Highly pure γ- and α-cyclodextrin, German Patent 3,446,080 (1985); *Chem. Abstr.*, 103 (1986) 194961c.
3316. Y. H. Lee, S. H. Lee, and I. K. Han, Continuous production of cyclodextrin in two-stage immobilized enzyme reactor coupled with ultrafiltration recycle system, *Sanop Misaengmul Hakhoechi*, 19 (1991) 171–178. *Chem. Abstr.*, 119 (1993) 26774.
3317. R. Yoshii, T. Okada, and T. Saiga, Enzymic manufacture of branched cyclodextrin, Jpn. Kokai Tokkyo Koho 02109990 (1990); *Chem. Abstr.*, 113 (1991) 57521.
3318. S. Suzuki, S. Kobayashi, and K. Kainuma, Cyclodextrin production, Jpn. Kokai Tokkyo Koho 77 38,038 (1975); *Chem. Abstr.*, 87 (1978) 20639q.
3319. T. Okada, T. Saiga, and R. Yoshii, Ultrafiltration membranes in enzymic manufacture of branched cyclodextrin, Jpn. Kokai Tokkyo Koho 02211890 (1990); *Chem. Abstr.*, 113 (1990) 189816.
3320. D. Wang, G. Yang, J. Li, and Q. Liu, Method for directly manufacturing maltosyl-β-cyclodextrin from starch, Chinese Patent 1850867 (2006); *Chem. Abstr.*, 145 (2006) 473363.
3321. J. A. Rendleman, Jr., Enhancement of cyclodextrin production through use of debranching enzymes, *Biotechnol. Appl. Biochem.*, 26 (1997) 51–61.
3322. Y. K. Kim and J. F. Robyt, Enzyme modification of starch granules: Formation and retention of cyclomaltodextrin inside starch granules by reaction of cyclomaltodextrin glucanosyltransferase with solid granules, *Carbohydr. Res.*, 328 (2000) 509–515.
3323. S. Wang, Y. Li, Y. Xu, and X.H. Yang, Enzymatic method for producing cycloamylose from natural starch, Chinese Patent 1817181 (2006); *Chem. Abstr.*, 145 (2006) 270203.
3324. H. O. S. Lima, F. F. De Moraes, and G. M. Zanin, β-Cyclodextrin production by simultaneous fermentation and cyclization, *Appl. Biochem. Biotechnol.*, 70–72 (1998) 789–804.
3325. S. Sakai, N. N. Yamamoto, M. Chiwa, H. Hashimoto, and K. Hara, Cyclodextrin glucanotransferase its immobilization and use for cyclodextrin manufacture, Jpn. Kokai Tokkyo Koho 63044886 (1988); *Chem. Abstr.*, 110 (1989) 191242.
3326. S. H. Choi, M. S. Kim, J. J. Ryoo, K. P. Lee, H. D. Shin, S. H. Kim, and Y. H. Lee, Immobilization of a cyclodextrin glucanotransferase (CGTase) onto polyethylene film with a carboxylic acid group and production of cyclodextrins from corn starch using CGTase-immobilized PE-film, *J. Appl. Polym. Sci.*, 85 (2002) 2451–2457.
3327. N. Tanaka, Y. Endo, and A. Miyzazaki, Enzymic manufacture of cyclodextrin, Jpn. Kokai Tokkyo Koho 62,130,688 (1987); *Chem. Abstr.*, 107 (1988) 174427n.
3328. R. P. Rohrbach and D.S. Scherl, Cyclodextrin enzymic manufacture from starch, yield enhancement by addition of organic solutes, US Patent US 4748237 (1988); *Chem. Abstr.*, 110 (1989) 133709.
3329. S. Sakai, N. Yamamoto, H. Hashimoto, and K. Hara, Cyclodextrin production with immobilized enzyme, Jpn. Kokai Tokkyo Koho 61,185,196 (1986); *Chem. Abstr.*, 105 (1987) 224630e.

3330. T. Okada, H. Ishizuka, H. Sahashi, T. Hibino, S. Inoue, and M. Ito, Membrane filter-immobilized enzymes for the manufacture of cycldextrins, Jpn. Kokai Tokkyo Koho, 01063390 (1989); *Chem. Abstr.*, 111 (1990) 213350.

3331. J. K. Hong and K. H. Youm, Enzymatic production and simultaneous separation of cyclodextrins using cross-flow membrane module, *Hwahak Konghak*, 38 (2000) 86–91. *Chem. Abstr.*, 133 (2000) 194853.

3332. H. Ishigami, K. Mikuni, and K. Hara, Continuous production of cyclodextrin. I. Continuous liquefaction of starch and continuous filtration, *Seito Gijutsu Kenkyu Kaishi*, 39 29–37. *Chem. Abstr.*, 116 (1992) 82112.

3333. H. P. Chen, C. P. Yang, and V. Z. Marmarelis, Kinetic studies on the production of cyclodextrin: Analysis of initial rate, inhibitory effect of glucose and degradation, *J. Chin. Inst. Chem. Eng.*, 20(4), (1989) 195–199.

3334. E. J. Wilson, Jr., T. J. Schoch, and C. S. Hudson, Action of macerans amylase of the fractions form starch, *J. Am. Chem. Soc.*, 65 (1943) 1380–1383.

3335. Japan Ministry of Agriculture, Forestry and Fishery, Cyclodextrin production, Jpn. Kokai Tokkyo Koho 57,202,298 (1982); *Chem. Abstr.*, 98 (1983) 141869a.

3336. J. Chiirin and S. Shenni, Manufacture of cyclodextrin with Bacillus, Jpn. Kokai Tokkyo Koho 03168096 (1991); *Chem. Abstr.*, 115 (1991) 157170.

3337. C. L. Jeang and H. Y. Sung, Production of cyclodextrin. I. The production of α-cyclodextrin, *Zhongguo Nongye Huaxue Huizhi*, 26 (1988) 525–535. *Chem. Abstr.*, 110 (1989) 191152.

3338. K. Yamamoto and S. Kobayashi, Manufacture of cyclodextrins from starch granules with cyclodextrin synthease, Jpn. Kokai Tokkyo Koho 09262091 (1997); *Chem. Abstr.*, 127 (1998) 318163.

3339. H. Hashimoto, K. Hara, S. Kobayashi, and K. Kainuma, Manufacture of cyclodextrins from starch, Jpn. Kokai Tokkyo Koho 62,104,590 (1987); *Chem. Abstr.*, 107 (1988) 174810g.

3340. T. J. Kim, B. C. Kim, and H. S. Lee, Production of cyclodextrins using moderately heat-treated cornstarch, *Enzyme Microb. Technol.*, 17 (1995) 1057–1061.

3341. S. Yano, T. Miyauchi, H. Hitaka, and M. Sawata, Preparation of cyclodextrins, *Jpn. Kokai Tokkyo Koho* 71 09,224 (1971).

3342. X. Yang, W. Liu, and Q. Ma, Process for producing β-cyclodextrin, Chinese Patent 101077895 (2007); *Chem. Abstr.*, 148 (2007) 56832.

3343. F. C. Armbruster, and S. A. Jacaway, Procedure for production of alpha-cyclodextrin, US Patent 3,640,847 (1972).

3344. K. H. Min, W. J. Chang, and Y. M. Koo, Increased production of β-cyclodextrin using aqueous two-phase system, *Biotechnol. Tech.*, 10 (1996) 395–400.

3345. K. C. M. Raja and S. V. Ramakrishna, Improved reaction conditions for preparation of β-cyclodextrin (β-CD) from cassava (*Manihot esculenta* Crantz) starch, *Starch/Stärke*, 46 (1994) 402–403.

3346. K. Nakanishi, N. Tomita, T. Fujushige, H. Akano, and K. Kawamura, Manufacture of cyclodextrin with cyclodextrin synthease, Jpn. Kokai Tokkyo Koho, 05103683 (1993); *Chem. Abstr.*, 119 (1993) 447620.

3347. G. Schmid and H. J. Eberle, Manufacture of cyclooctaamylose by enzymic digestion of starch in the presence of a complexing agent, EU. Patent EP 291067 (1988); *Chem. Abstr.*, 111 (1989) 55856.

3348. J. A. Rendleman, Jr., Enhanced production of cyclomaltooctaose (γ-cyclodextrin) through selective complexation with C_{12} cyclic compounds, *Carbohydr. Res.*, 230 (1992) 343–359.

3349. F. Shiraishi, K. Kawakami, H. Marushima, and K. Kusunoki, Effect of ethanol on formation of cyclodextrin from soluble starch by *Bacillus macerans* cyclodextrin glucanotransferase, *Starch/Stärke*, 41 (1989) 151–155.

3350. Toyo Jozo Co., Ltd, Improved method for the production of cyclodextrin, Jpn. Kokai Tokkyo Koho 60,120,998 (1984); *Chem. Abstr.*, 104 (1985) 4666k.
3351. R. N. Ammeraal, Preparation and isolation of cyclodextrins from gelatinized starch, German Patent 3,716,509 (1986); *Chem. Abstr.*, 108 (1987) 114587a.
3352. Y. D. Lee and H. S. Kim, Effect of organic solvents on enzymic production of cyclodextrins from unliquified corn starch in an attrition bioreactor, *Biotechnol. Bioeng.*, 39 (1992) 977–983.
3353. T. Takeshita, C. Hattori, and K. Kusunoki, Effect of polyether additives on enzymic synthesis of cyclodextrin, *Kyushu Sangyo Daigaku Kogakubu*, 33 (1996) 137–140. *Chem. Abstr.*, 127 (1997) 46738.
3354. I. K. Han and Y. H. Lee, Production of cyclodextrin from raw starch in the agitated bead reaction system and its reaction mechanism, *Sanop Misaengmul Hakhoechi*, 19 (1981) 163–170. *Chem. Abstr.*, 119 (1993) 26773.
3355. G. Seres and L. Barcza, Method development for the economic production of cyclodextrins by means of ternary complex formation, *Proc. 4th Int. Symp. Cyclodextrins*, pp. 81–86. *Chem. Abstr.*, 112 (1990) 160916.
3356. K. C. M. Raja, V. P. Sreedharan, P. Prema, and S. V. Ramakrishna, Cyclodextrin from cassava (*Manihot esculenta* Crantz) starch. Isolation and characterization as brmomobenzene and chloroform clathrates, *Starch/Stärke*, 42 (1990) 196–198.
3357. W. Shieh, Production of γ-cyclodextrin, German Patent 19545478 (1996); *Chem. Abstr.*, 125 (1996) 84818.
3358. F. C. Armbruster, and E. R. Kooi, Cyclodextrin, US Patent 3,425,910 (1969); *Chem. Abstr.*, 70 (1968) 79360u.
3359. J. L de Freitas Valle, Cyclodextrin complexing agents prepared from starch by enzyme treatment, Brazilian Patent 8704585 (1989); *Chem. Abstr.*, 111 (1989) 213353.
3360. K. Horikoshi, T. Ando, K. Yoshida, and N. Nakamura, Cyclodextrin, German Patent 2,453,860 (1975); *Chem. Abstr.*, 83 (1975) 99642j.
3361. H. S. Lee, B. K. Song, Y. H. Kim, Y. D. Lee, B. C. Kim, J. H. Kim, and T. J. Kim, Manufacture of cyclodextrin from starch with cyclodextrin glucanotransferase, S. Korean Patent 136363 (1998); *Chem. Abstr.*, 140 (2003) 6344.
3362. M. Fiedorowicz, G. Khachatryan, A. Konieczna-Molenda, and P. Tomasik, Formation of cyclodextrins with cyclodextrin glucotransferase stimulated with polarized light, *Biotechnol. Progr.*, 25 (2009) 147–150.
3363. A. Biwer and E. Heinzle, Process modeling and simulation can guide process development: Case study α-cyclodextrin, *Enzyme Microbiol. Technol.*, 34 (2004) 642–650.
3364. X. Cao and Z. Jin, Application of response surface methodology in enzymatic reaction using cyclodextrin glycosyltransferase, *Zhengzhou Gongcheng Xueyuan Xuebao*, 25 (2004) 58–61. *Chem. Abstr.*, 143 (2004) 76908.
3365. N. Nakamura, M. Matsuzawa, and K. Horikoshi, Cyclic-dextrin-containing starch-sugar powder, Jpn. Kokai Tokkyo Koho 79,44,045 (1979); *Chem. Abstr.*, 91 (1980) 54952j.
3366. Norin Suisansho Shokuhin Sogo Kenkyu Socho, Production of starch gel rich in cyclodextrin and maltotertaose, Jpn. Kokai Tokkyo Koho 58 81,744 (1981).
3367. P. Mattson, T. K. Korpela, and M. Makela, Manufacture α-, β- and/or γ-cyclodextrin-containing maltodextrin starch and maltose syrup, Finnish Patent FI 81116 (1990); *Chem. Abstr.*, 114 (1991) 99963.
3368. N. Kako, H. Kitagawa, T. Sato, H. Akano, H. Okumura, and K. Kawamura, Cyclodextrin-containing starch, its manufacture with cycodextrin synthetease and its use in food manufacture, Jpn. Kokai Tokkyo Koho 03083549 (1991); *Chem. Abstr.*, 115 (1992) 157601.
3369. A. A. Khakimzhanov and O. V. Fursov, Inhibition of rice grain α-amylase by β-cyclodextrin (in Russian), *Fiziol. Biokhim. Kult. Rast.*, 20 (1988) 157–162.

3370. I. Pocsi, N. Nogrady, A. Liptak, and A. Szentirmai, Cyclodextrins are likely to induce cyclodextrin glycosyltransferase production in *Bacillus macerans*, *Folia Microb. (Prague)*, 43 (1998) 71–74.
3371. L. M. Hamilton, C. T. Kelly, and W. M. Fogarty, Review: Cyclodextrins and their interaction with amylolytic enzymes, *Enzyme Microb. Technol.*, 26 (2000) 561–567.
3372. K. Mukai, H. Watanabe, K. Oku, T. Nishimoto, M. Kubota, H. Chaen, S. Fukuda, and M. Kurimoto, An enzymatically produced novel cyclic tetrasaccharide cyclo-{→6) -α-D-Glc$p$-(1-4)-α-D-Glc$p$-(1-6)-α-D-Glc$p$-(1-4)-α-D-Glc$p$-(1→}(cyclic maltosyl (1-6)-maltose) from starch, *Carbohydr. Res.*, 340 (2005) 1469–1474.
3373. H. Watanabe, H. Aga, N. Kubota, S. Fukuda, and T. Miyake, Cyclic pentasaccharide and its glycosyl derivatives, Jpn. Kokai Tokkyo Koho 2005162677 (2005); *Chem. Abstr.*, 143 (2005) 58930.
3374. V. Velusamy, S. Nithyanandham, and L. Palaniappyan, Ultrasonic study of adsorption in polysaccharide metabolism, *Main Group Chem.*, 6 (2007) 53-61; http:// www.informaworld.com/smpp/ftinterface˜Ñâcontent=a783637345˜Ñâfulltext=713240930; *Chem. Abstr.*, 148 (2008) 138008.
3375. R. J. Levin, Digestion and absorption of carbohydrates from molecules and membranes to humans, *Am. J. Clin. Nutr.*, 59(Suppl.), (1994) 690S–698S.
3376. G. M. Gray, Starch digestion and absorption in ruminants, *J. Nutr.*, 122 (1992) 172–177.
3377. D. L. Topping and P. M. Clifton, Short-chain fatty acids and human colonic function: Roles of resistant starch and nonstarch polysaccharides, *Physiol. Rev.*, 81 (2001) 1031–1064.
3378. P. J. Tomasik and P. Tomasik, Probiotics and prebiotics, *Cereal. Chem.*, 80 (2003) 113–117.
3379. D. C. Whitcomb and M. E. Lowe, Human pancreatic digestive enzyme, *Dig. Dis. Sci.*, 52 (2007) 1–17.
3380. C. F. Langworthy and A. T. Merrill, Digestibility of raw starches and carbohydrates, *US Dept. Agric. Bull.*, 1213 (1924) 1–15.
3381. M. T. Christian, C. A. Edwards, T. Preston, L. Johnston, R. Varlley, and L. T. Weawer, Starch fermentation by fecal bacteria of infants toddlers and adults, Importance of energy salvage, *Eur. J. Clin. Nutr.*, 57 (2003) 1486–1491.
3382. F. R. J. Bornet, Y. Bizais, S. Bruley des Varannes, B. Pouliquen, J. D. Lava, and J. P. Galmiche, α-Amylase (EC 3.2.1.1) susceptibility rather than viscosity or gastric emptying rate controls plasma response to starch in healthy humans, *Brit. J. Nutr.*, 63 (1990) 207–220.
3383. A. M. Aura, H. Harkonen, M. Fabritius, and K. Poutanen, Development of in vitro enzymic digestion method for removal of starch and protein and assessment of its performance using rye and wheat breads, *J. Cereal Sci.*, 29 (1999) 139–152.
3384. B. Držikova, G. Dongowski, E. Gebhardt, and A. Habel, The composition of dietary fibre-rich extrudates from oat affects bile acid binding and fermentation in vitro, *Food Chem.*, 90 (2005) 181–192.
3385. J. Larner and R. E. Gillespie, Gastrointestinal digestion of starch. II. Properties of the intestinal carbohydrates, *J. Biol. Chem.*, 223 (1956) 709–726.
3386. M. J. Werman and S. J. Bhathena, Fructose metabolizing enzymes in the rat liver and metabolic parameters. Interactions between dietary copper, type of carbohydrates and gender, *J. Nutr. Biochem.*, 6 (1995) 373–379.
3387. T. Doi, H. Yoshimatsu, I. Katsuragi, M. Kurokawa, H. Takahashi, A. Motoshio, and T. Sakata, Alpha-amylase inhibitor increases plasma 3-hydroxybutyric acid in food-restricted rats, *Experentia*, 51 (1995) 585–588.
3388. J. E. Moermann and H. R. Muehlemann, Influence of α-amylase and α-glucosidase inhibitors on oral starch metabolism, *Lebensm. Wiss. Technol.*, 14 (1981) 166–169.
3389. Z. Madar and Z. Omunsky, Inhibition of intestinal α-glucosidase activity and postprandial hyperglycemia by α-glucosidase inhibitors in fa/fa rats, *Nutr. Res.*, 11 (1991) 1035–1046.

3390. J. Zhang and S. Kashket, Inhibition of salivary amylase by black and green teas and their effects on the intraoral hydrolysis of starch, *Caries Res.*, 32 (1998) 233–238.
3391. H. W. Palat, P. Maletto, and D.A. Maiullo, Composition of amylase inhibitors and fibers to reduce the absorption of starch in human digestion from starchy foods and to reduce the glycemic index of starchy food, US Patent 2007020313 (2007); *Chem. Abstr.*, 146 (2007) 169387.
3392. J. C. Blake, Digestibility of bread. I. Salivary digestion in vitro, *J. Am. Chem. Soc.*, 38 (1916) 1245–1260.
3393. J. Effront, Influence of salts on the amylolysis of bread, *Mon. Sci.*, [5] 6 (1916) 5–12.
3394. A. S. Schultz and Q. Landis, The hydrolysis of starch in bread by four and malt amylase, *Cereal Chem.*, 9 (1932) 305–310.
3395. J. C. Favier, Digestibility of starches from various food plants of South Cameroun. Effect of technological transformations on manioc starch, *Ind. Aliment. Agric.*, 86 (1969) 9–13. *Chem. Abstr.*, 71 (1969) 121088.
3396. G. Haberlandt, The digestibility of potato starch, *Illustr. Landw. Ztg.*, 37 (1917) 107–108. *Chem. Abstr.*, 14 (1919) 7814.
3397. C. F. Langworthy and H. J. Deuel, Jr., Digestibility of raw corn, potato and wheat starches, *J. Biol. Chem.*, 42 (1920) 27–40.
3398. M. Pauletig, Verdaulichkeit der Stärke verschiedener pflanzlicher Futtermittel durch Malz, Pankreas und Speicheldiastase, *Z. Physiol. Chem.*, 100 (1917) 74–92.
3399. S. Rogols and S. Meites, Effect of starch species on α-amylase activity, *Starch/Stärke*, 20 (1968) 256–259.
3400. C. F. Boersting, K. E. B. Knudsen, S. Steenfeldt, H. Mejborn, and B. O. Eggum, The nutritive value of decorticated mill fractions of wheat.3. Digestibility experiments with boiled and enzyme treated fractions fed to mink, *Animal Feed Sci. Technol.*, 53 (1995) 317–336.
3401. J. J. Marshal and W. J. Whelan, Some aspects of dietary starch utilization in the mammalian digestive tract, *Proc. Congr. Biochem. Apects Nutr. Fed. Asian Oceanian*, 1977, pp. 125–130. *Chem. Abstr.*, 91 (1980) 190585u.
3402. F. Röhmann, Über die durch paranterale Rohzuckerinjectionen hervargelockten Fermente des Blutserum von trächtigen Kaninchen, *Biochem. Z.*, 84 (1917) 399–401.
3403. A. Bućko and D. Z. Kopeć, Adaptation of rat pancreatic α-amylase fermentation to changes on dietary starch, *Ernährungforsch.*, 13 (1968) 311–319. *Chem. Abstr.*, 70 (1969) 33583.
3404. F. Baker, H. Nasr, F. Morrice, and J. Bruce, Bacterial breakdown of structural starches and starch products in the digestive tract of ruminal and nomruminant mammals, *J. Path. Bact.*, 62 (1950) 617–638.
3405. S. F. Kotarski, R. D. Wanishka, and K. K. Thurn, Starch hydrolysis by the ruminal microflora, *J. Nutr.*, 122 (1992) 178–190.
3406. J. P. Jouany and C. Martin, Effect of protozoa in plant cell wall and starch digestion in the rumen, In: R. Onodera, (Ed.), *Rumen Microbes and Digestive Physiology in Ruminants,* Jpn. Sci. Soc. Press, Tokyo, 1997, pp. 11–24.
3407. D. C. Loper, C. O. Little, and G. E. Mitchell, Jr., In vitro procedure or studying starch digestion by rumen microorganisms, *J. Animal Sci.*, 25 (1966) 128–131.
3408. I. A. Dolgov and E. I. Kolenko, Amylolytic microflora of the sheep cecum and its enzymic activity at different sugar and starch levels in the diet (in Russian), *Byull. Vses. Naucghn.-Issled. Inst. Fiziol. Biokhim. Pitan. S-kh. Zivotn.*, 8(1), (1974) 47–49.
3409. R. H. Udall, The digestibility of starch in the cat family with special reference to the effect of freezing, *Cornell Vet.*, 33 (1943) 360–364.
3410. A. J. W. Hoorn and W. Scharloo, The functional significance of amylose polymorphism in *Drosophila melanogaster*. V. The effect of food components on amylase and α-glucosidase activity, *Genetica (The Hague)*, 49 (1978) 181–187.

3411. K. Hori, Physiological conditions in the midgut in relation to starch digestion and the salivary amylase of the bug *Lygus disponsi*, *J. Insect Physiol.*, 17 (1971) 1153–1167.
3412. K. Hori, Digestibility of insoluble starches by the amylases in digestive system of bug *Lygus disponsi* and the effect of Cl⁻ and NO_3^- on the digestion, *Entomol. Exp. Appl.*, 15 (1972) 13–22.
3413. K. Hori, Enzymes especially amylases in the digestive system of the bug *Lygus disponsi* and starch digestion in the system, *Obihiro Chikusan Daigaku Gakujutsu Kenkyu Hokoku Dai-1-Bu*, 8 (1973) 173–260.
3414. T. Doge, M. Obara, H. Okuda, T. Shiomi, and S. Onodera, Food, feed and sweeteners containing – glucosidase inhibitors, Jpn. Kokai Tokkyo Koho 09065836 (1997); *Chem Abstr.*, 126 (1997) 276663.
3415. A. C. Eliasson, Starch in Food: Structure, Function and Applications, (2004) Woodhead Publ, Cambridge.
3416. R. M. Faulks, S. Southon, and G. Livesey, The utilization of α-amylase resistant corn and pea starches in rats, *Roy. Soc. Chem., Special Publ.*, 72 (1989) 319–321.
3417. J. M. Gee, R. M. Faulks, and I. T. Johnson, Physiological effects of retrograded α-amylase-resistant cornstarch in rats, *J. Nutr.*, 121 (1991) 44–49.
3418. K. Vanhoof and R. De Schrijver, The influence of enzyme-resistant starch on cholesterol metabolism in rats fed on a conventional diet, *Brit. J. Nutr.*, 80 (1998) 193–198.
3419. K. Vanhoof and R. De Schrijver, Consumption of enzyme resistant starch and cholesterol metabolism in normo- and hypercholesterolemic rats, *Nutr. Res. (New York)*, 17 (1997) 1331–1340.
3420. G. Livesey, I. R. Davies, J. C. Brown, R. M. Faulks, and S. Southon, Energy balance and energy values of α-amylase (EC 3.2.1.1) –resistant maize and pea (*Pisum sativum*) starches in the rat, *Brit. J. Nutr.*, 63 (1990) 467–480.
3421. C. A. Reid, K. Hillman, and C. Henderson, Effect of retrogradation pancreatin digestion and amylose/amylopectin ratio on the fermentation of starch by *Clostridium butyricum* (NCIMB 7423), *J. Sci. Food Agric.*, 76 (1998) 221–225.
3422. M. Wacker, P. Wanek, E. Eder, S. Hylla, A. Gostner, and W. Scheppach, Effect of enzyme-resistant starch on formation of 1, N2-propanodeoxyguanosine adducts of trans-4-hydroxy-2-nonenal and cell proliferation in the colonic mucosa of healthy volunteers, *Cancer Epidem. Biomark. Prev.*, 11 (2002) 915–920.
3423. E. Horton, The use of taka-diastase to estimate starch, *J. Agr. Sci.*, 11 (1921) 240–257.
3424. H. Okazaki, Tripartite analysis of starch hydrolyzed by five types of molds (in Japanese), *J. Agr. Chem. Soc. Jpn.*, 24 (1950) 201–215.
3425. P. Thivend, C. Mercier, and A. Guilbot, Anwendung der Gluco-Amylase zur Stärkebestimmung, *Starch/Stärke*, 17 (1965) 278–283.
3426. H. Tsuge, E. Tatsumi, N. Ohtani, and A. Nakazima, Screening of α-amylase suitable for evaluating the degree of starch retrogradation, *Starch/Stärke*, 44 (1992) 29–32.
3427. Y. Takeda, Actions of various amylases to the vicinity of phosphate groups of potato starch and location of the phosphate groups in the starch (in Japanese), *Denpun Kagaku*, 34 (1987) 225–233.
3428. D. Bourdon, Determination of reducing sugars by the Michael Somogyi method, *Bull. Soc. Sci. Bretagne*, 43 (1968) 167–191. *Chem Abstr.*, 68 (1968) 3872.
3429. A. Astruc and A. Renaud, New specifications for starch intended for diastatic assays, *J. Pharm. Sci.*, 27 (1923) 333–337.
3430. C. H. Wie and S. M. McGarvey, A study of the effects of certain variations in preparation of a starch substrate in amylase viscosimetry, *J. Gen. Physiol.*, 16 (1932) 221–227.
3431. K. V. Giri, The differentiation of amylases and starches, *Science*, 81 (1935) 343–344.
3432. H. Ikeda, M. Gunji, Y. Takayama, and K. Tomita, Examination of degrading enzymes in measurement of gelatinization degree, *Kanzei Chuo Bunsekishoho*, 42 (2002) 41–47. *Chem. Abstr.*, 116 (1992) 104634.

3433. M. G. E. Wolters and J. W. Cone, Prediction of degradability of starch by gelatinization enthalpy as measured by differential scanning calorimetry, *Starch/Stärke*, 44 (1992) 14–18.
3434. H. O. Beutler, Enzymatische Bestimmung von Stärke in Lebensmitteln mit Hilfe der Hexokinase-Methode, *Starch/Stärke*, 30 (1978) 309–312.
3435. W. Ettel, Eine neue enzymatische Stärkebestimmung für Lebensmitteln Alimente, *Alimenta*, 20 (1981) 7–11.
3436. M. D. Morales, A. Escarpa, and M. C. Gonzalez, Simultaneous determination of resistant and digestible starch in foods and food products, *Starch/Stärke*, 49 (1997) 448–453.
3437. K. Brunt, P. Sanders, and T. Rozema, The dynamic determination of starch in food, feed and raw materials of the starch industry, *Starch/Stärke*, 50 (1998) 413–419.
3438. P. Aggarwal and D. Dollimore, Degradation of starchy food material studied using thermal analysis, *Proc. 26th Conf. N. Am. Thermal Anal. Soc., Cleveland*, pp. 320–323. *Chem. Abstr.*, 130 (1998) 65441.
3439. K. Brunt, Collaborative study concerning the enzymatic determination of starch in food, feed and raw materials of the starch industry, *Starch/Stärke*, 52 (2000) 73–75.
3440. H. O. Beutler, Starch, *Rep. Methods Enzym. Anal, 3rd Ed.*, 6 (1984) 2–10.
3441. R. D. Di Paola, R. Asis, and M. A. J. Aldao, Evaluation of the degree of starch gelatinization by a new enzymatic method, *Starch/Stärke*, 55 (2003) 403–409.
3442. L. ten Cate, Starch and amylose in meat products, *Fleischwirtschaft*, 15 (1963) 1021–1024. *Chem. Abstr.*, 60 (1963) 11767.
3443. G. Skrede, An enzymic method for the determination of starch in meat products, *Food Chem.*, 11 (1983) 175–185.
3444. F. Bauer and H. Stachelberger, Enzymatic determination of starch in meat products, *Lebensm. Wiss. Technol.*, 17 (1984) 160–162. *Chem. Abstr.*, 101 (1984) 128961.
3445. S. Hase and T. Yasui, Studies on determination of starch in agricultural products, *Shokuhin Sogo Kenkyusho Kenkyo Hokoku* (36), (1980) 98–103. *Chem. Abstr.*, 94 (1982) 45696p.
3446. J. C. Cuber, Study of the enzymatic determination of starch. 2. Comparative study of several methods for starch dispersion, *Sci. Aliments*, 2 (1982) 223–232.
3447. Z. Palacha, W. Pomarańska-Łazuka, D. Witrowa-Rajchert, and P. P. Lewicki, Use of enzymes as quality indicators of dried potatoes (in Polish), *Zesz. Nauk. Polit. Łódzka, Ser. Inz. Chem. Proces.* (25), (1999) 183–188.
3448. J. L. Batey, Starch analysis using thermostable alpha-amylases, *Starch/Stärke*, 34 (1982) 125–128.
3449. C. J. Cuber, Study of the enzymatic determination of starch. 2. Comparative study of several methods for starch dispersions, *Sci. Aliments*, 2 (1982) 233–244. *Chem. Abstr.*, 97 (1982) 116137.
3450. M. Huang, Determination of cereal starch content by amylase from *Aspergillus niger*, *Weishengwuxue Tongbao*, 8 (1981) 286–287. *Chem. Abstr.*, 97 (1982) 37564r.
3451. P. Weustink, Praktische Erfahrung mit der NIR-Reflexionspektroskopie beim enzymeinduzierten Abbau von Weizenstärke zu Glucose, *Starch/Stärke*, 38 (1986) 335–339.
3452. F. Alstin, Rapid method for starch determination in cereals using Fibertec E (in Italian), *Prodotto Chim. Aerosol Selezione*, 29(10), (1988) 28–29.
3453. B. V. McCleary, T. S. Gibson, V. Solah, and D. C. Mugford, Total starch measurement in cereal products. Interlaboratory evaluation of a rapid enzyme test procedure, *Cereal Chem.*, 71 (1994) 501–505.
3454. G. Sinnaeve, P. Dardenne, B. Weirich, and R. Agneessens, Online monitoring of enzymic degradation of wheat starch by near infrared, *Proc. 71th Int. Conf. Near Infrared Spectroscopy, Montreal*, pp. 290–294. *Chem. Abstr.*, 126 (1996) 225829.
3455. B. V. McCleary, T. S. Gibson, D. C. Mugford, and O. Lukow, Measurement of total starch in cereal products by amyloglucosidase – α-amylase method, *J. AOAC Int.*, 80 (1997) 571–579.

3456. H. M. Blasel, P. C. Hoffman, and R. D. Shaver, Degree of starch access. An enzymatic method to determine starch degradation potential of corn grain and corn silage, *Animal Feed Sci. Technol.*, 128 (2006) 96–107.
3457. A. Csiba, M. Szentgyoryi, and G. Lombai, A simple method for the determination of starch content in whole meal (in Hungarian), *Elemezesi Ipar*, 54(1), (2000) 23–25.
3458. E. T. Oakley, Enzymatic determination of starch in fresh green lyophilized green and cured tobacco, *J. Agric. Food Chem.*, 31 (1983) 902–905.
3459. B. E. Haissig and R. E. Dickson, Starch measurement in plant tissues using enzymatic hydrolysis, *Physiol. Plant.*, 47 (1979) 151–157.
3460. A. Schneider, S. Zuber, and G. Flachowsky, Untersuchungen zum Einsatz von α-amylase in der Zellwandanalytik, *Nahrung*, 35 (1991) 603–609.
3461. K. L. Tomlison, J. R. Lloyd, and A. M. Smith, Importance of isoforms of starch branching enzyme in determining the structure of starch in pea leaves, *Plant J.*, 11 (1997) 31–43.
3462. O. Theander, P. Åman, E. Westerlund, and H. Graham, Enzymic/chemical analysis of dietary fiber, *J. AOAC Int.*, 77 (1994) 703–709.
3463. R. A. Libby, Direct starch analysis using DMSO solubilization and glucoamylase, *Cereal Chem.*, 47 (1970) 273–281.
3464. M. Richter, Automatisierte enzymatische Bestimmung von Glucose, Maltose, und Stärke aur Mikrotiterplatten, *Starch/Stärke*, 46 (1994) 81–85.
3465. J. C. Cuber, Study on the enzymic determination of starch. 3. Assay for the measurement of initial and maximal rates of oxygen-consumption during the one-step procedure for starch determination, *Sci. Aliments*, 2 (1982) 207–222. *Chem. Abstr.*, 97 (1982) 161138.
3466. J. Karkalas, An improved enzymic method for the determination of native and modified starch, *J. Sci. Food Agric.*, 36 (1985) 1019–1027.
3467. P. Aaman, E. Westerlund, and O. Theander, Determination of starch using thermostable α-amylase, *Methods Carbohydr. Chem.*, 10 (1994) 111–115.
3468. P. R. Coulter and O. E. Potter, Measurement of the susceptibility of starch in hydrolysis by malt enzymes, *J. Inst. Brew. London*, 78 (1972) 444–449.
3469. P. M. Mathias, K. Bailey, J. J. Mcevoy, M. Cuffe, A. Savage, and A. Allen, Study on the amylolytic breakdown of damaged starch in cereal and non-ceeal flours using high performance liquid chromatography and scanning electron microscopy, *Roy. Soc. Chem., Spec. Publ.*, 205 (1997) 129–140.
3470. K. H. Meyer and W. F. Gonon, Recherches sur l'amidon. 49. Dosage de produits d'hydrolyse enzymatique de l'amidon par fermentation, *Helv. Chim. Acta*, 34 (1951) 290–294.
3471. W. Schmeider, Diastatic starch degradation and its possible use in analysis (in Roumanian), *Ind. Aliment.*, 19 (1968) 10–12.
3472. R. Haginoya, K. Sakai, T. Komatsu, S. Nagao, K. Yokoyama, T. Takeuchi, R. Matsukawa, and I. Karube, Determination of damaged starch and diastatic activity in wheat flour using a flow-injection analysis biosensor method, *Cereal Chem.*, 74 (1997) 745–749.
3473. T. S. Gibson, H. Al Qalla, and B. V. McCleary, An improved enzymic method for the measurement of starch damage, *J. Cereal Sci.*, 15 (1992) 15–27.
3474. W. C. Barnes, The rapid enzymic determination of starch damage in flours from sound and rain damaged wheat, *Starch/Stärke*, 30 (1978) 114–119.
3475. G. S. Ranhotra, J. A. Gelroth, and G. J. Eisenbraun, Correlation between Chopin and AACC methods of determining damaged starch, *Cereal Chem.*, 70 (1993) 236–237.
3476. Y. Liu, J. P. Gao, T. Lin, and C. F. Zhao, Establishment of the evaluation of resistant starch, *Tianjin Daxue Xuebao, Ziran Kexue Yu Gongcheng Jishuban*, 35 (2002) 775–777. *Chem. Abstr.*, 138 (2003) 384308.

3477. R. F. Jackson and J. A. Mathews, Determination of levulose in crude products, *Bur. Stand. J. Res.*, 8 (1932) 433–442.
3478. T. Sueno, Quality of sweet potato starch and liquefaction with enzyme (in Japanese), *Dempunto Gijutsu Kenkyu Kaiho*, 31 (1965) 1–19.
3479. H. Tsuge, M. Hishida, H. Iwasaki, S. Watanabe, and G. Goshima, Enzymatic evaluation for degree of starch retrogradation in foods, *Starch/Stärke*, 42 (1990) 213–216.
3480. X. F. Yu, H. L. Zhang, and G. Zhang, Microcalorimetric study on enzyme-catalyzed hydrolysis reaction of starch, *Yingyong Huaxue*, 19 (2002) 812–813. *Chem. Abstr.*, 137 (2002) 339246.
3481. Y. Fujio, M. Sambuichi, and S. Ueda, Rate of soluble starch consumption by *Aerobacter aerogenes* (in Japanese), *Hakko Kogaku Zasshi*, 50 (1972) 494–498.
3482. H. B. Sreerangachar, Dilatometric studies in the enzyme hydrolysis of polysaccharides. III. Hydrolysis of starch amylose and amylopectin by takadiastase, *Proc. Indian Acad. Sci.*, 2B (1935) 333–341.
3483. H. Brintzinger and W. Brintzinger, Messungen mit Hilfe der Dialysenmethode Untersuchungen über den enzymatischen Stärkeabbau sowie über die thermische Spaltung von Dextrin in Gegenwart von Säure, *Z. Anorg. Allgem. Chem.*, 196 (1930) 50–54.
3484. J. M. Wu, A rapid enzymic method for the quantitative determination of starch gelatinization, *Li Kung Hsueh Pao*, 14 (1977) 271–283. *Chem. Abstr.*, 90 (1978) 102027g.
3485. G. A. Mitchell, Methods of starch analysis, *Starch/Stärke*, 42 (1990) 131–134.
3486. D. T. Englis, G. T. Pfeifer, and J. L. Garby, Polarimetric reducing sugar relationships of starch hydrolytic products resulting from diastatic action, *J. Am. Chem. Soc.*, 53 (1931) 1883–1889.
3487. R. L. Gates and R. M. Sandstedt, A method of determining enzymic digestion of raw starch, *Cereal Chem.*, 30 (1953) 413–419.
3488. A. S. Schultz, R. A. Fisher, L. Atkin, and C. N. Frey, Sugar determination in starch hydrolyzates by yeast fermentation and chemical means, *Ind. Eng. Chem., Anal. Ed.*, 15 (1943) 496–498.
3489. T. Sabalitscka and R. Weidlich, Malzamylase. 5. Bestimmung der dextrinierenden und verzuckernden Wirkung der Amylase und Vergleich beider Wirkungen, *Biochem. Z.*, 207 (1929) 476–493.
3490. J. R. Broeze, Einwirkung der Ptyalins auf Stärke. I, *Biochem. Z.*, 204 (1929) 286–302.
3491. S. Rusznak, G. A. Gergely, and V. Komiszar, Polarographic chromatographic and oscillopolarographic estimation of the enzymic degradation of starch, *Abhandl. Deut. Akad. Wiss. Berlin, Kl. Chem. Geol. Biol.*, (1) (1964) 278–28; *Chem. Abstr.*, 62 (1964) 6663.
3492. V. Djordjevic, Relation between the specific rotation of starch hydrolysates and the extent of starch hydrolysis with respect to different hydrolysis methods (in Serbo-Croatian), *Hrana Ishrana*, 25 (1984) 221–223.
3493. K. Nishinari, R. K. Cho, and M. Iwamoto, Near infrared monitoring of enzymic hydrolysis of starch, *Starch/Stärke*, 41 (1989) 110–112.
3494. A. Lencses, Polarographic investigation of the products of enzyme starch hydrolyis (in Hungarian), *Soripar*, 14 (1967) 92–96.
3495. F. Svoboda and J. Šramek, Chromatographic and electrophoretic studies of the degradation of starch under the influence of various desizing agents, *Textil-Rundschau*, 20 (1965) 183–193. *Chem. Abstr.*, 63 (1966) 18434.
3496. P. Jackson, The use of polyacrylamide gel electrophoresis for the high resolution separation of reducing saccharides labeled with the fluorophore 8-aminonaphthalene-1,3,6-trisulfonic acid. Detection of picomolar quantities by an imaging system based on a cooled charge-coupled device, *Biochem. J.*, 270 (1990) 705–713.
3497. P. Jackson, Labeling reagent for analysis of carbohydrates by gel electrophoresis, EU Patent WO 9105256 (1991); *Chem Abstr.*, 115 (1991) 227800.
3498. Y. Otani, S. Takahashi, and Y. Fukuta, The identification and determination of the sugars in submerged koji by paper chromatography (in Japanese), *Hakko Kogaku Zasshi*, 36 (1958) 116–120.

3499. M. Ulmann and J. Seideman, Paper chromatographic characterization of degradation action of diastatic properties on potato starch, *Ernährungsforsch.*, 2 (1957) 83–96. *Chem. Abstr.*, 52 (1958) 609.
3500. J. Wohnlich, A simple technique for the enzymatic hydrolysis of poly- and oligosides combined with paper chromatography, *Bull. Soc. Chim. Biol.*, 43 (1961) 1121–1128.
3501. K. V. Giri, V. N. Nigam, and K. Saroya, Application of circular paper chromatography to the study of the mechanism of amylolysis, *Naturwissenschaften*, 40 (1953) 484.
3502. L. Vrbaski and Z. Lepojević, Thin-layer chromatographic separation of carbohydrates as possible products of microbial fermentation of starch (in Serbo-Croatian), *Zb. Matice Srpske Prirod. Nauke*, 77 (1989) 129–134.
3503. V. Ivanova, E. Emanuilova, M. Sedlak, and J. Pazlarova, HPLC study of starch hydrolysis products obtained with α-amylase from *Bacillus amyloliquefaciens* and *Bacillus licheniformis*, *Appl. Biochem. Biotechnol.*, 30 (1991) 193–202.
3504. E. Bertoft and R. Manelius, A method for the study of the enzymic hydrolysis of starch granules, *Carbohydr. Res.*, 227 (1992) 269–283.
3505. M. Fischl, T. Spies, and W. Praznik, Enzymic and chemical methods for the analysis of starch, *Acta Aliment. Pol.*, 17 (1991) 309–315.
3506. T. C. Sunarti, N. Yoshio, and M. Hisamatsu, Direct analysis of chains on outer layer of amylopectin through partial hydrolysis of normal starch by isoamylase, *J. Appl. Glycosci.*, 48 (2001) 123–130.
3507. E. Laszlo, Z. Toth, and J. Hollo, Amylase and amyloglucosidase–catalyzed starch hydrolysis, *Ernährungforsch.*, 13 (1968) 457–461. *Chem. Abstr.*, 70 (1969) 12807.
3508. K. Babor and V. Kalač, Characterization of enzymatically degraded starches (in Slovak), *Prum. Potravin*, 32 (1981) 538–541.
3509. M. R. Sarmidi and P. E. Baker, Saccharification of modified starch to maltose in a continuous rotating annular chromatograph, *J. Chem. Technol. Biotechnol.*, 57 (1993) 229–235.
3510. N. Torto, L. Gorton, G. Marko-Varga, J. Emneus, C. Akerberg, G. Zacchi, and T. Laurell, Monitoring of enzymic hydrolysis of starch by microdialysis sampling coupled online to anion exchange chromatography and integrated pulsed electrochemical detection using post-column switching, *Biotechnol. Bioeng.*, 56 (1997) 546–554.
3511. J. Emneus and L. Gorton, Flow system for starch determination based on consecutive enzyme steps and amperometric detection at a chemically modified electrode, *Anal. Chem.*, 62 (1990) 263–268.
3512. J. Emneus, G. Nilsson, and L. Gorton, A flow injection system for the determination of starch in starch from different origins with immobilized α-amylase and amyloglucosidase reactors, *Starch/Stärke*, 45 (1993) 264–270.
3513. J. C. Cuber, Fast method for enzymic determination of starch with oxygen probe, *Proc. 1st Eur. Conf. on Rec. Dev. Food Anal.*, 1981, pp. 286–291. *Chem. Abstr.*, 98 (1982) 16583q.
3514. J. Karkalas, Automated enzymic determination of starch by flow injection analysis, *J. Cereal Sci.*, 14 (1991) 279–286.
3515. E. F. Umoh and K. Schügerl, A flow injection system for simultaneous determination of starch and glucose concentrations, *J. Chem. Technol. Biotechnol.*, 61 (1994) 81–86.
3516. F. Scheller, D. Pfeiffer, and H. Weise, Enzyme electrodes for determining maltose, α-amylase and starch, East German Patent 211,425 (1980); *Chem. Abstr.*, 94 (1981) 153032c.
3517. D. Pfeiffer, F. Scheller, M. Jänchen, K. Betermann, and H. Weise, Bioenzyme electrodes for ATP, NAD^+, starch and disaccharides based on a glucose sensor, *Anal. Lett.*, 13(B13), (1980) 1179–1200.
3518. D. Pfeiffer, F. Scheller, M. Jänchen, and K. Betermann, Glucose oxidase bienzyme electroodes for ATP, NAD^+, starch and disaccharides, *Biochimie*, 62 (1980) 587–593.
3519. F. Zhou, J. Shi, and D. Feng, Rapid determination of starch content by enzyme electrode system with glucose oxidase, *Shipin Kexue*, 19(1), (1998) 42–44. *Chem. Abstr.*, 130 (1999) 89830.

3520. T. Komatsu, R. Oginoya, K. Sakai, and M. Karube, Enzyme digestion and oxygen electrode or luminal peroxidase system for starch determination, Jpn. Kokai Tokkyo Koho, JP 08242891 (1996); *Chem. Abstr.*, 126 (1996) 4196.
3521. E. Vrbova, J. Pečkova, and M. Marek, Biosensor for determination of starch, *Starch/Stärke*, 45 (1993) 341–344.
3522. L. Guo, X. Li, Q. Gao, C. Sun, and H. Xu, Glucose microelectrode based on polypyrrole, *Fenxi Huaxue*, 20 (1992) 828–830. *Chem. Abstr.*, 118 (1993) 5825.
3523. E. Watanabe, M. Takagi, S. Takei, M. Hoshi, and S. Cao, Development of biosensors for the simultaneous determination of sucrose and glucose lactose and glucose and starch and glucose, *Biotechnol. Bioeng.*, 38 (1991) 99–103.
3524. P. E. Buckle, R. J. Davies, and D. V. Pollard-Knight, Biosensors with a layer of dielectric material coupled to a porous matrix containing immobilized biochemicals, EU Patent. WO 9221976 (1992); *Chem. Abstr.*, 118 (1993) 55602.
3525. D. Raghavan, G. C. Wagner, and R. P. Wool, Aerobic biometer analysis of glucose and starch biodegradation, *J. Environ. Polym. Degr.*, 1 (1993) 203–211.
3526. D. Grebel, Quantitative Versuche zur enzymatischen Stärkespaltung, *Mikrokosmos*, 57 (1968) 183–186.
3527. D. Grebel, Spectrophotometric studies on enzyme starch splitting, *Prax Naturwiss. Biol.*, 27 (1978) 124–127. *Chem. Abstr.*, 89 (1978) 89645.
3528. S. Rogols and R. L. High, Microscopic evaluation for the enzyme conversion of starch. Laboratory technique and comparison, *Tappi*, 51(Pt. 1), (1968) 132A–135A.
3529. M. E. Amato, G. Ansanelli, S. Fisichella, R. Lamanna, G. Scarlata, A. P. Sobolev, and A. Segre, Wheat flour enzymatic amylolysis monitored by in situ ^{1}H NMR spectroscopy, *J. Agric. Food Chem.*, 52 (2004) 823–831.
3530. H. Anderson, H. Lubenow, W. Klingner, and K. Heldt, Equipments for modern thermal investigations, *Thermochim. Acta*, 187 (1991) 211–218.
3531. S. Chiba, A. Kimura, T. Kobori, and K. Saitoh, Quantitative determination of disaccharides produced from soluble starch through transglucosylation by buckwheat α-glucosidase, *Denpun Kagaku*, 32 (1985) 213–216.
3532. Y. Sekikawa and M. Deki, Analysis of several starch derivatives, *Kanzei Chuo Bunsekisho Ho*, 23 (1983) 101–108. *Chem. Abstr.*, 99 (1984) 40111n.
3533. J. W. Cone, A. C. Tas, and M. G. E. Wolters, Pyrolysis mass spectrometry and degradability of starch granules, *Starch/Stärke*, 44 (1992) 55–58.
3534. G. A. Grant, S. L. Frison, J. Yeung, T. Vasanthan, and P. Sporns, Comparison of MALDI-TOF mass spectrometric to enzyme colorimetric quantification of glucose from enzyme-hydrolyzed starch, *J. Agric. Food Chem.*, 51 (2003) 6137–6144.
3535. M. Marek and E.Vrbova, Enzymic determination of glycosides, oligosaccharides and polysaccharides in the presence of contaminating monosaccharides, Czech Patent 264586 (1992); *Chem. Abstr.*, 118 (1992) 76594.
3536. R. J. Henry, A. B. Blakeney, and C. M. R. Lance, Enzymic determination of starch in samples with high sugar content, *Starch/Stärke*, 42 (1990) 468–470.
3537. J. F. Kennedy and V. M. Cabalda, Evaluation of an enzymic method for starch purity determination, *Starch/Stärke*, 45 (1993) 44–47.
3538. J. W. Sullivan and J. A. Johnson, Measurement of starch gelatinization by enzyme susceptibility, *Cereal Chem.*, 41 (1964) 73–79.
3539. T. Toyama, Estimation of starch gelatinization by means of glucoamylase (in Japanese), *Denpun Kogyo Gakkaishi*, 13(3), (1966) 69–75.
3540. K. Kainuma, Determination of the degree of gelatinization and retrogradation of starch, *Methods Carbohydr. Chem.*, 10 (1994) 137–141.

3541. G. Goshima, K. Ohashi, and H. Tsuge, Measurement of the degree of starch gelatinization (in Japanese), *Gifu Daigaku Nogakubu Kenkyu Hokoku*, 40 (1977) 79–84. *Chem. Abstr.*, 88 (1979) 193169k.
3542. B. Y. Chiang and J. A. Johnston, Measurement of total and gelatinized starch by glucoamylase and o-toluidine reagent, *Cereal Chem.*, 54 (1977) 429–435.
3543. G. Tegge, Stärke und Stärkederivativen, 2nd edn. (1988) Behr's Verlag, Hamburg, p. 45.
3544. J. Blom, A. Bak, and B. Brae, Enzymatische Abbau der Stärke, *Z. Physiol. Chem.*, 241 (1936) 273–287.
3545. T. Kobayashi and M. Kadowaki, Determination of the end group in the polysaccharide produced by the action of isoamylase, *J. Agric. Chem. Soc. Jpn.*, 27 (1953) 599–602.
3546. W. J. Whelan, The contribution of enzymes to the structural analysis of glycogen and starch, *Biochem. J.*, 100 (1966) 1P–2P.
3547. D. French, Contribution of α-amylase to the structural determination of glycogen and starch, *Biochem. J.*, 100 (1966) 2P.
3548. D. J. Manners, Contribution of β-amylase to the structural determination of glycogen and starch, *Biochem. J.*, 100 (1966) 2P–3P.
3549. J. P. Robin, C. Mercier, F. Duprat, R. Charbonniere, and A. Guilbot, Amidons lintnerises etudes chromatographique et enzymatique des resides insolubles provenant de l'hydrolyse chlorhydrique d'amidons de cereals en particules de mais cireaux, *Starch/Stärke*, 27 (1975) 36–45.
3550. E. Bertoft, R. Manelius, and Z. Qin, Studies on structure of pea starches. 2. α-amylolysis of granular wrinkled pea starch, *Starch/Stärke*, 45 (1993) 258–263.
3551. W. Tuting, K. Wegemann, and P. Mischnick, Enzymatic degradation and electron spray tandem mass spectrometry as tools for determinig the structure of cationic starches prepared by wet and dry method, *Carbohydr. Res.*, 339 (2004) 637–648.
3552. J. Carter and E. Lee, An enzymatic method for determination of the average chain lengths of glycogens and amylopectins, *Anal. Biochem.*, 39 (1971) 373–386.
3553. J. Zeng, X. Li, and H. Gao, Influences of enzymolysis and fermentation on application quality of crosslinked starch, *Zhongguo Liangyou Xuebao*, 19(5), (2004) 29–32. *Chem. Abstr.*, 142 (2004) 265692.
3554. H. Rath, D. Keppler, and L. Rossling, The enzymic degradation of starch and evaluation of enzymic desizing agents, *Melliand Textilber.*, 25 (1943) 58–64. *Chem. Abstr.*, 38 (1944) 28238.
3555. H. Rath, The enzymic degradation of starch and evaluation of enzymic desizing agents, *Melliand Textilber.*, 25 (1944) 18–21. *Chem, Abstr.*, 38 (1944) 27224.
3556. J. Sun, R. Zhao, J. Zeng, G. Li, and X. Li, Characterization of dextrins with different dextrose equivalent, *Molecules*, 15 (2010) 5162–5173.
3557. K. Myrbäck and E. Willstädt, Starch degradation by α-amylases. II. Paper chromatography of low-molecular degradation products, *Arkiv Kemi*, 6 (1953) 417–425.
3558. R. Hayashi and M. Oka, Color indicators in food preservation, Jpn. Kokai Tokkyo Koho, 02257872 (1990); *Chem. Abstr.*, 114 (1991) 184069.
3559. M. Remiszewski, M. Grześkowiak, and R. Kaczmarek, Laboratory evaluation of enzymic preparations for production of glucose (in Polish), *Pr. Inst. Lab. Bad. Przem. Spoz.*, 25 (1975) 335–351.
3560. H. C. Lee and Y. S. Chen, Application of the β-amylase activity rapid test method on processing character of sweet potato, *Zhonghua Nongye Yaniju*, 53 (2004) 18–26. *Chem. Abstr.*, 144 (2005) 22032.
3561. S. Chaturved, B. Singh, L. Nain, S. K. Khare, A. K. Pandey, and S. Satya, Evaluation of hydrolytic enzymes in bioaugmented compost of *Jatropha* cake under aerobic and partly anaerobic conditions, *Ann. Microbiol.*, 60 (2010) 685–691.
3562. E. Podgórska and B. Achremowicz, Comparison of the methods of the estimation of α- and β-amylase activity in model systems (in Polish), *Proc. Conf. Post. Technol. Zywn.*, 60.

3563. B. Achremowicz and A. Wierbol, Evaluation of the enzyme activity in triticale grains (in Polish), *Folia Soc. Scient. Lublinensis Ser. Biol.*, 27 (1985) 69–73.
3564. W. Windisch, W. Dietrich, and A. Beyer, Stäerkeverflüssigende Funktion der Malzdiastase, *Wochenschr. Brau.*, 40 (1923) 49–50. 55–56, 61–63, 67–70.
3565. D. E. Evans, H. Collins, J. Eglinton, and A. Wilhelmson, Assesing the impact of the level of diastatic power enzymes and their thermostability on the hydrolysiss of starch during wort production to predict malt fermentability, *J. Am. Soc. Brew. Chemists*, 63 (2005) 185–198.
3566. I. D. Popov, Determination of the diastatic power of flours and some factors influencing it (in Bulgarian), *Ann. Univ. Sofia, V. Fac. Agronom. Sylvicult.*, 17 (1939) 79–137.
3567. C. S. O. Hagberg, Analysis of cereals, other starch-containing matter and products therefrom, German Patent 1,101,021 (1956).
3568. W. Schmidell and M. V. Fernandes, Comarison between acid and enzymic hydrolysis of starch for the determination of total reducing sugars (in Portuguese), *Rev. Microbiol.*, 8 (1977) 98–101.
3569. S. A. Waksman, A method of testing the amylolytic action of the diastase of *Aspergillus oryzae*, *J. Am. Chem. Soc.*, 42 (1920) 293–299.
3570. J. F. Robyt, R. J. Ackerman, and J. G. Keng, Reducing value method for maltodextrins. II. Automated methods of chain-length independence of alkaline ferricyanide, *Anal. Biochem.*, 45 (1972) 517–524.
3571. S. Hizukuri, Y. Takeda, M. Yasuda, and A. Suzuki, Multibranched nature of amylose and the action of debranching enzymes, *Carbohydr. Res.*, 94 (1981) 205–213.
3572. M. J. Somogyi, Notes on sugar determination, *Biol. Chem.*, 195 (1952) 19–23.
3573. N. J. Nelson, A photometric adaptation of the Somogyi method for the determination of glucose, *Biol. Chem.*, 153 (1944) 375–380.
3574. M. Sinner and J. Puls, Non-corrosive reagent for detection of reducing sugars in borate complex ion-exchange chromatography, *J. Chromatogr.*, 156 (1978) 197–204.
3575. P. Bernfeld, Enzymatic assay for α-amylase, *Methods Enzymol.*, 1 (1955) 149–158.
3576. H. Fuwa, A new method for microdetermination of amylose activity by the use of amylase as the substrate, *J. Biochem.*, 41 (1954) 583–603.
3577. M. John, J. Schmidt, and H. Kneifel, Iodine – Maltosaccharide complexes: Relation between chain length and colour, 119 (1983) 254–257.
3578. J. Effront, Method of determination of liquefying power of amylase, *Compt. Rend. Soc. Biol.*, 1 (1922) 7–9. *Chem. Abstr.*, 17 (1923) 10344.
3579. S. Jozsa and H. C. Gore, Determination of liquefying power of malt diastase, *Ind. Eng. Chem., Anal. Ed.*, 2 (1930) 26–28.
3580. T. Chrzaszcz, Eine neue Methode zur Bestimmung des Stärkeverflüsigungsvermogens, *Biochem. Z.*, 242 (1931) 130–136.
3581. T. Chrzaszcz and J. Janicki, Das Stärkeverflüssigungvermogen der Amylase und seine Bestimmung Methoden, *Biochem. Z.*, 256 (1932) 252–291.
3582. H. B. Sreerabgachar and M. Sreenivasaya, Dilatometric studies in the enzymic hydrolysis of polysaccharides. III. Hydrolysis of starch, amylose and amylopectin by takadiastase, *Proc. Indian Acad. Sci.*, 1B (1934) 43–47.
3583. F. Svoboda, Evaluation of enzymic desizing agents. Action of enzymic desizing agents (in Czekh), *Textil*, 17 (1962) 291–294, 424–426.
3584. T. Okano, Colorimetric determination of α-amylase in body fluids, Jpn. Kokai Tokkyo Koho JP 03065199 A; *Chem. Abstr.*, 115 (1991) 88308.
3585. E. Briones de Venegas and A. Gonzalez del Cueto, Spectrophotometric determination of α-amylase activity without the addition of β-amylase to the starch substrate suspension, *J. Am. Soc. Brew. Chemists.*, 50(2), (1992) 59–61.

3586. M. Marcazzan, F. Vianello, M. Scarpa, and A. Rigo, An ESR assay for α-amylase activity toward succinylated starch amylose and amylopectin, *J. Biochem. Biophys. Meth.*, 38 (1999) 191–202.
3587. V. D. Martin and J. M. Newton, Comparative rates of amylase action on starches, *Cereal Chem.*, 15 (1938) 456–462.
3588. H. M. El-Saied, Y. Ghali, and S. Gabr, Rate of starch hydrolysis by bacterial amylase, *Starch/Stärke*, 30 (1978) 96–98.
3589. I. Lovsin-Kumak, M. Zelenik-Blatnik, and V. Abram, Quantitative estimation of the action of α-amylase from *Bacillus subtilis* on native corn starch by HPLC and HPTLC, *Z. Lebensm. Untersuch Forsch.*, A 206 (1998) 175–178.
3590. H. Y. Kim and K. H. Park, Characterization of bacterial α-amylase by determination of rice starch hydrolysis product, *Hanguk Nonghwa Hakhoechi*, 29 (1986) 248–254. *Chem. Abstr.*, 106 (1986) 212781n.
3591. T. Sotoyama, Y. Nakai, Y. Hirata, K. Tanno, T. Hara, and Y. Fujio, Novel method of determining α-amylase by using starch gel as adsorbent (in Japanese), *Nippon Shokuhin Kagaku Gakkaishi*, 5 (1998) 197–200.
3592. M. Przybył, Determnation of glucoamylase activity with glucose electrode, *Starch/Stärke*, 43 (1991) 190–193.
3593. K. Inawa, N. Yamamoto, and H. Sasaki, Enzyme sensor and method for determination of α-amylase activity in body fluids, Jpn. Kokai Tokkyo Koho 10262645 (1998); *Chem. Abstr.*, 129 (1998) 241774.
3594. W. R. Johnston and S. Jozsa, General method for determining the concentration of enzyme preparations, *J. Am. Chem. Soc.*, 57 (1935) 701–706.
3595. A. Mukopadhyaya, A novel technique for determining half life of alpha amylase enzyme during liquefaction of starch, *Biotechnol. Techn.*, 6 (1992) 507–510.
3596. M. G. Goryacheva, Effect of different sources of soluble starch on the determination of amylase activity (in Russian), *Ferment. Spirt. Prom.* (8), (1973) 14–15.
3597. J. Seidemann, Charakterisierung von löslicher Stärke und anderer Stärkepolysaccharide zur Bestimmung der α-Amylaseaktivität, *Pharmazie*, 34 (1979) 79–85.
3598. T. Baks, A. E. M. Janssen, and R. M. Boom, The effect of carbohydrates on α-amylase activity measurements, *Enzyme Microb. Technol.*, 39 (2006) 114–119.
3599. G. Ritte, M. Steup, J. Kossman, and J. R. Lloyd, Determination of the starch-phosphorylating enzyme activity in plant extracts, *Planta*, 216 (2003) 798–801.
3600. K. R. Chandorkar and N. P. Badenhuizen, How meaningful are determinations of glucosyltransferase activities in starch – enzyme complexes? *Starch/Stärke*, 18 (1966) 91–95.
3601. R. H. McHale, A test for endo-amylase activity in the presence of exo-amylase, *Starch/Stärke*, 40 (1988) 422–426.
3602. L. Kuniak, J. Zemek, M. Tkač, and A. Taufel, Tablets for microbial α-amylase determination, Czechoslovak Patent 267157 (1990); *Chem. Abstr.*, 115 (1991) 251019.
3603. Y. She, B. Chen, M. Yan, and J. Hou, Preparation of the test kit for detecting oxido-reductase isoenzyme, Chinese Pat. CN 1381728 (2002); *Chem. Abstr.*, 139 (2002) 811154.
3604. B. V. McCleary and P. Sheehan, Measurement of cereal α-amylase. A new assay procedure, *J. Cereal Sci.*, 6 (1987) 237–251.
3605. H. Sheehan and B. V. McCleary, A new procedure for the measurement of fungal and bacterial α-amylase, *Biotechnol. Tech.*, 2 (1988) 289–292.
3606. B. V. McCleary and R. Sturgon, Measurement of α-amylase in cereal, food and fermentation products, *Cereal Foods World*, 47 (2002) 299–310.
3607. B. M. Brena, C. Pazos, L. Franco-Fraguas, and F. Batista-Viera, Chromatographic methods for amylases, *J. Chromatogr. B*, 684 (1996) 217–237.

3608. K. Okuda, K. I. Naka, and N. Shimoyo, *Handbook of Amylase and Related Enzymes*, (1988) Pergamon Press, Oxford, pp. 178–185.
3609. S. K. Bajpai and S. Saxena, Flow through diffusion cell method. A novel approach to study in vitro enzymatic degradation of a starch based ternary semi-interpenetrating network for gastrointestinal drug delivery, *J. Appl. Polym. Sci.*, 100 (2006) 2975–2984.
3610. S. Y. Chang, S. R. Delwiche, and N. S. Wang, Hydrolysis of wheat starch and its effect on the falling number procedure. Mathematical model, *Biotechnol. Bioeng.*, 79 (2002) 768–775.
3611. J. Zemek, L. Kuniak, and A. Matušova, Determination of alpha-, beta- and glucoamylase activities with corosslinked chromolytic starches, *Makromol. Chem. Suppl.*, 9 (1985) 219–225.
3612. Y. Ishige, T. Yoshid, and K. Yoritomi, Estimation of residual enzyme activities during enzymic hydrolysis of starch (in Japanese), *Dempunto Gijutsu Kenkyu Kaiho*, 35 (1967) 23–28.
3613. R. G. Ledder, T. Madhwan, P. K. Sreenivasan, W. De Vizio, and A. J. McBain, An in vitro evaluation of hydrolytic enzymes as dental plague control agents, *J. Med. Microbiol.*, 58 (2009) 482–491.

AUTHOR INDEX

Page numbers in roman type indicate that the listed author is cited on that page of an article in this volume; numbers in italic denote the reference number, in the list of references for that article, where the literature citation is given.

A

Aaman, P., 262, *3467*
Abarbri, M., 8, *42*
Abate, C., 78, *1086*, 79, *1086*
Abbas, I.R., 221, *2819*
Abd-Aziz, S., 226, *2897*
Abdel-Azize, R.M., 195, *2388*
Abdelfattah, A.F., 72, *662*
Abdelsalam, K., 78, *1079*
Abdulla, M., 109, *1117*
Abdullah, M., 98, *835*, 158, *1869*
Abe, A., 219, *2795*
Abe, J., 95, *618*, 118, *1233*, 118, *1234*, 129, *1419*, 150, *1233*
Abe, M., 216, *2715*
Abe, T., 156, *1785*
Abe, Y., 125, *1346*
Abee, T., 233, *2972*
Abelyan, V.A., 101, *895*, 124, *1327*, 252, *3275*
Abidin, Z., 226, *2891*, 226, *2892*, 226, *2893*
Abou Akkada, A.R., 125, *1334*
Aboul-Enein, A.M., 195, *2388*
Aboushadi, N.M., 112, *1174*
Aboutboul, H., 217, *2750*
Abouzied, M.M., 187, *2215*
Abraham, T.E., 107, *1009*, 174, *2037*, 219, *2782*, 242, *3091*, 243, *3091*
Abram, V., 267, *3589*
Abramov, Z.T., 73, *720*, 74, *720*, 76, *720*
Absar, N., 163, *1916*

Aburaki, S., 34, *122*, 37, *130*, 37, *132*
Acharya, C.N., 85, *412*
Achremowicz, B., 65, *302*, 266, *3562*, 266, *3563*
Ackerman, R.J., 73, *447*, 88, *447*, 119, *447*, 239, *447*, 266, *3570*
Adachi, M., 72, *647*, 79, *647*, 100, *875*, 176, *875*
Adachi, S., 137, *1483*, 139, *1523*, 139, *1524*
Adachi, T., 89, *491*, 156, *1784*, 205, *2509*
Adamczak, M., 219, *2792*
Adams, C.A., 81, *348*
Adams, F.R., 194, *2368*
Adams, G.A., 232, *2955*, 232, *2956*, 232, *2957*
Adams, M., 71, 82, *376*
Adams, M.W.W., 117, *1220*
Adeleye, A.I., 73, *710*
Adeni, A., 170, *2005*, 176, *2005*
Adeni, D.S.A., 226, *2898*
Adesogan, A.T., 207, *2587*, 236, *3027*
Adlercreutz, P., 247, *3187*
Afanaseva, V.P., 74, *745*, 77, *745*
Afanasyev, N.V., 207, *2585*
Afrikyan, E.G., 252, *3275*
Afrikyan, E.K., 101, *895*
Afschar, A., 231, *2949*
Aga, H., 258, *3373*
Aggarwal, N.K., 220, *2806*
Aggarwal, P., 262, *3438*
Aghajari, N., 64, *89*

AUTHOR INDEX

Aghani, N., 183, *2174*, 188, *2174*
Agneessens, R., 262, *3454*, 263, *3454*, 264, *3454*
Agnew, A.L., 83, *387*
Ago, K.I., 235, *2998*
Agrawal, M., 158, *1860*
Aguado, L., 27, *107*
Aguilar, G., 234, *2989*
Ah, J.S., 234, *2975*
Ahipo, E.D., 71, *638*, 78, *638*, 113, *638*, 122, *638*
Ahlborg, K., 63, *37*, 80, *317*, 80, *318*, 80, *328*, 80, *329*, 80, *330*, 94, *591*, 129, *37*, 151, *1709*, 214, *2673*
Ahmad, R., 101, *897*
Ahmed, S.H., 140, *1530*
Ahn, D.H., 167, *1979*, 218, *2770*, 218, *2775*, 224, *2770*
Ahn, E.K., 245, *3166*
Ahn, S., 163, *1926*
Ahn, Y.K., 101, *896*
Aibara, S., 76, *928*
Aikawa, K., 94, *580*
Aikawa, M., 252, *3274*, 254, *3310*
Aivasidis, A.C., 214, *2681*
Ajele, J.O., 76, *923*
Ajongwen, N.J., 191, *2319*
Akai, H., 158, *1873*
Akano, H., 255, *3346*, 257, *3368*
Akasawa, T., 71, *410*, 85, *410*
Akdogan, H., 190, *2307*
Akerberg, C., 142, *1538*, 236, *3019*, 264, *3510*
Akiba, T., 74, *727*, 75, *802*, 88, *456*, 88, *457*, 91, *539*, 126, *539*
Akimaru, K., 251, *3256*
Akino, T., 110, *1136*
Akinyanju, J.A., 72, *688*
Akiya, S., 88, *468*, 88, *469*, 88, *470*, 88, *471*, 249, *3226*, 249, *3227*
Akpan, I., 95, *619*, 105, *619*
Al Kazaz, M., 187, *2212*
Al Qalla, H., 263, *3473*
Alaa, M., 244, *3131*

Alam, R., 78, *1075*
Alazard, D., 74, *761*
Alcalde, M., 64, *176*
Aldao, M.A.J., 262, *3441*, 264, *3441*
Aldercreutz, P., 238, *3055*
Aleksandrov, V.V., 207, *2585*
Alemany, M., 81, *362*
Alexander, R.J., 205, *2516*
Alfin, R.B., 83, *386*, 83, *388*
Ali, S., 78, *1075*, 237, *3031*
Alissandrolos, A., 243, *3115*
Alken, C., 151, *1705*
Allain, E., 64, *83*, 92, *550*, 213, *2670*, 217, *2755*
Allam, A.M., 73, *722*
Allan, E., 147, *1587*
Allen, A., 263, *3469*, 264, *3469*
Allen, W.G., 89, *486*, 215, *2702*
Allerman, K., 75, *794*
Almeida-Dominguez, H.D., 192, *2332*
Alonso-Benito, M., 113, *1184*
Alpers, D.H., 77, *944*, 81, *370*, 82, *370*, 103, *944*
Alstin, F., 262, *3452*, 263, *3452*, 264, *3452*
Altaf, M., 234, *2988*, 235, *2988*
Alter, J.E., 88, *479*
Altieri, P., 206, *2534*
Altintas, M.M., 217, *2757*
Altman, C., 142, *1546*
Alverson, L.H., 158, *1839*
Alves-Prado, H.F., 251, *3247*
Amachi, T., 64, *104*
Åman, P., 177, *2098*, 262, *3462*, 263, *3462*
Amanlou, M., 187, *2208*
Amano, S., 104, *959*
Amante, R.E., 164, *1930*
Amarakone, S.P., 112, *1176*, 114, *1176*
Amato, M.E., 265, *3529*
Ambard, L., 127, *1382*, 180, *2142*
Amin, G., 227, *2920*
Amirul, A.A., 74, *756*, 78, *1064*
Ammeraal, R.N., 256, *3351*
Amorim, D.S., 67, *307*
Amsai, A., 64, *105*

AUTHOR INDEX

Amsal, A., 64, *95*
Amud, O.O., 73, *706*, 73, *707*
Amund, O.O., 77, *950*, 103, *950*
Amutha, R., 219, *2783*
An, S.C., 157, *1810*
Anastassiadis, S., 65, *277*, 237, *3045*, 237, *3046*
Andersen, C., 64, *78*, 64, *92*, 64, *97*, 91, *530*, 92, *548*, 121, *530*, 193, *2357*, 202, *2468*, 251, *3252*
Andersen, E., 155, *1772*
Andersen, K.V., 64, *75*
Andersen, L.N., 92, *550*
Anderson, H., 265, *3530*
Anderson, K.L., 64, *145*, 91, *533*
Anderson, L., 101, *888*, 139, *888*
Anderson, M.A., 75, *586*, 78, *586*, 94, *586*, 104, *586*, 107, *586*, 112, *586*, 114, *586*, 116, *586*
Anderson, W.A., 212, *2642*
Andersson, R., 177, *2098*
Ando, A., 78, *1088*
Ando, T., 256, *3360*
Andrade, C.M., 75, *797*, 117, *797*
Andrade, C.M.M., 74, *731*, 77, *731*
Andrade, M.C., 74, *731*, 77, *731*
Andreev, N.R., 158, *1872*
Andrews, M.R., 169, *1997*, 171, *1997*
Andrieux, C., 73, *701*
Andtfolk, H., 176, *2084*
Anfinsen, C.B., 75, *798*
Ang, D.C., 226, *2897*
Anger, H., 156, *1783*, 164, *1937*, 166, *1958*, 179, *2129*, 180, *2133*, 180, *2134*, 180, *2135*
Anikeeva, L.A., 165, *1943*
Anikieeva, L.A., 176, *2087*
Anindyawati, T., 95, *609*, 121, *609*, 127, *1385*
Ankrah, N.O., 208, *2601*
Ann, Y.G., 127, *1385*
Annous, B.A., 92, *554*
Ano, Y., 72, *654*
Anoop, C.V., 149, *1653*, 153, *1730*
Ansanelli, G., 265, *3529*

Anselmi, E., 8, *42*
Anslyn, E.V., 5, *27*
Anstaett, F.R., 182, *2166*
Antecki, Z., 238, *3049*
Antonanzas, J., 246, *3171*
Antranikian, G., 64, *112*, 64, *212*, 75, *797*, 75, *799*, 75, *801*, 77, *949*, 92, *552*, 103, *949*, 114, *1186*, 115, *1204*, 117, *797*, 117, *801*, 117, *1221*, 121, *1272*, 211, *2627*, 211, *2628*
Antrim, R.I., 244, *3142*
Antrim, R.L., 89, *484*, 151, *1704*, 188, *2234*
Anuradha, R., 235, *2997*
Anyanwu, C.U., 74, *753*, 76, *753*
Aoki, H., 251, *3242*
Aoki, K., 101, *900*
Apar, D.K., 125, *1343*, 133, *1462*, 149, *1650*, 165, *1462*, 168, *1986*
Arai, M., 74, *612*, 74, *613*, 74, *614*, 74, *764*, 75, *775*, 95, *612*, 95, *613*, 95, *614*
Arai, N., 104, *959*, 163, *1915*
Arai, T., 77, *1040*, 154, *1739*
Arai, Y., 110, *1135*, 116, *1135*, 150, *1690*
Araki, M., 74, *744*, 77, *744*
Arami, F., 149, *1646*
Arami, S., 95, *606*
Arami, T., 149, *1646*
Arasarantnam, V., 166, *1956*
Arasaratnam, V., 109, *1131*, 121, *1278*, 156, *1789*, 158, *1846*
Araujo, E.F., 217, *2748*, 217, *2749*
Araujo, N., 219, *2788*
Arbakariya, A., 170, *1999*, 207, *1999*
Archer, D.B., 129, *1415*, 129, *1426*
Ardá, A., 21, *92*, 21, *93*, 21, *94*
Arendt, E., 217, *2729*
Arendt, E.K., 147, *1595*
Argun, H., 245, *3164*
Arias, J.M., 73, *713*
Arica, M.Y., 190, *2293*
Ariff, A.B., 78, *1046*, 153, *1721*, 226, *2897*, 238, *3051*, 238, *3052*
Ariga, K., 190, *2298*
Ariga, O., 72, *654*

Arima, M., 204, *2505*
Arimura, T., 234, *2982*
Armbruster, F.C., 109, *1117*, 154, *1749*, 158, *1869*, 255, *3343*, 256, *3358*
Arpigny, J.L., 72, *642*
Arrien, N., 246, *3171*
Arroyo, J., 117, *1226*
Artjariyasripong, S., 220, *2796*
Artom, C., 121, *1284*, 138, *1495*
Artz, W.E., 190, *2278*
Asad, M.J., 89, *497*
Asai, S., 219, *2780*
Asano, K., 158, *1871*, 219, *2795*
Asano, M., 150, *1688*
Asano, S., 215, *2693*, 222, *2693*
Asayama, Y., 196, *2426*
Asbi, B., 170, *1999*, 207, *1999*
Aschengreen, N.H., 244, *3141*
Asgher, M., 89, *497*
Ashakumary, L., 78, *1053*, 78, *1054*, 78, *1055*, 78, *1056*, 78, *1057*, 78, *1058*
Asher, R.A., 67, *312*, 68, *312*
Ashikari, T., 64, *104*
Asis, R., 262, *3441*, 264, *3441*
Asp, N.-G., 202, *2466*
Astolfi-Filho, S., 72, *686*, 73, *686*, 74, *741*, 77, *741*
Astruc, A., 262, *3429*
Aten, J., 150, *1682*
Atichokudomchai, N., 80, *337*, 165, *337*
Atifah, N., 232, *2965*
Atkin, L., 88, *467*, 264, *3488*
Atkins, D.J., 161, *1898*
Atkins, D.P., 166, *1964*
Atomi, H., 246, *3172*
Atsumi, S., 231, *2953*, 232, *2953*
Attasampunna, P., 219, *2793*, 220, *2796*
Attia, R.M., 140, *1530*, 172, *2018*, 173, *2028*, 244, *3131*
Audebourg, A., 8, *42*
Auerman, L.Ya, 203, *2489*
Augustat, S., 149, *1635*, 150, *1656*
Augustin, J., 96, *632*, 137, *632*
Auh, J.H., 168, *1989*

Aunstrup, K., 95, *615*
Aura, A.M., 259, *3383*
Aurangzeb, M., 91, *537*, 217, *2736*, 218, *2736*
Auterinen, A.L., 89, *484*
Autio, K., 167, *1974*, 167, *1975*, 176, *2067*
Auwerx, J., 47, *154*, 48, *157*
Avakyants, S.P., 127, *1373*
Avall, A.K., 176, *2084*
Avram, A., 234, *2991*, 235, *2991*
Avramiuc, M., 86, *438*
Axen, R., 102, *916*
Ayano, Y., 179, *2118*
Aymard, C., 140, *1531*
Ayukawa, Y., 110, *1136*
Ayyanna, C., 110, *1150*
Ayyar, C.V.R., 188, *2246*
Azarova, G.G., 211, *2620*
Azemi, B.M.N.M., 173, *2031*
Azizan, M.N., 74, *756*, 78, *1064*
Azuma, M., 235, *2998*
Azzini, A., 227, *2918*, 227, *2919*

B

Baba, K., 206, *2553*
Baba, S., 223, *2850*
Baba, T., 99, *858*, 240, *3076*
Baba, Y., 194, *2379*, 194, *2380*
Babacan, S., 71, *398*, 83, *398*
Babenko, K., 145, *1566*
Babkina, M.V., 180, *2138*
Babor, K., 264, *3508*
Babu, K.R., 72, *658*
Bach Knudsen, K.E., 208, *2597*
Bachrach, E., 27, 62
Badenhuizen, N.P., 127, *1384*, 267, *3600*
Badocha, E., 224, *2869*, 230, *2939*
Badr, H.R., 231, *2950*
Bae, H., 123, *1316*
Bae, M., 220, *2804*
Baev, V.V., 195, *2398*
Bagai, R., 73, *702*
Bagby, M.O., 148, *1614*, 149, *1614*
Bagdasarian, M., 243, *3122*
Bai, F.W., 212, *2642*

Baig, M.A., 101, *897*
Baigori, M., 78, *1086*, 79, *1086*
Bailey, A.L., 83, *393*
Bailey, C.H., 98, *839*, 100, *839*
Bailey, K., 263, *3469*, 264, *3469*
Bailey, K.W., 11, *54*
Bailey, M.J., 120, *1253*, 248, *1253*
Bains, G.S., 198, *2444*, 206, *2443*
Bajpai, R., 235, *3010*
Bajpai, S.K., 268, *3609*
Bak, A., 80, *326*, 265, *3544*
Bakar, F.A., 79, *1154*
Bakar, J., 106, *1002*
Baker, E.P., 72, *641*
Baker, F., 260, *3404*
Baker, F.L., 165, *1952*
Baker, G.L., 204, *2502*, 204, *2504*
Baker, J., 100, *869*
Baker, J.C., 161, *1903*
Baker, J.L., 80, *334*, 86, *430*, 128, *1405*, 157, *1821*, 181, *2145*
Baker, P.E., 264, *3509*
Bakhle, S., 251, *3250*
Bak-Jensen, K.S., 64, *89*
Baks, T., 267, *3598*
Balagopalan, C., 220, *2805*
Balance, D.L., 176, *2055*
Balashova, G.B., 180, *2137*
Balasubramanian, K., 109, *1131*, 121, *1278*, 156, *1789*, 165, *1944*, 166, *1956*
Balauta, E., 222, *2833*
Balayan, A.M., 124, *1327*
Balaz, S., 64, *107*
Baldensperger, J.F., 74, *761*
Baldwin, T.L., 108, *1023*
Ballesteros, A., 64, *176*
Balls, A.K., 76, *927*, 122, *1287*
Baly, E.C.C., 135, *1468*
Balzarini, J., 2, *11*, 2, *12*, 2, *13*, 11, *11*, 11, *12*, 11, *13*, 11, *50*, 11, *52*, 11, *55*, 11, *61*, 11, *62*, 11, *63*, 11, *66*, 11, *67*, 11, *68*, 11, *70*, 11, *71*, 14, *75*, 27, *107*, 30, *68*, 46, *153*, 47, *11*, 47, *12*, 47, *13*, 47, *154*, 47, *155*, 47, *156*, 48, *52*, 48, *55*, 48, *157*

Ban, Q., 238, *3056*
Banasik, O.J., 147, *1590*
Bandaru, V.V.R., 226, *2889*
Bandmann, H., 16, *76*
Bandopadhyay, I.A., 72, *661*
Banerjee, M., 212, *2644*
Banerjee, R., 96, *804*
Banerjee, U.C., 93, *566*, 107, *566*
Banik, G., 194, *2382*
Bankole, M.O., 95, *619*, 105, *619*
Banks, W., 128, *1407*
Bannerjee, M., 215, *2699*
Baradarajan, A., 213, *2660*, 213, *2661*
Baraš, J., 222, *2840*
Baratti, J., 224, *2864*
Barba de la Rosa, A.P., 168, *1994*
Barbier, G., 112, *1173*, 113, *1173*
Barbu, E., 182, *2162*
Barcza, L., 256, *3355*
Baret, J.L.A.G., 227, *2911*
Barfoed, H., 62, *10*, 65, *10*, 150, *1661*
Barfoed, H.C., 64, *170*
Bargagliotti, C., 230, *2936*
Barghava, S., 156, *1786*
Barinova, S.A., 94, *593*
Barker, P.E., 191, *2319*
Barker, S.A., 64, *156*, 120, *1251*, 120, *1252*, 242, *1251*, 248, *1251*, 248, *1252*
Barmenkov, Ya.P., 128, *1404*, 129, *1404*
Barnard, D.L., 11, *54*
Barnes, W.C., 263, *3474*
Barnett, C.C., 196, *2423*
Barnett, E.A., 75, *796*
Barresi, F.W., 197, *2437*
Barreto di Menezes, T.J., 227, *2918*, 227, *2919*
Barta, M., 109, *1127*
Bartenev, S.P., 127, *1373*
Bartfay, J., 146, *1574*, 154, *1735*, 154, *1741*, 158, *1735*, 159, *1741*
Barton, L.L., 73, *695*, 77, *695*
Barton, N., 64, *82*
Barton, R.R., 154, *1748*
Baruah, J.N., 88, *465*, 99, *465*

Barwell, N.P., 24, *101*
Basappa, S.C., 78, *1087*, 213, *2663*, 217, *2663*, 219, *2786*, 225, *2786*
Basset, J., 182, *2162*
Basu, K.P., 168, *1984*
Bataus, L.A., 72, *686*, 73, *686*
Batey, J.L., 262, *3448*, 264, *3448*
Bath, K.S., 93, *566*, 107, *566*
Batista-Viera, F., 268, *3607*
Batsalov, I.P., 203, *2489*
Batsalova, M., 99, *861*, 159, *861*
Baudendistel, N., 243, *3115*
Bauer, E., 228, *2927*, 228, *2928*
Bauer, F., 262, *3444*, 265, *3444*
Baum, F., 193, *2359*
Baum, L.G., 12, *74*
Bauman, I., 200, *2457*
Baumgarten, G., 183, *2175*
Bayramoglu, G., 190, *2293*
Baysal, Z., 183, *2170*
Baytan, P., 206, *2552*
Bazin, H.G., 197, *2437*
Bazin, S.L., 210, *2612*
Bea, S., 225, *2874*
Bealin-Kelly, F., 88, *452*
Beall, D.S., 217, *2746*
Bean, S.R., 225, *2880*, 226, *2888*
Beazell, J.M., 208, *2593*
Bebić, Z.J., 165, *1942*, 222, *2840*
Beck, M., 208, *2598*
Becker, A., 177, *2101*
Becker, T., 96, *630*
Becker, U., 236, *3022*
Beckerich, R., 222, *2836*
Beckett, P., 203, *2490*
Beckord, O.C., 63, *33*, 267, *33*
Bednarski, W., 219, *2792*
Beech, S.C., 65, *274*
Beeftink, H.H., 180, *2139*
Beer, S.K., 246, *3173*
Behmenburg, H., 121, *1270*
Beier, L., 64, *75*, 64, *92*
Beishuizen, J.W., 150, *1678*
Bejar, S., 90, *509*, 102, *509*

Bejar, S.A., 72, *677*
Belafi-Bako, K., 126, *1371*, 190, *2284*
Belarbi, A., 64, *177*, 67, *177*, 75, *177*
Belayeva, L.I., 204, *2496*
Beldie, C., 110, *1134*
Beldman, G., 77, *947*, 103, *947*
Beleia, A., 85, *407*
Belenkii, D.M., 139, *1511*
Bellal, M., 110, *1142*
Belleville, M.P., 139, *1507*, 190, *2288*, 190, *2290*
Bellini, M.Z., 75, *790*, 94, *583*
Belov, E.M., 139, *1514*
Belshaw, N.J., 64, *132*, 129, *132*, 129, *1422*
Belsue, M., 246, *3171*
Ben Ali, M., 72, *677*, 90, *509*, 102, *509*
Ben Anmar, Y., 95, *609*, 121, *609*
Ben Elarbi, M., 71, *639*
Ben Hamida, J., 71, *639*
Ben-Bassat, A., 222, *2831*
Bende, P., 150, *1677*
Bender, H., 77, *952*, 103, *952*, 123, *1311*, 253, *3286*, 253, *3290*
Bender, H.F., 253, *3288*
Bendetskii, K.M., 141, *1534*
Benedetski, K.M., 187, *2209*
Benedito, C., 203, *2478*
Benedito De Barber, C., 166, *1966*
Beneditto de Barb, C., 198, *2454*
Benkova, M., 122, *1306*
Benner, S.A., 245, *3158*
Benson, C.P., 104, *957*
Beran, K., 126, *1369*
Berentzen, H., 55, *63*
Berg, H., 228, *2927*, 228, *2928*
Bergelin, R., 64, *85*
Berger, R., 236, *3022*
Bergerov, D., 74, *748*, 77, *748*
Berghgracht, M., 196, *2405*
Berghmans, E.F., 149, *1620*
Bergkvist, R.R., 203, *2488*
Berglund, K.A., 190, *2314*
Bergmann, F.W., 95, *618*
Bergmann, H., 217, *2729*

Bergsma, J., 180, *2139*, 241, *3078*
Bergthaller, W., 180, *2133*, 180, *2134*
Berka, E., 76, *933*
Berkson, J., 129, *1440*
Berliner, E., 167, *1969*
Bernat, J.A., 218, *2776*
Berner, A., 8, *39*
Bernfeld, P., 64, *116*, 98, *831*, 98, *846*, 100, *884*, 125, *1342*, 139, *1508*, 242, *3095*, 242, *3096*, 267, *3575*
Bernhardsdotter, E.C.M.J., 90, *526*
Berović, M., 65, *276*
Berry, J.W., 221, *2819*
Bertaux, C., 11, *52*, 48, *52*
Bertof, E., 176, *2066*
Bertoft, E., 146, *1584*, 167, *1977*, 172, *2012*, 174, *2039*, 174, *2040*, 176, *2078*, 176, *2084*, 183, *2172*, 264, *3504*, 265, *3550*
Bertoldo, C., 64, *112*
Bertolini, M.C., 74, *747*, 77, *747*
Bertoti, I., 196, *2404*
Bertrand, G., 71, *417*, 85, *417*
Bérubé, M., 9, *45*, 10, *48*, 12, *45*
Beschkov, V., 89, *482*, 251, *3251*
Beshkov, M., 158, *1851*
Best, R.W., 179, *2121*
Betermann, K., 264, *3517*, 264, *3518*
Beutler, H.O., 262, *3434*, 262, *3440*, 264, *3434*
Bevan, E.A., 186, *2191*
Bewley, C.A., 47, *155*
Beyer, A., 266, *3564*
Bhargava, S., 148, *1605*, 148, *1607*, 211, *2617*, 211, *2618*, 217, *2732*
Bhat, J.V., 93, *572*
Bhat, U.R., 176, *2063*, 177, *2063*, 177, *2095*
Bhathena, S.J., 259, *3386*
Bhatia, Y., 64, *179*, 119, *179*
Bhatnagar, L., 102, *910*, 243, *3122*
Bhatty, R.S., 192, *2327*
Bhella, R.S., 74, *752*
Bhide, N.V., 127, *1390*
Bhosale, S.H., 243, *3118*
Bhuiyan, M.Z.H., 99, *859*

Bhumibhamon, O., 75, *767*, 109, *767*
Bhuwapathanapun, S., 236, *3015*
Bi, Y., 149, *1647*
Białas, W., 80, *341*, 179, *2117*
Bidrawn, J.L., 151, *1712*, 154, *1752*, 178, *1752*
Bidziński, Z., 71, *431*, 86, *431*
Biedermann, W., 62, *17*, 62, *20*, 62, *25*, 62, *26*
Bielecki, M., 9, *47*
Bielecki, S., 119, *1247*, 208, *2589*, 208, *2590*, 266, *1247*
Bienkowski, P.R., 213, *2658*
Bijttebier, A., 81, *361*, 177, *2100*
Bilard, C., 241, *3088*
Billard, P.H., 235, *3005*
Billet, I., 191, *2315*
Billford, H.R., 63, *34*, 210, *34*
Binnema, D.J., 77, *947*, 103, *947*, 241, *3078*, 241, *3079*, 248, *3198*
Birkett, A.M., 206, *2532*
Birnbaum, D., 217, *2740*
Bisaria, V.S., 64, *179*, 119, *179*
Bisgard-Frantzen, H., 64, *91*, 89, *490*, 118, *1235*, 147, *1587*, 148, *1605*, 195, *2395*, 217, *2732*
Biswas, A.K., 217, *2737*
Biwer, A., 256, *3363*
Bizais, Y., 259, *3382*
Bjoerck, I., 166, *1965*
Bjoerling, C.O., 100, *870*
Bjorck, I., 177, *2098*
Bjork, I., 202, *2466*
Björndal, A., 11, *64*
Bjornvad, M., 64, *91*
Blachere, G.F.J., 243, *3108*
Blachere, H., 214, *2686*
Black, L.T., 148, *1614*, 149, *1614*
Black, P.T., 145, *1560*
Blake, J.C., 260, *3392*
Blakeney, A.B., 265, *3536*
Blakeslee, D., 11, *56*
Blanch, H.W., 218, *2763*
Blanshard, J.M.V., 64, *240*, 99, *859*

Blaschek, H.P., 73, *694*, 77, *694*, 92, *554*, 231, *2951*
Blasel, H.M., 262, *3456*
Blaszków, W., 187, *2220*
Blazłewicz, J., 210, *2607*, 210, *2608*
Błaszczyńska, A., 109, *1125*
Błaszkow, W., 64, *166*, 110, *1137*, 244, *3133*
Błażejczak, M., 224, *2868*
Blennow, A., 64, *113*
Blinnikova, E.I., 94, *589*
Blish, M.J., 71, *432*, 86, *432*
Blom, J., 80, *326*, 80, *342*, 83, *342*, 83, *389*, 161, *1907*, 265, *3544*
Blue, E.K., 197, *2431*, 197, *2432*, 197, *2433*
Blüher, A., 194, *2382*
Blum, H., 196, *2424*
Blum, P., 75, *800*, 113, *1183*, 114, *1188*, 114, *1189*
Boas, F., 126, *1360*
Bod, R., 217, *2740*
Boekestein, P.T., 176, *2092*
Boersting, C.F., 260, *3400*
Boese, R., 16, *79*
Bogdanescu, S., 234, *2991*, 235, *2991*
Bohnenkamp, C.G., 98, *834*
Böhnke, H., 5, *24*
Boidin, A., *40*, 63
Bok, S.H., 243, *3119*
Boki, K., 193, *2362*
Bolduan, G., 208, *2598*
Bölmstedt, A., 11, *63*, 11, *64*, 11, *67*, 11, *68*, 11, *69*, 30, *68*
Bolton, D.J., 72, *659*, 72, *660*
Boltyanskaya, E.V., 73, *712*
Bon, E.P.S., 78, *1047*
Boncoddo, B.N.F., 163, *1919*
Bondar, E.G., 108, *1031*
Bondareva, E.D., 127, *1393*, 137, *1486*, 174, *2046*
Bong, H.U., 105, *981*, 105, *982*
Bonissol, C.M.T.A., 243, *3108*
Bonse, D., 243, *3124*
Boom, R.M., 182, *2160*, 267, *3598*
Boone, D.R., 124, *1325*

Boons, G.J., 24, *101*
Borchardt, H., 188, *2237*
Borchert, T., 217, *2755*
Borchert, T.V., 64, *75*, 89, *490*, 195, *2395*
Borglum, G.B., 217, *2722*, 222, *2722*
Borhardt, H., 83, *390*, 163, *390*
Borissovskii, V., 186, *2205*
Bornet, F.R.J., 259, *3382*
Borochoff, E.H., 193, *2352*
Borodina, Z.M., 108, *1031*, 120, *1259*, 165, *1259*
Borovtsova, O.Yu., 189, *2263*, 189, *2264*
Borriss, R., 156, *1782*
Bort, B.R., 73, *737*, 74, *737*
Bortel, R., 97, *815*
Boruch, M., 179, *2124*, 244, *3151*, 245, *3151*, 245, *3161*, 245, *3162*
Bos, C., 149, *1639*, 155, *1772*
Bos, H., 123, *1313*, 241, *1313*
Bos, H.T., 241, *3079*
Bos, H.T.P., 241, *3078*
Bosch, B.J., 11, *55*, 48, *55*
Boskov, Z., 64, *226*, 79, *1027*, 108, *1027*, 114, *1027*, 115, *1027*, 154, *1747*, 155, *1766*
Bothast, R.J., 165, *1950*, 223, *2842*, 223, *2843*
Botos, I., 11, *51*
Bouchet, B., 160, *1893*
Bouchet, P., 70, *617*, 95, *617*, 104, *617*
Boudarel, M.J., 214, *2686*
Boudrant, J., 110, *1142*, 190, *2309*
Boulard, H., 63, *50*, 212, *50*
Boundy, J.A., 76, *934*, 97, *808*
Bouquelet, S.J.L., 129, *1428*
Bourdon, D., 262, *3428*
Bourke, A.C., 88, *450*, 160, *450*
Bourke, A.C.M., 88, *449*
Bourne, E.J., 120, *1251*, 120, *1252*, 127, *1378*, 145, *1565*, 242, *1251*, 248, *1251*, 248, *1252*
Bowen, W.H., 63, *69*
Bower, B.S., 108, *1023*
Boy, M., 145, *1568*
Boyd, D., 197, *2431*, 197, *2432*, 197, *2433*
Boyd, M.R., 11, *56*, 30, *108*

Boyer, E.W., 72, *646*
Boyes, P.N., 64, *239*
Boyle, D., 225, *2880*
Boze, H., 96, *634*, 105, *634*, 106, *1003*, 106, *1004*
Bozich, F.A. Jr., 195, *2394*
Braak, J.P., 83, *392*
Braathen, E., 198, *2450*
Brabender, C.W., 93, *571*
Brae, B., 80, *326*, 265, *3544*
Brandam, C., 210, *2616*, 211, *2616*
Bravo da Costa Ferreira, T., 164, *1932*
Brehm, S.D., 117, *1220*
Brekelmans, A.M., 72, *657*, 108, *657*
Brena, B.M., 268, *3607*
Brennan, L., 129, *1432*
Bretschneider, R., 204, *2493*
Breves, R., 196, *2415*
Brewer, C.F., 12, *74*
Bridger, S.L., 117, *1220*
Briess, E., 146, *1585*
Briggs, D.E., 121, *1269*
Bringhenti, L., 219, *2791*
Brinkner, F., 204, *2498*
Brintzinger, H., 264, *3483*
Brintzinger, W., 264, *3483*
Briones de Venegas, E., 267, *3585*
Britto, R.A., 133, *1461*, 134, *1461*, 142, *1543*
Brock, A., 83, *395*, 259, *395*
Broeze, J.R., 183, *2183*, 264, *3490*
Brooks, A.E., 207, *2587*
Brooks, J.R., 168, *1990*
Brouard, F.E.M.E., 227, *2911*
Brovarets, T.V., 107, *1006*
Brown, J.C., 262, *3420*
Brown, S.H., 114, *1185*, 115, *1212*, 117, *1212*
Bruce, J., 260, *3404*
Bruckmayer, P., 159, *1880*
Bruijn, J., 204, *2495*
Bruinenberg, P.M., 248, *3198*
Bruley des Varannes, S., 259, *3382*
Brumm, P.J., 72, *671*, 172, *2020*
Brunt, K., 262, *3437*, 262, *3439*
Bryce, J.H., 217, *2719*

Bryjak, J., 64, *161*, 64, *252*, 122, *1294*, 132, *1460*
Bryne, M., 8, *39*
Brzeziłski, S., 163, *1909*
Buchholz, K., 235, *2992*
Buckheit, R., 11, *56*
Buckheit, R.W. Jr., 30, *108*
Buckle, P.E., 264, *3524*
Bućko, A., 260, *3403*
Buckow, R., 100, *883*
Budryn, K., 108, *1032*
Buesa, C., 74, *737*, 73, *737*
Buethner, R., 217, *2740*
Bugatti, A., 46, *153*
Bui, D.M., 77, *1039*
Buitelaar, R.M., 72, *685*
Bujang, K., 170, *2005*, 176, *2005*, 226, *2898*
Buka, B., 214, *2674*
Buleon, A., 64, *155*, 64, *160*, 165, *1947*, 166, *1947*, 196, *2409*, 203, *2482*
Bull, V.J., 167, *1976*
Buranakarl, L., 92, *545*
Burchhardt, G.F., 217, *2746*
Bureau, G., 173, *2035*, 173, *2036*
Burel, C., 190, *2300*
Burger, J., 183, *2180*
Burger, M., 126, *1369*
Burgess, I.G.R., 204, *2497*
Burgess-Cassler, A., 73, *705*, 165, *1952*
Burhan, N., 251, *3251*
Burianek, J., 89, *496*, 149, *1642*, 190, *2310*
Burkhart, B.A., 99, *856*
Burle, J.A., 219, *2788*
Burmann, W.E., 78, *1072*
Bush, J.E., 73, *726*
Busk, P.K., 78, *1062*
Buthe, A.C., 19, *88*
Butler, D.P., 64, *141*
Butler, L.G., 242, *3100*
Butterworth, P.J., 127, *1375*, 138, *1496*, 178, *1375*, 181, *1375*
Buttrose, M.S., 175, *2051*
Buzzini, P., 30, *109*
Byrd, R.A., 43, *148*

Byrom, D., 114, *1193*, 116, *1193*
Byun, M.S., 239, *3059*
Byun, S.M., 79, *1014*, 88, *448*, 107, *1014*

C

Cabalda, V.M., 64, *213*, 265, *3537*
Cabello, C., 219, *2791*
Caboche, J.J., 157, *1800*
Cabral, J.M.S., 65, *268*, 189, *2269*, 189, *2270*, 189, *2274*, 190, *2274*
Cacciarini, M., 19, *89*, 20, *90*
Cacioppo, F., 243, *3111*
Cadmus, M.C., 155, *1777*
Cai, L., 78, *1079*
Cai, M., 136, *1479*
Cairns, P., 179, *2130*
Calderon, M., 234, *2990*, 237, *3036*
Calderon Santoyo, M., 236, *3020*
Caldini, G., 64, *102*
Caldwell, M.L., 71, *376*, 75, *774*, 80, *340*, 81, *359*, 82, *376*, 83, *386*, 83, *387*, 83, *388*, 94, *590*, 127, *340*, 138, *1491*, 138, *1492*, 188, *2242*, 188, *2244*, 188, *2245*, 258, *590*
Calleja, G.B., 93, *563*, 212, *2643*, 218, *2643*
Callen, W., 64, *76*, 64, *77*, 64, *82*
Calsavara, L.P.V., 189, *2271*, 189, *2272*
Camacho, F., 142, *1547*
Camacho Rubio, F., 167, *1971*
Camarasa, M.J., 27, *107*
Cameron, L.E., 150, *1675*
Camire, A.L., 190, *2312*
Camire, M.E., 190, *2312*
Campbell, G.L., 208, *2601*
Campbell, H.A., 189, *2252*
Campbell, L.D., 208, *2602*
Campbell, L.L., 72, *678*, 72, *679*, 72, *680*
Campbell, N., 233, *2971*
Cañada, F.J., 21, *92*, 21, *93*, 21, *94*
Canepa, P., 232, *2958*
Canganella, F., 75, *797*, 117, *797*
Cao, J., 192, *2335*
Cao, L., 146, *1573*
Cao, N., 178, *2108*
Cao, S., 264, *3523*
Cao, X., 123, *1307*, 256, *3364*
Capron, I., 241, *3079*
Caraban, A., 192, *2343*
Carabez Trejo, A., 168, *1994*
Caratti de Lima, D., 108, *1025*
Cardellina, J.H. II., 30, *108*
Cardin, C.E., 248, *3200*
Cardoso, J.P., 189, *2268*, 189, *2269*, 189, *2270*, 189, *2274*, 190, *2274*
Carnell, R., 73, *724*
Carr, M.E., 148, *1614*, 149, *1614*, 222, *2834*
Carrin, M.E., 204, *2501*
Carrol, J.O., 120, *1261*, 154, *1751*, 155, *1261*
Carroll, J.O., 133, *1461*, 134, *1461*, 142, *1543*
Carson, J., 7, *36*
Carter, J., 266, *3552*
Carter, J.E., 85, *403*
Carter, N.P., 81, *366*, 82, *366*
Casas, C., 65, *273*
Casey, J.P., 146, *1586*, 147, *1586*
Castilho, V.M., 75, *790*
Castillo, F.J., 96, *622*
Catapano, G., 190, *2285*
Cattell, G.S., 149, *1617*
Cattle, M., 129, *1447*, 187, *1447*
Caujoli, F., 188, *2230*
Caujolie, F., 188, *2229*, 188, *2231*
Caumon, J., 190, *2299*
Cavanna, C., 230, *2936*
Cave, L.E., 194, *2368*
Cavicchioli, R., 64, *79*
Cayle, T., 108, *1035*, 263, *1035*
Cazares, L.H., 8, *41*
Ceci, L.N., 65, *291*, 204, *2501*
Cenci, G., 64, *102*
Cereda, M.P., 218, *2766*
Cerletti, P., 202, *2465*
Cernigoj, F., 250, *3233*
Cha, J.H., 72, *669*
Cha, P.S.O.D.J., 226, *2885*
Cha, X., 244, *3154*
Chae, H.Y., 168, *1989*, 207, *2568*

Chaen, H., 64, *254*, 193, *2350*, 207, *2567*, 241, *3083*, 246, *3177*, 257, *3372*
Chakrabarti, A.C., 109, *1124*
Chakroborty, S., 73, *708*, 75, *785*, 242, *3089*, 242, *3090*
Chakraverty, R., 102, *914*
Chakravorty, R., 76, *941*
Chamberlain, N., 198, *2445*
Champagne, E.T., 206, *2543*
Champenois, Y., 203, *2482*
Chamy, R., 189, *2260*
Chan, C.T., 75, *779*
Chan, K.Y., 76, *932*
Chan, P.K., 11, *54*
Chandorkar, K.R., 267, *3600*
Chandra, A.K., 64, *87*, 88, *481*, 92, *549*
Chandraraj, K., 158, *1860*
Chandrokar, K.A., 127, *1384*
Chang, C.C., 245, *3163*
Chang, H.N., 158, *1844*, 167, *1979*
Chang, H.Y., 142, *1539*
Chang, J.E., 101, *896*
Chang, J.S., 246, *3169*
Chang, L., 64, *184*
Chang, L.W., 73, *700*, 79, *700*
Chang, S.M., 170, *2002*
Chang, S.S., 93, *560*
Chang, S.Y., 268, *3610*
Chang, W.J., 255, *3344*
Chang, Y.H., 177, *2103*, 177, *2104*
Chang, Y.L., 252, *3279*
Chaokasem, N., 96, *803*
Chapman, D.W., 85, *415*
Chapman, G., 76, *929*
Charbonniere, R., 265, *3549*
Chardonloriaux, I., 73, *701*
Charlebois, R.L., 112, *1172*, 114, *1172*
Charm, S.E., 186, *2192*
Chary, S.J., 96, *625*
Chaterjee, B., 78, *1074*
Chatterjee, B., 78, *1076*, 130, *1454*
Chatterjee, B.S., 105, *987*
Chaturvedi, S., 266, *3561*

Chaudhary, K., 79, *1109*, 217, *2743*, 217, *2744*, 217, *2745*
Chaudhary, V.K., 115, *1201*, 157, *1826*
Chaudhry, M.A., 179, *2119*
Chauhan, B., 64, *121*
Chedid, L., 206, *2552*
Cheftel, C., 110, *1142*
Cheftel, J.C., 190, *2309*, 191, *2315*
Cheirslip, B., 170, *2003*
Chelkowski, J., 129, *1445*, 160, *1445*
Chen, B., 213, *2662*, 268, *3603*
Chen, C., 158, *1836*
Chen, C.L., 72, *651*
Chen, C.Y., 246, *3169*
Chen, G., 247, *3184*
Chen, H., 74, *557*, 78, *1079*, 92, *557*, 207, *2573*
Chen, H.P., 255, *3333*
Chen, H.Y., 109, *1121*
Chen, J., 65, *287*, 244, *3136*, 247, *3184*
Chen, J.P., 97, *806*
Chen, K.C., 190, *2294*
Chen, K.H., 252, *3278*
Chen, L., 129, *1423*, 201, *2463*, 219, *2777*
Chen, L.F., 109, *1129*, 178, *2108*, 194, *2364*
Chen, P., 213, *2655*, 213, *2662*, 220, *2809*
Chen, Q., 193, *2345*
Chen, Q.H., 150, *1681*
Chen, S., 189, *2259*, 224, *2860*
Chen, S.C., 74, *557*, 92, *557*
Chen, S.D., 246, *3169*
Chen, W.J., 190, *2301*
Chen, X., 193, *2358*, 218, *2762*, 242, *3094*, 244, *3136*
Chen, Y., 179, *2116*, 216, *2713*, 216, *2714*, 218, *2762*
Chen, Y.H., 88, *474*
Chen, Y.S., 266, *3560*
Chen, Z., 149, *1649*, 193, *2347*
Cheng, C.Y., 72, *651*
Cheng, K., 232, *2959*
Cheng, S.C., 76, *932*
Cheng, Y., 3, *14*, 5, *14*, 7, *35*
Cheong, T.K., 72, *669*

Cherevko, N.G., 139, *1515*
Cherry, J.R., 64, *92*
Cheryan, M., 186, *2199*, 190, *2278*, 190, *2281*
Cheshkova, A.V., 195, *2399*
Cheung, C., 232, *2962*
Chi, Z., 222, *2829*, 222, *2835*, 246, *3181*
Chiang, B.Y., 265, *3542*
Chiang, H.Y., 239, *3064*
Chiang, J.P., 88, *479*
Chiarello, M.D., 79, *1100*
Chiba, M., 122, *1296*
Chiba, S., 106, *1005*, 107, *1015*, 111, *1156*, 111, *1157*, 111, *1160*, 111, *1161*, 111, *1163*, 112, *1165*, 112, *1180*, 113, *1156*, 113, *1157*, 113, *1160*, 113, *1161*, 113, *1165*, 265, *3531*
Chiba, T., 190, *2296*
Chicoye, E., 115, *1201*, 157, *1826*
Chien, H.R., 125, *1347*
Chiirin, J., 255, *3336*
Chikano, T., 124, *1328*
Chin, T.Z., 204, *2492*
Chino, M., 75, *784*
Chiou, B.S., 175, *2053*
Chiou, J.Y., 190, *2294*
Chiquetto, M.L., 78, *1045*
Chistyakova, E.A., 138, *1498*, 141, *1535*
Chithra, N., 213, *2660*, 213, *2661*
Chityala, A., 226, *2889*
Chiu, C.P., 239, *3064*
Chiu, C.W., 197, *2431*, 197, *2432*, 197, *2433*, 206, *2533*, 206, *2534*, 206, *2549*
Chiwa, M., 254, *3325*
Cho, K.H., 252, *3285*
Cho, R.K., 264, *3493*
Cho, S.H., 109, *1120*
Cho, W.I., 214, *2682*
Cho, Y.J., 244, *3156*
Cho, Y.K., 248, *3207*
Cho, Y.U., 214, *2679*, 226, *2679*
Chodat, F., 128, *1403*
Choe, J.S., 250, *3236*
Choi, C.R., 207, *2568*
Choi, J.S., 222, *2832*

Choi, J.Y., 252, *3284*, 252, *3285*
Choi, K.H., 252, *3284*
Choi, M.H., 219, *2785*, 226, *2785*
Choi, S.C., 213, *2657*
Choi, S.H., 89, *504*, 145, *1556*, 156, *1556*, 254, *3326*
Choi, S.K., 12, *73*
Chojecki, A., 73, *694*, 77, *694*
Chon.J.H., 250, *3236*
Chotani, G.K., 108, *1023*
Chou, C.J., 117, *1220*
Chou, W.P., 118, *1232*
Choukeatirote, E., 78, *1079*
Chouvel, H., 190, *2309*
Chowdary, G.V., 164, *1935*
Chrempinska, H., 188, *2241*
Christa, K., 126, *1363*, 176, *1363*
Christensen, T., 64, *89*
Christian, M.T., 259, *3381*
Chrząszcz, T., 71, *369*, 71, *431*, 81, *369*, 82, *369*, 86, *431*, 267, *3580*, 267, *3581*
Chu, D.H., 97, *806*
Chu, W.S., 118, *1232*
Chu, X., 213, *2655*
Chu, Y., 8, *41*
Chua, J.W., 220, *2808*
Chuenko, T.V., 109, *1123*, 189, *2264*
Chugunov, A.I., 204, *2496*
Chun, J., 234, *2976*
Chung, H.C., 252, *3273*
Chung, M.J., 90, *515*, 105, *991*, 106, *991*
Chung, W.C., 197, *2427*, 197, *2428*, 197, *2429*
Chung, Y.C., 74, *727*
Chunxi, T., 249, *3211*
Chyau, C., 72, *673*
Ciacco, C.F., 176, *2060*, 180, *2131*, 223, *2847*
Ciao, J., 65, *297*
Ciegler, A., 126, *1355*
Ciesielski, K., 122, *1294*
Cieślak, M., 224, *2868*
Cioglia, L., 121, *1284*
Cirsteanu, M., 99, *855*
Claesson, K.O., 203, *2488*

Clark, R., 182, *2165*
Clark, R.H., 145, *1560*
Clarkson, K.A., 64, *101*
Clary, J.J., 83, *391*
Claus, W.D., 139, *1510*
Cleerwerck, J., 237, *3041*
Clemens, R.M., 115, *1211*, 129, *1431*
Clements, K.D., 71, *401*, 85, *401*
Cleveland, P.D., 72, *680*
Clifford, W.M., 183, *2179*, 184, *2179*, 185, *2179*
Clifton, P.M., 259, *3377*
Closset, G.P., 189, *2255*
Clotet, R., 173, *2032*
Clow, K.G., 81, *366*, 82, *366*
Cobb, J.T., 108, *1033*, 189, *2255*, 189, *2256*
Cochet, N., 127, *1374*
Coduro, E., 181, *2155*, 202, *2470*
Coglia, L., 121, *1285*
Cojocaru, C., 158, *1852*
Coker, L.E., 149, *1643*
Colill, W., 244, *3142*
Colin, H., 64, *218*
Collazos Hernandez, G., 153, *1729*
Collins, B.S., 73, *723*, 82, *383*, 98, *383*
Collins, E.V. Jr., 158, *1864*
Collins, H., 266, *3565*
Collins, I.D., 95, *600*
Collins, S.H., 114, *1193*, 116, *1193*
Collins, T.H., 198, *2445*
Colliver, S., 64, *100*
Colman, P., 100, *877*
Colonna, P., 64, *155*, 64, *160*, 70, *617*, 74, *758*, 95, *617*, 104, *617*, 130, *1451*, 130, *1452*, 130, *1453*, 160, *1893*, 165, *1947*, 166, *1947*, 177, *2096*, 203, *2482*
Coma, V., 173, *2033*, 173, *2034*
Combes, D., 97, *807*
Comes, S.B., 117, *1220*
Comfort, D.A., 117, *1220*
Compagno, C., 218, *2761*
Compton, A., 71, *417*, 85, *417*
Concone, B.R.V., 191, *2321*
Conde-Petit, B., 203, *2477*

Condoret, J.S., 97, *807*
Cone, J.W., 262, *3433*, 265, *3533*
Cong, L., 97, *813*
Conil, N., 214, *2686*
Conn, J.F., 198, *2447*
Constable, C., 150, *1686*
Constantinides, A., 97, *821*
Constantino, H.R., 114, *1185*
Converti, A., 230, *2934*, 230, *2936*, 232, *2958*
Conway, R.L., 173, *2030*
Coogan, M., 123, *1315*
Cook, D.H., 63, *62*
Coolhaas, C., 233, *2968*
Cooper, C.S., 217, *2719*
Copeland, L., 186, *2200*
Čopikova, J., 204, *2493*
Copinet, A., 166, *1960*, 173, *2033*, 173, *2034*, 173, *2036*, 241, *3088*
Cordenunei, B.R., 99, *866*, 101, *866*
Cormier, E.G., 11, *52*, 48, *52*
Cornet, J.M.P., 158, *1841*
Coronado, M.J., 73, *703*
Corredor, D.Y., 225, *2874*
Costa, R.M., 143, *1551*
Costabile, T., 112, *1172*, 114, *1172*, 190, *2276*
Cote, G.L., 165, *1952*
Cotta, M.A., 67, *308*, 125, *1333*, 223, *2843*, 226, *2899*
Cottaz, S., 64, *89*
Coulter, P.R., 262, *3468*
Courtin, C.M., 196, *2411*
Coutinho, P.M., 64, *108*
Couturier, Y., 173, *2033*, 173, *2034*, 173, *2035*, 173, *2036*, 241, *3088*
Covan, D.A., 74, *730*
Cowie, J.M.G., 99, *849*, 119, *1242*
Crabb, G.A., 217, *2733*
Crabb, W.D., 64, *162*, 64, *181*
Crabbe, E., 236, *3016*
Craig, R., 173, *2025*
Craig, W.L., 65, *292*, 194, *2375*
Cramer, F., 64, *144*, 249, *144*, 250, *144*
Creedy, A.E., 126, *1372*
Cresswell, P., 11, *59*

Crichton, R.R., 97, *818*
Crispeels, M.J., 64, *88*
Cros, P., 125, *1338*
Croux, C., 73, *693*
Crowe, T.C., 186, *2200*
Crowell, C.D., 236, *3013*
Crowley, N., 93, *574*
Crump, M.P., 24, *100*, 24, *101*, 25, *103*
Cruz, R., 94, *583*
Cruz, V.D., 94, *583*
Crzellitzer, R., 186, *2186*
Csiba, A., 262, *3457*, 265, *3457*
Cuber, J.C., 262, *3446*, 262, *3449*, 262, *3465*, 264, *3446*, 264, *3449*, 264, *3513*
Cucolo, G.R., 251, *3247*
Cuffe, M., 263, *3469*, 264, *3469*
Cui, B., 158, *1836*
Cui, L., 149, *1648*
Cui, X., 206, *2532*
Čulik, K., 109, *1127*
Cunefare, J., 89, *484*
Curić, D., 191, *2317*, 200, *2457*
Curioni, A., 116, *1218*
Curran, A.D., 203, *2490*
Currens, M.J., 30, *108*
Curvelo-Santana, J.C., 128, *1408*
Cushing, M.L., 194, *2374*
Cynter, C.V.A., 248, 3209
Czarnecki, Z., 219, *2778*
Czupryński, B., 218, *2769*, 222, *2830*, 226, *2887*, 227, *2926*, 228, *2925*
Czwalinna, A., 74, *755*

D

da Costa, M.S., 74, *730*
da Mota, F.F., 253, *3293*
Da Silva, P.A., 251, *3257*
da Silva, R., 251, *3247*
Dadashev, M.N., 64, *233*
Dadswell, I.W., 164, *1939*
Daelemans, D., 11, *52*, 46, *153*, 48, *52*
Dahiya, J., 79, *1109*
Dahiya, J.S., 78, *1061*, 79, *1061*
Dahlems, U.M., 78, *1042*

Dai, C., 8, *41*
Dai, Z., 2, *9*, 167, *1972*
Dale, J.K., 205, *2514*
Dalmia, B.K., 103, *954*, 129, *954*
Dang, X., 219, *2790*
Dang, Y., 90, *524*
Daniel, R.M., 88, *461*, 114, *1191*, 115, *1210*, 115, *1211*, 116, *1210*, 129, *1431*
Danielsen, H.E., 8, *39*
Danielsen, S., 217, *2755*
Danilova, L.A., 136, *1482*
Danilyak, N.I., 229, *2930*
Dantal, L., 194, *2385*, 195, *2385*
Danzas, E., 243, *3110*
Daoud, I.S., 149, *1617*
D'Appolonia, B.L., 203, *2480*, 203, *2481*
Darabsett, D.B., 167, *1983*, 210, *1983*
Dardenne, P., 262, *3454*, 263, *3454*, 264, *3454*
Darii, E., 8, *42*
Darkanbaev, T.B., 161, *1900*, 176, *2087*
Darnoko, D., 190, *2278*
Das, A., 78, *1074*, 78, *1076*, 105, *987*, 130, *1454*
Das, D., 95, *607*, 104, *972*, 217, *2737*
Das, K., 89, *495*
Das, S., 66, *303*, 66, *304*
Dasinger, B.L., 65, *258*, 145, *258*
Daurat-Lerroque, S., 242, *3101*
David, F., 196, *2412*
David, M.H., 150, *1679*
Davies, D.R., 207, *2587*
Davies, H.V., 111, *1158*, 113, *1158*
Davies, I.R., 262, *3420*
Davies, R.J., 264, *3524*
Davies, W.A., 111, *1155*
Davila, A.M., 235, *3001*
Davis, A.P., 3, *16*, 15, *16*, 21, *16*, 21, *96*, 22, *97*, 23, *98*, 24, *100*, 24, *101*, 25, *103*, 26, *104*
Davis, B.R., 75, *798*
Davis, L., 227, *2904*
Davison, B.H., 213, *2658*, 222, *2839*
Davson, W.C., 93, *576*

Davydova, I.M., 248, *3202*
Dawod, M.A., 195, *2388*
Day, A., 89, *489*, 196, *2416*
Day, D.F., 93, *560*
Day, J.T., 150, *1662*
Day, R., 8, *43*
Dayton, A.D., 182, *2166*
De Almeida Carneiro, A.P., 195, *2396*
de Appolonia, F., 121, *1283*
de Araujo Tait, M.C., 230, *2938*
de Arauno Costa, F., 164, *1932*
De Barros, E.G., 217, *2748*, 217, *2749*
De Baynast de Septifontaines, R.J.M.P., 227, *2911*
de Burlet, G., 113, *1181*
de Castro, I.E., 164, *1932*
de Chatelperon, G.P.R., 243, *3108*
De Clercq, E., 11, *61*, 11, *63*, 11, *67*
de Clercq, E., 11, *69*
De Cordt, S., 177, *2105*, 182, *2167*
de Fabrizio, S.V., 235, *3007*
De Figuero Allieri, L.I.C., 78, *1086*, 79, *1086*
De Franca, F.P., 219, *2789*
de Freitas Valle, J.L., 256, *3359*
De Gooijer, C.D., 163, *1911*
de Haan, C.A.M., 11, *55*, 48, *55*
de Hoop, L., 63, *54*
De Jersey, J., 248, 3209
De la Cruz Medina, J., 140, *1529*
de Lima Camargo Giordano, R., 139, *1521*
de Lima, V.A., 75, *790*
De Mo, R., 74, *734*
De Moraes, F.F., 112, *1179*, 189, *2271*, 189, *2272*, 251, *3259*, 254, *3324*
De Moraes, L.M.P., 74, *741*, 77, *741*
de Mores, F.F., 109, *1128*
De Mot, R., 74, *733*, 74, *750*, 78, *750*, 96, *629*, 107, *629*, 227, *2920*
De Noord, K.G., 64, *241*
de Oliveira Barbosa, J., 164, *1932*
de Queiroz Araujo, N., 164, *1932*
De Rosa, M., 112, *1172*, 114, *1172*, 114, *1187*, 190, *2276*, 246, *3180*
De Schrijver, R., 261, *3418*, 262, *3419*

de Souza, E.L., 75, *790*, 94, *583*
de Souza Rocha, T., 195, *2396*
De Vizio, W., 268, *3613*
de Vos, P., 237, *3041*
De Wijk, R.A., 187, *2221*, 207, *2570*
De-Almeida, S.E.M., 72, *640*
Debnath, S., 212, *2644*, 215, *2699*
Debonne, I., 251, *3248*
Decker, J.M., 11, *65*
Declecq, C., 251, *3248*
Decordt, S., 72, *666*
Deeter, S., 4, *22*
Degn, P., 240, *3072*
DeGroot, H.S., 63, *73*
Deguchi, Y., 72, *644*
DeHaas, B.W., 85, *415*
Dehelean, T., 192, *2343*
Deinhammer, R., 193, *2357*
Deki, M., 265, *3532*
Dekker, K., 243, *3121*
Del Bianco, V., 219, *2788*
Del Borghi, A., 232, *2958*
Del Borghi, M., 230, *2934*, 230, *2936*
Del Rio, G., 238, *3053*, 239, *3053*
Del Vedovo, S., 176, *2079*
Delahaye, E.P., 73, *701*
Delcour, J.A., 81, *361*, 177, *2100*, 192, *2337*, 196, *2411*, 202, *2469*, 203, *2483*
Delecourt, R., 64, *224*, 217, *224*
Deleyn, F., 151, *1705*
Delgado, O., 78, *1086*, 79, *1086*
Della Valle, G., 203, *2482*
Delpeuch, F., 174, *2047*
Delwiche, S.R., 268, *3610*
Demirkan, E.S., 72, *647*, 79, *647*
DeMoraes, F.F., 251, *3259*
Denault, L.J., 145, *1561*, 153, *1561*, 183, *1561*
Denki Kagaku Kogyo, K.K., 158, *1862*
Dennis, J.W., 2, *4*
Dennis, M.V., 154, *1752*, 178, *1752*
Derez, F., 151, *1705*
Derjani, R., 244, *3137*
Derkanosova, N., 122, *1297*
Deroanne, C., 72, *643*

Derradji-Serghat, H., 173, *2035*
Deschamps, M., 8, *43*
Deshpande, S.S., 186, *2199*, 251, *3250*
Deshpande, V.V., 243, *3118*
Desmons, A., 8, *43*
Desseaux, V., 130, *1457*, 187, *2212*
Destexhe, A.M.P., 122, *1289*
Dettori-Campus, B.G., 90, *506*
Deuel, H.J. Jr., 260, *3397*
Devdariani, D., 182, *2156*, 182, *2157*, 182, *2158*
Devos, F., 154, *1745*, 154, *1746*
Dey, G., 96, *804*
Dey, N., 150, *1692*, 225, *2879*
Dhawale, M.R., 125, *1345*
Dhawale, R.M., 64, *103*
Di Lemia, I., 112, *1172*, 114, *1172*, 114, *1187*
Di Paola, R.D., 262, *3441*, 264, *3441*
Diaz Ruiz, G., 236, *3021*
Dickmann, R.S., 169, *1997*, 171, *1997*, 171, *1997*
Dickson, A.D., 99, *856*
Dickson, R.E., 262, *3459*
Diderichsen, B.K., 92, *547*
Diedering, P., 64, *194*, 64, *195*
Dien, B.S., 223, *2842*, 223, *2843*, 226, *2899*
Dierchen, W., 91, *544*
Dietrich, W., 266, *3564*
Dijkhuizen, L., 64, *182*, 65, *182*, 72, *685*, 123, *1313*, 129, *1429*, 241, *1313*, 251, *3252*
Dijkstra, B.W., 129, *1429*, 251, *3252*
Dijkstra, P., 150, *1682*
Dikhtyarev, S.I., 252, *3275*
Dilova, E.D., 197, *2430*
Ding, Y., 238, *3050*
Dinsdale, J., 157, *1824*
Dissing, U., 97, *813*
Dittmar, H.R., 63, *56*
Djordjevic, V., 264, *3492*
Do, J.H., 96, *628*
Do, M.H., 151, *1707*, 157, *1810*, 207, *2579*
Do Nascimento, A.J., 251, *3257*
Doane, W.M., 64, *235*

Dobreva, E., 63, *72*, 72, *480*, 72, *652*, 72, *653*, 72, *665*, 88, *480*, 89, *482*, 121, *72*
Dobriyan, V.M., 215, *2696*
Dodan, V., 234, *2991*, 235, *2991*
Dodziuk, H., 65, *283*, 65, *284*, 65, *288*, 257, *283*, 257, *284*
Doehlert, D.C., 101, *888*, 139, *888*
Doelle, H.W., 125, *1351*, 125, *1352*, 212, *2635*
Dogan, I.S., 198, *2442*
Doge, T., 261, *3414*
Dogo-Isonagie, C.I., 47, *155*
Dogru, M., 183, *2170*
Dohmen, J., 74, *749*, 77, *749*, 78, *1042*, 212, *2648*
Doi, T., 42, *142*, 43, *145*, 43, *149*, 259, *3387*
Doi, Y., 206, *2536*
Doin, P.P., 191, *2321*
Dolar, P., 145, *1563*
Doley, R., 89, *495*
Dolgov, I.A., 261, *3408*
Dollimore, D., 262, *3438*
Dominey, A.P., 23, *98*
Dominigues, C.M., 74, *759*
Dominiquez, J., 168, *1992*
Dominy, N.J., 81, *366*, 82, *366*
Dondero, M.L., 206, *2560*
Donelyan, V., 64, *249*, 243, *249*
Dong, C., 117, *1223*
Dong, J., 225, *2876*
Dong, Q., 149, *1629*, 217, *1629*, 220, *1629*
Dong, Y., 64, *137*
Dongowski, G., 206, *2539*, 259, *3384*
Donnini, C., 78, *1085*
Dos Mohd, A.M., 192, *2338*
Dos Santos, V.L., 217, *2748*, 217, *2749*
Doughty, R.A., 179, *2121*
Doumer, E., 233, *2974*
Douzals, J.P., 158, *1841*
Dowlut, M., 9, *44*, 9, *45*, 12, *44*, 12, *45*
Doyle, E.M., 73, *723*, 82, *383*, 88, *450*, 88, *452*, 90, *523*, 95, *620*, 98, *383*, 160, *450*
Dragic, T., 11, *52*, 48, *52*
Dragsdorf, R.D., 203, *2479*

Dragulescu, A., 225, *2882*
Drake, R.R., 8, *41*
Drapron, R., 181, *2153*, 181, *2154*
Dreimane, M., 104, *970*
Drews, B., 63, *35*, 126, *1366*
Driguez, H., 64, *89*
Držikova, B., 259, *3384*
D'Souza, S.F., 191, *2320*
Du, A., 179, *2114*
Du, G., 65, *287*
Du, L., 216, *2713*, 216, *2714*
du Preez, J.C., 213, *2666*, 217, *2754*, 225, *2666*
Duan, G., 65, *266*, 65, *267*, 207, *2583*
Duan, J., 64, *81*
Duan, K.J., 234, *2985*
Dubart, D.J., 142, *1543*
Dubreucq, E., 96, *634*, 105, *634*
Dubuisson, J., 11, *53*
Duchene, D., 65, *290*
Duchiron, F., 112, *1173*, 113, *1173*, 115, *1213*, 117, *1213*, 117, *1222*, 166, *1960*
Duckert, F., 82, *371*
Ducroo, P., 95, *616*, 153, *1727*
Dudkin, M.S., 163, *1922*, 171, *2011*, 172, *1922*
Duedahl-Olesen, L., 64, *96*, 65, *298*, 79, *96*, 111, *96*, 253, *3292*
Dugenet, Y., 194, *2365*
Dugum, J., 200, *2457*
Duhart, D.J., 123, *1309*, 133, *1461*, 134, *1461*
Duke, S.H., 101, *888*, 139, *888*
Dull, M.F., 179, *2127*
Duncan, G., 111, *1158*, 113, *1158*
Dunn-Coleman, N., 64, *101*, 108, *1023*, 207, *2583*, 213, *2671*
Dunnill, P., 64, *168*, 189, *2258*
Dunning, H.N., 193, *2352*
Duprat, F., 265, *3549*
Duran, E., 166, *1966*, 198, *2454*
Durmishidze, S.V., 108, *1028*, 115, *1028*
Duryea, C.B., 63, *32*, 179, *2123*
Duus, J.O., 240, *3072*
Dvadtsatova, E.A., 217, *2730*

Dwek, R.A., 2, *5*, 11, *59*
Dwiarti, L., 234, *2982*
Dwiggins, B.L., 155, *1767*
Dworschack, R.G., 244, *3126*
Dygert, S., 64, *111*
Dzulkifly, M.H., 170, *2004*

E

Eadie, G.S., 181, *2150*
Ebata, T., 216, *2710*
Eberle, H.J., 255, *3347*
Ebertova, H., 66, *306*, 74, *743*, 77, *743*
Ebisu, S., 72, *655*
Echtermann, K.W., 224, *2858*, 225, *2871*
Eckhard, M., 63, *53*
Eden, J.L., 196, *2406*
Eder, E., 262, *3422*
Edwards, C.A., 259, *3381*
Edwards, J.O., 4, *21*
Eegelseer, E.M., 72, *684*
Effront, J., 63, *40*, 260, *3393*, 267, *3578*
Efremov, P.S., 217, *2735*
Egas, M.C.V., 74, *730*
Egawa, S., 201, *2459*
Egberink, H.F., 11, *55*, 48, *55*
Eggert, H., 9, *46*, 9, *47*
Eggink, G., 72, *685*
Eggum, B.O., 260, *3400*
Egizabal, E., 246, *3171*
Eglinton, J., 266, *3565*
Egorova, M.I., 204, *2496*
Eid, B.M., 195, *2397*
Eisenbraun, G.J., 263, *3475*
Ejembi, O., 95, *621*
El-Aassar, S.A., 72, *662*
El-Hossamy, M., 195, *2397*
Eliasson, A.-C., 177, *2098*, 202, *2466*, 261, *3415*
Ellington, J.K., 100, *882*
Elliott, T., 11, *59*
Ellis, P.R., 127, *1375*, 138, *1496*, 178, *1375*, 181, *1375*
Ellis, R.P., 146, *1576*
Elms, J., 203, *2490*

El-SAhy, K.M., 140, *1530*
El-Saied, A.S.H., 173, *2028*
El-Saied, H.M., 267, *3588*
Emanuilova, E., 89, *482*, 264, *3503*
Emery, A., 244, *3130*
Emery, A.H., 191, *2322*
Emery, A.N., 72, *650*, 189, *2268*
Emneus, J., 264, *3510*, 264, *3511*, 264, *3512*
Enderle, G., 196, *2414*
Endo, I., 78, *1069*, 79, *1069*
Endo, K., 196, *2426*
Endo, S., 207, *2563*
Endo, Y., 115, *1198*, 116, *1198*, 190, *2295*, 197, *2435*, 244, *3157*, 254, *3327*
Enevoldsen, B.S., 214, *2683*
Engasser, J.M., 137, *1487*, 147, *1487*, 190, *2299*, 190, *2300*
Englis, D.T., 264, *3486*
Ensley, B., 73, *695*, 77, *695*
Epchtein, J.S., 221, *2816*
Eposito, E., 121, *1283*
Eriksson, E., 85, *420*
Ermoshina, G.K., 94, *594*
Ernandes, J.R., 74, *747*, 77, *747*
Ernst, H.A., 114, *1189*
Erokhin, I.I., 227, *2906*
Esaki, K., 179, *2126*
Escarpa, A., 262, *3436*, 263, *3436*
Escher, F., 203, *2477*
Escovar-Kousen, J., 211, *2625*
Eslick, R., 85, *416*
Ettel, W., 262, *3435*, 264, *3435*
Ettema, T.J., 123, *1313*, 241, *1313*
Euler, H., 187, *2211*
Eur-Aree, A., 220, *2796*
Euverink, G.J.W., 241, *3078*, 241, *3079*
Evans, D.E., 100, *882*, 266, *3565*
Evans, I.H., 186, *2191*
Evans, P., 203, *2490*
Evreinova, T.N., 99, *854*, 188, *2250*, 248, *3202*
Ewing, F.G., 63, *73*, 151, *1712*, 154, *1752*, 178, *1752*
Eynard, L., 202, *2465*

Ezeji, T.C., 231, *2951*
Ezeogu, L.I., 74, *753*, 74, *763*, 76, *753*, 76, *763*, 79, *763*, 168, *1991*
Ezhilvannan, M., 79, *527*, 91, *527*, 106, *527*, 119, *527*
Ezure, Y., 155, *1760*, 240, *3066*, 240, *3067*

F

Fabiano, B., 245, *3167*
Fabritius, M., 259, *3383*
Facciotti, M.C.R., 78, *1045*, 78, *1048*
Fagerstroem, R., 129, *1424*
Fahmy, A.H., 140, *1530*, 172, *2018*, 173, *2028*
Fairbarn, D.A., 93, *579*
Fajbergerówna, S., 129, *1439*
Falcao dos Reis, R., 164, *1932*
Fan, J., 2, *9*
Fan, M.L., 76, *925*
Fang, H., 5, *28*, 5, *33*
Fang, J., 207, *2573*, 221, *2822*
Fang, S., 212, *2632*, 221, *2822*
Fang, X., 193, *2346*
Fantl, P., 81, *363*
Faradzheva, E.D., 217, *2734*
Faraj, A., 167, *1973*
Farcasescu, I., 225, *2882*
Farezvidal, M.E., 73, *713*
Farhat, I.A., 177, *2101*
Faria, J.B., 217, *2750*
Farjon, B., 64, *126*
Farley, F.F., 100, *880*
Fassatiova, O., 96, *632*, 137, *632*
Fatima Grossi de Sa, M., 64, *88*
Faulks, R.M., 83, *393*, 261, *3416*, 261, *3417*, 262, *3420*
Favier, J.C., 174, *2047*, 260, *3395*
Fecht, R.J., 97, *808*
Fedorov, A.F., 120, *1265*, 212, *2629*
Fedorova, T.V., 139, *1514*
Feher, J., 205, *2515*
Felby, C., 149, *1628*, 222, *2836*, 227, *2921*
Felix, C.R., 72, *686*, 73, *686*
Fell, H., 97, *822*

AUTHOR INDEX

Feller, G., 72, *642*, 72, *643*, 183, *2174*, 188, *2174*
Feng, D., 264, *3519*
Feng, H., 219, *2787*
Feng, P., 215, *2692*
Feniksova, R.V., 94, *594*, 126, *1359*, 215, *2706*
Fenton, D.M., 65, *258*, 145, *258*
Fenyö, E.M., 11, *64*
Ferdes, M., 222, *2833*
Ferenczi, T., 108, *1030*, 129, *1432*
Ferenz, H.F., 79, *1092*, 220, *1092*
Fergus, C.L., 73, *704*, 75, *704*, 75, *796*
Ferguson, T.J., 124, *1325*
Fernandes, M.V., 266, *3568*
Fernandez-Lafuente, R., 110, *1149*
Fernandezvivas, A., 73, *713*
Fernando-Lobato, M., 113, *1184*
Feroza, B., 106, *993*, 215, *2704*
Ferraiolo, G., 230, *2934*
Ferrand, Y., 24, *101*, 25, *103*
Ferrari, E., 64, *84*
Ferrarotti, S.A., 251, *3246*
Ferreira de Castro, H., 164, *1932*
Ferrero, I., 78, *1085*
Fetzer, W.R., 179, *2120*
Fiedorowicz, M., 61, *9*, 136, *1470*, 136, *1471*, 136, *1472*, 136, *1473*, 136, *1474*, 256, *3362*, 257, *3362*, 258, *3362*
Fiedurek, J., 78, *1068*, 109, *1125*, 109, *1126*
Fiegler, H., 81, *366*, 82, *366*
Figueira, M.M., 74, *731*, 77, *731*
Figueroa, C., 235, *3001*, 243, *3123*, 244, *3123*
Fikkert, V., 11, *69*
Filipowitz, B., 76, *840*, 98, *840*, 100, *840*
Filippov, S.A., 127, *1373*
Finck, J.M., 213, *2670*
Finguerut, J., 221, *2816*
Finogenova, T.V., 65, *277*
Fischer, E.H., 82, *371*, 183, *2169*, 183, *2171*
Fischer, J., 120, *1257*, 186, *2187*
Fischer, T., 72, *641*
Fischl, M., 264, *3505*
Fisher, E.E., 158, *1838*
Fisher, R.A., 264, *3488*
Fisher, S.F., 87, *443*
Fisichella, S., 265, *3529*
Fitzsimon, R., 217, *2753*
Flachowsky, G., 208, *2596*, 262, *3460*
Flaschel, E., 253, *3289*
Flayeux, F., 137, *1487*, 147, *1487*
Fleming, I.D., 103, *946*, 119, *1242*
Fleming, I.T., 99, *849*
Fleming, L.D., 77, *951*, 79, *951*, 103, *951*
Fletcher, D., 169, *1997*, 171, *1997*
Flexner, L.B., 129, *1440*
Flint, V.M., 123, *1309*
Flitsch, S.L., 243, *3115*
Flor, P.Q., 78, *977*, 97, *817*, 105, *977*, 129, *1418*
Flores, D.M., 236, *3014*
Flores-Carreón, A., 27, *106*
Fodor, A., 192, *2343*
Fogarty, M., 88, *450*, 160, *450*
Fogarty, W.M., 64, *172*, 72, *477*, 72, *659*, 72, *660*, 72, *674*, 73, *717*, 73, *723*, 74, *739*, 76, *939*, 77, *739*, 78, *172*, 82, *383*, 88, *477*, 88, *478*, 88, *449*, 88, *452*, 90, *523*, 94, *582*, 95, *620*, 98, *383*, 104, *957*, 246, *172*, 257, *3371*
Foglietti, M.J., 73, *701*
Fomina, S., 122, *1297*
Fontana, J.D., 79, *1100*
Foolbrook, P., 64, *229*, 65, *229*
Ford, C., 64, *108*, 104, *971*, 129, *971*
Forney, L.J., 235, *3008*
Forsell, P.M., 176, *2068*
Forster, S., 77, *1039*
Foszczyńska, B., 210, *2608*
Fournier, I., 8, *43*
Fowles, F.L., 145, *1560*
Fox, D.L., 173, *2025*
Francesconi, O., 19, *89*, 20, *90*, 21, *91*, 21, *92*, 21, *93*, 21, *94*, 23, *99*, 30, *109*
Francis, G., 129, *1432*
Franco, C.M.L., 149, *1654*, 176, *2060*, 180, *2131*
Franco de Sarabia Rosado, P.M., 63, *65*

Franco, O.L., 71, *637*
Franco-Fraguas, L., 268, *3607*
François, K.O., 2, *13*, 11, *13*, 47, *13*, 47, *154*, 48, *157*
Frandsen, T.P., 64, *92*, 78, *1062*
Franke, G.T., 163, *1911*
Frankiewicz, J.R., 104, *966*, 126, *1364*
Frankova, L., 64, *139*
Frantzen, H.B., 64, *75*
Franzen, S., 5, *25*, 5, *26*
Fratkin, S.B., 232, *2957*
Frederix, S.A., 196, *2411*
Fredriksson, H., 177, *2098*
Freer, S.N., 73, *716*
Freiberger, S., 187, *2217*, 188, *2217*
Freire, D.M.G., 189, *2261*
Freitas, R.P., 218, *2765*
French, D., 9, *868*, 64, *110*, 76, *868*, 265, *3547*
Freudenberg, K., 249, *3216*, 249, *3217*
Freudenberger, J., 97, *821*
Frey, C.N., 88, *467*, 186, *2204*, 187, *2204*, 264, *3488*
Frey, G., 64, *76*, 64, *77*, 64, *82*
Freyberg, W., 225, *2871*
Freyer, S., 145, *1568*
Friedman, I., 83, *394*, 84, *394*
Friedrichs, O.v., 124, *1322*
Friend, B.A., 218, *2760*, 224, *2760*
Fris, J.J., 208, *2594*
Frisner, H., 148, *1605*, 156, *1786*, 217, *2732*
Frison, S.L., 265, *3534*
Froeyen, M., 11, *66*, 11, *67*
Fu, X., 174, *2041*, 193, *2344*
Fu, Y., 193, *2349*
Fucinos, P., 171, *2007*, 171, *2008*
Fuentes, M., 110, *1149*
Fuglsang, C.C., 149, *1628*
Fujii, K., 123, *1317*
Fujii, M., 74, *766*, 97, *812*, 121, *1279*, 121, *1281*, 176, *2076*, 212, *2645*, 214, *2687*, 214, *2688*, 236, *3023*
Fujii, N., 192, *2340*, 213, *2653*
Fujii, T., 78, *1088*
Fujikawa, K., 41, *141*

Fujimura-Ito, T., 146, *1581*
Fujio, Y., 79, *1097*, 79, *1104*, 103, *955*, 103, *956*, 150, *1695*, 219, *2793*, 219, *2794*, 264, *3481*, 267, *3591*
Fujita, Y., 251, *3258*
Fujkushi, O., 218, *2767*
Fujui, S., 75, *771*
Fujui, T., 78, *1083*
Fujushige, T., 255, *3346*
Fukada, S., 246, *3174*
Fukagawa, Y., 3, *19*, 37, *133*, 38, *135*, 38, *136*, 40, *136*, 40, *137*
Fukai, Y., 176, *2064*, 176, *2077*, 206, *2548*, 206, *2557*
Fuke, K., 136, *1476*
Fukimbara, F., 79, *1101*
Fukimbara, T., 146, *1578*, 146, *1579*, 210, *1578*
Fukuda, H., 64, *98*, 77, *1040*, 112, *1177*, 113, *1177*, 114, *1177*, 190, *2295*, 197, *2435*, 206, *2536*, 212, *2649*, 216, *2715*, 227, *2905*
Fukuda, K., 236, *3018*
Fukuda, S., 207, *2567*, 246, *3177*, 257, *3372*, 258, *3373*
Fukui, M., 141, *1536*
Fukui, N., 220, *2808*
Fukui, S., 77, *1041*
Fukui, T., 163, *1924*, 176, *2076*, 242, *3102*, 246, *3172*
Fukumoto, J., 74, *744*, 77, *744*, 88, *466*, 88, *472*, 88, *473*, 98, *836*, 99, *851*, 101, *891*, 120, *1263*, 127, *1392*, 148, *851*, 151, *851*
Fukusa, S., 241, *3085*
Fukushima, K., 234, *2986*
Fukushima, S., 214, *2689*
Fukushima, Y., 158, *1861*, 159, *1861*
Fukuta, Y., 264, *3498*
Fukuyama, S., 92, *550*
Fülbier, B., 64, *216*
Fuld, M., 168, *1987*
Funaguma, T., 243, *3114*
Funane, K., 240, *3076*
Furegon, L., 116, *1218*

Furio, Y., 191, *2318*
Furlanetto, L., 79, *1100*
Furniss, C.S.M., 64, *130*, 129, *130*, 129, *1415*
Fursov, O.V., 176, *2087*, 257, *3369*
Fursova, O.V., 165, *1943*
Furukawa, K., 129, *1425*, 217, *2752*
Furukawa, T., 158, *1850*, 159, *1850*
Furumai, T., 34, *119*, 34, *121*, 34, *122*, 34, *123*, 34, *124*
Furuta, H., 77, *1040*
Furuta, M., 204, *2503*
Furuta, S., 102, *918*
Furutani, S., 89, *483*
Furuyoshi, S., 94, *581*
Fusco, S., 112, *1172*, 114, *1172*, 114, *1187*
Fuse, M., 204, *2506*
Fuwa, H., 64, *99*, 64, *186*, 64, *188*, 64, *205*, 159, *186*, 160, *1890*, 174, *2043*, 176, *2043*, 176, *2061*, 176, *2062*, 176, *2070*, 188, *2226*, 267, *3576*

G

Gabanyi, M., 254, *3315*
Gabr, S., 267, *3588*
Gabriel, C.L., 231, *2944*
Gabrielli, G., 21, *91*, 21, *93*, 21, *94*, 30, *109*
Gago, F., 11, *68*, 30, *68*
Gaidenko, V.P., 107, *1006*
Gale, R.A., 194, *2370*
Galkin, A.V., 237, *3037*
Galkina, G.V., 148, *1603*, 150, *1603*, 150, *1659*, 153, *1720*
Gallagher, F.H., 63, *34*, 210, *34*
Gallant, D., 64, *160*
Gallant, D.J., 70, *617*, 95, *617*, 104, *617*, 160, *1893*
Galmiche, J.P., 259, *3382*
Galzy, P., 74, *738*, 75, *587*, 77, *738*, 94, *587*, 96, *634*, 104, *587*, 105, *634*, 106, *1003*, 106, *1004*, 217, *2739*
Gambacorta, A., 121, *1283*, 246, *3180*
Gan, X., 150, *1672*
Ganchev, I., 181, *2151*
Ganetsos, G., 191, *2319*

Gangadharan, D., 72, *649*
Gantelet, H., 115, *1213*, 117, *1213*, 117, *1222*
Gao, D.W., 176, *2072*, 176, *2080*, 176, *2081*, 176, *2082*
Gao, F., 207, *2582*
Gao, H., 266, *3553*
Gao, J., 227, *2907*
Gao, J.P., 263, *3476*, 265, *3476*
Gao, P., 220, *2807*, 226, *2883*
Gao, Q., 149, *1651*, 206, *2531*, 206, *2544*, 207, *2571*, 207, *2588*, 264, *3522*
Gao, S., 5, *32*, 7, *36*
Gao, X., 5, *32*, 5, *33*, 7, *36*
Gao, X.D., 64, *193*
Gaouar, O., 140, *1531*
Gaponenko, V., 43, *148*
Garbutt, J.T., 155, *1764*, 208, *1764*
Garby, J.L., 264, *3486*
Garcia, A. III., 226, *2884*
Garcia Alvarado, M., 140, *1529*
Garcia, J.L., 124, *1325*
Garcia Salva, T.J.G., 72, *687*
Garcia-Alonso, A., 65, *296*, 261, *296*
Gardner, J.F., 164, *1939*
Gareiss, P.C., 21, *95*
Garg, S.K., 125, *1351*, 125, *1352*
Gargova, S., 158, *1851*
Garrett, J.B., 112, *1174*
Garriott, O.K., 90, *526*
Gasdorf, H., 155, *1777*
Gashe, B.A., 72, *672*
Gąsiorek, E., 237, *3032*
Gasparik, H., 74, *746*, 77, *746*
Gasparyan, A.V., 101, *895*
Gassesse, A., 72, *672*
Gates, R.L., 264, *3487*
Gathman, A.C., 221, *2819*
Gaudix, E.M., 142, *1547*
Gaugadharan, D., 75, *778*, 80, *336*, 82, *336*
Gavrilova, V.P., 195, *2399*
Gawande, B., 253, *3287*
Gawande, B.N., 123, *1318*, 123, *1319*
Gazea, M., 158, *1852*
Gdala, J., 208, *2597*

Ge, S., 115, *1202*, 115, *1203*, 116, *1203*
Ge, X., 68, *313*, 68, *315*, 79, *1094*, 223, *2848*
Gebhardt, E., 259, *3384*
Gee, J.M., 261, *3417*
Geerink, M., 144, *1553*
Geessese, A., 78, *1077*
Geffert, C., 19, *86*
Gehman, H., 64, *189*
Gelroth, J.A., 263, *3475*
Gembicka, D., 110, *1139*
Generoso, M., 112, *1172*, 114, *1172*, 190, *2276*
Gentelet, H., 112, *1173*, 113, *1173*
Geoffroy, R., 101, *893*, 201, *2462*
Geol, A., 251, *3260*
Geonnotti, A.R., 14, *75*
Gerard, C., 165, *1947*, 166, *1947*
Gerday, C., 72, *642*, 72, *643*, 183, *2174*, 188, *2174*
Gergely, G.A., 264, *3491*
Gerges, U., 156, *1782*
Gerlach, S.R., 74, *755*
Gernat, C., 192, *2336*
Gerrits, N., 64, *94*
Gervais, P., 158, *1841*
Gessesse, A., 72, *476*, 88, *476*
Ghaemi, N., 187, *2208*
Ghalanbor, Z., 187, *2208*
Ghali, Y., 244, *3131*, 267, *3588*
Ghazali, H.M., 170, *2004*, 219, *2784*
Ghelardini, C., 30, *109*
Ghirlando, R., 47, *155*
Ghose, A., 105, *987*
Ghosh, A., 78, *1076*, 130, *1454*
Ghosh, A.B., 78, *1074*
Ghosh, S.B., 92, *549*
Ghozlan, D., 214, *2686*
Gibinski, M., 164, *1940*
Gibson, T.S., 262, *3453*, 262, *3455*, 263, *3473*, 264, *3455*
Giglio, J.R., 75, *789*
Giglio, R., 72, *640*
Gigras, P., 64, *121*
Gil, M.L., 27, *105*

Gilbert, R.D., 195, *2392*
Gill, G.E., 211, *2618*
Gillan, E.P., 195, *2391*
Gillespie, R.E., 259, *3385*
Ginsburg, S., 82, *378*, 163, *378*
Giordano, R.L.C., 213, *2654*, 230, *2937*
Giraud, F., 73, *709*
Giri, J.V., 183, *2173*
Giri, K.V., 100, *871*, 101, *892*, 262, *3431*, 264, *3501*
Giudicelli, J., 113, *1181*
Giurgiulescu, G., 225, *2882*
Gizycki, I., 63, *59*, 87, *59*
Gjörling, L.G., 250, *3234*
Gladilin, A.K., 188, *2222*
Gładkowski, J., 61, *3*, 82, *3*
Glemza, A., 216, *2712*
Glenn, G.M., 175, *2053*
Glimm, E., 63, *59*, 87, *59*
Glock, E.G., 81, *353*
Glock, G.E., 81, *365*
Glover, D.V., 64, *99*, 176, *2061*
Glymph, J.L., 73, *725*
Gocalek, M., 190, *2287*
Göck, K., 225, *2871*
Godia, F., 65, *273*
Goel, A., 123, *1318*
Goering, K.J., 85, *415*, 85, *416*
Goesaert, H., 81, *361*, 202, *2469*, 203, *2483*
Goetheer, E., 180, *2139*
Goff, D.J., 82, *375*
Goie, A., 225, *2882*
Golc-Wondra, A., 165, *1951*
Goldsbrough, A., 64, *100*
Goldstein, I.J., 2, *6*
Golubeva, L.A., 108, *1114*
Goma, G., 73, *693*
Gombin, D., 215, *2701*
Gome, L.M.S., 150, *1680*
Gomes, E.A., 251, *3247*, 253, *3293*
Gomez de Seruga, A., 64, *176*
Gomez, M.R.A., 182, *2165*
Gomi, K., 129, *1420*
Gomi, S., 34, *110*, 34, *112*, 34, *116*, 37, *110*

Gonçalves, J.E., 250, *3232*
Gonçalves, L.R.B., 213, *2654*, 230, *2937*
Gonçalves, R.A.C., 250, *3232*
Gong, C.R., 193, *2363*
Gong, W.Z., 108, *1111*
Goni, I., 65, *296*, 261, *296*
Gonon, W.F., 203, *2475*, 203, *2476*, 263, *3470*
Gonsalves, C.A., 221, *2816*
Gonzales Tello, P., 167, *1971*
Gonzalez, B., 113, *1184*
Gonzalez, C., 78, *1086*, 79, *1086*
Gonzalez del Cueto, A., 267, *3585*
Gonzalez, F., 73, *713*
Gonzalez, M.C., 262, *3436*, 263, *3436*
Gonzalez, M.P., 75, *777*
Gonzalez, P., 121, *1274*
Gonzalez-Tello, P., 142, *1547*
Goode, D.L., 147, *1595*
Goodman, A.H., 148, *1600*
Goossens, K., 182, *2167*
Gopalakrishnan, K.S., 79, *1108*
Gopalarathanam, B.A., 207, *2574*
Goranovic, D., 237, *3042*
Gordon, S.H., 165, *1952*, 175, *2053*
Gore, H.C., 64, *223*, 85, *405*, 85, *406*, 100, *223*, 100, *405*, 157, *1822*, 186, *2204*, 187, *2204*, 267, *3579*
Gorges, E., 156, *1782*
Gorinstein, S., 170, *2002*
Gortner, R.A., 176, *2093*
Gorton, L., 142, *1538*, 264, *3510*, 264, *3511*, 264, *3512*
Goryacheva, M.G., 267, *3596*
Gosh, A.K., 96, *631*
Goshima, G., 263, *3479*, 265, *3541*
Goss, H., 153, *1723*, 153, *1724*
Gosselin, L., 73, *709*
Gostner, A., 262, *3422*
Goswami, V.K., 64, *121*
Goto, C.E., 74, *760*
Goto, K., 91, *536*
Goto, M., 64, *134*, 92, *558*, 104, *973*, 104, *974*, 129, *973*, 129, *974*, 129, *1425*, 239, *3063*

Goto, S., 110, *1136*
Goto, Y., 207, *2563*
Gottlieb, K.F., 248, *3198*
Gottschalk, A., 63, *49*
Gottschalk, G., 92, *552*
Gottschalk, T.E., 64, *89*
Gottvaldova, M., 249, *3231*
Gou, T., 154, *1750*
Gouda, M.K., 72, *662*
Govindasamy, S., 170, *2001*
Gow, N.A.R., 27, *106*
Goyal, N., 90, *522*
Gozalbo, D., 27, *105*
Graber, M., 97, *807*
Grabovska, O.V., 105, *976*, 151, *1708*
Gracheva, I.M., 78, *1084*, 107, *1006*
Grafelman, D.D., 189, *2265*
Graham, H., 262, *3462*, 263, *3462*
Graham, P.R., 194, *2378*
Grajek, W., 80, *341*, 150, *1674*, 179, *2117*, 190, *2287*, 191, *2316*
Grampp, E., 223, *2855*
Grancea, I., 225, *2882*
Grant, G.A., 265, *3534*
Granum, P.E., 88, *475*
Gravesen, T., 208, *2592*
Gray, G.M., 259, *3376*
Gray, K., 64, *82*, 112, *1174*
Gray, M., 5, *27*
Gray, O.P., 152, *1715*
Grebel, D., 265, *3526*, 265, *3527*
Greenwood, C.T., 64, *167*, 66, *167*, 71, *408*, 85, *408*, 99, *849*, 119, *1242*, 128, *1407*
Gregory, S.R., 207, *2572*
Gregory, T.J., 11, *58*, 12, *58*
Grey, K., 64, *76*
Griffin, P., 203, *2490*
Griffin, V.K., 168, *1990*
Griko, Y.V., 75, *798*
Grilione, P., 226, *2884*
Griskey, R.G., 182, *2161*
Groesaert, H., 177, *2100*
Gromada, A., 78, *1068*
Gromov, M.A., 64, *201*

Gromov, S.I., 217, *2735*
Gross, R.A., 241, *3084*, 242, *3089*, 242, *3090*
Grosse, H.-H., 228, *2928*
Grossi-De-Sá, M.F., 71, *637*
Gryszkin, A., 127, *1391*
Gryuner, V.S., 207, *2585*
Grześkowiak, A., 190, *2287*, 190, *2289*, 253, *3296*
Grześkowiak, M., 120, *1254*, 266, *3559*
Grześkowiak-Przywecka, A., 168, *1996*, 170, *1996*, 190, *2291*, 190, *2302*
Grzybowska, B., 64, *175*
Gu, B.U., 88, *448*
Gu, M., 72, *664*
Gu, Z., 65, *287*
Guan, B., 215, *2691*
Guan, H., 104, *968*
Gubicza, L., 126, *1371*, 190, *2284*
Guenter, W., 208, *2602*
Guerra, N.P., 171, *2007*, 171, *2008*, 171, *2009*
Guerrieri, N., 202, *2465*
Guertler, P., 98, *831*, 100, *884*
Guerzoni, M.E., 62, *13*, 64, *13*
Guilbo, A., 179, *2111*
Guilbot, A., 181, *2153*, 181, *2154*, 262, *3425*, 265, *3549*
Guimaraes, W.V., 217, *2746*, 217, *2748*, 217, *2749*
Guisan, J.M., 110, *1149*
Gulakowski, R.J., 30, *108*
Gulyaev, S.P., 228, *2923*
Gulyuk, N.G., 108, *1031*, 153, *1719*
Gummadi, S.N., 158, *1860*
Gunasekaran, M., 73, *714*
Gunasekaran, P., 219, *2783*
Gunji, M., 262, *3432*
Gunkel, J., 146, *1571*, 210, *1571*
Gunkina, N.I., 217, *2734*
Gunning, A.P., 64, *130*, 129, *130*
Gunsekaran, P., 215, *2700*
Gunther, H., 150, *1679*
Guo, B., 216, *2713*
Guo, L., 148, *1598*, 264, *3522*
Guo, S., 136, *1479*

Gupta, A.K., 86, *437*
Gupta, G., 2, *10*
Gupta, J.K., 90, *522*, 224, *2866*
Gupta, M.N., 110, *1152*
Gupta, R., 64, *121*
Guraya, H.S., 206, *2543*
Gustafson, K.R., 11, *53*, 11, *56*, 30, *108*
Gut, I.G., 8, *42*
Guthke, J.A., 243, *3112*
Guttapadu, S., 207, *2574*
Guy, C.L., 64, *137*
Guyot, J.B., 106, *1004*
Guyot, J.P., 129, *1417*, 234, *2989*, 234, *2990*, 235, *3004*, 236, *3020*, 236, *3021*, 237, *3036*
Guzhova, E.P., 72, *656*, 73, *712*
Guzman-Maldonado, H., 64, *159*, 168, *1992*
Gzyl, P., 238, *3049*

H

Ha, J.U., 222, *2832*
Haase, G.W., 146, *1583*, 148, *1583*
Habel, A., 259, *3384*
Haberer, B., 208, *2596*
Haberland, D., 64, *216*
Haberlandt, G., 260, *3396*
Habibi-Rezaei, M., 187, *2208*
Hackl, W., 208, *2598*
Haeupke, K., 120, *1257*
Haeusler, G., 156, *1783*
Haga, R., 149, *1633*, 208, *2603*
Hagberg, C.S.O., 266, *3567*
Hagenimana, V., 71, *442*, 76, *442*, 87, *442*, 99, *442*, 100, *442*, 163, *1927*
Haginoya, R., 263, *3472*
Hagishihara, M., 102, *913*, 115, *913*, 116, *913*, 117, *913*
Hagita, H., 191, *2318*
Hagiwara, H., 201, *2460*
Hagiwara, S., 179, *2126*
Hahn, B.H., 11, *65*
Hähn, H., 63, *52*, 63, *55*
Hahn, W., 195, *2393*
Hahn-Hagerdal, B., 236, *3019*

Haijema, B.J., 11, *55*, 48, *55*
Haissig, B.E., 262, *3459*
Hakamata, S., 153, *1726*
Hakkarainen, L., 190, *2306*
Hakoda, M., 190, *2296*
Halim, M.A., 106, *993*
Hall, D.G., 9, *44*, 9, *45*, 10, *48*, 12, *44*, 12, *45*
Hall, G.M., 158, *1842*, 159, *1842*
Halling, P.J., 243, *3115*
Halpern, M., 129, *1434*, 129, *1435*, 129, *1436*
Hama, H., 77, *1040*
Hamada, H., 205, *2518*, 241, *2518*
Hamada, K., 93, *567*
Hamada, M., 34, *110*, 34, *116*, 37, *110*
Hamada, S., 197, *2441*
Hamakawa, S., 192, *2339*, 193, *2339*
Hamaker, B.R., 168, *1993*, 188, *2228*
Hamaki, H., 221, *2824*
Hamalainen, C., 176, *2093*
Hamar, K., 194, *2367*
Hamazaki, Y., 192, *2341*
Hambali, E., 232, *2965*
Hamdi, G., 174, *2038*
Hamer, R.J., 187, *2221*, 196, *2410*, 207, *2570*
Hamilton, L.M., 72, *477*, 72, *674*, 88, *477*, 88, *478*, 257, *3371*
Hamilton, R.M., 179, *2120*
Hamilton, S., 248, *3209*
Hammar, L., 242, *3101*
Hammershoj, M., 208, *2594*
Hammond, N., 217, *2718*
Han, I.K., 254, *3316*, 256, *3354*
Han, I.Y., 221, *2821*, 221, *2823*, 222, *2827*, 223, *2827*
Han, J., 148, *1609*
Han, J.C., 250, *3236*
Han, J.Y., 142, *1539*
Han, M.H., 220, *2801*, 226, *2890*, 226, *2891*, 226, *2892*
Han, Y.J., 151, *1714*, 163, *1912*, 163, *1913*
Han, Y.T., 220, *2801*
Hanai, T., 231, *2953*, 232, *2953*
Hanatani, T., 157, *1815*, 159, *1815*, 244, *3155*
Handa, T., 110, *1136*

Hanes, B., 243, *3115*
Hanes, C.S., 65, *256*, 98, *841*, 98, *847*, 129, *1438*, 243, *3106*
Hang, M., 248, 3209
Hang, M.T., 94, *581*
Hang, Y.D., 74, *757*
Hanna, T.G., 186, *2197*
Hanno, T., 206, *2537*, 206, *2538*
Hanno, Y., 149, *1623*
Hanrahan, V.M., 80, *340*, 94, *590*, 127, *340*, 258, *590*
Hansen, C.E., 202, *2473*
Hansen, G., 73, *721*
Hansen, T.T., 93, *575*, 155, *1757*
Hansley, D.E., 155, *1777*
Hanson, A.M., 155, *1764*, 208, *1764*
Hanson, M.A., 195, *2392*
Hantson, A., 11, *69*
Hanumatha Rao, G., 164, *1935*
Hao, J., 232, *2959*
Hao, X., 149, *1648*
Hao, Z., 68, *313*
Haq, I.U., 237, *3031*
Hara, A., 243, *3114*
Hara, H., 34, *110*, 37, *110*, 91, *532*
Hara, K., 90, *511*, 165, *1953*, 249, *3220*, 249, *3221*, 250, *3221*, 254, *3325*, 255, *3329*, 255, *3332*
Hara, S., 75, *791*, 180, *2132*
Hara, T., 267, *3591*
Hara, Y., 186, *2207*
Harada, A., 62, *11*, 161, *1899*
Harada, H., 127, *1376*
Harada, K., 183, *2176*
Harada, T., 62, *11*, 75, *782*, 102, *920*, 118, *1236*, 161, *1899*
Haraguchi, T., 188, *2225*
Harahan, V.M., 75, *774*
Harangi, J., 75, *772*
Harder, A., 72, *657*, 108, *657*
Hardiner, K.M., 198, *2444*, 206, *2443*
Hari Krishna, S., 164, *1935*
Hariantono, J., 212, *2646*, 222, *2646*
Hario, Y., 157, *1820*

Harjes, C.F., 154, *1749*
Harkonen, H., 259, *3383*
Harley, C.P., 87, *443*
Harley, J.P., 209, *2604*
Haros, M., 203, *2478*
Harris, D.W., 207, *2578*
Harris, J., 215, *2702*
Harris, R.J., 11, *58*, 12, *58*
Harry-O'kuru, R., 175, *2053*
Hart, G.W., 25, *102*
Hartman, P.A., 119, *1246*
Hartmann, A., 17, *82*, 18, *83*
Hartnell, G.F., 227, *2914*, 228, *2914*
Hase, S., 262, *3445*
Hasegawa, A., 8, *37*
Hasegawa, M., 73, *711*
Hasegawa, N., 193, *2351*
Hasegawa, T., 34, *121*, 34, *122*, 37, *130*
Haseltine, C., 75, *800*
Haseltine, C.W., 78, *1044*
Haser, R., 64, *89*, 90, *509*, 102, *509*, 183, *2174*, 188, *2174*
Hashimoto, H., 90, *511*, 106, *1001*, 112, *1178*, 114, *1178*, 165, *1953*, 249, *3220*, 249, *3221*, 250, *3221*, 254, *3325*, 255, *3329*
Hashimoto, K., 137, *1483*, 139, *1523*, 139, *1524*, 247, *3183*, 250, *3241*
Hashimoto, N., 163, *1916*, 226, *2886*
Hashimoto, S., 94, *580*, 159, *1883*, 159, *1884*, 205, *2511*
Haska, N., 170, *2000*, 188, *2240*, 226, *2896*
Haska, R., 75, *788*
Hassan, M.A., 238, *3052*, 247, *3183*, 250, *3241*
Hassan, Z., 106, *1002*
Hasselbeck, G., 65, *295*
Hassid, W.Z., 64, *191*, 247, *3189*
Hata, Y., 78, *979*, 105, *979*, 125, *1346*, 129, *1420*
Hatanaka, C., 188, *2225*
Hatanaka, Y., 99, *860*
Hatori, M., 34, *117*, 34, *118*, 34, *120*, 34, *122*, 34, *124*
Hatse, S., 11, *63*, 11, *66*, 11, *67*, 46, *153*

Hattori, C., 256, *3353*
Hattori, F., 149, *1624*
Hattori, H., 161, *1896*
Hattori, K., 62, *21*
Hattori, T., 72, *654*, 205, *2512*
Hattori, Y., 108, *1034*, 158, *1847*
Haworth, N., 145, *1565*
Haworth, W.N., 80, *335*
Hayakawa, I., 191, *2318*
Hayakawa, S., 150, *1666*, 174, *2048*, 178, *2048*, 186, *2201*
Hayakawa, Y., 148, *1610*
Hayasaka, I., 206, *2556*, 207, *2576*
Hayashda, S., 129, *1425*
Hayashi, J., 176, *2083*
Hayashi, K., 151, *1702*, 151, *1703*, 180, *2132*
Hayashi, R., 179, *2115*, 266, *3558*
Hayashi, S., 88, *446*, 89, *483*, 115, *1209*, 116, *1209*, 119, *446*, 246, *3170*
Hayashi, T., 88, *456*, 88, *457*, 92, *546*, 179, *2112*
Hayashibara, K., 248, *3203*
Hayashida, A., 179, *2115*
Hayashida, K., 150, *1690*
Hayashida, S., 72, *690*, 78, *977*, 78, *1051*, 89, *494*, 92, *558*, 102, *909*, 104, *973*, 104, *974*, 104, *975*, 105, *977*, 129, *973*, 129, *974*, 129, *1413*, 129, *1418*
Hayashida, Y., 97, *817*
Haye, B., 64, *177*, 67, *177*, 75, *177*
Hayes, L.P., 155, *1761*
Hazlewood, G., 80, *337*, 165, *337*
He, F., 219, *2777*
He, G., 146, *1573*
He, G.Q., 150, *1681*
He, P., 237, *3044*
He, W., 213, *2662*
He, X., 193, *2344*, 197, *2436*
He, Y., 104, *965*
Heady, R.E., 205, *2522*
Heathcote, J.G.M., 214, *2684*
Hebeda, R.E., 72, *671*, 148, *1616*, 149, *1620*, 158, *1868*, 217, *2759*, 244, *3127*, 244, *3129*, 244, *3143*

Hedge, M.V., 76, *876*, 100, *876*
Hedges, A., 254, *3314*
Hefter, J., 120, *1262*
Hehre, E.J., 241, *3077*
Heide, M., 224, *2867*
Heimsch, R.C., 96, *635*, 106, *635*
Heinath, A., 125, *1341*
Heinig, W., 156, *1782*
Heinzle, E., 256, *3363*
Heitmann, T., 139, *1502*, 161, *1502*
Heitz, F., 188, *2222*
Heldt, K., 265, *3530*
Heldt-Hansen, H.P., 95, *615*
Helen, A., 78, *1054*
Helenius, A., 11, *60*
Helle, F., 11, *53*
Hellebrandt, A., 196, *2415*
Helwig Nielsen, B., 244, *3141*
Hemdrickx, M., 182, *2167*
Henderson, C., 262, *3421*
Henderson, L., 150, *1686*
Henderson, L.E., 30, *108*
Henderson, W.E., 138, *1500*
Hendrickx, M., 72, *666*, 72, *667*, 177, *2105*
Hendriks, W.H., 208, *2595*
Hendriksen, H.V., 64, *90*
Hendrikx, P.J., 206, *2550*
Heneen, W.K., 87, *435*
Henis, Y.I., 102, *902*, 145, *902*, 148, *902*, 151, *902*
Henley, M., 206, *2533*, 206, *2534*
Henriksnas, H., 98, *832*, 146, *1584*, 176, *2066*
Henry, R.J., 265, *3536*
Hensley, D.E., 76, *934*
Henson, C.A., 85, *422*, 112, *1164*, 113, *1164*, 122, *422*, 147, *1591*, 244, *1164*
Heo, G.Y., 234, *2975*
Herbert, D., 77, *556*, 92, *556*
Herrman, H., 206, *2554*, 206, *2555*
Herrmann, H.A., 208, *2599*
Hersiczky, A., 150, *1677*
Herstad, O., 207, *2586*
Herzfeld, E., 127, *1398*
Hesse, A., 64, *227*

Hesse, O., 73, *721*
Hettwer, W., 120, *1257*
Heusch, E., 186, *2195*
Hibi, N., 111, *1156*, 113, *1156*
Hibino, K., 190, *2282*
Hibino, T., 110, *1138*, 150, *1685*, 190, *2279*, 190, *2303*, 255, *3330*
Hicks, K.B., 223, *2843*
Hidaka, H., 64, *140*, 117, *1224*, 207, *2566*, 205, *2509*
Hidaka, T., 192, *2339*, 193, *2339*
Hidaka, Y., 238, *3054*
Hidi, P., 204, *2497*
Higasa, T., 72, *647*, 79, *647*, 100, *875*, 176, *875*
Higashihara, M., 98, *837*, 98, *838*, 99, *860*
Higgins, T.J.V., 64, *88*
High, R.L., 150, *1658*, 247, *3188*, 265, *3528*
Higuchi, Y., 127, *1376*
Higuti, I.H., 251, *3257*
Hill, G.A., 80, *339*, 135, *339*, 166, *1961*, 176, *1961*
Hill, R.D., 186, *2193*
Hill, S.E., 177, *2101*
Hiller, P., 72, *650*
Hillman, K., 262, *3421*
Hincal, A.A., 174, *2042*
Hinman, C.W., 149, *1621*
Hino, K., 207, *2567*
Hirabayashi, J., 2, *8*
Hirakawa, E., 136, *1478*
Hirano, M., 34, *115*, 34, *119*, 34, *121*, 37, *125*, 37, *129*, 37, *134*
Hirano, P.C., 213, *2654*
Hirao, M., 154, *1740*, 157, *1809*, 157, *1812*, 205, *2520*, 205, *2521*, 205, *2526*, 242, *3104*
Hirata, J., 140, *1526*
Hirata, M., 93, *567*, 159, *1885*
Hirata, Y., 267, *3591*
Hirayama, M., 11, *57*
Hirayama, N., 227, *2917*
Hirayama, O., 245, *3165*
Hirayama, Y., 64, *221*

Hirohashi, T., 122, *1290*
Hiromi, K., 74, *762*, 76, *928*, 100, *886*, 106, *995*
Hirooka, S., 122, *1290*, 205, *2512*
Hirose, J., 246, *3170*
Hirose, S., 112, *1170*, 113, *1170*
Hirose, Y., 101, *890*
Hirschberg, F., 202, *2472*
Hirte, W., 224, *2858*
Hisada, H., 125, *1346*
Hisaka, S., 118, *1233*, 150, *1233*
Hisaku, S., 223, *2856*
Hisamatsu, M., 264, *3506*
Hisayasu, T., 215, *2693*, 222, *2693*
Hishida, M., 263, *3479*
Hitaka, H., 250, *3239*, 255, *3341*
Hixon, R.M., 98, *842*
Hizukuri, S., 95, *618*, 103, *943*, 118, *1234*, 165, *1949*, 167, *1968*, 244, *3153*, 266, *3571*
Hjort, C., 64, *83*, 64, *97*
H-Kittikun, A., 170, *2003*
Ho, C.T., 97, *811*
Ho, Y.C., 219, *2784*
Hobson, P.N., 67, *310*, 115, *1196*, 116, *1196*, 125, *1334*, 242, *3097*
Hock, A., 127, *1388*
Hockenhull, D.J.D., 77, *556*, 92, *556*
Hoekema, A., 89, *485*, 96, *485*
Hoeschke, A., 64, *146*, 64, *157*, 64, *190*
Hoescke, A., 64, *158*, 65, *158*
Hofemeister, J., 73, *703*
Hoff, T., 91, *530*, 121, *530*
Hoffman, C.L., 123, *1309*
Hoffman, P.C., 262, *3456*
Hoffmann, E.H.E., 75, *790*, 94, *583*
Hofreiter, B.T., 97, *808*
Hofvendahl, K., 236, *3019*
Hohaus, W., 120, *1257*, 236, *3022*
Hohne, W.E., 73, *721*
Hokari, S., 82, *372*
Holck, A., 207, *2586*
Holder, J.B., 169, *1997*, 171, *1997*
Holdgate, R.H., 204, *2497*
Holik, D.J., 148, *1616*, 149, *1620*, 158, *1868*
Hollenberg, C.P., 74, *749*, 77, *749*, 78, *1042*, 212, *2648*
Hollis, F. Jr., 189, *2252*
Hollo, J., 64, *119*, 64, *146*, 64, *157*, 64, *158*, 64, *185*, 64, *190*, 64, *203*, 65, *158*, 129, *1410*, 129, *1411*, 145, *1570*, 146, *1570*, 147, *1570*, 153, *1725*, 264, *3507*
Holm, J., 166, *1965*
Holmbergh, O., 129, *1444*, 187, *2210*
Holt, N.C., 149, *1639*
Holthuis, J.J., 145, *1569*
Home, S., 114, *1192*
Honda, H., 227, *2908*
Honda, K., 145, *1554*
Honda, M., 186, *2207*
Hong, J.K., 255, *3331*
Hong, S.C., 72, *651*
Honorio de Aranjo, E., 88, *444*
Hood, L.F., 173, *2030*
Hooker, B.S., 227, *2907*
Hooks, W.B., 7, *36*
Hoorelbeke, B., 27, *107*, 47, *154*
Hoorn, A.J.W., 261, *3410*
Hoover, R., 64, *163*, 82, *377*, 167, *1973*, 167, *1981*, 171, *163*, 176, *163*
Hope, P.M., 124, *1332*, 127, *1377*
Hopkins, R.H., 147, *1589*, 211, *2621*
Horai, J., 64, *106*
Hore, E.F., 145, *1561*, 153, *1561*, 183, *1561*
Hori, K., 11, *57*, 261, *3411*, 261, *3412*, 261, *3413*
Horie, Y., 71, *400*, 83, *400*
Horikoshi, K., 65, *279*, 75, *802*, 88, *456*, 88, *457*, 91, *534*, 115, *1215*, 116, *1215*, 150, *1683*, 246, *3178*, 251, *3243*, 251, *3244*, 251, *3254*, 251, *3255*, 252, *3254*, 252, *3270*, 252, *3276*, 252, *3277*, 253, *3300*, 254, *3308*, 256, *3360*, 257, *3365*
Horn, C.H., 213, *2666*, 225, *2666*
Horn, H.E., 173, *2027*
Horn, J.S., 215, *2707*
Horsley, M., 204, *2491*
Horstmann, C., 77, *1039*

Horton, E., 262, *3423*
Horvathova, V., 64, *198*, 107, *1019*, 200, *2458*
Horwath, R.O., 243, *3120*
Hoschke, A., 129, *1410*, 129, *1411*, 202, *2472*
Hoseney, R.C., 198, *2453*
Hoshi, H., 37, *130*, 37, *131*
Hoshi, M., 264, *3523*
Hoshino, H., 45, *151*
Hoshino, K., 64, *231*, 74, *766*, 75, *771*, 97, *812*, 212, *2645*, 214, *2687*, 214, *2688*, 236, *3023*
Hoshino, M., 101, *890*, 234, *2982*
Hoshino, Y., 34, *114*
Hosoya, K., 240, *3075*
Hossain, Z., 78, *1075*
Hossein, F., 106, *993*
Hosting, R., 225, *2875*
Hostinova, E., 74, *746*, 77, *746*
Hotakainen, M., 197, *2434*
Hou, J., 268, *3603*
Hou, L., 223, *2845*
Hou, W.N., 105, *991*, 106, *991*
Houben, H., 156, *1791*, 217, *1791*
Houng, J.Y., 190, *2294*
Howard, B.H., 125, *1334*
Howling, D., 64, *187*
Hrabova, H., 249, *3231*
Hsieh-Wilson, L.C., 2, *3*, 4, *3*
Hsiu, J., 183, *2169*
Hsu, P.H., 71, *397*, 83, *397*
Hsu, W.H., 72, *673*, 125, *1347*
Hu, J., 72, *666*, 225, *2876*
Hu, M., 41, *138*, 41, *139*, 41, *140*, 72, *664*
Hu, Q., 139, *1504*
Hu, W., 75, *786*, 97, *786*
Hu, Y., 164, *1929*, 218, *2772*
Hu, Z., 232, *2960*
Huang, C., 190, *2286*
Huang, C.M., 62, *23*
Huang, D., 206, *2545*
Huang, D.P., 206, *2549*, 206, *2552*
Huang, H., 247, *3184*
Huang, L.P., 141, *1537*, 149, *1651*, 153, *1728*, 193, *2353*, 234, *2983*
Huang, M.C., 232, *2963*, 262, *3450*, 264, *3450*
Huang, N., 149, *1651*, 193, *2353*
Huang, Q., 193, *2344*
Huang, S., 193, *2346*
Huang, T., 149, *1651*
Huang, T.I., 246, *3169*
Huang, X.W., 124, *1324*
Huang, Y., 215, *2692*
Huang, Y.L., 232, *2962*
Huchette, M., 154, *1745*, 154, *1746*
Hudson, C.S., 249, *3223*, 249, *3224*, 255, *3334*
Hughes, D.A., 65, *292*
Hughes, R.C., 2, *6*, 71, *440*, 87, *440*
Hug-Iten, S., 203, *2477*
Hui, H., 11, *65*
Huizing, H.J., 72, *685*
Hulleman, S.H.D., 176, *2068*
Hulseweh, B., 78, *1042*
Hultin, E., 71, *399*, 83, *399*
Hulton, H.F.E., 86, *430*, 128, *1405*, 181, *2145*
Hung, C.H., 245, *3163*, 250, *3240*
Hung, P.V., 164, *1936*, 216, *1936*
Huong, N., 123, *1316*
Hupfer, H., 188, *2237*
Hurst, T.L., 154, *1742*, 155, *1778*, 159, *1875*
Husemann, E., 63, *68*
Huske, E.A., 149, *1620*
Huskens, D., 47, *156*
Hussain, Z., 197, *2431*, 197, *2432*, 197, *2433*
Hussein, A.A., 150, *1691*
Hussein, A.M., 73, *722*
Hutchinson, J.B., 85, *403*
Hwang, J.B., 254, *3303*
Hwang, Y.I., 222, *2832*
Hyde, K.D., 78, *1079*
Hyldon, R.G., 164, *1938*
Hylla, S., 262, *3422*
Hyodo, H., 192, *2339*, 193, *2339*
Hyun, H.H., 76, *936*, 77, *936*, 77, *948*, 77, *1037*, 102, *911*, 103, *948*, 115, *1214*, 116, *1214*, 124, *936*, 216, *948*, 217, *2741*, 217, *2747*

I

Iancu, E., 234, *2991*, 235, *2991*
Iannotti, E., 235, *3010*
Ibrahim, N.A., 195, *2397*
Ichikawa, E., 78, *979*, 105, *979*
Ichikawa, K., 98, *836*, 101, *891*, 120, *1248*, 127, *1392*
Ichiki, M., 100, *874*
Ideie, S., 206, *2562*
Idgeev, B.K., 176, *2087*
Idikio, H.A., 8, *40*
Ienco, A., 19, *89*, 19, *99*, 21, *94*
Igarashi, S., 206, *2562*
Igarashi, T., 221, *2816*
Igarashi, Y., 11, *52*, 41, *138*, 42, *142*, 43, *145*, 43, *149*, 46, *153*, 47, *154*, 47, *155*, 48, *52*
Iguchi, R., 5, *31*
Iida, S., 158, *1847*
Iida, T., 223, *2850*
Iimura, S., 37, *129*, 37, *130*, 37, *131*
Iizuka, H., 77, *953*, 77, *1008*, 103, *953*, 107, *1008*
Iizuka, M., 95, *609*, 121, *609*, 127, *1385*
Ikasari, K., 79, *1103*
Ikawa, Y., 176, *2070*
Ikeda, D., 34, *116*
Ikeda, H., 262, *3432*
Ikeda, K., 100, *874*
Ikeda, T., 175, *2052*
Ikegami, Y., 79, *990*, 105, *986*, 105, *990*, 106, *990*, 129, *1421*
Iliescu, G., 194, *2377*
Iloki, I., 79, *1099*
Ilori, M.O., 73, *706*, 73, *707*, 77, *950*, 103, *950*
Imachi, H., 127, *1376*
Imada, K., 88, *446*, 89, *483*, 115, *1209*, 116, *1209*, 119, *446*
Imai, Y., 78, *1070*
Imam, S.H., 73, *705*, 165, *1952*, 175, *2053*
Imamura, T., 214, *2676*, 242, *3105*
Imanaka, H., 246, *3172*
Imanaka, T., 72, *525*, 90, *525*, 117, *1224*, 117, *1225*, 243, *3122*, 246, *3172*
Imanishi, A., 72, *681*
Imata, K., 158, *1871*
Imayasu, S., 78, *979*, 105, *979*
Imo, E., 174, *2043*, 176, *2043*
Imperiali, B., 2, *1*, 21, *1*
Imshenetskii, A.A., 124, *1323*
Inaba, T., 216, *2711*
Inada, K., 206, *2537*, 206, *2538*
Inage, M., 125, *1349*, 126, *1349*
Inawa, K., 267, *3593*
Ingale, S., 24, *101*
Ingle, M.B., 72, *646*
Ingledew, W.M., 64, *103*, 74, *561*, 77, *561*, 93, *561*, 106, *561*, 125, *1345*
Inglett, G.E., 90, *507*
Ingram, L.O., 217, *2746*
Inkemann, P.A., 248, 3209
Inlow, D., 222, *2831*
Inoe, K., 207, *2566*
Inokuchi, N., 105, *985*
Inoue, H., 121, *1280*, 128, *1409*, 155, *1756*
Inoue, S., 255, *3330*
Inoue, T., 72, *690*, 129, *1413*
Inoue, Y., 236, *3024*, 246, *3179*
Inui, M., 232, *2954*
Inui, T., 253, *3291*
Inuzuka, K., 75, *783*
Ionescu, A.D.D., 234, *2991*, 235, *2991*
Iqbal, J., 91, *537*, 101, *897*, 217, *2736*, 218, *2736*, 237, *3031*
Irbe, R.M., 243, *3120*
Ire, F.S., 74, *753*, 76, *753*
Irie, M., 79, *990*, 105, *985*, 105, *986*, 105, *990*, 106, *990*, 129, *1421*
Irie, R., 91, *528*
Irwin, C.L., 81, *370*, 82, *370*
Isaksen, M., 208, *2592*
Isawa, T., 186, *2203*
Isemura, T., 72, *681*
Ishibashi, H., 205, *2517*
Ishida, H., 78, *979*, 105, *979*, 235, *2993*
Ishida, M., 149, *1633*, 208, *2603*
Ishida, N., 176, *2083*, 179, *2118*
Ishidao, T., 110, *1135*, 116, *1135*

Ishido, T., 211, *2624*
Ishigami, H., 78, *996*, 78, *997*, 106, *996*, 106, *997*, 106, *1001*, 112, *1176*, 114, *1176*, 255, *3332*
Ishige, Y., 149, *1645*, 268, *3612*
Ishii, A., 193, *2356*
Ishii, K., 193, *2355*
Ishii, S., 140, *1526*
Ishii, T., 193, *2351*
Ishii, Y., 90, *510*, 145, *1555*, 157, *1820*, 178, *2109*
Ishikura, T., 252, *3272*
Ishimoto, C., 190, *2297*
Ishimoto, E., 73, *696*, 77, *696*
Ishizaki, A., 110, *1151*, 190, *2277*, 190, *2283*, 226, *2895*, 226, *2898*, 227, *2895*, 236, *3015*, 236, *3016*, 236, *3017*
Ishizuka, H., 110, *1138*, 150, *1685*, 158, *1866*, 190, *2279*, 190, *2282*, 190, *2303*, 255, *3330*
Ishizuka, Y., 41, *138*
Islam, M.N., 192, *2338*
Ismail, A.M., 72, *662*
Itagawa, M., 172, *2019*
Iten, L.B., 223, *2842*
Ito, A., 197, *2441*, 227, *2905*
Ito, H., 64, *95*, 64, *105*
Ito, I., 211, *2626*
Ito, J., 156, *1784*
Ito, K., 92, *545*, 95, *609*, 121, *609*, 127, *1385*
Ito, M., 110, *1138*, 150, *1685*, 158, *1861*, 158, *1866*, 159, *1861*, 190, *2279*, 190, *2303*, 255, *3330*
Ito, T., 71, *400*, 83, *400*, 88, *458*, 111, *1157*, 113, *1157*
Ito, Y., 3, *17*, 15, *17*, 42, *142*, 43, *145*, 43, *149*
Ivanov, N.N., 168, *1985*
Ivanov, V.V., 230, *2932*
Ivanova, L.A., 108, *1114*
Ivanova, V., 63, 72, 72, *480*, 72, *652*, 72, *665*, 88, *480*, 121, 72, 264, *3503*
Ivy, A.C., 208, *2593*
Iwai, Y., 110, *1135*, 116, *1135*
Iwamoto, A., 165, *1953*

Iwamoto, M., 264, *3493*
Iwasaki, H., 263, *3479*, 250, *3238*
Iwasaki, K., 190, *2280*
Iwase, K., 227, *2905*
Iwata, K., 74, *762*
Iwata, M., 240, *3065*
Iwata, T., 243, *3114*
Izaki, K., 73, *692*, 92, *545*
Izydorczyk, M.S., 167, *1982*

J

Jacaway, S.A., 255, *3343*
Jackel, S.S., 203, *2487*
Jackson, L.E., 158, *1849*
Jackson, P., 264, *3496*, 264, *3497*
Jackson, R.F., 263, *3477*
Jacob, H.-E., 228, *2927*, 228, *2928*
Jae, C.M., 88, *453*
Jafri, R.H., 101, *897*
Jahnke, M., 201, *2461*
Jakovlević, J.B., 64, *226*, 79, *1027*, 108, *1027*, 114, *1027*, 115, *1027*, 154, *1747*, 155, *1766*, 222, *2840*
Jalba, N., 194, *2377*
Jaleel, S.A., 220, *2802*, 226, *2802*
James, C., 206, *2543*
James, T.D., 5, *29*, 5, *30*, 5, *31*, 7, *34*, 7, *35*
James, W.O., 129, *1447*, 187, *1447*
Jamroz, J., 63, *71*, 176, *2074*, 176, *2075*, 177, *2074*, 180, *2140*
Jan, S.Y., 93, *577*
Jana, D., 73, *719*
Jana, S.C., 101, *899*, 188, *2227*
Janari, K.S., 237, *3034*
Jänchen, M., 264, *3517*, 264, *3518*
Jane, J., 80, *337*, 165, *337*
Janeček, S., 64, *107*, 64, *177*, 64, *198*, 65, *301*, 67, *177*, 75, *177*
Janes, B.J.H., 64, *93*
Jang, L.Y., 237, *3047*
Jang, S.Y., 72, *669*
Janicki, J., 110, *1139*, 267, *3581*
Janke, A., *43*, 63
Jankowski, T., 191, *2316*, 179, *2117*

Jansen, N.B., 109, *1119*
Janssen, A.E.M., 267, *3598*
Janusić, V., 223, *2844*
Jarai, M., 254, *3315*
Jarniewicz, J., 218, *2609*, 210, *2609*
Jaroslawski, L., 150, *1674*, 190, *2289*
Jarvinen, K., 197, *2434*
Jascanu, V., 192, *2343*
Jay, J.I., 12, *72*
Jayko, L.G., 155, *1777*
Jeang, C.L., 73, *700*, 79, *700*, 252, *3278*, 252, *3279*, 255, *3337*
Jeanloz, R., 160, *1892*
Jędrychowska, B., 68, *314*, 204, *2499*
Jeffcoat, R., 197, *2433*
Jeffreys, G.A., 145, *1562*
Jennings, R.P., 204, *2495*
Jenny, F.E. Jr., 117, *1220*
Jensen, B., 75, *794*, 75, *795*
Jensen, M.T., 64, *89*
Jeon, B.Y., 218, *2770*, 218, *2775*, 224, *2770*
Jeong, J.Y., 88, *448*
Jespersen, H., 129, *1412*
Jezek, D., 191, *2317*
Ji, G.E., 73, *691*
Ji, H.S., 151, *1707*, 157, *1810*, 207, *2579*
Ji, Y., 149, *1629*, 217, *1629*, 220, *1629*
Jian, H., 206, *2531*
Jiang, C., 213, *2655*
Jiang, G., 99, *857*, 165, *1948*
Jiang, J., 154, *1750*
Jiang, L., 64, *80*, 226, *2901*
Jiang, N., 237, *3044*, 249, *3211*
Jiang, S.T., 71, *402*, 85, *402*
Jiang, S.T.S.P., 234, *2984*, 235, *2984*
Jiang, X., 65, *266*
Jiang, Y., 226, *2901*
Jimenez, T., 198, *2449*, 199, *2455*, 200, *2455*, 201, *2455*
Jiménez-Barbero, J., 21, *92*, 21, *93*, 21, *94*, 24, *101*
Jimi, N., 115, *1200*
Jin, B., 234, *2983*
Jin, H., 223, *2845*
Jin, K.Y., 105, *981*, 105, *982*
Jin, S., 3, *14*, 5, *14*
Jin, Z., 256, *3364*
Jo, K.H., 177, *2106*, 189, *2266*
Jodal, I., 75, *772*
Joergensen, H., 227, *2921*
Joergensen, O.B., 147, *1592*, 147, *1596*
Joergensen, S.T., 91, *540*
Johal, M., 148, *1605*, 156, *1786*, 211, *2617*
Johannesen, P.F., 64, *81*
Johansen, H.N., 208, *2597*
Johansen, L.H., 202, *2467*
Johansson, N.O., 181, *2149*
Johari, D.P., 125, *1335*
John, M., 267, *3577*
Johnson, A.J., 198, *2447*
Johnson, D.E., 126, *1355*
Johnson, I.T., 261, *3417*
Johnson, J.A., 123, *1308*, 265, *3538*
Johnson, L.A., 192, *2333*
Johnson, M.R., 117, *1220*
Johnson, T.R., 142, *1543*
Johnston, D.B., 155, *1758*, 192, *2329*, 223, *2843*
Johnston, J.A., 265, *3542*
Johnston, L., 259, *3381*
Johnston, W.R., 267, *3594*
Johnston, W.W., 92, *555*
Jojima, T., 232, *2954*
Jolhiri, P., 170, *2005*, 176, *2005*
Jones, D.T., 65, *275*
Jones, P.G., 18, *83*
Joneukeu, R.N.D., 64, *86*
Jongboom, R.O.J., 97, *822*
Jonkers, J., 163, *1911*
Joo, C.Y., 105, *980*
Jore, J., 129, *1417*
Jorge, J.A., 67, *307*, 79, *1092*, 88, *584*, 94, *584*, 220, *1092*
Jorgensen, P.L., 114, *1186*, 117, *1221*
Jorgensen, T., 8, *39*
Josephson, K., 187, *2211*
Joshi, G., 3, *16*, 15, *16*, 21, *16*, 26, *104*
Joshi, P.N., 76, *876*, 100, *876*

Joszt, A., 86, *425*
Jouany, J.P., 260, *3406*
Joula, H., 63, *51*, 80, *51*
Joutsjoki, V.V., 108, *1022*
Jovanović, S.M., 110, *1140*
Jowett, P., 126, *1372*
Jozsa, S., 85, *405*, 85, *406*, 100, *405*, 267, *3579*, 267, *3594*
Jridi, T., 71, *639*
Ju, Y.H., 190, *2301*
Juge, N., 64, *130*, 129, *130*, 129, *1415*, 129, *1417*
Juhasz, O., 75, *780*
Julian, G.St., 222, *2834*
Juln, T., 78, *1088*
Jun, B.J., 244, *3156*
Jung, K.T., 250, *3236*
Junko Tomotani, E., 244, *3147*
Jurado, E., 142, *1547*
Jurcoane, S., 234, *2991*, 235, *2991*
Jurić, Z., 94, *588*
Jurisova, E., 190, *2310*
Juvekar, V.A., 149, *1653*, 153, *1730*

K

Kaalhus, O., 8, *39*
Kabaivanova, L., 72, *665*
Kabiri-Badr, A., 126, *1371*
Kaczmarek, M.J., 117, *1228*, 118, *1228*
Kaczmarek, R., 120, *1254*, 266, 3559
Kaczmarowicz, G., 238, *3049*
Kadowaki, K., 72, *655*
Kadowaki, M., 265, *3545*
Kaelin, A., *29*, 63
Kaetsu, I., 187, *2107*, 188, *2107*
Kago, N., 206, *2536*
Kai, M., 139, *1522*
Kainuma, K., 78, *997*, 88, *458*, 88, *459*, 99, *858*, 106, *997*, 106, *999*, 106, *1001*, 112, *1176*, 114, *1176*, 172, *2019*, 176, *2083*, 223, *2852*, 265, *3540*, 249, *3221*, 250, *3221*, 254, *3318*
Kainuma, R., 90, *511*
Kaji, A., 78, *1012*, 107, *1012*, 78, *1082*

Kakihara, K., 218, *2767*
Kakimi, M., 115, *1198*, 116, *1198*, 150, *1684*
Kakinuma, S., 34, *122*
Kako, K., *11*, 62
Kako, N., 257, *3368*
Kakushima, M., 34, *121*, 34, *122*, 34, *124*, 36, *127*, 36, *128*, 37, *125*, 37, *129*, 37, *134*
Kakutani, K., 221, *2812*
Kalać, V., 264, *3508*
Kalashnikova, N.A., 186, *2189*
Kalb, H., 164, *1937*
Kalinowska, H., 208, *2589*, 208, *2590*
Kalk, K.H., 129, *1429*
Kallai, B., 196, *2404*
Kalunyants, K.A., 136, *1482*, 167, *1980*
Kamachi, H., 36, *128*, 37, *130*, 37, *131*, 37, *132*, 40, *137*
Kamaeda, J., 94, *580*
Kamagata, Y., 96, *626*
Kamaruzaman, M., 74, *756*
Kamasaka, H., 150, *1673*
Kambara, L.M., 189, *2271*, 189, *2272*
Kamei, H., 34, *115*, 37, *115*, 37, *125*, 37, *132*, 37, *134*
Kamenan, A., 71, *638*, 78, *638*, 113, *638*, 122, *638*
Kamikubo, T., 79, *1102*, 109, *1102*
Kamimoto, F., 250, *3238*
Kamimura, M., 124, *1328*
Kamio, Y., 73, *692*
Kamitori, S., 120, *1248*
Kamogawa, A., 242, *3102*
Kamperman, A., 145, *1559*
Kamzolov, S.V., 65, *277*
Kan, J., 193, *2347*, 193, *2349*
Kanaeda, J., 159, *1883*, 159, *1884*, 205, *2511*
Kanai, H., 74, *727*
Kanai, T., 246, *3172*
Kanaya, K., 112, *1180*
Kanbara, I., 206, *2551*
Kandra, L., 75, *772*
Kaneko, J., 73, *692*
Kaneko, N., 147, *1594*

Kaneko, T., 79, *1153*, 164, *1933*, 244, *3139*, 251, *3255*, 253, *3300*
Kanenaga, K., *11*, 62, 161, *1899*
Kang, J., 181, *2148*, 183, *2148*
Kang, M., 171, *2010*
Kang, W., 164, *1929*
Kanlayakrit, W., 104, *974*, 129, *974*
Kannagi, R., 8, *38*, 8, *37*
Kanno, H., 78, *1089*
Kanno, M., 90, *510*, 90, *519*
Kano, S., 104, *969*, 105, *969*, 107, *969*
Kapdan, I.K., 245, *3164*
Kaper, F.S., 150, *1682*
Kaper, T., 241, *3079*
Kaper, T.N., 123, *1313*, 241, *1313*
Kaplan, F., 64, *137*
Kappes, J.C., 11, *65*
Kaprelyants, L.V., 166, *1957*
Kapustina, O.I., 119, *1244*
Kapustina, V.V., 163, *1922*, 172, *1922*
Karaichev, S.I., 217, *2730*
Karakatsanis, A., 121, *1276*, 121, *1277*, 189, *2262*
Karazhiya, V.F., 204, *2500*
Kargi, F., 245, *3164*
Karim, M.I.A., 153, *1721*, 226, *2897*, 238, *3052*, 247, *3183*, 250, *3241*
Karkalas, J., 64, *164*, 159, *1887*, 161, *1887*, 175, *164*, 262, *3466*, 264, *3466*, 264, *3514*, 265, *3466*
Karlović, D., 191, *2317*
Karmakaran, T., 215, *2700*
Karnati, V.V.R., 5, *32*, 7, *36*
Karpelenia, C.B., 147, *1591*
Karpenko, R.M., 153, *1719*
Karppelin, S., 89, *484*
Karpukhina, S.Ya., 72, *656*
Karrer, P., 172, *2014*
Karube, I., 159, *1882*, 263, *3472*
Karube, M., 151, *1702*, 151, *1703*, 240, *3074*, 264, *3520*
Karunanthi, T., 158, *1840*
Kas, S., 174, *2042*
Kashket, S., 259, *3390*

Kasica, J.J., 197, *2427*, 197, *2428*, 197, *2429*, 206, *2549*
Kasthuriban, M., 215, *2700*
Katagiri, M., 74, *766*
Katakura, Y., 212, *2650*
Katano, H., 139, *1516*, 139, *1517*, 175, *2052*, 267, *1516*
Kataoka, S., 95, *604*, 129, *1427*
Kataura, K., 157, *1820*
Katayama, M., 207, *2566*
Katchalski-Katzir, E., 102, *902*, 145, *902*, 148, *902*, 151, *902*
Kathrein, H.R., 150, *1660*
Katkocin, D.M., 107, *1018*, 123, *1309*, 253, *3294*
Kato, A., 158, *1850*, 159, *1850*
Kato, J., 126, *1361*
Kato, K., 79, *990*, 105, *990*, 106, *990*, 108, *1112*, 139, *1522*, 157, *1820*, 204, *2506*, 207, *2581*
Kato, M., 75, *784*, 100, *886*
Kato, T., 252, *3270*
Kato, Y., 90, *511*, 249, *3220*
Katsumata, N., 204, *2505*
Katsunori, K., 240, *3067*
Katsuragi, H., 205, *2518*, 241, *2518*
Katsuragi, I., 259, *3387*
Katsurayama, M., 149, *1633*, 208, *2603*
Katsuro, M., 193, *2351*
Katsuta, Y., 206, *2537*, 206, *2538*
Katz, D.F., 12, *72*
Kaufmann, H.A., 194, *2372*
Kaul, R., 97, *813*
Kaur, A., 224, *2866*
Kaur, G., 5, *26*, 5, *33*
Kaur, K., 86, *437*
Kaur, N., 86, *437*
Kaur, P., 79, *527*, 91, *527*, 106, *527*, 119, *527*
Kavitha, A.P., 74, *765*
Kawabata, Y., 247, *3185*
Kawaguchi, H., 34, *111*, 34, *113*, 42, *113*
Kawahara, H., 192, *2341*
Kawai, M., 112, *1170*, 113, *1170*

Kawakami, K., 102, *919*, 108, *1112*, 139, *1522*, 141, *1533*, 150, *1690*, 256, *3349*
Kawamaki, K., 139, *1512*
Kawamoto, M., 214, *2676*
Kawamura, K., 255, *3346*, 257, *3368*
Kawamura, S., 145, *1567*
Kawamura, Y., 121, *1279*, 121, *1281*
Kawano, M., 197, *2438*
Kawase, N., 154, *1737*
Kawashima, K., 179, *2112*
Kawashima, Y., 163, *1915*
Kawata, A., 78, *979*, 105, *979*
Kawato, A., 125, *1346*
Kayane, S., 242, *3105*
Kayashima, S., 102, *909*
Kazakova, S.I., 204, *2496*
Kazaoka, M., 64, *77*
Kele, A., 195, *2400*
Kelly, C.T., 64, *172*, 72, *449*, 72, *659*, 72, *660*, 72, *674*, 73, *717*, 73, *723*, 74, *739*, 76, *939*, 77, *739*, 78, *172*, 88, *449*, 88, *450*, 88, *452*, 88, *477*, 88, *478*, 90, *523*, 94, *582*, 95, *620*, 160, *450*, 257, *3371*
Kelly, J.J., 77, *944*, 103, *944*
Kelly, R.M., 114, *1185*, 115, *1212*, 117, *1212*, 117, *1220*
Kelly, S.J., 242, *3100*
Kelly, T.T., 82, *383*, 98, *383*
Kemler, M., 74, *748*, 77, *748*
Kempa, W., 120, *1257*
Kende, S., 186, *2202*
Keneres-Ursu, I., 194, *2373*
Keng, J.G., 266, *3570*
Kennedy, J.F., 64, *213*, 65, *268*, 89, *493*, 89, *505*, 98, *824*, 161, *1898*, 166, *1964*, 189, *2270*, 265, *3537*
Kensington, M.W., 173, *2031*
Keppler, D., 266, *3554*
Kerkhoven, P.H., 145, *1569*
Kerovuo, J.S., 64, *76*, 64, *82*
Kerr, R.W., 64, *189*, 150, *1668*, 194, *2371*, 221, *2820*, 249, *3225*, 249, *3229*
Kersten, E., 225, *2871*
Kertesz, Z., 19, 62

Kettlitz, B., 156, *1783*, 164, *1937*, 166, *1958*, 180, *2135*, 192, *2336*
Khachatryan, G., 136, *1470*, 136, *1471*, 136, *1473*, 136, *1474*, 256, *3362*, 257, *3362*, 258, *3362*
Khaidarova, Z.S., 165, *1943*
Khajeh, K., 187, *2208*
Khakimzhanov, A.A., 165, *1943*, 257, *3369*
Khalid, N.M., 244, *3148*
Khan, J., 101, *897*
Khare, S.K., 266, *3561*
Khaw, T.S., 212, *2650*
Khemakhem, B., 90, *509*, 102, *509*
Khemiri, H., 71, *639*
Khomov, V.V., 189, *2264*
Khomova, O.V., 189, *2263*, 189, *2264*
Khoo, S.L., 74, *756*, 78, *1064*
Khurana, A.L., 97, *811*
Khvorova, L.S., 158, *1872*
Kiba, Y., 65, *271*, 212, *271*, 212, *271*
Kida, K., 197, *2430*, 222, *2693*
Kiermeier, F., 181, *2155*, 202, *2470*
Kiersnowski, J., 210, *2608*
Kihlberg, B., 163, *1921*
Kikuchi, K., 254, *3310*
Kilby, J.M., 11, *65*
Kilian, S.G., 213, *2666*, 217, *2754*, 225, *2666*
Kilić Apar A.D., 181, *2146*
Kilkian, B.V., 78, *1045*
Kim, B.C., 253, *3295*, 255, *3340*, 256, *3361*
Kim, C.H., 101, *896*, 158, *1845*, 213, *2659*, 226, *2892*, 226, *2893*, 226, *2894*, 226, *2890*, 226, *2891*
Kim, C.J., 145, *1556*, 156, *1556*
Kim, D., 93, *560*
Kim, D.H., 218, *2770*, 218, *2775*, 224, *2770*
Kim, D.W., 93, *560*
Kim, E., 123, *1316*
Kim, H.C., 214, *2679*, 226, *2679*
Kim, H.H., 239, *3062*
Kim, H.J., 79, *1014*, 107, *1014*
Kim, H.O., 234, *2980*
Kim, H.S., 219, *2785*, 226, *2785*, 252, *3268*, 252, *3269*, 256, *3352*

Kim, H.Y., 267, *3590*
Kim, I.C., 72, *669*, 89, *487*
Kim, J., 90, *513*
Kim, J.C., 176, *2056*
Kim, J.H., 256, *3361*
Kim, J.I., 176, *2056*
Kim, J.R., 72, *669*
Kim, J.T., 209, *2605*, 252, *3269*
Kim, J.W., 89, *487*, 157, *1832*, 252, *3273*
Kim, K., 64, *74*, 125, *74*
Kim, K.S., 101, *896*, 239, *3062*
Kim, K.W., 151, *1707*, 157, *1810*
Kim, M., 234, *2976*
Kim, M.H., 101, *896*
Kim, M.J., 176, *2056*, 252, *3273*
Kim, M.S., 245, *3166*, 254, *3326*
Kim, N.M., 164, *1934*
Kim, S.D., 96, *628*
Kim, S.H., 254, *3303*, 254, *3326*
Kim, S.J., 218, *2775*
Kim, S.K., 239, *3062*
Kim, S.S., 94, *595*, 94, *596*
Kim, T.D., 151, *1707*
Kim, T.G., 240, *3071*
Kim, T.H., 105, *980*
Kim, T.J., 253, *3295*, 255, *3340*, 256, *3361*
Kim, T.K., 241, *3082*
Kim, T.U., 88, *448*
Kim, W.S., 219, *2785*, 226, *2785*
Kim, Y.B., 73, *691*
Kim, Y.H., 217, *2742*, 239, *3062*, 256, *3361*
Kim, Y.K., 177, *2097*, 178, *2097*, 254, *3322*
Kim, Y.R., 91, *541*, 123, *1320*, 168, *1989*, 207, *2568*
Kim, Z.U., 109, *1120*
Kimball, B.A., 173, *2027*
Kimura, A., 79, *1091*, 108, *1029*, 111, *1157*, 113, *1157*, 117, *1230*, 118, *1230*, 174, *2044*, 175, *2044*, 177, *2044*, 265, *3531*
Kimura, E., 205, *2518*, 241, *2518*
Kimura, I., 197, *2441*
Kimura, K., 129, *1427*, 254, *3302*
Kimura, T., 97, *816*, 115, *1215*, 116, *1215*, 122, *1290*, 163, *1915*, 205, *2529*, 240, *3075*
Kimura, Y., 73, *715*, 75, *784*, 234, *2986*
King, E.L., 136, *1481*
King, J.T., 138, *1491*
King, N.J., 78, *1080*
King, S., 206, *2558*
Kingma, W.G., 148, *1604*, 154, *1744*
Kinzburgskaya, F.M., 231, *2947*, 231, *2948*
Kinzinger, S.M., 148, *1600*
Kio, S., 197, *2435*
Kira, I., 89, *494*
Kiransree, N., 224, *2861*, 225, *2861*
Kirby, K.D., 213, *2667*
Kirdar, B., 217, *2756*, 217, *2757*
Kiribuchi, S., 120, *1255*
Kirimura, K., 237, *3033*
Kirsanova, V.A., 168, *1985*
Kirschner, M., 80, *325*, 82, *325*
Kirsh, D., 233, *2967*
Kirste, K., 225, *2871*
Kiser, P.F., 12, *72*, 14, *75*
Kishimoto, M., 78, *1070*, 216, *2715*
Kislaya, L.V., 224, *2863*
Kislykh, V.I., 189, *2273*
Kiso, M., 8, *37*
Kistler, H., 206, *2558*
Kitagawa, H., 257, *3368*
Kitagawa, M., 243, *3117*
Kitahara, K., 75, *783*, 156, *1785*, 235, *2993*
Kitahata, S., 65, *281*, 91, *538*, 112, *1178*, 114, *1178*, 205, *2517*, 251, *3263*, 251, *3264*, 253, *3298*, 253, *3299*
Kitamoto, K., 79, *1091*, 129, *1420*
Kitamura, S., 179, *2126*
Kitamura, Y., 240, *3076*
Kitano, T., 64, *123*, 81, *345*, 81, *346*, 81, *347*, 95, *601*, 138, *1499*, 181, *2143*, 181, *2144*, 186, *2196*
Kitao, S., 119, *1245*
Kitazawa, T., 196, *2407*
Kitchen, H., 80, *335*
Kittikusolthum, K., 236, *3015*
Kiyofuji, T., 244, *3153*
Klamar, G., 109, *1133*, 109, *1132*, 215, *2701*
Klein, D.A., 209, *2604*

Klein, E., 24, *100*, 24, *101*, 190, *2285*
Kleindienst, A., 86, *425*
Kleinman, M., 186, *2191*
Klenk, H.P., 124, *1324*
Klich, M., 75, *770*
Klimovskii, D.N., 95, *608*, 99, *852*, 104, *608*, 145, *852*
Klingenberg, P., 93, *565*, 166, *1958*
Klinger, R., 127, *1398*
Klingner, W., 265, *3530*
Kłosowski, G., 218, *2769*, 222, *2830*, 226, *2887*, 227, *2926*, 228, *2925*
Knap, I.H., 208, *2597*
Knapik, H.P.G., 244, *3149*
Knauber, H., 249, *3216*, 249, *3217*
Kneck, C., 78, *1071*
Kneen, E., *33*, 63, *33*, 267
Knegtel, R.M.A., 129, *1429*
Kneifel, H., 267, *3577*
Knittel, D., 195, *2400*
Knorr, D., 100, *883*, 138, *1501*
Knowles, R., 112, *1174*
Knox, A.M., 217, *2754*
Knudsen, F.E., 159, *1887*, 161, *1887*
Knudsen, K.E.B., 260, *3400*
Knutson, C.A., 86, *436*, 176, *436*
Ko, J.H., 101, *896*
Ko, S.H., 212, *2647*
Koba, Y., 110, *1151*, 120, *1260*, 176, *2094*, 190, *2283*, 215, *2704*, 217, *2720*, 220, *2720*, 220, *2800*, 221, *2800*, 222, *2720*, 225, *2881*
Kobayashi, G., 190, *2277*, 236, *3017*, 236, *3016*, 236, *3014*
Kobayashi, H., 97, *816*
Kobayashi, I., 112, *1178*, 114, *1178*
Kobayashi, J., 140, *1526*
Kobayashi, M., 78, *1012*, 107, *1012*, 78, *1082*, 119, *1245*, 207, *2576*, 240, *3076*, 247, *3185*
Kobayashi, N., 157, *1811*
Kobayashi, S., 78, *997*, 88, *458*, 90, *511*, 106, *997*, 117, *1229*, 129, *1427*, 157, *1805*, 159, *1879*, 165, *1953*, 172, *2019*, 176, *2064*, 176, *2077*, 240, *3069*, 247, *3182*, 249, *3221*, 250, *3221*, 255, *3338*, 249, *3228*, 254, *3318*, 249, *3220*
Kobayashi, T., 74, *727*, 75, *802*, 78, *1069*, 79, *1069*, 114, *1194*, 114, *1195*, 116, *1194*, 116, *1195*, 117, *1227*, 118, *1227*, 247, *3192*, 247, *3193*, 247, *1227*, 247, *3194*, 265, *3545*
Kobelev, K.V., 167, *1980*
Kobori, T., 265, *3531*
Koch, R., 75, *799*, 115, *1204*, 195, *2393*
Kochanowski, P., 97, *815*
Kodaka, A., 125, *1346*
Kodet, J., 206, *2547*
Koellreutter, B., 176, *2079*
Koenraad, K.F.P., 122, *1289*
Koh, G.R., 89, *491*
Kohl, S., 148, *1607*
Kohno, A., 76, *930*
Koishi, M., 193, *2351*
Koivikko, H., 234, *2979*
Koizumi, K., 240, *3068*, 252, *3271*
Koizumi, S., 157, *1811*
Kojima, I., 65, *285*, 89, *491*
Kojima, M., 240, *3067*, 240, *3066*, 234, *2982*
Kojima, T., 102, *919*, 141, *1533*
Kojima, Y., 78, *979*, 105, *979*, 194, *2381*
Kok, J., 233, *2972*
Kokonashvili, G.N., 108, *1028*, 115, *1028*
Kokubun, F., 97, *819*
Kolachov, P.J., *34*, 63, 210, *34*
Kolawole, A.O., 76, *923*
Kolbach, P., 146, *1582*, 146, *1583*, 148, *1583*
Kolcheva, R.A., 167, *1980*
Kolenko, E.I., 261, *3408*
Kolkman, M., 64, *84*
Kolpakchi, A., 136, *1482*
Komai, Y., 163, *1925*, 173, *2026*
Komaki, T., 64, *200*, 148, *1611*, 148, *1615*, 150, *1611*, 150, *1667*, 180, *2136*
Komarova, N.L., 11, *65*
Komatsu, T., 263, *3472*, 264, *3520*
Komine, S.I., 82, *372*
Komiszar, V., 264, *3491*

Komoda, T., 82, *372*
Komolprasert, V., 135, *1463*, 139, *1463*, 190, *2308*, 190, *2314*
Komova, O.V., 109, *1123*
Konda, T., 112, *1178*, 114, *1178*
Kondo, A., 64, *98*, 77, *1040*, 212, *2649*, 212, *2650*, 216, *2715*
Kondo, H., 106, *995*
Kondo, K., 75, *781*
Kondo, S., 34, *110*, 34, *112*, 34, *116*, 37, *110*
Konerth, H., 194, *2377*
Kong, B.W., 176, *2056*
Kong, D.M., 226, *2901*
Konieczna-Molenda, A., 136, *1470*, 136, *1472*, 136, *1473*, 256, *3362*, 257, *3362*, 258, *3362*
Konieczny-Janda, G., 104, *961*, 206, *2554*, 208, *2599*, 217, *2724*
Konieczny-Molenda, A., 97, *815*
Konishi, H., 157, *1813*
Konishi, M., 34, *111*, 34, *113*, 34, *114*, 34, *117*, 36, *127*, 36, *128*, 37, *125*, 37, *129*, 37, *130*, 37, *131*, 37, *132*, 42, *113*
Kono, T., 64, *140*
Konsula, Z., 89, *498*
Konsuola, Z., 97, *809*
Koo, Y.M., 255, *3344*
Kooi, E.R., 154, *1749*, 256, *3358*
Kooryama, T., 197, *2441*
Kopec, D.Z., 260, *3403*
Kopplin, M.J., 221, *2819*
Koreeda, A., *11*, 62
Korhola, M., 64, *85*, 92, *553*
Korhonen, I., 149, *1627*
Korner, D., 166, *1958*
Körner, D., 73, *721*, 93, *565*
Korotkii, V.M., 64, *233*
Korotkyj, B., *60*, 63
Korpela, T., 251, *3249*
Korpela, T.K., 251, *3253*, 257, *3367*
Korus, R.A., 96, *635*, 106, *635*
Kosakai, Y., 234, *2981*
Koshijima, T., 240, *3073*
Kosicki, Z., 64, *246*

Kossman, J., 267, *3599*
Kostyukova, T.N., 161, *1900*
Kosyuk, I.P., 189, *2273*
Kotaka, T., 88, *462*, 91, *536*
Kotarska, K., 218, *2769*
Kotarski, S.F., 260, *3405*
Kotlarska, K., 226, *2887*
Kottwitz, B., 196, *2415*, 196, *2417*, 196, *2418*, 196, *2419*, 196, *2420*, 196, *2421*, 196, *2422*, 196, *2424*, 204, *2422*
Kou, X., 206, *2545*
Kouame, F.A., 71, *638*, 78, *638*, 113, *638*, 122, *638*
Kouame, L.P., 71, *638*, 78, *638*, 113, *638*, 122, *638*
Koukiekolo, R., 130, *1457*
Kourio, K., 253, *3291*
Koutinas, R.A., 126, *1371*
Kovago, A., 205, *2515*
Kovalenko, G.A., 109, *1123*, 189, *2263*, 189, *2264*
Kovalenko, V.I., 136, *1482*
Kovaleva, T.A., 142, *1544*
Kovolevskaya, I.D., 73, *712*
Koyama, I., 82, *372*
Koyama, S., 75, *776*
Koyano, T., 74, *614*, 75, *775*, 95, *614*
Kozaki, Y., 128, *1409*
Koziak, I., 89, *496*
Koziol, A., 132, *1458*, 143, *1549*
Kozuma, T., 118, *1234*
Kraetz, H., 164, *1937*, 166, *1958*
Kragh, K., 64, *142*, *142*, 245, 208, *2592*
Kragh, K.M., 64, *96*, 79, *96*, 79, *96*, 111, *96*, 203, *2485*
Kraic, J., 122, *1306*
Kramhoft, B., 64, *89*, 64, *130*, 129, *130*, 129, *1415*
Krasnobajew, V., 158, *1865*
Krause, A., 71, *431*, 86, *431*
Krauze, D.R., 107, *1021*
Krauze, J., 110, *1146*
Krebs, J., 154, *1754*, 198, *2448*
Kreppel, L.K., 25, *102*

Kricłka, T., 223, *2844*
Krieger, N., 79, *1100*
Krishnan, M.S., 222, *2839*
Krishnan, T., 64, *87*
Krishnaswamy, C.I., 219, *2782*
Krivopalov, F.G., 146, *1580*
Kroeyer, K., 192, *2330*
Kroeyer, K.K.K., 153, *1718*, 155, *1768*, 157, *1806*
Królikowski, L.J., 132, *1460*
Kroll, G.W.F., 248, *3206*
Krome Sander, P.A., 149, *1638*
Krone, P.A., 150, *1663*
Kroumov, A.D., 230, *2938*
Kroyer, K.K.K., 154, *1736*
Kruger, J.E., 101, *889*, 167, *1967*
Krutsch, H.C., 75, *798*
Kruus, K., 247, *3186*
Krymkiewicz, N., 251, *3246*
Krzechowska, M., 78, *1078*
Krzewinski, F.S., 129, *1428*
Krzyżaniak, W., 80, *341*, 179, *2117*, 191, *2316*
Kubal, B.S., 191, *2320*
Kuboi, J., 145, *1554*
Kubomura, S., 94, *580*, 159, *1883*, 159, *1884*, 205, *2511*
Kubota, A., 247, *3183*, 250, *3241*
Kubota, K., 219, *2779*, 219, *2781*
Kubota, M., 119, *1237*, 204, *2508*, 240, *3070*, 241, *3085*, 246, *3176*, 257, *3372*
Kubota, N., 258, *3373*
Kubota, Y., 240, *3068*
Kučera, J., 109, *1127*, 140, *1528*, 249, *3231*
Kučerova, J., 224, *2870*
Kudo, K., 180, *2132*
Kudo, T., 74, *727*, 75, *802*, 91, *539*, 126, *539*
Kudryashova, V.E., 188, *2222*
Kuehl, H., 161, *1901*
Kuge, T., 75, *776*, 179, *2126*
Kugimiya, M., 146, *1581*
Kuhn, R., *57*, 87, 63, 87, 71, 87
Kujawski, M., 154, *1738*
Kukharenko, A.A., 64, *233*

Kulaev, I.S., 74, *736*
Kulka, D., 211, *2621*
Kull, J.F., 82, *375*
Kulp, K., 198, *2451*
Kumada, Y., 245, *3165*
Kumagai, C., 129, *1420*
Kumagaya, T., 100, *874*
Kumaki, Y., 11, *54*
Kumakura, M., 187, *2107*, 188, *2107*
Kumar, A., 242, *3099*, 243, *3107*
Kumar, E.V., 234, *2988*, 235, *2988*
Kumar, J.K., 158, *1840*
Kumar, S., 79, *527*, 79, *1106*, 91, *527*, 106, *527*, 119, *527*
Kume, T., 179, *2113*
Kumundinu, B.S., 74, *765*
Kundu, A.K., 66, *303*, 66, *304*
Kung, J.T., 80, *340*, 127, *340*
Kung, L. Jr., 75, *786*, 97, *786*, 236, *3025*, 236, *3026*
Kuniak, L., 96, *632*, 96, *627*, 137, *632*, 268, *3602*, 268, *3611*
Kunimoto, K., 150, *1690*
Kunisaki, S., 129, *1418*, 217, *2725*
Kunitake, T., 190, *2298*
Kunz, M., 123, *1312*
Kunze, I., 77, *1039*
Kupfer, B., 236, *3022*
Kurahashi, Y., 180, *2132*
Kurakake, M., 180, *2136*
Kurgatnikov, M.M., 168, *1985*
Kuriki, T., 117, *1224*, 117, *1225*, 129, *1430*, 150, *1673*, 253, *3297*
Kurimoto, M., 155, *1775*, 155, *1776*, 157, *1804*, 246, *3174*, 246, *3177*, 257, *3372*
Kurimura, T., 244, *3157*
Kuroda, M., 112, *1177*, 113, *1177*, 114, *1177*
Kurokawa, M., 259, *3387*
Kurokawa, T., 119, *1245*
Kurosaki, T., 242, *3105*
Kurosawa, H., 213, *2651*, 213, *2652*
Kurosawa, K., 78, *1081*, 96, *626*, 107, *1013*
Kursheva, N.G., 120, *1265*, 212, *2629*
Kuruma, A., 72, *644*

Kurushima, M., 156, *1785*
Kurushima, N., 206, *2553*
Kusai, K., 149, *1624*
Kusakabe, M., 156, *1792*, 217, *1792*
Kuschel, M., 17, *80*
Kushimoto, T., 80, *316*
Kusnadi, A.R., 104, *971*, 129, *971*
Kusubara, E., 157, *1823*
Kusunoki, K., 102, *919*, 108, *1112*, 139, *1512*, 139, *1522*, 141, *1533*, 256, *3349*, 256, *3353*
Kuwabara, M., 240, *3073*
Kuwatani, N., 163, *1915*
Kuzin, A., 148, *1597*
Kuznetsov, E.V., 227, *2906*
Kuzovlev, V.A., 165, *1943*
Kvesitadze, G.I., 108, *1028*, 115, *1028*
Kwak, Y.S., 75, *802*, 91, *539*, 126, *539*
Kwan, H.S., 76, *932*
Kweon, K.S., 252, *3273*
Kweon, M.R., 179, *2128*, 180, *2128*
Kwon, I.B., 252, *3284*, 252, *3285*
Kwong, P.D., 11, *65*
Kyotani, Y., 240, *3066*

L

Laane, N.C., 145, *1559*
Labarre, J., 83, *385*
Labbe, R.G., 73, *697*, 77, *697*
Labout, J.J.M., 95, *616*, 108, *1024*
Łączyński, B., 224, *2868*
Laddee, M., 236, *3015*
Laderman, K.A., 75, *798*
Ladrat, C., 112, *1173*, 113, *1173*
Ladur, T.A., 108, *1031*, 120, *1259*, 153, *1719*, 165, *1259*
Lagoda, A.L., 76, *934*
Lagrain, B., 202, *2469*
Lahtela-Kakkonen, M., 197, *2434*
Lai, B.E., 12, *72*
Lai, J.T., 117, *1231*
Lai, T.S., 97, *810*
Lajolo, F.M., 99, *866*, 101, *866*
Laluce, C., 74, *747*, 77, *747*
Lama, L., 121, *1283*, 246, *3180*
Lamanna, R., 265, *3529*
Lambertini, P., 62, *13*, 64, *13*
Lambrechts, M.G., 64, *183*
Lamed, R., 102, *902*, 145, *902*, 148, *902*, 151, *902*
Lampe, B., 35, *63*
Lan, M.J., 97, *810*
Lance, C.M.R., 265, *3536*
Land, C.E. Jr., 154, *1748*
Landert, J.P., 253, *3289*
Landi Franco, C.M., 195, *2396*
Landis, B.H., 151, *1712*, 154, *1752*, 178, *1752*
Landis, Q., 260, *3394*
Landvik, S., 92, *550*
Lang, X., 80, *339*, 135, *339*
Lange, N.K., 127, *1400*, 129, *1400*
Langheinrich, K., 12, *72*
Langlois, D.P., 151, *1697*, 158, *1839*, 205, *2514*, 205, *2519*
Langridge, P., 100, *882*
Langworthy, C.F., 259, *3380*, 260, *3397*
Lant, P., 234, *2983*
Lantero, O., 65, *266*
Lantero, O.J., 212, *2633*, 213, *2671*
Lantero, O. Jr., 108, *1023*
Lantz, S.E., 64, *101*, 108, *1023*, 213, *2671*
Lapidus, T.V., 158, *1872*
Lapointe, P., 83, *385*
Lappalainen, A., 116, *1208*, 167, *1975*
Larinova, T.I., 248, *3202*
Larkin, A., 2, *1*, 21, *1*
Larlsson, R., 87, *435*
Larner, J., 259, *3385*
Larsen, B., 64, *96*, 79, *96*, 111, *96*
Larsen, J., 227, *2921*
Larsen, K.L., 65, *298*, 240, *3072*, 253, *3292*
Larsen, S., 114, *1189*
Larson, M., 214, *2685*
Larson, R.F., 151, *1696*, 158, *1839*
Larsson, J., 238, *3055*
Larsson, M., 158, *1846*
Larsson-Razinkiewicz, M., 87, *435*
Łasik, M., 219, *2778*

Laszlo, E., 64, *119*, 64, *203*, 64, *146*, 64, *190*, 129, *1410*, 129, *1411*, 145, *1570*, 146, *1570*, 147, *1570*, 153, *1725*, 202, *2472*, 264, *3507*
Laszlo, I.F., 99, *864*
Laszlo, L., 109, *1133*
Lau, K.S., 2, *4*
Laube, K., 224, *2867*
Lauer, M., 5, *24*
Laurell, T., 264, *3510*
Laurencot, C.M., 30, *108*
Laurenz, A., 206, *2558*
Lauro, M., 114, *1192*, 167, *1974*, 167, *1975*, 167, *1976*, 176, *2067*, 176, *2068*, 176, *2069*
Lauterback, G.E., 161, *1905*
Lava, J.D., 259, *3382*
Law, D.K., 206, *2560*
Lazareva, A.N., 156, *1793*, 211, *2622*, 212, *2640*, 215, *2694*, 217, *2622*, 217, *2721*, 218, *2768*, 217, *2721*
Lazič, M.L., 218, *2771*
Le Corvaisier, H., 176, *2057*
Le Gal-Coeffet, M.F., 64, *130*, 129, *130*, 129, *1415*, 129, *1426*
Le Thanh, J., 241, *3087*
Leach, H.W., 148, *1616*, 149, *1620*, 158, *1868*, 160, *1895*, 176, *1895*, 177, *1895*, 244, *3127*, 244, *3129*, 244, *3143*
Leahu, A., 86, *438*
Leal Martins, M.L., 72, *676*
Lebeault, J.M., 79, *1096*, 127, *1374*
Lebedeva, V.I., 195, *2399*
Leclerc, C., 227, *2911*
Lecollinet, G., 23, *98*
Lecoq, R., 81, *360*, 86, *429*, 154, *1734*
LeCureux, L.W., 115, *1205*, 116, *1205*
Ledder, R.G., 268, *3613*
Ledward, D.A., 182, *2165*
Lee, A., 227, *2924*
Lee, A.S., 81, *366*, 82, *366*
Lee, B.H., 164, *1934*
Lee, C., 81, *366*, 82, *366*, 243, *3122*
Lee, C.C., 64, *243*, 68, *243*

Lee, C.G., 226, *2894*
Lee, C.K., 190, *2301*
Lee, D.D., 108, *1113*, 109, *1118*, 158, *1864*, 158, *1870*
Lee, D.H., 222, *2832*
Lee, D.S., 72, *669*
Lee, E., 76, *828*, 98, *828*, 98, *835*, 100, *828*, 139, *828*, 266, *3552*
Lee, G.M., 220, *2801*, 226, *2890*, 226, *2892*, 226, *2891*
Lee, G.O., 93, *560*
Lee, H., 158, *1844*
Lee, H.C., 266, *3560*
Lee, H.G., 123, *1320*, 207, *2568*
Lee, H.S., 91, *541*, 117, *1220*, 158, *1844*, 158, *1845*, 252, *3273*, 253, *3295*, 255, *3340*, 256, *3361*
Lee, I.W., 252, *3273*
Lee, J.H., 93, *560*, 212, *2636*, 252, *3284*, 252, *3285*
Lee, J.I., 239, *3062*
Lee, J.M., 220, *2804*
Lee, J.S., 164, *1934*, 234, *2975*
Lee, J.W., 234, *2975*
Lee, K.I., 207, *2579*
Lee, K.J., 220, *2801*, 226, *2890*
Lee, K.P., 254, *3326*
Lee, K.Y., 123, *1320*
Lee, M.J., 252, *3273*
Lee, S., 123, *1316*
Lee, S.C., 222, *2832*
Lee, S.H., 254, *3316*
Lee, S.J., 89, *487*
Lee, S.K., 73, *691*, 145, *1556*, 156, *1556*
Lee, S.M., 101, *896*
Lee, S.O., 93, *560*, 238, *3057*
Lee, S.P., 72, *525*, 90, *525*
Lee, S.W., 214, *2678*, 214, *2679*, 216, *2710*, 213, *2657*, 226, *2679*
Lee, T.H., 238, *3057*
Lee, W.C., 125, *1344*, 235, *2999*
Lee, W.G., 158, *1844*
Lee, W.S., 209, *2605*
Lee, Y.C., 2, *2*, 4, *2*, 41, *141*

Lee, Y.D., 252, *3268*, 252, *3269*, 256, *3352*, 256, *3361*
Lee, Y.E., 102, *910*, 243, *3122*
Lee, Y.H., 73, *700*, 79, *700*, 177, *2106*, 189, *2266*, 189, *2267*, 239, *3064*, 240, *3071*, 241, *3082*, 254, *3307*, 254, *3313*, 254, *3316*, 256, *3354*, 254, *3326*
Lee, Y.K., 89, *504*
Lee, Y.S., 252, *3284*
Lee, Y.Y., 108, *1113*, 109, *1118*
Leech, J.G., 145, *1558*, 191, *2323*
Leemhuis, H., 64, *182*, 65, *182*
Lefuji, H., 75, *784*
Legg, D.A., 231, *2945*
Legge, R.L., 89, *497*
Legin, E., 112, *1173*, 113, *1173*, 166, *1960*, 173, *2034*
Legin-Copinet, E., 114, *1190*
Legisa, M., 65, *276*
Legov, M.D., 63, *72*, 121, *72*
Lehmbeck, J., 107, *1016*, 107, *1017*
Lehmussari, A., 192, *2328*
Lehtonen, U.R., 149, *1627*
Lei, G.D., 249, *3222*
Leibowitz, J., 63, *63*, 81, *351*, 129, *1436*
Leibowitz, Y., 129, *1434*, 129, *1435*
Leisola, M., 166, *1959*
Leite, S.G.F., 219, *2789*
Lelievre, J., 186, *2197*
Leloir, L.F., 248, *3200*, 248, *3201*
Leloup, V., 130, *1453*, 179, *2130*
Leloup, V.M., 130, *1451*, 130, *1452*
Lema, J.M., 139, *1518*, 139, *1519*, 189, *2260*
Lemaire, R., 8, *43*
Leman, P., 177, *2100*, 202, *2469*, 203, *2483*
Lemke, K., 75, *799*
Lencses, A., 264, *3494*
Lenders, J.P., 97, *818*
Leon, A., 166, *1966*
Leon, A.E., 198, *2454*
Leonard, C.K., 11, *58*, 12, *58*
Leonard, G., 114, *1189*
Leopold, H., 214, *2674*
Lepojevic, Z., 264, *3502*

Lercker, G., 62, *13*, 64, *13*
Lerminet, P., 224, *2862*
Les Usines de Melle, 154, *1743*
Leslie, J.D., 232, *2956*
Lesniak, W., 237, *3030*, 237, *3032*
Leszčzynski, W., 127, *1391*
Lettinga, G., 144, *1553*
Levashov, A.V., 188, *2222*
Levchik, A.P., 211, *2619*, 211, *2620*, 215, *2694*, 215, *2695*
Levchik, L.A., 211, *2619*, 215, *2695*
Leveque, E., 64, *177*, 67, *177*, 75, *177*
Levin, M.S., 94, *597*
Levin, R.J., 259, *3375*
Levitskaya, M.V., 155, *1771*, 163, *1910*
Levy, R., 83, *390*, 163, *390*
Levy-Rick, S.R., 93, *563*
Lewandowicz, G., 126, *1363*, 176, *1363*, 241, *3087*
Lewandowska, M., 219, *2792*
Lewen, K., 169, *1997*, 171, *1997*
Lewicki, P.P., 262, *3447*, 263, *3447*, 264, *3447*
Lewis, M.S., 75, *798*
Lewis, S.M., 213, *2670*, 227, *2912*
Li, A., 150, *1657*, 206, *1657*, 237, *3043*
Li, C., 126, *1370*, 206, *2545*, 213, *2655*, 220, *2809*
Li, D., 65, *278*
Li, F., 150, *1657*, 206, *1657*
Li, G., 154, *1750*, 237, *3043*, 266, *3556*
Li, H., 5, *33*, 126, *1370*, 190, *2286*, 218, *2773*
Li, H.M., 64, *209*
Li, J., 254, *3320*
Li, J.F., 222, *2838*
Li, J.H., 167, *1981*, 226, *2900*
Li, J.K.-K., 11, *54*
Li, K., 215, *2697*, 217, *2697*
Li, L., 136, *1479*, 193, *2344*, 201, *2463*, 219, *2777*, 250, *3237*
Li, M., 3, *14*, 5, *14*, 157, *1832*, 212, *2633*
Li, P., 64, *80*
Li, Q., 65, *297*, 206, *2541*, 218, *2772*
Li, S.P., 193, *2363*

Li, S.Z., 109, *1121*
Li, T.S., 215, *2697*, 217, *2697*
Li, W., 244, *3140*, 244, *3154*
Li, W.J., 124, *1324*
Li, X., 201, *2463*, 219, *2777*, 219, *2790*, 230, *2935*, 264, *3522*, 266, *3556*, 266, *3553*
Li, X.J., 234, *2984*, 235, *2984*
Li, X.X., 197, *2435*
Li, Y., 64, *108*, 65, *266*, 104, *965*, 136, *1479*, 171, *2010*, 254, *3323*
Li, Z., 65, *287*, 149, *1647*, 190, *2286*, 207, *2573*
Liakopoulou-Kyriakides, M., 89, *498*, 97, *809*, 121, *1276*, 121, *1277*, 189, *2262*
Liang, G., 193, *2348*
Liang, S., 206, *2531*
Liang, X., 211, *2623*
Liang, Y., 176, *2072*, 176, *2080*, 176, *2081*, 176, *2082*, 174, *2041*
Liang, Y.T., 182, *2166*
Liao, J.C., 231, *2953*, 232, *2953*
Libby, R.A., 262, *3463*, 264, *3463*
Libioulle, C., 72, *643*
Libranz, A.F., 128, *1408*
Liebl, W., 74, *729*
Liekens, S., 47, *154*
Lietz, P., 156, *1782*, 215, *2708*, 216, *2709*, 224, *2867*
Ligget, R.W., 158, *1839*
Lii, C.Y., 170, *2002*, 191, *2326*
Likola, M., 197, *2434*
Liliensiek, B., 72, *641*
Liljeberg, H., 177, *2098*
Lilly, M.D., 64, *168*, 189, *2258*
Lim, D.J., 149, *1634*
Lim, H.C., 244, *3130*
Lim, H.J., 159, *1882*
Lim, J.W., 166, *1955*
Lim, S.H., 250, *3236*
Lima da Rocha, L., 64, *236*
Lima, D.C., 110, *1141*
Lima, H.O.S., 254, *3324*
Limones Limones, G., 63, *65*
Limonov, P., 146, *1575*

Lin, C.F., 75, *779*
Lin, C.S., 72, *651*
Lin, C.Y., 245, *3163*, 246, *3169*
Lin, F.P., 74, *557*, 92, *557*
Lin, G., 193, *2345*
Lin, J.H., 177, *2103*, 177, *2104*
Lin, K.F., 158, *1834*
Lin, K.T., 157, *1818*, 157, *1819*
Lin, L.L., 72, *673*, 125, *1347*, 181, *2148*, 183, *2148*, 197, *2435*, 244, *3136*
Lin, R., 232, *2959*
Lin, S.F., 232, *2963*
Lin, T., 263, *3476*, 265, *3476*
Lin, T.Y., 204, *2492*
Linardi, V.R., 74, *731*, 77, *731*
Lindeman, L.R., 218, *2764*
Lindemann, E., 63, *68*
Linderer, M., 196, *2414*
Lindet, L., 228, *2922*
Lindroos, A., 167, *1978*, 244, *3132*
Line, W.F., 115, *1201*, 157, *1826*
Lineback, D.R., 78, *1072*, 104, *958*
Ling, A.R., 153, *1732*, 153, *1733*
Linke, B., 114, *1186*
Linko, M., 150, *1664*
Linko, P., 139, *1505*, 166, *1959*, 167, *1978*, 176, *2067*, 190, *2313*, 190, *2306*, 244, *3132*
Linko, Y.Y., 150, *1664*, 166, *1959*, 167, *1978*, 190, *2313*, 244, *3132*
Linnecar, K.E.M., 64, *240*
Lintner, C.J., 80, *325*, 82, *325*
Liptak, A., 257, *3370*
Lis, H., 2, *7*
Lisbonne, M., 128, *1406*
Lisitsyn, D.I., 36, *63*
Little, C.O., 83, *391*, 260, *3407*
Little, J.A., 207, *2578*
Liu, A., 193, *2345*
Liu, C.Q., 150, *1681*
Liu, D., 206, *2540*, 232, *2959*
Liu, H., 193, *2349*, 207, *2573*, 232, *2959*
Liu, H.S., 117, *1231*
Liu, J., 64, *80*, 68, *313*, 92, *548*, 222, *2835*

Liu, Q., 72, *664*, 82, *377*, 165, *1948*, 254, *3320*
Liu, S., 139, *1503*, 139, *1520*, 193, *2346*, 245, *3168*
Liu, W., 255, *3342*
Liu, W.H., 91, *535*, 237, *3047*
Liu, X., 193, *2347*, 193, *2349*
Liu, X.D., 72, *675*
Liu, Y., 2, *9*, 65, *297*, 92, *550*, 97, *823*, 123, *1307*, 189, *2259*, 263, 265, *3476*
Liu, Y.C., 216, *2710*
Liu, Y.Q., 215, *2697*, 217, *2697*
Liu, Z., 149, *1649*, 150, *1676*, 222, *2829*, 232, *2960*
Livesey, G., 261, *3416*, 262, *3420*
Lloyd, J.R., 262, *3461*, 267, *3599*
Lloyd, N.E., 123, *1308*, 244, *3126*
Łobarzewski, J.L., 109, *1125*, 109, *1126*
Lo Leggio, L., 114, *1189*
Lo, Y.C., 246, *3169*
Lockwood, L.B., 65, *259*, 209, *259*, 232, *2961*
Loes, M., 160, *1891*
Loftsson, T., 65, *290*
Loginova, L.G., 72, *656*, 73, *712*, 73, *720*, 74, *720*, 76, *720*
Logue, S.J., 100, *882*
Lohmar, R.L. Jr., 80, *320*, 161, *1905*
Lohnienne, T., 72, *643*
Lohscheidt, M., 145, *1568*
Loiseau, G., 236, *3020*, 237, *3036*
Lombai, G., 262, *3457*, 265, *3457*
Lonner, C., 202, *2466*
Lonsane, B.K., 72, *668*, 73, *668*, 72, *670*, 127, *1397*, 183, *1397*, 213, *2668*
Looten, P.J., 129, *1428*
Loper, D.C., 260, *3407*
Lopez, C., 171, *2007*, 171, *2008*, 171, *2009*
Lopez, E., 246, *3171*
Lopez-Munguia, A., 89, *492*, 238, *3053*, 239, *3053*
López-Romero, E., 27, *106*
Lopez-Ulibarri, R., 158, *1842*, 159, *1842*
Lorand, J.P., 4, *21*
Lorenzen, S., 202, *2467*

Loseva, A., 146, *1575*
Losman, B., 11, *64*
Lovgren, T., 98, *832*
Lovsin-Kumak, I., 267, *3589*
Lowe, M.E., 259, *3379*
Lozano, J.E., 65, *291*, 204, *2501*
Lozano, V., 27, *107*
Lozińska, E., 168, *1996*, 170, *1996*
Lu, D., 237, *3044*
Lu, D.M., 142, *1542*
Lu, H., 126, *1370*
Lu, J., 223, *2845*
Lu, P.C., 100, *881*
Lu, X., 136, *1477*
Lubenow, H., 265, *3530*
Lubiewski, Z., 241, *3087*
Lucca, M.E., 217, *2753*
Ludikhuyze, L., 182, *2167*
Ludvig, L., 64, *230*, 108, *1030*, 109, *1132*, 148, *1612*
Luenser, S.J., 64, *248*
Luis, R., 206, *2558*
Lukačova, V., 75, *780*
Lukin, N.D., 158, *1872*
Lukow, O., 262, *3455*, 264, *3455*
Lundberg, B., 80, *324*
Lundblad, G., 71, *399*, 83, *399*
Lunden, R., 82, *380*, 98, *380*
Luo, F., 191, *2324*, 193, *2344*, 197, *2436*
Luo, H., 211, *2623*
Luo, J., 153, *1728*
Luo, Z., 197, *2436*, 206, *2544*
Lupandina, I.B., 74, *745*, 77, *745*
Lupashin, V.V., 74, *736*
Lusena, C.V., 93, *563*, 218, *2643*, 212, *2643*
Lushchik, T.A., 78, *1084*
Lützen, N.W., 212, *2631*
Luzon Morris, D., 81, *364*
Lvov, Y., 190, *2298*
Lynch, V.M., 5, *27*
Lynn, A., 64, *238*, 176, *238*
Lysyuk, F.A., 176, *2086*

M

Ma, D., 226, *2883*
Ma, G.T., 222, *2838*
Ma, H., 245, *3168*
Ma, K.J., 125, *1347*
Ma, Q., 255, *3342*
Ma, S.C., 100, *881*
Ma, W., 238, *3056*
Ma, Y.F., 100, *882*
MacAllister, R.V., 189, *2252*, 244, *3126*
MacDonald, D.G., 80, *339*, 135, *339*, 166, *1961*, 176, *1961*
Macey, A., 145, *1565*
MacGregor, A.W., 64, *208*, 71, *408*, 85, *408*, 86, *424*, 112, *1166*, 113, *1166*, 175, *2049*, 176, *2055*, 186, *2193*, 210, *2612*
MacGregor, E.A., 65, *301*, 86, *424*, 129, *1412*
Macheboeuf, M., 182, *2162*
Machek, F., 237, *3029*, 250, *3029*
Maclean, D.J., 107, *1021*
Macovschi, E., 99, *855*
MacPherson, M., 67, *310*
Mączyński, M., 150, *1674*, 238, *3049*
Madamwar, D., 73, *702*
Madamwar, M., 78, *1052*
Madar, Z., 259, *3389*
Madejon Seiz, A., 63, *65*
Madgavkar, A.M., 108, *1033*
Madhwan, T., 268, *3613*
Madicherla, N.R., 226, *2889*
Madsen, G.B., 88, *464*, 148, *1613*
Maeda, H., 109, *1129*, 155, *1759*
Maeda, I., 64, *151*, 115, *1199*, 115, *1200*, 116, *1199*, 120, *1255*
Maeda, S., 227, *2903*
Maesmans, G., 72, *666*, 177, *2105*
Maezawa, T., 148, *1610*, 150, *1666*, 174, *2048*, 178, *2048*, 186, *2201*
Magara, M., 207, *2581*
Mah, R.A., 124, *1325*
Mahalingam, A., 12, *72*, 14, *75*
Mahendran, S., 165, *1944*
Maheshwari, R., 91, *543*, 92, *543*, 107, *543*
Mahmood, S., 78, *1075*

Mahomed, N., 93, *572*
Maidanets, O.M., 105, *976*, 151, *1708*
Maiorella, B.L., 218, *2763*
Maiullo, D.A., 259, *3391*
Majumdar, S.K., 105, *983*, 111, *983*, 215, *2699*, 235, *2995*
Majumdar, S.L., 212, *2644*
Makarious, A.G., 197, *2439*
Makeev, D.M., 217, *2721*
Makela, H., 190, *2313*
Makela, M., 251, *3249*, 257, *3367*
Makela, M.J., 251, *3253*
Makesh, B., 74, *765*
Makino, A., 219, *2780*
Makino, S., 159, *1883*, 159, *1884*
Makoto, S., 83, *396*
Makrinov, I.A., 124, *1330*
Maksimova, E.A., 215, *2696*
Maksimovič, V.M., 110, *1140*
Malcata, F.X., 143, *1551*
Maldakar, N.K., 221, *2818*
Maldakar, S.N., 221, *2818*
Maletto, P., 259, *3391*
Malfait, M.H., 217, *2739*
Maliarik, M.J., 109, *1115*, 157, *1818*, 157, *1819*
Malkin, V.A., 217, *2734*
Malkki, V., 139, *1505*
Malloch, J.G., 167, *1970*
Malloy, T.P., 109, *1115*, 157, *1818*, 157, *1819*
Malyi, G.D., 108, *1031*
Malzahn, R.C., 150, *1662*
Mamaril, F.P., 188, *2233*
Mammem, M., 12, *73*
Mamo, G., 72, *476*, 72, *672*, 78, *1077*, 88, *476*
Man, J., 72, *664*
Mandai, T., 246, *3174*
Mandavilli, S.N., 218, *2774*
Manelius, R., 167, *1977*, 172, *2012*, 174, *2039*, 174, *2040*, 176, *2078*, 176, *2084*, 183, *2172*, 264, *3504*, 265, *3550*
Mangallum, A.S.K., 79, *1108*
Mangels, C.E., 95, *602*

Maninder, K., 198, *2444*, 206, *2443*
Manjunath, P., 105, *984*, 78, *1066*
Manna, S., 66, *303*
Mannelli, L.D.C., 30, *109*
Mannen, S., 187, *2216*
Manners, D.J., 64, *150*, 64, *245*, 99, *849*, 119, *1242*, 265, *3548*
Manning, G.B., 72, *678*, 72, *679*
Mano, D.S., 219, *2789*
Mansfeld, H.W., 120, *1257*
Manstala, P., 80, *338*
Manyailova, I.I., 107, *1007*
Marashi, S.A., 187, *2208*
Marc, A., 137, *1487*, 147, *1487*, 190, *2299*, 190, *2300*
Marcazzan, M., 267, *3586*
Marchal, L.M., 64, *242*, 129, *1443*, 163, *1911*, 180, *2139*
Marchetti, R., 62, *13*, 64, *13*
Marchis-Mouren, G., 130, *1451*, 130, *1457*, 187, *2212*
Marchylo, B.A., 167, *1967*
Marcipar, A., 127, *1374*
Marco, J.L., 72, *686*, 73, *686*
Marcussen, J., 64, *255*
Mardon, C.J., 213, *2667*
Marek, M., 264, *3521*, 265, *3535*
Margo, A.A., 63, *45*, 124, *45*
Marin Alberdi, M.D., 63, *65*
Marin, B., 73, *709*, 79, *1096*, 79, *1099*
Marinchenko, V.A., 212, *2641*, 224, *2863*
Marion, D., 158, *1841*
Markakis, P., 244, *3148*
Markosyan, L.S., 124, *1327*
Marko-Varga, G., 264, *3510*
Marlida, Y., 79, *1154*, 106, *1002*
Marmarelis, V.Z., 255, *3333*
Marmiroli, N., 78, *1085*
Marotta, M., 112, *1172*, 114, *1172*, 190, *2276*
Marques da Silva, I., 158, *1853*
Marrant, K., 192, *2337*
Marseille, J.P., 196, *2409*, 196, *2410*
Marshal, J.J., 260, *3401*

Marshall, J.J., 64, *133*, 64, *169*, 78, *585*, 94, *585*, 101, *898*, 107, *585*, 112, *585*, 115, *585*, 115, *1217*, 116, *585*, 116, *1217*, 189, *2257*
Martelli, H.L., 158, *1853*
Martens, F.B., 74, *749*, 77, *749*, 212, *2648*
Martens, I.S.H., 204, *2494*
Mårtensson, K., 102, *915*, 102, *917*, 143, *1550*
Martensson, K.B., 157, *1796*
Martin, C., 260, *3406*
Martin, J.D., 72, *648*
Martin, J.J. Jr., 95, *602*
Martin, M.L., 198, *2453*
Martin, M.T., 64, *176*
Martin, V.D., 98, *842*, 267, *3587*
Martin-Carron, N., 65, *296*, 261, *296*
Martinez, A., 235, *3007*
Martinez, J., 73, *737*, 74, *737*
Martinez, R.G., 147, *1596*
Martinez-Anaya, M.A., 198, *2449*, 199, *2455*, 200, *2455*, 201, *2455*
Martini, A., 74, *747*, 77, *747*
Martini, A.V., 74, *747*, 77, *747*
Martinkova, L., 237, *3029*, 250, *3029*
Martino, A., 112, *1172*, 114, *1172*, 190, *2276*
Martynenko, N.S., 180, *2138*
Marumoto, H., 214, *2687*, 214, *2688*, 236, *3023*
Marumoto, K., 80, *316*
Maruo, B., 114, *1195*, 116, *1195*, 117, *1227*, 118, *1227*, 247, *1227*, 247, *3192*, 247, *3193*, 247, *3194*, 248, *3197*
Maruo, S., 155, *1760*, 240, *3066*
Maruo, T., 78, *1012*, 99, *860*, 107, *1012*
Marushima, H., 256, *3349*
Maruta, I., 180, *2132*
Maruta, K., 246, *3176*
Maruyama, A., 207, *2569*
Maruyama, Y., 88, *453*, 90, *515*, 90, *517*, 90, *518*
Marvanova, L., 96, *627*
Masaki, H., 254, *3304*
Masaki, K., 180, *2136*
Masamoto, M., 78, *1070*

AUTHOR INDEX

Mase, T., 252, *3266*, 252, *3281*, 252, *3282*
Masenko, L.V., 150, *1669*
Maslova, O.P., 211, *2619*
Maslow, H.L., 93, *576*
Maslyk, E., 127, *1391*
Masuda, H., 77, *945*, 103, *945*
Masuda, K., 157, *1808*
Masuda, Y., 42, *142*, 43, *145*
Masumoto, Y., 196, *2426*
Masure, M.P., 87, *443*
Masuyoshi, S., 37, *130*, 37, *132*, 37, *134*
Mateo, C., 110, *1149*
Mather, E.J., 64, *76*
Matheson, N.K., 242, *3103*
Mathew, S., 247, *3187*
Mathews, J.A., 263, *3477*
Mathias, P.M., 263, *3469*, 264, *3469*
Mathur, E.J., 64, *77*
Mathur, S.N., 88, *465*, 99, *465*
Matin, A., 223, *2844*
Matioli, G., 250, *3232*, 251, *3259*
Matos da Silva, J.C., 72, *676*
Matsubara, K., 111, *1157*, 113, *1157*
Matsubara, T., 95, *609*, 121, *609*
Matsubayashi, T., 167, *1968*
Matsuda, I., 206, *2537*, 206, *2538*
Matsuda, K., 79, *1095*, 145, *1567*
Matsuda, M., 206, *2542*
Matsudaira, M., 149, *1645*, 166, *1954*, 193, *2354*
Matsui, H., 111, *1163*, 112, *1165*, 113, *1165*
Matsui, T., 92, *550*
Matsukawa, R., 263, *3472*
Matsumo, R., 106, *995*
Matsumoto, K., 137, *1485*
Matsumoto, M., 218, *2767*
Matsumoto, N., 217, *2725*, 223, *2849*
Matsumoto, S., 95, *605*, 95, *606*
Matsumoto, T., 182, *2163*, 182, *2164*, 216, *2715*, 227, *2903*
Matsumura, M., 140, *1526*
Matsumura, Y., 219, *2781*, 221, *2812*
Matsunaga, K., 227, *2916*
Matsunaga, M., 79, *1104*

Matsunaga, S., 172, *2019*
Matsuno, R., 79, *1102*, 109, *1102*
Matsusue, T., 202, *2474*
Matsuura-Endo, C., 163, *1916*
Matsuzawa, M., 195, *2389*, 252, *3276*, 257, *3365*
Matsuzawa, T., 206, *2548*, 206, *2557*
Matthews, B., 100, *877*
Mattiasson, B., 97, *813*, 158, *1846*, 214, *2685*
Mattoon, J.R., 64, *74*, 125, *74*
Mattson, P., 251, *3249*, 257, *3367*
Mattson, P.T., 251, *3253*
Matu, S.D., 225, *2882*
Matušova, A., 268, *3611*
Maurer, H.W., 153, *1723*, 153, *1724*
Maurer, K.H., 196, *2415*, 196, *2417*, 196, *2418*, 196, *2419*, 196, *2420*, 196, *2421*, 196, *2422*, 196, *2424*, 204, *2422*
Maurer, W., 65, *293*
Mauro, D., 151, *1705*
Mauui, H.E.H., 107, *1015*
Mazik, M., 3, *15*, 15, *15*, 16, *76*, 16, *77*, 16, *78*, 16, *79*, 16, *87*, 17, *80*, 17, *81*, 17, *82*, 18, *83*, 18, *84*, 18, *85*, 19, *86*, 19, *88*
Mazo, J.C., 243, *3123*, 244, *3123*
Mazur, N.S., 119, *1244*, 186, *2189*, 217, *2723*
Mba, C.L., 74, *763*, 76, *763*, 79, *763*
McBain, A.J., 268, *3613*
McBurney, L.J., 163, *1918*
McCleanhan, W.S., 249, *3224*
McCleary, B.V., 75, *586*, 78, *586*, 94, *586*, 104, *586*, 107, *586*, 112, *586*, 114, *586*, 116, *586*, 262, *3453*, 262, *3455*, 263, *3473*, 264, *3455*, 268, *3604*, 268, *3605*, 268, *3606*
McClure, F.J., 183, *2182*
McConiga, R.E., 194, *2383*
McCord, J.D., 127, *1395*, 190, *2292*, 239, *3060*
McCready, R.M., 247, *3189*
McDermott, E.E., 198, *2445*
Mcevoy, J.J., 263, *3469*, 264, *3469*
McGarvey, S.M., 262, *3430*
McGhee, J.E., 222, *2834*

McGill, L.A., 206, *2560*
McGuigan, C., 11, *70*, 11, *71*
McGuigan, H., 71, *367*, 81, *367*
McHale, R.H., 267, *3601*
McHugh, J.J., 73, *695*, 77, *695*
McKain, N., 227, *2914*, 228, *2914*
McKenzie, E.H.C., 78, *1079*
McMahon, H.E.M., 73, *717*, 94, *582*
McMahon, J., 11, *69*
McMahon, J.B., 30, *108*
McMullan, W.H., 64, *171*
McRae, D., 111, *1158*, 113, *1158*
McRae, J., 222, *2831*
McTigue, M.A., 90, *523*
Meahger, M.M., 189, *2265*
Mecham, D.K., 71, *432*, 86, *432*
Medda, S., 88, *481*, 130, *1455*
Mednokova, A.P., 74, *745*, 77, *745*
Medvecki, D., 149, *1642*
Meertens, L., 11, *52*, 48, *52*
Meeuse, B.J.D., 64, *115*, 176, *2054*
Meites, S., 260, *3399*
Mejborn, H., 260, *3400*
Mekvichitsaeng, P., 236, *3015*
Melasniemi, H., 72, *645*, 92, *553*
Meliawati, R., 232, *2965*
Mello, T., 30, *109*
Melnichenko, L.A., 108, *1114*
Melnikov, B.N., 195, *2399*
Melo, F.R., 71, *637*
Mencinicopshi, G., 222, *2833*
Mendicino, J., 76, *929*
Mendow, E., 236, *3022*
Mendu, D.R., 226, *2889*
Meng, X., 208, *2602*
Meng, Y., 179, *2116*
Mercer, G.D., 72, *646*
Mercier, C., 64, *155*, 262, *3425*, 265, *3549*
Merrill, A.T., 259, *3380*
Mersmann, A., 139, *1502*, 161, *1502*
Meschonat, B., 206, *2554*, 208, *2599*
Meyer, F., 77, *949*, 103, *949*
Meyer, H., 158, *1859*
Meyer, K., 64, *222*

Meyer, K.H., 82, *371*, 98, *831*, 98, *846*, 125, *1342*, 139, *1508*, 139, *1509*, 160, *1892*, 168, *1987*, 203, *2475*, 203, *2476*, 242, *3095*, 242, *3096*, 263, *3470*
Meyer, X.M., 211, *2616*, 210, *2616*
Meyerhof, O., 234, *2977*, 234, *2978*
Meyers, G.E., 124, *1329*
Mezghani, M., 72, *677*
Miah, M.N., 78, *1067*, 79, *1067*
Miceli, A., 219, *2788*
Miceli, M.C., 12, *74*
Michael, J., 239, *3061*
Michalak, S., 219, *2792*
Michelena, V.V., 96, *622*
Michelin, M., 79, *1092*, 220, *1092*
Middlehoven, W.J., 79, *1110*
Middlehover, W.J., 75, *793*, 96, *623*
Miecznikowski, A., 224, *2868*
Miecznikowski, A.H., 235, *3002*, 235, *3009*
Mifune, K., 173, *2026*
Mikami, B., 72, *647*, 79, *647*
Mikami, R., 100, *875*, 176, *875*
Mikulikova, D., 122, *1306*
Mikuni, K., 90, *511*, 223, *2852*, 249, *3220*, 255, *3332*
Miles, M.J., 176, *2088*, 179, *2130*
Milewski, J., 224, *2868*
Millan, S., 111, *1158*, 113, *1158*
Miller, B.L., 21, *95*
Miller, B.S., 198, *2447*
Miller, C., 64, *77*, 64, *82*
Miller, C.A., 208, *2600*
Miller, F.D., 150, *1693*, 150, *1694*, 214, *1693*, 216, *2717*, 217, *1694*, 217, *2731*
Miller, L.M., 242, *3089*, 242, *3090*
Milne, E., 64, *167*, 66, *167*
Milne, E.A., 71, *408*, 85, *408*
Milner, J., 151, *1706*
Milner, J.A., 72, *648*
Milosavič, N.B., 110, *1140*
Milovskaya, V.F., 161, *1897*
Mimura, K., 159, *1883*, 159, *1884*
Min, K.H., 255, *3344*
Minagawa, T., 247, *3190*

AUTHOR INDEX

Minakawa, T., 192, *2334*
Minamiura, N., 95, *609*, 121, *609*, 127, *1385*
Mindell, F.M., 83, *387*
Minegishi, I., 72, *654*
Mineki, S., 77, *953*, 77, *1008*, 103, *953*, 107, *1008*
Mino, Y., 164, *1933*
Minoda, Y., 74, *611*, 74, *612*, 74, *613*, 74, *614*, 74, *764*, 75, *775*, 95, *611*, 95, *612*, 95, *613*, 95, *614*
Miranda, M., 121, *1274*, 139, *1518*
Miron, J., 121, *1274*, 227, *2902*
Mironova, N.I., 119, *1243*, 127, *1393*, 137, *1486*, 174, *2046*
Miroshnik, V.O., 151, *1708*
Mirovich, A.I., 119, *1243*, 127, *1393*, 137, *1486*, 180, *2138*
Misawa, M., 251, *3242*
Mischnick, P., 265, *3551*
Mishra, R.S., 91, *543*, 92, *543*, 107, *543*
Mishra, S., 64, *179*, 119, *179*
Mis'kiewicz, T., 110, *1137*, 110, *1137*, 219, *2778*, 244, *3133*
Misra, R., 81, *366*, 82, *366*
Misra, U., 9, *868*, 76, *868*
Mitchell, D.A., 79, *1103*
Mitchell, G.A., 264, *3485*
Mitchell, G.E. Jr., 83, *391*, 260, *3407*
Mitchinson, C., 64, *181*, 149, *1632*, 196, *2416*, 196, *2423*
Mitchinson, G., 188, *2234*
Mitsugi, K., 101, *890*
Mitsuhashi, M., 154, *1740*, 157, *1808*, 157, *1812*, 205, *2521*, 205, *2526*
Mitsui, T., 71, *410*, 85, *410*
Mitsuiki, S., 129, *1413*, 217, *2752*
Mitsutake, T., 188, *2225*
Miura, S., 234, *2982*, 234, *2986*
Miwa, N., 110, *1148*
Miyado, S., 155, *1759*
Miyake, T., 34, *111*, 34, *114*, 34, *117*, 34, *118*, 112, *1175*, 207, *2567*, 241, *3080*, 241, *3081*, 241, *3085*, 246, *3175*, 258, *3373*
Miyamoto, H., 206, *2551*

Miyamoto, T., 150, *1688*, 157, *1813*
Miyanaga, M., 218, *2767*
Miyashita, Y., 122, *1290*
Miyauchi, T., 250, *3239*, 255, *3341*
Miyazaki, K., 240, *3066*
Miyazaki, N., 252, *3280*
Miyazaki, S., 223, *2849*
Miyoshi, S., 98, *837*, 98, *838*, 102, *913*, 115, *913*, 116, *913*, 117, *913*, 246, *3179*
Miyzazaki, A., 254, *3327*
Mizerak, R.J., 115, *1201*, 157, *1826*
Mizukura, T., 110, *1136*
Mizuno, H., 156, *1784*
Mizuno, K., 79, *1104*, 240, *3076*
Mizuno, M., 120, *1248*
Mizuta, K., 72, *640*, 75, *789*
Mo, X., 225, *2880*
Mochizuki, Y., 149, *1623*
Modenes, A.N., 230, *2938*
Moermann, J.E., 259, *3388*
Mohamed, A., 175, *2053*
Mohamed, M.I.A., 238, *3051*
Mohamed, R., 238, *3051*
Mohan, R., 64, *178*, 65, *178*
Mohapatra, H., 64, *121*
Moheno-Perez, J.A., 192, *2332*
Mohsin, S.M., 150, *1691*
Mojovič, L., 222, *2841*
Molenda-Konieczny, A., 136, *1474*
Molendijk, L., 89, *485*, 96, *485*
Molhant, A., 214, *2672*
Molin, D.R., 149, *1652*
Molinier, J., 188, *2229*, 188, *2230*, 188, *2231*
Moll, M., 137, *1487*, 147, *1487*
Möller, E., 64, *216*
Monceaux, P., 223, *2854*
Moneti, G., 19, *89*, 20, *90*, 23, *99*
Monma, M., 106, *999*, 106, *1000*, 105, *988*, 106, *988*, 223, *2852*
Monma, N., 88, *459*
Monsigny, M., 2, *6*
Monteiro, C.D., 220, *2799*
Monteiro, N. Jr., 126, *1362*
Montero, C.I., 117, *1220*

Montesinos, T., 224, *2857*, 224, *2865*
Montet, D., 235, *3005*
Montgomery, M.W., 206, *2560*
Monti, R., 67, *307*
Moon, T.W., 207, *2568*
Moore, G.R.P., 164, *1930*
Moo-Young, M., 212, *2642*
Moracci, M., 112, *1172*, 114, *1172*
Moradian, A., 245, *3158*
Moraes, I.O., 72, *687*, 150, *1680*
Moraes, L.A., 79, *1092*, 220, *1092*
Morales, M.D., 262, *3436*, 263, *3436*
Morales Villena, L.J., 167, *1971*
Mora-Montes, H.M., 27, *106*
Morana, A., 114, *1187*
Moranelli, F., 212, *2643*, 218, *2643*
Morcombe, C.R., 43, *148*
Moreau, R.A., 223, *2843*
Moreau, Y., 130, *1457*
Morehouse, A.L., 149, *1638*, 150, *1662*, 150, *1663*
Morgan, A., 64, *142*, 245, *142*
Morgan, F.J., 72, *663*
Morgan, H.W., 88, *461*, 115, *1210*, 115, *1211*, 116, *1210*, 129, *1431*
Morgan, J.E., 175, *2049*, 186, *2193*
Morgan, K., 172, *2013*
Morgumov, I.G., 65, *277*
Morgunov, I.G., 237, *3046*
Mori, H., 64, *89*, 106, *1005*, 111, *1157*, 113, *1157*
Mori, M., 72, *655*
Mori, S., 112, *1170*, 113, *1170*, 122, *1296*, 239, *3063*, 252, *3266*, 252, *3281*, 252, *3282*
Mori, T., 11, *56*, 30, *108*
Mori, Y., 216, *2711*
Moriarty, M.E., 74, *739*, 77, *739*
Morikawa, M., 72, *525*, 90, *525*
Morimoto, K., 11, *57*, 226, *2886*
Morishita, H., 125, *1337*
Morishita, M., 242, *3105*
Morita, H., 79, *1097*, 79, *1104*
Morita, M., 236, *3011*
Morita, N., 164, *1936*, 198, *2446*, 203, *2484*, 216, *1936*
Morita, S., 153, *1726*
Morita, T., 151, *1702*, 151, *1703*, 159, *1882*, 240, *3074*
Morita, Y., 76, *928*, 100, *886*
Moriwaki, C., 250, *3232*
Moriyama, H., 227, *2916*
Moriyama, M., 201, *2460*
Moriyama, S., 79, *1102*, 109, *1102*
Moriyama, T., 37, *132*
Morlon-Guyot, J., 129, *1417*, 234, *2989*, 236, *3021*
Morohashi, S., 64, *231*
Morrice, F., 260, *3404*
Morrill, J.L., 182, *2166*
Morris, V.J., 64, *130*, 129, *130*, 179, *2130*
Morsi, M.K., 150, *1691*
Morsy, M.S., 195, *2397*
Morton-Guyot, J., 234, *2990*
Moruya, S., 237, *3034*
Morya, V.K., 78, *1050*
Morzillo, P., 246, *3180*
Mose, W.L., 146, *1577*, 210, *1577*
Moskalova, E.N., 151, *1713*
Mosu, C., 222, *2833*
Motoshio, A., 259, *3387*
Moukamnerd, C., 212, *2650*
Moulin, G., 74, *738*, 75, *587*, 77, *738*, 94, *587*, 96, *634*, 104, *587*, 105, *634*, 106, *1003*, 106, *1004*, 217, *2739*
Mounfort, D.O., 71, *401*, 85, *401*
Mountain, J.L., 81, *366*, 82, *366*
Mountfort, D.O., 67, *309*, 67, *312*, 68, *312*
Mozhayev, V.V., 188, *2222*
Mracec, M., 225, *2882*
Mu, D., 171, *2010*
Mücke, J., 224, *2867*
Mudrak, T.E., 224, *2863*
Muehlemann, H.R., 259, *3388*
Mueller, F., 217, *2728*
Mueller, H., 217, *2728*, 244, *3134*
Mueller, J., 231, *2946*
Mueller, R., 235, *3010*

Mueller, W.C., 217, *2731*
Mueller, W.H., 244, *3149*
Mueller-Stoll, W.R., 193, *2360*
Mugford, D.C., 262, *3453*, 262, *3455*, 264, *3455*
Mugibayashi, N., 101, *887*
Muhammed, S.K.S., 170, *2004*
Mukai, K., 65, *286*, 137, *1483*, 241, *3085*, 246, *3177*, 257, *3372*
Mukaiyama, F., 71, *400*, 83, *400*
Mukerjea, R., 188, *2223*
Mukerji, B.K., 125, *1335*
Mukherjee, A.K., 89, *495*
Mukherjee, G., 235, *2995*
Mukherjee, S., 168, *1984*
Mukherjee, S.K., 79, *1093*
Mukopadhyaya, A., 267, *3595*
Mulder, J., 145, *1569*
Müller, H., 244, *3135*, 244, *3146*
Muller, R., 210, *2615*
Muller, W.C., 150, *1693*, 150, *1694*, 214, *1693*, 217, *1694*, 216, *2717*
Mullertz, A., 208, *2594*
Mun, S.H., 166, *1955*
Mura, K., 197, *2438*
Murad, M.A., 121, *1274*
Murado, M.A., 75, *777*, 139, *1518*, 139, *1519*, 227, *2902*
Murakami, H., 213, *2651*
Murakami, S., 141, *1536*
Murakami, T., 35, *126*, 37, *133*, 38, *126*
Muraki, N., 79, *1091*
Muralikrishna, G., 86, *433*, 86, *434*, 176, *434*
Muramatsu, K., 146, *1578*, 210, *1578*
Muramatsu, M., 89, *499*
Muramatsu, Y., 74, *728*
Murao, T., 78, *1082*
Murata, M., 77, *945*, 103, *945*
Murata, S., 111, *1163*
Murata, Y., 252, *3274*
Murayama, W., 157, *1811*
Muroi, N., 129, *1421*
Murthy, V.S., 75, *768*, 75, *769*, 95, *610*
Murugapoopthy, T., 109, *1131*

Murukami, H., 161, *1896*
Musaev, D.S., 74, *736*
Mushnikova, L.N., 237, *3037*, 237, *3038*
Muslin, E.H., 147, *1591*
Mussulman, W.C., 155, *1778*
Muto, N., 241, *3081*
Muylwijk, C.M., 96, *623*
Myashita, Y., 91, *536*
Myers, D.K., 242, *3103*
Myhara, R.M., 78, *1073*
Myllarinen, P., 167, *1977*
Myrbäck, K., 63, *37*, 63, *66*, 64, *118*, 64, *147*, 64, *148*, 64, *192*, 64, *196*, 64, *219*, 71, *418*, 80, *317*, 80, *318*, 80, *319*, 80, *321*, 80, *322*, 80, *323*, 80, *324*, 80, *327*, 80, *328*, 80, *329*, 80, *330*, 80, *331*, 80, *332*, 80, *333*, 81, *368*, 82, *368*, 82, *373*, 82, *374*, 82, *380*, 82, *381*, 85, *418*, 86, *423*, 93, *568*, 94, *591*, 94, *592*, 98, *829*, 98, *833*, 98, *380*, 98, *381*, 99, *850*, 99, *862*, 100, *872*, 120, *1258*, 127, *1387*, 129, *37*, 145, *862*, 145, *1564*, 151, *1709*, 161, *1904*, 163, *1921*, 172, *2021*, 172, *2022*, 181, *2149*, 186, *2206*, 188, *2249*, 214, *2673*, 250, *3234*, 250, *3235*, 266, *3557*
Myrbäck, S., 188, *2249*
Myrrey, H.E., 2, *3*, 4, *3*
Myszka, D.G., 12, *72*

N

Na, B.G., 105, *989*
Na, B.K., 218, *2770*, 218, *2775*, 224, *2770*
Nabetani, H., 190, *2280*
Nachtergaele, W., 194, *2369*
Nacu, A., 194, *2373*
Naczk, M., 78, *1073*
Naele, R., 177, *2101*
Nagahama, T., 75, *783*
Nagai, K., 127, *1386*
Nagai, M., 150, *1688*
Nagamoto, H., 121, *1280*
Nagamura, T., 78, *1069*, 79, *1069*
Naganawa, H., 34, *110*, 37, *110*
Nagano, H., 252, *3272*

Naganum, M., 232, *2952*
Nagao, S., 263, *3472*
Nagasaka, Y., 78, *1081*, 79, *1091*
Nagata, T., 105, *988*, 106, *988*, 176, *2058*
Nagatomo, S., 195, *2389*
Nagatsuka, K., 197, *2435*
Nagayama, T., 89, *503*
Nagayasu, K., 78, *1070*
Nagle, N., 223, *2843*
Nagpal, V., 96, *804*
Nagulwar, V., 251, *3250*
Nagura, M., 72, *654*
Nahar, S., 106, *993*
Nahete, P.N., 110, *1143*
Naigai Shokuhin Koki, K.K., 225, *2877*
Nain, L., 266, *3561*
Naito, H., 159, *1885*
Naito, T., 34, *113*, 42, *113*
Naka, K.I., 268, *3608*
Nakada, T., 119, *1237*, 204, *2508*, 240, *3070*, 246, *3177*
Nakagawa, R., 117, *1229*
Nakagawa, Y., 3, *17*, 15, *17*, 42, *142*, 43, *145*, 43, *149*
Nakai, H., 111, *1157*, 113, *1157*
Nakai, T., 149, *1624*
Nakai, Y., 267, *3591*
Nakajima, A., 246, *3172*
Nakajima, E., 218, *2767*
Nakajima, I., 126, *1361*
Nakajima, M., 190, *2280*, 240, *3073*
Nakajima, R., 93, *567*
Nakajima, T., 38, *135*, 90, *511*, 150, *1688*, 157, *1813*, 249, *3220*
Nakajima, Y., 156, *1784*
Nakakuki, T., 88, *459*, 89, *499*, 97, *816*, 119, *1239*, 204, *2506*, 205, *2529*
Nakamura, A., 129, *1427*, 254, *3304*
Nakamura, H., 95, *599*, 188, *2235*, 188, *2236*
Nakamura, K., 190, *2296*, 223, *2849*
Nakamura, L.K., 236, *3013*
Nakamura, M., 90, *515*, 99, *867*, 115, *1200*, 115, *1199*, 116, *1199*, 120, *1255*, 149, *1624*, 248, *3195*, 248, *3196*, 248, *3199*

Nakamura, N., 64, *104*, 65, *279*, 74, *728*, 91, *534*, 91, *542*, 118, *1233*, 118, *1234*, 119, *1239*, 119, *1240*, 150, *1233*, 150, *1683*, 165, *1949*, 195, *2389*, 246, *3178*, 251, *3243*, 251, *3244*, 251, *3255*, 252, *3276*, 253, *3300*, 254, *3308*, 256, *3360*, 257, *3365*
Nakamura, S., 43, *146*, 43, *147*
Nakamura, T., 126, *1357*, 141, *1536*, 176, *2089*, 205, *2509*
Nakamura, Y., 97, *819*, 217, *2751*
Nakanishi, H., 41, *138*
Nakanishi, K., 126, *1353*, 223, *2856*, 228, *2929*, 229, *2929*, 230, *2929*, 247, *3183*, 250, *3241*, 251, *3258*, 255, *3346*
Nakanishi, N., 238, *3054*
Nakano, H., 89, *503*, 112, *1178*, 114, *1178*
Nakao, E., 105, *986*
Nakao, H., 106, *1005*
Nakao, M., 129, *1418*
Nakashima, H., 45, *150*
Nakata, Y., 92, *558*
Nakatani, H., 106, *995*, 143, *1548*
Nakatani, N., 174, *2043*, 176, *2043*
Nakaya, T., 91, *528*
Nakazima, A., 262, *3426*, 265, *3426*
Nakhapetyan, L.A., 107, *1007*
Nakhmanovich, B.M., 139, *1514*, 211, *2619*, 211, *2620*, 212, *2639*, 215, *2695*, 217, *2730*, 230, *2933*, 230, *2940*, 231, *2940*
Nam, K.D., 219, *2785*, 220, *2797*, 226, *2785*
Nam, M.S., 89, *504*
Nampathiri, K.M., 80, *336*, 82, *336*
Nampoorthiri, K.M., 72, *649*
Nampoothiri, K.M., 75, *778*
Nanakui, T., 73, *715*
Nanakuki, T., 204, *2505*
Nanasi, P., 75, *772*
Nanda, G., 73, *719*, 101, *899*, 188, *2227*
Nandi, R., 79, *1093*
Nandini, B.N., 74, *765*
Nanji, D.R., 153, *1732*, 153, *1733*
Nanmori, T., 76, *930*, 90, *513*, 90, *514*, 99, *514*, 101, *900*

Napier, E.J., 77, *951*, 79, *951*, 103, *951*
Nara, P.L., 30, *108*
Narayanamurti, D., 188, *2246*
Narayanan, A.S., 74, *735*
Narinx, E., 72, *642*
Narita, J., 64, *98*
Narita, K., 75, *781*
Narita, Y., 122, *1288*
Naruse, Y., 245, *3165*
Nascimento, H.J., 78, *1047*
Nashimoto, J., 157, *1813*
Nasim, A., 93, *563*
Nasr, H., 260, *3404*
Nativi, C., 19, *89*, 20, *90*, 21, *91*, 21, *92*, 21, *93*, 21, *94*, 23, *99*, 30, *109*
Natsui, K., 235, *3000*
Navarro, J.M., 213, *2668*, 224, *2857*, 224, *2865*
Naveena, B.J., 234, *2988*, 235, *2988*
Navez, A.E., 135, *1469*
Navikul, O., 249, *3210*
Navrodskaya, L.I., 215, *2696*
Nawa, H., 236, *3018*
Naylor, N.M., 98, *842*, 100, *880*, 188, *2232*
Nazalan, M.N., 78, *1064*
Nazalan, N., 74, *756*
Nazarets, E.P., 237, *3037*
Neale, R.J., 64, *240*
Nebelung, G., 225, *2871*
Nebesny, E., 121, *1282*, 122, *1293*, 161, *1902*, 161, *1906*, 161, *1908*, 162, *1902*, 163, *1909*, 166, *1962*, 166, *1963*, 186, *2198*, 244, *3151*, 245, *3151*, 245, *3161*, 245, *3162*
Negrut, G., 225, *2882*
Nelson, G.E.N., 126, *1355*
Nelson, J.M., 221, *2819*
Nelson, N.J., 266, *3573*
Nelson, N.M., 11, *54*
Nelson, R.P., 65, *258*, 145, *258*
Nelson, W., 149, *1621*
Nelson, W. Jr., 244, *3126*
Nemestothy, N., 190, *2284*
Nene, S., 251, *3260*
Nene, S.N., 123, *1318*
Nesiewicz, R.J., 196, *2406*
Nestyuk, M.N., 99, *854*, 188, *2250*
Netsu, Y., 97, *812*, 212, *2645*
Neugebauer, W., 183, *2178*
Neuman, S., 191, *2325*
Neumuller, G., 98, *829*
Neurnberger, H., 164, *1937*
Newton, J.M., 100, *880*, 267, *3587*
Neyts, J., 11, *61*
Ng, J.D., 90, *526*
Ngee, C.C., 226, *2893*
Nghiem, N.P., 213, *2658*, 222, *2839*
Nguyen, C.H., 75, *773*
Nguyen, Q., 223, *2843*
Nguyen, T.K., 237, *3029*, 250, *3029*
Ni, J., 178, *2108*
Ni, W., 5, *25*, 5, *26*, 5, *32*, 7, *36*
Niamke, S.L., 71, *638*, 78, *638*, 113, *638*, 122, *638*
Nichols, N.N., 223, *2842*, 223, *2843*, 226, *2899*
Nicholson, R.I., 204, *2491*
Nicol, D., 96, *634*, 105, *634*
Nicola, L., 234, *2991*, 235, *2991*
Nicolas, P., 202, *2473*
Nicolaus, B., 121, *1283*, 246, *3180*
Nicolaus, G., 121, *1283*
Nicolella, C., 230, *2936*
Niedbach, J., 192, *2331*
Niekamp, C.W., 155, *1767*
Nielsen, B.H., 127, *1400*, 129, *1400*, 149, *1630*, 223, *1630*
Nielsen, B.R., 107, *1016*, 107, *1017*
Nielsen, H.K., 123, *1310*
Nielsen, J.B., 206, *2530*
Nielsen, J.E., 64, *75*
Nielsen, R.I., 107, *1016*, 107, *1017*
Nielsen, T., 63, *70*
Nierle, W., 150, *1670*
Nieto, J.J., 73, *703*
Nigam, P., 64, *178*, 64, *180*, 65, *178*, 72, *180*, 73, *180*, 76, *180*, 78, *180*, 78, *1061*, 79, *180*, 79, *1061*, 113, *180*, 114, *180*, 116, *180*, 118, *180*, 217, *2743*, 220, *2806*

Nigam, V.N., 264, *3501*
Niimi, M., 153, *1716*, 157, *1820*
Nikiforova, T.A., 237, *3035*, 237, *3037*, 237, *3038*
Nikolaev, Z.L., 104, *971*, 129, *971*
Nikolič, S., 222, *2841*
Nikolov, Z.L., 103, *954*, 129, *954*
Nikuni, J., 176, *2076*
Nikuni, Z., 78, *1083*, 115, *1199*, 116, *1199*, 163, *1924*, 176, *2062*
Niku-Paavola, M.L., 116, *1208*, 247, *3186*
Nilsson, G., 264, *3512*
Nimi, T., 155, *1765*
Nimmagadda, A., 64, *122*
Ninomiya, K., 212, *2650*
Ninomiya, M., 95, *604*
Ninomyia, H., 172, *2016*, 188, *2239*
Nipkow, A., 76, *938*, 77, *938*
Nirmala, M., 86, *433*, 86, *434*, 176, *434*
Nishida, K., 101, *900*
Nishida, M., 242, *3098*
Nishikawa, A., 120, *1248*
Nishikawa, K., 95, *598*, 101, *900*
Nishikawa, U., 74, *762*
Nishimoto, T., 65, *257*, 65, *286*, 207, *2567*, 257, *3372*
Nishimura, S., 126, *1356*, 172, *2015*, 188, *2247*, 188, *2248*, 247, *3191*
Nishimura, Y., 243, *3109*
Nishina, N., 234, *2981*
Nishinari, K., 264, *3493*
Nishio, M., 3, *18*, 34, *111*, 34, *113*, 34, *114*, 34, *117*, 34, *118*, 36, *127*, 37, *125*, 37, *129*, 37, *130*, 40, *137*, 42, *113*
Nishira, H., 101, *900*
Nishiyama, K., 179, *2126*
Nithyanandham, S., 258, *3374*
Nitsch, C., 196, *2418*, 196, *2419*, 196, *2420*, 196, *2421*, 196, *2422*, 196, *2424*, 204, *2422*
Niu, Y.L., 193, *2363*
Noda, T., 102, *918*, 105, *988*, 106, *988*, 163, *1916*, 176, *2058*
Nodiya, M., 182, *2158*

Noelting, G., 98, *831*
Nogrady, N., 257, *3370*
Noguchi, T., 158, *1850*, 159, *1850*
Nohr, J., 129, *1415*
Nojiri, K., 206, *2562*
Nojiri, M., 149, *1624*
Nolasco-Hipolito, C., 190, *2277*, 236, *3016*, 236, *3017*
Nomura, G., 246, *3179*
Nomura, J., 249, *3220*
Nomura, M., 145, *1555*, 178, *2109*
Nomura, N., 213, *2652*
Nomura, Y., 235, *3000*
Noor, B.M., 192, *2338*
Noordam, B., 72, *657*, 95, *616*, 108, *657*, 159, *1878*
Noorwez, M., 79, *527*, 91, *527*, 106, *527*, 119, *527*
Norjehan, N.R., 170, *1999*, 207, *1999*
Norman, B.E., 64, *90*, 64, *247*, 88, *464*, 92, *547*, 101, *894*, 116, *1206*, 157, *1798*
Norrild, J.C., 9, *46*, 9, *47*
Norton, G., 64, *240*
Novais, J.M., 189, *2269*, 189, *2270*, 189, *2274*, 190, *2274*
Novellie, L., 81, *348*
Novichkova, Yu.V., 195, *2398*
Nowacka, I., 110, *1139*
Nowak, J., 213, *2656*, 219, *2778*
Nowak, M.A., 11, *65*
Nowak, S., 125, *1341*, 224, *2867*
Nowakowska, K., 155, *1780*, 205, *2523*, 205, *2524*, 205, *2525*
Noworyta, A., 132, *1458*, 143, *1549*
Nowotny, F., 127, *1389*, 154, *1738*
Nozawa, Y., 37, *133*
Nsereko, V.L., 227, *2914*, 228, *2914*
Nugey, A.L., 148, *1606*, 149, *1606*
Numata, K., 3, *19*, 35, *126*, 36, *127*, 38, *126*, 38, *135*, 40, *137*
Nunez, M.J., 189, *2260*
Nurmi, K., 174, *2039*, 174, *2040*
Nwosu, V.C., 244, *3145*
Nyachoti, C.M., 208, *2602*

Nybergh, P.M.A., 64, *85*
Nycander, G., 80, *322*
Nyman, M., 202, *2466*
Nyunina, N.N., 158, *1872*

O

Oakley, E.T., 262, *3458*, 263, *3458*, 265, *3458*
Oates, C.G., 170, *1998*, 170, *2001*, 170, *2002*, 176, *1998*
Obara, M., 261, *3414*
Obata, K., 95, *618*, 223, *2856*
Oberon, X.S., 238, *3053*, 239, *3053*
O'Brian, P.J., 223, *2842*
Oda, Y., 219, *2795*
O'Dea, K., 206, *2546*
Odibo, F.J.C., 75, *792*, 76, *940*, 117, *792*
Odibo, J.F.C., 74, *753*, 76, *753*
O'Donoghue, E., 64, *82*, 112, *1174*
Oensan, Z.I., 217, *2757*
Oezbek, B., 125, *1343*, 168, *1986*
Ofoli, R.Y., 135, *1463*, 139, *1463*, 190, *2308*, 190, *2314*
Ogasahara, K., 72, *681*, 72, *683*
Ogasawara, H., 201, *2459*
Ogasawara, J., 157, *1809*, 205, *2520*, 205, *2521*
Ogata, M., 73, *715*, 97, *816*, 219, *2794*
Ogawa, H., 240, *3066*
Ogawa, K., 192, *2340*, 239, *3058*
Ogawa, T., 207, *2569*
Oginoya, R., 264, *3520*
Oguma, T., 119, *1245*
Ogura, M., 149, *1623*
Oh, B.H., 89, *487*, 252, *3273*
Oh, Y.J., 234, *2975*
Ohashi, A., 127, *1376*
Ohashi, K., 265, *3541*
Ohata, K., 118, *1236*
Ohba, R., 104, *969*, 105, *969*, 107, *969*
Ohdan, K., 129, *1430*
Ohisa, N., 79, *1153*
Ohkouchi, Y., 236, *3024*

Ohkuma, H., 34, *113*, 34, *114*, 37, *129*, 40, *137*, 42, *113*
Ohlsson, E., 81, *350*, 187, *2218*, 187, *2219*
Ohmori, K., 8, *37*
Ohnishi, H., 77, *1036*
Ohnishi, M., 64, *215*, 74, *762*, 129, *215*, 190, *2297*
Ohno, N., 78, *1088*
Ohno, T., 79, *1153*
Ohnuma, T., 37, *132*
Ohta, A., 37, *134*
Ohta, K., 217, *2746*
Ohta, S., 37, *129*, 204, *2503*
Ohta, T., 77, *1036*
Ohta, Y., 73, *718*, 75, *788*, 170, *2000*, 188, *2240*, 226, *2896*
Ohtani, N., 110, *1135*, 116, *1135*, 262, *3426*, 265, *3426*
Ohya, T., 64, *152*, 252, *3282*
Oi, S., 73, *696*, 77, *696*, 124, *1331*
Oishi, K., 97, *816*
Oiwa, H., 232, *2952*
Oka, H., 136, *1476*
Oka, M., 38, *136*, 40, *136*, 266, *3558*
Oka, S., 192, *2342*, 193, *2355*, 193, *2356*
Okabe, M., 234, *2981*, 234, *2982*
Okada, G., 89, *499*
Okada, M., 34, *110*, 37, *110*, 190, *2305*
Okada, S., 65, *281*, 79, *1098*, 91, *538*, 98, *837*, 98, *838*, 117, *1224*, 117, *1225*, 129, *1430*, 150, *1673*, 205, *2517*, 207, *2566*, 248, *3208*, 251, *3263*, 251, *3264*, 253, *3297*, 253, *3298*, 253, *3299*
Okada, T., 110, *1138*, 150, *1685*, 158, *1866*, 190, *2279*, 190, *2303*, 254, *3319*, 254, *3317*, 255, *3330*
Okada, Y., 240, *3068*
Okafor, B.U., 75, *792*, 117, *792*
Okafor, N., 75, *792*, 76, *940*, 117, *792*
Okamoto, A., 122, *1288*
Okamoto, T., 207, *2569*
Okamoto, Y., 227, *2905*
Okano, K., 64, *98*
Okano, T., 267, *3584*

Okazaki, H., 93, *573*, 102, *903*, 102, *904*, 102, *905*, 102, *906*, 102, *907*, 102, *908*, 120, *1264*, 122, *1298*, 122, *1299*, 141, *1532*, 181, *2147*, 262, *3424*
O'Keefe, B.R., 11, *56*, 11, *69*, 30, *108*
Okemoto, H., 91, *532*
Okemoto, T., 165, *1953*
Oki, T., 3, *19*, 11, *52*, 34, *111*, 34, *113*, 34, *114*, 34, *115*, 34, *117*, 34, *118*, 34, *119*, 34, *120*, 34, *121*, 34, *122*, 34, *123*, 34, *124*, 35, *126*, 36, *127*, 36, *128*, 37, *115*, 37, *125*, 37, *129*, 37, *130*, 37, *131*, 37, *132*, 37, *133*, 37, *134*, 38, *126*, 38, *135*, 38, *136*, 40, *136*, 40, *137*, 41, *138*, 41, *141*, 42, *113*, 45, *150*, 45, *152*, 46, *153*, 47, *154*, 48, *52*, 213, *2653*
Okkerse, C., 61, *1*
Okolo, B.N., 74, *753*, 74, *763*, 76, *753*, 76, *763*, 79, *763*, 168, *1991*
Oku, K., 65, *286*, 257, *3372*
Okuba, M., 186, *2201*
Okubo, M., 148, *1610*, 150, *1666*, 174, *2048*, 178, *2048*
Okuda, H., 261, *3414*
Okuda, K., 268, *3608*
Okumura, H., 257, *3368*
Okumura, K., 140, *1525*
Okumura, N., 221, *2824*
Okura, T., 207, *2567*
Okuyama, K., *11*, 62
Okuyama, M., 106, *1005*, 111, *1157*, 113, *1157*
Okuyama, S., 11, *57*, 36, *128*, 37, *130*, 37, *131*
Olered, R., 64, *234*
Olesienkiewicz, A., 179, *2117*
Oliveira de Nascimento, J.R., 99, *866*, 101, *866*
Oliver, S.G., 217, *2756*
Olkku, J., 139, *1505*, 190, *2306*
Ollivier, B., 124, *1325*
Oloffson, S., 11, *64*
Olsen, H.S., 64, *114*, 212, *2634*, 213, *2669*, 222, *2836*
Olsen, J., 75, *795*
Olson, J., 75, *794*
Olson, W.J., 99, *856*
Olsson, U., 186, *2194*
Olutiola, P.O., 74, *754*
Omaki, T., 158, *1858*
Omar, S.H., 72, *662*
Omemu, A.M., 95, *619*, 105, *619*
Omidiji, O., 73, *706*, 73, *707*, 77, *950*, 103, *950*
Omori, T., 127, *1381*
Omori, Y., 246, *3172*
Omunsky, Z., 259, *3389*
Onda, M., 190, *2298*
One, T., 141, *1536*
O'Neill, S.P., 64, *168*, 189, *2258*
Oner, E.T., 213, *2664*, 217, *2756*
Ongen, G., 97, *822*
Onishi, H., 74, *732*
Onishi, K., 89, *503*, 160, *1890*
Ono, K., 78, *1065*
Ono, N., 75, *771*
Onodera, S., 261, *3414*
Onoe, A., 207, *2566*
Onteniente, J.P., 173, *2033*, 241, *3088*
Onuma, S., 232, *2952*
Ookuma, K., 206, *2537*, 206, *2538*
Oosterbaan, R.A., 64, *214*
Ooya, R., 240, *3069*, 252, *3266*, 252, *3281*
Oparin, A.I., 188, *2250*, 248, *3202*
Opwis, K., 195, *2400*
Ordorica Falomir, C., 168, *1994*
Oren, A., 77, *1038*
Orestano, G., 71, *409*, 81, *349*, 85, *409*, 138, *1493*, 138, *1495*, 174, *2045*
Orlando, P., 121, *1283*
Orlova, L.A., 211, *2620*, 212, *2639*
Örtenblad, B., 63, *37*, 80, *327*, 80, *328*, 80, *329*, 80, *331*, 93, *568*, 99, *862*, 129, *37*, 145, *862*, 145, *1564*, 214, *2673*
Ortez, C., 110, *1149*
Orts, W.J., 64, *243*, 68, *243*, 175, *2053*
Osawa, T., 2, *6*
Ostaszewicz, D., 193, *2361*
Östdal, H., 202, *2468*

AUTHOR INDEX

Ostergaard, J., 244, *3141*
Ostern, P., 243, *3112*
Ostwald, G., 156, *1791*, 217, *1791*
O'Sullivan, J., 127, *1399*
Otani, Y., 120, *1268*, 125, *1349*, 126, *1349*, 125, *1350*, 126, *1350*, 147, *1594*, 153, *1717*, 154, *1755*, 172, *2017*, 264, *3498*
Otero, R.R.C., 217, *2758*
Otomo, K., 110, *1148*
Otsubo, K., 193, *2339*, 192, *2339*
Ottinger A.F., 194, *2378*
Otto, E., 211, *2625*
Ottombrino, A., 114, *1187*
Otzen, D., 64, *75*
Ou, J., 207, *2580*
Ou, Y., 150, *1657*, 206, *1657*
Outtrup, H., 101, *894*, 92, *547*
Ouyang, S., 138, *1497*
Owen, W.L., 215, *2705*
Oya, R., 122, *1296*
Oya, T., 156, *1794*, 135, *1466*
Oyeka, C.A., 76, *940*
Ozawa, H., 74, *614*, 95, *614*
Ozawa, J., 78, *1049*, 79, *1049*
Ozbek, B., 133, *1462*, 149, *1650*, 165, *1462*, 181, *2146*
Oztekin, R., 245, *3164*

P

Paavilainen, S., 251, *3249*
Padmasree, D.V.S., 110, *1150*
Pagan, R.J., 212, *2636*
Pagenstedt, B., 93, *571*
Pak, S.P., 93, *577*
Pal, A., 10, *48*
Pal, S.C., 72, *661*
Palacha, Z., 262, *3447*, 263, *3447*, 264, *3447*
Palacios, H.R., 203, *2480*, 203, *2481*
Palaniappyan, L., 258, *3374*
Palas, J.S., 78, *1090*
Palasinski, M., 164, *1940*
Palat, H.W., 259, *3391*
Palde, P.B., 21, *95*
Pallas, J., 76, *929*

Pan, C., 225, *2876*
Pan, L.H., 234, *2984*, 235, *2984*
Pan, L.J., 234, *2984*, 235, *2984*
Pan, W.Z., 124, *1324*
Pan, Y.C., 125, *1344*
Pan, Y.Z., 71, *402*, 85, *402*
Panasyuk, T.E., 171, *2011*
Panda, S.H., 235, *3003*, 235, *3006*
Pandey, A.K., 64, *131*, 64, *178*, 65, *178*, 72, *649*, 75, *778*, 78, *131*, 78, *1053*, 78, *1054*, 78, *1055*, 78, *1056*, 78, *1057*, 78, *1058*, 78, *1059*, 78, *1060*, 78, *1063*, 79, *131*, 80, *336*, 82, *336*, 103, *131*, 266, *3561*
Pandey, M., 224, *2859*
Pandit, A.B., 97, *805*
Pang, Y., 201, *2463*
Pankiewicz, U., 190, *2289*
Pankratov, A.Ya., 211, *2620*
Pannecouque, C., 11, *69*
Pannell, L.K., 30, *108*
Paolucci-Jeanjean, D., 139, *1507*, 190, *2288*, 190, *2290*
Paquet, V., 73, *693*
Para, A., *9*, 61
Parada, J.L., 73, *709*, 235, *3007*
Paramahans, S.V., 103, *942*, 177, *2063*, 176, *2063*
Paraschiv, M., 234, *2991*, 235, *2991*
Paredes Lopez, O., 168, *1994*
Paredes-Lopez, O., 64, *159*, 168, *1992*
Parekh, R.S., 224, *2860*
Parekh, S.R., 223, *2851*, 224, *2860*
Paris, G.M.S., 243, *3108*
Parisi, F., 230, *2934*
Park, B.C., 149, *1634*
Park, C.S., 64, *74*, 91, *541*, 125, *74*
Park, D.C., 240, *3071*, 254, *3307*, 254, *3313*
Park, D.H., 218, *2770*, 218, *2775*, 224, *2770*
Park, J.A., 234, *2975*
Park, J.H., 207, *2568*
Park, J.S., 189, *2267*
Park, J.T., 137, *1488*, 137, *1489*
Park, J.Y., 238, *3057*, 239, *3062*

Park, K.H., 72, *669*, 89, *487*, 91, *541*, 123, *1320*, 168, *1989*, 207, *2568*, 248, *3207*, 252, *3273*, 267, *3590*,
Park, S.K., 213, *2657*, 214, *2679*, 226, *2679*
Park, S.W., 158, *1844*
Park, T.H., 245, *3166*
Park, Y.H., 234, *2975*, 234, *2981*, 251, *3261*
Park, Y.K., 75, *787*, 104, *967*, 108, *1025*, 110, *1141*, 122, *1292*, 204, *2494*, 220, *2799*, 223, *2847*, 251, *3261*
Parker, K., 244, *3145*
Parkkinen, E.E.M., 108, *1022*
Parmanick, M., 235, *3006*
Parsiegla, G., 254, *3305*
Pasari, A.B., 96, *635*, 106, *635*
Pashchenko, L.P., 139, *1513*
Pastrana, L.M., 75, *777*, 142, *1541*, 171, *2007*, 171, *2008*, 171, *2009*, 227, *2902*
Paszczyński, A., 109, *1126*
Patel, B., 248, 3209
Patel, B.K.C., 88, *461*
Patel, S., 73, *702*
Patil, N.B., 121, *1273*
Patil, V.B., 121, *1273*
Patkar, A., 253, *3287*
Patkar, A.Y., 123, *1318*, 123, *1319*
Patnaik, P.R., 93, *566*, 107, *566*
Pattaragulwanit, S., 236, *3015*
Patwardhan, N.V., 85, *414*, 87, *414*
Pauletig, M., 255, *3398*, 260, *3398*
Paulson, B.A., 150, *1687*
Pavgi-Upadhye, S., 243, *3107*
Payen, A., *14*, 62
Payre, N., 64, *89*
Pazlarova, J., 264, *3503*
Pazos, C., 268, *3607*
Pazur, J.H., 64, *199*, 79, *1098*, 82, *382*
Pearce, J., 227, *2904*
Peat, S., 80, *335*, 99, *848*, 115, *1196*, 115, *1197*, 116, *1196*, 119, *848*, 119, *1240*, 119, *1241*, 120, *1251*, 120, *1252*, 120, *1253*, 127, *1378*, 145, *1565*, 172, *2013*, 242, *1251*, 242, *3097*, 248, *1251*, 248, *1252*, 248, *1253*, 248, *3205*, 248, *3206*
Peckova, J., 264, *3521*

Pedersen, B.O., 240, *3072*
Pedersen, L.H., 65, *298*, 253, *3292*
Pedersen, S., 64, *91*, 64, *97*, 91, *530*, 121, *530*, 157, *1827*, 207, *2584*, 212, *2634*, 222, *2836*
Peeples, T.L., 157, *1832*
Peglow, K., 224, *2858*
Peiris, P., 227, *2904*
Peitonen, S., 197, *2434*
Peixoto, S.C., 88, *584*, 94, *584*
Pelissari, F.M., 250, *3232*
Pellicano, T., 112, *1172*, 114, *1172*
Pen, J., 89, *485*, 96, *485*
Peng, M., 68, *313*, 68, *315*
Penner, D.W., 157, *1818*, 157, *1819*, 158, *1834*
Penninga, D., 129, *1429*
Pepe de Moraes, L.M., 217, *2753*
Pepsin, M.J., 108, *1023*, 213, *2671*
Peralta, R.M., 74, *759*
Perego, P., 230, *2934*, 232, *2958*, 245, *3167*
Perevozchenko, I.I., 74, *751*
Pérez-Pérez, M.J., 27, *107*
Perlstein, H., 148, *1599*
Perminova, L.V., 189, *2263*, 189, *2264*, 109, *1123*
Perry, G.H., 81, *366*, 82, *366*
Persia, M.E., 75, *786*, 97, *786*
Person, B., 186, *2206*
Persoz, J.F., *14*, 62
Persson, B., 161, *1904*
Perttula, M., 114, *1192*
Peruffo, A.D.B., 116, *1218*
Pessela, B.C., 110, *1149*
Petach, H.H., 114, *1191*
Peterson, W.H., 124, *1321*
Petit, P., 63, *38*, 127, *1380*, 144, *1380*, 182, *1380*
Petroczy, J., 205, *2515*
Pettijohn, O.G., 232, *2961*
Petzel, J.P., 119, *1246*
Peumans, W., 11, *61*, 11, *63*, 11, *66*, 11, *67*, 11, *68*, 30, *68*
Peumans, W.J., 11, *52*, 48, *52*
Pfau, K., 72, *641*
Pfeifer, G.T., 264, *3486*

Pfeiffer, D., 264, *3516*, 264, *3517*, 264, *3518*
Pfost, H.D., 182, *2166*
Philia, M., 128, *1403*
Philip, R.G., 244, *3143*
Piao, J., 223, *2845*
Pickens, C.E., 155, *1767*
Pielecki, J., 64, *126*
Pieper, H.J., 65, *269*, 147, *1593*, 210, *2610*, 210, *2614*
Pierzgalski, T., 121, *1275*, 127, *1394*, 155, *1780*, 163, *1909*, 166, *1962*, 179, *2124*, 205, *2523*, 205, *2524*, 205, *2525*
Pietkiewicz, J., 237, *3030*
Pietschke, B.I., 243, *3116*
Pigman, W.W., 122, *1286*
Pikhalo, D.M., 108, *1031*
Pilar Gonzalez, M., 227, *2902*
Pilgrim, C.E., 64, *84*, 64, *101*
Piller, K., 114, *1191*
Pinchukova, E.E., 78, *1084*
Pincussen, L., 135, *1466*
Pingaud, H., 210, *2616*, 211, *2616*
Pinheiro, L.I., 125, *1348*, 126, *1354*
Pintado, J., 129, *1417*, 235, *3004*
Pinto da Soilva Bon, E., 126, *1362*
Pires, E.M.V., 74, *730*
Pirkotsch, M., 195, *2393*
Pirt, S.J., 99, *848*, 119, *848*, 119, *1241*
Pirttijarvi, T.S.M., 112, *1168*
Pisareva, M.S., 163, *1923*, 186, *2189*
Pisheto, I., 75, *521*, 90, *521*
Pishtiyski, I., 252, *3267*
Pitcher, W.H. Jr., 109, *1118*
Pittner, F., 204, *2498*
Piukovich, S., 254, *3315*
Pivarnik, L.F., 71, *398*, 83, *398*
Piyachamkwon, K., 249, *3210*
Plaksin, G.V., 109, *1123*
Planchot, C., 74, *758*
Planchot, V., 64, *160*, 70, *617*, 95, *617*, 104, *617*, 129, *1415*, 165, *1947*, 166, *1947*, 177, *2096*, 203, *2482*
Plankenhorn, E., 249, *3216*, 249, *3217*
Plant, A.R., 88, *461*, 115, *1210*, 115, *1211*, 116, *1210*, 129, *1431*

Plou, F.J., 64, *176*
Pocsi, I., 257, *3370*
Podgórska, E., 266, *3562*
Podgorski, W., 237, *3030*
Pohjonen, E., 64, *85*
Polak, F., 85, *419*
Polanyi, L., 196, *2415*
Poli, A., 121, *1283*
Polikashin, Yu.V., 217, *2735*
Polizeli, M.L., 67, *307*, 79, *1092*, 220, *1092*
Polizeli, M.T.M., 88, *584*, 94, *584*
Pollaczek, H., 163, *1917*
Pollard-Knight, D.V., 264, *3524*
Polo, A., 113, *1184*
Polyakov, V.A., 217, *2735*
Pomarańskązuka, W., 262, *3447*, 263, *3447*, 264, *3447*
Pomeranz, V., 160, *1894*, 177, *1894*
Pomeranz, Y., 93, *570*, 188, *2233*
Pomeroy, D., 158, *1853*
Pompejus, M., 145, *1568*
Ponce-Noyola, P., 27, *106*
Ponchel, G., 174, *2038*, 174, *2042*
Pongjanta, J., 249, *3210*
Pongsawasdi, P., 249, *3219*
Ponte, J.G. Jr., 198, *2451*
Popa, J., 110, *1134*
Popadich, I.A., 119, *1243*, 127, *1393*, 137, *1486*, 167, *1980*, 174, *2046*, 180, *2138*
Popadich, J.A., 176, *2086*
Popov, I.D., 266, *3566*
Popovic, M., 235, *3010*
Porodko, F.M., 178, *2110*
Porro, D., 218, *2761*
Poso, A., 197, *2434*
Posternak, T., 163, *1917*
Postier, P.H.E., 149, *1618*
Pothiraj, C., 227, *2909*
Potter, O.E., 262, *3468*
Potter, R.S., 186, *2185*
Pouliquen, B., 259, *3382*
Poulivong, S., 78, *1079*
Poulsen, P.B., 63, *70*
Pourquie, J., 235, *3001*, 235, *3005*

Poutanen, K., 116, *1208*, 167, *1974*, 167, *1975*, 167, *1976*, 176, *2067*, 197, *2440*, 259, *3383*
Powell, A.D., 170, *1998*, 176, *1998*
Power, D.S., 150, *1687*
Power, S.D., 196, *2423*
Pozdnyakova, T.A., 237, *3037*, 237, *3038*,
Pozen, M.A., 146, *1572*
Pozerski, E., 22, 62, *31*, 63, *2122*, 179, *2122*
Prabhu, K.A., 109, *1130*, 244, *3152*
Pradeep, S., 158, *1860*
Prado, F.C., 237, *3039*
Pradyumnarai, V.P., 207, *2574*
Prasad, V.S., 242, *3091*, 243, *3091*
Prasnjak, V.B., 151, *1713*
Pratt, K.L., 215, *2702*
Praznik, W., 264, *3505*
Preece, I.A., 64, *117*, 120, *1267*
Preignitz, R.D., 208, *2591*
Preiswerk, E., 160, *1892*
Prema, P., 256, *3356*
Prescott, L.M., 209, *2604*
Press, J., 139, *1509*
Preston, T., 259, *3381*
Pretorius, I.S., 64, *93*, 64, *183*, 217, *2758*
Prevost, J., 217, *2718*
Pries, F.G., 93, *579*
Priest, F.G., 72, *663*, 90, *506*,
Prieto, J.A., 73, *737*, 74, *737*, 117, *1226*
Primarin, D., 73, *718*
Pringsheim, H., 63, *61*, 63, *63*, 63, *64*, 81, *351*, 82, *378*, 83, *390*, 163, *378*, 163, *390*, 188, *2237*
Privalov, P.L., 75, *798*
Prodanov, E., 187, *2212*
Prodanović, R.M., 110, *1140*
Proess, G., 8, *43*
Pronin, S., 86, *426*
Pronin, S.I., 94, *589*, 137, *1490*, 138, *1494*, 180, *2137*, 180, *2141*
Prosvetova, T.I., 217, *2730*
Proth, J., 211, *2616*, 210, *2616*
Protzmann, J., 177, *2101*
Provost, D., 151, *1705*

Prudhomme, J.C., 173, *2034*, 173, *2035*
Przybył, M., 267, *3592*
Przybylski, D.K., 149, *1620*
Przylas, I., 123, *1317*
Puglisi, P.P., 78, *1085*
Puls, J., 266, *3574*
Punnett, P.W., 81, *358*, 87, *358*
Punpeng, B., 92, *558*
Purgatto, E., 99, *866*, 101, *866*
Purr, A., 214, *2680*
Pusey, M.L., 90, *526*
Pye, E.K., 135, *1464*
Pykhova, S.V., 211, *2622*, 212, *2640*, 215, *2694*, 217, *2622*, 217, *2721*, 217, *2730*, 217, *2721*
Pyun, Y.R., 234, *2975*

Q

Qadee, M.A., 101, *897*
Qadeer, M.A., 91, *537*
Qi, B., 193, *2345*
Qi, Q., 254, *3306*
Qi, X., 64, *164*, 175, *164*
Qian, R., 136, *1480*
Qian, Y., 65, *266*, 207, *2583*
Qiao, M., 217, *2738*
Qin, F., 72, *664*
Qin, X.X., 222, *2838*
Qin, Z., 172, *2012*, 176, *2078*, 176, *2084*, 265, *3550*
Qiu, F., 206, *2535*
Quadeer, M.A., 101, *897*, 217, *2736*, 218, *2736*
Quan, Y., 97, *823*
Quareshi, N., 231, *2951*
Quarles, J.M., 205, *2516*
Quax, W.J., 89, *485*, 96, *485*
Queiroz, M.C.R., 78, *1048*
Qui, S., 2, *9*
Quilez, M.A., 173, *2032*
Quivey, R.G. Jr., 63, *69*

R

Raabe, E., 138, *1501*
Rachlitz, K.V., 149, *1639*

Radaeli, G., 183, *2181*
Radhakrishnan, S., 78, *1059*, 78, *1060*
Radosavljevic, M., 165, *1942*
Radosta, S., 150, *1656*, 180, *2135*, 192, *2336*
Radu, S., 79, *1154*, 106, *1002*
Radunz, W., 16, *78*, 16, *79*
Rafaeli, G., 129, *1446*
Ragab, A.M., 73, *722*
Raghavan, D., 264, *3525*
Raghavandra, M.R.R., 78, *1066*
Rahman, S.U., 89, *497*
Raibaud, P., 73, *701*
Raimbault, M., 73, *709*, 79, *1099*, 125, *1348*, 126, *1354*, 213, *2668*, 235, *3004*
Rainey, F.A., 112, *1168*
Raitaru, G., 234, *2991*, 235, *2991*
Raja, K.C.M., 255, *3345*, 256, *3356*
Rajan, A., 174, *2037*, 242, *3091*, 243, *3091*
Raju, K.J., 110, *1150*
Rakin, M., 222, *2841*
Ram, K.A., 150, *1689*
Ramachandran, N., 75, *768*, 75, *769*
Ramaiya, P., 76, *933*
Ramakrishna, S.V., 107, *1009*, 219, *2782*, 255, *3345*, 256, *3356*
Ramakrishnan, C.V., 64, *149*
Raman, N., 227, *2909*
Ramaseh, N., 95, *610*
Ramazanov, Yu.A., 189, *2273*
Ramesh, M.V., 72, *668*, 72, *670*, 73, *668*, 127, *1397*, 183, *1397*
Ramesh, V., 102, *922*
Ramirez, A., 214, *2686*
Ramstad, P.E., 177, *2099*, 179, *2099*, 203, *2486*
Rand, A.G., 71, *398*, 83, *398*
Randez-Gil, F., 73, *737*, 74, *737*
Ranhotra, G.S., 263, *3475*
Rani, R.R., 101, *899*
Ranjbar, B., 187, *2208*
Ranjit, N.K., 236, *3025*, 236, *3026*
Ranzi, B.M., 218, *2761*
Rao, J.L.U.M., 79, *527*, 91, *527*, 106, *527*, 119, *527*

Rao, K.R.S.S., 64, *122*
Rao, L.V., 224, *2861*, 225, *2861*
Rao, M.B., 243, *3118*
Rao, M.R.R., 105, *984*
Rao, M.V., 104, *972*
Rao, P.S., 101, *892*
Rao, P.V.S., 79, *1107*
Rao, R.J., 110, *1150*
Rao, S.S., 110, *1150*
Rao, V.B., 79, *1107*
Rasap, R., 216, *2712*
Rashba, O.Ya., 96, *624*
Rasković, S., 218, *2771*
Rasmussen, P.B., 64, *96*, 64, *250*, 79, *96*, 111, *96*
Rasmussen-Wilson, S.J., 78, *1090*
Raspor, P., 237, *3042*
Ratanakhanokchai, K., 73, *692*
Rath, F., 146, *1571*, 210, *1571*
Rath, H., 266, *3554*, 266, *3555*
Rausch, K.D., 155, *1758*
Ravi, R., 158, *1840*
Ravindran, V., 208, *2595*
Ray, L., 105, *983*, 111, *983*, 235, *2995*
Ray, R.C., 235, *3003*, 235, *3006*
Ray, R.R., 64, *127*, 73, *719*, 76, *941*, 102, *914*, 188, *2227*
Rayson, G.D., 42, *144*
Razip, M.S., 78, *1064*
Reaimbault, M., 79, *1096*
Recondo, E.F., 248, *3201*
Reczey, K., 153, *1725*
Reddy, C.A, 235, *3008*
Reddy, C.A., 187, *2215*
Reddy, G., 234, *2987*, 234, *2988*, 235, *2988*
Reddy, L.V.A., 213, *2663*, 217, *2663*
Reddy, N.S., 64, *122*
Reddy, O.V.S., 78, *1087*, 213, *2663*, 217, *2663*, 219, *2786*, 225, *2786*
Reddy, P.R.M., 76, *937*, 77, *937*
Reddy, S.M., 96, *625*
Redon, R., 81, *366*, 82, *366*
Reece, J., 233, *2971*
Rees, W.R., 248, *3205*

Reeve, A., 157, *1824*
Regmer, F.E., 244, *3130*
Rehm, H.J., 237, *3045*
Reichel, M., 214, *2680*
Reid, C.A., 187, *2214*, 262, *3421*
Reid, S., 3, *14*, 5, *14*
Reilly, P.J., 64, *153*, 98, *834*, 158, *1864*, 205, *2528*, 244, *2528*
Reinders, M.A., 150, *1682*
Reinikainen, P., 139, *1505*
Reinus, J.F., 11, *52*, 48, *52*
Remesar, X., 81, *362*
Remiszewski, M., 64, *246*, 120, *1254*, 166, *1959*, 218, *2776*, 238, *3049*, 266, 3559
Ren, Y., 249, *3211*
Renaud, A., 262, *3429*
Renders, M., 27, *107*, 47, *154*
Rendleman, J.A. Jr., 173, *2024*, 177, *2024*, 254, *3321*, 256, *3348*
Renken, A., 253, *3289*
Rentshler, D.F., 158, *1839*
Requadt, C., 196, *2416*
Requin, E., 154, *1745*
Rey, M., 76, *933*
Rhee, S.K., 213, *2659*, 220, *2801*, 226, *2890*, 226, *2891*, 226, *2892*, 226, *2893*, 226, *2894*
Ribeiro Correa, T.L., 72, *676*
Ribeiro Coutinho de Oliverira Mansur, L., 72, *676*
Ribeiro do Santos, M.G.G., 217, *2750*
Ribeiro, L., 190, *2284*
Richardson, T., 64, *76*, 64, *77*, 64, *82*
Richey, D.D., 125, *1336*
Richter, G., 159, *1888*, 243, *3124*
Richter, K., 236, *3022*
Richter, M., 64, *211*, 93, *565*, 108, *1026*, 115, *1026*, 137, *1484*, 149, *1635*, 150, *1656*, 156, *1783*, 158, *1859*, 164, *1937*, 166, *1958*, 180, *2135*, 192, *2336*, 262, *3464*, 263, *3464*
Richter, R., 244, *3138*
Richter, T., 182, *2161*
Riddiford, C.L., 89, *505*

Riddle, L., 11, *58*, 12, *58*
Riegels, M., 195, *2393*
Rietved, K., 89, *485*, 96, *485*
Rigden, D.J., 71, *637*
Rigo, A., 267, *3586*
Rimbault, M., 202, *2471*
Ring, S., 130, *1453*
Ring, S.G., 130, *1452*, 167, *1976*, 176, *2088*, 179, *2130*
Ringer, C., 196, *2416*
Rios, G.M., 139, *1507*, 140, *1531*, 190, *2288*, 190, *2290*
Rios, R., 243, *3123*, 244, *3123*
Rishpon, J., 102, *902*, 145, *902*, 148, *902*, 151, *902*
Ritte, G., 267, *3599*
Rivera, M.H., 89, *492*
Rivera, M.I., 30, *108*
Rizzeti, A.C., 67, *307*
Robb, D., 72, *688*
Roberts, F.F., 65, *258*, 145, *258*
Roberts, I., 111, *1158*, 113, *1158*
Roberts, S.M., *16*, 62
Robertson, G.H., 64, *243*, 68, *243*
Robin, J.P., 265, *3549*
Robinson, R., 124, *1325*
Robles Medina, A., 167, *1971*
Robyt, J.F., 73, *447*, 88, *447*, 108, *1029*, 117, *1230*, 118, *1230*, 119, *447*, 174, *2044*, 175, *2044*, 177, *2044*, 177, *2097*, 178, *2097*, 182, *2159*, 188, *2223*, 188, *2224*, 239, *447*, 254, *3322*, 266, *3570*
Rocchiccioli, C., 218, *2764*
Roche, J., 242, *3092*, 243, *3110*
Rodda, E.D., 64, *232*, 156, *232*
Rodrigues de Canto, L., 164, *1930*
Rodrigues-Leon, L.A., 237, *3039*
Rodriguez, M.E., 238, *3053*, 239, *3053*
Rodriguez Sanoja, R., 129, *1417*, 236, *3020*
Rodriguez-Sanoja, R., 129, *1414*, 129, *1417*
Rodwell, V.W., 244, *3130*
Rodzevich, V.I., 95, *608*, 99, *852*, 104, *608*, 119, *1244*, 145, *852*, 163, *1923*, 186, *2189*
Rodzevich, V.L., 217, *2723*

Roelens, S., 19, *89*, 20, *90*, 21, *91*, 21, *92*, 21, *93*, 21, *94*, 23, *99*, 30, *109*
Roels, J.A., 140, *1527*
Roff, M., 203, *2490*
Rogalski, J., 63, *71*
Roger Carcassonne-Leduc, P.C., 149, *1619*
Rogers, P., 227, *2904*
Rogers, P.L., 212, *2636*
Rogols, S., 150, *1658*, 164, *1938*, 247, *3188*, 260, *3399*, 265, *3528*
Röhmann, F., 260, *3402*
Rohrbach, R.P., 109, *1115*, 157, *1818*, 157, *1819*, 255, *3328*
Roick, T., 156, *1783*
Roig, P., 27, *105*
Rojas, L.F., 243, *3123*, 244, *3123*
Rolfsmeier, M., 75, *800*, 114, *1188*
Rollings, J.E., 137, *1488*, 137, *1489*
Romaszeder, K., 65, *294*
Rome, M.N., 81, *363*
Romenskii, N.V., 150, *1669*
Romijn, C., 80, *344*
Romouts, F.M., 97, *814*
Rona, P., 120, *1262*
Rongine de Fekete, M., 248, *3200*
Roos, J., 80, *344*
Ropp, T., 196, *2416*
Rosell, C.M., 203, *2478*
Rosen, O., 81, *350*
Rosenberg, M., 64, *107*
Rosenberger, A., 65, *272*
Rosenblum, J.L., 81, *370*, 82, *370*
Rosendal, P., 127, *1400*, 129, *1400*, 149, *1630*, 223, *1630*, 244, *3141*
Rosenfeld, E.L., 178, *2110*
Rosenmund, H., 117, *1228*, 118, *1228*
Rosenthal, F.R., 176, *2089*
Rosicka, J., 166, *1962*, 166, *1963*, 186, *2198*
Ross, A.B., 207, *2587*
Ross, H.A., 111, *1158*, 113, *1158*
Rossi, A.L., 251, *3246*
Rossi, C., 78, *1085*
Rossi, M., 112, *1172*, 114, *1172*, 114, *1187*
Rossling, L., 266, *3554*
Rossnagel, B.G., 167, *1981*, 208, *2601*
Rosso, A.M., 251, *3246*
Rosted, C.O., 80, *342*, 83, *342*
Roszak, H., 213, *2656*
Roth, D.L., 213, *2670*
Rothenbach, E., 126, *1366*
Rotsch, A., 165, *1946*
Roushdi, M., 244, *3131*
Roussel, L., 191, *2315*
Roussos, S., 213, *2668*
Rovinskii, L.A., 230, *2931*, 230, *2932*
Rowe, A., 99, *863*, 101, *863*
Rowe, A.W., 99, *865*
Roy, D.K., 210, *2611*
Roy, F., 76, *876*, 100, *876*
Roy, I., 110, *1152*
Roy, J., 95, *607*
Roy, S.K., 75, *785*
Royal, C.L., 155, *1770*
Rozema, T., 262, *3437*
Rozenboom, H.J., 129, *1429*
Rozfarizan, M., 238, *3052*
Rozie, H., 97, *814*
Ruadze, I.D., *12*, 62, 110, *1144*, 129, *1433*
Ruadze, I.D., 110, *1145*
Ruan, H., 150, *1681*
Rubenstein, B.B., 135, *1469*
Rudd, P.M., 11, *59*
Rudina, N.A., 109, *1123*
Rudlof, I., 8, *43*
Ruediger, A., 114, *1186*, 117, *1221*
Ruessbuelt, I., *46*, 63, *46*, 233
Ruff, J., 150, *1677*
Rugh, S., 63, *70*, 155, *1757*
Rugsaseel, S., 237, *3033*
Rühlemann, I., 236, *3022*
Ruille, P., 74, *729*
Ruiz, B., 129, 1414
Ruiz, M.J., 149, *1652*
Ruiz Teran, F., 236, *3021*
Ruller, E., 79, *1092*, 220, *1092*
Rupp, P.L.C., 164, *1928*
Rusek, J.J., 227, *2913*
Rusek, M-L.R., 227, *2913*

Rusendi, D., 232, *2966*
Rusnati, M., 46, *153*
Russell, R.R.B., 123, *1315*
Russina, H., 71, *428*, 86, *428*
Rusznak, S., 264, *3491*
Ruter, R., 167, *1969*
Ruttloff, H., 74, *742*, 77, *742*, 93, *565*, 156, *1782*, 158, *1859*
Ruttloff, N.A., 215, *2708*, 216, *2709*
Ryoo, J.J., 254, *3326*
Ryu, B.H., 219, *2785*, 220, *2797*, 226, *2785*
Ryu, S., 123, *1316*
Ryu, Y.W., 158, *1845*, 212, *2647*
Ryun, H.W., 234, *2980*
Rzędowski, W., 193, *2361*
Rzepka, E., 224, *2869*, 230, *2939*

S

Saab, G., 238, *3053*, 239, *3053*
Saab-Rincon, G., 89, *492*
Saag, M.S., 11, *65*
Saari, N., 79, *1154*, 106, *1002*, 170, *2004*
Saarinen, P., 150, *1664*
Sabalitscka, T., 186, *2186*, 264, *3489*
Sa-Correia, I., 65, *268*, 92, *559*
Sada, H., 252, *3280*
Sadhukhan, R., 75, *785*
Sadova, A.I., 107, *1006*
Saeki, F., 154, *1737*
Saenger, W., 123, *1317*
Safonov, V.V., 195, *2398*
Saganuma, T., 75, *783*
Saha, B.C., 64, *135*, 102, *910*, 102, *912*, 103, *956*, 104, *962*, 104, *963*, 104, *964*, 105, *964*, 106, *994*, 115, *1205*, 116, *1205*, 156, *1788*, 165, *1950*, 120, *1260*, 130, *1455*, 176, *2094*, 190, *2314*, 205, *2527*, 221, *2811*, 243, *3122*
Sahar, E., 102, *902*, 145, *902*, 148, *902*, 151, *902*
Sahara, H., 125, *1346*
Saharan, B., 224, *2859*
Sahashi, H., 110, *1138*, 158, *1866*, 190, *2279*, 190, *2303*, 255, *3330*

Sahashi, Y., 190, *2282*
Sahashiand, H., 150, *1685*
Sahasrabuddhe, D.L., 127, *1390*
Sahidi, F., 78, *1073*
Sahlstrøm, S., 176, *2059*, 198, *2450*, 207, *2586*
Sahoo, B., 242, *3089*, 242, *3090*
Sai Ram, M., 76, *935*
Saiga, T., 150, *1685*, 158, *1866*, 190, *2279*, 190, *2303*, 254, *3317*, 254, *3319*
Saikai, S., 119, *1237*
Saileh, M.M., 238, *3051*
Saito, K., 244, *3139*
Saito, N., 72, *451*, 88, *451*
Saito, S., 141, *1532*
Saitoh, K., 34, *111*, 34, *114*, 34, *119*, 34, *120*, 34, *123*, 34, *124*, 37, *131*, 265, *3531*
Saitsu, A., 246, *3170*
Saka, M., 217, *2752*
Sakaguchi, K., 243, *3121*
Sakai, H., 77, *1036*
Sakai, K., 263, *3472*, 264, *3520*
Sakai, S., 156, *1790*, 158, *1873*, 204, *2507*, 204, *2508*, 241, *3083*, 254, *3325*, 255, *3329*
Sakai, T., 64, *152*, 216, *2716*
Sakai, Y., 159, *1879*
Sakakibara, M., 213, *2653*
Sakamoto, K., 191, *2318*
Sakamoto, R., 163, *1915*, 240, *3075*
Sakano, Y., 120, *1248*, 240, *3075*
Sakata, T., 259, *3387*
Sakurai, A., 213, *2653*
Sakurai, K., 73, *711*
Salama, J.J., 195, *2390*
Salas, M., 244, *3145*
Salawu, M.B., 207, *2587*, 236, *3027*
Salazar, M.G., 11, *65*
Salazar-Gonzalez, J.F., 11, *65*
Salkinoja-Salonen, M.S., 112, *1168*
Salmanova, I.S., 212, *2640*
Salmanova, L.S., 215, *2694*
Salo, M.L., 149, *1627*
Salomon, N., 158, *1854*, 159, *1854*

Salomonsson, L., 87, *435*
Salwani, D., 170, *2005*, 176, *2005*
Salzet, M., 8, *43*
Sambuichi, M., 264, *3481*
Samchez, C.I., 149, *1652*
Samec, M., 63, *28*, 76, *878*, 81, *28*, 82, *379*, 98, *843*, 100, *878*, 129, *1437*, 145, *1563*, 163, *1920*, 243, *3113*, 250, *3233*
Sampathkumar, S., 2, *10*
Sams, C., 203, *2490*
San Martin, E.M., 223, *2847*
Sanchez, C., 243, *3123*, 244, *3123*
Sanchez, S., 129, *1414*
Sandanayake, K.R.A.S., 5, *29*, 5, *31*
Sander, V., 206, *2554*, 208, *2599*
Sanders, P., 262, *3437*
Sanders, W.T.M., 144, *1553*
Sandhu, I.K., 93, *566*, 107, *566*
Sandiford, C.P., 76, *926*
Sandra, S.W., 150, *1687*
Sandstet T.M., 264, *3487*
Sandstedt, R.M., 63, *33*, 71, *432*, 82, *382*, 86, *432*, 130, *1450*, 176, *2085*, 267, *33*
San-Félix, A., 27, *107*
Sangan, P., 73, *714*
Sani, A., 72, *688*
Sanjay, G., 109, *1122*
Sankar, S., 5, *32*
Sano, K., 101, *890*
Sano, Y., 111, *1157*, 113, *1157*
Sanroman, A., 139, *1518*, 139, *1519*, 189, *2260*
Sansen, W., 72, *667*
Santamaria, R.I., 238, *3053*, 239, *3053*
Sant'anna, G.L. Jr., 189, *2261*
Santimone, M., 130, *1457*
Sanwal, G.G., 242, *3099*
Sanz, P., 73, *737*, 74, *737*, 117, *1226*
Sanz-Aparicio, J., 113, *1184*
Sapronov, A.R., 138, *1498*, 141, *1535*
Sara, M., 72, *684*
Sargeant, J.G., 86, *439*
Sarikaya, E., 100, *875*, 176, *875*
Saris, P.E.J., 112, *1168*

Sarmidi, M.R., 264, *3509*
Saroya, K., 264, *3501*
Sarzana, G., 243, *3111*
Sasaibe, M., 88, *462*, 91, *536*
Sasaki, H., 96, *626*, 107, *1013*, 267, *3593*
Sasaki, I., 72, *683*
Sasaki, K., 252, *3274*, 254, *3310*
Sasaki, S., 201, *2460*, 236, *3018*
Sasaki, T., 125, *1339*, 125, *1340*, 176, *2083*
Sasakura, T., 74, *766*, 64, *231*
Sastry, N.V.S., 79, *1107*
Sata, H., 90, *518*
Sato, F., 106, *1005*
Sato, H.H., 122, *1292*, 204, *2494*, 223, *2847*, 251, *3261*
Sato, K., 153, *1726*, 223, *2849*
Sato, M., 75, *791*, 78, *1012*, 78, *1082*, 107, *1012*, 252, *3271*, 252, *3272*
Sato, S., 192, *2342*
Sato, T., 64, *136*, 75, *773*, 236, *3018*, 257, *3368*
Sato, Y., 11, *57*, 242, *3104*
Satoh, E., 212, *2649*
Satya, S., 266, *3561*
Satyanarayana, T., 79, *527*, 79, *1106*, 91, *527*, 106, *527*, 119, *527*
Satynarayana, T., 72, *658*
Saucedo, C.J., 11, *56*
Saucedo-Castaneda, G., 213, *2668*
Sauder, M.E., 96, *633*, 103, *633*, 106, *633*, 107, *633*
Sauer, J., 64, *89*
Sauter, J.J., 121, *1271*
Savage, A., 263, *3469*, 264, *3469*
Savage, B.J., 211, *2618*
Savin, S.I., 217, *2734*
Sawada, M., 107, *1013*
Sawada, T., 217, *2751*
Sawada, Y., 34, *117*, 34, *118*, 34, *120*, 35, *126*, 37, *125*, 37, *133*, 38, *126*, 38, *135*, 40, *137*
Sawaguchi, Y., 252, *3274*, 254, *3310*
Sawai, T., 241, *3077*
Sawata, M., 250, *3239*, 255, *3341*

Sawicka-Zukowska, R., 68, *314*, 204, *2499*
Sawicki, J.E., 142, *1545*
Saxena, S., 268, *3609*
Scamanna, A.F., 218, *2765*
Scarlata, G., 265, *3529*
Scarpa, M., 267, *3586*
Schaafsma, G., 233, *2972*
Schaeder, W.E., 203, *2487*
Schaefer, B., 63, *43*
Schafhauser, D.Y., 110, *1147*
Schäller, K., 231, *2949*
Schamala, T.R., 235, *2996*
Schardinger, F., 249, *3213*, 249, *3214*, 249, *3215*
Scharloo, W., 261, *3410*
Schechtlowna, Z., 71, *369*, 81, *369*, 82, *369*
Scheerens, J.C., 221, *2819*
Schellart, J.A., 75, *793*, 79, *1110*
Schellenberger, A., 120, *1257*
Scheller, F., 264, *3516*, 264, *3517*, 264, *3518*
Schenberg, A.C.G., 217, *2750*
Schepers, H.J., 88, *460*, 124, *460*
Scheppach, W., 262, *3422*
Scherl, D.S., 255, *3328*
Schierbaum, F., 64, *211*, 93, *565*, 149, *1635*, 150, *1656*, 158, *1859*, 166, *1958*, 192, *2336*
Schierz, V., 217, *2729*
Schild, E., 146, *1582*
Schilling, C.H., 61, *6*, 61, *7*, 61, *8*, 176, *2065*
Schimmelpennink, E.B., 180, *2139*
Schink, N.F., 194, *2383*, 221, *2820*
Schiraldi, C., 190, *2276*
Schirner, R., 156, *1783*
Schleyer, M., 72, *641*
Schmedding, D.J.M., 145, *1559*
Schmeider, W., 65, *299*, 263, *3471*
Schmid, G., 254, *3301*, 255, *3347*
Schmidell, N.W., 78, *1045*, 78, *1048*, 88, *444*, 139, *1521*, 213, *2654*, 217, *2750*, 266, *3568*
Schmidt, C.R., 208, *2593*
Schmidt, D., 156, *1782*
Schmidt, I., 196, *2415*

Schmidt, J., 239, *3061*, 267, *3577*
Schmiedl, D., 180, *2133*, 180, *2134*
Schmith, T., 83, *389*, 161, *1907*
Schmitt, R., 198, *2448*
Schneider, A., 262, *3460*
Schneider, D.R., 244, *3130*
Schnutzenberger, F., 73, *724*
Schnyder, B.J., 244, *3142*
Schoch, T.J., 160, *1895*, 177, *1895*, 176, *1895*, 255, *3334*
Schocher, J., 72, *684*
Schoenebaum, U., 123, *1312*
Schollmeyer, E., 195, *2400*
Schols, D., 11, *52*, 11, *61*, 11, *63*, 11, *66*, 11, *67*, 11, *68*, 27, *107*, 30, *68*, 46, *153*, 47, *154*, 47, *156*, 48, *52*, 48, *157*
Schönbohm, D., 194, *2382*
Schønning, K., 11, *64*
Schoonees, B.M., 135, *1465*
Schoonover, F.D., 203, *2487*
Schopmeyer, H.H., 194, *2372*
Schotting, J.B., 248, *3198*
Schraldi, C., 112, *1172*, 114, *1172*
Schroeder, H., 193, *2360*
Schroeder, S.W., 210, *2612*
Schroh, I., 251, *3246*
Schubert, T.A., 188, *2250*
Schugerl, K., 74, *755*
Schügerl, K., 264, *3515*
Schulein, M., 64, *91*, 92, *547*
Schultz, A.S., 88, *467*, 203, *2487*, 260, *3394*, 264, *3488*
Schulz, A., 93, *569*
Schulz, E., 208, *2596*
Schulz, F.N., 62, *24*
Schulz, G., 224, *2858*
Schulz, G.E., 254, *3305*
Schulz, P., 164, *1937*, 166, *1958*
Schuurman, N.M.P., 11, *55*, 48, *55*
Schwachula, G., 120, *1257*
Schwartz, B., 161, *1907*
Schwartz, S.J., 164, *1928*
Schwarz, P.B., 203, *2480*, 203, *2481*
Schweizerische Ferment A-G, 194, *2376*

Schwengers, D., 155, *1772*, 158, *1843*
Schwermann, B., 72, *641*
Schwimmer, S., 122, *1287*, 122, *1291*
Scott, S.W., 124, *1321*
Sedlak, M., 264, *3503*
Seenayya, G., 73, *698*, 73, *699*, 76, *937*, 77, *937*, 77, *698*, 234, *2987*
Segard, E., 223, *2854*
Segraves, J.M., 215, *2702*
Segre, A., 265, *3529*
Seichert, L., 237, *3029*, 250, *3029*
Seideman, J., 264, *3499*
Seideman, M., 243, *3119*
Seidemann, J., 80, *343*, 267, *3597*
Seidman, M., 155, *1770*, 158, *1849*
Seiji, M., 89, *501*, 89, *502*, 120, *1250*
Seitz, A., 195, *2393*
Seki, J., 45, *151*
Sekikawa, Y., 265, *3532*
Seldin, L., 253, *3293*
Selek, H., 174, *2042*
Selemenev, V.F., 110, *1144*
Selemenev, V.F., 110, *1145*
Seligman, S.A., 186, *2200*
Selitrennikoff, C.P., 78, *1090*
Selmi, B., 158, *1841*
Selvakumar, P., 78, *1053*, 78, *1054*, 78, *1055*, 78, *1056*, 78, *1057*, 78, *1058*
Semimaru, T., 129, *1425*
Semmens, E., 135, *1467*
Semmens, E.S., 135, *1468*, 136, *1475*
Semmes, O.J., 8, *41*
Sen, G.P., 95, *607*
Sen, S., 73, *708*
Senayya, G., 76, *935*
Sene, T., 219, *2795*
Sengupta, S., 96, *631*
Senise, J.T., 191, *2321*
Senkeleski, J.L., 206, *2550*
Senn, T., 65, *269*, 147, *1593*, 210, *2610*, 210, *2614*, 215, *2703*
Sensen, C.W., 112, *1172*, 114, *1172*
Seo, B.C., 72, *669*
Seo, C.W., 239, *3062*

Seo, E.S., 93, *560*
Seong, J.H., 212, *2647*
Serbulov, Yu., 122, *1297*
Serebrinskaya, R.A., 64, *237*
Seregina, L.M., 73, *712*
Seres, G., 249, *3230*, 254, *3315*, 256, *3355*
Serghat-Derradji, H., 173, *2036*
Seri, S.K., 72, *661*
Serna-Saldivar, S.O., 192, *2332*
Serrano, P.O., 149, *1654*
Seto, J., 190, *2297*
Seu, J.H., 217, *2742*
Sewaki, T., 64, *98*
Sezaki, M., 34, *110*, 34, *112*, 37, *110*
Shabbir, S.H., 5, *27*
Shadaksharaswamy, M., 120, *1267*
Shah, A.R., 78, *1052*
Shah, H., 197, *2431*, 197, *2432*, 197, *2433*
Shah, Y.T., 108, *1033*, 189, *2255*, 189, *2256*
Shahani, K.M., 218, *2760*, 224, *2760*
Shahzad-ul-Hussan, S., 47, *155*
Shambe, T., 95, *621*
Shamrin, V.F., 217, *2730*
Shankar, V., 110, *1143*
Shanmugasundaram, E.R.B., 74, *735*
Shann-Tzong, 71, *402*, 85, *402*
Shapiro, E., 63, *61*
Sharma, R., 76, *931*
Sharma, S., 224, *2859*
Sharon, N., 2, *7*, 2, *6*
Sharova, N.Yu., 206, *2559*, 237, *3035*, 237, *3038*
Shatangeeva, N.I., 151, *1708*
Shaver, R.D., 262, *3456*
Shaw, G.M., 11, *65*
Shaw, J.F., 74, *557*, 92, *557*
She, Y., 268, *3603*
Sheehan, H., 268, *3604*, 268, *3605*
Shellenberger, J.A., 160, *1894*, 177, *1894*
Shen, A., 237, *3044*
Shen, D., 179, *2116*, 190, *2311*
Shen, G., 76, *938*, 77, *938*
Shen, G.J., 102, *910*, 102, *912*, 115, *1205*, 116, *1205*

Shen, J., 244, *3154*
Shen, N., 216, *2714*
Shen, X., 136, *1477*
Shen, Y., 232, *2960*
Shenni, S., 255, *3336*
Shepardson, S., 176, *2065*
Sheppard, J.D., 232, *2966*
Sherman, H.C., 71, *376*, 81, *358*, 81, *359*, 82, *376*, 87, *358*, 161, *1903*, 186, *2190*, 188, *2232*, 188, *2242*, 188, *2243*, 188, *2244*, 188, *2245*
Shetty, J.K., 64, *162*, 65, *266*, 89, *486*, 108, *1023*, 150, *1687*, 207, *2583*, 212, *2633*, 213, *2671*
Sheu, D.C., 234, *2985*
Shewale, S.D., 97, *805*
Shewry, P.R., 176, *2088*
Shi, G., 221, *2815*, 222, *2815*
Shi, J., 148, *1598*, 264, *3519*
Shi, Y., 138, *1497*
Shi, Y.C., 196, *2406*, 206, *2532*
Shibamoto, N., 247, *3185*
Shibano, Y., 64, *104*
Shibata, K., 236, *3014*
Shibata, S., 154, *1737*
Shibuya, N., 247, *3182*
Shibuya, S., 242, *3098*
Shibuya, T., 193, *2350*, 241, *3083*, 246, *3174*, 246, *3175*, 246, *3177*
Shieh, M.T., 191, *2319*
Shieh, W., 256, *3357*
Shierbaum, F., 115, *1026*, 108, *1026*
Shigechi, H., 216, *2715*
Shih, E.K.Y., 147, *1590*
Shih, N.J., 73, *697*, 77, *697*
Shikata, S., 78, *1065*
Shilova, A.A., 126, *1359*
Shim, K.H., 168, *1989*
Shima, M., 118, *1236*
Shimada, M., 97, *819*
Shimada, S., 206, *2548*, 206, *2557*
Shimashita, M., 252, *3280*
Shimazaki, T., 75, *791*
Shimazu, K., 207, *2581*

Shimbori, F., 150, *1666*, 186, *2201*
Shimizu, A., 155, *1756*
Shimizu, H., 238, *3052*
Shimizu, K., 236, *3023*
Shimobayashi, Y., 91, *529*
Shimodi, K., 205, *2518*, 241, *2518*
Shimomura, T., 111, *1156*, 111, *1160*, 111, *1161*, 111, *1162*, 111, *1163*, 112, *1165*, 112, *1180*, 113, *1160*, 113, *1161*, 113, *1165*, 113, *1182*, 113, *1156*
Shimomura, Y., 73, *696*, 77, *696*, 124, *1331*
Shimoyo, N., 268, *3608*
Shin, H.D., 240, *3071*, 241, *3082*, 254, *3326*
Shin, H.J., 244, *3156*
Shin, J.H., 245, *3166*
Shin, K., 254, *3304*
Shin, M., 163, *1926*, 166, *1955*
Shin, M.S., 180, *2128*, 179, *2128*, 207, *2568*
Shin, Y.C., 88, *448*, 239, *3059*
Shinbori, F., 174, *2048*, 178, *2048*
Shinkai, S., 5, *29*, 5, *31*
Shinke, R., 76, *930*, 90, *513*, 90, *514*, 99, *514*, 101, *887*, 101, *900*
Shinoda, M., 37, *131*, 37, *134*
Shinoda, O., 75, *521*, 90, *521*
Shinohara, H., 88, *446*, 119, *446*
Shinohara, S., 158, *1850*, 159, *1850*
Shinoyama, H., 78, *1088*
Shintani, K., 78, *1065*
Shiomi, N., 77, *1040*
Shiomi, T., 261, *3414*
Shiotsu, S., 159, *1883*, 159, *1884*
Shioya, S., 153, *1721*, 170, *2003*, 212, *2650*, 226, *2897*, 238, *3052*
Shirai, Y., 247, *3183*, 250, *3241*
Shiraishi, F., 102, *919*, 102, *921*, 108, *1112*, 139, *1512*, 139, *1522*, 141, *1533*, 150, *1690*, 256, *3349*
Shiratori, S., 157, *1803*
Shirkhande, J.G., 183, *2173*
Shkutina, I.V., 110, *1145*, 110, *1144*
Shneider, P., 78, *1062*
Shobsngob, S., 96, *803*
Shockley, K.R., 117, *1220*

Shoemaker, R.H., 30, *108*
Shon, B.S., 213, *2657*
Shorniko, V.G., 108, *1031*
Short, J.M., 64, *76*, 64, *77*
Shtangeeva, N.I., 105, *976*
Shub, I.S., 174, *2046*, 176, *2086*
Shubert, T.A., 99, *854*
Shukla, G.L., 244, *3152*
Shulman, A.H., 167, *1977*
Shulman, M.S., 130, *1448*, 130, *1449*, 172, *1488*
Shut, G.J., 117, *1220*
Si, A., 244, *3140*
Sialyl Lewis, X., 8, *40*
Sicard, P.J., 129, *1428*
Sicking, W., 16, *76*, 16, *77*, 16, *78*
Siddiqui, K.S., 64, *79*
Sidorenko, I.A., 195, *2398*
Sidorin, V.N., 217, *2735*
Siedenberg, D., 74, *755*
Siedich, A.S., 63, *39*
Sieh, W., 254, *3314*
Sierks, M.R., 105, *978*, 129, *1412*
Siezen, R.J., 233, *2972*
Sigurskjold, B.W., 64, *89*
Sihlbom, E., 98, *833*
Sijmons, P.C., 89, *485*, 96, *485*
Silberstein, O., 198, *2452*
Silhol-Bernere, M.J., 242, *3092*
Siljestrom, M., 202, *2466*
Sillen, L.G., 64, *196*, 172, *2021*
Sillinger, V., 249, *3231*
Sills, A.M., 93, *564*, 96, *633*, 103, *633*, 106, *633*, 107, *633*
Silva, J.G., 78, *1047*
Silverio, J., 177, *2098*
Sim, J.S., 245, *3166*
Sim, P.C.X., 238, *3051*
Simakov, A.V., 189, *2263*, 189, *2264*
Simard, R.E., 71, *442*, 76, *442*, 87, *442*, 99, *442*, 100, *442*, 163, *1927*
Simmons, G., 11, *54*
Simoes-Mendes, B., 93, *562*, 106, *562*
Simone, I.D., 30, *109*

Sims, K.A., 190, *2281*
Sin, S.U., 93, *577*
Sinclair, P.M., 158, *1848*
Sinelnikova, L.E., 146, *1580*
Sineriz, F., 217, *2753*
Singer, P.A., 148, *1599*
Singh, B., 266, *3561*
Singh, C., 102, *922*
Singh, D., 64, *178*, 64, *180*, 65, *178*, 72, *180*, 73, *180*, 78, *1061*, 79, *1061*, 76, *180*, 78, *180*, 79, *180*, 113, *180*, 114, *180*, 116, *180*, 118, *180*, 217, *2743*, 220, *2806*
Singh, R.P., 238, *3048*
Singh, V.O., 155, *1758*, 192, *2329*, 223, *2843*, 238, *3048*
Sinnaeve, G., 262, *3454*, 263, *3454*, 264, *3454*
Sinner, M., 266, *3574*
Sippel, A., 149, *1631*
Sippel, G.J., 148, *1602*
Siqueira, E.M.D., 75, *789*
Sirdeshmukh, R., 76, *923*
Sirisansaneeyakul, S., 236, *3015*
Sirotti, D.A., 191, *2322*
Siso, I.G., 121, *1274*
Siso, M.I.G., 97, *807*
Sissons, M.J., 112, *1166*, 113, *1166*
Siswoyo, T.A., 203, *2484*
Sitch, D.A., 127, *1378*
Sivuramakrishnan, S., 72, *649*, 75, *778*, 80, *336*, 82, *336*
Sjöberg, K., 64, *217*
Sjoeberg, K., 48, 63, 85, *420*
Sjoeholm, C., 211, *2627*, 211, *2628*
Sjoholm, C., 75, *801*, 117, *801*
Skarka, B., 75, *780*
Skea, G.L., 71, *401*, 85, *401*
Skeen, R.S., 227, *2907*
Škoda, F., 89, *496*
Skrede, A., 207, *2586*
Skrede, G., 207, *2586*, 262, *3443*, 264, *3443*
Skreekantiah, K.R., 75, *768*, 75, *769*
Slade, J., 196, *2425*
Slajsova, K., 107, *1019*

Slaughter, S.L., 127, *1375*, 138, *1496*, 178, *1375*, 181, *1375*
Slepokurova, Yu.I., 110, *1144*
Slepokurova, Yu.I., 110, *1145*
Sleyter, U.B., 72, *684*
Slinde, E., 207, *2586*
Slocum, G., 188, *2223*
Slolheim, B.A., 89, *484*
Słominska, L., 64, *246*, 65, *260*, 120, *1256*, 122, *1295*, 150, *1674*, 151, *1711*, 152, *1711*, 168, *1996*, 170, *1996*, 190, *2287*, 190, *2289*, 190, *2291*, 190, *2302*, 192, *2331*, 218, *2776*, 253, *3296*
Slominski, B.A., 208, *2602*
Slott, S., 88, *464*, 148, *1613*
Słowiński, W., 218, *2776*
Slupska, M., 64, *76*, 64, *82*
Smaley, K.L., 76, *934*
Smee, D.F., 11, *54*, 11, *56*
Smeekens, J.C.M., 64, *94*
Smeraldi, C., 218, *2761*
Smiley, K.L., 97, *808*, 155, *1777*, 158, *1863*, 158, *1867*
Smirnov, V.B., 110, *1145*
Smirnova, M.I., 176, *2073*
Smith, A., 72, *648*
Smith, A.E., 11, *60*
Smith, A.J., 11, *54*
Smith, A.M., 262, *3461*
Smith, B.N., 176, *2054*
Smith, I.A., 88, *463*
Smith, J.A., 104, *966*, 126, *1364*
Smith, J.S., 104, *958*
Smith, M.D., 163, *1918*
Smith, M.P.T., 217, *2733*
Smith, M.R., 64, *243*, 68, *243*
Smith, S.M., 65, *282*
Snow, P., 206, *2546*
So, K.H., 76, *932*
Soares, V.F., 78, *1047*
Soberon, X., 89, *492*
Sobolev, A.P., 265, *3529*
Sobri, M.A., 153, *1721*
Sobue, H., 137, *1485*

Soccol, C.R., 64, *178*, 65, *178*, 75, *778*, 79, *1096*, 79, *1099*, 79, *1100*, 125, *1348*, 126, *1354*, 237, *3039*
Soccol, V.T., 64, *178*, 65, *178*
Socol, C.R., 80, *336*, 82, *336*
Soetaert, W., 235, *2992*
Soga, T., 126, *1353*, 228, *2929*, 229, *2929*, 230, *2929*
Sogo, K., 234, *2986*
Sohn, C.B., 101, *896*, 104, *967*
Sokhansanj, S.R.T., 208, *2601*
Sokołowski, A., 238, *3049*
Sola, C., 65, *273*
Solah, V., 262, *3453*
Soldi, V., 164, *1930*
Solheim, L., 89, *484*
Solheim, L.P., 149, *1632*, 196, *2416*, 188, *2234*
Soloducha, A., 149, *1655*
Solomon, G.L., 72, *650*
Solomon, J., 74, *765*
Solomon, J.D., 179, *2127*
Soloveichik, I.Ya., 178, *2110*
Soltuzu, F., 234, *2991*, 235, *2991*
Somalanka, S.R., 226, *2889*
Somchai, P., 220, *2796*
Somers, W., 97, *814*
Somkuti, G.A., 96, *636*
Sommer, M., 206, *2554*, 208, *2599*
Sommerville, M.D., 130, *1456*, 181, *2152*
Somogyi, M., 82, *384*
Somogyi, M.J., 266, *3572*
Song, B.K., 256, *3361*
Song, C.H., 105, *981*, 105, *982*
Song, C.P., 193, *2363*
Song, F.B., 109, *1121*
Song, S., 78, *1088*
Soni, R., 150, *1692*, 225, *2879*
Soni, S.K., 90, *522*, 93, *566*, 104, *972*, 107, *566*, 150, *1692*, 224, *2866*, 225, *2879*
Sonnenberg, C., 18, *85*
Sonomoto, K., 190, *2277*, 236, *3017*, 236, *3014*, 236, *3016*
Sonoyama, K., 194, *2366*

Soong, C.L., 92, *550*
Soral-Smietana, M., 126, *1363*, 176, *1363*
Sorey, S., 5, *27*
Sorimachi, K., 129, *1426*
Sorkin, S.Z., 163, *1919*
Soto Ibanez, R., 164, *1931*
Sotoyama, T., 267, *3591*
Soucaille, P., 73, *693*
Sousa, N.O., 158, *1853*
Southon, S., 261, *3416*, 262, *3420*
Souza, A.V.V., 74, *731*, 77, *731*
Souza e Silva, P.C., 219, *2788*
Sowder, R.C. II., 30, *108*
Spannagel, R., 208, *2599*
Spannägl, R., 206, *2554*, 206, *2555*
Spassova, D., 72, *652*, 72, *653*
Speakman, E.L., 195, *2387*
Speakman, H.B., 231, *2942*, 231, *2943*
Specht, H., 126, *1366*
Specka, U., 77, *949*, 103, *949*
Speckmann, H.D., 196, *2417*, 196, *2418*, 196, *2419*, 196, *2420*, 196, *2421*, 196, *2422*, 196, *2424*, 204, *2422*
Spellman, M.W., 11, *58*, 12, *58*
Spencer-Martins, I., 64, *207*, 66, *305*, 72, *689*, 74, *689*, 77, *689*
Spendler, T., 202, *2468*, 206, *2530*
Spiers, H.M., 233, *2969*
Spies, T., 264, *3505*
Spiesser, D., 253, *3289*
Spiridonova, V.A., 252, *3273*
Sporns, P., 265, *3534*
Spouge, J.W., 217, *2719*
Spradlin, J., 64, *111*
Spreinat, A., 75, *799*, 121, *1272*
Springsteen, G., 4, *20*, 4, *22*, 5, *26*
Sprockhoff, M., 87, *441*
Sproessler, B., 198, *2448*
Sprössle, B., 223, *2855*
Šramek, J., 264, *3495*
Sreedharan, V.P., 107, *1009*, 256, *3356*
Sreekantiah, K.R., 95, *610*, 220, *2802*, 226, *2802*, 235, *2996*
Sreenath, H.K., 176, *2071*

Sreenivasan, A., 179, *2125*
Sreenivasan, P.K., 268, *3613*
Sreenivasaya, M., 267, *3582*
Sreerabgachar, H.B., 267, *3582*
Sreerangachar, H.B., 264, *3482*
Srikanta, S., 220, *2802*, 226, *2802*
Sringam, S., 127, *1396*
Srivastava, K.C., 115, *1205*, 116, *1205*
Srivastava, R.A.K., 72, *508*, 76, *508*, 77, *508*, 88, *465*, 90, *508*, 90, *512*, 99, *465*
Sroczynski, A., 108, *1032*, 121, *1275*, 127, *1394*, 155, *1780*, 179, *2124*, 205, *2523*, 205, *2524*, 205, *2525*
St Denis, E., 166, *1961*, 176, *1961*
Stachelberger, H., 262, *3444*, 265, *3444*, 265, *3444*
Staley, A.E., 149, *1640*, 179, *2121*
Stalin, M.O., 151, *1705*
Stamatatos, L., 11, *69*
Stamatoudis, M., 189, *2262*
Stamatoutis, M., 121, *1277*
Stamberg, O.E., 100, *879*
Stamberg, O.F., 98, *839*, 100, *839*
Stanasel, O., 192, *2343*
Stanković, M., 218, *2771*
Starace, C.A., 64, *171*
Stark, J.R., 64, *238*, 90, *506*, 93, *579*, 176, *238*
Stark, W.H., 63, *34*, 210, *34*
Starka, J., 93, *578*
Starnes, R.L., 123, *1309*, 252, *3283*, 253, *3294*
Starobinskaya, A.B., 141, *1535*
Starogardzka, G., 120, *1256*, 122, *1295*
Staroszczyk, H., 136, *1470*
Starzyk, F., 191, *2326*
Stauber, J., 8, *43*
Stecka, K.M., 224, *2868*, 224, *2869*, 230, *2939*, 235, *3002*, 235, *3009*
Steeneken, P.A.M., 64, *141*
Steenfeldt, S., 208, *2594*, 260, *3400*
Steenken, P.A.M., 241, *3079*
Stefanova, M., 72, *652*, 72, *653*, 72, *665*
Steffen, P., 156, *1782*, 215, *2708*, 216, *2709*
Stein, E.A., 183, *2169*, 183, *2171*

Stein, I., 129, *1441*
Steinberg, D.H., 96, *636*
Steinberg, M.P., 64, *232*, 156, *232*, 221, *2821*, 221, *2823*, 222, *2827*, 223, *2827*
Steinbök, H., 233, *2973*
Steinbüchel, A., 232, *2964*
Steingröver, A., 63, *64*
Steinke, J.D., 192, *2333*
Steinle, D., 64, *144*, 249, *144*, 250, *144*
Stelmaschuk, S., 92, *551*
Stemplinger, I., 74, *729*
Stendera, L., 241, *3087*
Stenholm, K., 114, *1192*
Stepanov, A.I., 74, *745*, 77, *745*
Stepanov, A.V., 148, *1597*
Stepanov, V.M., 73, *720*, 74, *720*, 76, *720*
Šterba, S., 206, *2547*
Stern, E., 158, *1837*
Sternberg, M., 88, *479*
Stertz, S.C., 79, *1100*, 125, *1348*, 126, *1354*
Steup, M., 267, *3599*
Stevens, A., 73, *714*
Stevnebo, A., 176, *2059*
Stewart, D., 111, *1158*, 113, *1158*
Stewart, G.G., 93, *564*, 96, *633*, 103, *633*, 106, *633*, 107, *633*, 217, *2719*
Stoffer, B., 78, *1062*
Stolbikova, O.E., 119, *1243*
Stone, A.C., 81, *366*, 82, *366*
Stoof, G., 179, *2129*, 180, *2133*, 180, *2134*, 206, *2539*
Storey, K.B., 109, *1124*
Story, K.B., 110, *1147*
Stouffs, R.H.M., 150, *1675*, 155, *1762*
Stoyanova, O.F., 110, *1144*, 110, *1145*
Straeter, N., 123, *1317*
Strasser, A.W.M., 74, *749*, 77, *749*, 78, *1042*, 212, *2648*
Stredansky, M., 64, *107*
Strehaiano, P., 210, *2616*, 211, *2616*
Stretton, S., 129, *1432*
Strickberger, M.W., 188, *2251*
Strohm, B.A., 108, *1023*
Strumeyer, D.H., 97, *821*

Stryer, L., 209, *2606*
Stuart, R.D., 63, *44*
Sturdik, E., 64, *198*, 107, *1019*
Sturgon, R., 268, *3606*
Stutzenberger, F.J., 73, *725*, 73, *726*
Styngach, I.V., 166, *1957*
Styrlund, C.R., 217, *2759*
Su, C.S., 254, *3311*
Su, Q., 104, *965*
Su, Y.C., 232, *2963*, 237, *3047*
Subbramaiah, K., 76, *931*
Subrahmanyam, A., 79, *1108*
Subramanian, T.V., 144, *1552*
Sucharzewska, D., 166, *1963*
Suda, S., 102, *918*
Suda, T., 225, *2878*
Sudaka, P., 113, *1181*
Suenaga, H., 204, *2503*
Sueno, T., 263, *3478*
Suetsugu, N., 75, *776*
Suga, K., 140, *1525*
Suga, M., 204, *2506*
Sugano, T., 211, *2626*
Suganuma, T., 76, *928*
Sugawara, S., 77, *945*, 103, *945*
Sugihara, T., 153, *1716*, 155, *1765*
Sugimot, T., 246, *3174*
Sugimoto, K., 157, *1812*, 205, *2521*
Sugimoto, T., 64, *253*, 202, *2474*, 246, *3175*, 246, *3176*
Sugimoto, Y., 64, *99*, 64, *186*, 159, *186*, 174, *2043*, 176, *2043*, 176, *2061*, 176, *2062*, 176, *2090*, 188, *2226*
Suginami, K., 78, *979*, 105, *979*
Sugita, H., 72, *644*
Sugiyama, M., 206, *2542*, 240, *3067*
Sugunan, S., 109, *1122*
Suh, H.W., 226, *2885*
Suhas, R., 74, *765*
Sujka, M., 176, *2074*, 176, *2075*, 177, *2074*, 180, *2140*
Suk, K.K., 125, *1336*
Sukhinin, S.V., 189, *2263*, 189, *2264*
Sullivan, J.W., 265, *3538*

Sumerwell, W.N., 183, *2171*
Sumiyoshi, M., 235, *3000*
Summers, M.F., 42, *143*
Sun, C., 65, *266*, 65, *267*, 213, *2665*, 226, *2883*, 264, *3522*
Sun, H., 68, *313*, 68, *315*, 79, *1094*
Sun, J., 213, *2665*, 266, *3556*
Sun, J.W., 107, *1011*
Sun, M.Y., 213, *2658*
Sun, W., 249, *3211*
Sun, X., 120, *1249*
Sun, X.S., 225, *2880*
Sun, Y., 250, *3237*
Sun, Y.M., 97, *806*
Sun, Z., 85, *422*, 112, *1164*, 113, *1164*, 122, *422*, 220, *2803*, 221, *2803*, 244, *1164*
Sunarti, T.C., 264, *3506*
Sung, A., 207, *2583*
Sung, C.K., 213, *2657*
Sung, H.Y., 252, *3279*, 255, *3337*
Sung, M.H., 64, *98*
Sung, N.K., 213, *2657*, 214, *2679*, 226, *2679*
Sung, S.J., 91, *541*
Suortti, M.T., 116, *1208*, 176, *2068*
Suortti, T., 114, *1192*, 139, *1505*, 167, *1974*, 167, *1975*, 176, *2067*
Suresh, A.K., 149, *1653*, 153, *1730*, 224, *2861*, 225, *2861*, 235, *2997*
Surolia, A., 2, *10*
Suryanarayaba, M., 71, *413*, 85, *413*
Suryani, A., 232, *2965*
Suterska, A.M., 235, *3002*, 235, *3009*
Sutra, R., 173, *2023*
Suvee, A.J., 150, *1682*
Suyanadona, P., 219, *2793*
Suye, S., 213, *2653*
Suzuki, A., 266, *3571*
Suzuki, H., 89, *491*, 155, *1759*
Suzuki, K., 34, *119*, 34, *121*, 193, *2351*, 197, *2435*
Suzuki, M., 78, *1070*
Suzuki, N., 227, *2916*
Suzuki, S., 64, *244*, 88, *458*, 154, *1737*, 227, *2908*, 242, *3098*, 254, *3318*

Suzuki, T., 153, *1726*
Suzuki, W., 240, *3069*
Suzuki, Y., 65, *285*, 78, *1049*, 79, *1049*, 89, *503*, 107, *1020*, 112, *1020*, 112, *1167*, 112, *1175*, 113, *1167*, 114, *1020*, 160, *1890*, 160, *1890*, 250, *3238*
Suzzi, P., 164, *1932*
Svanborg, K., 86, *423*
Svedsen, A., 64, *75*
Svedsen, I., 78, *1062*
Svendsen, A., 64, *78*, 64, *92*, 89, *490*, 118, *1235*, 195, *2395*, 202, *2468*
Svenson, K.B., 105, *978*
Svensson, B., 64, *89*, 64, *130*, 65, *301*, 78, *1062*, 129, *130*, 129, *1412*, 129, *1415*
Svensson, D., 238, *3055*
Sverdlova, N.I., 195, *2399*
Svihus, B., 176, *2059*
Svoboda, F., 264, *3495*, 267, *3583*
Swain, E.W., 78, *1044*
Swamy, M.V., 73, *698*, 73, *699*, 76, *935*, 76, *937*, 77, *698*, 77, *937*
Swanson, B., 89, *489*
Swanson, C.L., 97, *808*
Swanson, R.G., 175, *2050*, 179, *2127*
Swanson, T.R., 120, *1261*, 133, *1461*, 134, *1461*, 154, *1751*, 155, *1261*
Swings, J., 72, *642*
Syamsu, K., 232, *2965*
Sym, E.A., 127, *1379*, 183, *1379*
Syniewski, V., 81, *352*
Synowiecki, J., 64, *143*, 64, *175*, 112, *1171*, 113, *1171*
Syu, M.J., 88, *474*
Szajani, B., 109, *1132*, 109, *1133*, 215, *2701*
Szatmar, E., 64, *230*
Szczepaniak, E., 80, *341*
Szczodrak, J., 78, *1068*
Szejtli, J., 65, *280*, 65, *289*, 75, *772*, 249, *3230*, 254, *3315*, 257, *280*
Szentgyoryi, M., 262, *3457*, 265, *3457*
Szentirmai, A., 257, *3370*
Szerman, N., 251, *3246*
Szewczyk, K.W., 149, *1655*

Szigetvari, G., 254, *3315*
Szostek, A., 253, *3296*
Szyli, O., 73, *701*

T

Tabet, J.C., 8, *42*
Table, L., 225, *2882*
Tabuchi, A., 246, *3174*, 246, *3177*
Tachauer, E., 189, *2256*
Tachibana, Y., 180, *2136*
Tachikake, N., 155, *1760*
Tada, S., 100, *874*
Tadokoro, T., 125, *1339*, 125, *1340*
Taeufel, A., 156, *1782*
Taft, C.S., 78, *1090*
Tagaki, E., 206, *2548*, 206, *2557*
Tagubase, J.L., 236, *3014*
Taguchi, H., 220, *2808*
Taguchi, Y., 243, *3117*
Tahara, N., 75, *783*
Tahara, T., 93, *567*
Taira, R., 64, *83*
Taji, N., 65, *285*, 148, *1615*, 149, *1624*
Takabane, T., 253, *3297*
Takada, A., 8, *37*
Takada, H., 253, *3297*
Takada, Y., 204, *2503*
Takagi, H., 72, *655*
Takagi, M., 72, *525*, 90, *525*, 243, *3122*, 264, *3523*
Takagi, S., 64, *83*
Takaha, T., 123, *1317*
Takahara, Y., 101, *901*
Takahash, S., 125, *1349*, 126, *1349*, 244, *3139*
Takahashi, C., 159, *1883*, 159, *1884*, 205, *2511*
Takahashi, F., 159, *1879*
Takahashi, H., 91, *532*, 92, *545*, 207, *2565*, 259, *3387*
Takahashi, K., 74, *728*, 119, *1239*, 156, *1792*, 217, *1792*, 235, *2998*
Takahashi, M., 105, *985*, 193, *2362*
Takahashi, N., 111, *1161*, 111, *1162*, 113, *1161*, 113, *1182*

Takahashi, S., 120, *1268*, 125, *1350*, 126, *1350*, 147, *1594*, 153, *1717*, 154, *1755*, 172, *2017*, 216, *2715*, 252, *3280*, 264, *3498*
Takahashi, T., 65, *282*, 77, *945*, 79, *990*, 103, *945*, 105, *986*, 105, *990*, 106, *990*, 129, *1421*, 149, *1645*, 206, *2562*, 247, *3185*
Takahashi, Y., 91, *542*, 118, *1233*, 118, *1234*, 150, *1233*, 223, *2846*
Takahasi, N., 64, *174*
Takahata, Y., 105, *988*, 106, *988*, 176, *2058*
Takaki, E., 176, *2064*, 176, *2077*
Takami, K., 197, *2441*
Takamine, J. Jr., 63, *47*
Takanayagi, Y., 149, *1623*
Takano, R., 180, *2132*
Takano, T., 129, *1427*
Takao, F., 188, *2226*
Takao, S., 96, *626*, 107, *1013*, 107, *1015*, 212, *2646*, 222, *2646*
Takaoka, K., 91, *528*
Takaoka, Y., 214, *2676*
Takasaki, Y., 64, *125*, 64, *173*, 64, *204*, 88, *445*, 88, *446*, 88, *455*, 89, *483*, 89, *488*, 90, *520*, 101, *901*, 115, *1209*, 115, *1216*, 116, *1207*, 116, *1209*, 116, *1216*, 119, *446*, 119, *1238*, 122, *1300*, 122, *1301*, 122, *1302*, 122, *1303*, 122, *1304*, 122, *1305*, 123, *1314*, 151, *1701*, 154, *1753*, 157, *1799*, 157, *1828*, 157, *1829*, 157, *1830*, 157, *1831*, 159, *1874*, 159, *1881*, 183, *2177*, 239, *1238*, 244, *3144*, 246, *3170*
Takasugi, N., 125, *1339*
Takata, H., 65, *285*, 117, *1225*, 129, *1430*
Takaya, T., 160, *1890*, 160, *1890*, 174, *2043*, 176, *2043*, 176, *2062*
Takayama, Y., 262, *3432*
Takeda, M., 228, *2929*, 229, *2929*, 230, *2929*
Takeda, S., 106, *992*, 128, *992*
Takeda, Y., 107, *1015*, 167, *1968*, 244, *3153*, 262, *3427*, 266, *3571*
Takegoshi, K., 42, *142*, 43, *145*, 43, *146*, 43, *147*, 43, *149*

Takei, S., 264, *3523*
Takeo, K.I., 75, *776*
Takeshi, S., 202, *2474*
Takeshita, T., 256, *3353*
Takeuchi, M., 75, *773*
Takeuchi, T., 34, *110*, 34, *116*, 37, *110*, 45, *151*, 263, *3472*
Taki, A., 190, *2305*, 239, *3058*
Takigami, M., 64, *95*, 64, *105*
Takigawa, S., 163, *1916*
Talik, B., 123, *1313*, 241, *1313*
Talwar, G.P., 182, *2162*
Tamaya, H., 153, *1726*
Tambourgi, E., 128, *1408*
Tams, J.W., 217, *2732*
Tamura, G., 129, *1420*
Tamura, N., 179, *2113*
Tamura, S., 37, *131*
Tamura, T., 64, *228*, 154, *1737*
Tamuri, M., 90, *510*
Tan, H., 139, *1503*
Tan, X., 218, *2772*
Tanabe, A., 45, *150*
Tanabe, K., 204, *2505*
Tanabe, O., 64, *124*, 126, *1357*, 126, *1358*
Tanabe, S., 194, *2366*
Tanabe-Tochikura, A., 45, *152*
Tanaka, A., 106, *992*, 128, *992*, 212, *2649*, 216, *2715*
Tanaka, H., 112, *1177*, 113, *1177*, 114, *1177*, 213, *2651*, 213, *2652*, 216, *2710*
Tanaka, K., 207, *2581*
Tanaka, M., 64, *99*, 176, *2061*, 207, *2563*, 214, *2678*
Tanaka, N., 115, *1198*, 116, *1198*, 150, *1684*, 197, *2435*, 203, *2484*, 206, *2536*, 254, *3327*
Tanaka, R., 95, *603*
Tanaka, T., 73, *696*, 77, *696*, 124, *1331*, 251, *3258*
Tanaka, Y., 64, *104*, 89, *499*, 93, *567*, 159, *1885*, 179, *2112*
Tanba, T., 159, *1885*
Tang, G., 190, *2286*

Tang, L., 64, *81*
Tang, M.S., 75, *779*
Tang, S., 150, *1657*, 206, *1657*
Tang, S.K., 124, *1324*
Tang, Z., 218, *2762*
Tanida, M., 96, *626*, 107, *1015*
Tanigaki, M., 242, *3105*
Tanigawa, K., 104, *974*, 129, *974*
Taniguchi, H., 88, *453*, 88, *454*, 90, *515*, 90, *516*, 90, *517*, 90, *518*, 115, *1199*, 115, *1200*, 116, *1199*, 176, *516*
Taniguchi, M., 64, *231*, 73, *696*, 74, *766*, 77, *696*, 97, *812*, 124, *1331*, 212, *2645*, 214, *2687*, 214, *2688*, 236, *3023*
Tanimichi, H., 35, *126*, 38, *126*
Tanimura, W., 197, *2438*
Tanino, T., 64, *98*
Tanizawa, S., 111, *1157*, 113, *1157*
Tanner, R., 64, *85*
Tanno, K., 267, *3591*
Tao, C., 206, *2535*
Tao, W.T., 85, *411*, 86, *427*
Tapodo, J., 202, *2472*
Tarakhtii, L.V., 166, *1957*
Tareen, J.A.K., 103, *942*
Targoński, Z., 64, *126*
Tarnovski, G., *46*, 63, *46*, 233
Tarr, H.L.A., *41*, 63
Tarytsa, V.F., 204, *2500*
Tas, A.C., 265, *3533*
Tateno, T., 64, *98*
Tateno, Y., 207, *2581*
Tatham, A.S., 176, *2088*
Tatsinkou Fossi, B., 64, *86*
Tatsumi, E., 262, *3426*, 265, *3426*
Tatsumi, H., 139, *1516*, 139, *1517*, 175, *2052*, 267, *1516*
Taufel, A., 74, *742*, 77, *742*, 268, *3602*
Taufel, F., 215, *2708*, 216, *2709*
Tauro, P., 79, *1109*
Tavares, D.Q., 180, *2131*
Tavea, F., 64, *86*
Taylor, A.J., 76, *926*
Taylor, C.C., 236, *3026*

Taylor, M.A., 111, *1158*, 113, *1158*
Taylor, P.M., 77, *951*, 79, *951*, 103, *951*
Teague, W.M., 72, *671*, 138, *1500*
Tearney, R.L., 65, *293*
Tee, R.D., 76, *926*
Tegge, G., *15*, 62, 155, *1769*, 159, *1888*, 164, *1941*, 244, *3138*, 265, *3543*
Teixera, C.G., 220, *2798*
Teller, G.L., 85, *404*
ten Cate, L., 262, *3442*, 265, *3442*
Teniola, O.D., 95, *619*, 105, *619*
Tenmyo, O., 34, *115*, 34, *123*, 37, *115*, 45, *150*
Terada, Y., 123, *1317*
Terai, T., 91, *529*
Teramoto, Y., 89, *494*, 72, *690*, 92, *558*, 129, *1413*
Terao, T., 43, *146*, 43, *147*
Teraoka, I., 242, *3089*, 242, *3090*
Terasawa, K., 240, *3076*
Terenti, H.F., 67, *307*
Terenzi, H.F., 88, *584*, 94, *584*
Terris, B., 8, *42*
Tester, R.F., 64, *164*, 130, *1456*, 175, *164*, 181, *2152*
Teuerkauf, H.D., 158, *1859*
Textor, S.D., 166, *1961*, 176, *1961*
Thammarutwasik, P., 225, *2881*
Thanh, V.N., 74, *740*, 77, *740*
Thannhauser, S.J., 163, *1919*
Tharanathan, R.N., 103, *942*, 176, *2063*, 177, *2063*
Tharathan, R.N., 177, *2095*
Thayanathan, K., 156, *1789*
Theander, O., 262, *3462*, 262, *3467*, 263, *3462*
Thevenot, G.D., 153, *1731*, 210, *2613*
Thiery, G., 8, *42*
Thilo, L., 197, *2430*
Thisted, T., 64, *78*
Thivend, P., 262, *3425*
Thoai, N.V., 242, *3092*, 243, *3110*
Thoma, J., 64, *111*
Thomas, G.J., 115, *1197*, 119, *1240*

Thomas, J.N., 11, *58*, 12, *58*
Thomas, L., 210, *2610*
Thompson, G., 158, *1834*
Thompson, G.J., 157, *1818*, 157, *1819*
Thompson, K.N., 123, *1308*
Thompson, P.W., 156, *1781*
Thompson, R., 76, *927*
Thompson, W.R., 83, *394*, 84, *394*
Thompson, W.S., 124, *1321*
Thomson, N.H., 176, *2088*
Thorsell, W., 99, *850*, 99, *851*, 148, *851*, 151, *851*
Thorup, J.H., 214, *2684*
Thoth, A., 190, *2284*
Thurn, K.K., 260, *3405*
Tian, Y., 64, *80*
Tilden, E.B., 249, *3223*, 249, *3224*
Tischler, M., 224, *2858*
Tkač, M., 268, *3602*
Tkaczyk, M., 186, *2198*
Tobback, P., 72, *666*, 177, *2105*
Tobe, K., 119, *1245*
Tochikura, T.S., 45, *152*
Toda, H., 75, *781*, 159, *1884*, 205, *2511*
Toda, S., 40, *137*
Toeda, K., 247, *3185*
Tokida, M., 196, *2426*
Tokimura, K., 100, *874*
Tokuda, H., 126, *1353*, 228, *2929*, 229, *2929*, 230, *2929*
Tokuda, M., 112, *1178*, 114, *1178*
Tokuoka, Y., 100, *873*
Toldra, F., 109, *1119*
Tollier, M.T., 179, *2111*
Tom, M.U., 76, *940*
Tomar, M., 109, *1130*
Tomasik, J., 219, *2792*
Tomasik, P., 61, *2*, 61, *3*, 61, *4*, 61, *5*, 61, *6*, 61, *7*, 61, *8*, 61, *9*, 82, *3*, 97, *815*, 136, *1470*, 136, *1472*, 136, *1474*, 164, *1940*, 176, *2065*, 178, *5*, 191, *2326*, 256, *3362*, 257, *3362*, 258, *3362*, 259, *3378*
Tomatsu, K., 34, *111*, 34, *115*, 37, *115*
Tominaga, K., 91, *529*

Tominaga, T., 153, *1716*, 155, *1765*, 163, *1925*
Tominaga, Y., 174, *2043*, 176, *2043*
Tomita, F., 78, *1081*, 79, *1091*, 212, *2646*, 222, *2646*
Tomita, G., 94, *595*, 94, *596*
Tomita, K., 34, *111*, 34, *114*, 34, *120*, 251, *3258*, 262, *3432*
Tomita, N., 255, *3346*
Tomita, T., 74, *762*
Tomlison, K.L., 262, *3461*
Tomoo, K., 123, *1317*
Tonda, H., 159, *1883*
Tonkova, A., 72, *652*, 72, *653*, 72, *665*
Tonoue, H., 100, *874*
Tonozuka, T., 120, *1248*
Topping, D.L., 259, *3377*
Torigoe, Y., 74, *612*, 95, *612*
Torkkeli, T.K., 108, *1022*
Torok, G., 205, *2515*
Torrado, A., 75, *777*, 171, *2007*, 171, *2008*, 171, *2009*
Torres, E.F., 224, *2864*
Torres, R., 110, *1149*
Torres, R.G., 149, *1652*
Torto, N., 142, *1538*, 264, *3510*
Tosaar, I., 74, *752*
Toshihiko, I., 146, *1579*
Toth, A., 126, *1371*
Toth, M., 215, *2701*
Toth, T.A., 196, *2404*
Toth, Z., 264, *3507*
Totsuka, A., 119, *1239*, 204, *2505*, 204, *2506*
Towprayoon, S., 103, *955*, 103, *956*
Toyama, N., 192, *2340*
Toyama, T., 265, *3539*
Trackman, P.C., 120, *1261*, 123, *1309*, 154, *1751*, 155, *1261*, 253, *3294*
Tramper, J., 129, *1443*, 163, *1911*, 180, *2139*
Tran, H.T., 218, *2775*
Traubenberg, S.E., 176, *2086*
Trautmann, S., 127, *1382*, 180, *2142*
Tregubov, N.N., 127, *1393*, 137, *1486*
Trejgel, A., 190, *2289*
Trejo-Aguilar, B., 234, *2989*
Trimde, G., 225, *2871*
Trincone, A., 246, *3180*
Tripalo, B., 191, *2317*
Tripetchkul, S., 226, *2895*, 227, *2895*
Trippier, P.C., 11, *70*, 11, *71*
Troesch, W., 88, *460*, 124, *460*
Trojaborg, S., 147, *1596*
Trotta, F., 64, *102*
Troyer, D.A., 8, *41*
Trubiano, P.C., 197, *2431*, 197, *2432*, 197, *2433*, 197, *2439*
Truesdell, S.J., 65, *258*, 145, *258*
Truter, P.A., 197, *2430*
Tsai, H.T., 239, *3064*
Tsai, Y.C., 232, *2963*
Tsao, C.Y., 71, *402*, 85, *402*
Tsao, G.T., 108, *1113*, 109, *1118*, 109, *1119*, 109, *1129*
Tsaplina, I.A., 73, *712*, 73, *720*, 74, *720*, 76, *720*
Tsekova, K., 104, *960*, 181, *2151*
Tseng, C.C., 65, *261*
Tsiomenko, A.B., 74, *736*
Tskhvediani, L., 182, *2156*
Tsuchiya, H., 241, *3080*
Tsuge, H., 262, *3426*, 263, *3479*, 265, *3426*, 265, *3541*
Tsuji, H., 112, *1170*, 113, *1170*, 122, *1296*
Tsujisaka, Y., 64, *129*, 64, *251*, 74, *744*, 77, *744*, 246, *3177*
Tsukagoshi, H., 72, *655*
Tsukagoshi, N., 75, *521*, 90, *521*
Tsukamoto, K., 243, *3109*
Tsukamoto, Y., 41, *141*
Tsukano, Y., 117, *1227*, 118, *1227*, 247, *1227*, 247, *3194*
Tsunakawa, M., 34, *111*, 34, *113*, 42, *113*
Tsuno, T., 34, *113*, 34, *120*, 34, *124*, 42, *113*
Tsuru, H., 126, *1357*, 126, *1358*
Tsuruhisa, M., 88, *446*, 119, *446*
Tsurumizu, R., 75, *773*
Tsuruta, T., 115, *1209*, 116, *1209*
Tsusaki, K., 241, *3085*

Tsutsumi, N., 118, *1235*
Tsuyama, N., 156, *1790*
Tsuyuoka, K., 8, *37*
Tsyperovich, A.S., 74, *751*
Tucker, M.P., 223, *2843*
Tuener, N., *16*, 62
Tumbleson, M.E., 155, *1758*
Tuovinen, L., 197, *2434*
Tur, W., 80, *341*
Turchetti, B., 30, *109*
Turchi, S.L., 96, *630*
Turecek, P.L., 204, *2498*
Turhan, M., 149, *1650*
Turk, S.C.H.J., 64, *94*
Turkiewicz, M., 208, *2589*, 208, *2590*
Turner, A.W., 151, *1696*, 155, *1778*
Turner, C.W., 194, *2374*
Turner, M.K., 62, *16*
Turpin, J.A., 11, *56*
Turvey, J.R., 71, *440*, 87, *440*, 172, *2013*
Tuting, W., 265, *3551*
Tveter, K.J., 8, *39*
Tychowski, A., 85, *419*, 85, *421*, 98, *844*, 98, *845*
Tyler, R.T., 208, *2601*
Tyofuku, H., 72, *654*
Tyrsin, Y.A., 78, *1084*
Tzekova, A., 104, *960*

U

Uchida, H., 246, *3170*
Uchida, K., 112, *1167*, 113, *1167*
Uchida, T., 126, *1357*
Uchiyama, K., 226, *2897*
Uchiyama, S., 78, *1088*
Udagawa, H., 64, *83*, 92, *550*
Udaka, S., 72, *655*, 243, *3121*
Udaka, Y., 75, *521*, 90, *521*
Udall, R.H., 261, *3409*
Udeh, K.O., 180, *2140*
Ueda, A., 125, *1346*
Ueda, M., 212, *2649*, 212, *2650*, 216, *2715*, 240, *3073*
Ueda, R., 227, *2917*
Ueda, S., 64, *197*, 64, *206*, 65, *262*, 65, *263*, 65, *271*, 78, *1067*, 79, *1067*, 99, *853*, 101, *898*, 102, *909*, 103, *955*, 103, *956*, 104, *962*, 104, *963*, 104, *964*, 104, *969*, 105, *964*, 105, *969*, 106, *994*, 107, *969*, 120, *1260*, 120, *1266*, 126, *1365*, 126, *1367*, 126, *1368*, 127, *1383*, 130, *1450*, 130, *1455*, 150, *1695*, 176, *2094*, 180, *1365*, 212, *271*, 212, *271*, 214, *2677*, 215, *2704*, 217, *2720*, 220, *2677*, 220, *2720*, 219, *2793*, 219, *2794*, 220, *2799*, 220, *2800*, 221, *2800*, 221, *2811*, 222, *2677*, 222, *2720*, 225, *2881*, 264, *3481*
Uehara, K., 187, *2216*
Uejima, O., 239, *3058*
Ueki, T., 3, *19*, 37, *133*, 38, *135*, 38, *136*, 40, *136*, 40, *137*
Uelgen, K.O., 217, *2757*
Uenakai, K., 221, *2812*
Ueno, F., 149, *1623*
Ueshima, O., 190, *2305*
Ueyama, M., 139, *1523*, 139, *1524*
Uguru, G.C., 72, *688*
Ugwuanyi, K.E., 168, *1991*
Uhlig, H., 198, *2448*, 223, *2855*
Uitdehaag, J.C.M., 64, *182*, 65, *182*
Ulhao, C.J., 72, *686*, 73, *686*, 74, *741*, 77, *741*
Ulmann, M., 80, *343*, 137, *1484*, 264, *3499*
Ulme, H.M., 147, *1595*
Umeda, K., 179, *2112*
Umeki, K., 89, *500*
Umezawa, H., 34, *110*, 37, *110*
Umoh, E.F., 264, *3515*
Umschweif, B., 243, *3112*
Underkoefler, L.A., 145, *1561*, 153, *1561*, 183, *1561*
Underkofler, L.A., 65, *259*, 209, *259*
Unno, T., 204, *2505*, 204, *2506*
Uno, K., 195, *2389*
Uno, T., 91, *529*
Uotsu, N., 240, *3075*
Uozumi, T., 254, *3304*
Upadhye, K., 251, *3250*
Uppalanchi, A.K., 150, *1672*

Urbanek, H., 78, *1078*
Urbano, L.H., 219, *2791*
Usami, S., 237, *3033*
Ushiro, S., 165, *1949*
Ustinnikov, B.A., 156, *1793*, 211, *2622*, 212, *2640*, 215, *2694*, 215, *2696*, 215, *2708*, 216, *2709*, 217, *2622*, 217, *2721*, 217, *2723*, 217, *2730*, 217, *2721*, 218, *2768*
Ustinnikov.B.A., 217, *2735*
Usui, Y., 239, *3058*
Uta, Y., 163, *1924*
Utaipatanacheep, A., 249, *3210*
Utamura, T., 240, *3068*, 252, *3271*
Utsumi, S., 72, *647*, 79, *647*
Uttapap, D., 110, *1151*, 190, *2283*
Uwamori, K., 246, *3172*
Uyama, K., 216, *2715*
Uyar, F., 183, *2170*

V

Vaboe, B., 64, *229*, 65, *229*
Vacca, A., 19, *89*, 21, *91*
Vacca-Smith, A.M., 63, *69*
Vadehra, D., 222, *2828*
Vadlamani, K.R., 169, *1997*, 171, *1997*
Vainer, L.M., 130, *1448*, 130, *1449*, 172, *1488*
Vakurov, A.V., 188, *2222*
Valachova, K., 200, *2458*
Valencia, F.F., 72, *686*, 73, *686*
Valero, J.M., 246, *3171*
Valetudie, J.C., 160, *1893*
Valies-Pamies, B., 177, *2101*
Valjakka, T.T., 198, *2451*
Vallee, B.L., 183, *2171*
van Arem, E.J.F., 79, *1110*
van Bekkum, H., 61, *1*
Van Beynum, G.M.A., 140, *1527*
Van Boekel, M.A.J.S., 79, *1110*
Van Camp, C., 8, *43*
Van Damme, E., 11, *61*, 11, *63*, 11, *66*, 11, *67*, 11, *68*, 30, *68*
Van Damme, E.J., 11, *52*, 48, *52*
Van de Laar, A.M.J., 180, *2139*

Van den Elzen, P.J.M., 89, *485*, 96, *485*
Van der Burgh, L.F., 179, *2121*
Van der Goot, A.J., 182, *2160*
Van der Ham, W., 192, *2328*
van der Maarel, M.J.E.C., 64, *141*, 64, *182*, 65, *182*, 123, *1313*, 124, *1326*, 241, *1313*, 241, *3078*, 241, *3079*
van der Meer, F.J.U.M., 11, *55*, 48, *55*
Van der Meeren, J., 194, *2369*
Van der Merwe, T.L., 197, *2430*
Van der Oost, J., 112, *1172*, 114, *1172*
Van der Osten, C., 64, *78*, 251, *3252*
van der Veen, B., 64, *182*, 65, *182*
van der Veen, B.A., 129, *1429*
Van der Veen, M.E., 182, *2160*
Van Dijk, K., 227, *2920*
Van Dyk, J.W., 138, *1491*, 138, *1492*
van Hilum, S.A.F.T., 129, *1429*
Van Hung, P., 198, *2446*
van Klinkenberg, G.A., 98, *830*
van Laer, J.A., 63, *54*
Van Laethem, K., 11, *63*, 11, *66*, 11, *67*, 11, *68*, 30, *68*, 46, *153*, 47, *154*
Van Lancker, F.R.G.M., 122, *1289*
Van Ooteghem, S.A., 246, *3173*
Van Ooyen, A.J.J., 89, *485*, 96, *485*
Van Oudendijck, E., 74, *733*
Van Rensburg, P., 217, *2758*
Van Solingen, P., 64, *101*
Van Tilburg, R., 140, *1527*
Van Twisk, P., 155, *1769*
Van Uden, N., 66, *305*, 72, *689*, 74, *689*, 77, *689*, 92, *559*
Vanbrrumrn, J., 72, *643*
Vandamme, A.M., 11, *63*
Vandamme, E.J., 235, *2992*, 251, *3248*
Vandemeulebroucke, E., 47, *156*
Vandeputte, G.E., 177, *2100*, 192, *2337*, 202, *2469*
Vanderberghe, L.P.S., 237, *3039*
Vandor, J., 99, *864*
Vang Hendricksen, H., 157, *1827*, 207, *2584*
Vanhemelrijck, B., 151, *1705*
Vanhoof, K., 72, *666*, 261, *3418*, 262, *3419*

VanLoey, A., 72, *667*
Vann, W.P., 109, *1118*
Vannier, C., 113, *1181*
Van't Riet, K., 97, *814*
Varalakshmi, K.N., 74, *765*
Varavinit, S., 96, *803*
Varbanets, I.D., 96, *624*
Varella, V.L., 191, *2321*
Varga, M., 148, *1612*, 154, *1735*, 154, *1741*, 158, *1735*, 159, *1741*
Vargas, C., 73, *703*
Varlan, A.R., 72, *667*
Varlley, R., 259, *3381*
Varriano-Marston, E., 85, *407*, 203, *2479*
Vasanthan, T., 167, *1973*, 167, *1981*, 265, *3534*
Vasileva, N.Ya., 217, *2730*
Vasu, S.S., 99, *855*
Vecher, A.S., 163, *1910*
Vecher, C.A., 155, *1771*
Veeder, G.T., 125, *1336*
Veelaert, S., 182, *2160*
Veen, A., 77, *947*, 103, *947*, 124, *1326*
Vehrmaanpera, J., 64, *85*
Veit, C., 149, *1628*, 222, *2836*
Vekjović, V.B., 218, *2771*
Velasco, T., 23, *98*
Velazquez, J.A., 227, *2902*
Veliky, I.A., 212, *2643*, 218, *2643*
Velusamy, V., 258, *3374*
Venkataraman, K., 78, *1066*
Venkatasubramanian, K., 149, *1643*, 150, *1689*
Venkatesh, K.V., 235, *2997*
Venkateshwar, M., 234, *2988*, 235, *2988*
Venker, H.P., 155, *1762*
Venkitaraman, A.R., 63, *69*
Ventosa, A., 73, *703*
Venturi, C., 21, *92*, 21, *93*
Verachtert, H., 74, *733*, 74, *734*, 74, *750*, 78, *750*, 96, *629*, 107, *629*, 227, *2920*
Verheije, M.H., 11, *55*, 48, *55*
Verma, G., 217, *2743*, 217, *2744*, 217, *2745*
Vermeire, K., 11, *63*, 47, *156*

Verplaetse, A.R.J.R., 122, *1289*
Verwimp, T., 192, *2337*
Vest, Z.G., 11, *54*
Vetsesol, A., *42*, 63
Vezina, L.P., 71, *442*, 76, *442*, 87, *442*, 99, *442*, 100, *442*, 163, *1927*
Vianello, F., 267, *3586*
Vibe-Pedersen, J., 227, *2921*
Vicent, C., 24, *101*
Vicheva, A., 104, *960*
Viebrock, F., 108, *1035*, 263, *1035*
Vieille, A., 191, *2315*
Vieira, C.R., 94, *583*
Vieira de Carvalho, R., 72, *676*
Vielle, C., 117, *1223*
Vihinen, M., 80, *338*
Viikari, L., 247, *3186*
Vijaylakshmi, K.S., 78, *1057*
Viksoe-Nielsen, A., 64, *97*, 91, *530*, 92, *548*, 121, *530*, 202, *2468*
Vikso-Nielsen, A., 64, *83*, 92, *550*, 148, *1605*
Viljava, T., 234, *2979*
Villagómez-Castro, J.C., 27, *106*
Villamon, E., 27, *105*
Villanea, F.A., 81, *366*, 82, *366*
Villette, J.R., 129, *1428*
Villiers, A., 249, *3212*
Vilpoux, O.F., 218, *2766*
Vinayak, S.A., 207, *2574*
Vishnu, C., 234, *2987*
Visigalli, C., 147, *1596*
Visser, F.M.W., 75, *793*
Visser, J., 97, *814*
Visvanath, B., 71, *413*, 85, *413*
Vitez, L., 165, *1951*
Vitolo, M., 244, *3147*
Vlad, M., 234, *2991*, 235, *2991*
Voča, N., 223, *2844*
Voetz, M., 146, *1571*, 210, *1571*
Voisset, C., 11, *53*
Vojcišek, V., 109, *1127*
Volker, H., 100, *883*
Vollek, V., 75, *780*
Vollu, R.E., 253, *3293*

Volz, F.E., 177, *2099*, 179, *2099*, 203, *2486*
von Falkenhausen, F., 242, *3093*
von Klinkerberg, V.A., 129, *1442*
Von Tigerstrom, R.G., 92, *551*
Vonk, H.J., 83, *392*
Voorpostel, A.M.B., 196, *2409*
Vorlop, K.D., 123, *1312*
Voronova, I.N., 237, *3037*
Voss, J., 100, *885*
Vrbaski, L., 264, *3502*
Vrbova, E., 264, *3521*, 265, *3535*
Vretblad, P., 102, *916*
Vu-Dac, N., 11, *53*
Vujcič, Z.M., 110, *1140*
Vukasinovich, M., 222, *2841*
Vulquin, E., 128, *1406*
Vvedenskii, N., 186, *2205*

W

Wacher, C., 236, *3021*
Wachman, J., 148, *1602*
Wacker, M., 262, *3422*
Wadetwar, R., 251, *3250*
Wagner, G.C., 264, *3525*
Wagner, P., 208, *2597*
Wagschal, K., 64, *243*, 68, *243*
Wahl, G., *18*, 62, 120, *1257*, 158, *1859*
Wahlstrom, G., 112, *1168*
Wakabayashi, H., 207, *2569*
Wakabayashi, S., 159, *1879*
Wakabayashi, Y., 220, *2808*
Wako, K., 94, *580*, 159, *1883*, 159, *1884*
Waksman, S.A., 233, *2967*, 266, *3569*
Walaszek, A., 97, *815*
Walden, C.C., 201, *2464*
Walden, M., 76, *927*
Waldschmidt-Leitz, E., 82, *379*
Walisch, S., 238, *3049*
Waliszewski, K.N., 140, *1529*
Walker, D.B., 3, *16*, 15, *16*, 21, *16*
Walker, F., 81, *359*, 188, *2243*
Walker, G.J., 67, *311*, 124, *1332*, 127, *1377*
Walker, J.A., 186, *2190*
Walker, T.K., 126, *1372*

Walker, T.S., 86, *439*
Walkup, J.H., 145, *1558*, 191, *2323*
Wallace, R.J., 227, *2914*, 228, *2914*
Wallace, S.J.A., 227, *2914*, 228, *2914*
Wallerstein, L., 152, *1715*
Wallerstein, M., 183, *2168*
Walon, R.G.P., 149, *1620*, 149, *1636*, 149, *1637*, 156, *1787*, 157, *1797*, 157, *1824*, 244, *3129*, 245, *3159*, 245, *3160*, 244, *3125*, 244, *3127*
Walsh, D.R., 65, *275*
Walsh, J.F., 148, *1600*
Walton, J.H., *56*, 63
Walton, M.T., 231, *2945*
Wandersee, M.K., 11, *54*
Wandrey, A., 214, *2681*
Wanek, P., 262, *3422*
Wang, B., 3, *14*, 4, *20*, 4, *22*, 5, *14*, 5, *25*, 5, 26, 5, *28*, 5, *32*, 5, *33*, 7, *36*, 8, *41*
Wang, D., 164, *1929*, 225, *2874*, 225, *2880*, 226, *2888*, 254, *3320*
Wang, F., 65, *265*, 65, *287*, 217, *265*, 225, *2876*
Wang, H.K., 7, *35*, 78, *1044*, 108, *1023*, 142, *1539*, 149, *1629*, 193, *2353*, 217, *1629*, 220, *1629*, 233, *2970*
Wang, J., 64, *128*, 65, *128*, 150, *1676*, 197, *2437*, 223, *2853*
Wang, L., 8, *41*, 68, *315*, 150, *1657*, 158, *1836*, 179, *2116*, 206, *1657*, 223, *2848*, 244, *3140*
Wang, M., 65, *265*, 65, *287*, 217, *265*
Wang, N.S., 268, *3610*
Wang, P., 155, *1758*, 168, *1988*, 221, *2814*, 254, *3306*
Wang, P.M., 118, *1232*
Wang, Q., 237, *3044*
Wang, R., 65, *278*
Wang, S., 11, *65*, 64, *154*, 149, *1648*, 168, *1988*, 206, *2545*, 223, *2853*, 254, *3323*
Wang, S.L., 8, *41*
Wang, W.J., 170, *1998*, 176, *1998*
Wang, X., 120, *1249*, 149, *1629*, 189, *2253*, 217, *1629*, 220, *1629*, 246, *3181*

Wang, X.D., 189, *2275*
Wang, Y., 158, *1836*, 163, *1914*, 179, *2114*, 244, *3140*
Wang, Y.S., 107, *1011*
Wang, Z., 167, *1972*, 196, *2413*, 206, *2561*, 208, *2561*, 238, *3050*, 254, *3306*
Wang, Z.Y., 108, *1111*
Wang,W.F., 193, *2363*
Wanishka, R.D., 260, *3405*
Wankat, P.C., 244, *3130*
Ward, D.E., 150, *1687*
Ward, G.E., 232, *2961*
Ward, M., 64, *101*
Ward, R.J., 79, *1092*, 220, *1092*
Wardsack, C., 193, *2359*
Wareham, R.S., 22, *97*
Warren, R.A.J., 64, *165*, 65, *165*
Warth, F.J., 167, *1983*, 210, *1983*
Wartmann, T., 77, *1039*
Wary, S., 154, *1734*
Wase, D.A.J., 72, *650*
Washino, K., 240, *3065*
Watanabe, A., 190, *2280*, 204, *2505*
Watanabe, E., 264, *3523*
Watanabe, H., 93, *567*, 257, *3372*, 258, *3373*
Watanabe, J., 194, *2366*
Watanabe, K., 79, *1101*
Watanabe, M., 194, *2366*, 207, *2564*
Watanabe, N., 240, *3069*
Watanabe, S., 263, *3479*
Watanabe, T., 79, *1095*, 145, *1567*, 240, *3073*, 249, *3226*, 249, *3227*
Waters, E., 112, *1174*
Watson, S.D., 243, *3116*
Watson, T.G., 81, *348*
Watts, R., 79, *1109*
Wawrzyniak, B., 110, *1146*
Wayman, M., 223, *2851*
Weakley, F.B., 161, *1905*
Weawer, L.T., 259, *3381*
Webb, C., 78, *1046*, 126, *1371*, 190, *2284*
Webb, O.F., 213, *2658*
Weber, A., 196, *2415*
Weber, H., 88, *460*, 124, *460*

Wee, Y.J., 234, *2980*
Weegels, P.L., 196, *2409*, 196, *2410*
Weemanes, G., 182, *2167*
Weenen, H., 187, *2221*, 207, *2570*
Weetall, H.H., 109, *1118*
Wegemann, K., 265, *3551*
Wegner, H., 196, *2408*
Wei, C., 72, *664*, 205, *2513*
Wei, J., 190, *2286*, 190, *2304*
Wei, J.Q., 108, *1111*
Wei, K., 222, *2838*
Wei, W., 195, *2401*
Wei, X., 11, *65*, 225, *2876*
Wei, Y., 138, *1497*
Weibel, M.K., 64, *171*
Weidenbach, G., 243, *3124*
Weidenhagen, R., 63, *60*
Weidlich, R., 264, *3489*
Weill, C.E., 99, *863*, 99, *865*, 101, *863*
Weirich, B., 262, *3454*, 263, *3454*, 264, *3454*
Weisbeek, P.J., 64, *94*
Weise, H., 264, *3516*, 264, *3517*
Weitz, R., 86, *429*
Weizmann, C., 230, *2941*, 233, *2969*
Weller, C., 64, *232*, 156, *232*
Welz, G., 217, *2729*
Welzmüller, F., 63, *30*
Wen, Q., 153, *1728*, 193, *2353*
Wenge, K.S., 147, *1587*
Wenger, K.S., 213, *2670*
Wenzig, E., 139, *1502*, 161, *1502*
Werman, M.J., 259, *3386*
Werner, J., 81, *366*, 82, *366*
Wertheim, M., 98, *846*
Weselake, R.J., 186, *2193*
Wesenberg, J., 125, *1341*, 224, *2867*
Wessenberg, J., 225, *2871*
Westerlund, E., 262, *3462*, 262, *3467*, 263, *3462*
Westin, C., 11, *64*
Weston, B., 7, *36*
Weustink, P., 262, *3451*, 264, *3451*
Wheeler, H.R., 151, *1712*, 154, *1752*, 178, *1752*

Whelan, W.J., 64, *220*, 65, *300*, 98, *835*, 99, *848*, 115, *1196*, 115, *1197*, 116, *1196*, 119, *848*, 119, *1240*, 119, *1241*, 120, *1253*, 189, *2257*, 242, *3097*, 242, *3101*, 248, *1253*, 248, *3205*, 248, *3206*, 260, *3401*, 265, *3546*
Whelen, W.J., 172, *2013*
Whitcomb, D.C., 259, *3379*
White, C.A., 64, *213*, 65, *268*, 89, *493*, 89, *505*, 98, *824*
Whitesides, G.M., 12, *73*
Wickerham, L.J., 232, *2961*
Wie, C.H., 262, *3430*
Wie, K.B., 124, *1324*
Wieczorek, J. Jr., 196, *2406*
Wiejak, S., *4*, 61
Wierbol, A., 266, *3563*
Wierciński, J., 190, *2289*
Wierzchowski, Z., 111, *1159*, 113, *1159*
Wigner, N., 117, *1220*
Wijbeng, D.J., 77, *947*, 103, *947*
Wijbenga, D.J., 124, *1326*
Wild, G.M., 63, *67*, 80, *67*
Wildbrett, G., 196, *2414*
Wilhelmson, A., 266, *3565*
Wilke, C.R., 218, *2763*, 218, *2765*
Wilkinson, I.A., 120, *1251*, 120, *1252*, 242, *1251*, 248, *1251*, 248, *1252*
Willemoes, M., 114, *1189*
Willetts, A.J., *16*, 62
Williamson, G., 64, *130*, 64, *132*, 129, *130*, 129, *132*, 129, *1415*, 129, *1422*, 129, *1426*
Williamson, M.P., 129, *1426*
Willstätter, E., 81, *368*, 82, *368*, 82, *374*, 250, *3235*, 266, *3557*
Wilson, A.L., 158, *1857*
Wilson, E.J. Jr., 255, *3334*
Wilson, I.A., 11, *59*
Wilson, J.J., 74, *561*, 77, *561*, 93, *561*, 106, *561*
Wilson, J.P., 226, *2888*
Wilson, L.A., 169, *1997*, 171, *1997*, 171, *1997*

Wind, R.D., 72, *685*
Windisch, W., 128, *1401*, 128, *1402*, 146, *1582*, 159, *1889*, 186, *2188*, 266, *3564*
Wingard, L.B., 135, *1464*
Wisniewska, D., 192, *2331*
Wisztorski, M., 8, *43*
Witrowa-Rajchert, D., 262, *3447*, 263, *3447*, 264, *3447*
Witt, P.R., 149, *1644*
Witt, W., 121, *1271*
Wittenberg, G., 114, *1186*
Witvrouw, M., 11, *69*
W.Jülicher, 194, *2384*
Wlodarczyk, Z., 210, *2609*, 218, *2609*
Wlodawer, A., 11, *51*
Wnuk, W., 238, *3049*
Wohnlich, J., 264, *3500*
Wojciechowski, A.L., 237, *3039*
Wojciechowski, P., 143, *1549*
Wojciechowski, P.M., 132, *1458*, 132, *1459*
Wójcik, A., 109, *1125*
Wójcik, W., 65, *302*
Wolf, V.J., 78, *1090*
Wölk, H.U., 64, *225*
Wolska, M., 222, *2830*, 226, *2887*
Wolski, T., 109, *1126*
Wolters, M.G.E., 262, *3433*, 265, *3533*
Wong, B.L., 186, *2192*
Wong, C.W., 170, *2004*
Wong, D., 150, *1672*
Wong, D.W.S., 64, *109*, 64, *243*, 68, *243*
Wong, H.A., 170, *2001*
Woo, G.J., 127, *1395*, 190, *2292*, 239, *3060*
Wood, B.E., 217, *2746*
Wood, C.J., 107, *1021*
Woodams, E.E., 74, *757*
Woods, A.S., 8, *43*
Wool, R.P., 264, *3525*
Woolhouse, A.D., 64, *202*
Wootton, M., 179, *2119*
Wopet, P.W., 243, *3119*
Word, N.S., 107, *1018*
Wrede, H., *58*, 63
Wright, F., 111, *1158*, 113, *1158*

Wu, C.Y., 234, *2985*
Wu, H., 190, *2286*
Wu, J., 65, *287*, 238, *3056*
Wu, J.M., 264, *3484*
Wu, L., 30, *108*, 238, *3050*
Wu, Q., 247, *3184*
Wu, S., 150, *1676*, 211, *2623*
Wu, S.C., 117, *1231*
Wu, W., 64, *81*, 92, *550*
Wu, X., 11, *65*, 226, *2888*
Wu, X.H., 170, *2006*
Wu, Y., 220, *2803*, 221, *2803*
Wu, Y.B., 208, *2595*
Wu, Y.I., 150, *1681*
Wu, Y.V., 226, *2899*
Wu, Z., 232, *2962*
Wuersch, P., 176, *2079*
Wuester-Botz, D., 214, *2681*
Wulff, G., 5, *23*, 5, *24*
Wunderly, C., 188, *2238*
Wychowski, C., 11, *53*
Wyman, M., 224, *2860*
Wyne, A.M., 92, *555*

X

Xavier, X., 90, *509*, 102, *509*
Xia, H., 42, *144*
Xia, Y., 68, *313*
Xiang, H., 201, *2463*
Xiao, D., 216, *2713*, 216, *2714*
Xie, B., 244, *3136*, 254, *3312*
Xie, H., 150, *1671*, 189, *2254*
Xie, J., 151, *1710*, 244, *3137*
Xie, Z.I., 170, *2006*
Xing, Z., 7, *35*
Xiong, L., 193, *2353*
Xiong, S., 148, *1598*
Xiong, X., 218, *2772*
Xiong, Y., 225, *2876*
Xu, B., 72, *664*
Xu ,C., 249, *3211*
Xu, G., 149, *1649*
Xu, H., 65, *266*, 264, *3522*
Xu, K., 216, *2713*

Xu, P., 222, *2835*
Xu, Q., 178, *2108*, 215, *2692*
Xu, S., 153, *1728*, 206, *2561*, 208, *2561*
Xu, X., 237, *3043*
Xu, Y., 64, *120*, 72, *675*, 104, *965*, 193, *2349*, 254, *3323*
Xu, Y.S., 145, *1557*
Xu, Z., 151, *1714*, 153, *1728*, 163, *1912*, 163, *1913*
Xu, Z.F., 206, *2549*, 206, *2550*
Xue, H., 155, *1758*
Xue, W., 219, *2790*

Y

Yabuki, M., 75, *771*
Yabuki, S., 244, *3155*
Yabuki, T., 157, *1815*, 159, *1815*
Yadav, B.S., 220, *2806*
Yadov, D., 78, *1050*
Yagi, F., 76, *928*
Yagi, T., 94, *581*, 251, *3256*
Yagi, Y., 252, *3271*, 252, *3272*, 253, *3291*
Yagisawa, M., 249, *3219*
Yago, N., 197, *2435*
Yahiro, K., 234, *2981*
Yajima, M., 214, *2678*
Yakovenko, V.A., 150, *1669*
Yakovle, A., 122, *1297*
Yakovlev, A.N., 129, *1433*
Yaku, H., 136, *1476*
Yakushijin, M., 82, *372*
Yamada, B., 217, *2752*
Yamada, H., 194, *2379*, 194, *2380*
Yamada, K., 42, *142*, 74, *611*, 74, *612*, 74, *613*, 74, *614*, 74, *764*, 75, *775*, 95, *611*, 95, *612*, 95, *613*, 95, *614*, 204, *2505*, 206, *2542*
Yamada, M., 211, *2624*
Yamada, N., 117, *1227*, 118, *1227*, 207, *2569*, 247, *1227*, 247, *3194*
Yamada, S., 244, *3155*
Yamada, T., 157, *1815*, 159, *1815*
Yamada, Y., 141, *1536*, 193, *2362*, 237, *3040*
Yamadaki, M., 215, *2693*, 222, *2693*

Yamade, K., 214, *2689*
Yamagata, H., 72, *655*, 243, *3121*
Yamagishi, G., 76, *356*, 81, *354*, 81, *355*, 81, *356*, 81, *357*, 100, *356*, 100, *357*
Yamagishi, H., 187, *2213*
Yamaguchi, J., 71, *410*, 85, *410*
Yamaguchi, K., 217, *2751*
Yamaguchi, Y., 126, *1357*
Yamakawa, S., 100, *874*
Yamamori, M., 198, *2446*
Yamamoto, E., 125, *1350*, 126, *1350*
Yamamoto, H., 34, *114*, 34, *117*, 34, *118*, 34, *121*, 240, *3066*
Yamamoto, I., 241, *3081*
Yamamoto, K., 248, *3208*, 249, *3228*, 255, *3338*
Yamamoto, M., 65, *285*, 91, *542*, 118, *1234*, 165, *1949*, 251, *3255*
Yamamoto, N., 11, *57*, 45, *150*, 45, *152*, 255, *3329*, 267, *3593*
Yamamoto, N.N., 254, *3325*
Yamamoto, R., 95, *603*, 95, *604*, 95, *605*, 95, *606*
Yamamoto, S., 34, *121*, 34, *123*, 35, *126*, 38, *126*, 91, *531*, 94, *581*, 95, *609*, 121, *609*, 164, *1933*, 251, *3256*
Yamamoto, T., 64, *210*, 89, *500*, 98, *836*, 101, *891*, 111, *1157*, 113, *1157*, 127, *1392*, 219, *2781*
Yamamoto, Y., 221, *2812*
Yamanaka, K., 244, *3150*
Yamane, K., 101, *900*, 129, *1427*, 254, *3302*
Yamanobe, T., 122, *1300*, 122, *1301*, 122, *1302*, 122, *1303*, 122, *1304*, 122, *1305*
Yamanouchi, K., 114, *1194*, 116, *1194*
Yamasaki, Y., 76, *924*, 78, *1049*, 78, *1089*, 79, *1049*, 107, *1020*, 112, *1020*, 112, *1175*, 114, *1020*
Yamashita, H., 37, *132*, 76, *928*, 240, *3066*
Yamashita, I., 77, *1041*
Yamashita, M., 145, *1554*
Yamauchi, H., 163, *1916*
Yamazaki, I., 102, *909*
Yamazaki, N., 205, *2512*

Yan, H., 221, *2817*, 222, *2837*
Yan, J., 4, *22*, 5, *28*, 64, *120*, 235, *3010*
Yan, M., 268, *3603*
Yan, Z., 104, *968*
Yanase, M., 117, *1225*
Yang, C.H., 91, *535*, 105, *989*
Yang, C.P., 249, *3222*, 250, *3240*, 254, *3311*, 255, *3333*
Yang, G., 169, *1997*, 171, *1997*, 171, *1997*, 254, *3320*
Yang, H., 108, *1111*
Yang, J., 165, *1945*, 226, *2900*
Yang, K., 244, *3136*
Yang, L., 174, *2041*, 176, *2080*, 191, *2324*, 195, *2401*, 206, *2544*
Yang, L.S., 142, *1542*, 176, *2072*, 176, *2081*, 176, *2082*
Yang, L.X., 145, *1557*
Yang, M., 207, *2582*
Yang, R., 208, *2561*, 206, *2561*
Yang, R.D., 218, *2765*
Yang, S., 115, *1202*, 237, *3028*
Yang, S.J., 115, *1203*, 116, *1203*
Yang, S.S., 107, *1018*
Yang, S.T., 232, *2962*
Yang, W., 5, *32*, 7, *36*
Yang, X., 193, *2353*, 255, *3342*
Yang, X.H., 254, *3323*
Yang, Y., 154, *1750*, 207, *2571*, 207, *2588*, 242, *3094*
Yankov, D., 89, *482*
Yano, S., 250, *3239*, 255, *3341*
Yao, H., 142, *1540*, 193, *2348*
Yao, W., 142, *1540*, 193, *2348*
Yaoi Kagaku Kogyo, K.K., 196, *2403*
Yaron, T., 102, *902*, 145, *902*, 148, *902*, 151, *902*
Yarovenko, V.L., 139, *1514*, 141, *1534*, 156, *1793*, 163, *1923*, 187, *2209*, 211, *2619*, 211, *2620*, 212, *2639*, 212, *2640*, 215, *2694*, 215, *2695*, 215, *2708*, 216, *2709*, 217, *2721*, 217, *2730*, 217, *2721*, 218, *2768*, 230, *2931*, 230, *2933*, 230, *2940*, 231, *2940*

Yarovenko, V.V., 212, *2639*
Yasuda, M., 266, *3571*
Yasui, T., 262, *3445*
Ye, D.J., 92, *550*
Yebra, M.J., 117, *1226*
Yelchits, S.V., 108, *1114*
Yeung, J., 265, *3534*
Yiang, S., 165, *1945*
Yilmaz, G., 97, *822*
Yilmaz, M., 190, *2293*
Yim, D.G., 251, *3261*
Yin, J., 163, *1914*
Yin, P., 234, *2981*
Yin, Y., 167, *1972*
Yokoi, H., 246, *3170*
Yokoi, N., 64, *152*, 156, *1794*
Yokota, A., 78, *1081*, 79, *1091*, 94, *580*, 212, *2646*, 222, *2646*
Yokota, Y., 236, *3011*
Yokoyama, K., 263, *3472*
Yokoyama, M., 235, *2994*
Yomo, T., 145, *1555*, 178, *2109*
Yoneda, T., 8, *37*
Yoo, S.H., 89, *487*, 168, *1989*
Yoo, S.K., 93, *560*
Yook, C., 182, *2159*
Yoon, J.H., 245, *3166*
Yoon, S.H., 252, *3273*
Yoon, S.L., 194, *2381*
Yoritomi, K., 157, *1825*, 159, *1825*, 254, *3309*, 268, *3612*
Yoshid, T., 268, *3612*
Yoshida, E., 206, *2556*
Yoshida, H., 118, *1234*
Yoshida, K., 161, *1896*, 200, *2456*, 256, *3360*
Yoshida, M., 73, *715*, 97, *816*, 246, *3178*, 251, *3255*
Yoshida, O., 45, *150*, 45, *152*
Yoshida, S., 125, *1337*
Yoshida, T., 149, *1645*, 157, *1825*, 159, *1825*, 166, *1954*, 193, *2354*, 220, *2808*
Yoshida, T.T., 254, *3309*
Yoshigi, N., 88, *453*, 124, *1328*
Yoshii, R., 254, *3317*, 254, *3319*

Yoshikawa, K., 117, *1224*, 207, *2566*, 248, *3208*
Yoshimatsu, H., 259, *3387*
Yoshimoto, M., 100, *874*
Yoshimura, S., 95, *603*
Yoshimura, Y., 253, *3298*, 253, *3299*
Yoshinaga, T., 228, *2929*, 229, *2929*, 230, *2929*
Yoshino, E., 78, *1051*
Yoshino, Z., 156, *1792*, 157, *1817*, 217, *1792*
Yoshio, N., 264, *3506*
Yoshioka, K., 226, *2886*
Yoshizawa, A., 196, *2407*
Yoshizawa, K., 221, *2824*, 223, *2849*
Yoshizume, H., 64, *104*
Yoshizumi, H., 65, *264*
Yoshizumi, M., 218, *2767*
You, S., 167, *1982*
Youm, K.H., 255, *3331*
Young, D.C., 158, *1835*
Youssef, A.M., 195, *2388*
Yu, A., 64, *233*
Yu, B., 221, *2814*
Yu, C., 234, *2985*
Yu, E.K.C., 251, *3242*
Yu, J.B., 109, *1121*
Yu, J.H., 252, *3284*
Yu, L., 10, *49*, 201, *2463*
Yu, P., 189, *2273*
Yu, R., 193, *2353*
Yu, S., 64, *142*, 64, *255*, 174, *2041*, 245, *142*
Yu, T.S., 105, *980*
Yu, W., 218, *2762*
Yu, X.F., 264, *3480*
Yu, Y., 207, *2582*
Yu, Z., 217, *2738*
Yuan, X., 150, *1676*, 164, *1929*
Yuasa, A., 102, *919*, 141, *1533*
Yue, P.C., 246, *3173*
Yukawa, H., 232, *2954*
Yukhi, A., 79, *1095*
Yukhpan, P., 237, *3040*
Yuki, H., 195, *2402*
Yukimasa, T., 136, *1476*

Yun, J.S., 234, *2980*
Yunusov, T., 207, *2572*
Yurkov, A.M., 74, *748*, 77, *748*
Yusaku, F., 215, *2704*
Yusof, H.M., 226, *2897*
Yusof, Z.A.M., 220, *2801*, 226, *2890*
Yutani, K., 72, *682*
Yutani, Y., 72, *683*
Yvon, G.A.J.M., 227, *2911*

Z

Zablowski, P., 92, *552*
Zacchi, G., 142, *1538*, 236, *3019*, 264, *3510*
Zaidul, I.S.M., 163, *1916*
Zaitseva, G.V., 74, *745*, 77, *745*
Zając, A.M., 168, *1995*
Zakhia, M.P., 190, *2290*
Zakhia, N., 139, *1507*, 140, *1531*
Zakirova, M.R., 147, *1588*
Zalesskaya, M.I., 231, *2947*, 231, *2948*
Zallie, J.P., 206, *2550*
Zandstva, T., 75, *793*
Zanin, G.M., 109, *1128*, 112, *1179*, 189, *2271*, 189, *2272*, 251, *3259*, 254, *3324*
Zapata, E., 235, *3007*
Zapletal, J., 89, *496*, 149, *1642*, 190, *2310*
Zaranyika, M.F., 5, 61, 178, *5*
Zbieć, M., 224, *2869*
Zbycinski, I., 122, *1294*
Zdziebo, A., 64, *143*, 112, *1171*, 113, *1171*
Zeeman, G., 144, *1553*
Zegar, T., 129, 160, *1445*
Zeikus, J.G., 64, *135*, 76, *936*, 76, *938*, 77, *936*, 77, *938*, 77, *948*, 77, *1037*, 102, *910*, 102, *911*, 102, *912*, 103, *948*, 115, *1205*, 115, *1214*, 116, *1205*, 116, *1214*, 117, *1223*, 124, *936*, 156, *1788*, 205, *2527*, 217, *2741*, 216, *948*, 243, *3122*
Zekhini, Z., 72, *642*
Zelder, O., 145, *1568*
Zelenik-Blatnik, M., 267, *3589*
Zelinka, J., 74, 77, *746*
Zelinskaya, E.V., 224, *2863*
Zeltsin, R., 104, *970*

Zemek, J., 96, *627*, 96, *632*, 137, *632*, 268, *3602*, 268, *3611*
Zeng, A., 150, *1676*
Zeng, J., 266, *3553*, 266, *3556*
Zeng, R., 193, *2358*
Zeng, X., 174, *2041*
Zenin, C.T., 75, *787*, 220, *2799*
Zeretsov, N.A., 110, *1145*
Zhan, D., 207, *2575*
Zhan, G., 90, *524*
Zhan, X., 225, *2880*
Zhang, B., 174, *2041*, 176, *2080*
Zhang, B.S., 176, *2072*, 176, *2081*, 176, *2082*
Zhang, C., 163, *1914*
Zhang, C.M., 124, *1324*
Zhang, F., 221, *2814*
Zhang, G., 65, *270*, 138, *1497*, 168, *1993*, 188, *2228*, 250, *270*, 264, *3480*
Zhang, H., 136, *1477*, 151, *1714*, 163, *1912*
Zhang, H.L., 264, *3480*
Zhang, H.W., 163, *1913*
Zhang, J., 64, *80*, 124, *1324*, 259, *3390*
Zhang, K., 217, *2738*, 219, *2787*, 221, *2815*, 222, *2815*
Zhang, K.Q., 124, *1324*
Zhang, L.M., 64, *120*, 139, *1504*, 139, *1520*, 207, *2580*, 221, *2815*, 222, *2815*, 232, *2962*
Zhang, M., 136, *1477*
Zhang, P., 168, *1988*, 193, *2363*
Zhang, Q., 223, *2845*
Zhang, R., 218, *2775*
Zhang, S., 99, *857*, 104, *968*, 115, *1202*
Zhang, S.Z., 107, *1010*, 107, *1011*, 115, *1203*, 116, *1203*
Zhang, T., 164, *1929*, 193, *2347*
Zhang, W., 79, *1094*, 223, *2848*, 245, *3168*
Zhang, X., 120, *1249*, 233, *2970*
Zhang, Y., 189, *2253*, 241, *3086*
Zhang, Z., 148, *1598*, 150, *1657*, 206, *1657*
Zhang, Z.Z., 249, *3228*
Zhao, C., 97, *823*
Zhao, C.F., 263, 265, *3476*
Zhao, H., 64, *193*, 216, *2713*

Zhao, J., 7, *35*, 211, *2623*, 244, *3136*
Zhao, M., 153, *1728*
Zhao, P., 68, *313*, 68, *315*
Zhao, R., 266, *3556*
Zhao, S., 148, *1598*, 216, *2713*, 225, *2876*
Zhao, W., 149, *1647*, 171, *2010*, 206, *2535*
Zhao, X., 64, *80*
Zhao, Y., 215, *2692*, 218, *2773*
Zhao, Z., 97, *820*
Zhekova, B., 252, *3267*
Zheng, C., 179, *2114*
Zheng, G.H., 192, *2327*
Zheng, W.W., 226, *2900*
Zheng, X., 225, *2876*
Zheng, Y., 221, *2814*, 232, *2960*
Zheng, Z., 234, *2984*, 235, *2984*
Zherebtsov, N., 122, *1297*
Zherebtsov, N.A., *12*, *1144*, 62, *1144*, 110, *1144*, 139, *1513*
Zhong, B., 150, *1672*
Zhong, G., 230, *2935*
Zhong, S.S., 139, *1504*, 139, *1520*, 142, *1539*
Zhou, D.H., 226, *2900*
Zhou, F., 264, *3519*, 220, *2807*
Zhou, H., 65, *266*, 218, *2772*, 244, *3154*
Zhou, H.B., 64, *193*
Zhou, J., 2, *9*, 141, *1537*
Zhou, J.P., 108, *1111*
Zhou, L., 171, *2010*
Zhou, M., 238, *3050*
Zhou, Q., 193, *2347*
Zhou, W., 217, *2738*
Zhou, X., 207, *2580*

Zhou, Y., 11, *54*, 64, *163*, 82, *377*, 153, *1728*, 171, *163*, 176, *163*, 227, *2915*
Zhrebtsov, N.A., 129, *1433*
Zhu, C., 7, *35*
Zhu, L., 5, *27*, 244, *3137*
Zhu, M., 212, *2632*
Zhyryanov, A.I., 214, *2675*
Zickler, F., 74, *742*, 77, *742*
Ziegler, P., 64, *138*
Zielińska, K.J., 68, *314*, 235, *3002*, 235, *3009*
Zielonka, R., 190, *2289*
Zienkiewicz, E., 143, *1549*
Ziese, W., 173, *2029*
Zilli, M., 230, *2934*
Zilm, K.W., 43, *148*
Zimare, U., 225, *2871*
Zimmermann, W., 64, *96*, 79, *96*, 111, *96*, 253, *3292*, 240, *3072*
Ziulkowsky, J.D., 227, *2913*
Zlatarov, A., 186, *2184*
Zmeeva, N.N., 227, *2906*
Zou, S., 193, *2358*
Zubak, T.A., 204, *2500*
Zubchenko, A.V., 139, *1513*
Zuber, S., 262, *3460*
Zuchner, L., 196, *2424*
Zucker, F.J., 156, *1791*, 217, *1791*
Zummo, C., 71, 85, *409*
Zuo, D., 237, *3043*
Zuo, Y., 206, *2540*
Zygora, P.S.J., 93, *564*

SUBJECT INDEX

Note: Page numbers followed by "*f*" indicate figures, "*t*" indicate tables and "*s*" indicate schemes.

A

Acetic acid (vinegar) fermentation, 237
Acetone-butanol fermentation, 230–231
Adhesives, 196
Admixed inorganic salts, amylolytic starch conversions
   alkaline media, 183
   autolysis, potato starch, 183–186
   calcium ion, 183
   cations, alkaline media, 183
   pancreatin, 183
   ptyalin and pancreatin stimulation
      bromides, 185*f*
      chlorides, 184*f*
      fluorides, 184*f*
      iodides, 185*f*
Adsorption
   active-site domains, enzymes, 129
   alpha amylase, *B. subtilis*, 130
   amylase decay, 127
   beta amylase, 130
   DE, 127–128
   glucoamylases, 130
   glycosidic bond, 129
   histidine residues, 129
   hydrolytic action, diastase, 128
   liquefaction, starch, 127
   low temperature and high pH, 127
   mathematical models, 128
   Michaelis–Menten relationship, 129
   model, amylolysis, 128
   pH dependence, optical activity, 129–130
   provenance, 130
   single-chain attack, 128
   "starch dialdehyde", 127
   starch–enzyme complex, 128
Alcohol and alcohol–acetone fermentation
   acetone–butanol, 230–231
   ethanol (*see* Ethanol fermentation)
   isopropyl alcohol, 231–232
   1,3-propanediol and 2,3-butanediol, 232
Alfalfa tap root, ginseng, mango, and canna starches, 164
Alpha amylases
   bacterial, 88–92
   degree of multiple attack (DMA), 81
   divisions, 80
   enzymes, 65
   evolutionary tree, 65, 66*f*
   fungal alpha amylases, 93–96
   human and animal, 81–85
   hydrolysis, purified and nonpurified enzymes, 81
   immobilized alpha amylases, 96–98
   liquefaction, 81
   pancreatin, 80
   plant alpha amylases, 85–87
   sources, molecular weights, and optimum temperature and pH, 71*t*
   yeast alpha amylases, 92–93
Amaranth starch, 168
Amylolytic starch conversions
   admixed inorganic salts, 183–186
   botanical origin, 159–174

Amylolytic starch conversions (cont.)
   copper(II) sulfate, 145
   elevated pressure, 182
   engineering problems, 189–191
   enzymatic process applications, 191–208
   hydrolysis stimulators, 188
   inhibitors, 186–187
   liquefaction, 148–151
   malting, 146
   mashing, 146–147
   pH, 182–183
   pretreatment, 174–180
   pulping, 145
   saccharification (*see* Saccharification, starch)
   substrate concentration, 181
   temperature, 180–181
   water, 181–182
Amylopectin, 172–173
Amylose, 172–173
Antibiotic activity
   in vitro, PRM-A and BMY-28864, $35t$
   against yeast and yeast-like microorganisms, $31t$, $33t$

B

Bacterial alpha amylases
   Bacillus mesentericus group, 88
   B. amyloliquefaciens, 88
   B. circulans F-2, 90
   chimeric alpha amylases, 92
   extracellular alpha amylase, 92
   liquefying amylases, 89
   Lysobacter brunescens, 92
   maltohexaose-forming alpha amylase, 90
   meso- and thermophilic anaerobic nonspore-forming bacteria, 88
   purple photosynthetic bacterium, 92
   Streptococcus bovis PCSIR-7B, 91–92
   thermostable alpha amylase, 89
   Zooglea ramigera, 91
Bacterial polyester formation, 247
Bakery production and enzymatic processes
   alpha amylase, 198
   amylase-containing flour preparations, 203
   bread
      dry yeast, 200–201
      staling and retrogradation, 203
      workability and storability, 203
   commercial enzyme characteristics, 199, $199t$
   denaturation, alpha amylase, 201–202
   emulsifiers, 203
   gelatinization, 202
   RS, 198
   sweet products, 202
Bananomicins, 33–40
Barley starch
   gelatinized and waxy, 167
   liquefaction and saccharification, 167
   waxy, normal and hylon, 167
Beta amylases
   affinity constant, 99
   vs. alpha amylase, 99
   animal, 100
   bacterial
      adsorption, 101–102
      B. stearothermophilus, 102
      NCA26, 89–90
      NP33 and NP54, 91
      saccharifying/liquefying activity, 101
   corn starch, 99
   cyanides, 99
   fungal, 102
   immobilized, 102
   maltotetraose, 98
   plant
      corn-starch granules digestion, 100
      flours, 101
      nongerminated rice, 100–101
      optimum pH, malt, 100
      potato beta amylase, 100
      wheat, 100
   sources and characteristics, $76t$
Boronic acid-dependent lectin mimics
   antiviral potential
      benzoboroxole-functionalized polymers, 12–14, $13f$
      biocompatibility evaluations, 14–15

bisphenylboronic acid compounds,
11–12, 13f
envelope glycans, 11
high mannose-type and complex-type
N-glycans, 12–14, 14f
monophenyl boronic acid compounds,
11–12, 12f
multivalent interactions, 12–14
molecular architecture
arylboronic acid/diol association, 5
benzoboroxole binding, 9–10, 9s
binding constants, 6t
carbohydrate binding, 4–5
chirality, 7
dimer analogues, 5–7
fluorescent probe, 7–8
hexopyranosides binding constants,
9–10, 9t
o-(N,N-dialkylaminomethyl)-
phenylboronic acid binding, 5s
phenylboronic acid binding, diol, 4–5, 4s
tetrasaccharide sLe^x, 7–8, 8f
TF disaccharide, 10, 10f
TIMS, 7–8
Boronic acid-independent lectin mimics
antiviral and antimicrobial potential
aminopyrrolic tripod-type receptor,
30–33, 32f
analogues, aminopyrrolic tripod-type
receptor, 30–33, 32f
antibiotic activities, 30–33, 31t, 33t
anti-HIV activities, 27–30, 30t
dimers, 1,3,5-triazine compounds,
27–30, 28f
monomers, 1,3,5-triazine compounds,
27–30, 28f
trimers, 1,3,5-triazine compounds,
27–30, 29f
tripod-type receptors, 30–33
molecular architecture
amidopyridine moieties replacement,
15–16
binding affinities, Oct-β-Gal, 18–19
binding constants, n-octyl D-glycosides,
17t, 22t

cage type, 21–22, 22f
cis-1,3,5-trisubstituted cyclohexane
scaffold, 21, 21f
H-NMR analysis, 16–18
intrinsic median binding concentrations,
n-octyl D-glycosides, 19–21, 20t
molecular-modeling, 16–18, 22–23
pyrrolic tripodal receptors, 19–21, 20f
tetrapod-type, 18–19, 19f
tripod-type receptors, 15–18, 15f, 18f
Botanical origin effect, amylolytic starch
conversions
amylose and amylopectin, 172–173
beta amylase, 160
cereal starches, 164–169
chemically modified starches, 173–174
chestnut starch, 170–171
enzymatic hydrolysis, starch, 161, 162t
hydrolysis, 159, 160t
legume starch, 171–172
lichen and millet starch, 172
sago starch, 170
solubilization, 160
synthetic starch, 173
tuber and root starches, 161–164
Branching, starch, 247

C

Carbohydrate recognition, PRMs
complex-forming equilibrium, 41s
2D-DARR spectra, 43, 44f
derivatives, BMY-28864, 40–41, 41f
Man-binding conformation, 43, 45f
Man-binding model, 43, 45f
solid-state CP/MAS ^{113}Cd-NMR spectra, 42f
UV-visible spectrophotometric analysis,
40–41
Carboxylic acid fermentations
acetic acid, 237
citric acid, 236–237
gluconic acid, 237–238
glutamic acid, 238
kojic acid, 238
lactic acid, 233–236
pyruvic acid, 237

Cassava starch, 164
CDs. *See* Cyclodextrins (CDs)
Cerealpha method, 267–268
Cereal starches
    barley starch, 167
    corn starch, 164–166
    oat starch, 168–169
    rice starch, 167–168
    saccharification, 211–212
    sorghum, amaranth, and triticale starches, 168
    wheat starch, 166–167
Chemically modified starches
    acetylation, starch, 173–174
    deuterated starch, 173
    epichlorohydrin-treated starch microspheres, 174
    etherification, corn starch, 173
Chestnut starch, 170–171
Citric acid fermentation, 236–237
Corn starch
    A. fumigatus alpha amylase, 165
    genetic modifications, maize, 166
    hydrolysis, 164–166
    malt amylases, 165
    synergism, alpha amylase intrinsic enzymes, 165
    thermostable alpha amylase, 165–166
Cyclodextrins (CDs)
    aliphatic alcohols
        C_1-C_6, 254–255
        C_8-C_{16}, 255–256
    Arthrobacter globiformis bacterium, 257–258
    branched, 252, 253
    CGTase
        Bacillus coagulans, 251
        Bacillus firmus, 251
        Bacillus megaterium, 251–252
        Bacillus subtilis, 252
        Clostridium thermoamylolyticum, 252–253
        isolation, 250–251
        Micrococcus varians, 253
        thermophilic, 252
        thermostable, 253
    cyclomaltononaose (δ-CD), 253
    illumination time, α-, β-and γ-CDs concentrations, 257*f*
    inclusion complexes, 257
    kinetics, formation, 256, 258*f*
    limit dextrins, 255
    mononbranched- and dibranched, 253
    organic solvents, 256
    preparation, 249
    reducing sugars, 255
    taka amylase, 258
Cyclomaltoses. *See* Cyclodextrins (CDs)

D

Dextrinogenic amylase. *See* Alpha amylases
Dextrins. *See also* Cyclodextrins (CDs)
    amylodextrins, 148
    branched, 151, 167, 168
    degree of polymerization (DP), 146
    hydrolysis, 194
    β-limit, 89, 111, 166–167
    limit dextrins (*see* Limit dextrins)
    low-molecular-weight, 94, 198
    potato starch, 82
    pullulanase type I, 114
    wort and beer, 211
Digestible starch
    bread and cereals, 259–260
    felines, 261
    human body fluids, 259
    hydrolysis, corn starch, 260
    insects, 261
    intestinal carbohydrates digestion, 259
    mammals, 260–261
    starch-containing diet, 259

E

Elevated pressure, amylolytic starch conversions, 182
Engineering problems, amylolytic starch conversions
    continuous starch-liquefying ejector, 191

SUBJECT INDEX

electro-ultrafiltration bioreactor, 190
fixed-bed reactors, 189
hollow-fiber reactors, 190
hydrolysis, extruders, 190–191
immobilized enzyme catalyst, 191
membrane reactors, 190
Enzymatic cocktails
  amylase/glucoamylase ratio, 121
  cooperative hydrolysis, 122
  degree of polymerization (DP), 120–121
  glucoamylase and mutarotase, 122
  α-glucosidase, 121–122
  hydrolysis of starch, 122
  isoamylase cocktail, 123
  poplar wood, 121
  preparation, 123
  starch conversion, 122
  synergism, 120–121
  thermophilic amylase, 121
  transglucosidase admixture, 122
Enzymatic conversions, starch
  amylolytic starch conversions
    (*see* Amylolytic starch conversions)
  analytics
    enzyme evaluation, 266–268
    evaluation and analysis, 262–266
  enzymes
    alpha amylases, 80–98
    beta amylases, 98–102
    decomposing raw starch, 68, 68*t*
    degrading ability, porcine pancreatin and alpha amylase, 70*t*
    dextrins and isomaltose, 119
    enzymatic cocktails, 120–123
    glucan 1,4-α-maltohydrolase, 119
    glucan 1,4-α-maltotetraohydrolase, 119
    β-glucanase, 119
    α-1,4-glucan lyase, 120
    glucoamylase, 103–111
    glucodextrinase, 119–120
    α-glucosidase, 111–113
    β-glucosidase, 119
    β-glucosiduronase, 119
    glycosyltransferases, 123
    hydrolase hydrolysis, 67, 67*f*

isoamylase, 117–118
neopullulanase, 117
nonmaltogenic exoamylases, 111
pullulanase, 114–116
thermophilic amylolytic enzymes, 67
xylanases, 67
feedstock, fermentations
  (*see* Fermentation)
hydrolysis pathways
  adsorption, 127–130
  inhibition mechanism, 130–132
  kinetics, 136–144
  light, microwaves and external electric field, 135–136
  mathematical models, 132–135
metabolism
  digestible starch, 258–261
  RS (*see* Resistant starch (RS))
microorganisms
  A. awamori, 126
  A. awamori and *A. oryzae*, 127
  A. niger activity, 126
  Aspergillus oryzae, 126
  bacteria and yeasts, 124
  brewery yeast, 125
  CO_2 and hydrogen, 124
  Rhizopus species, 125–126
  Schwanniomyces castelli R68 mutant, 125
  sheep rumen, 124–125
  strains, *Rhizobium* species, 125
nonamylolytic starch conversions
  (*see* Nonamylolytic starch conversions)
Esterification and hydrolysis
  acylation of starch, 241
  amylase hydrolyzed starch, 243
  lipase-catalyzed acylation, 243
  "nanostarch", 241–242
  phosphatase, potato, 242
  phosphorylases, 242–243
  Typha latifolia pollen, 243
Ethanol fermentation
  coimmobilized microorganisms, 212–213
  description, 209–210

Ethanol fermentation (cont.)
    fructose and glucose, 210
    liquefaction, 210–211
    malting, 210
    mashing, 210, 211f
    potatoes, 218–219
    production
        amaranthus, 226
        barley, 226
        buffalo gourd, 221
        cassava, 219–220
        corn, 221–223
        dc and ac current, 228–229, 229f
        enzymes, 214–230
        field pea, 226–227
        kinetic studies, 228
        organic acids and aromas, 230t
        pearl millet, 226
        potatoes, 218–219
        rice, 225
        rye, 224
        sago, 226
        sorghum, 225–226
        starch–cellulose substrates, 229
        stimulators, 228
        substrates, 217–218
        sweet potatoes, 220–221
        time-dependent unimolecular process, 230
        triticale, 224–225
        waste and biomass, 227–230
        wheat, 223–224
        yam, 221
    saccharification, 211–212
    Saccharomyces cerevisiae, 212
    Schwanniomyces alluvis, 212
    Schwanniomyces castellii, 213
    solid-phase fermentation, 213
    sterility, 213–214

F

Fermentation
    alcohol and alcohol–acetone, 209–232
    carboxylic acid, 232–238
    definition, 208–209
    glycolysis pathway, 209
Food industry
    animal feed, 208
    bakery production, 198–203
    fruit and vegetable juices and pomace processing, 204
    RS and dietary fiber, 205–206
    starch processing and enzymes, 206–208
    sweeteners, 204–205
Fungal alpha amylases
    A. niger, 95
    Aspergillus fumigatus (K-27), 95
    Corticium rolfsii, 96
    and human and bacterial alpha amylases, 93
    Pichia burtoni Boldin, 94
    sources, molecular weights and optimum temperature and pH, 71t
    Streptomyces strains, 93–94
    takadiastase, 94–95
Fungal glucoamylases
    A. awamori, 104–105
    A. candidus and *A. foetidus*, 105
    A. oryzae, 105
    Chalara paradoxa, 106
    hydrolysis, wheat starch, 104
    Rhizopus sp., 105–106

G

Glucoamylases
    bacterial, 103
    fungal, 103–106
    granular starch digestion, 103
    immobilized, 108–111
    sources, molecular weights, and optimum temperature and pH, 77t
    yeast, 106–108
Gluconic acid fermentation, 237–238
α-Glucosidase
    B. subtilis secreted, 112
    corn, 111
    covalent immobilization, 112
    description, 111
    isoform, 111–112
    Mucor javanicus fungus, 112

plants, 111
  sources and characteristics, 112, 113t
Glutamic acid fermentation, 238
Glycosylation
  alpha amylolysis, 240
  candida transglucosyl amylase, 241
  CGTase, 239–240
  ethyl α-D-glucopyranoside, 238–239
  α-glucosylation, 238–239
  glycyrrhizin, capsaicin and ascorbic acid, 241
  maltotetraose, 239
  transglycosylation, 240
Glycosyltransferases
  cyclodextrin, 123
  cyclomaltosaccharides, 123
  thermostable α-glucanotransferase, 123

H

Human and animal alpha amylase
  isolation, 81
  pancreatin, 82–83
  ptyalin, 82
  silkworm *Bombycx mori*, 83–85
Hydrogen production, nonamylolytic starch conversions
  anaerobic *Enterobacteriaceae* SO5B, 245
  sucrose, 245–246
Hydrolysis stimulators, amylolytic starch conversions
  aliphatic amines, 188
  organic solvents, 188
  polyethylene glycol (PEG), 188

I

Immobilized alpha amylases
  B. licheniformis, 96–97
  B. subtilis, 97–98
Immobilized glucoamylase
  Amberlite IRA 93, 110
  A. niger, 109–110
  granular poly(acrylonitrile) resin, 110
  mineral supports, 109
  operational and storage stability, 108
  pullulanase, 110–111

synthetic organic polymers, 109
Inhibition mechanism, enzymatic conversions
  competitive inhibition, 132
  DE, 152f
  dry substance yield, dextrose, 134f
  noncompetitive inhibition, 131
  nonstarchy polysaccharides, 130
  oat starch slurry concentration, 169f
  random-type noncompetitive inhibition, 131
  three-dimensional response surface, 134f
  uncompetitve inhibition, 132
Inhibitors, amylolytic starch conversions
  acarbose, 187
  heparin, 186–187
  lower saccharides, 187
  phenols, 186
  phytic acid, 186
Isoamylase
  DSM 4252, 118
  Flavobacterium odoratum, 118
  Pseudomonas amyloderamosa, 117–118
  sources, 118, 118t
Isolation and purification, starch
  alpha amylase, 193–194
  fiber-degrading enzymes, 192
  microporous granular starch, 193
  porous starch powder, 193
  sweet potato, 192–193
  wet-milling process, 192
Isomerization
  1,5-anhydro-D-fructose, 245
  glucose isomerase, 243, 244
  glucose 6-phosphate isomerase, 244
  glucose-to-fructose, 245
  immobilization, glucoamylase, 244
  syrups, 245
Isopropyl alcohol fermentation, 231–232

K

Kinetics
  alpha amylase-catalyzed depolymerization, 137
  debranching enzyme, 141
  dependence on temperature, 137–138

Kinetics (cont.)
  endo- and exo-enzymes, 143–144
  enzyme concentration, 136–137
  glucoamylase, 139–140
  glucose, rate of formation, 141–142
  hydrolysis, 138–139
  Michaelis A and B constants, 140–141
  Michaelis–Menten kinetics, 136
  multisubstrate model, 143
  parameters, 137, 137t
  potato and barley starches, hydrolysis, 141
  potato starch, *Bacillus* sp. IIA, 143
  rate constant, hydrolysis, 137
  retrogradation, starch, 142
  saccharification, corn starch, 142
  soluble and liquefied cassava starch, 140
  soybean amylase, 138
  waxy corn starch, 138
Kojic acid fermentation, 238

L

Lactic acid fermentation
  bacteria, 236
  fermented hard cheeses, 233–234
  Lactobacillus, 233
  molecule, 234
  polysaccharides, 235
  pretreatment, 236
  sago and cassava starches, 236
  sauerkraut, 233–234
  Streptococcus bovis 148, 234–235
  synthesis, muscles, 234
  yogurt, 233
Lectin mimics
  carbohydrate-binding molecules, 2–3
  molecular-design principles, 3
  naturally occurring, 33–48
  synthetic
    boronic acid-dependent lectin mimics, 4–10, 11–15
    boronic acid-independent lectin mimics, 15–33
Legume starch, 171–172
Lichen starch, 172

Limit dextrins
  alpha amylase generated, 151
  CDs, 255
  decomposition, 181
  glucose, 85–86
  α-limit dextrins, 260
  β-limit dextrins, 89, 103, 166–167
  maltotriose, 129
Liquefaction
  amylodextrins, 148
  bacterial alpha amylase, 150
  branched dextrins and oligosaccharides, 151
  corn starch, 148–149, 150
  heat-stable enzymes, 149
  malting and mashing stages, 148
  patented procedures, hydrolysis, 149
  pH optimum, 148
  rate, amylolytic liquefaction, 149
  rice starch, 150
  soluble calcium compounds, 148–149
  starch, 150
  thinned, nonretrograding hydrolyzates, 151
  three-step, 149
  two-step, 148

M

Malting
  dextrins, 146
  diastase, 146
  germination, 146
Mashing
  acidification, 147
  gelatinization, 146–147
  β-glucanase, 147
  α-glucosidase, 147
  unmalted millet, 146
  viscosity, gelatinized starch, 147
Mathematical models, enzymatic hydrolysis
  coefficients, Taylor equation, 133, 133t
  depolymerization, starch, 132
  enzyme concentration, 135
  glucoamylase/pullulanase saccharification, 133
  glucose, 135

SUBJECT INDEX

hydrolyzate, 135
impeller speed, 134
maltose, 135
pH effect, 133
processing time, 135
three-dimensional response surface, 133
viscosity, 135
Methanogenic and biosulfidogenic conversions, 243
Michaelis–Menten kinetics, 136
Millet starch, 172

N

Naturally occurring lectin mimics. *See* Pradimicins (PRMs)
Neopullulanase, 117
Nonamylolytic starch conversions
  bacterial polyester formation, 247
  branching, starch, 247
  cyclodextrins, 249–258
  esterification and hydrolysis, 241–243
  glycosylation, 238–241
  hydrogen production, 245–246
  isomerization, 243–245
  methanogenic and biosulfidogenic conversions, 243
  oxidation, 247
  polymerization, 247–249
  trehalose, 246–247

O

Oat starch, 168–169
Oxidation, nonamylolytic starch conversions, 247

P

Pancreatin
  erythrodextrins, 82
  nonbranched amylose component, 82
  purified hog, 82
  spontaneous loss, amylolytic potence, 84f
Pharmaceutical industry, 197
Plant alpha amylases
  apple tissues, 87
  cereals
    barley, 85
    maize, 86
    ragi, 86
    reducing sugars, 86, 87f
    rice, 85
    subsite model, 86
  salicinase/emulsin, 87
  tubers, 86–87
Polymerization, nonamylolytic starch conversions
  cellulases, 248
  dextrin dextranase, 248
  high-amylose rice starch debranching, 249
  potato phosphorylase, 248
  Q-enzyme, 248
  starch gels, 248
  synthetic starch, 247
Potato starch
  beta amylase, 161–163
  dextrose equivalent (DE), 161
  liquefaction, 161–163
  phosphate ester groups, 163
Pradimicins (PRMs)
  antimicrobial and carbohydrate-binding profiles
    benzo[*a*]naphthacenequinone, 33–36, 34f
    BMY-28864, 37–40, 38f
    carbohydrate-binding ability, 33–36
    in vitro antibiotic activities, 33–36, 35t
    in vivo activities, 33–36, 36t
    D-mannose, 37–40, 40f
    precipitation, BMY-28864, 37–40, 39t
    structural requirements, antifungal activity, 36f
    structure-microbial activity relationship, 36–37
  antiviral profile and mode of action
    anti-HIV activities, 46–47, 46t
    HIV infection, 45–46
    PRM-A, 47–48
    PRM-S, 47–48
    time-of-drug-addition studies, 46–47
  carbohydrate recognition
    (*see* Carbohydrate recognition, PRMs)

Pretreatment, amylolytic starch conversions
  autoclaving starch, 177
  barley starch, 176
  corn granules, 174
  digestion, starch granules, 177
  extensive erosion, granules, 177
  fungal and bacterial alpha amylases, 174–175
  granular starch
    crushing, 178
    glucoamylase, 174, 175f
    preheating, 178–179
  hydrothermal treatment, 179–180
  mechanical damaging starch, 177
  porcine pancreatic alpha amylase, 176
  retrogradation, 177
  swelling capacity, granules, 179
  time-course, glucose formation, 177, 178f
  ultramicroscopic studies, 176–177
PRMs. See Pradimicins (PRMs)
1,3-Propanediol and 2,3-butanediol fermentations, 232
Ptyalin
  crude, 82
  dextrins, 82
  pH ranges, 82
Pullulanase
  bacterial, 115
  hyperthermophilic, 115
  molecular weights, and optimum temperature and pH, 116t
  potato and broad bean, 115
  sources, 114, 115, 116t
  type I, 114
Pulp industry, 194
Pulping, 145
Pyruvic acid fermentation, 237

R

Resistant starch (RS)
  fermentation, 261–262
  oxidative stress, 262
Rice starch
  high-protein rice flour, 168
  hydrolysis, 168

liquefaction, 167
RS. See Resistant starch (RS)

S

Saccharification, starch
  glucose
    fungi, 158
    immobilized glucoamylase, 158
    isolation, 159
    production procedures, 158
    pullulanase, 158–159
  maltohexaose, 159
  maltopentaose, 159
  maltose, 157–158
  syrups
    alpha amylase, 151–152, 154
    beta amylase, 153
    cassava starch, 155–156
    cationic surfactants, 152–153
    fungal alpha amylases, 153
    glucoamylase, 155
    glucoamylase-saccharified starch and alpha amylase, 155–156
    hydrolysis, liquefied starch, 155
    limit dextrins, 151
    liquefied starch, 156
    mild acid-catalyzed hydrolysis, 153–154
    one-step and two-step process, 153
    pullulanase, 156–157
    saccharification power (SP), 156
    sago starch, 153
    sliced cassava roots, 153
Sago starch, 170, 247
Schardinger dextrins. See Cyclodextrins (CDs)
Sorghum starch, 168
Starch
  acid-catalyzed "saccharification", 62
  description, 61, 62
  enzymatic conversions (see Enzymatic conversions, starch)
  enzymolysis, 63, 64–65
Starch-based washing and cleaning, 196
Sweet-potato starch, 163–164
Synthetic lectin mimics

boronic acid-dependent, 4–10, 11–15
boronic acid-independent, 15–33
Synthetic starches, 173

T

Temperature, amylolytic starch conversions, 180–181
Textile sizing and desizing
 cellulose-containing fabrics, 195
 enzymatic reactions, 194–195
 glucoamylase and beta amylase, 195
 pancreatic amylase, 194–195
Thomsen–Friedenreich (TF) disaccharide, 10, 10f
Trehalose, 246–247
Triticale starche, 168
Tuber and root starches
 alfalfa tap root, ginseng, mango, and canna starches, 164
 cassava starch, 164
 potato starch, 161–163
 sweet-potato starch, 163–164

W

Water, amylolytic starch conversions, 181–182
Wheat starch
 alpha amylase isoenzymes, 166–167
 A. oryzae alpha amylase, 167
 hydrolysis, 166
 proteolytic and alpha amylase activity, 167
 thermal pretreatment, 166

Y

Yeast alpha amylases, 92–93
Yeast glucoamylase
 Colletotricum gloelosporioides, 107–108
 commercial preparations, 108
 optimum pH, *Endomycopsis* species, 107
 Schwanniomyces alluvius, 106
 sources and characteristics, 77t, 108

Z

Z-enzyme, 98–99, 119